# Signaling and Communication in Plants



## Series Editor

František Baluška
Department of Plant Cell Biology, IZMB, University of Bonn, Kirschallee 1, D-53115
Bonn, Germany

For further volumes:
http://www.springer.com/series/8094

P. Vidhyasekaran

# PAMP Signals in Plant Innate Immunity

## Signal Perception and Transduction

Springer

P. Vidhyasekaran
Plant Pathology
Tamil Nadu Agricultural University
Coimbatore, Tamil Nadu, India

ISSN 1867-9048 ISSN 1867-9056 (electronic)
ISBN 978-94-024-0755-6 ISBN 978-94-007-7426-1 (eBook)
DOI 10.1007/978-94-007-7426-1
Springer Dordrecht Heidelberg New York London


Softcover reprint of the hardcover 1st edition 2014


Printed on acid-free paper

Springer is part of Springer Science+Business Media (www.springer.com)

# Contents

# Chapter 1
# Introduction

**Abstract** Innate immunity is the first line of defense against invading microorganisms in plants. Pathogen-associated molecular patterns (PAMPs) are the classical activators of immune responses. These are alarm signal molecules are perceived as 'nonself' by plant pattern recognition receptors (PRRs) to switch on the plant immune responses. PAMPs are not only detected in pathogens, but also detected in nonpathogens and even in saprophytes. The PAMPs are often called as microbe-associated molecular patterns (MAMPs). MAMPs are molecular signatures typical of whole classes of microbes and their recognition by PRRs activates the plant innate immunity. Most of the PRRs are receptor-like kinases (RLKs) and RLKs are proteins with a "receptor" and a "signaling domain" in one molecule. The extracellular domains of RLKs bind directly to legands to perceive extracellular signals, whereas the cytoplasmic kinase domains transduce these signals into the cell. PRRs interact with additional transmembrane proteins which act as "signaling amplifiers". PAMPs induce autophosphorylation of the kinase domain of PRRs and the autophosphorylated PRRs are translocated to endosomes. The biogenesis of transmembrane PRRs occurs through endoplasmic reticulum (ER) with the aid of ER-resident chaperones. The PRR in ER is transported from ER to plasma membrane and *N*-glycosylation of PRRs is required for the transport of PRRs. Second messengers deliver the information generated by the PAMP/PRR signaling complex to the proteins which decode/interpret signals to initiate defense gene expression. Calcium ion is a ubiquitous intracellular second messenger involved in various defense signaling pathways. $Ca^{2+}$ is a master regulator of gene expression in plants. Calcium signatures are recognized by calcium sensors to transduce calcium-mediated signals into downstream events. Guanosine triphosphate (GTP)-binding proteins (G-proteins) act as molecular switches in signal transduction system. Mitogen-activated protein kinase (MAPK) cascades transduce extracellular stimuli into intracellular responses in plants. Reactive oxygen species is a second messenger in transmitting the PAMP signal. Nitric oxide (NO) is a diffusible second messenger acting in cellular signal transduction through stimulus-coupled S-nitrosylation of cysteine residues. The plant hormones salicylic acid, jasmonate,

P. Vidhyasekaran, *PAMP Signals in Plant Innate Immunity: Signal Perception and Transduction*, Signaling and Communication in Plants 21,
DOI 10.1007/978-94-007-7426-1_1, 

ethylene, abscisic acid, auxin, cytokinin, gibberellins, and brassinosteroids play important role in immune response signaling. Plant hormones activate different signaling pathways inducing distinctly different defense genes. These signaling pathways can crosstalk with each other and this crosstalk helps the plant to "decide" which defensive strategy to follow, depending on the type of attacker it is encountering. Potential pathogens produce several effectors to nullify the defense responses induced by the innate immune system. Pathogens may also hijack some signaling systems to cause disease. The war between the plant and pathogen appears to be in fine-tuning the signaling systems to cause disease or to enhance host defense response. Recent advances in our understanding of the molecular basis of plant innate immunity have opened new era in developing potential tools in management of crop diseases.

**Keywords** Pathogen-associated molecular patterns (PAMPs) • Microbe-associated molecular patterns (MAMPs) • Plant pattern recognition receptors (PRRs) • Endocytosis of PRR proteins • PAMP-triggered immunity (PTI) • PAMP-PRR signaling complex

## 1.1 Classical PAMPs

Innate immunity is the first line of defense against invading microorganisms in vertebrates and the only line of defense in invertebrates and plants (Silipo et al. 2010; Zamioudis and Peterse 2012). Several elicitors of microbial origin have been identified as primary danger/alarm signal molecules to switch on the plant immune systems culminating in activation of defense genes (Aziz et al. 2003; D'Ovidio et al. 2004; Cavalcanti et al. 2006; Vidhyasekaran 2007; Thomma et al. 2011). The classical general elicitors reported in plant pathogens resemble the pathogen-associated molecular patterns (PAMPs), the classical activators of innate immune responses in mammals (Nürnberger and Brunner 2002; Nürnberger et al. 2004; Nürnberger and Lipka 2005). These historically termed general elicitors have been renamed as PAMPs (Jones and Dangl 2006; Bent and Mackey 2007). PAMPs are often vital for microbial survival and are therefore not subject to mutational variation (Gust et al. 2007; Zhang and Zhou 2010). PAMPs are defined as evolutionarily conserved building blocks of microbial surfaces that directly bind to plant pattern recognition receptors (PRRs) and induce defense responses (Nürnberger and Brunner 2002; Qutob et al. 2006; Nicaise et al. 2009; Tsuda and Katagiri 2010; Thomma et al. 2011). The molecular signatures in PAMPs are not present in the host and these are perceived as 'non-self' by plant pattern recognition receptors (Mackey and McFall 2006).

PAMPs that trigger innate immune responses in various vertebrates and non-vertebrate organisms include eubacterial flagellin, elongation factors, lipopolysaccharides (LPS) from gram-negative bacteria, viral and bacterial nucleic acids, fungal cell wall-derived chitins, glucans, mannans, or proteins and peptidoglycans from gram-positive bacteria (Zipfel and Felix 2005; Jones and Dangl 2006). Similar

PAMPs have been detected in a wide range of plant pathogens (Shinya et al. 2007; Boller and Felix 2009; Silipo et al. 2010; Tsuda and Katagiri 2010; Nürnberger and Kufner 2011). One of the common features of PAMPs is their presence in a broad range of microbial species (Brunner et al. 2002). The general structure of lipopolysaccharides (LPS) is shared by all gram-negative bacteria (Medzhitov 2001) and the protein PAMP flagellin is highly conserved among bacterial taxa (Felix et al. 1999). Chitin is the widespread, conserved, and intrinsic structure detected in fungi (Thomma et al. 2011). CBEL (for Cellulose-Binding Elicitor Lectin) is a glycoprotein PAMP and it occurs widely in the oomycete *Phytophthora* species (Khatib et al. 2004). The PAMP double-stranded RNA is a structural signature of several groups of viruses (Medzhitov 2001; Ding 2010).

PAMPs are exclusively recognized as the molecules involved in triggering innate immunity. PAMPs are actually defined as the molecules, which bind to plant PRRs and induce defense responses (Nicaise et al. 2009; Tsuda and Katagiri 2010). However, most of the PAMPs also have virulence functions besides eliciting defense responses (Thomma et al. 2011). The well characterized PAMP flagellin also has a role in virulence. Glycosylation of the flagellin molecule has been shown to be required for virulence in *Pseudomonas syringae* pv. *tabaci* (Taguchi et al. 2010). *P. syringae* pv. *tabaci* flagellin mutants affected in their elicitor activity also showed reduced virulence in plants due to reduced motility (Naito et al. 2008; Taguchi et al. 2010). The bacterial lipopolysaccharide (LPS) generally acts as PAMP inducing defenses (Tellström et al. 2007; Aslam et al. 2008; Silipo et al. 2008; Thomma et al. 2011). However, changes in composition of LPS affect bacterial virulence, suggesting a role for LPS in virulence of pathogens (Newman et al. 2007). When the PAMP chitin synthesis was disrupted in the fungal pathogen *Botrytis cinerea*, virulence of the pathogen was drastically reduced (Soulie et al. 2006). Mutation of peptidoglycan (PGN) genes reduces the virulence of *Ralstonia solanacearum* and of *Erwinia amylovora* (Cloud-Hansen et al. 2006), suggesting the role of the PAMP peptidoglycan in virulence of pathogens.

PAMPs are detected not only in pathogens, but also in several nonpathogens, and saprophytes. Since the PAMPs are detected in all microbes, the PAMPs are better called as microbe-associated molecular patterns (MAMPs) (Viterbo et al. 2007; Zhang et al. 2007; Denoux et al. 2008; Aslam et al. 2009; Jeworutzki et al. 2010; Thomma et al. 2011; de Freitas and Stadnik 2012). MAMPs are molecular signatures typical of whole classes of microbes, and their recognition plays a key role in innate immunity (Boller and Felix 2009).

## 1.2 Plant Pattern Recognition Receptors (PRRs)

PAMPs are perceived as alarm/danger signals by cognate plant pattern recognition receptors (PRRs) and the PAMP-PRR complex activates the plant immune system (Takakura et al. 2004; Jones and Dangl 2006; Altenbach and Robatzek 2007; He et al. 2007; Wan et al. 2008; Iriti and Faoro 2009). Several receptors for the PAMPs have been recognized in plasma membrane of plant cells (Nicaise et al. 2009; Petutschnig et al. 2010; Shinya et al. 2010; Schulze et al. 2010; Segonzac and Zipfel 2011). The PRRs identified to date are modular proteins harbouring an extracellular domain consisting of leucine-rich repeat (LRR) or lysine motifs (LysM) (Saijo 2010; Segonzac and Zipfel 2011). Most of the PRRs are receptor-like kinases (RLKs) and the sensors for extracellular molecules consisting of an extracellular ligand-binding domain, a single transmembrane domain, and a cytosolic protein kinase domain are called RLKs (Seifert and Blaukopf 2010). RLKs are proteins with a "receptor" and a "signaling domain" in one molecule. The extracellular domains of RLKs bind directly to legands to perceive extracellular signals (PAMPs), whereas the cytoplasmic kinase domains transduce these signals into the cell (Bi et al. 2010).

PRRs interact with additional transmembrane proteins which act as signaling amplifiers to achieve their functionality (Zipfel 2009). PAMPs bind with PRRs and induce conformational alteration in PRRs leading to their activation (Ali et al. 2007). PAMPs trigger increased transcription of PRR genes and accumulation of PRR proteins (Qutob et al. 2006; Lohmann et al. 2010). Most of the PRRs are receptor kinases and the PAMPs induce autophosphorylation of the kinase domain of PRRs (Kanzaki et al. 2008; Xiang et al. 2008).

The plasma membrane resident autophosphorylated PRRs are translocated to endosomes and it helps to extend the signaling surface ensuring a robust and efficient cellular signaling system (Geldner and Robatzek 2008). PAMPs induce ubiquitin-proteasome- or clathrin-mediated endocytosis of PRR proteins (Robatzek et al. 2006; Aker and de Vries 2008). PAMP-induced PRR-induced endocytosis has been shown to be dependent on phosphorylation of the PRR (Robatzek et al. 2006). PAMP-induced internalization of PRRs from the plasma membrane is closely correlated with their immune function (Bar et al. 2009; Saijo 2010). The biogenesis of trans-membrane PRRs may occur through the endoplasmic reticulum (ER) with the aid of ER –resident chaperones (Dodds and Rathjen 2010; Popescu 2012). After biosynthesis of PRR in ER, it is transported from the ER to the plasma membrane. *N*-glycosylation of PRRs is required for transport of PRRs from ER to plasma membrane (Häweker et al. 2010). Sustained activation of plasma membrane–resident PRR signaling is important for mounting robust PAMP-triggered immunity (Saijo 2010).

## 1.3 Second Messengers in PAMP Signal Transduction

The plant immune system uses several second messengers to encode information generated by the PAMP/PRR signaling complex and deliver the information downstream of PRRs to proteins which decode/interpret signals and initiate defense gene expression (van Verk et al. 2008; Mersmann et al. 2010; Boudsocq et al. 2010; Hwang and Hwang 2011). It is still not known how the PAMP signals are transmitted downstream of PRR. In plant cells, the calcium ion is a ubiquitous intracellular second messenger involved in numerous signaling pathways (Luan 2009; McAinsh and Pittman 2009: Abdul Kadar and Lindsberg 2010; DeFalco et al. 2010; Hamada et al. 2012; Stael et al. 2012).

Guanosine triphosphate (GTP)-binding proteins (G-proteins) are the regulatory GTPases, which act as molecular switches in signal transduction system (Yalowsky et al. 2010; Zhang et al. 2011, 2012). Two classes of signaling G-proteins, heterotrimeric G-proteins and small monomeric G-proteins (Ras/Ras-like small GTPases), have been reported. In the Ras superfamily of small GTPases, only the Ras and Rho families have been shown to transmit extracellular signals (Gu et al. 2004). Ras superfamily is named the Ras superfamily because the founding members are encoded by human Ras genes initially discovered as cellular homologs of the viral *ras* oncogene. Plants do not possess a true Ras GTPase such as those that are pivotal signaling in animals. Instead, they have a unique subfamily of Rho-family GTPases, called ROPs (Rho-related GTPase of plants). ROP is the sole subfamily of Rho GTPase in plants. ROPs are also referred to as RAC (for Ras [rat sarcoma oncogene product] related C3 botulinum toxin substrate) proteins (Gu et al. 2004; Kiirika et al. 2012). RAC/ROP small GTPases share a common ancestor with Rho, cdc42 and Rac and they are the only Rho-like GTPases in plants (Gu et al. 2004).

$Ca^{2+}$ is a master regulator of gene expression in plants (Galon et al. 2010). Calcium ion acts as a signal carrier (Allen et al. 2000). Calcium signaling is modulated by specific calcium signatures. $Ca^{2+}$ signatures are generated in the cytosol, and in noncytosolic locations including the nucleus and chloroplast through the coordinated action of $Ca^{2+}$ influx and efflux pathways (McAinsh and Pittman 2009). Specific calcium signatures are recognized by different calcium sensors to transduce calcium-mediated signals into downstream events (Reddy et al. 2011; Wang et al. 2012; Hashimoto et al. 2012).

Mitogen-activated protein kinase (MAPK) cascades are major pathways downstream of sensors/receptors that transduce extracellular stimuli into intracellular responses in plants (Hettenhausen et al. 2012; Zhang et al. 2012). A typical MAPK signaling module consists of three interconnected protein kinases: a MAP kinase kinase kinase (MAPKKK or MEKK [for MAPK/Extracellular signal-regulated kinase Kinase Kinase]), a MAP kinase kinase (MAPKK or MKK), and a MAP kinase (MAPK or MPK) (Mészáros et al. 2006). MAP kinase cascade involves sequence of phosphorylation events (Hirt 2000). MAPKs function at the bottom of the three-kinase cascade and are activated by MAPKKs through phosphorylation on the Thr and Tyr residues in their activation motif between the kinase subdomain VII and VIII. The activity of MAPKKs is, in turn, regulated by MAPKKKs via phosphorylation of two Ser/Thr residues in the activation loop of MAPKKs. MAPKKKs receive signals from upstream receptors/sensors (Ichimura et al. 2002; Li et al. 2012).

The oxidative burst involving rapid and transient production of reactive oxygen species (ROS) is a very rapid response, occurring within seconds (Bolwell et al. 1995) or within a few minutes (Arnott and Murphy 1991) of PAMP treatment, suggesting that the oxidative burst may not require *de novo* protein synthesis but involves the activation of pre-existing enzymes. NADPH oxidase (Bae et al. 2006), peroxidases (Halliwell 1978; Lehtonen et al. 2012), and xanthine oxidase (Allan and Fluhr

1997; Ori et al. 1997) have been shown to be involved in triggering ROS production. ROS is a messenger in transmitting the PAMP signal. Nitric oxide (NO) has been identified as a gaseous second messenger (Besson-Bard et al. 2008; Bellin et al. 2013). NO is a diffusible molecular messenger that plays an important role in the plant immune response signal transduction system (Grennan 2007). PAMPs trigger NO burst within minutes in plant cells (Foissner et al. 2000; Lamotte et al. 2004; Tischner et al. 2010). NO acts substantially in cellular signal transduction through stimulus-coupled S-nitrosylation of cysteine residues (Benhar et al. 2008). It serves as a key redox-active signal for the activation of various defense responses (Klessig et al. 2000).

## 1.4 Plant Hormone Signals in Plant Immune Signaling System

The plant hormones salicylic acid (Mukherjee et al. 2010; Dempsey et al. 2011; Liu et al. 2011a, b), jasmonate (Wang et al. 2009; Sheard et al. 2010; Bertoni 2012), ethylene (Boutrot et al. 2010; Laluk et al. 2011; Nie et al. 2011; Nambeesan et al. 2012), abscisic acid (Yazawa et al. 2012), auxin (Fu and Wang 2011), cytokinin (Choi et al. 2011), gibberellins (Qin et al. 2013), and brassinosteroids (De Vleeschauwer et al. 2012) play important role in defense signaling against various pathogens. It has been demonstrated that specific plant hormone signaling pathways should be activated to confer resistance against specific pathogens. JA and SA signaling systems may differentially contribute for resistance against specific pathogens. JA-mediated pathway effectively confers resistance against necrotrophic fungal pathogens (Berrocal-Lobo and Molina 2004; McGrath et al. 2005; Zheng et al. 2006), while SA- mediated pathway confers resistance against biotrophic fungal pathogens and also against virus and bacterial diseases in some plants (Thomma et al. 1998, 2001; Thaler and Bostock 2004; Nie 2006; De Vos et al. 2006; Spoel et al. 2007; Zheng et al. 2006, 2007). Two forms of induced resistance, systemic acquired resistance (SAR) and induced systemic resistance (ISR), have been recognized based on the induction of specific plant hormone signaling systems (Li et al. 2008). SAR refers to a distinct signaling pathway mediated by SA (Oostendorp et al. 2001), while ISR refers to the signaling pathway mediated by JA and ET. SA signaling system activates not only local resistance, but also systemic acquired resistance (SAR) observed in distal (systemic) tissues. SAR is a SA-dependent heightened defense to a broad spectrum of pathogens that is activated throughout a plant following local infection (Liu et al. 2011a). SAR is associated with priming of defense (Kohler et al. 2002; Jung et al. 2009; Luna et al. 2011) and the priming results in a faster and stronger induction of defense mechanisms after pathogen attack (Conrath 2011). The priming can be inherited epigenetically from disease-exposed plants (Pastor et al. 2012) and descendants of primed plants exhibit next-generation systemic acquired resistance (Slaughter et al. 2012; Luna et al. 2011). The transgenerational SAR has been recently reported (Luna et al. 2011). Thus, SA signal is transduced not only within the plant tissues, but also transferred even to the next generations.

Plant hormones activate different signaling pathways inducing distinctly different defense genes (Spoel et al. 2007; Zhang et al. 2007; Mitsuhara et al. 2008). These signaling pathways are not simple linear and isolated cascades, but can crosstalk with each other. Both antagonism and synergism between the signaling systems have been reported. Cross-talk between defense signaling pathways is thought to provide the plant with a powerful regulatory potential, which helps the plant to "decide" which defensive strategy to follow, depending on the type of attacker it is encountering (De Vos et al. 2005). It may also allow pathogens to manipulate plants to their own benefit by shutting down induced defense through influences on the signaling network.

## 1.5 War Between Host Plants and Pathogens and the Winner Is ------- ?

Plant innate immune systems have high potential to fight against viral, bacterial, oomycete, and fungal pathogens and protect the crop plants against wide range of diseases (Knecht et al. 2010; Lacombe et al. 2010; Hwang and Hwang 2011; Alkan et al. 2012). However, potential pathogens produce several effectors to nullify the defense responses induced by the innate immune system (Wu et al. 2011; Akimoto-Tomiyama et al. 2012; Cheng et al. 2012). To avoid or suppress or delay the expression of the defense gene-activating signaling systems, the pathogens secrete several effectors into the host cell (Göhre et al. 2008; Kim et al. 2010; Wu et al. 2011; Cheng et al. 2012). Pathogens may also hijack some signaling systems to cause disease (de Torres-Zabala et al. 2007; Thatcher et al. 2009; El Rahman et al. 2012). It has also been demonstrated that the virulent pathogen may suppress the particular defense signaling system which induce the expression of specific defense genes conferring resistance against the particular pathogen (van Verk et al. 2008; Koornneef and Pieterse 2008; Makandar et al. 2010). Activation of some signaling systems may induce susceptibility, rather than resistance (Atsumi et al. 2009; Yazawa et al. 2012). To overcome antiviral RNA silencing immunity, plant viruses express silencing-suppressor proteins which can counteract the host silencing-based antiviral process (Qu and Morris 2005; Ding and Voinnet 2007; Lewsey et al. 2010).

The war between the plant and pathogen appears to be in fine-tuning the signaling systems to cause disease or enhance host defense. Fast and strong activation of the plant immune responses aids the host plants to win the war against the pathogens (Großkinsky et al. 2011). Overexpression or suppression of some specific signaling systems in the plant immune system has been shown to help the plants to win in the arms race between plants and pathogens (Cheung et al. 2007; Zhang et al. 2008; Hwang and Hwang 2010, 2011; Wu et al. 2010).

Engineering durable nonspecific resistance to phytopathogens is one of the ultimate goals of plant breeding. However, most of the attempts to reach this goal fail as a result of rapid changes in pathogen populations and the sheer diversity of pathogen infection mechanisms. Recently several bioengineering and molecular manipulation technologies have been developed to activate the 'sleeping' plant innate immune system, which has potential to detect and suppress the development of a wide range of plant pathogens in economically important crop plants (Lacombe et al. 2010). Enhancing disease resistance through altered regulation of plant immunity signaling systems would be durable and publicly acceptable (Yamamizo et al. 2006; Shao et al. 2008; Gust et al. 2010; Lacombe et al. 2010). Strategies for activation and improvement of plant immunity aim at enhancing host capacities for recognition of potential pathogens, at boosting the executive arsenal of plant immunity, and interfering with virulence strategies employed by microbial pathogens (Gust et al. 2010). Major advances in our understanding of the molecular basis of plant immunity and of microbial infection strategies have opened new ways for engineering durable resistance in crop plants (Gust et al. 2010; Huffaker et al. 2011). This book describes the most fascinating PAMP-PRR signaling complex and signal transduction systems. It discusses the highly complex networks of signaling pathways involved in transmission of the signals to induce distinctly different defense-related genes to mount offence against different pathogens.

## References

Abdul Kadar M, Lindsberg S (2010) Cytosolic calcium and pH signaling in plants under salinity stress. Plant Signal Behav 5:233–238

Aker J, de Vries SC (2008) Plasma membrane receptor complexes. Plant Physiol 147:1560–1564

Akimoto-Tomiyama C, Furutani A, Tsuge S, Washington EJ, Nishizawa Y, Minami E, Ochiai H (2012) XopR, a type III effector secreted by *Xanthomonas oryzae* pv. *oryzae*, suppresses microbe-associated molecular pattern-triggered immunity in *Arabidopsis thaliana*. Mol Plant Microbe Interact 25:505–514

Ali GS, Prasad KVSK, Day I, Reddy ASN (2007) Ligand-dependent reduction in the membrane mobility of FLAGELLIN SENSITIVE2, an Arabidopsis receptor-like kinase. Plant Cell Physiol 48:1601–1611

Alkan N, Fluhr R, Prusky D (2012) Ammonium secretion during *Colletotrichum coccodes* infection modulates salicylic and jasmonic acid pathways of ripe and unripe tomato fruit. Mol Plant Microbe Interact 25:85–96

Allan AG, Fluhr R (1997) Two distinct sources of elicited reactive oxygen species in tobacco epidermal cells. Plant Cell 9:1559–1572

Allen GJ, Chu SP, Schumacher K, Shimazaki CT, Vafeados D, Kemper A, Hawke SD, Tallman G, Tsien RY, Harper JF, Chory J, Schroeder JI (2000) Alteration of stimulus-specific guard cell calcium oscillations and stomatal closing in *Arabidopsis det3* mutant. Science 289:2338–2342

Altenbach DD, Robatzek S (2007) Pattern recognition receptors: from the cell surface to intracellular dynamics. Mol Plant Microbe Interact 20:1031–1039

Arnott T, Murphy TM (1991) A comparison of the effects of a fungal elicitor and ultraviolet radiation on ion transport and hydrogen peroxide synthesis by rose cells. Environ Exp Bot 31:209–216

Aslam SN, Newman MA, Erbs G, Morissey KL, Chinchilla D, Boller T, Jensen TT, De Castro C, Iearno T, Molinaro A, Jackson RW, Knight MR, Cooper RM (2008) Bacterial polysaccharides suppress induced innate immunity by calcium chelation. Curr Biol 18:1078–1083

Aslam SN, Erbs G, Morrissey KL, Newman M-A, Chinchilla D, Boller T, Molinaro A, Jackson RW, Cooper RM (2009) Microbe-associated molecular pattern (MAMP) signatures, synergy, size, and charge: influences on perception or mobility and host defense responses. Mol Plant Pathol 10:375–387

Atsumi G, Kagaya U, Kitazawa H, Nakahara KS, Uyeda I (2009) Activation of the salicylic acid signaling pathway enhances *Clover yellow vein virus* virulence in susceptible pea cultivars. Mol Plant Microbe Interact 22:166–175

Aziz A, Poinssot B, Daire X, Adrian M, Bezier A, Lambert B, Joubert JM, Pugin A (2003) Laminarin elicits defense responses in grapevine and induces protection against *Botrytis cinerea* and *Plasmopara viticola*. Mol Plant Microbe Interact 16:1118–1128

Bae H, Kim MS, Sicher RC, Bae H-J, Bailey BA (2006) Necrosis- and ethylene-inducing peptide from *Fusarium oxysporum* induces a complex cascade of transcripts associated with signal transduction and cell death in *Arabidopsis*. Plant Physiol 141:1056–1067

Bar M, Sharfman M, Schuster S, Avni A (2009) The coiled-coil domain of EHD2 mediates inhibition of LeEix2 endocytosis and signaling. PLoS One 4:e7963

Bellin D, Asai S, Delledonne M, Yoshioka H (2013) Nitric oxide as a mediator for defense responses. Mol Plant Microbe Interact 26:271–277

Benhar M, Forrester MT, Hess DT, Stamler JS (2008) Regulated protein denitrosylation by cytosolic and mitochondrial thioredoxins. Science 320:1050–1054

Bent AF, Mackey D (2007) Elicitors, effectors, R genes: the new paradigm and a lifetime supply of questions. Annu Rev Phytopathol 45:399–436

Berrocal-Lobo M, Molina A (2004) Ethylene response factor 1 mediates Arabidopsis resistance to the soilborne fungus *Fusarium oxysporum*. Mol Plant Microbe Interact 17:763–770

Bertoni G (2012) Oxylipins and plant palability. Plant Cell 24:1305

Besson-Bard A, Courtois C, Gauthier A, Dahan J, Dobrowolska G, Jeandroz S, Pugin A, Wendehenne D (2008) Nitric oxide in plants: production and cross-talk with $Ca^{2+}$ signaling. Mol Plant 1:218–228

Bi D, Cheng YT, Li X, Zhang Y (2010) Activation of plant immune responses by a gain-of-function mutation in an atypical receptor-like kinase. Plant Physiol 153:1771–1779

Boller T, Felix G (2009) A renaissance of elicitors: perception of microbe-associated molecular patterns and danger signals by pattern-recognition receptors. Annu Rev Plant Biol 60:379–406

Bolwell GP, Butt VS, Davies DR, Zimmerlin A (1995) The origin of the oxidative burst in plants. Free Radic Res 23:517–532

Boudsocq M, Willmann MR, McCormack M, Lee H, Shan L, He P, Bush J, Cheng SH, Sheen J (2010) Differential innate immune signalling via $Ca^{2+}$ sensor protein kinases. Nature 464:418–422

Boutrot F, Segonzac C, Chang KN, Qiao H, Ecker JR, Zipfel C (2010) Direct transcriptional control of the *Arabidopsis* immune receptor FLS2 by the ethylene-dependent transcription factors EIN3 and EIL1. Proc Natl Acad Sci USA 107:14502–14507

Brunner F, Rosahl S, Lee J, Rudd JJ, Geller C, Kauppinen S, Rasmussen G, Scheel D, Nürnberger T (2002) Pep-13, a plant defense-inducing pathogen-associated pattern from *Phytophthora* transglutaminases. EMBO J 21:6681–6688

Cavalcanti FR, Resende MLV, Lima J, Silveira JAG, Oliveira TA (2006) Activities of antioxidant enzymes and photosynthetic responses in tomato pre-treated by plant activators and inoculated by *Xanthomonas vesicatoria*. Physiol Mol Plant Pathol 68:198–208

Cheng B, Yu X, Ma Z, Dong S, Dou D, Wang Y, Zheng X (2012) *Phytophthora sojae* effector Avh331 suppresses the plant defence response by disturbing the MAPK signalling pathway. Physiol Mol Plant Pathol 77:1–9

Cheung M-Y, Zeng N-Y, Tong S-W, Li FW-Y, Zhao K-J, Zhang Q, Sun SS-M, Lam H-M (2007) Expression of a RING-HC protein from rice improves resistance to *Pseudomonas syringae* pv. *tomato* DC3000 in transgenic *Arabidopsis thaliana*. J Exp Bot 58:4147–4159

Choi J, Choi D, Lee R, Ryu CM, Hwang I (2011) Cytokinins and plant immunity: old foes or new friends. Trends Plant Sci 16:388–394

Cloud-Hansen KA, Brook Petersen S, Stabb EV, Goldman WE, McFall-Ngai MJ, Handelsman J (2006) Breaching the great wall: peptidoglycan and microbial interactions. Nat Rev Microbiol 4:710–716

Conrath U (2011) Molecular aspects of defence priming. Trends Plant Sci 16:524–531
D'Ovidio R, Mattei B, Roberti S, Bellincampi D (2004) Polygalacturonases, polygalacturonase-inhibiting proteins and pectic oligomers in plant-pathogen interactions. Biochim Biophys Acta Proteins Proteomics 1696:237–244
de Freitas MB, Stadnik MJ (2012) Race-specific and ulvan-induced defense responses in bean (*Phaseolus vulgaris*) against *Colletotrichum lindemuthianum*. Physiol Mol Plant Pathol 78:8–13
de Torres-Zabala M, Truman W, Bennett MH, Lafforgue G, Mansfield JW, Rodriguez Egea P, Bogre L, Grant M (2007) *Pseudomonas syringae* pv. *tomato* hijacks the *Arabidopsis* abscisic acid signalling pathway to cause disease. EMBO J 26:1434–1443
De Vleeschauwer D, Van Buyten E, Satoh K, Balidion J, Mauleon R, Choi I-R, Vera-Cruz C, Kikuchi S, Höfte M (2012) Brassinosteroids antagonize gibberellin- and salicylate-mediated root immunity in rice. Plant Physiol 158:1833–1846
De Vos M, Van Oosten VR, Van Poecke RMP, Van Pelt JA, Pozo MJ, Mueller MJ, Buchala AJ, Métraux J-P, Van Loon LC, Dicke M, Pieterse CMJ (2005) Signal signature and transcriptome changes of *Arabidopsis* during pathogen and insect attack. Mol Plant Microbe Interact 18:923–937
De Vos M, Van Zaanen W, Koornneef A, Korzelius JP, Dicke M, Van Loon LC, Pieterse CMJ (2006) Herbivore-induced resistance against microbial pathogens in *Arabidopsis*. Plant Physiol 142:352–363
DeFalco TA, Bender KW, Snedden WA (2010) Breaking the code: $Ca^{2+}$ sensors in plant signalling. Biochem J 425:27–40
Dempsey DA, Vlot AC, Wildermuth MC, Klessig DF (2011) Salicylic acid biosynthesis and metabolism. Arabidopsis Book 9:e0156. doi:10.1199/tab.0156
Denoux C, Galletti R, Mammarella N, Gopalan S, Werck D, De Lorenzo G, Ferrari S, Ausubel FM, Dewdney J (2008) Activation of defense response pathways by OGs and Flg22 elicitors in Arabidopsis seedlings. Mol Plant 1:423–445
Ding S-W (2010) RNA-based antiviral immunity. Nat Rev Immunol 10:632–644
Ding S-W, Voinnet O (2007) Antiviral immunity directed by small RNAs. Cell 130:413–426
Dodds PN, Rathjen JP (2010) Plant immunity: towards an integrated view of plant-pathogen interactions. Nat Rev Genet 11:539–548
El Rahman TA, El Oirdi M, Gonzalez-Lamothe R, Bouarab K (2012) Necrotrophic pathogens use salicylic acid signaling pathway to promote disease development in tomato. Mol Plant Microbe Interact 25:1584–1593
Felix G, Duran JD, Volko S, Boller T (1999) Plants have a sensitive perception system for the most conserved domain of bacterial flagellin. Plant J 18:265–276
Foissner L, Wendehenne D, Langebartels C, Durner J (2000) In vivo imaging of an elicitor-induced nitric oxide burst in tobacco. Plant J 23:817–824
Fu J, Wang S (2011) Insights into auxin signaling in plant-pathogen interactions. Front Plant Sci 2:74. doi:10.3389/fpls.2011.0074, Article 2:pp7
Galon Y, Aloni R, Nachmias D, Snir O, Feldmesser E, Scrase-Field S, Boyce JM, Bouché N, Knight MR, Fromm H (2010) Calmodulin-binding transcription activator 1 mediates auxin signaling and responds to stresses in Arabidopsis. Planta 232:165–178
Geldner N, Robatzek S (2008) Plant receptors go endosomal: a moving view on signal transduction. Plant Physiol 147:1565–1574
Göhre V, Spallek T, Häweker H, Mersmann S, Mentzel T, Boller T, de Torres M, Mansfield JW, Robatzek S (2008) Plant pattern-recognition receptor FLS2 is directed for degradation by the bacterial ubiquitin ligase AvrPtoB. Curr Biol 18:1824–1832
Grennan AK (2007) Protein *S*-nitrosylation: protein targets and roles in signal transduction. Plant Physiol 144:1237–1239
Großkinsky DK, Naseem M, Abdelmohsen UR, Plickert N, Engelke T, Griebel T, Zeier J, Novák O, Strnad M, Pfeifhofer H, Graaff EVD, Simon U, Roitsch T (2011) Cytokinins mediate resistance against *Pseudomonas syringae* in tobacco through increased antimicrobial phytoalexin synthesis independent of salicylic acid signaling. Plant Physiol 157:815–830
Gu Y, Wang Z, Yang Z (2004) ROP/RAC GTPase: an old new master regulator for plant signaling. Curr Opin Plant Biol 7:527–536

Gust AA, Biswas R, Lenz HD, Rauhut T, Ranf S, Kemmerling B, Götz F, Glawischnig E, Lee J, Felix G, Nurnberger T (2007) Bacteria-derived peptidoglycans constitute pathogen-associated molecular patterns triggering innate immunity in *Arabidopsis*. J Biol Chem 282:32338–32348

Gust AA, Brunner F, Nürnberger T (2010) Biotechnological concepts for improving plant innate immunity. Curr Opin Biotechnol 21:204–210

Halliwell B (1978) Superoxide-dependent formation of hydroxyl radicals in the presence of iron chelates. Is it a mechanism for hydroxyl radical production in biochemical systems? FEBS Lett 92:321–326

Hamada H, Kurusu T, Okuma E, Nokajima H, Kiyoduka M, Koyano T, Sugiyama Y, Okada K, Koga J, Saji H, Miyao A, Hirochika H, Yamane H, Murata Y, Kuchitsu K (2012) Regulation of a proteinaceous elicitor-induced $Ca^{2+}$ influx and production of phytoalexins by a putative voltage-gated cation channel, OsTPC1, in cultured rice cells. J Biol Chem 287:9931–9939

Hashimoto K, Eckert C, Anschūtz U, Scholz M, Held K, Waadt R, Reyer A, Hippler M, Becker D, Kudla J (2012) Phosphorylation of calcineurin B-like (CBL) calcium sensor proteins by their CBL-interacting protein kinases (CIPKs) is required for full activity of CBL-CIPK complexes toward their target proteins. J Biol Chem 287:7956–7968

Häweker H, Rips S, Koiwa H, Salomon S, Saijo Y, Chinchilla D, Robatzek S, von Schaewen A (2010) Pattern recognition receptors require *N*-glycosylation to mediate plant immunity. J Biol Chem 285:4629–4636

He P, Shan L, Sheen J (2007) Elicitation and suppression of microbe-associated molecular pattern-triggered immunity in plant-microbe interactions. Cell Microbiol 9:1385–1396

Hettenhausen C, Baldwin IT, Wu J (2012) Silencing *MPK4* in *Nicotiana attenuata* enhances photosynthesis and seed production but compromises abscisic acid-induced stomatal closure and guard cell-mediated resistance to *Pseudomonas syringae* pv. *tomato* DC3000. Plant Physiol 158:759–776

Hirt H (2000) Connecting oxidative stress, auxin, and cell cycle regulation through a plant mitogen-activated protein kinase pathway. Proc Natl Acad Sci USA 97:2405–2407

Huffaker A, Dafoe NJ, Schmelz EA (2011) ZmPep1, an ortholog of Arabidopsis elicitor peptide 1, regulates maize innate immunity and enhances disease resistance. Plant Physiol 155:1325–1338

Hwang IS, Hwang BK (2010) The pepper 9-lipoxygenase gene *CaLOX1* functions in defense and cell death responses to microbial pathogens. Plant Physiol 152:948–967

Hwang IS, Hwang BK (2011) The pepper mannose-binding lectin gene *CaMBL1* is required to regulate cell death and defense responses to microbial pathogens. Plant Physiol 155:447–463

Ichimura K, Shinozaki K, Tena G, Sheen J, Henry Y (2002) Mitogen-activated protein kinase cascades in plants: a new nomenclature. Trends Plant Sci 7:301–308

Iriti M, Faoro F (2009) Chitosan as a MAMP, searching for a PRR. Plant Signal Behav 4:66–68

Jeworutzki E, Roelfsema MR, Anschütz U, Krol E, Elzenga JT, Felix G, Boller T, Hedrich R, Becker D (2010) Early signaling through the *Arabidopsis* pattern recognition receptors FLS2 and EFR involves $Ca^{2+}$-associated opening of plasma membrane anion channels. Plant J 62:367–378

Jones JDG, Dangl JL (2006) The plant immune system. Nature 444:323–329

Jung HW, Tschaplinski TJ, Wang L, Glazebrook J, Greenberg JT (2009) Priming in systemic plant immunity. Science 324:89–91

Kanzaki H, Saitoh H, Takahashi Y, Berberich T, Ito A, Kamoun S, Terauchi R (2008) NbLRK1, a lectin-like receptor kinase protein of *Nicotiana benthamiana*, interacts with *Phytophthora infestans* INF1 elicitin and mediates INF1-induced cell death. Planta 228:977–987

Khatib M, Lafitte C, Esquerré-Tugayé M-T, Bottin A, Rickauer M (2004) The CBEL elicitor of *Phytophthora parasitica* var. *nicotianae* activates defence in *Arabidopsis thaliana* via three different signalling pathways. New Phytol 162:501–510

Kiirika LM, Bergmann HF, Schikowsky C, Wimmer D, Korte J, Schmitz U, Niehaus K, Colditz F (2012) Silencing of the Rac1 GTPase *MtROP9* in *Medicago truncatula* stimulates early mycorrhizal and oomycete root colonizations but negatively affects rhizobial infection. Plant Physiol 159:501–516

Kim Y-T, Oh J, Kim K-H, Uhm J-Y, Lee B-M (2010) Isolation and characterization of *NgRLK1*, a receptor-like kinase of *Nicotiana glutinosa* that interacts with the elicitin of *Phytophthora capsici*. Mol Biol Rep 37:717–727

Klessig DF, Durner J, Noad R, Navarre DA, Wendehenne D, Kumar D, Zhou JM, Shah J, Zhang S, Kachroo P, Trifa Y, Pontier D, Lam E, Silva H (2000) Nitric oxide and salicylic acid signalling in plant defense. Proc Natl Acad Sci USA 97:8849–8855

Knecht K, Seyffarth M, Desel C, Thurau T, Sherameti I, Lou B, Oelmüller R, Cai D (2010) Expression of *BvGLP-1* encoding a germin-like protein from sugar beet in *Arabidopsis thaliana* leads to resistance against phytopathogenic fungi. Mol Plant Microbe Interact 23:446–457

Kohler A, Schwindling S, Conrath U (2002) Benzothiadiazole-induced priming for potentiated responses to pathogen infection, wounding, and infiltration of water into leaves requires the *NPR1/NIM1* gene in *Arabidopsis*. Plant Physiol 128:1046–1056

Koornneef A, Pieterse CMJ (2008) Cross talk in defense signaling. Plant Physiol 146:839–844

Lacombe S, Rougon-Cardoso A, Sherwood E, Peeters N, Dahlbeck D, van Esse HP, Smoker M, Rallapalli G, Thomma BPHJ, Stakawicz B, Jones JDG, Zipfel C (2010) Interfamily transfer of a plant pattern-recognition receptor confers broad-spectrum bacterial resistance. Nat Biotechnol 28:365–369

Laluk K, Luo H, Chai M, Dhawan R, Lai Z, Mengiste T (2011) Biochemical and genetic requirements for function of the immune response regulator BOTRYTIS-INDUCED KINASE1 in plant growth, ethylene signaling, and PAMP-triggered immunity in *Arabidopsis*. Plant Cell 23:2831–2849

Lamotte O, Gould K, Lecourieux D, Sequeira-Legrand A, Lebrun-Garcia A, Durner J, Pugin A, Wendehenne D (2004) Analysis of nitric oxide signaling functions in tobacco cells challenged by the elicitor cryptogein. Plant Physiol 135:516–529

Lehtonen NT, Akita M, Frank W, Reski R, Valkonen JPT (2012) Involvement of a class III peroxidase and the mitochondrial protein TSPO in oxidative burst upon treatment of moss plants with a fungal elicitor. Mol Plant Microbe Interact 25:363–371

Lewsey MG, Murphy AM, MacLean D, Dalchau N, Westwood JH, Macaulay K, Bennett M, Moulin M, Hanke DE, Powell G, Smith AG, Carr JP (2010) Disruption of two signaling pathways by a viral RNA silencing suppressor. Mol Plant Microbe Interact 23:835–845

Li YM, Zhang ZK, Jia YT, Shen YM, He HM, Fang RX, Chen XY, Hao XJ (2008) 3-acetonyl-3-hydroxy-oxindole: a new inducer of systemic acquired resistance in plants. Plant Biotechnol J 6:301–308

Li G, Meng X, Wang R, Mao G, Han L, Liu Y, Zhang S (2012) Dual-level regulation of ACC synthase activity by MPK3/MPK6 cascade and its downstream WRKY transcription factor during ethylene induction in Arabidopsis. PLoS Genet 8(6):e1002767. doi:10.1371/journal.pgen.1002767

Liu P-P, von Dahl CC, Klessig DF (2011a) The extent to which methyl salicylate is required for signaling systemic acquired resistance is dependent on exposure to light after infection. Plant Physiol 157:2216–2226

Liu P-P, von Dahl CC, Park S-W, Klessig DF (2011b) Interconnection between methyl salicylate and lipid-based long-distance signaling during the development of systemic acquired resistance in *Arabidopsis* and tobacco. Plant Physiol 155:1762–1768

Lohmann GV, Shimoda Y, Nielsen W, Jørgensen FG, Grossmann C, Sandal N, Sørensen K, Thirup S, Madsen LH, Tabata S, Sato S, Stougaard J, Radutoiu S (2010) Evolution and regulation of the *Lotus japonica* LysM receptor gene family. Mol Plant Microbe Interact 23:510–521

Luan S (2009) The CBL-CIPK network in plant calcium signaling. Trends Plant Sci 14:37–42

Luna E, Bruce TJA, Roberts MR, Flors V, Ton J (2011) Next generation systemic acquired resistance. Plant Physiol 158:844–853

Mackey D, McFall AJ (2006) MAMPs and MIMPs: proposed classification for inducers of innate immunity. Mol Microbiol 61:1365–1371

Makandar R, Nalam V, Chaturvedi R, Jeannotte R, Sparks AA, Shah J (2010) Involvement of salicylate and jasmonate signaling pathways in *Arabidopsis* interaction with *Fusarium graminearum*. Mol Plant Microbe Interact 23:861–870

McAinsh MR, Pittman JK (2009) Shaping the calcium signature. New Phytol 181:275–294
McGrath KC, Dombrecht B, Manners JM, Schenk PM, Edgar CI, Maclean DJ, Scheible W-R, Udvardi MK, Kazan K (2005) Repressor- and activator-type ethylene response factors functioning in jasmonate signaling and disease resistance identified via a genome-wide screen of Arabidopsis transcription factor gene expression. Plant Physiol 139:949–959
Medzhitov R (2001) Toll-like receptors and innate immunity. Nat Rev Immunol 1:135–145
Mersmann S, Bourdais G, Rietz S, Robatzek S (2010) Ethylene signaling regulates accumulation of the FLS2 receptor and is required for the oxidative burst contributing to plant immunity. Plant Physiol 154:391–400
Mészáros T, Helfer A, Hatzimasoura E, Magyar Z, Serazetdinova L, Rios G, Bardóczy V, Teige M, Koncz C, Peck S, Bögre L (2006) The Arabidopsis MAP kinase kinase MKK1 participates in defence responses to the bacterial elicitor flagellin. Plant J 48:485–495
Mitsuhara I, Iwai T, Seo S, Yanagawa Y, Kawahigasi H, Hirose S, Ohkawa Y, Ohashi Y (2008) Characteristic expression of twelve *PR1* family genes in response to pathogen infection, wounding, and defense-related signal compounds. Mol Genet Genomics 279:415–427
Mukherjee M, Larrimore KE, Ahmed NJ, Bedick TS, Barghouthi NT, Traw MB, Barth C (2010) Ascorbic acid deficiency in *Arabidopsis* induces constitutive priming that is dependent on hydrogen peroxide, salicylic acid, and the *NPR1* gene. Mol Plant Microbe Interact 23:340–351
Naito K, Taguchi F, Suzuki T, Inagaki Y, Toyoda K, Shiraishi TT, Ichinose Y (2008) Amino acid sequence of bacterial microbe-associated molecular pattern flg22 is required for virulence. Mol Plant Microbe Interact 21:1165–1174
Nambeesan S, AbuQamar S, Laluk K, Mattoo AK, Mickelhart MV, Ferruzzi MG, Mengiste T, Handa AK (2012) Polyamines attenuate ethylene-mediated defense responses to abrogate resistance to *Botrytis cinerea* in tomato. Plant Physiol 158:1034–1045
Newman M-A, Dow JM, Molinaro A, Parrilli M (2007) Priming, induction and modulation of plant defense responses by bacterial lipopolysaccharides. Plant Immun 13:69–84
Nicaise V, Roux M, Zipfel C (2009) Recent advances in PAMP-triggered immunity against bacteria: pattern recognition receptors watch over and raise the alarm. Plant Physiol 150:1638–1647
Nie X (2006) Salicylic acid suppresses *Potato virus Y* isolate N:O-induced symptoms in tobacco plants. Phytopathology 96:255–263
Nie H, Wu Y, Yao C, Tang D (2011) Suppression of *edr2*-mediated powdery mildew resistance, cell death and ethylene-induced senescence by mutations in *ALD1* in *Arabidopsis*. J Genet Genomics 38:137–148
Nürnberger T, Brunner F (2002) Innate immunity in plants and animals: emerging parallels between the recognition of general elicitors and pathogen-associated molecular patterns. Curr Opin Plant Biol 5:318–324
Nürnberger T, Kűfner I (2011) The role of the plant plasma membrane in microbial sensing and innate immunity: the plant plasma membrane. Plant Cell Monogr 19:471–483
Nürnberger T, Lipka V (2005) Non-host resistance in plants – new insights into an old phenomenon. Mol Plant Pathol 6:335–345
Nürnberger T, Brunner F, Kemmerling B, Piater L (2004) Innate immunity in plants and animals: striking similarities and obvious differences. Immunol Rev 198:249–266
Oostendorp M, Kunz W, Dietrich B, Staub T (2001) Induced disease resistance in plants by chemicals. Eur J Plant Pathol 107:19–28
Ori N, Eshed Y, Paran I, Presting G, Aviv D, Tanksley S, Zamir D, Fluhr R (1997) The *I2C* family from the wilt disease resistance locus *I2* belongs to the nucleotide binding, leucine-rich repeat super family of plant resistance genes. Plant Cell 9:521–532
Pastor V, Luna E, Mauch-Mani B, Ton J, Flors V (2012) Primed plants do not forget. Environ Exp Bot 94:46–56
Petutschnig EK, Jones AME, Serazetdinova L, Lipka U, Lipka V (2010) The lysin motif receptor-like kinase (LysM-RLK) is a major chitin binding protein in *Arabidopsis thaliana* and subject to chitin-induced phosphorylation. J Biol Chem 285:28902–28911

Popescu SC (2012) A model for the biosynthesis and transport of plasma membrane-associated signaling receptors to the cell surface. Front Plant Sci 3:7

Qin X, Liu JH, Zhao WS, Chen XJ, Guo ZJ, Peng YL (2013) Gibberellin 20-oxidase gene *OsGA20ox3* regulates plant stature and disease development in rice. Mol Plant Microbe Interact 26:227–239

Qu F, Morris TJ (2005) Suppressors of RNA silencing encoded by plant viruses and their role in viral infection. FEBS Lett 579:5958–5964

Qutob D, Kemmerling B, Brunner F, Küfner I, Engelhardt S, Gust AA, Luberacki B, Seitz HU, Stahl D, Rauhut T, Glawischnig E, Schween G, Lacombe B, Watanabe N, Lam E, Schlichting R, Scheel D, Nau K, Dodt G, Hubert D, Gijzen M, Nürnberger T (2006) Phytotoxicity and innate immune responses induced by NEP1-like proteins. Plant Cell 18:3721–3744

Reddy ASN, Ali GS, Celesnik H, Day IS (2011) Coping with stresses: roles of calcium- and calcium/calmodulin-regulated gene expression. Plant Cell 23:2010–2032

Robatzek S, Chinchilla D, Boller T (2006) Ligand-induced endocytosis of the pattern recognition receptor FLS2 in *Arabidopsis*. Genes Dev 20:537–542

Saijo Y (2010) ER quality control of immune receptors and regulators in plants. Cell Microbiol 12:716–724

Schulze B, Mentzel T, Jehle A, Mueller K, Beeler S, Boller T, Felix G, Chinchilla D (2010) Rapid heteromerization and phosphorylation of ligand-activated plant transmembrane receptors and their associated kinase BAK1. J Biol Chem 285:9444–9451

Segonzac C, Zipfel C (2011) Activation of plant pattern-recognition receptors by bacteria. Curr Opin Microbiol 14:54–61

Seifert GJ, Blaukopf C (2010) Irritable walls: the plant extracellular matrix and signaling. Plant Physiol 153:467–478

Shao M, Wang J, Dean RA, Lin Y, Gao X, Hu S (2008) Expression of a harpin-encoding gene in rice confers durable nonspecific resistance to *Magnaporthe grisea*. Plant Biotechnol J 6:73–81

Sheard LB, Tan X, Mao H, Withers J, Ben-Nissan G, Hinds TR, Kobayashi Y, Hsu EF, Sharon M, Browse J, He SY, Rizo J, Howe GA, Zheng N (2010) Jasmonate perception by inositol-phosphate-potentiated COI1-JAZ co-receptor. Nature 468:400–405

Shinya T, Galis I, Narisawa T, Sasaki M, Fukuda H, Matsuoka H, Saito M, Matsuoka K (2007) Comprehensive analysis of glucan elicitor-regulated expression in tobacco BY-2 cells reveals a novel MYB transcription factor involved in the regulation of phenylpropanoid metabolism. Plant Cell Physiol 48:1404–1413

Shinya T, Osada T, Desaki Y, Hatamoto M, Yamanaka Y, Hirano H, Takai R, Che F S, Kaku H, Shibuya N (2010) Characterization of receptor proteins using affinity cross-linking with biotinylated ligands. Plant Cell Physiol 51:262–270

Silipo A, Sturiale L, Garozzo D, Erbs G, Tandrup Poulsen T, Lanzetta R, Dow JM, Parrilli M, Newman M-A, Molinaro A (2008) The acylation and phosphorylation pattern of lipid A from *Xanthomonas campestris* strongly influence its ability to trigger the innate immune response in *Arabidopsis*. Chembiochem 9:896–904

Silipo A, Erbs G, Shinya T, Dow JM, Parrilli MT, Lanzetta Shibuya N, Newman M-A, Molinaro A (2010) Glyco-conjugates as elicitors or suppressors of plant innate immunity. Glycobiology 20:406–419

Slaughter A, Daniel X, Flors V, Luna E, Hohn B, Mauch-Mani B (2012) Descendants of primed Arabidopsis plants exhibit resistance to biotic stress. Plant Physiol 158:835–843

Soulie MC, Perino C, Piffeteau A, Choquer M, Malfatti P, Cimerman A, Kunz O, Boccara M, Vidal-Cros A (2006) *Botrytis cinerea* virulence is drastically reduced after disruption of chitin synthesis class III gene (*Bcchs3a*). Cell Microbiol 8:1310–1321

Spoel SH, Johnson JS, Dong X (2007) Regulation of tradeoffs between plant defenses against pathogens with different lifestyles. Proc Natl Acad Sci USA 104:18842–18847

Stael S, Wurzinger B, Mair A, Mehlmer N, Vothknecht UC, Teige M (2012) Plant organellar calcium signaling: an emerging field. J Exp Bot 63:1525–1542

Taguchi F, Yamamoto M, Ohnishi-Kameyama M, Iwaki M, Yoshida M, Ishii T, Ichinose Y (2010) Defects in flagellin glycosylation affect the virulence of *Pseudomonas syringae* pv. *tabaci* 6605. Microbiology 156(Pt 1):72–80

Takakura Y, Ishida Y, Inoue Y, Tsutsumi F, Kuwata S (2004) Induction of a hypersensitive response-like reaction by powdery mildew in transgenic tobacco expressing harpin$_{Pss}$. Physiol Mol Plant Pathol 64:83–89

Tellström V, Usadel B, Thimm O, Stitt M, Küster H, Niehaus K (2007) The lipopolysaccharide of *Sinorhizobium meliloti* suppresses defense-associated gene expression in cell cultures of the host plant *Medicago truncatula*. Plant Physiol 143:825–837

Thaler JS, Bostock RM (2004) Interactions between abscisic acid-mediated responses and plant resistance to pathogens and insects. Ecology 85:48–58

Thatcher LF, Manners JM, Kazan K (2009) *Fusarium oxysporum* hijacks COI1-mediated jasmonate signaling to promote disease development in *Arabidopsis*. Plant J 58:927–939

Thomma BPHJ, Eggermont K, Penninckx IAMA, Mauch-Mani B, Vogelsang R, Cammue BPA, Broekaert WF (1998) Separate jasmonate-dependent and salicylate-dependent defense-response pathways in Arabidopsis are essential for resistance to distinct microbial pathogens. Proc Natl Acad Sci USA 95:15107–15111

Thomma B, Penninckx I, Broekaert WF, Cammue BPA (2001) The complexity of disease signaling in *Arabidopsis*. Curr Opin Immunol 13:63–68

Thomma BPHJ, Nürnberger T, Joosten MHAJ (2011) Of PAMPs and effectors: the blurred PTI-ETI dichotomy. Plant Cell 23:4–15

Tischner R, Koltermann M, Haesse H, Plath M (2010) Early responses of *Arabidopsis thaliana* to infection by *Verticillium longisporum*. Physiol Mol Plant Pathol 74:419–427

Tsuda K, Katagiri P (2010) Comparing signaling mechanisms engaged in pattern-triggered and effector-triggered immunity. Curr Opin Plant Biol 13:459–465

van Verk MC, Pappaaioannou D, Neeleman L, Bol JF, Linthorst HJM (2008) A novel WRKY transcription factor is required for induction of *PR-1a* gene expression by salicylic acid and bacterial elicitors. Plant Physiol 140:1983–1995

Vidhyasekaran P (2007) Fungal pathogenesis in plants and crops: molecular biology and host defense mechanisms, IIth edn. CRC Press/Taylor & Francis Group, Boca Raton, p 510

Viterbo A, Wiest A, Brotman Y, Chet I, Kenerley C (2007) The 18mer peptaibols from *Trichoderma virens* elicit plant defense responses. Mol Plant Pathol 8:737–746

Wan J, Zhang X-C, Neece D, Ramonell KM, Clough S, Kim S-Y, Stacey MG, Stacey G (2008) A LysM receptor-like kinase plays a critical role in chitin signaling and fungal resistance in *Arabidopsis*. Plant Cell 20:471–481

Wang Z, Mao H, Dong C, Ji R, Cai L, Fu H, Liu S (2009) Overexpression of *Brassica napus* MPK4 enhances resistance to *Sclerotinia sclerotiorum* in oilseed rape. Mol Plant Microbe Interact 22:235–244

Wang Z, Tan X, Zhang Z, Gu S, Li G, Shi H (2012) Defense to *Sclerotinia sclerotiorum* in oilseed rape is associated with the sequential activations of salicylic acid signaling and jasmonic acid signaling. Plant Sci 184:75–82

Wu T, Guo A, Zhao Y, Wang X, Wang Y, Zhao D, Li X, Ren H, Dong H (2010) Ectopic expression of the rice lumazine synthase gene contributes to defense responses in transgenic tobacco. Phytopathology 100:573–581

Wu S, Lu D, Kabbage M, Wei H-L, Swingle B, Records AR, Dickman M, He P, Shan L (2011) Bacterial effector HopF2 suppresses *Arabidopsis* innate immunity at the plasma membrane. Mol Plant Microbe Interact 24:585–593

Xiang T, Zong N, Zou Y, Wu Y, Zhang J, Xing W, Li Y, Tang X, Zhu L, Chai J, Zhou J-M (2008) *Pseudomonas syringae* effector AvrPto blocks innate immunity by targeting receptor kinases. Curr Biol 18:74–80

Yalowsky S, Baluska F, Jones A (2010) Integrated G proteins signaling in plants. Springer, Heidelberg

Yamamizo C, Kuchimura K, Kobayashi A, Katou S, Kawakita K, Jones JDG, Doke N, Yoshioka H (2006) Rewiring mitogen-activated protein kinase cascade by positive feedback confers potato blight resistance. Plant Physiol 140:681–692

Yazawa K, Jiang C-J, Kojima M, Sakakibara H, Takatsuji H (2012) Reduction of abscisic acid signaling in rice during the early phase of *Magnaporthe oryzae* infection decreases its susceptibility to the fungus. Physiol Mol Plant Pathol 78:1–7

Zamioudis C, Peterse CMJ (2012) Modulation of host immunity by beneficial microbes. Mol Plant Microbe Interact 25:139–150

Zhang J, Zhou J-M (2010) Plant immunity triggered by microbial molecular signatures. Mol Plant 3:783–793

Zhang J, Shao F, Li Y, Cui H, Chen L, Li H, Zou Y, Long C, Lan L, Chai J, Chen S, Tang X, Zhou J-M (2007) A *Pseudomonas syringae* effector inactivates MAPKs to suppress PAMP-induced immunity in plants. Cell Host Microbe 1:175–185

Zhang W, Chen J, Zhang H, Song F (2008) Overexpression of a rice diacylglycerol kinase gene *OsBIDK1* enhances disease resistance in transgenic tobacco. Mol Cells 26:258–264

Zhang L, Xi D, Luo L, Meng F, Li Y, Wu CA, Guo X (2011) Cotton GhMPK2 is involved in multiple signaling pathways and mediates defense responses to pathogen infection and oxidative stress. FEBS J 278:1267–1378

Zhang H, Gao Z, Zheng X, Zhang Z (2012) The role of G-proteins in plant immunity. Plant Signal Behav 7:1284–1288

Zheng Z, Qamar SA, Chen Z, Mengiste T (2006) Arabidopsis WRKY33 transcription factor is required for resistance to necrotrophic fungal pathogens. Plant J 48:596–605

Zheng Z, Mosher SL, Fan B, Klessig DF, Chen Z (2007) Functional analysis of Arabidopsis WRKY25 transcription factor in plant defense against *Pseudomonas syringae*. BMC Plant Biol 7:2

Zipfel C (2009) Early molecular events in PAMP-triggered immunity. Curr Opin Plant Biol 12:414–420

Zipfel C, Felix G (2005) Plants and animals: a different taste for microbes? Curr Opin Plant Biol 8:353–360

# Chapter 2
# PAMP Signaling in Plant Innate Immunity

**Abstract** Plant innate immunity is a potential basal defense system existing in plant kingdom. This system provides powerful weapons to the host plants to fight against viral, bacterial, fungal, and oomycete pathogens and serves as a surveillance system against invasion of pathogens. It is not active in normal healthy plants and it requires specific signals to get activated. Pathogen-associated molecular patterns (PAMPs) act as alarm/danger signals to trigger the plant innate immune responses. When pathogens land on the plant's surface, plants read the molecular fingerprints/signatures of pathogens (PAMPs) by binding the PAMPs with cognate pattern-recognition receptors (PRRs) residing in plant cell plasma membrane and trigger several defense signaling systems. Pathogens contain a wide array of PAMPs of diverse chemical structures and every pathogen contains or secretes multiple PAMPs. Each PAMP may regulate induction of different defense genes. The time of induction, intensity of induction, and duration of induction of the defense signals may vary depending on PAMPs. Amount of PAMP available in the plant-pathogen interaction site may determine the intensity of induced gene expression. Each PAMP may regulate distinctly different signaling pathway(s). Sometimes different PAMPs may induce the same signaling system, but the intensity of the defense signaling gene expression may differ. The same PAMP may behave differently in different plant system. A single PAMP may not be able to activate all the defense signaling-related genes and several PAMPs may be required to activate the complex signaling systems. PAMPs may act synergistically or antagonistically in inducing defense signaling. Some PAMPs have additive effect, while others show antagonistic effect between them. The PAMPs are perceived as danger signals by PRRs and the PAMP-PRR complex activates the plant innate immunity. PAMPs trigger phosphorylation of PRRs. Fine control of membrane-resident PRR activity is essentially achieved by a combination of proper endoplasmic reticulum (ER) folding, degradation and trafficking of PRRs. Strict elimination of the misfolded PRR occurs in the absence of the identified ER folding machineries, which would avoid precocious immune activation. Pre-recognition membrane traffic of PRRs from the ER to their functional sites, together with post-recognition internalization

P. Vidhyasekaran, *PAMP Signals in Plant Innate Immunity: Signal Perception and Transduction*, Signaling and Communication in Plants 21,
DOI 10.1007/978-94-007-7426-1_2, 

is crucial for PRR function. The signals generated by PAMPs are perceived by PRRs and several second messengers are involved in transmission of the signals downstream of the PRRs. Highly complex networks of signaling pathways are activated by the PAMP-PRR signaling system.

**Keywords** Innate immunity • PAMPs • PRRs • PAMP-PRR signaling complex • Second messengers • Trafficking of PRRs

## 2.1 Classical PAMPs as Alarm Signals

Plants are endowed with innate immune system, which acts as a surveillance system against possible attack by pathogens (Vidhyasekaran 2004, 2007a; Akira et al. 2006; Chisholm et al. 2006; Boller and He 2009; Chen et al. 2010c; Dodds and Rathjen 2010; Tsuda and Katagiri 2010; Silipo et al. 2010; Drutskaya et al. 2011; Zamioudis and Peterse 2012). Discrimination between self and non-self is a fundamental ability of the innate immune systems (Sanabria et al. 2008; Takken and Tameling 2009; Ronald and Beutler 2010; Saijo 2010; Segonzac and Zipfel 2011; Dubery et al. 2012). The immune system is activated on perception of the pathogen-associated molecular patterns (PAMP; the pathogen's signature) of invading pathogens (Dodds and Rathjen 2010; Keinath et al. 2010; Park et al. 2010a, b; Ronald and Beutler 2010; Shimizu et al. 2010; Boutrot et al. 2010; Nürnberger and Kufner 2011; Segonzac and Zipfel 2011). PAMPs are the new name given to the classical general elicitors identified in various plant pathogens (Jones and Dangl 2006; Bent and Mackey 2007). The classical general elicitors reported in plant pathogens resemble the PAMPs reported in mammals (Nürnberger and Brunner 2002; Nürnberger et al. 2004; Nürnberger and Lipka 2005; Zipfel and Felix 2005; Jones and Dangl 2006). PAMPs are evolutionarily conserved building blocks of microbial surfaces that directly bind to the PRRs and induce defense responses (Nürnberger and Brunner 2002; Mackey and McFall 2006; Qutob et al. 2006; Gust et al. 2007; Nicaise et al. 2009; Tsuda and Katagiri 2010; Zhang and Zhou 2010; Thomma et al. 2011). The PAMPs are vital for microbial survival (Gust et al. 2007; Zhang and Zhou 2010). The molecular signatures in PAMPs are not present in the host (Mackey and McFall 2006). Several PAMPs have been detected in fungal, oomycete, bacterial and viral plant pathogens (Felix et al. 1999; Medzhitov 2001; Brunner et al. 2002; Boller and Felix 2009; Silipo et al. 2010; Ding 2010; Tsuda and Katagiri 2010; Thomma et al. 2011; Nürnberger and Kufner 2011).

The informations generated by the PAMP alarm/danger signals are perceived by the PRRs (Nürnberger and Brunner 2002; Qutob et al. 2006; Nicaise et al. 2009; Petutschnig et al. 2010; Shinya et al. 2010; Schulze et al. 2010; Tsuda and Katagiri 2010; Segonzac and Zipfel 2011; Thomma et al. 2011). PAMPs are actually defined as the molecules, which bind to plant PRRs and induce plant immune responses (Nicaise et al. 2009; Tsuda and Katagiri 2010). PAMPs are detected not only in pathogens, but also in saprophytes, probably in all microbes. Hence the PAMPs are also called as microbe-associated molecular patterns (MAMPs) (Bittel and Robatzek 2007; Viterbo et al. 2007; Denoux et al. 2008; Aslam et al. 2009; Boller and Felix 2009; Erbs and Newman 2009; Jeworutzki et al. 2010; Thomma et al. 2011).

The plant immune system uses several second messengers to encode information generated by the PAMPs and deliver the information downstream of PRRs to proteins which decode/interpret signals and initiate defense gene expression (Snedden and Fromm 2001; Lecourieux et al. 2006; van Verk et al. 2008; Mersmann et al. 2010; Boudsocq et al. 2010; Hwang and Hwang 2011). Highly complex networks of signaling pathways are involved in transmission of the signals to induce distinctly different defense-related genes to mount offence against different biotrophic, hemibiotrophic, and necrotrophic pathogens (Zheng et al. 2007; Koornneef and Pieterse 2008; Gaige et al. 2010; Gfeller et al. 2010; Leon-Reyes et al. 2010; Perchepied et al. 2010; Katagiri and Tsuda 2010; Ahmad et al. 2011; Choi and Hwang 2011; Fernández-Calvo et al. 2011; Kobeasy et al. 2011; Zhu et al. 2011; Alkan et al. 2012; Cheng et al. 2012).

Several studies have provided evidences that plant innate immune systems have high potential to fight against viral, bacterial, oomycete, and fungal pathogens and protect the crop plants against wide range of diseases (Mandal et al. 2008; Zipfel 2008; Pitzschke et al. 2009a, b; Véronési et al. 2008; D'Amelio et al. 2011; Knecht et al. 2010; Lacombe et al. 2010; Molloy 2010; Hwang and Hwang 2011; Alkan et al. 2012). Potential pathogens contain several PAMPs and they serve as alarm signals to activate the plant innate immunity.

## 2.2 Effector-Like PAMPs

Pathogens have additional pathogen-only molecules called effectors, besides PAMPs (Kwon 2010). Effectors are pathogen molecules that manipulate host cell structure and function thereby facilitating infection and/or triggering defense responses (Hogenhout et al. 2009). Effectors are double-edged swords; they enhance virulence of pathogens and also trigger resistance in plants carrying cognate defense receptors (Zong et al. 2008).

Effectors induce susceptibility, mostly by suppressing PAMP-induced immune responses. The effector proteins target basic innate immunity in plants (Block et al. 2008; Bartetzko et al. 2009; Boller and He 2009; Song and Yang 2010; Szczesny et al. 2010). The effector AvrBsT from *Xanthomonas campestris* pv. *vesicatoria* has been identified as a suppressor of specific plant defense in pepper plants (Szczesny et al. 2010). The effector $XopZ_{Pxo99}$ from the rice bacterial blight pathogen interferes with host innate immunity during the pathogen infection process in rice (Song and Yang 2010). The *Pseudomonas syringae* pv. *tomato* DC3000 effector HopF2 intercepts PAMP signaling at the plasma membrane immediately of signal perception. It acts as a potent suppressor of early immune-response gene transcription and mitogen-activated protein kinase signaling activated by multiple PAMPs (Wu et al. 2011). LysM domain-containing effector protein Ecp6 of the fungal plant pathogen *Cladosporium fulvum* mediates virulence through perturbation of chitin-triggered host immunity. During infection, Ecp6 sequesters chitin oligosaccharides that are released from the cell walls of invading hyphae to prevent elicitation of host immunity. Since LysM effectors are widely conserved in the fungal kingdom, this type of effector action may be a common strategy of fungal pathogens in suppressing host immune systems (de Jonge et al. 2010).

The PWL proteins are the effectors detected in the rice blast pathogen *Magnaporthe oryzae* (Thomma et al. 2011) and the effectors were shown to have virulence function and suppress host defense responses (Khang et al. 2010; Valent and Khang 2010). The effector ATR13 of the downy mildew pathogen *Hyaloperonospora arabidopsidis* suppresses callose deposition triggered by the bacterial pathogen *P. syringae* (Sohn et al. 2007). Circumvention of innate immunity is crucial for pathogenesis in plants and effectors play important role in suppression of plant immunity (Block et al. 2008; Schornack et al. 2009).

While PAMPs are essential for microbial fitness and survival, effectors specifically contribute to virulence by targeting host plant innate immunity (Thomma et al. 2011). However, this type of differentiation of effectors from PAMPs is only blurred one, as most of the effectors are also known to trigger innate immunity (Thomma et al. 2011; Gassmann and Bhattacharjee 2012). The effector AvrPto from *P. syringae* pv. *tomato* inhibits immune responses in *Arabidopsis* but triggers immune responses in some tomato plants carrying resistance proteins Pto, a serine/threonine kinase, and Prf, a nucleotide-binding leucine-rich repeat (LRR) protein (Zong et al. 2008). Plant innate immunity triggered by the effectors is called "effectors-triggered immunity (ETI)" (Nürnberger and Kemmerling 2009; Gassmann and Bhattacharjee 2012), whereas the immunity triggered by PAMP is called "PAMP-triggered immunity (PTI)" (Tsuda et al. 2009; Tsuda and Katagiri 2010; Thomma et al. 2011). It is difficult to distinguish between PAMPs and effectors based on virulence and elicitor functions as both of them have the dual virulence and elicitor functions (Cloud-Hansen et al. 2006; Soulie et al. 2006; Newman et al. 2007; Naito et al. 2008; Taguchi et al. 2010; Thomma et al. 2011). The bacterial PAMPs flagellin (Naito et al. 2008; Taguchi et al. 2010), lipopolysaccharide (LPS) (Newman et al. 2007), and peptidoglycan (Cloud-Hansen et al. 2006) and the fungal PAMP chitin (Soulie et al. 2006) have been reported to have a function in virulence of the pathogens.

Other differences between PAMPs and effectors include the wide occurrence of PAMPs in microbes as against narrow occurrence of effectors in specific pathogens. PAMPs are generally considered as molecules which are widely conserved across genera of microbes, while effectors are specific to single or a few related species of pathogens (Jones and Dangl 2006; Erbs et al. 2008). However, it is now known that several groups of effector proteins are also widespread (Thomma et al. 2011). LysM effectors widely occur in the fungal kingdom (Bolton et al. 2008; de Jonge and Thomma 2009; Thomma et al. 2011). The effectors harpins are produced by several Gram-negative bacteria (Tampakaki et al. 2010). The Nep1 – like proteins are the effectors that are conserved among oomycetes, fungi, and bacteria (Gijzen and Nürnberger 2006; Kamoun 2006). In contrast with groups of widely conserved effectors, some PAMPs are only narrowly conserved (Lee et al. 2009). The PAMP AxYS22 has been detected only in a few *Xanthomonas* species (Lee et al. 2009). The PAMP Pep-13 is conserved only in *Phytophthora* species (Brunner et al. 2002).

These observations suggest that it may be difficult to distinguish effectors from PAMPs based on their elicitor function or wide/narrow occurrence.

Another major difference between PAMPs and effectors is in induction of hypersensitive response (HR). Effectors often induce HR (Wei et al. 1992; Ron and Avni 2004; Zhang and Zhou 2010), while the PAMPs do not (Mishina and Zeier 2007; Thomma et al. 2011). However, some of the PAMPs do induce HR. It has been demonstrated that the PAMP flg22 induces an HR in *Arabidopsis*, rice, and tobacco (Naito et al. 2008; Taguchi et al. 2003; Hann and Rathjen 2007). CBEL, the glycoprotein PAMP from *Phytophthora parasitica var. nicotianae* induces HR in tobacco and *Arabidopsis* (Khatib et al. 2004).

To call an elicitor as a PAMP, it should be present in a broad range of microbial species but not in plants, directly bind to plant pattern recognition receptor, and activate defense responses (Nicaise et al. 2009; Tsuda and Katagiri 2010). Based on this definition of PAMPs, several elicitors previously classified as effectors may have to be reclassified as PAMPs. These include harpins produced by bacterial pathogens (Tampakaki et al. 2010; Boureau et al. 2011), Nep1-like proteins from bacterial, fungal, and oomycete pathogens (Gijzen and Nürnberger 2006; Kamoun 2006), crinkers produced by oomycete pathogens (Haas et al. 2009), avrXa21 detected in *Xanthomonas* species (Lee et al. 2009), Avr4 protein from *Cladosporium fulvum* (Sterigopoulos et al. 2010; Thomma et al. 2011), and Ecp2 from different fungal pathogens (Sterigopoulos et al. 2010). All these effector proteins have been now reclassified as PAMPs (Thomma et al. 2011).

## 2.3 PAMPs Found Within Effectors

Some of the PAMPs may be contained within effectors. The fungal effector ethylene inducing xylanase (EIX) is an important factor for the success of *Trichoderma viride* as an invasive pathogen (Rotblat et al. 2002). EIX is not recognized by its enzyme activity as an elicitor. Instead a PAMP composed of five amino acids of a surface-exposed β- strand of EIX is essential for its defense response triggering activity (Rotblat et al. 2002). EIX is a fungal effector that contains a PAMP that is recognized by PAMP-receptors (Mackey and McFall 2006). The fungal effector 'AvrPita' from the rice blast pathogen *Magnaporthe oryzae* contains a PAMP that interacts directly with the LRRs of the rice R-protein, Pi-ta (Jia et al. 2000). AvrL567 effectors from the flax rust fungus *Melampsora lini* contain a PAMP (Dodds et al. 2006). A PAMP is also found within a cell wall transglutaminase from the oomycete *Phytophthora sojae* (Brunner et al. 2002). A mutation was identified that disrupted the defense eliciting activity without altering the enzymatic activity of the protein. A 13-aminoacid peptide, called Pep-13, contains the PAMP from within the transglutaminase (Brunner et al. 2002). A bacterial effector from *Ralstonia solanacearum*, POP2, contains a PAMP that directly interacts with the host R-proteins RRS1-RRS1-S (Deslandes et al. 2003).

The RXLR effector AVR3a of *Phytophthora infestans* has dual function; it activates immune system and also suppresses the defense responses induced by the PAMP

INF1 secreted by the pathogen *P. infestans* in potato (Bos et al. 2009, 2010). Mutagenesis of the conserved C-terminal tyrosine residue at position 147 in AVR3a retained the ability to trigger defense, but lost the ability to suppress the defense responses. The results suggest that tyrosine residue at position 147 in the AVR3a is the critical factor in suppression of defense responses. Distinct amino acids appear to condition the two effector activities (induction and suppression of defense responses) in AVR3a (Bos et al. 2009). In the effector protein AvrPtoB of *P. syringae* pv. *tomato*, the two activities (induction and suppression of plant immunity) have been shown to be carried out by distinct domains within the protein (Abramovitch et al. 2003; Janjusevic et al. 2006).

## 2.4 Toxins Acting as PAMPs

Microbial toxins constitute a separate class of elicitors produced by oomycete, fungal, and bacterial pathogens. These cytolytic compounds function as key virulence determinants of pathogens (van't Slot and Knogge 2002; Glazebrook 2005; Gijzen and Nurnberger 2006) and these toxins can be called effectors. The same toxic compounds also function as PAMPs by acting as nonself recognition determinants for the activation of plant innate immune responses (Gijzen and Nürnberger 2006).

Nep1 (for Necrosis and ethylene-inducing peptide1) from *Fusarium oxysporum* f. sp. *erythroxyli* and Nep1-like proteins (NLPs) detected in several oomycetes, fungi, and bacteria are host nonselective toxins (Mattinen et al. 2004; Pemberton and Salmond 2004; Qutob et al. 2006; Staats et al. 2007; Kufner et al. 2009; Cabral et al. 2012). The NLPs exert cytolytic activity that causes cell maceration and cell death in dicotyledonous plants in a manner that is similar to disease symptom development (Kufner et al. 2009). NLPs act as positive virulence factors (effectors) during infection of plants (Mattinen et al. 2004; Pemberton et al. 2005; Ottmann et al. 2009). The NLPs can also activate defense-related responses (Bae et al. 2006; Qutob et al. 2006; Ottmann et al. 2009). The NLPs trigger immune responses similar to that of classic PAMPs. They mediate the activation of MAPKs, induction of ion fluxes, production of reactive oxygen species, induction of defense-related genes, production of phytoalexins and callose deposition. These responses resemble to a great extent those triggered by classical PAMPs (Kufner et al. 2009). The NLPs act like PAMPs in many instances, as they are detected in a wide range of pathogens and not in plants and recognize nonself triggering defense responses (Qutob et al. 2006). However, NLPs differ from classical PAMPs in that the elicitor-active minimal motif has not been detected in NLPs and the NLPs are transiently expressed proteins (Qutob et al. 2006; Kufner et al. 2009; Ottmann et al. 2009).

The maize pathogen *Fusarium moniliforme* produces a phytotoxin, fumonisin (FB1) that elicits cytolysis of plant cells (Gilchrist et al. 1995). FBI also triggers accumulation of reactive oxygen species (ROS), deposition of callose, phytoalexin synthesis, and defense-related genes (Asai et al. 2000). *Fusarium graminearum* and *F. culmorum* produce trichothecene family phytotoxins (Nishiuchi et al. 2006). The type B trichothecene, deoxynivalenol (DON), is considered as virulence factor in infection of plants (Bai et al. 2002). The toxin also acts as an elicitor and triggers generation of hydrogen peroxide, deposition of callose, accumulation of salicylic acid, activation of mitogen-activated protein kinases, and expression of *PR-1* and *PR-2* genes (Nishiuchi et al. 2006). Another plant pathogen *Alternaria alternata* f. sp. *lycopersici* produces a phytotoxin, AAL toxin. The AAL toxin triggers cytolysis and also triggers expression of defense genes (Gechev et al. 2004). The host-selective toxin victorin produced by *Cochliobolus victoriae*, the maize victoria blight pathogen, induces defense-related responses such as extracellular alkalinization, generation of ROS and nitric oxide (NO), and production of phytoalexin (Tada et al. 2005). Collectively these studies suggest that the cytolytic toxins play dual roles in plant-pathogen interactions as virulence determinants (effectors) and as nonself recognition determinants (PAMPs) for the activation of plant innate immune responses (Qutob et al. 2006).

## 2.5 PAMP-Induced HAMPs (DAMPs/MIMPs/PAMP Amplifiers/Endogenous Elicitors)

Polygalacturonases and cellulases are produced by a wide range of pathogens and they act as effectors and also function as general elicitors (Rotblat et al. 2002; Boudart et al. 2003; Poinssot et al. 2003). During host-pathogen interaction, many pathogens secrete these cell-wall-degrading enzymes (Vidal et al. 1998; Furman-Matarasso et al. 1999; Boudart et al. 2003; Poinssot et al. 2003). These enzymes can themselves function as elicitors (Rotblat et al. 2002; Poinssot et al. 2003), but their enzymatic products are also known to be general elicitors of plant defense responses (Shibuya and Minami 2001). These enzymes degrade the plant cell wall structure and some of the degradation products such as pectin-derived oligogalacturonides (OGs) and cellodextrins act as potent elicitors of innate immunity. These host-derived elicitors function almost in the same fashion as the PAMPs function in plant innate immunity.

The host-derived elicitors are called by different names by different authors. They are called host-associated molecular patterns (HAMPs) as they are of host origin (Galletti et al. 2009), damage-associated molecular patterns (DAMPs) as they are also induced by cellular damage (Zipfel 2009), or endogenous/internal elicitors (Ryan et al. 2007; Huffaker et al. 2011). It was also suggested that these plant cell-wall degradation products can be called microbe-induced molecular patterns (MIMPs) recognized through receptors as 'pathogen-induced modified self' (Mackey and McFall 2006; Aziz et al. 2007). It may be better to call the MIMPs as pathogen-induced molecular patterns (PIMPs) as these oligosaccharides are formed during pathogenesis by the production of cell-wall-degrading enzymes (Vidal et al. 1998; Furman-Matarasso et al. 1999; Boudart et al. 2003; Poinssot et al. 2003). Most of the endogenous elicitors have the property of inducing expression of their own genes to initiate a feedback mechanism to the

**Table 2.1** PAMPs/pathogens induced expression of the HAMP/endogenous elicitor PROPEP2 in *Arabidopsis*

| PAMPs/pathogens | Fold change in *PROPEP2* gene expression |
|---|---|
| PAMP flg22 | 21.8 |
| PAMP HrpZ | 40.8 |
| PAMP NPP1 | 26.9 |
| Fungal pathogen *Botrytis cinerea* | 27.9 |
| Oomycete pathogen *Phytophthora infestans* | 31.2 |
| Bacterial pathogen *Pseudomonas syringae* | 3.2 |

Adapted from Huffaker et al. (2006)

original PAMP signals and therefore they can also be called "PAMP amplifiers" (Huffaker and Ryan 2007).

Several PAMPs including Flg22, NPP1, and HrpZ and pathogens including *Botrytis cinerea*, *Phytophthora infestans*, and *Pseudomonas syringae* trigger expression of *PROPEP2* gene encoding PROPEP2, the precursor for the endogenous elicitor AtPep2 (Table 2.1; Huffaker et al. 2006). Similar induced expression of genes encoding PROPEP3 and PROPEP1 due to PAMP and pathogens treatment has also been reported (Huffaker et al. 2006). Treatment with the bacterial PAMP flg22 upregulates the transcription of genes encoding PROPEP family precursors for the endogenous elicitors AtPeps and PEPR receptors (Zipfel et al. 2004; Ryan et al. 2007). These endogenous elicitors are also induced by fungal pathogen infection (Huffaker et al. 2011).

*At*Pep family elicitors and the classical PAMPs activate similar downstream responses using many of the same molecular components (Ryan et al. 2007; Krol et al. 2010; Postel et al. 2010; Qi et al. 2010; Yamaguchi et al. 2010; Huffaker et al. 2011). Both the PAMP flg22 and the endogenous elicitors AtPeps bind specific LRR receptors and both activate the same downstream signaling events (Yamaguchi et al. 2006; Huffaker and Ryan 2007; Krol et al. 2010). The endogenous elicitor AtPep1 treatment induces the transcription of FLS2, the PRR for the PAMP flg22 (Ryan et al. 2007). The receptors for both flg22 and AtPep1 associate with the interacting receptor partner, BAK1 (Ma et al. 2009; Postel et al. 2010). Collectively these studies suggest that the endogenous elicitors are functionally similar to classical PAMPs and may act as amplifiers of PAMP-induced signals.

## 2.6 Bacterial PAMPs

### *2.6.1 PAMPs from Various Bacterial Structures*

Several PAMPs have been isolated and characterized from wide range of bacterial pathogens. These PAMPs have been purified from bacterial flagella structure and bacterial cell envelope lipopolysaccharides and peptidoglycans. PAMPs have also

**Table 2.2** PAMPs detected in bacterial pathogens

| Bacterial PAMPs | Elicitor-active domain(s) |
|---|---|
| Flagellin | flg22, flg15 |
| Lipopolysaccharides | Lipid A part of lipopolysaccharides, O-antigen polysaccharides |
| Peptidoglycans | Muropeptide |
| Elongation factors | elf18 |
| Cold shock proteins | CSP22 |
| Harpins | C-terminal fragment |
| Ax21 | Tyrosine-sulfated protein |
| Bacterial DNA | Nonmethylated CpG sequence |
| Rhamnolipids | – |

been detected in several bacterial proteins including elongation factor proteins, cold-shock proteins, harpin proteins, sulfated proteins, and the bacterial superoxide dismutase enzyme. PAMPs have also been detected in bacterial rhamnolipids and bacterial DNA structure (Table 2.2).

## 2.6.2 *PAMPs Detected in Flagella*

### 2.6.2.1 Flagellin Proteins May Contain Several Distinct PAMPs

Bacterial flagella have been found to be potential sources for the PAMPs, which trigger innate immunity in both plants and animals. The bacterial extracellular flagella are involved in motility of bacteria. The structural components of flagella include a basal body capable of rotary motion, a hook apparatus, and thousands of flagellin monomers that polymerize to form the flagellar filament (Schuster and Khan 1994). Flagellin is the structural protein that forms the major portion of flagellar filaments. The protein flagellin contains PAMPs that can be recognized by some plants, leading to activation of defense responses (Felix et al. 1999; Che et al. 2000; Gómez-Gómez and Boller 2002; Boller and Felix 2009; Nicaise et al. 2009). The flagellin consists of a conserved elicitor-active domain that is widespread in bacterial species. Synthetic peptides comprising 15–22 aminoacids of the highly conserved domain within N terminus of flagellin acted as elicitors of defense responses at sub-nanomolar concentrations in cells of tomato and several other plant species. Peptides comprising only the central 8–11 amino acids of the active domain had no elicitor activity but acted as specific, competitive inhibitors of flagellin elicitor activity in tomato cells (Felix et al. 1999). These antagonists suppressed the plant's response to flagellin, crude bacterial extracts and living bacterial cells (Felix et al. 1999).

A peptide flg22, the stretch of 22 amino acids in the N terminus of bacterial flagellin has been identified as the bacterial PAMP epitope (Naito et al. 2007; Boller and Felix 2009). The peptide flg22 elicits responses in most plant species and is active as the full-length flagellin. However, flg22 is recognized by rice but this response was weaker than with full-length flagellin (Takai et al. 2008). The results suggest that additional PAMP epitopes, besides flg22 may be present in the full-length flagellin. A shortened version of flg22 epitope derived from *Escherichia coli*, flg15, was highly active in triggering innate immunity in tomato, but not in *Arabidopsis* (Robatzek et al. 2007). In contrast, flg22 derived from *Pseudomonas syringae* is active in both *Arabidopsis* and tomato (Meindl et al. 2000; Bauer et al. 2001), suggesting existence of other forms of epitopes similar to flg22.

The flg22 region of flagellin is normally buried in the assembled polymer's tertiary structure (MacNab 1996). Hence it may be difficult for the plant pattern recognition receptors (PRRs) to recognize the buried flg22 epitope in flagellin. It is not known whether plants that do detect flagellin recognize assembled flagella shed from the bacterium, free flagella, or fragments of degraded flagellin (Pfund et al. 2004). It is suggested that alternative epitopes of flagellin may be displayed by shed flagella compared with intact flagellin (Pfund et al. 2004).

#### 2.6.2.2 Flagellin from Different Bacteria May Differ in Their Action

Flagellin purified from the incompatible *Acidovorax avenae* N1141 strain induced immune responses, whereas flagellin from the compatible K1 strain induced no responses (Takai et al. 2008). Flagellin purified from the K1 strain was identical to that of the N141 flagellin, suggesting that N1141 flagellin has an epitope in addition to the flg22 region capable of eliciting immune responses (Takai et al. 2008). Flagellins purified from *P. syringae* pv. *glycinea*, an incompatible pathogen for tobacco, induced immune responses in tobacco, whereas flagellin from *P. syringae* pv. *tabaci*, a compatible pathogen, does not, despite complete amino acid identity (Taguchi et al. 2003b). Flagellins derived from nonadapted bacteria but having identical protein sequences differentially induce strong defense responses in nonhost plants, suggesting that other domains and/or posttranslational modifications of flagellin may be involved in triggering immune responses (Taguchi et al. 2003a, b, 2006; Takeuchi et al. 2003, 2007). The major difference between various flagellins has been suggested to be in the glycosylation sites in flagellin (Ishiga et al. 2005; Taguchi et al. 2006; Takai et al. 2008).

#### 2.6.2.3 Flg22 Upregulates Several Signals and Signaling Systems Involved in Plant Immune Responses

Flg22 trigger the upregulation of *Arabidopsis thaliana* genes involved in signal perception (*FLS2*), $Ca^{2+}$ influx (*CNGC4*, *DND1*), calmodulin-mediated signaling (*CML41*), mitogen-activated kinase (*MKS1*, *MPK3*, *EDR1*) signaling, phosphatase (*VSP1*) signaling, reactive oxygen species (ROS) signaling (*RbohC*, *RbohD*, *RbohF*), nitric oxide (NO) signaling (*NOS1*), salicylic acid (SA) signaling (*NPR1*, *SID2*, *PAD4*, *EDS1*, *EDS5*), jasmonate signaling (*LOX3*, *OPR3*, *ERF4*, *CYP81F2*, *ACX1*), and ethylene (ET) signaling (*ACS2*, *ACS7*, *ACS8*, *CTR1*, *ETR1*, *ETR2*, *EIN2*, *EIN4*, *EBF1*) systems (Denoux et al. 2008). PAMPs generally do not induce hypersensitive resistance. However, it has been shown that flg22, as well as flagellin, induces the hypersensitive response (Naito et al. 2008).

#### 2.6.2.4 Some Peptides Derived from Flagellin May Not Have Elicitor Activity

Several studies suggest that the N-terminal flg22 region may not be the sole determinant of flagellin recognition by plants (Che et al. 2000; Taguchi et al. 2003b; Tanaka et al. 2003). A peptide derived from *Ralstonia solanacearum* flg22 region contains at least one significant change from a consensus sequence derived from the flagellin of many bacteria and it did not trigger defense responses (Pfund et al. 2004). The Gly to Ala change at position 18 has been shown to reduce the elicitor activity of flg22 in tomato cells by 96 % (Felix et al. 1999). Felix et al. (1999) demonstrated that peptides derived from the flagellins of *Agrobacterium* spp. also had no elicitation activity.

#### 2.6.2.5 Flagellin May Not Be a Major Defense Elicitor in Some Bacteria

Flagellin may not be a major defense elicitor in *R. solanacearum* cells applied to *Arabidopsis thaliana*. Flagellin also was not the primary elicitor of responses in tobacco (Pfund et al. 2004). Boiled extracts from *R. solanacearum* contained a strong elicitor of defense-associated responses. However, *R. solanacearum* flagellin was not that elicitor, because extracts from wild-type bacteria and mutants defective in flagellin production all elicited similar plant responses. The primary eliciting activity in boiled *R. solanacearum* extracts applied to *Arabidopsis* was attributable to one or more proteins other than flagellin (Pfund et al. 2004).

### *2.6.3 Lipopolysaccharide Components Acting as PAMPs*

Liposaccharides (LPS) of bacterial pathogens act as PAMPs triggering immune responses in both dicots and monocots (Dow et al. 2000; Newman et al. 2001, 2002, 2007; Gerber et al. 2004, 2006; Desaki et al. 2006; Desender et al. 2006; Munford and Varley 2006; Nicaise et al. 2009; Erbs et al. 2010). LPS induces local and systemic resistance and mobilization/translocation occurs through the xylem in *Arabidopsis* (Zeidler et al. 2010). LPS from Gram-negative bacterial pathogens are a great source of novel monosaccharides with unusual and occasionally astounding chemical structures, never found in plants and hence qualify to be recognized as PAMPs (Molinaro et al. 2009). LPS are amphiphilic macromolecules composed of a hydrophilic heteropolysaccharide (comprising the core oligosaccharide and

*O*-specific polysaccharide or *O*-chain) covalently linked to a lipophilic moiety termed lipid A. LPS not possessing the *O*-chain are termed rough LPS or lipooligosaccharides (LOSs) (Silipo et al. 2005). The polysaccharide moiety contains a long-chain polysaccharide, called *O*-antigen, which is highly variable with respect to composition, length, and the branching of its carbohydrate subunits (Knirel 2009). In contrast, the oligosaccharide core and the lipid A, which form the sheet of the membrane, are highly conserved in different bacteria (Holst and Molinaro 2009). LPS are present in the outer monolayer of the external membrane of almost all Gram-negative bacteria. LPS contribute to the structural properties of the cell envelope and play a vital role for bacterial growth (Silipo et al. 2010).

The lipid A part of LPS has been considered as the PAMP epitope in LPS. The lipid A part of LPS is as effective as intact LPS in inducing defense response in *Arabidopsis* (Zeidler et al. 2004). Phosphorylation and acylation of the lipid A moiety seem to influence LPS elicitor activity (Silipo et al. 2008). The structure of LPS of *Xanthomonas campestris* pv. *campestris* (*Xcc*) shows a strong accumulation of negatively charged groups in the lipid A inner-core region and has a number of novel features, including a galacturonyl phosphate attached at a 3-deoxy-D-manno-oct-2-ulosonic acid residue and a unique phosphoramide group in the inner core region. Dephosphorylated LPS molecule, which retains a single negative charge on the inner core, does not induce any defense response in *A. thaliana.* It suggests a key role for the charged phosphate, phosphoramide, and galacturonic residues in LPS signaling (Silipo et al. 2005).

The lipid A moiety is not solely responsible for all of the effects of LPS in plants: core oligosaccharide and *O*-antigen components can elicit specific responses (Newman et al. 2007). *O*-chain in LPS may also act as a PAMP, besides the lipid part. Synthetic oligorhamnans, which are common components of *O*-chain in LPS, can trigger innate immune responses in *Arabidopsis* (Bedini et al. 2005). Besides activating defenses, LPS can suppress defense responses, probably by chelating calcium ions (Newman et al. 2007; Tellstrom et al. 2007; Aslam et al. 2008).

The *O*-antigen of the LPS from many phytopathogenic bacteria comprises a rhamnan backbone with the trisaccharide repeating unit [α-L-Rha-(1 → 3)-α-L-Rha-(1 → 2)-α-L-Rha-(1 → 3)]. This trisaccharide was synthesized and oligomerized to obtain hexa-and nona-saccharides. These rhamnans were effective in elicitation of transcription of the defense-related genes *PR1* and *PR2* (Bedini et al. 2005). The results suggest that the coil structure containing *O*-antigen polysaccharides may also be a plant-recognizable PAMP.

Both LPS and LOS have been described in *Xcc*, with LOS being the predominant form in some strains (Dow et al. 1995). Both lipid A and the core oligosaccharide of *Xcc* LOS were able to trigger defense responses in *A. thaliana* (Silipo et al. 2005). LOS induced defense responses in two temporal phases, while the core oligosaccharide induced only the earlier phase and lipid A induced only the later phase (Table 2.3; Silipo et al. 2005). The results suggest that plant cells can recognize lipid A and core oligosaccharide structures within LPS to trigger defense responses and that this may occur via two distinct recognition events (Silipo et al. 2005) (Table 2.3).

**Table 2.3** Relative ability of LOS and lipid A and core oligosaccharide structures within LPS of *Xanthomonas campestris* pv. *campestris* in triggering *PR1* gene expression in *Arabidopsis thaliana*

| | *PR* gene expression (fold regulated) | | |
|---|---|---|---|
| Time after treatment | LOS | Core oligosaccharide | Lipid A |
| 12 h | 131 | 144 | 11 |
| 20 h | 5 | 17 | 8 |
| 24 h | 1192 | 11 | 448 |

Adapted from Silipo et al. (2005)

### 2.6.4 *Muropeptides and Sugar Backbone Structure PAMPs in Peptidoglycans*

Peptidoglycan, not found in eukaryotes, is an essential and unique component of the bacterial envelope that provides rigidity and structure to the bacterial cell. Peptidoglycan is found as thick outer layer in the cell wall of Gram-positive bacteria, whereas a relatively thin layer is present in the cell wall of Gram-negative bacteria, where it is overlaid with lipopolysaccharides (Erbs et al. 2008; Vollmer and Born 2009). Peptidoglycan from both Gram-positive and Gram-negative bacteria is composed of a network of glycan strands that are interlinked by short peptides. The glycan chains are formed by alternating *N*-acetylmuramic acid (MurNAc) and *N*-acetylglucosamine (GLcNAc) linked by β-(1 → 4)-glycosidic bonds (Cloud-Hansen et al. 2006; Erbs et al. 2008; Cirillo et al. 2010; Silipo et al. 2010). The presence of the lactyl group of the muramic acid allows for the covalent attachment of a short peptide stem that typically contains alternating L- and D-amino acids. The structure of the carbohydrate backbone is generally conserved in all bacteria, but different degrees of acetylation are the major variations among the bacteria (Silipo et al. 2010). Glycan strands are frequently deacetylated and/or *O*-acetylated in bacterial species (Volmmer 2008). The peptide moiety also displays considerable diversity among the Gram-positive and Gram-negative bacteria. In general, the third-position amino acid in Gram-positive bacteria is L-lysine (Lys), whereas Gram-negative bacteria possess the *meso*-2,6-diaminopimelic (DAP) as the third amino acid (McDonald et al. 2005). Gram-positive bacteria have peptide stems that are usually cross-linked through an interpeptide bridge (generally glycine), whereas gram-negative bacteria peptide stems are usually directly crosslinked (Erbs et al. 2008).

Peptidoglycan is located on most bacterial surfaces, which constitute excellent targets for recognition by the innate immune system. Peptidoglycan is considered as a typical PAMP, because it is widely found in bacteria, structurally stable, displayed on the cell surface and not found in plant cells (Gust et al. 2007). Peptidoglycans from both gram-positive and gram-negative bacteria have been reported to be PAMPs (Gust et al. 2007; Erbs et al. 2008). Perception of gram-positive peptidoglycans mostly depends on their sugar backbones (Gust et al. 2007), whereas

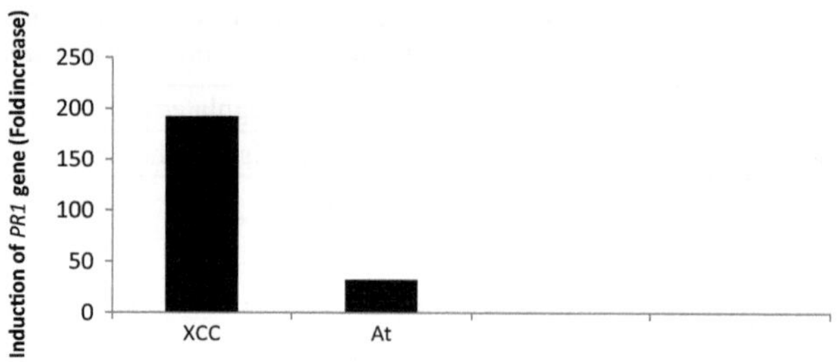


**Fig. 2.1** Induction of *PR1* gene in *Arabidopsis* after treatment with *Xanthomonas campestris* pv. *campestris* (*Xcc*) and *Agrobacterium tumefaciens* (*At*) muropeptides of peptidoglycans (Adapted from Erbs et al. 2008)

muropeptides derived from gram-negative peptidoglycans are more potent elicitors than intact peptidoglycans (Erbs et al. 2008). The Gram-positive bacteria-derived peptidoglycan triggered immune responses and the peptidoglycan-mediated immunity in *Arabidopsis* has been found to be based upon recognition of the sugar backbone in the peptidoglycan (Gust et al. 2007). The purified muropeptides of the Gram-negative bacterial pathogens (*Xanthomonas campestris* pv. *campestris* and *Agrobacterium tumefaciens*) show higher elicitor activity than the peptidoglycan preparations from those bacteria, suggesting that the PAMP epitope may reside in the muropeptide moiety of the peptidoglycan (Erbs et al. 2008). Peptidoglycans in the bacterial surface may be degraded to muropeptides by host lysozyme activities. The released muropeptides are highly mobile, while the peptidoglycan diffuses only slowly (Erbs et al. 2008).

The structure of muropeptides may differ in different bacterial pathogens. Differences in the structures of *X. campestris* pv. *campestris* (*Xcc*) and *Agrobacterium tumefaciens* muropeptides include the presence of a Gly residue replacing Ala in the case of *A. tumefaciens* peptidoglycan and by the lack of an acetyl group in the case of *Xcc* peptidoglycan (Erbs et al. 2008). The differences observed in the muropeptides of the two pathogens would have contributed to the differences in their elicitor activity. The elicitor activity of muropeptide of *X. campestris* pv. *campestris* is very high when compared with that of *A. tumefaciens* (Fig. 2.1, Erbs et al. 2008). The studies suggest that structure and activity of peptidoglycans may vary widely (Fig. 2.1).

Peptidoglycan is associated with inner membrane and, in Gram-negative bacteria, is shielded by the LPS-containing outermembrane. Peptidoglycans may be released during growth process of the bacteria (Cloud-Hansen et al. 2006). It is also suggested that degradation of bacterial cells by host defenses may contribute to release of peptidoglycan (Erbs et al. 2008). Some plants possess peptidoglycan-modifying lysozymes (Brunner et al. 1998) and these enzymes may release muropeptides which may sense the pattern recognition receptors (PRRs) in host plants and activate the innate immunity (Erbs et al. 2008).

### 2.6.5 *elf18 PAMP Epitope in Elongation Factor Tu (EF-Tu)*

Elongation factor Thermo unstable (EF-Tu) is the abundant bacterial protein and is involved in translation of bacterial mRNAs (Zipfel 2008). EF-Tu is recognized as a PAMP in *Arabidopsis* and other members of the family Brassicaceae (Kunze et al. 2004). EF-Tu possesses all the characteristics of a typical PAMP; highly abundant, high sequence conservation over thousands of bacterial species and vital for microbial survival (Zipfel 2008). The PAMP epitope has been detected in the N terminus of bacterial EF-Tu (Kunze et al. 2004; Zipfel et al. 2006). A highly conserved *N*-acetylated 18 amino acid peptide, elf18, is sufficient to trigger those responses induced by full-length EF-Tu. Peptides derived from mitochondrial or plastid EF-Tu are inactive as PAMPs, revealing that this perception is specific to the infectious non-self (Zipfel 2008).

EF-Tu is mostly intracellular and surface localized in bacterial cell. It lacks classical signal and transport sequences for secretion (Zipfel 2008). It is still not known how EF-Tu inside the bacterial cell is recognized by the plant. Lysis of dying bacteria in the plant cell during plant colonization may release sufficient EF-Tu to stimulate the receptor (Zipfel 2008).

### 2.6.6 *Cold-Shock Protein (CSP22) as PAMP*

Elicitation activity in some bacterial species was attributed to a cold-shock protein rather than flagellin (Felix and Boller 2003). An elicitor of defense responses found in extracts of *Micrococcus lysodeikticus* was a member of the cold-shock protein family (Felix and Boller 2003). The highly conserved RNA- binding motif RNP-1 of bacterial cold shock proteins (CSPs) acts as a PAMP. It triggers defense responses in Solanaceous plants. The 22-amino acid core of RNP-1 named CSP22 is recognized as elicitor by Solanaceous plants (Felix and Boller 2003).

### 2.6.7 *Harpins with PAMP and Protein Secretion Structure*

#### 2.6.7.1 Several Different Harpins Are Produced by Various Phytopathogenic Bacteria

Harpins are acidic, glycine rich, protease sensitive, and heat stable proteins that are encoded by *hrp* genes present in several phytopathogenic bacteria including members of the genera *Erwinia*, *Pantoea*, *Pseudomonas*, *Xanthomonas*, and *Ralstonia* (Kvitko et al. 2007; Chen et al. 2008; Engelhardt et al. 2009; Tampakaki et al. 2010; Boureau et al. 2011). Harpins are structurally unrelated proteins that are produced and secreted by many bacterial pathogens and that share a number of biochemical features (Engelhardt et al. 2009). *Pseudomonas syringae* pv. *tomato* DC3000 produces two harpins, HrpZ1 and HrpW1 (Kvitko et al. 2007). The harpins $HrpN_{Ea}$ (Wei et al. 1992), $HrpZ_{Pss}$ (He et al. 1993), $HrpZ_{Psph}$, (Tampakaki et al. 2000), $HrpG_{Xoo}$ (Wen and Wang 2001), and $HpaG_{Xooc}$ (Chen et al. 2008) are produced by *Erwinia amylovora*, *Pseudomonas syringae* pv. *syringae*, *P. syringae* pv. *phaseolicola*, and *Xanthomonas oryzae* pv. *oryzae*, and *X. oryzae* pv. *oryzicola*, respectively.

#### 2.6.7.2 PAMP May Reside Within the Harpin Structure

It has been demonstrated that harpins act as PAMPs triggering plant immune responses in several plants (Alfano and Collmer 2004; Wu et al. 2011). The harpin proteins HrpZ1 from *P. syringae*, HrpN from *Erwinia amylovora*, and PopA from *Ralstonia solanacearum*, elicit innate immune responses in a non-cultivar-specific manner in various plants (Wei et al. 1992; He et al. 1993; Lee et al. 2001a; Racapé et al. 2005; Wu et al. 2011). Transgenic *Nicotiana benthamiana* and sugar beet plants expressing *hrpG* gene of *P. syringae* pv. *phaseolicola* triggered the activation of several defense signaling genes (Pavli et al. 2011). The HrpN of *E. amylovora* contributes directly or indirectly to callose elicitation on apple leaves (Boureau et al. 2011). The harpin HrpZ1 triggers several defense signaling systems and hypersensitive response (HR) in various plant species (Nürnberger et al. 2004; Grant et al. 2006).

A C-terminal fragment of the HrpZ1 protein retained the ability of the harpin to trigger plant immunity. Random insertion mutagenesis of HrpZ1 further revealed that the C- terminus is important for the PAMP activity of the protein (Engelhardt et al. 2009). The 24-amino-acid HrpG fragment found in the C-terminal regions showed the PAMP activity (Haapalainen et al. 2011). These studies suggest that a PAMP resides within in the harpin structure.

#### 2.6.7.3 Harpin-Binding Sites in Plant Membranes

The harpin HrpZ1 binds to plant membranes with high affinity and specificity, suggesting that the activation of plant immunity-associated responses by HrpZ1 is receptor-mediated. The binding site found in the microsomes was protease- and heat-resistant, suggesting that the binding site may therefore not be a protein at all (Engelhardt et al. 2009). HrpN harpin from *E. amylovora* has been shown to bind to a small 6.5-kDa plasma membrane-associated protein from apple (HrpN-interacting protein from *Malus*, HIPM) and *Arabidopsis* (AtHIPM) (Oh and Beer 2007). These studies suggest that harpins possess all important characters of PAMPs: wide occurrence in various bacterial species, binding with PRRs and triggering innate immune responses.

#### 2.6.7.4 A Specific Region in Harpin with Pore Formation Function May Be Involved in Delivery of the PAMP Residing Within Harpin into Plant Cells

Many bacterial pathogens use type-III protein secretion systems (TTSS) to infect plants. TTSS are molecular conduits that facilitate the injection of bacterial effectors into plant cells to manipulate host physiology. Harpins from various *P. syringae* pathovars form ion-conducting pores, suggesting a role of the harpin proteins in effector delivery during infection (Lee et al. 2001b; Fu et al. 2006). The harpin HrpZ of *P. syringae* showed membrane-binding and pore-forming activities *in vitro*, suggesting that it could be targeted to the host cell plasma membrane (Haapalainen et al. 2011). HrpZ was found to interact with the lipid phosphatidic acid and pore-formation by HrpZ in artificial lipid vesicles was found to be dependent on the presence of phosphatidic acid. In addition, HrpZ was able to form pores in vesicles prepared from *Arabidopsis thaliana* plasma membrane. These studies suggest that the harpin HrpZ is targeted to the host cell plasma membrane and it binds with the lipid layer.

HrpG forms dimers and higher order oligomers. The oligomerization was mainly mediated by a region near the C-terminus of the protein and the same region was also found to be essential for membrane pore formation. Phosphatidic acid binding appears to be mediated by two regions separate in the primary structure. A 24-amino-acid HrpG fragment found in the region was shown to be indispensable for the oligomerization and pore formation functions (Haapalainen et al. 2011). The pore formation activity of harpins may facilitate translocation of the PAMP found within the harpin structure into plant cells and trigger the expression of genes involved in defense signaling systems.

### *2.6.8 Ax21 Sulfated Protein as a PAMP*

Ax21 (activator of XA21-mediated immunity) isolated from the rice bacterial blight pathogen *Xanthomonas oryzae* pv. *oryzae* (*Xoo*) has been identified as an elicitor and it triggers hypersensitive reaction (HR) in rice cultivars expressing the R protein XA21 (Lee et al. 2006b). *Xa21*was cloned in 1995 as a disease resistance gene expressing resistance against wide range of *Xoo* strains and it was the first disease resistance gene cloned from rice (Song et al. 1995). The sequence of the predicted protein encoded by the *Xa21* gene carried both a leucine-rich repeat motif and a serine-threonine kinase-like domain, suggesting a role in cell surface recognition of a pathogen ligand and subsequent activation of an intracellular defense response (Song et al. 1995). After 14 years, Lee et al. (2009) have shown that XA21 is a pattern recognition receptor (PRR) and it recognizes a 194-amino acid protein designated Ax21 as the pathogen ligand and as a pathogen-associated molecular pattern (PAMP). A tyrosine-sulfated 17-amino acid synthetic peptide corresponding to the N-terminus of Ax21 was fully active in eliciting

XA21-mediated resistance. Cross-linking experiments suggested that Ax21 directly binds XA21. Ax21 is conserved in most species of *Xanthomonas* and the tyrosine sulfation is required for its recognition by XA21 (Shen et al. 2002; Lee et al. 2006b, 2009).

The PAMP Ax21 protein carries two predicted tyrosine sulfation sites. An Ax21-derived synthetic peptide (17-amino acid) containing a sulfated tyrosine-22 (axY$^{s}$22) is sufficient for Ax21 activity, whereas peptides lacking tyrosine sulfation and peptide variants carrying alanine in place of the tyrosine are inactive (Lee et al. 2009). The peptide axY$^{s}$22 directly binds to XA21 (Lee et al. 2009). Although all *Xanthomonas oryzae* pv. *oryzae* (*Xoo*) strains tested carry ax21 (Lee et al. 2009), *Xoo* strains lacking the sulfation and/or secretion systems can no longer elicit the XA21-mediated defense response (de Silva et al. 2004). These results suggest that sulfation on the axY$^{s}$22 peptide is critical for XA21/Ax21 recognition in rice.

The sulfated protein Ax21 is secreted by *Xoo* through type I secretion system (Zhang and Zhou 2010). The genes *raxA*, *raxB*, and *raxC* encoding components of a bacterial type I secretion system have been detected in *Xoo*. *Xoo* mutants carrying knockouts in any of these genes lose the ability to trigger XA21-mediated immunity and are no longer able to secrete Ax21 (Lee et al. 2006b). The genes involved in sulfation, *raxST*, *raxR*, and *raxP*, were also detected in *Xoo*. Tha *raxST* encodes tyrosine sulfotransferase and the *raxR* and *raxP* genes are involved in synthesis of 3′-phosphoadenosine 5′-phosphosulfate (PAPS). RaxST may utilize PAPS to transfer a sulfuryl group to Ax21 (Lee et al. 2006b). These results suggest Ax21 is sulfated and secreted through the bacterial type 1 secretion system (Park et al. 2010b).

Ax21 is present in all sequenced *Xanthomonas* species, in *Xyllella fastidiosa*, the causal agent of Pierce's disease on grapes, and in the human pathogen, *Stenotrophomonas maltophila* (Lee et al. 2009). Thus, Ax21 satisfies a key aspect of the definition of PAMPs. A *Xoo* mutant strain lacking Ax21 was unable to trigger XA21-mediated immunity (Park et al. 2010b). It shows that Ax21 is a PAMP triggering defense responses.

### 2.6.9 Rhamnolipids as PAMPs

Rhamnolipids derived from *Pseudomonas aeruginosa*, an opportunistic pathogen of plants, were identified as PAMPs recognized by grapevine (Varnier et al. 2009). They trigger the early signaling events including $Ca^{2+}$ influx, mitogen-activated protein kinase activation and reactive oxygen species production, which are the characteristic components in PAMP-triggered immunity (Varnier et al. 2009). The rhamnolipids efficiently protected grapevine against *Botrytis cinerea* (Varnier et al. 2009). Rhamnolipids potentiate defense responses induced by another PAMP chitosan in grapevine (Vatsa et al. 2010). Rhamnolipids were able to stimulate defense genes in tobacco, wheat, and *Arabidopsis thaliana*, besides grapevine (Vatsa et al. 2010).

### *2.6.10 Superoxide Dismutase (SOD) as a PAMP*

A protein of the extracellular proteome of *Xanthomonas campestris* pv. *campestris* was identified as a PAMP. The protein PAMP was a superoxide dismutase (SodM). It elicited $H_2O_2$ production in tobacco cell cultures. The amino acid sequence of this protein was found to be responsible for the elicitation of the oxidative burst reaction (Watt et al. 2006).

### *2.6.11 Bacterial DNA as PAMP*

Elicitor activity of bacterial DNA has been demonstrated on the model plant *Arabidopsis thaliana*. *Eco*RI-digested plasmid DNA induced generation of ROS generation and deposition of callose, whereas *Sma*1- and *Hap*II-digested plasmid DNA and *Eco*RI-digested herring DNA did not remarkably induce these responses (Yakushiji et al. 2009). Further, methylation of the CpG sequence of plasmid DNA and *Escherichia coli* DNA reduced the level of the defense responses (Yakushiji et al. 2009). These results suggest that the nonmethylated CpG DNA is a PAMP/MAMP. The non-methylated DNA of the bacteria seems to be translocated into the cytoplasm of plant cells by endocytosis, as the endocytosis inhibitors significantly inhibited DNA-induced defense responses (Yakushiji et al. 2009).

### *2.6.12 NEP1-Like Proteins as Bacterial PAMPs*

Nep1-like proteins (NLPs) have been detected in some bacterial pathogens, including *Pectobacterium carotovorum* subsp. *carotovorum* (Mattinen et al. 2004; Kufner et al. 2009) and *P. carotovorum* subsp. *atrosepticum* (Pemberton et al. 2005). These proteins were shown to be virulence factors (Ottmann et al. 2009) and these NLPs also trigger immune responses. The NLPs have dual functions, acting both as triggers of immune responses and as toxin-like virulence factors (Ottmann et al. 2009).

## 2.7 Fungal PAMPs

### *2.7.1 Chitooligosaccharides as PAMPs*

Several PAMPs have been detected in various plant fungal pathogens. Chitin, ($\beta$-1 $\rightarrow$ 4-linked polymer of *N*-acetylglucosamine; GlcNAc), is a major component of fungal cell walls. It is not found in plants; however the plants secrete chitin-degrading enzymes (Vidhyasekaran 2007a). Fungal infection induces the expression of chitinases in plant cells, and these chitin-degrading enzymes accumulate at the sites of invasion. The chitinases release chitin fragments (chitin oligomers or chitooligosaccharides) from fungal cell walls (Eckardt 2008).

The chitooligosaccharides are the classical PAMPs detected in fungi (Miya et al. 2007; Hamel and Beaudoin 2010; Lizasa et al. 2010) and they are known to trigger the plant innate immune responses in a wide range of plants including both monocots and dicots (Silipo et al. 2010; Shimizu et al. 2010; Son et al. 2012). Plant cells perceive the chitin fragments and the elicitor activity of chitin fragments increases with the increase of degree of polymerization up to the octasaccharide. *N*-acetylchitoheptaose generated from cell walls of the rice blast pathogen *Magnaporthe oryzae* has been identified as a potential elicitor triggering defense responses in rice cells (Yamaguchi et al. 2002). Chitooctaose has been reported to be the most potential chitin fragment in eliciting defense responses in plants (Wan et al. 2008a, b). However, Petutschnig et al. (2010) provided evidences that insoluble polymeric chitin may also be a potential elicitor. They showed that the chitin receptor CERK1 (for Chitin Elicitor Receptor Kinase 1) binds to polymeric chitin more strongly than to chitin oligomers. It suggests that the polymeric chitin is potentially an active molecule in chitin signaling and generation of short chitooligomers by apoplastic chitinases might not be an absolute prerequisite for chitin recognition.

Chitin treatment of rice induced transient membrane depolarization (Kuchitsu et al. 1993a), ion efflux, cytoplasmic acidification (Kuchitsu et al. 1997), transient generation of ROS (Kuchitsu et al. 1995), protein phosphorylation (Kuchitsu et al. 1993b), and jasmonic acid biosynthesis (Nojiri et al. 1996). Chitin oligomers induced medium alkalinization and ROS generation in suspension-cultured soybean cells (Day et al. 2001). Chitin treatment caused an immediate oxidative burst in *Physcomitrella patens* (Lehtonen et al. 2012). The chitin-induced oxidative burst was associated with the induction of alternative oxidase (AOX), lipoxygenase (LOX), and NADPH oxidase (Lehtonen et al. 2012).

Chitin elicits phosphorylation of various proteins. Peck et al. (2001) identified a number of proteins that were phosphorylated within minutes after chitin treatment of *Arabidopsis* tissue culture cells. Calcium-dependent protein kinase (CDPK) was transiently induced upon chitin elicitation (Zhang et al. 2002a). The phosphorylation event necessary for transmission of the chitin signal was completed within the first 20 min of chitin addition in *Arabidopsis* (Zhang et al. 2002a).

Mitogen-activated protein kinases play important role in chitin signaling (Zhang et al. 2002a; Wan et al. 2004). Perception of PAMPs by receptors leads to the rapid activation of MAP kinases including MPK3, MPK4, and MPK6. In particular, MAP kinase 3 and 6 (MPK3/MPK6) were shown to be rapidly activated by chitin in *Arabidopsis* and their activation depended on upstream MAPK kinases (MKK4 and MKK5) (Wan et al. 2004).

Transcription factors (TF) are critical in reprogramming gene expression in plant cells in response to various stimuli. Plant cells reprogram gene expression in response to chitin elicitation and 118 TF genes have been shown to be induced by chitin in *Arabidopsis* (Wan et al. 2008b).

### 2.7.2 *β-Glucan PAMPs Isolated from Fungal Cell Wall*

Glucans are important components of cell walls of various fungi and oomycetes. Several β-glucan poly- and oligosaccharides seem to be generated from fungal and oomycete cell walls at the site of infection through the action of plant β-1,3-glucanases (Silipo et al. 2010). Most of these β-glucan wall components have been recognized as PAMPs (Yamaguchi et al. 2000a, b; Klarzynski et al. 2000; Shibuya and Minami 2001; Silipo et al. 2010). These fragments may act differently in different plants. A glucan fragment, tetraglucosyl glucitol from *Magnaporthe oryzae* is active in rice cells, but not in soybean (Yamaguchi et al. 2000a, b). β-1,3 glucan oligomers with a degree of polymerization (DP) 4 and DP 6 trigger defense responses (Inui et al. 1997).

### 2.7.3 *Fungal Cell Wall Ergosterol PAMP*

Ergosterol is a typical fungal sterol, which is absent in plants. It has been detected in cell walls of most fungi (Kasparovsky et al. 2003, 2004; Laquitaine et al. 2006; Lochman and Mikes 2006). Ergosterol is perceived by plant cells. It triggers production of ROS even at nano-molar concentrations in tobacco and tomato cells. It also activates mitogen-activated protein kinases on alfalfa cells (Lochman and Mikes 2006). The ergosterol induced oxidative burst and it enhanced NADPH oxidase and superoxide dismutase activities (Rossard et al. 2010). Based on these elicitor activities and wide occurrence in fungal species, and its absence in plants, ergosterol has been recognized as a PAMP (Naito et al. 2008).

### 2.7.4 *EIX Protein as PAMP*

The fungal protein ethylene-inducing xylanase (EIX) is a PAMP inducing defense reactions in tobacco and tomato (Bailey et al. 1990; Avni et al. 1994). EIX is a potent elicitor of plant defense responses in specific cultivars of tobacco and tomato (Bar et al. 2010). EIX was shown to specifically bind to the plasma membrane of both tomato and tobacco responsive cultivars (Hanania and Avni 1997). The response to EIX in tobacco and tomato cultivars is controlled by a leucine-rich-repeat receptor-like-protein (LRR-RLP) encoded by a single locus, termed LeEix (Ron and Avni 2004).

### 2.7.5 *Cerebrosides as PAMPs*

Cerebrosides A, B. and C. categorized as sphingolipids are novel elicitors detected in a wide range of pathogens (Koga et al. 1998; Umemura et al. 2000). Sphingolipid is a structural component of many eukaryotic cell membranes and has been shown to be essential for normal development of fungi (Levery et al. 2002). Several fungal pathogens including several forme speciales of *Fusarium oxysporum* (f. sp. *lycopersici*, f. sp. *melonis*, f. sp. *cucumerinum*, and f. sp. *lactucae*), *Pythium graminicola*, *Glomerella cingulata*, and *Sclerotinia cepivorum* contain cerebroside elicitor (Umemura et al. 2004). The cerebroside elicitor induced defense signaling systems and induced resistance against pathogens (Umemura et al. 2004). These results suggest that cerebroside elicitors are general elicitors and considered as PAMPs (Nakashima et al. 2008).

### *2.7.6 NEP1-Like Proteins as PAMPs*

A 24-kDa necrosis- and ethylene-inducing protein (Nep1) isolated from culture filtrates of *Fusarium oxysporum* f. sp. *erythroxyli* has been found to trigger both ethylene production and necrosis in numerous dicotyledonous plants (Bailey 1995; Bailey et al. 1997). Nep1 acts as a general elicitor and induces $Ca^{2+}$-dependent signaling system, changes in $K^+$ and $H^+$ channel fluxes, accumulation of ROS, production of ethylene, enhanced transcription of *PR* genes and callose deposition (Jennings et al. 2001; Fellbrich et al. 2002; Keates et al. 2003; Bae et al. 2006). Nep1-like proteins (NLPs) have been isolated from other fungal pathogens including *Verticillium dahliae* (Wang et al. 2004), *Botrytis elliptica* (Statts et al. 2007), *Botrytis cinerea* (Schouten et al. 2008; Arenas et al. 2010) and *Magnaporthe grisea*, *Fusarium graminearum* and *Mycosphaerella graminicola* (Motteram et al. 2009).

## 2.8 Oomycete PAMPs

### *2.8.1 PEP-13 as an Oomycete PAMP*

Several different PAMPs have been detected in various oomycete pathogens. A peptide fragment (Pep-13), within an abundant cell wall glycoprotein (GP42) from *Phytophthora sojae*, has been identified as a PAMP in oomycete pathogens (Halim et al. 2004). It was found to be necessary and sufficient for receptor-mediated defense gene expression and synthesis of antimicrobial phytoalexins in parsley (Nürnberger et al. 1994; Hahlbrock et al. 1995). GB 42 is a $Ca^{2+}$-dependent TGase (*R-glutaminyl-peptide*:amine-γ-glutamyltransferase), which catalyzes an acyl transfer reaction between peptide-bound glutamine residues and primary amines including the ε-amino group of peptide-bound lysine residues, forms intra- or intermolecular isopeptide bonds resulting in irreversible protein cross-linking (Brunner et al. 2002). The PAMP epitope of GP42 has been identified as a surface exposed 13 amino acid spanning domain (Pep-13), which is also essential for TGase activity (Brunner et al. 2002). Mutational analysis within Pep-13 identified the same amino acids indispensable for both TGase and defense-eliciting activity. Several species of *Phytophthora* including *P. cactorum*, *P. capsici*, *P. cinnamomi*, *P. cryptogea*, *P. drechsleri*, *P. infestans*, *P. nicotianae*, *P. palmivora*, *P. parasitica*, and *P. sojae* possess a gene family encoding GP42 TGase-related proteins. GP42 TGase homologs have been detected in all the 10 *Phytophthora* species tested. The peptide fragment Pep-13 was found to be highly conserved among all *Phytophthora* species tested (Brunner et al. 2002).

Pep-13 treatment triggered the accumulation of defense-related transcripts encoding lipoxygenase, 4-coumarate::CoA ligase and PR protein 1 in potato cells and in intact potato leaves (Brunner et al. 2002). Plants possess neither orthologs of *Phytophthora* TGase nor proteins containing peptide with Pep-13 elicitor activity (Brunner et al. 2002). Collectively these studies reveal that Pep-13 is a PAMP found in *Phytophthora* spp. Pep-13 binds to its receptor, which is a protease- and heat-sensitive 100-kDa protein (Nürnberger et al. 1994).

### 2.8.2 Elicitins as Oomycete PAMPs

Elicitins are small lipid binding proteins secreted by the oomycetes *Phytophthora* and *Pythium* (Bonnet et al. 1996; Bourque et al. 1999; Baillieul et al. 2003; Qutob et al. 2003; Kawamura et al. 2009; Kim et al. 2010). Elicitin genes have been cloned from *Phytophthora infestans* (Kamoun et al. 1997a, b), *P. sojae* (Mao and Tyler 1996), *P. parasitica* (Kamoun et al. 1993a, b), *P. cinnamomi* (Duclos et al. 1998), *P. cryptogea* (Panabieres et al. 1995), and *P. capsici* (Kim et al. 2010). Elicitins are highly conserved 10-kDa proteins that are secreted in culture by all tested *Phytophthora* and *Pythium* species (Kamoun et al. 1993a, b; Pernollet et al. 1993a, b). All the elicitins share a conserved elicitin domain from amino acids 1 to 98.

The elicitins are grouped into five classes based on their primary structure (Vidhyasekaran 2007a). Class I-A and class I-B comprise 98 amino acid-long elicitin proteins. Class I-A elicitins have an acidic pI, while class I-B elicitins have basic pI. Class II contains highly acidic elicitins, which possess a short (5–6 amino acids long) hydrophilic C-terminal tail. Class III comprises elicitins with a long (65–101 amino acids long) amino acid C-terminal domain rich in Ser, Thr, Ala, and Pro. Elicitins from *Pythium* spp. are classified as a distinct group called Py class (Ponchet et al. 1999). They trigger innate immunity in a narrow range of plants, including *Nicotiana* species in the Solanaceae and some radish and rape cultivars in the Brassicaceae (Ponchet et al. 1999). However, they trigger a wide range of defense responses in most *Nicotiana* species and this response is sufficient to protect against infection not only by *Phytophthora* but also by bacteria, fungi, and viruses (Tyler 2002).

### 2.8.3 Oomycete Cell Wall Glucans as PAMPs

A glucan fragment β-1,6-1,3 heptaglucan from *Phytophthora sojae* induced defense responses in soybean and the minimal structural requirements for the elicitation of defense responses by this glucan were established as a succession of five β-1,6-linked glucosyl residues with two side branches of β-1,3 glucan (Cheong et al. 1991). A glucan fragment, hexa-β-glucopyranosyl-D-glucitol isolated from the cell walls of *Phytophthora megasperma* f. sp. *glycinea*, acted as PAMP triggering defense responses (Sharp et al. 1984). A doubly-branched hepta-β-glucoside generated from *P. megasperma* f. sp. *glycinea* glucan is a highly active PAMP (Cheong et al. 1991); however its elicitor action is restricted to the plants belonging to the family Fabaceae (Cote et al. 2000). β-1,3 glucans induced a variety of defense reactions in a wide range of host plants (Klarzynski et al. 2000; Ménard et al. 2004; Aziz et al. 2003). The biological activity of β-1,3 glucans is dependent on the degree of polymerization (DP) and decorations carried by the sugar backbone (Inui et al. 1997; Ménard et al. 2004).

### 2.8.4 Cell Wall Glycoprotein CBEL with CBD Motifs as PAMPs

A cell wall glycoprotein named CBEL (for Cellulose-Binding Elicitor Lectin) isolated from the root rot pathogen *Phytophthora parasitica* var. *nicotianae* has been identified as an elicitor of defense responses in tobacco (Séjalon-Delmas et al. 1997; Villalba Mateos et al. 1997). It shows cellulose-binding and lectin-like activities (Villalba Mateos et al. 1997). CBEL binds to crystalline cellulose and isolated plant cell walls, but it is devoid of enzyme activity on cellulose and various glucans. CBEL was found to be localized in the inner and outer layers of the cell walls of the oomycete, and it was present in close contact to the host cell wall during infection (Séjalon et al. 1995).

The protein portion of CBEL is composed of two cysteine-rich domains, each containing subdomain homologous to cellulose-binding domains (CBDs) of fungal glucan hydrolases. CBDs belong to the Carbohydrate Binding Module 1 family, which is found exclusively in fungi and oomycetes (Gaulin et al. 2006). CBDs of CBEL have been shown to be essential and sufficient to stimulate defense responses and CBDs are considered as a novel class of PAMPs (Gaulin et al. 2006).

CBEL triggers the expression of various defense genes encoding lipoxygenase, peroxidase, sesquiterpene cyclase, and anthranilate synthase and accumulation of defense proteins including PR-1, PDF1.2, and hydroxyproline-rich glycoprotein (HRGP) (Villalba Mateos et al. 1997; Khatib et al. 2004). CBEL has been shown to be necessary for the structure of the hyphal cell walls (Gaulin et al. 2002). CBEL homologues have been detected in various *Phytophthora* species and these homologues contain highly conserved CBD motifs (Khatib et al. 2004).

### 2.8.5 NEP1-Like Proteins as Oomycete PAMPs

Nep1-like proteins (NLPs) have been detected in a wide range of oomycete pathogens (*Pythium* spp., *Phytophthora* spp., and *Hyaloperonospora arabidopsidis*) (Veit et al. 2001; Fellbrich et al. 2002; Qutob et al. 2002, 2006; Gijzen and Nürnberger 2006; Kanneganti et al. 2006; Kufner et al. 2009; Ottmann et al. 2009; Cabral et al. 2012). They share a high degree of sequence similarity (Pemberton and Samond 2004; Qutob et al. 2006). The NLPs induce defense responses in both susceptible and resistant plants (Kanneganti et al. 2006). NLPs show important characters of PAMPs: they trigger nonself immunity in plants, functional orthologs of NLPs occur in a wide range of pathogens, and they are absent in plant genome. The NLPs are also required for virulence of the pathogens. NLPs evoke complex defense responses in diverse range of dicotyledonous plant species, such as tomato, tobacco, soybean, parsley, and *Arabidopsis* (Fellbrich et al. 2002; Qutob et al. 2002; Kanneganti et al. 2006; Kufner et al. 2009).

NLP isolated from the oomycete *Phytophthora parasitica* ($NLP_{Pp}$) induced genes involved in signal perception (*AtLECRK*, *RKF1*, *RPK1*, *BRL3*, *WAKL7*, *RFO1/WAKL* and genes encoding receptor Ser/Thr kinase and receptor protein kinase), mitogen-activated kinase (*MPK3*, *MPK11*), ROS signaling (*RbohD*, *RbohF*), SA signaling (*ICS/SID2*, *EDS5/SID1*) and ethylene signaling (*ACS2*, *ACS7*, *EAT1*, *ERF5*) systems (Qutob et al. 2006).

## 2.9 Viral Elicitors

### 2.9.1 Several Viral Components Show Elicitor Function

Microbial nucleic acids have been recognized as classical PAMPs in mammals (Kawai and Akira 2009). Viral proteins and/or nucleic acid products comprise an array of PAMP signatures that can engage specific PRRs, including Toll-like receptors (TLRs) or RIG-1-like receptors (RLRs) in mammals (Liu et al. 2010; Drutskaya et al. 2011). However, no classical PAMPs have been detected in plant viruses. Several elicitors have been detected in viruses. A *Tobacco mosaic virus* (TMV) coat protein (CP) gene has been reported to code for an elicitor, which induced disease resistance in *Nicotiana sylvestris* plants (Culver and Dawson 1991; Culver et al. 1994). The coat protein of *Potato virus X* (PVX) is an elicitor (Bendahmane et al. 1995). The coat proteins of *Paprika mild mottle virus* (PaMMV) and *Pepper mild mottle virus* (PMMoV) act as elicitors inducing host defense genes in *Capsicum* (Gilardi et al. 2004; Matsumoto et al. 2008). A nucleocapsid protein (N) has been identified as an elicitor of *Tomato spotted wilt virus* (Lovato et al. 2008).

A TMV replicase gene product acts as elicitor and it activates defense genes in tobacco (Padgett et al. 1997). The *Potato virus Y* Nlb-replicase protein has been reported to be the elicitor of a hypersensitive response in tobacco (Fellers et al. 2002). The helicase domain of the TMV replicase proteins act as elicitors and induce the defense response in tobacco (Erikson et al. 1998). A 29 K-movement protein (MP) of *Tobacco rattle virus* is considered as an elicitor (Ghazala and Varrelmann 2007). The P3 protein involved in virus multiplication has been found to be an elicitor in *Turnip mosaic virus* (Jenner et al. 2002, 2003) and in *Soybean mosaic virus* (Hajimorad et al. 2005). Virion structural proteins have been shown to act as elicitor in *Cauliflower mosaic virus* (Love et al. 2005).

### *2.9.2 Viral Double-Stranded RNAs May Be PAMPs*

Small RNA-directed RNA silencing is a major immune system targeting foreign nucleic acids of invading pathogens (Ding and Voinnet 2007; Jaubert et al. 2011). In small RNA-based antiviral immunity, viral double-stranded RNA (vdsRNA) has been identified as a PAMP (Ding 2010). The RNA silencing pathway in plants presents a formidable defense against viral pathogens (Qu and Morris 2005). RNA silencing is triggered by dsRNA which are commonly generated during plant virus replication (Willmann et al. 2011). In case of single-stranded RNA (ssRNA) viruses, the viral RNA-dependent RNA polymerase (RdRP) encoded by the plant copies a plus-sense ssRNA generating a dsRNA molecule (Qi et al. 2009; Garcia-Ruiz et al. 2010; Wang et al. 2010b). In case of other RNA viruses, the two strands do not anneal but can fold into highly structured molecules that have dsRNA regions (Alvarado and Scholthof 2009). In case of geminiviruses, the RNAs transcribed from their circular genomes act as a source of dsRNA (Chellappan et al. 2005). Viroids form hairpin structures, which contain intervals of dsRNA (Papaefthimiou et al. 2001).

The plant innate immune system (defense surveillance system) detects the presence of dsRNA as aberrant RNA molecule (Wypijewski et al. 2009) and generates small RNAs. The generated small RNAs direct the antiviral machinery to cleave and destroy the invading viral genome (Alvarado and Scholthof 2009). The results suggest that the viral dsRNA may be a PAMP triggering in antiviral immune response in plants.

## 2.10 Host-Associated Molecular Patterns as Endogenous Elicitors

### *2.10.1 Oligogalacturonides as HAMPs*

Besides elicitors of pathogen origin (PAMPs), several elicitors of host origin (endogenous elicitors) have also been reported to trigger immune responses. These endogenous elicitors are called host-associated molecular patterns

(HAMPs). Fragments of pectic polysaccharide homogalacturonan (HGA) called oligogalacturonides (OGs) function as danger signals activating plant innate immune responses and these OGs are called endogenous elicitors/host-associated molecular patterns (HAMPs) (Ryan et al. 2007; Caffali and Mohnen 2009; Galletti et al. 2009; Zipfel 2009; Huffaker et al. 2011). Oligogalacturonides are generated by the action of endopolygalacturonases (PGs) secreted by pathogens during the infection process. The PGs generate OGs by hydrolyzing pectin. The activity of OGs depends on the degree of polymerization (DP). The OG oligomers with a degree of polymerization (DP) between 9 and 15 have been shown to be most potent inducers of defense responses in several plants (Poinssot et al. 2003; Aziz et al. 2004; Moscatiello et al. 2006; Huang et al. 2007). However, in some reports, small-sized oligomers have also been shown to induce the accumulation of protease inhibitor proteins and ethylene synthesis in tomato involved in defense responses, while larger oligomers were ineffective (Simpson et al. 1998).

The formation of a $Ca^{2+}$-dependent "egg box" conformation of OGs has been shown to be necessary for the biological activity of oligogalacturonide fragments (Cabrera et al. 2008). The term "egg box" is used to designate the calcium-induced association between chains of homogalacturonides of DP > 8 (Cabrera et al. 2008). In the egg box model, dimerization of two homogalacturonide rigid chains occurs by cooperative bridging of parallel facing chains through $Ca^{2+}$ ions. The binding of a first calcium cation by two pectin chains facilitates their alignment with respect to each other, which in turn allows the easier binding of a next calcium ion, and so on along the sequence (Cabrera et al. 2008). It has been shown that a minimum DP of 9 is critical for OGs to form stable egg boxes (Liners et al. 1992). At least five calcium ions are needed between two pectin chains to allow egg box dimer formation (Liners et al. 1989).

Once the generated OGs are of right size, they dimerize, and bind putative receptors, which trigger production of extracellular polygalacturonase inhibiting proteins (PGIP) that specifically recognize and inhibit PGs produced by pathogens (Di Matteo et al. 2003; Cabrera et al. 2008). PGIPs are plant extracellular leucine-rich repeat proteins that specifically bind and inhibit fungal polygalacturonases. The interaction with PGIP limits the destructive potential of polygalacturonases and might trigger the plant defense responses induced by oligogalacturonides (Di Matteo et al. 2006).

PG inhibition by PGIP delays OG hydrolysis by 24 h (D'Ovidio et al. 2004a, b). The egg box formation by OGs progressively increases with time and it needs about 10 h for OGs to form fully associated egg boxes (Cabrera et al. 2008). The major function of PGIP may therefore be to delay OGs hydrolysis enough for the oligomers not only to accumulate but also to mature into more bioactive egg boxes (Cabrera et al. 2008).

The egg box conformation of OGs may be specifically recognized by the OG perception system (Messiaen and Van Cutsem 1999). It has been shown that the extracellular domain of the transmembrane receptor-like wall-associated kinase 1 (WAK1) could ionically bind calcium-associated homogalacturonans (Decreux and Messiaen 2005). Any condition that impaired the formation of egg boxes also impaired the interaction between WAK1 and homogalacturonides (Decreux and Messiaen 2005). The results suggest that egg box conformation of OGs is the critical factor recognized by the PRR WAK1 to activate the innate immune system.

Cabrera et al. (2008) suggested that there may be at least two different perception systems for egg box dimers. One binds egg box junctions and the other binds egg box ends. Perception system 1 may be able to bind OGs in a size-and conformation-dependent way. The egg box junctions may be the ligand of this perception system 1. Perception system 2 may also bind OGs in a size- and conformation-dependent way but the ends of the egg boxes may constitute the ligand. Many of the early defense responses induced by OGs in *Arabidopsis* cells would depend on this type 2 perception system. According to the hypothesis of Cabrera et al. (2008), the modification of the reducing end of the OGs does not hinder egg box formation but it prevents egg box binding to the type 2 perception system.

### 2.10.2 *Cellodextrins as HAMPs*

Cellulose-derived oligosaccharides may also play an important role in triggering innate immune responses (Vidal et al. 1998). Cellodextrins (CD), water-soluble derivatives of cellulose composed of β-1,4 glucoside residues, have been shown to induce a variety of defense responses in grapevine cells (Aziz et al. 2007). Degree of polymerization determines efficacy of the CD in inducing defense responses. Oligomers of DP 3 and 4 induced a slight production of $H_2O_2$, those of DP5 and 6 were unable to induce any significant response, while CD of DP>6 were active. DP7 oligomer was 2-fold more active than CD 8 and 9 (Aziz et al. 2007). The CD oligomers stimulated chitinase and β-1,3-glucanase activities in grapevine cells. Overall, the high values were obtained with CD of DP 7–9 (Aziz et al. 2007).

### 2.10.3 **Arabidopsis At*Pep Peptides***

Besides the oligosaccharides, some peptides of host plant origin have been reported as endogenous elicitors, which are also called HAMPs. *At*Pep1 (*Arabidopsis thaliana* elicitor peptide 1), a 23-aa peptide, was identified in soluble extracts of *Arabidopsis* leaves as an endogenous elicitor (Huffaker et al. 2006; Yamaguchi et al. 2006; Huffaker and Ryan 2007; Krol et al. 2010; Shinya et al. 2010). Two other endogenous elicitors, *At*Pep2 and *At*Pep3, have also been identified in *Arabidopsis* and they are distant homologues of *At*Pep1. All three elicitors are recognized by the PRRs PEPR1 and PEPR2 (Krol et al. 2010). Several other homologs of *At*Pep1, including *At*Pep3, *At*Pep4, *At*Pep5, *At*Pep6, and *At*Pep7 have been identified in *Arabidopsis* (Huffaker et al. 2006).

*At*Pep1 and its homologs, *At*Pep2-7, are derived from the C-terminal portion of their precursor proteins PROPEP1-7, respectively (Huffaker et al. 2006). PROPEP1 is a 92-aa precursor protein. *PROPEP1* belongs to a seven-member gene family

(*PROPEP2*, *PROPEP3*, *PROPEP4*, *PROPEP5*, *PROPEP6* and *PROPEP7*). PROPEP1 orthologs have been found in rice, maize, wheat, barley, canola, potato, soybean, *Medicago*, and poplar plants (Huffaker et al. 2006). Transcripts of *PROPEP1*, *PROPEP2* and *PROPEP3* genes were induced by pathogen infection and also by PAMPs (Huffaker et al. 2006; Huffaker and Ryan 2007).

*At*Pep1 activates the expression of its own precursor gene *PROPEP1* (Huffaker et al. 2006). Because Pep peptides induced the transcription of their own precursor genes, it is likely that Pep peptides, which are initially induced by PAMPs, feed back into the signaling pathways to generate additional processed peptides to further upregulate downstream defense responses (Ryan et al. 2007). Thus, *PROPEP* genes are components of a feedback signaling system that is mediated by the PEPR1 receptor to amplify the innate immune response of *Arabidopsis*. *At*Pep1 and its homologs (*At*Pep2 to *At*Pep7) are endogenous amplifiers of innate immunity of *Arabidopsis thaliana*.

## 2.10.4 Soybean GmSubPep Peptide

In soybean, a 12-aa peptide was found to activate transcription of defense genes. The HAMP peptide was named G*lycine* m*ax* Subtilase Peptide (GmSubPep) (Pearce et al. 2010). The amino acid sequence of the peptide was determined and was found to be derived from a member of the subtilisin-like protease (subtilase) family. The sequence of the peptide was located within a region of the protein that is unique to subtilases in legume plants and not found within any other plant subtilases. GmSubPep peptide is processed from a unique region of an extracellular subtilisin-like protease (subtilase) (Pearce et al. 2010). It was capable of producing a pH change within 10 min and a maximal alkalinizing response in 15 min (Pearce et al. 2010). Among the suspension-cultured cells tested from a wide array of species, only the suspension cells produced from *Glycine max* (soybean) were capable of producing an alkalinizing response to GmSubPep (Pearce et al. 2010). The gene encoding this peptide, *Glyma18g48580*, has been cloned. The peptide was active at extremely low concentrations. The receptor for this peptide has not yet been isolated (Pearce et al. 2010).

## 2.10.5 Maize ZmPep1 Peptide

A pathogen-inducible gene orthologous to the Arabidopsis *AtPROPEP1*gene, which encodes the precursor protein of elicitor peptide 1 (*At*Pep1), has been identified in maize (*Zea mays*) and called *ZmPROPEP1* (Huffaker et al. 2011). Both *At*PROPEP1 and *Zm*PROPEP1 do not have a conventional signal sequence for export through the secretory pathway and both are predicted to localize to the cytosol (Huffaker et al. 2011). *ZmPROPEP1* encodes a peptide, *Zm*Pep1, which is an endogenous elicitor. The native length of *Zm*Pep1 is predicted to be 23 amino acids similar to the *At*Pep peptides of *Arabidopsis thaliana*.

The *ZmPROPEP1* gene is expressed in response to fungal infection and jasmonic acid. Treatment of maize leaves with *Zm*Pep1 induces the expression of *ZmPROPEP1* gene (Huffaker et al. 2011). Pretreatment with *Zm*Pep1 prior to infection enhances resistance to the southern leaf blight pathogen *Cochliobolus heterostrophus* and the stalk rot pathogen *Colletotrichum gramoinicola* in maize (Huffaker et al. 2011).

## 2.11 Plant Pattern Recognition Receptors

### *2.11.1 Structure of PRRs*

PAMPs are perceived by plants as danger signals and these signals trigger a network of signaling systems activating defense responses. Specific receptors for the recognition of PAMPs have been identified in the plant cell plasma membrane and these receptors are called 'Pattern Recognition Receptors (PRRs)'. PAMPs are perceived as alarm/danger signals by cognate pattern recognition receptors (PRRs) and the PAMP-PRR complex activates the plant immune system (Takakura et al. 2004; Jones and Dangl 2006; Altenbach and Robatzek 2007; He et al. 2007; Wan et al. 2008a, b; Iriti and Faoro 2009). Several receptors for the PAMPs have been recognized in plasma membrane of plant cells (Table 2.4; Mithöfer et al. 2000; Montesano et al. 2003; Fliegmann et al. 2004; Ron and Avni 2004; Fritz-Laylin et al. 2005; Altenbach and Robatzek 2007; He et al. 2007). The PRRs identified to date are modular proteins harbouring an extracellular domain consisting of leucine-rich repeat (LRR) or lysine motifs (LysM) (Table 2.4; Saijo 2010; Segonzac

**Table 2.4** Structure of plant pattern recognition receptors (PRRs)

| PAMP/endogenous elicitor | PRR | Structure |
|---|---|---|
| flg22 | FLS2 | LRR-RLK |
| EFTu | EFR | LRR-RLK |
| Ax21 | XA21 | LRR-LK |
| Pep1 | PEPR1 | LRR-RLK |
| PEP2 | PEPR2 | LRR-RLK |
| EIX1 | LeEIX1, LeEIX2 | LRR-RLP |
| Chitin | CERK1 | LysM-RLK |
| Chitin | CEBiP | LysM-RLP |
| β-glucans | GBP | Glycoside hydrolases |
| Mannose | MBL | Lectin receptor kinase |
| Oligogalacturonides | WAK1 | Wall-associated RLK |
| INF1 elicitin | NbLRK1 | Lectin-like receptor kinase |
| Capsicein elicitin | NgRLK1 | PR5 protein kinase |
| Lipopolysaccharides | Receptor | RLK |
| Peptidoglycan | Receptor | LysM-receptor kinase |

and Zipfel 2011). The PRRs are largely divided into receptor-like kinases (RLKs) and receptor-like proteins (RLPs) on the basis of the presence or absence of an intracellular kinase domain (Wang et al. 2010a).

Most of the PRRs identified are LRR-RLKs (Goff and Ramonell 2007). LRR-RLKs are single-pass transmembrane proteins composed of an LRR ectodomain (eLRR), a transmembrane domain and a Ser/Thr protein kinase domain related to *Drosophila* Pelle (Shiu and Bleecker 2001a, b). The sensors for extracellular molecules consisting of an extracellular ligand-binding domain, a single transmembrane domain, and a cytosolic protein kinase domain have been termed RLKs (Seifert and Blaukopf 2010). Kinases are classified as arginine-aspartate (RD) or non-RD kinases. RD kinases carry a conserved arginine (R) immediately preceding the catalytic aspartate (D), while non-RD kinases typically carry a cysteine or glycine in place of the arginine (Dardick and Ronald 2006). RD kinases are regulated by autophosphorylation of the activation loop, a centrally located domain that is positioned close to the catalytic centre. In contrast, non-RD receptor kinases, the activation loop is not autophosphorylated. It suggests that the non-RD kinases use alternative mechanisms for activation (Dardick and Ronald 2006).

The PRRs typically carry or associate with non-RD (non-arginine-aspartate) kinases to control early events of innate immunity signaling (Chen et al. 2010a). For example, Arabidopsis the PRRs FLS2 and EFR, and the rice PRR XA21 contain an intracellular non-RD Ser/Thr kinase (Gómez-Gómez and Boller 2000; Zipfel et al. 2006; Chen et al. 2010a). However, some PRRs have been identified as RD kinases. The PRR for the PAMP elicitin INF1 of *Phytophthora infestans* has been identified as a lectin-like receptor kinase and it was designated NbLRK1. NbLRK1 is a typical RD kinase (Kanzaki et al. 2008). The Arabidopsis BRI1-associated receptor kinase 1 (BAK1) that associates with FLS2 and EFR is an RD kinase (Li et al. 2002; Chinchilla et al. 2007a, b).

The extracellular domains of RLKs are believed to bind directly to legands to perceive extracellular signals (PAMPs), whereas the cytoplasmic kinase domains transduce these signals into the cell (Bi et al. 2010). Thus, RLKs are proteins with a "receptor" and a "signaling domain" in one molecule (Shiu and Bleecker 2003). Some of the PRRs are RLPs and these include LeEIX1 and LeEIX2, the PRRs for the PAMP EIX proteins detected in tomato (He et al. 2007). Both RLKs and RLPs are characterized by an extracellular domain and a membrane-spanning domain. However, RLPs lack an intracellular activation domain and hence, require interaction with adaptor molecules for signal transduction (Altenbach and Robatzek 2007).

### 2.11.2 FLS2, the PRR for the PAMP Flagellin (flg22)

The PRR responsible for the PAMP flagellin (flg22) recognition in *Arabidopsis thaliana* is FLS2 (for FLAGELLIN-SENSING 2), a leucine-rich repeat receptor-like kinase (LRR-RLK) (Chinchilla et al. 2006; Sun et al. 2012). FLS2 belongs to a subfamily XII of LRR-RK and consists of an extracellular domain with 28 LRR motifs, a transmembrane domain, and a cytoplasmic Ser/Thr kinase domain (Gómez-Gómez and Boller 2000). The *A. thaliana* FLS2 (AtFLS2) LRR domains 9–15 contribute significantly to flg22 binding (Dunning et al. 2007). The exact flg22-binding site is unknown. Flg22 directly binds to FLS2 and contributes to recognition specificity (Chinchilla et al. 2006). FLS2 physically interacts with the flg22 epitope and determines ligand specificity (Chinchilla et al. 2006).

The genes similar to *Arabidopsis FLS2* genes have been detected in tomato (Robatzek et al. 2007), tobacco (Hann and Rathjen 2007; Hann et al. 2010), and *Brassica* species (Dunning et al. 2007). FLS2 with extracellular LRR domain has been detected in barley, tomato, tobacco, and *Arabidopsis* (Chinchilla et al. 2006, 2007a; Shen et al. 2007). The LRR kinase FLS2, which is homologous to *Arabidopsis* FLS2 has been detected in rice (Takai et al. 2008; Shinya et al. 2010) and an *FLS2-like* gene has also been identified in rice (Takai et al. 2008). These results suggest that both monocotyledonous and dicotyledonous plants may possess a flg22 perception system.

Flagellin perception is required for full immunity against bacteria, because plants deficient in FLS2 are more susceptible to adapted and nonadapted bacterial pathogens (Zipfel 2009). In *Arabidopsis*, pretreatment with flg22 restricts growth of the pathogenic bacterium *Pseudomonas syringae* pv. *tomato* DC3000 and *fls2* mutants are more susceptible to this bacterium (Zipfel et al. 2004). Lack of flagellin recognition allows more growth of the nonadapted bacteria *P. syringae* pv. *tabaci* and *P. syringae* pv. *phaseolicola* (Li et al. 2005b; de Torres et al. 2006). Successful bacterial pathogens need to avoid or suppress PAMP-triggered immunity induced by flagellin (Zipfel 2009). Some virulence effectors from phytopathogenic bacteria do so by directly targeting FLS2 (Göhre et al. 2008).

### *2.11.3 EFR, the PRR for the PAMP EF-Tu*

EFR (for EF-Tu RECEPTOR) is the PRR detected in *Arabidopsis* for binding the PAMP EF-Tu. It belongs to the same subfamily (LRR-RK XII) as FLS2 (Zipfel et al. 2006). EFR is a receptor-like kinase (Albert et al. 2010) and its structure is highly similar to FLS2, with a 21-LRR extracellular domain, a transmembrane domain, and a cytoplasmic Ser/Thr kinase domain (Zipfel et al. 2006; Albert et al. 2010). ERF physically interacts with the first 18 amino acids of the N-terminus of EF-Tu, the elf peptide (Zipfel et al. 2006). No binding of ligand was found with the ectodomain lacking the transmembrane domain or with EFR lacking the first 5 of its 21 LRRs (Albert et al. 2010). It suggests that the transmembrane domain and LRR ectodomain of EFR are essential for ligand binding.

It is suggested that the EFR receptor, which presumably resides in the plasma membrane of the plant cells, is exposed to bacterial EF-Tu during infection. It is still not known how EF-Tu gets released from the bacterial cells; however, EF-Tu has been detected in the secretome of *Xanthomonas campestris* and *Erwinia chrysanthemi* (Kazemi-Pour et al. 2004; Watt et al. 2005).

EF-Tu responsiveness was found only in Brassicaceae species (Kunze et al. 2004). Transient heterologous expression of AtEFR in *Nicotiana benthamiana*, a plant that normally lacks elf18 responsiveness, restores elf18 binding and responses (Zipfel et al. 2006). It suggests that downstream signaling components are conserved between Brassicaceae and Solanaceae (Nicaise et al. 2009). *Arabidopsis efr* mutants are more susceptible to *Agrobacterium tumefaciens* (Zipfel et al. 2006). EF-Tu is recognized by EFR at the host plasma membrane (Zipfel 2008). EFR autophosphorylation has also been reported, suggesting that EFR carries active kinase domain (Xiang et al. 2008).

### 2.11.4 XA21, the PRR for the PAMP Ax21

XA21 is a receptor kinase, which consists of LRR, transmembrane, juxtamembrane (JM) and intracellular kinase domains (Song et al. 1995). XA21 belongs to subfamily XII of the LRR-RKs and is highly similar to EFR. Similar to FLS2 and EFR, XA21 possesses a non-RD kinase, whose presence has been correlated with a role in innate immunity across kingdoms (Dardick and Ronald 2006).

### 2.11.5 CERK1, the PRR for the PAMP Chitin

Miya et al. (2007) identified a receptor-like kinase, designated CERK1 (for Chitin Elicitor Receptor Kinase 1) in *Arabidopsis* as a PRR for the fungal PAMP chitooligosaccharides. Unlike FLS2, EFR, and XA21, CERK1 possesses three extracellular Lysine Motif (LysM) domains instead of LRRs (Miya et al. 2007; Wan et al. 2008a). The Lysine motif is a ubiquitous protein module found in prokaryotes as well as eukaryotes. LysM proteins were first described in bacteria and shown to have binding capacity for peptidoglycan (PGN), a linear form of alternatively β-1,4-linked N-acetyl-muramic acid and GlcNAc (β-1,4 linked N-acetyl-glucosamine) (Zhang et al. 2009a).

CERK1 is a plasma membrane protein containing three LysM motifs in the extracellular domain and an intracellular Ser/Thr kinase domain with autophosphorylation/myelin basic protein (MBP) kinase activity, suggesting that CERK1 plays a critical role in fungal PAMP perception in plants (Miya et al. 2007; Iriti and Faoro 2009; Petutschnig et al. 2010). The CERK1 ectodomain binds chitin and partially de-acetylated chitosan directly without any requirement for interacting proteins (Petutschnig et al. 2010). The three LysM domains have been shown to be necessary for chitin binding (Petutschnig et al. 2010).

CERK1 contains an intracellular serine/threonine kinase domain, which makes it an excellent candidate for the *Arabidopsis* chitin receptor. CERK1 may be involved in the perception of the chitin oligosaccharide elicitor at the cell surface and the transduction of the signal into the cytoplasm via its intracellular serine/threonine kinase activity (Miya et al. 2007). These serine/threonine kinases have been considered good candidates for playing a role in fungal chitin reception (Eckardt 2008; Lohmann et al. 2010).

CERK1 has a higher affinity for chitin having a longer residue of *N*-acetyl glucosamine (Lizasa et al. 2010). CERK1 is autophosphorylated *in vitro* and chitin does not affect the phosphorylation of CERK1 (Lizasa et al. 2010). CERK1 binds specifically and directly to chitin. LysM RLK1 was shown to bind only to chitin, and not to colloidal chitosan and peptidoglycan (PGN), although all of them have a common backbone, GlcNAc. The results suggest that LysM RLK1 may recognize the acetyl group of N-acetylglucosamine residues of chitin and it may be inhibited by the bulky peptide group cross-linked to N-acetyl-muramic acid residues of peptidoglycan (Lizasa et al. 2010).

The knock-out mutants for *CERK1* completely lost the ability to respond to the chitin elicitor, including MAPK activation, ROS generation, and gene expression (Miya et al. 2007). The complete loss of the gene responses induced by the chitin elicitor indicates that CERK1 serves as the 'master switch' of the signaling cascade. The mutation in *CERK1* gene blocked the induction of almost all chitooligosaccharide-responsive genes and led to more susceptibility to fungal pathogens but had no effect on infection by a bacterial pathogen (Wan et al. 2008b). *Arabidopsis cerk1* mutants are more susceptible to fungal pathogens (Miya et al. 2007; Wan et al. 2008b). Exogenously applied chitooligosaccharides enhanced resistance against both fungal and bacterial pathogens in the wild-type plants but not in the mutant. These results suggest that CERK1 (LysM RLK1) is essential for chitin signaling in plants as part of the receptor complex and is involved in chitin-mediated plant innate immunity (Wan et al. 2008a).

CERK1 was also involved in bacterial recognition, as *cerk1* mutants are more susceptible to *P. syringae* pv. *tomato* DC3000 (Gimenez-Ibanez et al. 2009a). *cerk1* mutants, however, were not impaired in their responsiveness to flg22, elf18, LPS, or PGN (Gimenez-Ibanez et al. 2009a), suggesting that CERK1 is involved in the recognition of yet unknown bacterial PAMP. Gimenez-Ibanez et al. (2009b) reported reduced activation of a PAMP-induced defense response on plants lacking the CERK1 gene after treatment with crude extracts of the bacterial pathogen *P. syringae* pv. *tomato* DC3000. This strengthens the earlier findings that CERK1 mediates perception of an unknown bacterial PAMP in *Arabidopsis*.

A LysM receptor-like kinase similar to *Arabidopsis* CERK1 has been detected in rice. It was designated OsCERK1 and it showed high homology with *Arabidopsis* CERK1 (Shimizu et al. 2010). OsCERK1 encoded a receptor-like kinase consisting of 624 amino acid residues, containing a signal peptide, an extracellular domain, a transmembrane region and an intracellular Ser/Thr kinase domain. Motif analysis indicated the presence of one LysM in the OsCERK1 extracellular domain, while CERK1 contained three LysM motifs in its extracellular domain (Shimizu et al. 2010). The expression of *OsCERK1* was up-regulated by elicitor treatment (Shimizu et al. 2010).

### 2.11.6 CEBiP, the Second PRR for the PAMP Chitin

Another PRR protein for the perception of chitin has been recognized in rice and designated chitin elicitor-binding protein (CEBiP) (Kaku et al. 2006; Shinya et al. 2010). It is a receptor-like protein (RLP), unlike the PRR OsCERK1, which is a receptor-like kinase (RLK). CEBiP contains extracellular LysM motifs for chitin-binding but lacks an intracellular kinase domain, which is characteristically present in OsCERK1. CEBiP is a transmembrane protein with two extracellular LysM domains and a short cytoplasmic tail. It is predicted to have the two LysM domains in its extracellular part and a single transmembrane domain. It directly binds the fungal PAMP chitin (Kaku et al. 2006).

CEBiP specifically binds the chitin oligosaccharide. Its knock-down transformants exhibited the suppression of chitin-induced defense responses, suggesting that CEBiP functions as receptor for the PAMP (Kaku et al. 2006). RNAi experiment showed that CEBiP is required for chitin-induced defenses in rice (Kaku et al. 2006). CEBiP functions as a cell surface receptor for chitin elicitor in rice (Miya et al. 2007). The predicted structure of CEBiP does not contain any intracellular domains, suggesting that an additional component(s) is required for signaling through the plasma membrane into the cytoplasm (Miya et al. 2007). Although CeBiP possesses two LysM domains and a transmembrane region, it does not have any domain that could function as a signal transduction module (Kaku et al. 2006). Since CEBiP lacks a significant intracellular domain, it likely is only a part of the chitin receptor complex in rice (Kaku et al. 2006). An obvious partner for CEBiP would be a membrane-associated receptor-like kinase (Wan et al. 2008a).

Several studies have indicated that rice requires both the types of plasma membrane PRRs CEBiP and OsCERK1 for chitin signaling. The extracellular domain of OsCERK1 can interact with that of CEBiP. It appears that CEBiP plays a major role in chitin elicitor binding and that OsCERK1 functions as a signal transducer through its Ser/Thr kinase activity in rice (Shimizu et al. 2010). In the absence of chitin oligosaccharides, CEBiP and OsCERK1 mostly exist separately from each other, although a major portion of CEBiP appears to exist as homo-oligomers. CERK1 may form a heterodimer with CEBiP to bind chitin (Shimizu et al. 2010).

### 2.11.7 NbLRK1, the PRR for the PAMP INF1 Elicitin

The PRR for the PAMP elicitin INF1 of *Phytophthora infestans* has been identified as a lectin-like receptor kinase and it was designated NbLRK1. NbLRK1 is a typical RD kinase (Kanzaki et al. 2008). The gene encoding this receptor, *NbLRK1*, has been isolated from *Nicotiana benthamiana* (Kanzaki et al. 2008). The structure of lectin-like receptor kinases (LRKs) is similar to other plant receptor-like kinases with an N-terminal targeting signal, an extracellular domain, a single transmembrane (TM) spanning helix, and a highly conserved cytosolic kinase domain. The extracellular domain shows homology to lectin proteins known to bind carbohydrates (van Damme et al. 1998). NbLRK1 belongs to the class B lectin-like receptor kinases (Kanzaki et al. 2008).

The 31 amino acids fragment of NbLRK1 kinase domain within VIb subdomain has been shown to interact with INF1 *in vitro*. The VIb subdomain of Ser/Thr kinase is known to contain the catalytic loop with an invariant Asp serving as the catalytic base necessary for the kinase function. This site is close to the VII and VIII domains where the activation loop is located, which is necessary for autophosphorylation of kinases (Dardick and Ronald 2006; Kanzaki et al. 2008). It is suggested that INF1 binding to the VIb subdomain of NbLRK1 alters its kinase activity presumably by autophosphorylation (Kanzaki et al. 2008). NbLRK1 contains a conserved arginine (R) at immediately preceding the invariant aspartate (D) in subdomain VIb. It suggests that NbLRK1 is a typical RD kinase (Kanzaki et al. 2008). It does not belong to the non-RD kinases known to harbor many kinases involved in pathogen recognition receptors signaling (Dardick and Ronald 2006).

INF1 and NbLRK1 proteins also interact *in vitro*. INF1 treatment induced autophosphorylation of NbLRK1 *in vivo*. Virus-induced gene silencing of NbLRK1 delayed INF1-mediated defense responses in *N. benthamiana* (Kanzaki et al. 2008). These results suggest that NbLRK1 recognizes INF1 elicitor and transduces the defense signals.

NbLRK1 has been found to be localized at plasma membrane in tobacco (Kanzaki et al. 2008). INF1 is known to be secreted by *P. infestans* through its N-terminal signal peptide and was suggested to localize at the extracellular space of plant tissue (Kamoun et al. 1997a). Tyler (2002) suggested that plant recognition of elicitins takes place inside the plant cells and elicitins would have been transported inside plant cells by receptor-mediated endocytosis. The elicitin quercinin of *Phytophthora quercina* was reported to be localized inside the cells of host oak plants (Brummer et al. 2002). Collectively these studies suggest that INF1 protein initially localizes in the apoplast but then trafficks inside plant cells by endocytosis, where it interacts with kinase domain of NbLRK1 (Kanzaki et al. 2008).

### *2.11.8 NgRLK1, the PRR for the PAMP Elicitin Capsicein*

A PRR for the elicitin capsicein has been identified in tobacco (*Nicotiana glutinosa*) and it was designated NgRLK1 (Kim et al. 2010). NgRLK1 has a domain structure similar to that of all plant RLKs. The extracellular domain of NgRLK1 contains both lectin-like and S-locus glycoprotein domains, in addition to a PAN AP domain, which is known to mediate protein-protein or protein-carbohydrate interactions. Extracellular NgRLK1 was found to interact with the elicitin capsicein. Capsicein was found to bind to the intracellular kinase domain of NgRLK1 (Kim et al. 2010). NgRLK1 was more closely related to the protein kinase homologous to PR5 K from *Arabidopsis thaliana* than to the lectin-like receptor kinases. It has been suggested that plant recognition of capsicein occurs in the extracellular and intracellular spaces (Kim et al. 2010). NgRLK1 is a new type of plant RLK that recognizes capsicein (Kim et al. 2010).

## *2.11.9 LeEIX1 and LeEIX2, the PRRs for the PAMP EIX Proteins*

### 2.11.9.1 Two Different EIX Receptors Exist for Perception of the PAMP EIX Signals

Two PRRs have been identified in tomato for the perception of the fungal PAMP EIX (Ethylene-Inducing Xylanase elicitor). These include LeEIX1 and LeEIX2, which contain a leucine zipper, an extracellular LRR domain, a transmembrane domain, and a C-terminal domain with an endocytosis signal (Ron and Avni 2004). These PRRs have been identified as cell-surface receptors without kinase domain in plants (Ron and Avni 2004; Kaku et al. 2006). The two EIX proteins are highly similar to each other and have extracellular domains of 31 LRRs. The structure of these EIX receptors is similar to a family of receptor-like proteins (RLPs) (He et al. 2007). These EIX receptors belong to a superclade of leucine-rich repeat receptor-like proteins with a signal for receptor-mediated endocytosis, which was shown to be essential for proper induction of defense responses (Bar et al. 2010). LeEIX2 contains the conserved endocytosis signal YxxΦ within the short cytoplasmic domain, and mutation in this endocytosis motif resulted in abolishment of hypersensitive response (HR) induction in response to the PAMP EIX, suggesting that endocytosis plays a key role in mediating the signal generated by EIX that leads to hypersensitive response induction (Ron and Avni 2004).

### 2.11.9.2 LeEIX2 Transmits EIX-Induced Signals, Whereas LeEIX1 Attenuates EIX-Signaling of LeEIX2

The tomato PRRs LeEIX1 and LeEIX2 appear to act in distinctly different ways in the PAMP-PRR signaling complex. It has been observed that both the PRRs are able to bind the PAMP EIX, but only LeEIX2is involved in triggering defense responses (Bar et al. 2011). LeEIX1 heterodimerizes with LeEIX2 upon application of the EIX elicitor. LeEIX1 attenuates EIX-induced internalization and signaling of the LeEIX2 receptor. The brassinosteroid co-receptor, BAK1, binds LeEIX1 but not LeEIX2. In BAK1-silenced plants, LeEIX2 was no longer able to attenuate plant responses to EIX, indicating that BAK1 is required for this attenuation. It is suggested that LeEIX1 functions as a decoy receptor for LeEIX2, a function which requires BAK1 (Bar et al. 2010). For effective signal transduction, the effective PRR LeEIX2 requires action of the co-receptor BAK1, which may nullify the negative action of LeEIX1 in signal transduction. These studies suggest the potential role of BAK1 in signal transduction through the PAMP-PRR signaling net work.

## *2.11.10 Glucan—Binding Proteins*

### 2.11.10.1 Glucan-Binding Proteins in Soybean and *Medicago truncatula*

Many β-glucan components of fungal and oomycete cell walls act as PAMPs (Shibuya and Minami 2001; Silipo et al. 2010). A β-glucan-binding protein (GBP) has been identified as a PRR in soybean. It lacks a transmembrane domain and predominantly localizes to the cytoplasmic face of the plant cell wall (Fliegmann et al. 2004). In *Medicago truncatula*, a high-affinity β-glucan-binding site was characterized biochemically. Four full-length clones encoding putative β-glucan-binding proteins from *M. truncatula*, *Mt*GBP1, 2, 3, and 4, composing a multigene family encoding GBP-related proteins have been identified (Leclercq et al. 2008). The GBP has been detected in vesicles at the plasma membrane and in the cytoplasm and it indicates that GBP may interact with a transmembrane RLK or RLP (Fliegmann et al. 2004).

### 2.11.10.2 Glucan-Binding Protein Contains Two Different Activities: Releases β Glucan from β Glucan Polysaccharides and Also Acts as a Receptor of β Glucan Signaling

GBP represents a soluble extracellular binding protein. It is a member of family 81 glycoside hydrolases (Fliegmann et al. 2005). GBP has been shown to contain two different activities. As part of the plasma membrane-localized pathogen receptor complex, it binds the PAMP β-glucan, triggering the activation of defense responses. Additionally, the GBP is able to hydrolyze β-1,3-glucans present in the cell walls of potential pathogens (Fliegmann et al. 2005). GBP initially acts on oomycete/fungus-derived heptaglucosides as a glucan hydrolase, releasing β-glucans that subsequently are perceived by a different domain of GBP (Fliegmann et al. 2004).

## *2.11.11 Mannose-Binding Lectin Receptors*

Some mannose-binding lectin (MBL) receptor kinases have been identified as PRRs for perception of PAMPs. Function of MBL appears to be pattern recognition. MBL recognizes carbohydrate patterns, found on the surface of a large number of pathogenic microorganisms. MBL binds to carbohydrates (specifically mannose residues). The lectin receptor kinases (LecRKs) contain an extracellular domain of mannose specific binding lectin and an intracellular Ser/Thr kinase catalytic domain (Barre et al. 2002). A mannose-binding lectin (MBL) gene, *CaMBL1*, has been isolated from pepper (*Capsicum annuum*) leaves (Hwang and Hwang 2011). The *CaMBL1* gene contains a predicted *Galanthus nivalis* agglutinin-related lectin domain responsible for the recognition of high-mannose *N*-glycans. The CaMBL1 protein exhibits binding specificity for mannose and is mainly localized to the plasma membrane. Mannose has been recognized as a PAMP found in fungal pathogens (Meyer-Wentrup et al. 2007).

### 2.11.12 RLK Receptor for the PAMP Lipopolysaccharides

The bacterial PAMP lipopolysaccharide (LPS) was found to bind to tobacco cells and become internalized into endocytic vesicles, suggesting a receptor-mediated process (Gross et al. 2005). It has been suggested that the receptor of the LPS may be a receptor-like kinase (RLK) (Sanabria and Dubery 2006). Further studies are needed to characterize the PRR for the PAMP LPS.

### 2.11.13 Peptidoglycan-Binding Proteins

The PRR of the PAMP peptidoglycan is still not known. The LysM motif present in several receptor kinases and transmembrane proteins in plants can bind peptidoglycan (Guan and Mariuzza 2007; Zhang et al. 2007b; Buist et al. 2008). It is likely that these proteins may function as PRRs for the carbohydrate PAMPs (Nicaise et al. 2009).

### 2.11.14 Pep Receptors for the HAMPs Pep Proteins

Two receptors for the perception of *At*Pep1 have been recognized in *Arabidopsis*. The Pep1 receptor, PEPR1, and the gene encoding the receptor have been isolated from *Arabidopsis* suspension-cultured cells (Yamaguchi et al. 2006). PEPR1 is a typical LRR receptor kinase, having an extracellular LRR domain and an intracellular protein kinase domain, and belongs to the LRR XI subfamily of the 15 LRR-RLK subfamilies (Shiu et al. 2004; Qi et al. 2010). *At*PEPR1 has guanylyl cyclase activity, generating cGMP from GTP, and the cGMP can activate CNGC2-dependent cytosolic $Ca^{2+}$ elevation (Qi et al. 2010).

The second Pep1 receptor,PEPR2, has been identified by Yamaguchi et al. (2010).It is a plasma membrane LRR receptor kinase and has 76 % amino acid similarity to PEPR1 (Yamaguchi et al. 2010). PEPR1 has been identified as a receptor for the HAMPs *At*Pep1, *At*Pep2, *At*Pep3, *At*Pep4, *At*Pep5, and *At*Pep6 and PEPR2 is a receptor for *At*Pep1 and *At*Pep2 (Yamaguchi et al. 2010).

The *pepr1* and *pepr2* mutants affected in PEPR1 and PEPR2 and the wild type *Arabidopsis* plants were sensitive to AtPep1, but the double mutant *pepr1/pepr2* was completely insensitive. *At*Pep1 triggers a receptor-dependent transient depolarization through activation of plasma membrane anion channels. This effect was absent in the double mutant *pepr1/pepr2* (Krol et al. 2010). These results suggest that a receptor complex consisting of two PRRs (PEPR1 and PEPR2) acts as receptor for *At*Pep1. The double mutant also fails to respond to *At*Pep2 and *At*Pep3, the homologues of *At*Pep1, suggesting that the receptor complex PEPR1 and PEPR2 is responsible for the perception of all three HAMPs (Krol et al. 2010).

Both PEPR1 and PEPR2 were transcriptionally induced by the HAMP Pep peptides (Yamaguchi et al. 2010). Methyl jasmonate induced transcription of both *PEPR1* and *PEPR2* expression within 30 min, whereas methyl salicylate and 1-aminocyclopropan-1-carboxylic acid, an ethylene precursor, did not induce either PEPR1 or PEPR2 (Yamaguchi et al. 2010). The results suggest that JA signaling system may be involved in activation of PEPRs. The receptors may be involved in amplification of the innate immune response in *Arabidopsis* induced by Pep peptides (Yamaguchi et al. 2006).

### *2.11.15 WAK1 as a Receptor for the HAMP Oligogalacturonides*

Oligogalacturonides (OGs) released from the plant cell wall are active as damage-associated molecular patterns (DAMPs) or host-associated molecular patterns (HAMPs) for activation of the plant immune response (Brutus et al. 2010). The wall-associated kinase 1 (WAK1) has been identified as a receptor of OGs (Brutus et al. 2010). WAK1 belongs to the huge family of 610 receptor-like kinases identified in the *Arabidopsis thaliana* genome (Shiu and Bleecker 2001a, b). WAKs display a typical plant Ser/Thr kinase signature and an extra cytoplasmic domain (ectodomain) containing several EGF (Epidermal Growth Factor)-like repeats. WAK1 binds *in vitro* to OGs through the N-terminal non-EGF portion of the ectodomains (Decreux et al. 2006; Cabrera et al. 2008). WAK1 is induced by wounding, bacterial infection, and salicylic acid treatment (He et al. 1998, 1999; Wagner and Kohorn 2001). WAK1 is up-regulated in response to OGs, whereas it is slightly down-regulated by flg22 (Denoux et al. 2008). WAK1 was capable to sense OGs *in vivo* and trigger a defense response (Brutus et al. 2010). The role of WAK1 as a receptor of the HAMP OGs was demonstrated by constructing chimeric receptors carrying EFR and WAK1. Upon stimulation with OGs, the WAK1 ectodomain was capable of activating the EFR kinase domain. Transgenic plants over expressing WAK1 are more resistant to *Botrytis cinerea* (Brutus et al. 2010).

## 2.12 Transmembrane Proteins Interacting with PRRs in PAMP-PRR Signaling Complex

### 2.12.1 *Signaling Adapters/Amplifiers in PAMP-PRR Signaling Complex*

Plant pattern recognizing receptors (PRRs) interact with additional transmembrane proteins that act as signaling adapters or amplifiers to achieve full functionality (Zipfel 2009). These transmembrane proteins include BAK1 (for BRI1 (BRASSINOSTEROID INSENSITIVE1)-ASSOCIATEDKINASE1), BIK1 (BOTRYTIS-INDUCED KINASE1), and BIR1 (BRANCHING INHIBITING RECEPTOR1) (Postel et al. 2010; Zhang and Zhou 2010; Zhang et al. 2010a). BAK1 is also called SERK3 (for Somatic EMBRYO RECEPTOR KINASE3) and it belongs to the LRR-receptor-like kinase (RLK). BAK1 belongs to the LRR type II SERK subfamily (Chinchilla et al. 2007b; Heese et al. 2007; Zipfel 2008). BAK1 is a component of diverse processes, including brassinosteroid signaling, light responses, cell death, and plant innate immunity (Chinchilla et al. 2009). BAK1 was originally identified as a BRI1-associated receptor kinase mediating brassinosteroid signaling (Li et al. 2002; Nam and Li 2002). Brassinosteroids (BR), a class of plant hormone with essential roles in plant growth and development, are perceived by LRR-RK BRI1, which is structurally similar to the PRR FLS2 (Belkhadir et al. 2006). BAK1 is a positive regulator of PAMP-triggered plant immunity and it acts as an adaptor of multiple LRR-RKs that act in defense signaling, including the PRRs FLS2, EFR, PEPR1 and PEPR2 (Chinchilla et al. 2007a, b; Ryan et al. 2007; Gao et al. 2009a; Postel et al. 2010; Schulze et al. 2010). It also acts as an adaptor of the receptor kinases BIR1 and SOBIR1, which seem to act as part of a presumed PRR complex(es) and/or at a downstream step in the signaling cascade (Saijo 2010).

BAK1 forms a complex with the RLK BIR1 to negatively regulate defense responses (Gao et al. 2009a). BIK1 is a receptor-like cytoplasmic kinase (RLCK). The PAMP flg22 interacts with BAK1 to phosphorylate BIK1, which seems to act as positive regulator of the PAMP signaling pathway (Lu et al. 2010; Zhang et al. 2010a). BIK1 is an essential component in PAMP signal transduction, which links the PAMP receptor complex to downstream intracellular signaling (Lu et al. 2010).

Zhang et al. (2010a) identified a number of PBS1-like (PBL) RLCKs (cytoplasmic receptor-like kinases), including BIK1 and several other PBLs, as components in PAMP- signaling pathways. PBL proteins belong to the subfamily VII of cytoplasmic receptor-like protein kinases (Zhang et al. 2010a). PBL1, PBL2, and PBS1 appear to additively contribute to PAMP- signaling (Zhang et al. 2010a). AtPHOS32, AtPHOS34, and AtPHOS43 are the other signaling components in PAMP-triggered immunity (Peck et al. 2001; Merkouropoulos et al. 2008). Another LRR-RLK, ERECTA has been shown to interact with PRRs (Godiard et al. 2003; Llorente et al.

2005). Identification of extensive engagement of membrane-localized receptors and regulators in signaling suggests that these membrane proteins represent a key aspect of plant immunity.

### 2.12.2 BAK1 Is Required for Proper Functionality of PRRs

BAK1 appears to be a key component in plant innate immune system and suppression of *BAK1* gene expression results in suppression of the plant innate immunity. *BAK1*-silenced plants were found to be more susceptible to adapted and non-adapted *Pseudomonas* and to the oomycete *Hyaloperonospora parasitica* (Heese et al. 2007). Arabidopsis *bak1* mutants are also extremely susceptible to necrotrophic fungi *Botrytis cinerea* and *Alternaria brassicicola* (Kemmerling et al. 2007). These results demonstrate the importance of BAK1 in immune signaling.

BAK1and/or its homologs have been shown to be required for full function of PAMP-PRR signaling complex to activate plant immune responses (Schulze et al. 2010). Chinchilla et al. (2007b) reported that plants mutated in *BAK1* were strongly affected in the PAMP elf18 responses. Arabidopsis *bak1* mutants are largely impaired in the PAMP flg22-induced defense responses (Chinchilla et al. 2007b; Heese et al. 2007). Silencing of *NbBAK1*in *Nicotiana benthamiana* results in decreased responsiveness to the PAMPs CSP22 and INF1 (Heese et al. 2007). BAK1 is required for proper functionality of several PRRs including FLS2, EFR, CERK1, PEPR1, and PEPR2 (Postel et al. 2010; Zhang and Zhou 2010). BAK1 was identified as a positive regulator of the PRRs FLS2, EFR, and CSP22 (Chinchilla et al. 2007a, b; Heese et al. 2007; Zipfel 2008). It has been demonstrated that BAK1 forms a complex with FLS2 in a ligand-dependent fashion and is necessary for FLS2 signaling (Chinchilla et al. 2007b; Heese et al. 2007). BAK1 interacts with PRRs and acts downstream of PRRs. BAK1 has been shown to be dispensable for the PAMP flg22 binding, but it interacts with the PRR FLS2 in a ligand-dependent manner shortly after elicitation (Chinchilla et al. 2007b; Heese et al. 2007).

### 2.12.3 BAK1 Acts Downstream of PRR Perception by PAMP

The BAK1 kinase activity is required for FLS2-mediated signaling but not flg22 induced association of FLS2 and BAK1 (Schulze et al. 2010). It suggests that BAK1 acts downstream of FLS2 perception to regulate signaling (Chinchilla et al. 2007b). BAK1 is also required for the activation of another PRR CERK1 in chitin signaling (Zhang and Zhou 2010).

Flg22 perception by FLS2 triggers an interaction between FLS2 and BAK1 (Chinchilla et al. 2007b; Heese et al. 2007). The flg22 induced FLS2-BAK1 association occurs within seconds and is accompanied by increased phosphorylation on both FLS2 and BAK1 (Schulze et al. 2010).

### *2.12.4 BAK1 Functions as an Adapter or Signaling Partner for Regulation of PRRs*

BAK1 is a central regulator of innate immunity in plants (Heese et al. 2007). PAMP-PRR binding leads to the formation of a molecular platform that involves the recruitment of adaptor proteins to trigger downstream signaling (Heese et al. 2007). BAK1 probably acts as a co-activator of the receptor complex to enhance various signaling pathways (Wang et al. 2008a). Upon PAMP perception, PRR rapidly associates with BAK1, thereby initiating downstream signaling. BAK1 acts downstream of several PRRs (Nicaise et al. 2009). BAK1 acts not only as an adapter of multiple PRRs including the PRRs FLS2, EFR, PEPR1 and PEPR2, but also as an adapter of other transmembrane proteins BIR1 and SOBIR1. Unlike the direct binding of the PAMP to PRR (Chinchilla et al. 2007b; Kinoshita et al. 2005), BAK1 more likely functions as an adapter or signaling partner for the regulation of PRRs. Furthermore, BAK1 is required for the immune responses triggered by multiple PAMPs including flagellin, the bacterial elongation factor EF-Tu, peptidoglycans, lipopolysaccharides, cold-shock protein, HrpZ (harpin), and the oomycete elicitor INF1 in *Arabidopsis* and tobacco (Chinchilla et al. 2007a; Heese et al. 2007; Shan et al. 2008). A BAK1 ortholog has been detected in rice (Li et al. 2009a) and it may be involved in the PAMP Ax21-mediated immunity (Segonzac and Zipfel 2011). Thus, BAK1 appears to associate with multiple PRRs to integrate specific PAMP perception into convergent downstream signaling. It is still not known how the PAMP signal is transmitted from the BAK1-associated receptor complexes at the plasma membrane to intracellular events (Lu et al. 2010).

### *2.12.5 Rapid Heteromerization and Phosphorylation of PRRs and Their Associated Kinase BAK1*

PRRs form tight complexes with the receptor kinase BAK1 instantaneously after ligand binding. FLS2-BAK1 heteromerization occurs almost instantaneously after perception of the PAMP flg22. Flg22 can induce formation of a stable FLS2-BAK1 complex in microsomal membrane preparations *in vitro* (Schulze et al. 2010). However, the kinase inhibitor K-252a does not prevent complex formation. The results suggest that kinase activity of BAK1 is essential for FLS2 signaling, but not for flg22 induced association of FLS2 and BAK1 (Schulze et al. 2010).

Schulze et al. (2010) detected *de novo* phosphorylation of both FLS2 and BAK1 within 15 s of stimulation with flg22. Similarly, brassinolide induces BAK1 phosphorylation within seconds. Bacterial EF-Tu and AtPep1 induce rapid formation of heterocomplexes consisting of *de novo* phosphorylated BAK1 and proteins representing the ligand-specific binding receptors EF-Tu receptor and Pep1 receptor, respectively. It is suggested that several LRR-RKs form tight complexes with BAK1 almost instantaneously after ligand binding and the subsequent phosphorylation events are key initial steps in signal transduction (Schulze et al. 2010).

BAK1 may be a signal "amplifier" rather than an integral component of downstream signaling pathways (Nicaise et al. 2009). BRI1-BAK1 interaction leads to the transphorylation of their respective kinase domains and the subsequent enhancement of BRI1 signaling output (Wang et al. 2008a), suggesting that BAK is a signal amplifier rather than an integral component of downstream signaling pathways (Nicaise et al. 2009). The PRRs for both flg22 and AtPep1 associate with the interacting receptor partner, BAK1, and likely activate cyclic nucleotide-gated calcium channels via receptor guanylyl cyclase activity (Ma et al. 2009; Postel et al. 2010).

### *2.12.6 BIK1*

Downstream of PAMP-PRR-BAK1 signaling complex, several receptor-like cytoplasmic kinases (RLCKs) play important role in regulation of the signaling pathways. Lacking an apparent extracellular domain, RLCKs more likely function in signal transduction rather than in signal perception (Lu et al. 2010). An RLCK member BIK1 (Botrytis-induced kinase 1) plays an important role in mediating early flagellin signaling from the FLS2/BAK1 receptor complex (Lu et al. 2010). Flg22-induced oxidative burst has been shown to be reduced in *bik1* mutant *Arabidopsis* plants, suggesting the importance of BIK1 in plant innate immune system (Zhang et al. 2010a). BIK1 was originally identified as a component in plant defense against necrotrophic fungal pathogens (Veronese et al. 2006).

Both FLS2 and BAK1 are able to interact with and phosphorylate the receptor-like cytoplasmic kinase BIK1, which seems to act as positive regulator of the FLS2 signaling pathway (Lu et al. 2010; Zhang et al. 2010a). BIK1 forms a complex with unstimulated FLS2 in plants, and flg22 induces a rapid phosphorylation of BIK1 in both an FLS2- and BAK-dependent manner (Zhang et al. 2010a). BIK1 is phosphorylated within 1 min upon flagellin perception (Wu et al. 2011). BIK1 is rapidly phosphorylated by flg22 within the first minutes after stimulation, which may happen instantaneously with the formation of FLS2/BAK1 complex (Lu et al. 2010).

*In vivo* and *in vitro* data suggest that BIK1 associates with both FLS2 and BAK1. BIK1 is a substrate of BAK1, whereas BAK1 and FLS2 are also substrates of BIK1, suggesting transphosphorylation events between BIK1 and the FLS2/BAK1 complex. BIK1 is phosphorylated by BAK1, and BIK1 also directly phosphorylates BAK1 and FLS2 *in vitro*. The flagellin phosphorylation site $Thr^{287}$ of BIK1 is required for its phosphorylation on BAK1 and FLS2, suggesting that BIK1 is likely first phosphorylated upon flagellin perception and subsequently BIK1 transphosphorylates FLS2/BAK1 to propagate flagellin signaling (Lu et al. 2010). BIK1 appears to function downstream of FLS2/BAK1 complex formation and phosphorylation because BIK1 phosphorylation requires not only the presence of both FLS2 and BAK1, but also their kinase activity (Lu et al. 2010).

In addition to phosphorylation, flg22 induces a dissociation of BIK1 from FLS2 (Zhang et al. 2010a; Lu et al. 2010). The flg22-induced BAK1-FLS2 association is not affected by the ATP-binding site mutant form of BIK1, which does not dissociate from FLS2. In contrast, the flg22-induced phosphorylation of BIK1 and BIK1-FLS2 dissociation requires BAK1. Together, these results support the proposal that BIK1 acts downstream of FLS2 and BAK1. ATP-binding site and phosphorylation site mutant forms of BIK1 dominantly inhibit PAMP triggered immunity (PTI), indicating that the activated BIK1 kinase positively regulates PTI signaling. It is possible that the dissociation of the phosphorylated BIK1 and PBL1 proteins from FLS2 allows the activation of other components downstream of BIK1 and PBL1 (Zhang et al. 2010a). BIK1 is also required for the PAMPs elf18 and chitin-induced responses. BIK1 interacts with the PRRs CERK1 and EFR in protoplasts (Zhang et al. 2010a). BAK1 and/or its homologs are required for the activation of the receptor kinase CERK1 in chitin signaling (Zhang and Zhou 2010). The ligand-induced EFR-BAK1 interaction has been reported in the PAMP EF-Tu signaling (Schulze et al. 2010) and BIK1 has been shown to be required for interaction with the EFR (Zhang et al. 2010a). These studies suggest that upon PAMP binding, a complex forms between PRR, BAK1, and BIK1. An Arabidopsis *bik1* mutant is severely compromised in defense responses induced by flg22, elf18, and chitin, indicating that BIK1 plays a critical role in the integration of signals from multiple PRRs (Lu et al. 2010). The results demonstrate that BIK1 mediates PAMP-triggered immunity signal transduction from multiple PAMP receptor complexes. The BIK1 is an essential component in PAMP signal transduction, which links the PAMP receptor complex to downstream intracellular signaling (Lu et al. 2010). BIK1 may play a central role in signal integration from multiple surface-localized receptors (Zhang et al. 2010a).

### *2.12.7 PBL Proteins*

Zhang et al. (2010a) identified a number of the *Arabidopsis* resistance gene PBS1-like (PBL) cytoplasmic receptor-like kinases (RLCKs) as components in PAMP-signaling pathways. PBL proteins belong to the subfamily VII of cytoplasmic receptor-like protein kinases. One of the PBL proteins, BIK1, is required for signaling elicited by flg22, elf18, and chitin and is essential for PAMP-induced resistance to *P. syringae.* Other members including PBL1, PBL2, and PBS1 also contribute to PAMP-triggered defenses. BIK1 is localized to the plasma membrane (Veronese et al. 2006). PBS1 and many other PBL proteins may also localize to the plasma membrane, because these proteins possess putative myristoylation and palmitoylation sites at the N-terminus (Zhang et al. 2010a). PBL1, PBL2, and PBS1 appear to additively contribute to PAMP- signaling as their corresponding mutants showed slightly reduced PAMP-induced responses (Zhang et al. 2010a).

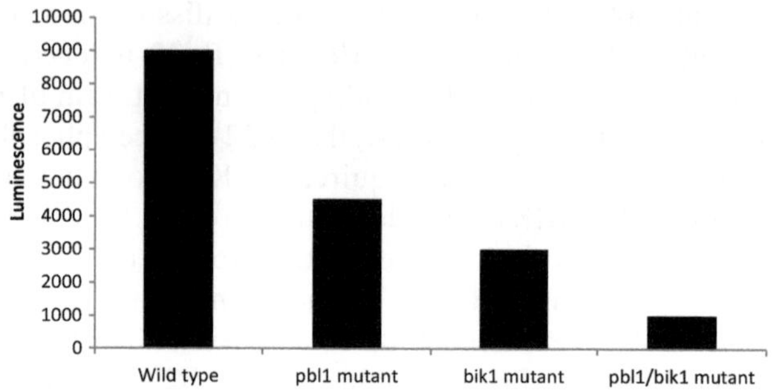


**Fig. 2.2** Flg22-induced oxidative burst in *bik1*, *pbl1*, and *bik1/pbl1* mutant *Arabidopsis* plants (Adapted from Zhang et al. 2010a)

Flg22-induced oxidative burst has been shown to be reduced in *pbl1* mutant *Arabidopsis* plants and the reduction was more in *pbl1* and *bik1* double mutant plants, suggesting that PBL1 acts additively with BIK1 in plant innate immune system (Fig. 2.2; Zhang et al. 2010a).

PBL1 interacts with unstimulated FLS2 and is phosphorylated upon flg22 stimulation (Zhang et al. 2010a). In addition to phosphorylation, flg22 induces a dissociation of PBL1 from FLS2 (Zhang et al. 2010a). The dissociation of the phosphorylated PBL1 and also phosphorylated BIK1 from FLS2 allows the activation of other components downstream of BIK1 and PBL1 (Lu et al. 2010; Zhang et al. 2010a). BIK1 and, to a lesser extent, PBL1, PBL2, and PBS1, are required for signaling from multiple PAMPs. BIK1, and likely PBS1 and other PBL proteins, directly act downstream of FLS2, EFR, CERK1 to trigger immune responses. Thus, the PBL1 and PBS1 proteins are key components that integrate signaling from multiple immune receptors. It appears that BIK1, PBL1, PBL2, and PBS1, integrate immunc signaling from multiple PRRs (Zhang et al. 2010a).

### 2.12.8 ERECTA Protein

ERECTA, a LRR- receptor-like protein kinase, is encoded by *erecta* (*er*) gene identified in *Arabidopsis* Landsberg *erecta* ecotype (Godiard et al. 2003). It is a member of the RLK family of transmembrane proteins with a conserved cytoplasmic serine/threonine kinase domain and a divergent extracellular LRR domain. ERECTA has been shown to interact with LRR-RLKs involved in plant innate immunity, potentially PRRs (Godiard et al. 2003; Llorente et al. 2005). It is suggested that ERECTA may function in signal perception and/or in transduction (Godiard et al. 2003). It is involved in triggering resistance against the necrotrophic fungus *Plectosphaerella cucumerina* (Llorente et al. 2005) and the bacterium *Ralstonia solanacearum* in *Arabidopsis* (Godiard et al. 2003). The leucine-rich repeat and the kinase domains of ERECTA were specifically required for resistance to *P. cucumerina*, as *er* mutant alleles impaired in any of these domains showed enhanced susceptibility to this fungus (Llorente et al. 2005).

### *2.12.9 AtPHOS32, AtPHOS34, and AtPHOS43 Proteins*

AtPHOS32, AtPHOS34, and AtPHOS43 are the other signaling components in PAMP-triggered immunity. These proteins were shown to be rapidly phosphorylated upon flg22 or chitin treatment (Peck et al. 2001; Merkouropoulos et al. 2008). AtPHOS43 and related proteins in tomato and rice are phosphorylated within minutes after treatment with flagellin or chitin fragments. Phosphorylation of AtPHOS43 after flagellin treatment was dependent on FLS2 (Peck et al. 2001). AtPHOS32 and AtPHOS34 show similarity to bacterial universal stress protein A.AtPHOS32 has been shown to be a substrate of the mitogen-activated kinases MPK3 and MPK6 (Merkouropoulos et al. 2008). The target phosphorylation site in AtPHOS32 is conserved in AtPHOS34 and among orthologues from many plant species (Merkouropoulos et al. 2008).

### *2.12.10 BIR1*

Upon recognition of the PAMP flg22, the PRR FLS2 heterodimerizes with BAK1 and activates the plant immune responses. Because constitutive activation of defense responses is detrimental, plant defense signaling pathways must be negatively controlled (Gao et al. 2009a). BAK1 forms a complex with BIR1 (for BRANCHING INHIBITING RECEPTOR 1) to negatively regulate defense responses (Tang et al. 2008; Gao et al. 2009a).

BIR1 is a BAK-1 interacting receptor-like kinase. Knocking out *BIR1* leads to activation of constitutive defense responses. A mutant, which suppresses the activity of *BIR1* has been obtained. The gene *SOBIR1* (suppressor of *bir1*) encodes another receptor-like kinase whose over expression activates defense responses (Gao et al. 2009a). SOBIR1 functions as a specific regulator of resistance activated by *bir1* mutation. SOBIR1 is not required for flg22-mediated defense responses. SOBIR1 and BIR1 did not interact with each other. SOBIR1 has been shown to be a positive regulator of innate immunity and it activates plant immune responses (Gao et al. 2009a). Transgenic plants over expressing *SOBIR1* showed enhanced expression of both *PR-1* and *PR-2* genes. BIR1 appears to negatively regulate SOBIR1 signaling pathways in *Arabidopsis* (Gao et al. 2009a).

## 2.13 PAMP Triggers Increased Transcription of PRR Gene and Accumulation of PRR Protein

### *2.13.1 PAMPs Activate Expression of Genes Encoding PRRs*

PAMPs activate the expression of genes encoding various PRRs. The PAMPs $NLP_{Pp}$ (from oomycete) and Flg22 (from bacteria) induced expression of the receptor genes in plants. They induced several folds the expression of receptor-like kinases such as lectin RLKs, S-locus RLKs, LRR-RLKs, wall-associated kinases, receptor protein kinase, and receptor Ser/Thr kinase in *A. thaliana* (Qutob et al. 2006). The LPS from *Burkholderia cepacia* up-regulated the gene encoding a receptor-like protein kinase (Sanabria and Dubery 2006). Many *Arabidopsis* genes encoding RLK and RLP were found to be induced by Flg22 or EF-Tu treatment (Zipfel et al. 2004). Chitin treatment activated the expression of LysM receptor kinase gene family (17 genes) in *Lotus japonica* (Lohmann et al. 2010). WAK1 was up-regulated upon perception of its own ligand, similarly to what has been shown for FLS2 and EFR (Brutus et al. 2010). Out of a total 216 LRR-RLK in *Arabidopsis*, 27 were found to be transcriptionally induced upon treatment with Flg22 or EF-Tu (Zipfel et al. 2006). These results suggest that PAMPs trigger enhanced expression of several PRRs (Table 2.5).

### *2.13.2 PRRs Are Activated by Widely Varying PAMPs*

FLS2 and EFR are induced also by bacterial LPS, fungal chitin, and the oomycete-derived NPP1 (Zipfel et al. 2006). Overall, these different PAMPs seem to trigger changes in a common set of genes, indicating that plants do not distinguish bacteria, fungi, and oomycetes on the basis of the signaling signature of their PAMPs. Rather,

**Table 2.5** PAMP-triggered upregulation of expression of PRR genes

| PAMP | Upregulated gene | Fold increase in expression | References |
|---|---|---|---|
| Flg22 | *FLS2* | 2.0 | Denoux et al. (2008) |
| | Receptor kinase gene | 3.3 | Qutob et al. (2006) |
| | *AtLECRK* | 9.8 | Qutob et al. (2006) |
| | *PROPEP* | 21.2 | Huffaker et al. (2006) |
| HrpZ | *PROPEP* | 40.8 | Huffaker et al. (2006) |
| Chitin | *CEBiP* | 2.6 | Kaku et al. (2006) |
| NPP1 | *PROPEP1* | 26.9 | Huffaker et al. (2006) |
| CBEL | *WAK1* | – | Khatib et al. (2004) |
| $NLP_{pp}$ | *AtLECRK* | 7.4 | Qutob et al. (2006) |
| | *WAK L7* | 19.5 | Qutob et al. (2006) |

presence of one type of PAMP seems to serve as an indicator of injury or danger in general (Zipfel et al. 2006). Treatment with flg22 upregulates the transcription of genes encoding PROPEP family precursors and both PEPR receptors, and AtPep1 treatment induces the transcription of FLS2, the flg22 receptor (Zipfel et al. 2004; Ryan et al. 2007).

### 2.13.3 PRRs May Act Additively in Perception of PAMP

Arabidopsis *efr1*, *fls2* and *fls2 efr1* mutants displayed more severe disease symptoms than the wild-type plants and allowed more bacterial growth when spray-inoculated with *P. syringae* pv. *tomato* compared with wild-type plants. It was also observed that *efr1* and *fls2 efr1* mutants were more susceptible to *P. syringae* pv. *tomato* compared with wild type and *fls2*, respectively, suggesting that FLS2 and EFR can act additively in perception of the PAMPs of this bacterium and triggering innate immunity (Nekrasov et al. 2009).

### 2.13.4 PRRs Bind with PAMPs for Their Activation

The PRRs bind with PAMPs. CERK1 binds polymeric chitin oligomers (Petutschnig et al. 2010). FLS2 binds with the PAMP flg22 (Chinchilla et al. 2006; Boutrot et al. 2010). PRRs such as FLS2 and EFR contain conserved cysteine residues flanking the LRR domain, which could form intermolecular disulfide bridges allowing stable homo- and heterodimerization or coupling to signaling molecules (van der Hoorn et al. 2005; Kolade et al. 2006). Chinchilla et al. (2006) showed that FLS2 itself is sufficient to mediate flg22 binding and there is no evidence for the involvement of additional proteins on flg22 binding. However, a reduction in the fluidity of FLS2 upon stimulation by flg22 was observed, indicative of the formation of larger complexes (Ali et al. 2007). The binding of an extracellular ligand (PAMP) induces a conformational alteration in PRRs leading to their activation (Ali et al. 2007).

### 2.13.5 Ethylene Regulates Transcription of PRRs on PAMP Perception

It is still not fully understood how the PAMP recognition by PRR leads to increased transcription of PRR genes. Boutrot et al. (2010) showed that ethylene is an integral part of PAMP-triggered immunity. Plants mutated in the key ethylene-signaling protein EIN2 are impaired in all FLS2-mediated responses, correlated with reduced FLS2 transcription and protein accumulation. The EIN3 and EIN3-like transcription factors, which depend on EIN2 activity for their accumulation, directly control FLS2 expression. The results suggest a direct role for ethylene in transcription of the PRR FLS2 (Boutrot et al. 2010).

Ethylene perception and signaling are crucial for *FLS2* gene transcription (Boutrot et al. 2010). FLS2 promoter revealed the presence of nine potential EIN3/EILs binding sites (Boutrot et al. 2010), suggesting that EIN3 may bind to the promoter of the *FLS2* gene to influence its transcription. EIN3 binds to two positions in FLS2 promoter in *Arabidopsis* seedlings treated with ethylene. The results suggest that endogenous ethylene controls FLS2 expression transcriptionally through direct binding of the transcription factor EIN3 and potentially, EIL1 to the FLS2 promoter (Boutrot et al. 2010).

Flg22 induces MAP kinases, which phosphorylate the ethylene biosynthetic enzymes ACC synthases 2 and 6 as well as EIN3, leading to its stabilization (Liu and Zhang 2004; Yoo et al. 2008). It is suggested that in the absence of flagellin, endogenous ethylene may ensure a constitutive level of FLS2 expression. On flg22 binding, FLS2 may activate MPK6 that, in turn may phosphorylate ACS2/6 and further leads to EIN3 stabilization, resulting in increased ethylene production and signaling (Boutrot et al. 2010). These studies suggest that endogenous ethylene may play an important role in PAMP-triggered expression of PRR genes in plants.

## 2.14 PAMPs Induce Phosphorylation of PRRs

### *2.14.1 PAMP-Induced Autophosphorylation of PRRs*

Most of the PRRs identified are receptor kinases and these protein kinase PRRs are known to be activated by PAMPs (Segonzac and Zipfel 2011). Before activation, the protein kinases are frequently autophosphorylated (Schlessinger 2000; Gómez-Gómez et al. 2001; Wang et al. 2005; Kanzaki et al. 2008). The PAMP flg22 induces autophosphosphorylation of the PRR FLS2 and the PRR receptor kinase is phosphorylated by its own serine/threonine kinase (Gómez-Gómez et al. 2001; Wang et al. 2001). Mutation of the threonine residue 867 in FLS2 hampers the autophosphorylation response, suggesting that autophosphorylation of the FLS2 occurs at threonine residue 867 (Robatzek et al. 2006). Autophosphorylation of the PRR EFR has also been reported, suggesting that EFR carries active kinase domain (Xiang et al. 2008).

The rice PRR, XA21, recognizes the PAMP, Ax21 (Activator of XA21-mediated immunity), which is highly conserved in all sequenced genomes of *Xanthomonas* and in *Xylella* (Lee et al. 2006a, 2009). It has been shown that the intracellular non-RD cytoplasmic kinase domain of XA21 contains intrinsic kinase activity (Liu et al. 2002). Chen et al. (2010b) showed that XA21 juxtamembrane (JM) domain is required for kinase autophosphorylation. Threonine 705 in the XA21 JM domain is essential for XA21 autophosphorylation *in vitro* and XA21-mediated innate immunity *in vivo*. The replacement of $Thr^{705}$ by an alanine or glutamic acid abolishes XA21 autophosphorylation (Chen et al. 2010d, e). Threonine residues analogous to $Thr^{705}$ of XA21 are present in the JM domains of most RD and non-RD plant receptor-like kinases (Chen et al. 2010d).

XA21 autophosphorylation occurs on multiple residues, some of which stimulate XA21 function and others of which inhibit XA21 function. Phosphorylation of certain residues on XA21 negatively regulates XA21 function, whereas phosphorylation on other residues may be required for activation of XA21 function. Autophosphorylation of the XA21 JM residues Ser686, Thr688 and Ser689 is important for stabilization of the XA21 protein (Xu et al. 2006a). The Thr705 residue in the XA21 JM region is required for binding to XA21 binding proteins (XBs) including XB3, XB10, XB15 and XB24 (Park et al. 2008; Chen et al. 2010d). The replacement of Thr705 residue by an alanine or a glutamic acid abolishes XA21 autophosphorylation and eliminates the interactions between XA21 and XB3, XB10, XB15 and XB24 in rice. These results suggest that after being autophosphorylated, Thr705 may transfer its phosphoryl group to another XA21 residue, which would activate XA21 (Chen et al. 2010a; Park et al. 2010b).

The *Phytophthora infestans* PAMP INF1 treatment of *Nicotiana benthamiana* results in autophosphorylation of the PRR NbLRK1. The autophosphorylation signal was stronger at 10 min after INF1 treatment (Kanzaki et al. 2008). Ser/Thr kinase domain of the tobacco PRR NgRLK1 shows autophosphorylation activity (Kim et al. 2010). NgRLK1 undergoes a conformational change upon enzymatic activation (Kim et al. 2010). The PAMP chitin oligomers and chitosan rapidly induce autophosphorylation of the PRR CERK1 at multiple residues in the juxtamembrane and kinase domain (Petutschnig et al. 2010). Kinase activity of CERK1 has been shown to be required for its chitin-dependent *in vivo* phosphorylation. The PRR CERK1 binds polymeric chitin oligomers. Subsequently, ligand binding leads to phosphorylation of CERK1 in the juxtamembrane and kinase domain (Petutschnig et al. 2010).

### *2.14.2 PRR Autophosphorylation Is Essential for PRR to Bind to Its Negative Regulators*

The activity of PRRs may be negatively regulated by some PRR binding proteins. Autophosphorylation of the rice PRR XA21 has been shown to be essential for XA21 to bind to its negative regulators including XB10 (OsWRKY62) (Peng et al. 2008), XB15 (a PP2C phosphatase) (Park et al. 2008) and XB24 (an ATPase) (Chen et al. 2010e). The replacement of $Thr^{705}$ residue by an alanine or glutamic acid abolishes XA21 autophosphorylation and eliminates interactions between XA21 and the three XA21-binding proteins (XB10, XB15, and XB24) in rice (Chen et al. 2010d). It suggests that PRR phosphorylation is important in binding these negative regulators.

Upon perception of PAMP, the PRR may be dissociated from the negative regulators. This may result in nullifying the function of the negative regulators and activating the function of PRR. One of the negative regulators in rice, XB24, is an ATPase (Chen et al. 2010e). The activity of the PRR XA21 is negatively regulated by XB24. Rice lines silenced for *xb24* display enhanced XA21-mediated immunity (Chen et al. 2010e). Association between XB24 and XA21 is compromised upon inoculation of the rice bacterial blight pathogen *Xanthomonas oryzae* pv. *oryzae*, which secretes the PAMP Ax21 (Chen et al. 2010e). The PRR XA21 protein is present on the plasma membrane after transit from the endoplasmic reticulum (ER) (Park et al. 2010a, b), where it recognizes the PAMP Ax21. XB24 physically associates with XA21 and uses ATP to promote phosphorylation of certain Ser/Thr sites on XA21, keeping the XA21 protein in an inactive state (Chen et al. 2010e). Upon recognition of the PAMP Ax21, XA21 may be dissociated from the negative regulator XB24 and/or XB24-promoted autophosphorylation may be removed (Chen et al. 2010e). When the function of XB24 is nullified, the XA21 kinase becomes activated, triggering downstream of defense responses (Chen et al. 2010e). These results suggest that PAMP may activate PRR by removing the action of negative regulators of PRR.

## 2.15 Negative Regulation of PRR Signaling

Improperly regulated plant immune responses can lead to the overexpression of defense-related genes and cell death (Park et al. 2008; Schweswinger and Zipfel 2008). It is therefore necessary that the PRR signaling components, as well as the PRRs themselves, are tightly regulated. Although PRR-mediated immune responses are clearly essential for innate immunity in plants, sustained or highly induced immune response can be harmful. It is therefore necessary that PRR signaling through non-RD kinases be under tight negative regulation.

One important class of negative regulators is protein phosphatase 2Cs (PP2Cs). *Arabidopsis* PP2C, kinase-associated protein phosphatase (KAPP) interacts with BAK1 and the PRR FLS2 (Gómez-Gómez et al. 2001). In *Arabidopsis*, FLS2 is negatively regulated by KAPP, which blocks activated FLS2 signaling and attenuates the downstream innate immune response (Gómez-Gómez et al. 2001). Overexpression of KAPP in *Arabidopsis* results in loss of sensitivity to flagellin treatment, suggesting that KAPP negatively regulates the FLS2-mediated immune response (Gómez-Gómez et al. 2001).

XB15, a PP2C phosphatase, is a PRR binding protein in rice. It dephosphorylates autophosphorylated XA21 and negatively regulates the PRR XA21-mediated innate immune responses (Wang et al. 2006; Park et al. 2008). Phosphorylation of certain residues is required for activation of XA21 function. These residues may be dephosphorylated by XB15 to down-regulate XA21 activity (Park et al. 2008). XB15, another rice protein, negatively regulates the XA21-mediated innate immune response (Park et al. 2008). Overexpression of the *Xb15* in an XA21 rice line compromised resistance to *X. oryzae* pv. *oryzae* (Park et al. 2008).

Another rice PRR binding protein, XB24, is an ATPase and it regulates XA21-mediated immunity (Chen et al. 2010e). XB24 displays significant ATP hydrolysis activity, while XB24 mutant containing a single aminoacid change Ser154 with Ala had only negligible ATPase activity, indicating that amino acid Ser154 is essential for its ATPase activity (Chen et al. 2010e). XB24 promotes autophosphorylation of the XA21 protein *in vitro*. XB24 is not transphosphorylated by the XA21 protein in the absence or presence of X. *oryzae* pv. *oryzae* expressing AX21 (Chen et al.

2010e). XB24 enhances XA21 autophosphorylation and its ATPase activity is required for this function. In plant a silencing of *Xb24* expression enhances XA21-mediated disease resistance (Chen et al. 2010e). Association between XB15 and XA21 is compromised while the association between XB15 and XA21 is enhanced upon Ax21 triggering (Park et al. 2008; Chen et al. 2010e). It is possible that the regulation by XB24 occurs before Ax21 recognition while regulation by XB15 occurs after Ax21 recognition (Park et al. 2010a, b).

## 2.16 Translocation of PRRs from Plasma Membrane to Endocytic Compartments

### *2.16.1 Endocytosis of PRRs*

Receptor endocytosis appears to be a common phenomenon in plant defense signaling system (Altenbach and Robatzek 2007; Chinchilla et al. 2007a; Robatzek 2007; Geldner and Robatzek 2008; Groen et al. 2008; Chen et al. 2010a). The endocytic machinery regulates the space and the time of signal transduction and processing in the cell (Irani and Russinova 2009). Plasma membrane resident receptors may be translocated into endosomes and it helps to extend the signaling surface ensuring a robust and efficient cellular signaling system (Geldner and Robatzek 2008). Translocation of PRRs from plasma membrane to endocytic compartments has been widely reported (Fliegmann et al. 2004; Gross et al. 2005; Robatzek et al. 2006; Leborgne-Castel et al. 2008).

The PRR FLS2 is found localized at the plasma membrane. When activated by the PAMP flg22, FLS2 is translocated to endocytic compartments. The induced FLS2 endocytosis is dependent on cytoskeleton and proteasome function. FLS2 lacks a YxxΦ motif (Y = Tyr, x = any amino acid, Φ = hydrophobic residue), which is known to play a role in clathrin-dependent endocytosis, but contains a PEST-like motif, which is reported to mediate receptor endocytosis via mono-ubiquitination. Single mutations in the PEST-like motif or at a conserved, potentially phosphorylated, residue in the JM region impaired FLS2 endocytosis (Robatzek et al. 2006). Endocytosis of the PRR FLS2 has been shown to be important for the PAMP flg22-induced defense signaling system (Chinchilla et al. 2007b).

The tomato PRR LeEIX2 belongs to a superclade of leucine-rich repeat receptor-like proteins (RLP) with a signal for receptor-mediated endocytosis (Bar et al. 2010). It carries the YxxΦ motif, which is involved in endocytosis, in its C-terminal part. Mutation in this endocytosis motif resulted in abolishment of HR induction in response to the PAMP EIX, suggesting that endocytosis plays a key role in mediating the signal generated by EIX that leads to HR induction (Ron and Avni 2004). EIX triggers internalization of the LeEIX2 receptor on endosomes, which is dependent on an intact cytoskeleton. LeEIX2 is internalized on highly motile endosome 15–20 min after EIX application (Bar and Avni 2008). Similar swift endocytic process has been reported in flg22-activated FLS2 internalization (Robatzek et al. 2006).

EFR is a glycosylated transmembrane protein and therefore needs to enter the secretory pathway to mature and to reach its final plasma membrane destination (Li et al. 2009b). It is localized at the PM, albeit with significant amounts of the receptor in the endo-membrane compartments (Nekrasov et al. 2009; Saijo et al. 2009; Häweker et al. 2010). The bacterial PAMP LPS was found to bind to the plasma membrane and to become internalized into vesicles in tobacco suspension-cultured cells (Gross et al. 2005). LPS uptake was abolished by amantadine, an inhibitor of receptor-mediated endocytosis (Gross et al. 2005). It suggests that the receptor of LPS may be translocated to the vesicle.

The β-glucan binding protein (GBP) in soybean is devoid of any transmembrane domain. Electron microscopy unraveled localization of GBP at the cytoplasmic face of the cell wall and to vesicles at the plasma membrane. It is suggested that GBP interacts with a receptor-like protein or RLK that is targeted for receptor-mediated endocytosis (Fliegmann et al. 2004). The PRR XA21 is localized in plasma membrane and may be endocytosed to initiate resistance responses during pathogen infection in rice (Chen et al. 2010a). XA21 is primarily localized to the ER but also to the PM (Park et al. 2010a, b). The chaperone complex interacts with rice CERK1 in the endoplasmic reticulum (ER) and mediates its maturation and transport to the PM (Chen et al. 2010c).

## 2.16.2 Clathrin-Mediated Endocytosis

The PRRs are translocated to endosomes through two different pathways. The first one is clathrin-mediated while the other one is ubiquitination-mediated (Robatzek et al. 2006). PRRs containing a YxxΦ motif show clathrin-dependent endocytosis, whereas the PRRs containing PEST-like motif undergo endocytosis via mono-ubiquitination (Gómez-Gómez and Boller 2000; Ron and Avni 2004; Robatzek et al. 2006; Bar and Avni 2008).

Clathrin is a basketlike network of protein molecules that forms on the cell membrane in response to the attachment of ligands (PAMPs) to receptors (PRRs). It coats the endocytotic vesicles which bud off from the membrane and it becomes inside surface of the coated vesicle during endocytosis. Clathrin-mediated endocytosis of PRR has been shown to be dependent on reactive oxygen species (ROS) production.

The oomycete PAMP cryptogein has been shown to stimulate endocytosis in tobacco cells. Internalization of the lipophilic dye FM4-64, which is a marker of endocytosis, was stimulated a few minutes after addition of cryptogein to tobacco cells. The cryptogein increased clathrin-mediated endocytosis. In the presence of cryptogein, increases in both FM4-64 internalization and clathrin-mediated endocytosis are specifically blocked upon treatment with tyrphostin, a receptor-mediated endocytosis inhibitor (Leborgne-Castel et al. 2008). The kinetics of the transient increase in clathrin-coated pits at the plasma membrane coincided with that of transitory reactive oxygen species (ROS) production occurring within the first 15 min after elicitation. In tobacco cells expressing the NADPH oxidase gene *NtrbohD* antisense cDNA, which are unable to produce ROS when treated with cryptogein, the clathrin-coated pits stimulation was inhibited. These results indicate that the very early endocytic process induced by cryptogein in tobacco is due to clathrin-mediated endocytosis and is dependent on ROS production (Leborgne-Castel et al. 2008).

### *2.16.3 Ubiquitin-Proteasome System May Be Involved in PRR Endocytosis*

Ubiquitin- and proteasome-mediated degradation of proteins plays an important role in endocytosis of PRRs (Robatzek et al. 2006; Aker and de Vries 2008). The main function of the proteasome is to degrade unneeded or damaged proteins by proteolysis. Proteasomes regulate the concentration of particular proteins and degrade misfolded proteins. Proteins are tagged for degradation by a small protein called ubiquitin. Proteins are targeted for degradation by the proteasome (Dreher and Callis 2007; Goritschnig et al. 2007). Proteasome inhibitors, such as MG132, which could deplete the cell's pool of freely available ubiquitin moieties, prevented flg22-induced internalization of FLS2 (Robatzek et al. 2006). It suggests that ubiquitin-proteasome system may be involved in endocytosis of the PRR.

### *2.16.4 Phosphorylation of PRR May Be Involved in PRR Endocytosis*

PAMP-induced PRR-mediated endocytosis has been shown to be dependent on phosphorylation of the PRR. Only those flg22 peptides that could activate the FLS2 receptor were able to target FLS2 for endocytosis (Robatzek et al. 2006). Mutation of a conserved potentially phosphorylated residue within the juxtamembrane region abolished FLS2 internalization. Addition of the protein kinase inhibitor K252a also abolished FLS2 internalization (Robatzek et al. 2006). The kinase-associated protein phosphatase interacts with FLS2 in vitro (Gómez-Gómez et al. 2001) and the protein phosphatase 2A inhibitor cantharidin affected FLS2 subcellular trafficking (Serrano et al. 2007). These results suggest that phosphorylation of PRR is involved in PRR endocytosis.

### *2.16.5 Involvement of EHD in Endocytosis*

Study of mammalian systems revealed that endocytosis depends on a large number of protein-protein interactions mediated by specific modules. One such module is the EH (Esp15 Homology) domain. The EHD (EHDomain) structure generally consists of two EF-hands and a helix-loop-helix structure that binds calcium, connected to an anti-parallel beta-sheet (Bar et al. 2010). The proteins containing the EHD have also been detected in plants (Bar et al. 2008). The EHDs isolated from *Arabidopsis* (AtEHD1 and AtEHD2) contain an EH domain with two EF calcium binding hands, a P-loop (GQYSTGKT) and DTPG with a predicted ATP/GTP binding site, a bipartite NLS (Nuclear localization signal) and a coiled-coil or helical domain, as well as a Dynamin –N motif (Dynamin like GTPase domain) (Bar et al. 2010).

The EHD has been shown to be involved in endocytosis in plants (Bar et al. 2008; Bar and Avni 2008). The plant EHD binds to the PRR and this binding is an important factor in internalization of the PRR (Bar and Avni 2008). EHD has an inhibitory effect on endocytosis in plant cells (Bar et al. 2008).

It has been shown that *Arabidopsis* EHD2 binds the cytoplasmic domain of the LeEix2 receptor and inhibits its internalization and signaling (Bar and Avni 2008). The ability of plant EHD2 protein to bind the LeEIX2 receptor was mediated by the EHD2 coiled-coil. Truncated EHD2 lacking the coiled-coil lost most of the ability to attenuate LeEix2 signaling (Bar et al. 2009). P-loop in EHDs was required for proper membranal localization of AtEHD2 (Bar et al. 2009). The results suggest that P-loop in EHDs is involved in proper membranal localization of AtEHD2, while the coiled-coil mediates the binding to target proteins thereby enabling the inhibitory function on endocytosis (Bar et al. 2009).

EHD2 was found to be specific to certain endocytic systems, in particular, in internalization of receptor-like proteins possessing a YxxΦ motif. It does not inhibit the internalization or signaling of FLS2, a receptor lacking this motif (Bar and Avni 2008). EFR, the receptor from *Arabidopsis* for the bacterial PAMP EF-Tu also carries YxxΦ motif (Zipfel et al. 2006). EHD2 appears to exert its inhibitory effect on endocytosis through the actin skeleton (Bar et al. 2009). *AtEHD2* expression is induced in response to the *Pseudomonas syringae* elicitor syringolin (Michel et al. 2006). The PAMP would have triggered expression of the endocytosis inhibitory protein in order to more tightly control the resultant HR (Bar and Avni 2008).

*NtEHD2* expression is rapidly induced upon EIX treatment in *Nicotiana tabacum* leaves, with a peak of 1.6–1.8 times the basal *NtEHD2* expression at 4 h after EIX application. However, the level of *NtEHD2* returned to normal at 8 h after EIX application (Bar and Avni 2008). EIX application triggers *NtEHD2* expression, upon which NtEHD2 acts to inhibit the defense response in the short term. Longer exposure to the PAMP leads to a 'full-blown' defense response including the HR, free of the inhibitory influence of EHD2, suggesting that a control mechanism based on the interplay of different proteins may be at work (Bar and Avni 2008).

At least two endocytic mechanisms for pattern recognition receptors may exist in plant cells, and EHD2 is involved in one such mechanism. EHD2 inhibits signaling of LeEix2, probably by inhibiting the endocytosis. The endocytosis mechanism of FLS2 appears not to require EHD2 involvement (Bar and Avni 2008). Flg22 does not induce *NtEHD2* expression (Bar and Avni 2008). EHD inhibits PAMP-induced endocytosis of PRR lacking a kinase domain, but not the PRR possessing a kinase domain (Bar and Avni 2008).

### 2.16.6 *What Is the Role for Endocytosis in PAMP-PRR Signaling?*

PAMP-induced internalization of PRRs from the plasma membrane (PM) is closely correlated with their immune function (Robatzek et al. 2006; Bar and Avni 2008). It is still not known whether this endocytosis leads to signal activation or attenuation of the PRRs (Saijo 2010). Endocytosis may help the exogenous ligands to provoke plant responses that are rapid, but transient, to ensure proper defense while preventing harm for the host cell by clearing the host cell of exogenous ligand (Geldner and Robatzek 2008). Translocation of activated cell surface receptors is associated with an attenuation of ligand-stimulated responses and also contributes to activate downstream signaling cascades (von Zastrow and Sorkin 2007).

## 2.17 ER-QC (for Endoplasmic Reticulum Quality Control) Pathways in Biogenesis of PRRs

### 2.17.1 *ERQC Mechanisms Monitor Protein Folding in ER*

In plants, pattern recognition receptors (PRRs) are known to reside in cell surface plasma membrane and no cytoplasmic PRRs have been reported so far (Zipfel 2009; Saijo 2010). In animals, extracellular PRRs are translated on the ER membrane, enter the ER lumen, and then are transported to plasma membrane (Akashi-Takamura and Miyake 2008). Similar mechanism of biogenesis and transport of proteins may exist in plants (Li et al. 2009b; Nekrasov et al. 2009; Park et al. 2010a, b). The biogenesis of trans-membrane PRRs may occur through the endoplasmic reticulum (ER) with the aid of ER-resident chaperones (Dodds and Rathjen 2010; Popescu 2012). After synthesis, proteins must rapidly fold to perform their biological activities (Anelli and Sitia 2008). ER is highly specialized for folding proteins and it greatly enhances protein folding efficiency (Kleizen and Braakman 2004). In the ER lumen, chaperones and folding enzymes are abundant and these folding factors in general prevent aggregation and thereby allow more efficient folding of a large variety of proteins (Kleizen and Braakman 2004). Folding status of client proteins is monitored during their folding and maturation by the process called ER-QC (for Endoplasmic Reticulum Quality Control) (Anelli and Sitia 2008). Unfolded proteins are retained in the ER until they are properly folded, or ultimately destroyed by ER-associated degradation (ERAD) in the cytosol (Vembar and Brodsky 2008). Plant cells are equipped with several ERQC mechanisms to monitor protein folding, allowing export of only correctly folded proteins to their final destinations but retaining misfolded proteins in the ER (Jin et al. 2007).

### 2.17.2 *Calnexin (CNX)/Calreticulin (CRT)/UGGT System*

ER-QC depends on two main different pathways involving different chaperones. The first pathway is specific to glycoproteins and is dependent on the calnexin (CNX)/calreticulin (CRT) cycle that relies on specific interaction between CNX/CRT, two ER-resident lectin-like chaperones, and Asn-linked monoglucosylated glycans. The CNX/CRT-glycan interaction depends on the availability of the terminal glucose residue, which is generated through sequential removal of two glucose residues of the core glycans on nascent proteins by glucosidases I and II. Eliminating the remaining glucose residue by glucosidase II releases the glycoproteins from the ER lectins. The released glycoprotein that has successfully acquired its native conformation can exit the ER to continue its secretory journey. By contrast, a deglucosylated glycoprotein with an incompletely/improperly folded conformation is recognized by the luminal enzyme UDP-glucose:glycoprotein glucosyltransferase (UGGT), which specifically functions as a folding sensor. UGGT transfers a glucose residue from UDP-glucose to glycans (Jin et al. 2007). This UGGT-catalyzed reglucosylation promotes its reassociation with CNX/CRT lectins to initiate another round of CNX/CRT-mediated folding (Williams 2006; Jin et al. 2007). CNX and CRT need assistance of α-glucosidase II as well as UGGT to release and re-bind substrate glycoprotein, respectively (Kleizen and Braakman 2004). The alternate action of glucosidase II and UGGT drives cycles of glycoprotein release from and binding to CNX/CRT until the glycoprotein is correctly folded. Terminally misfolded proteins are retrotranslocated into the cytosol for proteasome-mediated ER-associated degradation (ERAD) in the cytosol (Jin et al. 2007).

### 2.17.3 *BiP/ERdj/SDF2 System*

The second pathway in ERQC system involves Binding Protein (BiP), also called Glucose-Related Protein 78 (GRP78), which is a member of the heat shock protein70 (Hsp70) family of chaperones. It activates an adaptive signaling pathway termed the "unfolded protein response" (Kleizen and Braakman 2004). BiP consists of approximately 45 kDa domain at the N-terminus that is predicted to carry adenosine triphosphatase (ATPase) activity and a domain of 25 kDa at the C-terminus having a predicted substrate-binding domain (Mayer et al. 2003). BiP is localized to the ER (Park et al. 2010a, b). BiP interacts with the growing nascent chain of substrates containing N-linked glycans, facilitating their translocation into the ER (Molinary and Helenius 2000). In addition, it is involved in the ER-QC system by which misfolded or unassembled proteins are selectively retained in the ER (Kleizen and Braakman 2004; Park et al. 2010a). BiP targets permanently misfolded proteins for ER-associated degradation (ERAD) (Kleizen and Braakman 2004).

BiP ATPase cycle is controlled by a number of cofactors that regulate either ATP hydrolysis or nucleotide exchange. These include Hsp40 proteins, which act as co-chaperones in the ER. All Hsp 40s contain a J domain, named for a conserved about 70 amino acid motif in DnaJ, and are often referred to as J domain containing proteins or J proteins (Buck et al. 2007). DnaJ is a type I Hsp40, and contains an N-terminal J domain, a glycine/phenylalanine-rich domain and a cysteine-rich domain. DnaJ proteins specifically interact with the ATP-bound form of Hsp70s (Buck et al. 2007).

The ER-localised co-chaperone Hsp40 protein ERdj (ER DnaJ like protein) first directly binds to the misfolded substrate. ERdj then recruits BiP and activates BiP ATPase activity present in its N-terminus, leading to interaction of the C-terminal region of BiP with the substrate and the release of ERdj (Jin et al. 2008, 2009). The BiP retention system acts independently of, or subsequent to, the CNX/CRT cycle (Buck et al. 2007).

Another protein SDF2 (for Stromal Derived Factor 2) is also required for PRR biogenesis (Nekrasov et al. 2009). SDF2 resides in ER protein complex with the Hsp40 ERdj and the Hsp70 BiP, which are components of the ER-QC. Loss of SDF2 results in ER retention and degradation of PRR, suggesting a role for the BiP/ERdj/SDF2 in ER-QC system (Nekrasov et al. 2009). ER protein complex comprising stromal-derived factor-2 (SDF2), Erdj3B and BiP is required for the proper biogenesis of the PRR EFR (Nekrasov et al. 2009). *AtSDF2* is a single copy gene in *Arabidopsis* and orthologs exist in all eukaryotes. AtSDF2 is a small protein of 218 amino acids (24 kDa) consisting of a 23 amino-acid (aa) predicted N-terminal signal peptide and three repeats of the MIR domain. MIR domain is named after three of the proteins in which it occurs: protein mannosyltransferase, inositol 1,4,5-trisphosphate receptor (IP3R) and the ryanodine receptor (RyR). Although many eukaryotic proteins contain MIR domains, SDF2 is the only MIR domain-containing protein in plants (Nekrasov et al. 2009). SDF2 seems to be required for the function of EFR (the PRR for the PAMP elf18) and FLS2 (the PRR for the PAMP flg22). However, the requirement of SDF2 for the function of FLS2 is only to a lesser extent and SDF2 doesn't seem to be required for the function of CERK1, the PRR for the fungal PAMP chitin (Nekrasov et al. 2009). The oxidative burst induced by elf18 was strongly diminished in *sdf2* mutant *Arabidopsis* plants and it was less reduced after flg22 treatment. In contrast, the oxidative burst triggered by the fungal PAMP chitin was not impaired at all in *sdf2* mutant plants (Nekrasov et al. 2009).

### *2.17.4 Function of ERD2 in ER-QC*

Another protein involved in ER-QC is ERD2 (for ER Retention Defective2), the ER lumen protein-retaining receptor (Semenza et al. 1990). ERD2 binds the ER-escaped proteins and retrieves them back to the ER (Li et al. 2009b). The ERD2b protein is highly homologous to the yeast HDEL receptor. Five ERD2 paralogs (ERD2-likeproteins or ERPs) have been detected in *Arabidopsis*. ERD2b is essential for the calreticulin CRT3 accumulation. The retention of soluble ER protein relies mainly of the recognition of its C-terminal sorting signal (i.e. HDEL;

His-Asp-Glu-Leu tetrapeptide) by ERD2 (Semenza et al. 1990). *Arabidopsis erd2b* mutants show insensitivity to EFR, suggesting the importance of ERD2 in EFR function (Li et al. 2009b). The *erd2b* mutation specifically affects CRT3, which carries the C-terminal HDEL signal and CRT3 may be a likely substrate for ERD2b (Li et al. 2009b).

### 2.17.5 ER Quality Control Components Required for Biogenesis of the Pattern Recognition Receptor EFR

EFR is the PRR detected in *Arabidopsis* for binding the PAMP EF-Tu (Zipfel et al. 2006; Albert et al. 2010). It is a transmembrane glycoprotein and it needs to transit through the secretory pathway to mature and reach their final destination at the plasma membrane. The protein is secreted into the ER and undergoes quality control during folding and maturation in the ER. Several chaperones and enzymes resident in the ER take part in the ER-QC process. *Arabidopsis* genes encoding glucosidase II, UGGT, CRT3, ERdj3b, and ERD2b have been shown to be required for the EFR function and accumulation. SDF2 and STT3A, a subunit of the oligosaccharyltransferase complex are also necessary for EFR biogenesis (Li et al. 2009b; Nekrasov et al. 2009; Dodds and Rathjen 2010).

In the ER, the PRR proteins may be modified at glycosylable Asn residues by an oligosaccharyltransferase complex (OST). This function specifically depends on *STT3A* (*Staurosporin and temperature sensitive-3A*) coding for a component of the OST complex involved in N-glycosylation of nascent proteins (Nekrasov et al. 2009). Loss of STT3A-containing OST complex markedly decreases accumulation and signaling activity of the PRR EFR (Saijo et al. 2009; Häweker et al. 2010), suggesting the importance of the OST complex in ER-QC. The OST complex is involved in N-glycosylation of nascent proteins (Nekrasov et al. 2009). It covalently attaches a complex polysaccharide containing three terminal glucose residues. The glucose moieties may be subsequently trimmed by glucosidases I and II (Dodds and Rathjen 2010). A single glucose residue is added back by UGGT near regions of protein disorder. Monoglucosylated proteins interact with the lectin calreticulin (CRT) to retain misfolded substrates in the ER. In this way, UGGT acts as a folding sensor, and glycosylation is intimately related to protein maturation. Terminally misfolded proteins are degraded (Dodds and Rathjen 2010). UGGT mutant alleles that compromise EFR signaling have been identified (Li et al. 2009b). The allele *uggt-3* and a null insertion line (*uggt-4*) showed the PAMP elf18 insensitivity confirming that UGGT is required for EFR function (Li et al. 2009b).

Arabidopsis carries three *CRT* genes, including *CRT1*, *CRT2*, and *CRT3*. The *crt3-1* mutant was completely insensitive to elf18, as measured by oxidative burst, callose deposition, ethylene production, mitogen-activated protein kinase activation, and defense gene activation. Loss of *CRT1* together with *CRT2* compromises EFR function to a certain extent, while loss of *CRT3* alone abrogates EFR function completely, suggesting that *CRT3* contributes more to the ER-QC function than the other two CRT genes (Li et al. 2009b). CRT3 has been show n to be an ER-localized protein and it is required for EFR protein accumulation (Li et al. 2009b). A *crt3* null mutant did not accumulate EFR protein, suggesting that EFR is a substrate for CRT3 (Li et al. 2009b). The *Erd2b* mutant did not accumulate CRT3 protein. ERD2B seems to be HDEL receptor for CRT3 that allows its retro-translocation from the Golgi to the ER (Li et al. 2009b). The *crt3* mutants were more susceptible to *P. syringae* pv. *tomato* strains than *efr* mutants. It suggests that EFR is not the only PRR whose function is compromised by CRT3 mutations (Li et al. 2009b).

Nekrasov et al. (2009) have demonstrated the requirement of the soluble luminal protein Hsp40 ERdj3 for elf18 responses. ERDj3B is an ER-localized member of the HSP40 co-chaperone family. Arabidopsis *erdj3b-1* mutant plants were strongly affected in the bacterial PAMP elf18-triggered oxidative burst and MAP kinase activation (Nekrasov et al. 2009). SDF2 has also been shown to be required for EFR biogenesis (Nekrasov et al. 2009). *Sdf2* mutants are strongly impaired in EFR protein accumulation. The *sdf2* mutant plants were strongly affected in the PAMP elf18-triggered oxidative burst and MAP kinase activation. Loss of SDF2 results in ER retention and degradation of EFR (Nekrasov et al. 2009). However, the *sdf2* or *erdj3b* mutants are not completely insensitive to elf18, suggesting that BiP retention is less critical than CRT-based ER-QC for EFR proper folding and protein accumulation (Nekrasov et al. 2009).

SDF2 exists in a complex with ERdj3B and BiP3, in which ERdj3B may act as a bridge between SDF2 and BiP-3. As both SDF2 and ERdj3B lack an ER retention signal, their ER localisation might be due to interaction with BiPs (Nekrasov et al. 2009). BiP and CRT exist in an abundant large complex in tobacco (Crofts et al. 1998). CRT3, SDF2, Erdj3B, BiP, and potentially UGGT may exist in the same complex to regulate proper EFR folding (Li et al. 2009b). EFR biogenesis may require the SDF2/ERdj3B/BiP complex, in addition to ER-QC mediated by CRT3 and UGGT.

The other lectin component calnexin has been shown to be not necessary for biogenesis of EFR. A double mutant in two calnexin genes, *CNX1* and *CNX2* showed no impairment in elf18-triggered oxidative burst, or defense gene induction (Li et al. 2009b). These results suggest that calnexins may not be involved in the biogenesis of EFR.

### 2.17.6 ER Quality Control Components Required for Biogenesis of the PRR FLS2

The differential requirement of EFR and FLS2 for ER-QC and glycosylation components has been reported (Nekrasov et al. 2009). Both *crt3* and *uggt* mutants were unaltered in oxidative burst triggered by flg22, suggesting that UGGT and CTR may not have any significant role in biogenesis and accumulation of FLS2 (Li et al. 2009b). Like CRT3 and UGGT, ERD2b is also not required for FLS2 function (Li et al. 2009b). ERD2b is Golgi-localized and it is required for CRT3 protein accumulation (Li et al. 2009b). However, the *erdj3b-1* mutant was impaired in flg22 responses. ERDj3B is an ER-localized member of the HSP40 co-chaperone family and it may be involved in flg22 biogenesis (Nekrasov et al. 2009). It has also been reported that in the ER-QC mutants, EFR levels are greatly reduced while FLS2 levels remain unaffected, suggesting that ER-QC may not be involved in FLS2 signaling (Saijo 2010).

### *2.17.7 Role of BiP3 in ERQC of the Rice PRR XA21*

XA21 is glycosylated and is primarily localized to the ER and also to the plasma membrane (Park et al. 2010a). BiP3, the ER-localized chaperone HSP70, regulates XA21 processing and stability. BiP may serve as a PRR chaperone, and it may be involved in processing and degradation of XA21 (Park et al. 2010a). The rice *BiP3* gene encodes a 666 amino acid protein with an approximately 45 kDa domain at the N-terminus that is predicted to carry ATPase catalytic activity and a domain of approximately 25 kDa at the C-terminus having a predicted substrate-binding domain (Park et al. 2010a). In *BiP3*-overexpressing rice plants, XA21-mediated immunity is down-regulated and XA21 stability is significantly decreased. The results indicate that BiP3 regulates XA21 protein stability and processing and this regulation is critical for resistance to the bacterial pathogen *Xanthomonas oryzae* pv. *oryzae* (Park et al. 2010a).

BiPs are known to be involved in targeting unfolded glycoproteins for ER-associated degradation (ERAD) activity (Kleizen and Braakman 2004). If glycoproteins are not able to acquire their native performance within an appropriate time, misfolded or unassembled proteins are retained due to the ER-QC system. If unfolded and/or misfolded proteins may over accumulate in ER after PRR signaling and the cells may either initiate ER-associated cell death or attenuate the signal transduction pathway (Park et al. 2010a). BiP is involved in degradation of these proteins and these proteins are ultimately destroyed by ERAD (Kleizen and Braakman 2004; Meusser et al. 2005).

The rice PRR XA21 is highly glycosylated and this N-glycosylation is important for correct protein folding and ERAD (Kleizen and Braakman 2004). It has been shown that BiP3 accumulation drives glycosylated XA21 to the ERAD system, inhibiting its further processing (Park et al. 2010a). Fine control of membrane-resident PRR activity is essentially achieved by a combination of proper ER folding, degradation and trafficking of PRRs. Strict elimination of the misfolded proteins may occur by the action of BiP, which would avoid precocious immune activation (Saijo 2010).

## 2.18 *N*-Glycosylation of PRRs

### 2.18.1 Glycosylation of PRRs Is Required for Binding PAMP Ligands

*N*-glycans attached to ectodomains of plasma membrane PRRs constitute likely initial contact sites between plant cells and PAMPs. Under-glycosylated EFR and non-glycosylated FLS2 were not able to form functional ligand-binding sites. It suggests that LRR glycosylation of these PRRs is required for stably binding their elf18 and flg22 peptide ligands, an essential function for plant immunity (Häweker et al. 2010). Correct glycosylation also seems to be essential for the PRR EFR function. EFR accumulation was significantly reduced when synthesized without *N*-glucans. $EFR^{N143Q}$ lacking a single conserved *N*-glycosylation site from the EFR ectodomain accumulated to reduced levels, lost ability to bind to its ligand, and that to mediate elf18-mediated oxidative burst. However, $EFR^{N143Q}$ in wild type cells correctly targeted to the plasma membrane via the Golgi apparatus (Häweker et al. 2010).

EFR requires at least an *N* residue (N143) for stable accumulation and ligand binding (Häweker et al. 2010). This indicates that proper glycosylation on a particular site(s) is crucial for EFR function, despite extensive N-glycosylation of the receptor. It appears that N143 is located on the convex surface in the middle of the LRR domain. It is suggested that N143-glycosylation may mediate interactions with ER folding machineries, durable LRR folding, ligand binding and/or combinations there of (Saijo 2010). Ligand binding seems to occur in central LRRs 9–15 of FLS2 (Dunning et al. 2007).

### 2.18.2 N-Glycosylation Is Required for Transport of PRRs from Endoplasmic Reticulum to Plasma Membrane

*N*-glycans attached to PRRs seem to be critical for the export of proteins from the endoplasmic reticulum (ER) to the plasma membrane. The glycosylation of asparagines residues (*N*-glycosylation) is an essential, highly conserved co-translational modification of secreted proteins occurring in plant cells. The oligosaccharyltransferase (OST) complex in the endoplasmic reticulum (ER) controls transfer of *N*-glycans from dolicholpyrophosphate-linked lipid anchors to nascent polypeptide chains. *N*-glycans attached to polypeptides within ER lumen monitor correct folding of proteins. Only successfully folded proteins are exported from the ER and therefore, *N*-glycans are crucial for the transport of glycoproteins from ER to plasma membrane (Häweker et al. 2010).

PRRs are subject to *N*-glycosylation (an enzymatic process that attaches glycans to proteins), which in turn could be essential for the function of PRRs in triggering plant immunity. PRRs require *N*-glycosylation to mediate plant immunity (Häweker et al. 2010). The PRR EFR accumulation was significantly reduced when synthesized without *N*-glycans (Häweker et al. 2010). A single *N*-glycan plays a critical role for receptor abundance and ligand recognition during plant-pathogen interactions at the cell surface (Häweker et al. 2010).

PRRs such as EFR, FLS2, and the co-receptor BAK1 carry multiple putative *N*-glycosylation sites in their ectodomains. Successful folding in the ER and migration through the Golgi apparatus occur by *N*-glycosylation. *N*-glycosylation is important for folding and subsequent transport of the PRRs EFR and FLS2 to the cell surface (Häweker et al. 2010).

## 2.19 Significance of PRRs in Innate Immunity

Mutations in PRRs often compromise PAMP-induced defense responses and overall resistance to pathogens. For example, plants lacking *FLS2* are completely defective in flg22-induced ROS accumulation, MAPK activation, and defense gene expression (Gómez-Gómez et al. 1999; Asai et al. 2002). *Arabidopsis* plants mutated in FLS2 are more susceptible to the pathogen *Pseudomonas syringae* pv. *tomato* (Zipfel et al. 2004). The FLS2-mediated resistance to this strain is largely attributed to PAMP-induced guard cell closure that limits bacterial entry into the leaf tissue (Melotto et al. 2006). The flagellin gene *fliC*-induced defenses partially account for *Arabidopsis* non-host resistance to *P. syringae* pv. *tabaci* strain, a non-adapted pathogen on *Arabidopsis* (Li et al. 2005b).

The *efr* mutants are completely abolished in all responses to elf18 and show enhanced susceptibility to *Agrobacterium tumefaciens* (Zipfel et al. 2006). *Cerk1* mutants not only are insensitive to chitin treatment and display enhanced susceptibility to fungal pathogens (Miya et al. 2007; Wan et al. 2008b), but also are more susceptible to *P. syringae* bacteria (Gimenez-Ibanez et al. 2009a). Sustained activation of PRR signaling is important for mounting robust PAMP-triggered immunity (Saijo 2010). Collectively these studies demonstrate that PRR function is essential in triggering immune responses.

## 2.20 PAMPs-Induced Early Signaling Events Downstream of PRRs

### 2.20.1 *PAMPs Trigger Complex Networks of Signaling Pathways*

The plant immune system uses several second messengers to encode information generated by the PAMPs and deliver the information downstream of PRRs to proteins which decode/interpret signals and initiate defense gene expression (Snedden and Fromm 2001; Lecourieux et al. 2006; van Verk et al. 2008; Mersmann et al.

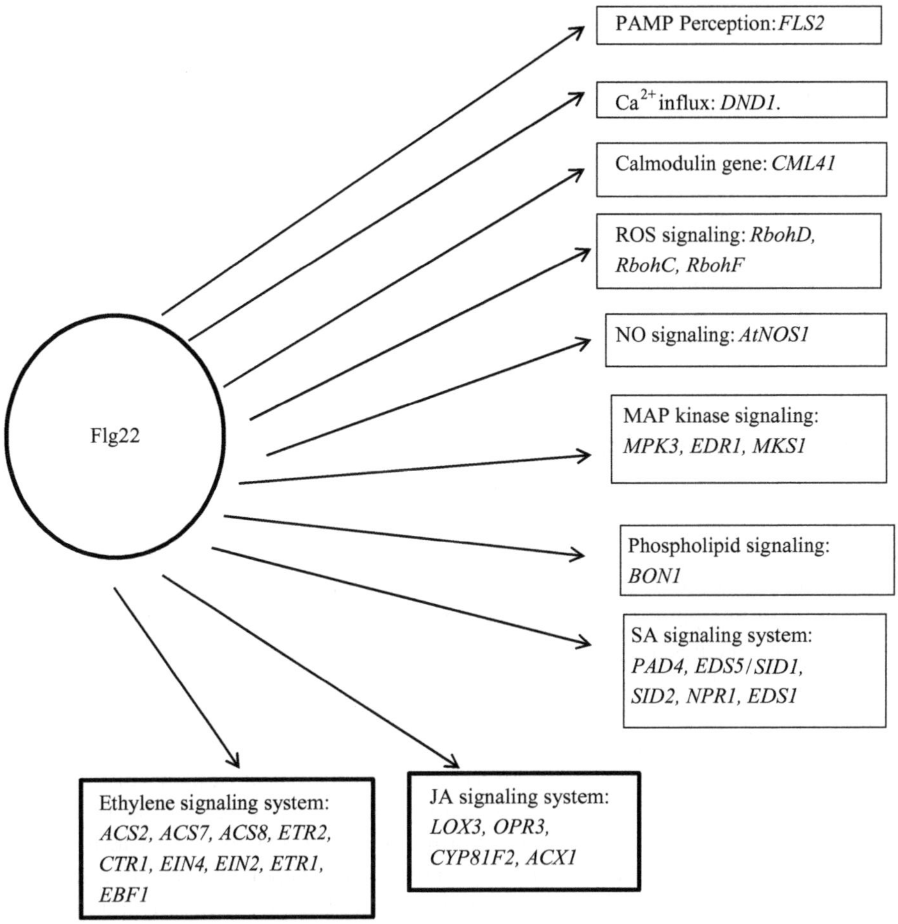


**Fig. 2.3** Flg22-induced expression of genes involved in various immune response signaling systems (Adapted from Denoux et al. 2008)

2010; Boudsocq et al. 2010; Hwang and Hwang 2011). It is still not known how the PAMP signals are transmitted downstream of PRR. The genes upregulated or downregulated by each PAMP are too many to find out their function in the downstream signaling events. Analysis of the *Arabidopsis* transcriptome revealed that more than 1,000 genes were significantly upregulated or downregulated within 30 min after the PAMP flg22 treatment (Zipfel et al. 2004). Denoux et al. (2008) observed 4,413 genes with altered expression in response to the PAMP flg22 in *Arabidopsis* seedlings. These genes are involved in activation of several distinct signaling systems in *Arabidopsis thaliana* (Fig. 2.3; Denoux et al. 2008).

A typical array of early defense responses induced by PAMPs includes distinctly different signaling systems and several second messengers. Second messengers are molecules that are used by plants to encode information and deliver it downstream to proteins which decode/interpret signals and initiate cellular responses (Snedden and Fromm 2001). Highly complex networks of signaling pathways are involved in transmission of the PAMP signals to induce the plant immune responses (Koornneef and Pieterse 2008; Gfeller et al. 2010; Leon-Reyes et al. 2010; Perchepied et al. 2010; Katagiri and Tsuda 2010; Choi and Hwang 2011; Fernández-Calvo et al. 2011). These signaling pathways are not simple linear and isolated cascades, but can crosstalk with each other (McGrath et al. 2005; Flors et al. 2008; Koornneef and Pieterse 2008). Both antagonism (Balbi and Devoto 2008; Flors et al. 2008) and synergism (Mur et al. 2006; De Vos et al. 2006; Mao et al. 2007) between the signaling systems have been reported.

### *2.20.2 $Ca^{2+}$ Signaling System*

In plant cells, the calcium ion is a ubiquitous intracellular second messenger involved in numerous signaling pathways (Lecourieux et al. 2006; Zhu et al. 2009; Ma et al. 2009). Calcium ion acts as a signal carrier and the calcium signaling is modulated by specific "calcium signatures" (Lecourieux et al. 2006). Spatial and temporal changes in cytosolic calcium ($[Ca^{2+}]_{cyt}$) are called "calcium signatures" (Luan et al. 2002). These changes may proceed as single calcium transients, oscillations, or repeated spikes/waves (Lecourieux et al. 2006). Specific calcium signatures are recognized by different calcium sensors to transduce calcium-mediated signals into downstream events (Harmon et al. 2000; Sanders et al. 2002; Reddy and Reddy 2004). The $Ca^{2+}$ signature controls diverse cellular processes via $Ca^{2+}$ sensors which include calmodulins (CaM), CaM-like and CaM-related proteins, calcineurin B-like (CBL) proteins, $Ca^{2+}$-dependent protein kinases (CDPKs) and $Ca^{2+}$-binding proteins without EF hands (Snedden and Fromm 2001; Luan et al. 2002; Tomsig et al. 2003; Kang et al. 2006; Kobayashi et al. 2007; Takabatake et al. 2007).

Several $Ca^{2+}$-permeable channels have been found in plant plasma membranes and they have been implicated in plant immune signaling system (White and Broadley 2003). Calcium ion channels are integral membrane proteins that are involved in transport of solutes across the cell membrane in plants (Maathuis et al. 1997). Cyclic nucleotide-gated ion channels (CNGCs) have been found in plant cell plasma membrane (Kaplan et al. 2007; Baxter et al. 2008). CNGCs are involved in $Ca^{2+}$-dependent signaling pathways (Talke et al. 2003; Yoshioka et al. 2003).

PAMPs elicit calcium ion influx within 15–30 min after PAMP treatment in plant cells, resulting in an immediate increase in $Ca^{2+}$ concentration in the cytosol (Lecourieux-Ouaked et al. 2000; Aslam et al. 2008). PAMP perception leads to membrane potential depolarization and an increase in cytoplasmic $Ca^{2+}$ concentration (Lecourieux et al. 2006; Aslam et al. 2008; Jeworutzki et al. 2010). Various PAMPs elicit an immediate increase in $[Ca^{2+}]_{cyt}$ (cytoplasmic calcium ion) concentration in plant cells (Lecourieux et al. 2002). The PAMP-induced $[Ca^{2+}]_{cyt}$ elevations predominantly result from a continuous $Ca^{2+}$ influx through the plasma membrane (Hu et al. 2004; Vandelle et al. 2006).

The cytoplasmic $Ca^{2+}$ spikes (oscillations and waves) result from two opposing reactions, $Ca^{2+}$ influx through channels and $Ca^{2+}$ efflux through pumps and transport systems (Hwang et al. 2000). Different messages can be encoded by changing a $Ca^{2+}$ spike's magnitude, duration, location, or frequency (Sanders et al. 1999). $Ca^{2+}$ signal is presented by the concentration of $Ca^{2+}$ (Trewavas 1999). PAMPs may activate $Ca^{2+}$ influx and the different signals may induce different $Ca^{2+}$ concentrations in the cytosol. The changes in $[Ca^{2+}]_{cyt}$ concentrations are monitored by the $Ca^{2+}$ sensors and the $Ca^{2+}$ signals are subsequently decoded and propagated downstream to activate plant defense responses.

Activation of $Ca^{2+}$-signaling system by different PAMPs has been demonstrated (Lecourieux et al. 2006; Aslam et al. 2009). The PAMP Flg22 recognized in several bacterial pathogens has been shown to induce $Ca^{2+}$ influx in *Arabidopsis*. The PAMP activates calmodulin-like gene *CML41* within 1 h after treatment and the peak expression of the gene was observed at 12 h (Denoux et al. 2008). Cyclic nucleotide-gated ion channels (CNGCs) are involved in the $Ca^{2+}$-dependent signaling pathways (Talke et al. 2003; Yoshioka et al. 2003). Arabidopsis *DND1* codes for a cyclic nucleotide-gated channel 2 (CNGC2) (Clough et al. 2000) and Flg22 induces the expression of *DND1* in *Arabidopsis* (Denoux et al. 2008). Flg22 also activates cyclic nucleotide gated channel 4 encoding gene *CNGC4* (Denoux et al. 2008). These channels have been found to be calmodulin (CaM)-binding proteins (Borsics et al. 2007).

The flagellin upregulated the gene encoding $Ca^{2+}$-dependent protein kinase in rice cells (Fujiwara et al. 2004). Calcium-dependent protein kinases CDPK4, CDPK5, CDPK6, and CDPK11 were shown to mediate the PAMP flg22-triggered defense responses, including defense gene expression and ROS production (Boudsocq et al. 2010). The pathway involving the calcium-dependent protein kinases (CDPK) 4/5/6/11 has been proposed to act in parallel to the MAPK pathways to control flg22-dependent gene expression (Boudsocq et al. 2010). Flg22 also induces the expression of BON1 gene which encodes a calcium-dependent phospholipid binding protein (Denoux et al. 2008). It also activated *CCD1* in cultured rice cells (Fujiwara et al. 2004). *CCD-1* encodes a $Ca^{2+}$-binding protein that shares homology with the C-terminal half domain of centrin and centrins are involved in $Ca^{2+}$ signaling (Takezawa 2000).

Several PAMPs are known to trigger $Ca^{2+}$ influx, as one of the earliest signaling systems (Aslam et al. 2009). The oomycete PAMP CBEL induces calcium ion fluxes in tobacco cells (Gaulin et al. 2006). Each PAMP may elicit different calcium signatures. A comparison of calcium influx patterns revealed that Flg22 induced a rapid (about 2 min) response, usually with two or three minor, decreasing peaks, whereas the elf18-induced calcium influx pattern was less defined and broader (Aslam et al. 2009). Eighteen transcripts involved in calcium sensing were up-regulated by Nep1 treatment. These genes encoded calcium-binding EF hand family proteins, CAM, CAM-binding, CAM-related proteins, and $Ca^{2+}$-ATPases (Bae et al. 2006). The induction of a transcript encoding CAM-related protein

(*CmCAL-1*) was reported in spotted knapweed and dandelion within 15 min in response to Nep1 (Keates et al. 2003).

The signaling cascade initiated by the HAMP AtPep1 leads to expression of defense genes in a $Ca^{2+}$-dependent manner (Qi et al. 2010). The endogenous elicitor AtPep1 after binding with its PRR AtPepR1 activates plant membrane inwardly conducting $Ca^{2+}$ permeable channels in mesophyll cells, resulting in cytosolic $Ca^{2+}$ elevation. This activity is dependent on the PRR AtPepR1 as well as a cyclic nucleotide-gated channel (CNGC2). The PRR AtPepR1 has guanylyl cyclase activity and this activity generates cGMP from GTP. The cGMP activates CNGC2-dependent cytosolic $Ca^{2+}$ elevation (Qi et al. 2010). AtPep-dependent expression of defense genes such as *PDF1.2*, *MPK3*, and *WRKY33*, is mediated by the $Ca^{2+}$ signaling pathway associated with AtPep peptides and their receptor (Qi et al. 2010). These studies indicate that downstream from AtPep and AtPepR1 in a signaling cascade, the cGMP-activated channel CNGC2 is involved in AtPep- and AtPepR1- dependent inward $Ca^{2+}$ conductance and resulting cytosolic $Ca^{2+}$ elevation.

Although several PAMPs trigger $Ca^{2+}$ influx, some PAMPs are capable of eliciting plant immune responses without triggering $Ca^{2+}$ influx (Lecourieux et al. 2002, 2006; Garcia-Brugger et al. 2006). Peptidoglycan and muropeptides are virtually ineffective at inducing $Ca^{2+}$ influx, yet are capable of eliciting other defense responses (Erbs et al. 2008). The role of $Ca^{2+}$ signaling system in plant innate immunity is further described in Chap. 4.

### *2.20.3 $H^+$ Fluxes and Extracellular Alkalinization*

$K^+/H^+$ exchange response is another important component in the plant immune signaling system (Orlandi et al. 1992; Felix et al. 1999). The PAMPs have been shown to trigger apoplastic alkalinization combined with cytosolic acidification in plant cells (Sakano 2001; Felle et al. 2004). Flagellin purified from *P. syringae* pv. *tabaci* is able to induce medium alkalinization in plant cell cultures from tomato, tobacco, potato, and Arabidopsis (Felix et al. 1999). The $Ca^{2+}$-dependent protein kinase, activated by increased cytosolic $Ca^{2+}$ concentration may evoke $H^+$ fluxes that lead to extracellular alkalinization and depolarization of the plasma membrane (Schaller and Oecking 1999).

The extracellular alkalinization is one of the earliest responses of suspension-cultured cells to OGs. Alkalinization results from the inhibition of the plasma membrane proton ATPases (PMA) via a signaling pathway that involves calcium ions and a calcium-dependent protein kinase (Schaller and Oecking 1999). In plants, PMA is the main electrogenic pump that generates the proton motive force across the plasma membrane and PMA inhibition reduces ATP consumption. OG elicitor significantly controls proton pumps, $K^+$ channels, and $H_2O_2$ production (Lecourieux et al. 2005, 2006).

### 2.20.4 G-Proteins

G-proteins (guanosine triphosphate-binding proteins) act as molecular switches in signal transduction system (Cabrera-Vera et al. 2003; Zeng et al. 2007). Several studies using inhibitors and agonists of G-proteins in different plant species have suggested that G-proteins are involved in activation of various defense signaling systems initiated by PAMPs (Beffa et al. 1995; Gelli et al. 1997; Ono et al. 2001; Park et al. 2000). G-proteins trigger changes in cytosolic $Ca^{2+}$ concentrations (Schultheiss et al. 2003). The G-proteins induce $Ca^{2+}$ channel opening in plants through the action of PAMPs (Gelli et al. 1997). G-proteins are involved in PAMP-activated ROS-mediated signaling system (Park et al. 2000; Suharsono et al. 2002). The PAMP flg22 induces G-protein-activated ROS signaling systems. The gene *AGB1*, encoding the β-subunit of G protein in *Arabidopsis*, is highly induced after flg22 treatment (Zipfel et al. 2006). The *agb1* mutants are impaired in the oxidative burst triggered by flg22, suggesting the importance of G-proteins in ROS signaling system (Ishikawa 2009). G-proteins are also involved in salicylate signaling system (Beffa et al. 1995), jasmonate signaling system (Zhao and Sakai 2003), ethylene signaling system (Fujiwara et al. 2006), and abscisic acid signaling system (Liu et al. 2007). G-protein OsRac1 induces biosynthesis of the important second messenger polyamine (Fujiwara et al. 2006). The G-protein may also be involved in generation of phospholipid second messengers (Viehweger et al. 2006). G-proteins may be involved in $Ca^{2+}$ channel opening (Gelli et al. 1997). Protein phosphorylation precedes $Ca^{2+}$ influx in tobacco cells treated with a PAMP isolated from the oomycete pathogen *Phytophthora cryptogea* (Tavernier et al. 1995). The G-proteins modulate the phosphorylation/dephosphorylation system in the plasma membrane of tomato cells and transduce the signal (Vera-Estrella et al. 1994a). Phosphorylation of proteins involved in G-protein coupled signaling has been reported in tobacco cells treated with a bacterial elicitor (Gerber et al. 2006). Heterotrimeric G proteins are involved in many diverse physiological processes in plants (Temple and Jones 2007; Chen 2008; Gao et al. 2008b; Oki et al. 2009). The role of G-proteins in signaling system in plant immune responses is further described in Chap. 3.

### 2.20.5 ROS Signaling System

The oxidative burst involving rapid and transient production of reactive oxygen species (ROS) is one of the most rapid defense responses observed in plants (Faize et al. 2004; Asada 2006; Sagi and Fluhr 2006; Vidhyasekaran 2007a, b). The oxidative burst is a very rapid response, occurring within seconds (Bolwell et al. 1995) or within a few minutes (Arnott and Murphy 1991) of PAMP treatment, suggesting that the oxidative burst may not require *de novo* protein synthesis but involves the activation of pre-existing enzymes. NADPH oxidase (Bae et al. 2006), peroxidases (Halliwell 1978; Lehtonen et al. 2012), and xanthine oxidase (Allan and Fluhr 1997;

Ori et al. 1997) have been shown to be involved in triggering ROS production. Different PAMPs may induce ROS production by different types of enzymes (Allan and Fluhr 1997). Cultured cells of rose treated with a PAMP derived from *Phytophthora* spp. produced ROS by a NADPH oxidase, whereas the cultured cells of French bean treated with a PAMP from *Colletotrichum lindemuthianum* produced ROS by the action of a cell wall peroxidase (Bolwell et al.1995, 1998). The transcript encoding NADPH oxidase (*AtrbohD*) was up-regulated (7.2 fold) by the PAMP Nep1 treatment and it generated ROS in *Arabidopsis* (Bae et al. 2006). Flagellin treatment triggers the expression of *RbohD* and *RbohC* genes encoding NADPH oxidases involved in generation of ROS (Denoux et al. 2008).

G-proteins regulate the production of ROS by activating the NADPH oxidase (Joo et al. 2005; Moeder et al. 2005; Wong et al. 2007). MAP kinases may also be involved in generation of ROS by activating the NADPH oxidase (Asai et al. 2008). Calcium-dependent protein kinase (CDPK) has also been shown to phosphorylate NADPH oxidase and trigger ROS production (Xing et al. 1997; Blumwald et al. 1998). Accumulation of ROS requires both $Ca^{2+}$ influx and protein kinase activity (Bolwell et al. 1995; Romeis et al. 1999).

ROS plays a central role in launching the defense response (Vandenabeele et al. 2003). It interacts with various defense signaling systems and activates $Ca^{2+}$ signaling system, NO signaling system, salicylic acid signaling system, jasmonate signaling system and ethylene signaling system (León et al. 1995; Desikan et al. 2001; Vranová et al. 2002; Vandenabeele et al. 2003; Fedoroff 2006; Hancock et al. 2006; Torres et al. 2006). The functions of ROS in the signaling network are described in Chap. 5.

### 2.20.6 Nitric Oxide Signaling System

PAMPs are known to trigger nitric oxide (NO) burst within minutes in plant cells (Foissner et al. 2000; Lamotte et al. 2004). The bacterial PAMP harpin induces NO generation in *Arabidopsis* cells (Krause and Durner 2004). The bacterial PAMP lipopolysaccharide (Lipid A) generates a rapid burst of NO production in *Arabidopsis* cells (Zeidler et al. 2004). Treatment of tomato cell cultures with the fungal PAMP xylanase resulted in a rapid NO accumulation (Laxalt et al. 2007).

NO is a gaseous readily diffusible free radical which acts as a messenger in plant immune signaling system (Besson-Bard et al. 2008; Wilson et al. 2008). NO may be synthesized through different pathways (Planchet et al. 2006; Yamasaki and Cohen 2006; Zhao et al. 2007; Zottini et al. 2007). NO is synthesized predominantly by nitric oxide synthase (NOS) (Zhao et al. 2007; Zottini et al. 2007). The PAMP flg22 triggered the expression of *AtNOS1* encoding nitric oxide synthase (NOS) (Denoux et al. 2008). NOS catalyses nitric oxide (NO) production (Crawford et al. 2006), suggesting that the PAMP activates NO signaling system. Polyamines may also be involved in NO synthesis (Tun et al. 2006; Besson-Bard et al. 2008) and polyamine oxidases or some unknown enzymes may be involved in the generation of NO from polyamines (Yamasaki and Cohen 2006; Besson-Bard et al. 2008). NO may be formed also from nitrite by the action of nitrate reductase (Rockel et al. 2002; Bethke et al. 2004; Wilson et al. 2008). Nitrate reductase transcript and protein levels increase in response to a PAMP in potato tubers, suggesting a role for nitrate reductase in the synthesis of NO during the plant immune response (Delledonne 2005). NO may also be synthesized from nitrite in a nonenzymatic manner (Yamasaki 2000).

$Ca^{2+}$ signaling system may be involved in activation of NOS-dependent NO generation (Lamotte et al. 2004). Application of a bacterial PAMP induced NO generation that was downstream from an influx of extracellular $Ca^{2+}$ (Ali et al. 2007). NO synthesis is regulated by the signaling cascade including cyclic nucleotide gated channel (CNGC)-mediated $Ca^{2+}$ currents with a concomitant increase in calmodulin (CaM) or calmodulin-like proteins (CML), and phosphorylation events (Ali et al. 2007; Ma and Berkowitz 2007). MAP kinase signaling cascades may also participate in NO production (Asai et al. 2008). NO acts in signal transduction through stimulus-coupled S-nitrosylation of cysteine residues (Benhar et al. 2008).

NO plays an important role in redox signaling system. It induced increased expression of catalase, peroxidase, glutathione S-transferase, glutathione-S-reductase, glutathione peroxidase, superoxide dismutase, thioredoxin, and glutaredoxin, which are involved in redox signaling system (Clarke et al. 2000; Polverari et al. 2003; Lindermayr et al. 2005) NO and ROS signaling systems appear to operate together in triggering innate immune responses (Grennan 2007; Asai and Yoshioka 2009). NO also acts in SA, ethylene, and jasmonate signaling systems. NO induces ACC synthase involved in ethylene biosynthesis (Lamotte et al. 2004). It induces the key enzymes of the JA biosynthesis pathway (del Rio et al. 2004; Palmieri et al. 2008). NO also triggers production of salicylic acid (Chamnongpol et al. 1998; Durner et al. 1998; Zago et al. 2006). SA in turn, activates nitric oxide synthesis in *A. thaliana* (Zottini et al. 2007). The role of NO in innate immune responses is further described in Chap.6.

### 2.20.7 Mitogen-Activated Protein Kinase (MAPK) Cascades

Mitogen-activated protein kinase (MAPK) cascades are major pathways downstream of PAMP/PRR signaling complex that transduce extracellular stimuli into intracellular responses in plants (Liu et al. 2003; Pedley and Martin 2005). Different PAMPs, including bacterial flagellin, elongation factor Tu, peptidoglycan, lipopolysaccharide, HrpZ1 harpin, and fungal chitin activate MAP kinase signaling system (Wu et al. 2011; Bethke et al. 2012). A typical MAPK signaling module consists of three protein kinases: a MAP kinase kinase kinase (MAPKKK or MEKK [for MAPK/Extracellular signal-regulated kinase Kinase Kinase]), a MAP kinase kinase (MAPKK or MKK), and a MAP kinase (MAPK or MPK) (Mészáros et al. 2006). MAPKKKs can be activated by various PAMPs (Teige et al. 2004). MAPKKK phosphorylates MAPKK, and MAPKK phosphorylates MAPK (Teige et al. 2004; Mészáros et al. 2006).

PAMP-triggered immunity (PTI) requires a signal transduction from receptors to downstream components via the MAPK cascade and many of the known PAMPs were shown to activate MAP kinases (Pitzschke et al. 2009a, b). MAPK signaling is activated by flagellin in *Arabidopsis* (Wu et al. 2011). The MAPK module MEKK1-MKK4/MKK5-MPK3/MPK6 has been proposed to be responsible for flg22 signal transmission (Asai et al. 2002). The involvement of MEKK1 in flg22-induced MKK4/MKK5-MPK3/MPK6 signaling is unlikely, since *mekk1* mutant plants are compromised in flg22-triggered activation of MPK4, but show normal activation of MPK3 and MPK6 (Suarez-Rodriguez et al. 2007). The PAMP-induced FLS2 activation leads to activation of the MAP kinase pathways MEKK4/5 and MPK3/6 and MEKK1-MKK1/2-MPK4, leading to transcription of defense-related genes through the WRKY transcription factors WRKY22/29 and WRKY25/33 (Nicaise et al. 2009).

The flagellin derived peptide flg22 triggers a rapid and strong activation of MPK3, MPK4 and MPK6 (Droillard et al. 2004). MPK3, MPK6, MKK4, and MKK5 form a cascade that positively regulates plant defenses (Pitzschke et al. 2009a). Flg22 activates the expression of *MPK3* in *Arabidopsis* (Denoux et al. 2008). MPK3 has been shown to be required for camalexin accumulation upon *Botrytis cinerea* infection (Ren et al. 2008). MPK3 phosphorylates a plant VirE2-interacting protein 1 (VIP1), a bZIP transcription factor (Liu et al. 2010; Zhang and Zhou 2010). VIP1 is a direct target of the PAMP-induced MPK3. Upon phosphorylation by MPK3, VIP1 relocalizes from the cytoplasm to the nucleus and regulates the expression of the *PR1* pathogenesis-related gene (Djamei et al. 2007). The MPK3 pathway is also involved in JA/ET signaling system (Takahashi et al. 2007). Inactivation of MPK3 and MPK6 by the *Pseudomonas syringae* effector HopA/1 and inactivation of MKKs by the *P. syringae* effector HopF2 severely impair PAMP-induced defenses and render plants highly susceptible to nonpathogenic *P. syringae* bacteria (Zhang et al. 2007a; Wang et al. 2010c). The results suggest that activation of MPK3 and MPK6 by the PAMP is an important component in plant immune responses. A recent study has shown that flg22 treatment increases expression of another *MPK* gene, *MPK11* in *Arabidopsis* (Bethke et al. 2012).

Flg22 induces the MAP kinase gene *MPK4* (Suarez-Rodriguez et al. 2007; Denoux et al. 2008). The MPK4 suppresses SA accumulation but induces the JA pathway (Brodersen et al. 2006). MPK4 interacts with its substrate MKS1; the latter interacts with WRKY transcription factors WRKY25 and WRKY33 (Andreasson et al. 2005). MPK4, MKS1, and WRKY33 form a complex in the nucleus, and the flg22-induced MPK4 activation releases WRKY33 from the complex. This enables WRKY33 to directly activate transcription of PAD3, which encodes a cytochrome P450 involved in the phytoalexin camalexin biosynthesis (Qiu et al. 2008b). MPK4, its upstream MAP kinase kinases MKK1 and MKK2, and the MAP kinase kinase kinase MEKK1 form a cascade that negatively regulates defenses in *Arabidopsis* because loss-of-function mutations in this cascade result in constitutive activation of defenses (Mėszáros et al. 2006; Suarez-Rodriguez et al. 2007; Gao et al. 2008a; Qiu et al. 2008a; Pitzschke et al. 2009b).

The PAMP-induced MAPKs may act either at upstream or downstream of various signaling systems in different plants. ROS (Nakagami et al. 2006), NO (Kumar and Klessig 2000), JA (Takahashi et al. 2007), SA (Zhang and Klessig 1997; Zhang and Liu 2001; Uppalapati et al. 2004), and abscisic acid (Uppalapati et al. 2004; Xiong and Yang 2003) have been reported to activate MAPK signaling cascade.

Although several studies have shown that MAPK cascades are important components in flagellin signaling system, there are also reports that flg22 may trigger the plant immune responses independent of MAPK cascades. The *bik1* mutant is significantly compromised in PAMP-induced resistance, but not the flg22-induced MAPK activation (Zhang and Zhou 2010). Transgenic plants expressing *AvrPphB*, which is capable of cleaving BIK1 and several PHL proteins, also show intact Flg22-induced MAPK activation (Zhang and Zhou 2010). The Arabidopsis *dde2/ein2/pad4/sid2*-quadruple mutant is largely impaired in flg22-induced resistance, but flg22-induced MAPK activation is comparable to wild-type plants (Tsuda et al. 2009). These results suggest that Flg22 may trigger immune responses against specific pathogens through a pathway other than MAPK signaling system. The role of MAPKs in plant innate immunity is further described in Chap. 7.

### *2.20.8 Salicylate Signaling System*

Salicylic acid (SA) is an important signal induced by PAMPs in plant innate immune system (Anand et al. 2008; Garcion et al. 2008; Mukherjee et al. 2010; Makandar et al. 2012). The PAMP Flg22 induces production of salicylic acid in plant cells (Mishina and Zeier 2007; Tsuda et al. 2008). It induces accumulation of SA in *Arabidopsis* (Mishina and Zeier 2007). SA is synthesized via both the isochorismate pathway and phenylalanine pathway in *Arabidopsis* (Wildermuth et al. 2001; Ferrari et al. 2003; Dempsey et al. 2011), while it is synthesized predominantly via the phenylalanine pathway in tobacco (Ogawa et al. 2006). Several regulatory proteins including SID2, EDS1, EDS4, EDS5, and PAD4are involved in triggering SA production in *Arabidopsis*. SID2 is an isochorismate synthase (Wildermuth et al. 2001), whereas EDS5exhibits homology to multidrug and toxin extrusion (MATE) transporter proteins from animals (Nawrath et al. 2002). EDS5 is involved in the transport of precursors for SA biosynthesis (Nawrath et al. 2002). EDS1 is another regulatory protein (Moreau et al. 2012) and it controls SA production to amplify defense signals (Eulgem et al. 2004; Song et al. 2004). PAD4 displays similarity to triacyl glycerol lipases and other esterases (Jirage et al. 1999). EDS1 forms distinct complexes with PAD4 (Wiermer et al. 2005; Xing and Chen 2006) and EDS1 and PAD4 complex transduces ROS-derived signals leading to SA production (Mateo et al. 2004; Wiermer et al. 2005). SA signaling consists of a linear pathway in which EDS1, PAD4 and EDS4 activate EDS5 and SID2, which produce SA (Glazebrook et al. 2003).

Another regulatory protein, flavin-dependent monooxygenase1 (FMO1) may also be involved in SA signal amplification (Mishina and Zeier 2006). FMO1 may contribute to a signal amplification loop involving ROS, SA, and NPR1 that is required to potentiate innate immune system (Mishina and Zeier 2006). Another protein, GH3.5, has been shown to be involved in SA accumulation in *A. thaliana* and it possesses adenylation activity on SA (Zhang et al. 2007c). The protein CDR1 has also been shown to take part in SA signaling (Xia et al. 2004). The CDR1 may be an aspartic protease. It induces accumulation of SA and also induces oxidative burst, suggesting that ROS-mediated SA accumulation is mediated by CDR1 (Xia et al. 2004). Several signaling systems, including $Ca^{2+}$- signaling network system (Garcia-Brugger et al. 2006), G-proteins (Fujiwara et al. 2006), MAPK signaling systems (Zhang et al. 2007a), ROS signaling system (Torres et al. 2006; Ahn et al. 2007), and NO signaling system (Durner et al. 1998) act upstream of SA synthesis.

Several SA-binding (SAB) proteins have been shown to be involved in accumulation of SA in plants. The first SAB protein identified is the cytosolic (peroxisomal) tobacco catalase that reversibly binds SA (Chen et al. 1993). SA inhibits $H_2O_2$-degrading activity of catalase and the SA-mediated inhibition of catalase may generate $H_2O_2$, which may activate the ROS signaling system (Chen et al. 1993). A second specific high-affinity SA-binding protein, SABP2, has been identified as a methyl salicylate esterase whose function is to convert biologically inactive methyl salicylate to active SA (Kumar and Klessig 2008; Vlot et al. 2008; Manosalva et al. 2010; Liu et al. 2011a, b). The third SA-binding protein, SAB3, identified in tobacco is the chloroplast carbonic anhydrase (Slaymaker et al. 2002). It shows antioxidant activity and SA may inhibit the antioxidant activity by binding with SABP3. The inhibition of antioxidant enzymes may enhance ROS levels (Slaymaker et al. 2002). Azelaic acid is a long-distance priming signal (Parker 2009). It primes plants to accumulate SA upon infection by pathogens (Jung et al. 2009).

SA triggers ROS and NO signaling systems (Blee et al. 2004; Zottini et al. 2007; Kobeasy et al. 2011). It also activates MAPK signaling cascade (Uppalapati et al. 2004; Brodersen et al. 2006). SA elevates NPR1 (for *non-expresser of PR gene1*), which is a master regulator of SA-mediated defense responses (Chern et al. 2008). SA induces increased expression of several WRKY and ERF transcription factors in *Arabidopsis* (Knoth et al. 2007; Mao et al. 2007; Miao and Zentgraf 2007; Zheng et al. 2006, 2007; Zhang et al. 2007d; Grennan 2008; Moreau et al. 2012). These transcription factors are necessary for the inducible expression of several defense genes (Yu et al. 2001; Grennan 2008).

Flg22 activates the transcription factor WRKY7, which is a negative regulator of SA-mediated responses in *Arabidopsis* (Denoux et al. 2008). Probably the transcription factor would have suppressed the SA- induced defense response. The transgenic *Arabidopsis* plants over expressing *WRKY7* showed reduced expression of defense-related genes, including *PR1* (Kim et al. 2006).

Although most of the PAMPs activate SA signaling system, there are also reports that SA signaling system may not be necessary to activate the plant immune systems against a particular pathogen. Ferrari et al. (2007) showed that resistance to *Botrytis cinerea* induced in *Arabidopsis* by flg22 was independent of SA signaling.

### 2.20.9 Jasmonate Signaling System

PAMPs have been shown to trigger accumulation of jasmonic acid (Wang et al. 2000; Fabro et al. 2008). The fungal PAMP chitosan treatment induced accumulation of jasmonic acid in tomato leaves (Doares et al. 1995). Several enzymes including lipoxygenase, allene oxide synthase, allene oxide cyclase, OPDA reductase 3 (OPR3) and acyl-CoA oxidase (ACX) are involved in biosynthesis of jasmonic acid (JA) (Mei et al. 2006; Schilmiller et al. 2006; Balbi and Devoto 2008; Delker et al. 2007; Vidhyasekaran 2007a). Flg22 induced the activation of JA signaling system. It enhanced the expression of *LOX3* and *LOX4* genes encoding lipoxygenases (LOX), *OPR3* gene encoding 12-oxophytodienoate reductase (OPR) and *ACX1* gene encoding acyl-CoA oxidase (ACX) (Denoux et al. 2008). The fungal PAMP chitosan also activates lipoxygenase, the key enzyme in JA-mediated signaling system (Bohland et al. 1997; Rakwal et al. 2002). Lipoxygenase activity significantly increased in chitosan-treated carrot plants (Jayaraj et al. 2009). The PAMP β-1,3-glucan induces expression of *LOX* gene encoding lipoxygenase in grapevine (Aziz et al. 2003; Balbi and Devoto 2008) and tobacco cells (Klarzynski et al. 2000). The PAMP Nep1 rapidly induces genes involved in JA biosynthesis (Bae et al. 2006). It triggered genes encoding lipoxygenases (*LOX*), 12-oxophytodienoate reductase, and allene oxide cyclase (*AOC2*), which are involved in JA biosynthesis (Bae et al. 2006). The HAMP oligogalacturonates triggered OPR3 and ACX1, the key enzymes involved in biosynthesis of JA (Denoux et al. 2008).

JA can be metabolized to several derivatives and some of them are involved in defense signaling system. Methyl jasmonate is one of these JA derivatives, which trigger the immune signaling system (Seo et al. 2001). JA is converted to methyl jasmonate (MeJA) by the action of jasmonic acid methyl transferase (Wasternack 2007). The JA amino acid conjugate JA-Ile (jasmonoyl-isoleucine) also has been shown to be involved in defense signaling (Kang et al. 2006; Katsir et al. 2008). JA-amino synthetase activates conjugation of JA to an amino acid and this enzyme may be involved in JA-Ile biosynthesis (Staswick and Tiryaki 2004). In addition to Ile, the JAR family of related GH3 enzymes has the potential to conjugate other amino acids, such as Trp, Val, and Leu in tobacco. The JA-Trp, JA-Val, and JA-Leu may also participate in JA signaling pathway (Wang et al. 2008c).

G-proteins activated $Ca^{2+}$ influx and the subsequent $Ca^{2+}$ wave may initiate calmodulin-dependent protein kinase cascade, ROS production, and eventually the jasmonate biosynthesis (Zhao et al. 2004; Trusov et al. 2006). MAPK cascades may also be involved in JA biosynthesis (Lee et al. 2004; Kandoth et al. 2007). NO is involved in induction of biosynthesis of JA (Xu et al. 2005; Palmieri et al. 2008). NO induces the key enzymes of the JA biosynthesis pathway (del Rio et al. 2004; Grün et al. 2006; Zago et al. 2006). Several PAMPs are known to activate plant innate immunity by triggering the action of several components in the JA signaling system. The oomycete PAMP cryptogein induced an increase in lipoxygenase activity and the accumulation of JA-responsive proteinase inhibitors in tobacco suggesting the role of JA signaling system in the PAMP-mediated disease resistance (Bottin et al. 1994).

### 2.20.10 Ethylene Signaling System

Ethylene (ET) signaling system is another important component in plant's innate immune system (Cao et al. 2006; Xu et al. 2007; Lin et al. 2008; Gaige et al. 2010; Al-Daoud and Cameron 2011; Sun et al. 2010; Zhu et al. 2011). Methionine (L-Met) is the precursor of ethylene. It is converted to *S*-adenosylmethionine (*S*-AdoMet) by the action of *S*-AdoMet synthetase (SAM synthetase) (Wang et al. 2002). Nitric oxide (NO) induces SAM synthetase, which catalyzes the conversion of ATP and methionine into *S*-adenosyl-Met (Zago et al. 2006). S-AdoMet is converted to 1-aminocyclopropane −1-carboxylic acid (ACC) by ACC synthase (ACS) (Chae et al. 2003; Peng et al. 2005). Accumulation of ACS isozymes leads to increased synthesis of ACC (Liu and Zhang 2004) and the ACC is oxidized by ACC oxidase (ACO) to form ET (Wang et al. 2002; Vidhyasekaran 2007a). Both $Ca^{2+}$and NO signaling systems are involved in ethylene biosynthesis. The $Ca^{2+}$ influx activates ACC oxidase (Gallardo et al. 1999), whereas NO induces ACC synthase, resulting in accumulation of ethylene (Lamotte et al. 2004).

Downstream components of ET signal transduction system in *Arabidopsis* include activation of ET receptors. In *Arabidopsis*, ethylene is perceived by a family of five membrane-bound receptors (ETR1, ERS1, ETR2, EIN4, and ERS2), which transmit the signal to downstream effectors (O'Malley et al. 2005; Wang et al. 2006; Qu et al. 2007; Grefen et al. 2008). Ethylene receptor ETR1 has been shown to mediate ROS signaling in *Arabidopsis* (Desikan et al. 2005). ETR1 functions as an ROS sensor. ROS up-regulates four ethylene-responsive element-binding proteins (EREBPs), the ethylene-responsive transcription factor (ERF1), and a CEO1-like protein, which is a potential cofactor of EREBP transcription factors in tobacco (Vandenabeele et al. 2003). NO signaling activates *EIN3*, which is involved in activation of transcription of ethylene-responsive genes (Chang and Stadler 2001).

The HAMP oligogalacturonides induced several plant cell membrane-bound ethylene receptors such as ETR1, EIN2, ERF1 and ERF4 (Denoux et al. 2008). Flg22 induced expression of the ethylene receptors ETR1 and EIN4 (Denoux et al. 2008). In *Arabidopsis*, ethylene is perceived by membrane-bound receptors such as ETR1 and EIN4, which transmit the signal to downstream effectors (Qu et al. 2007; Grefen et al. 2008).

A MAPK cascade, MAPKKK (CTR1) – MKK9 – MPK3/MPK6, has been shown to be an important downstream component in ET signaling system induced by PAMPs (Yoo et al. 2008). The MAPKKK CTR1 is a negative regulator of defense responses. Both the inhibition of CTR1 and activation of MKK9 are induced by ethylene signaling (Yoo et al. 2008). ET activates JA biosynthesis (O'Donnell et al. 1996) and also activates JA-inducible defense gene expression (Gu et al. 2002; Brown et al. 2003; Tournier et al. 2003; McGrath et al. 2005). The PAMP-induced ET also induces $Ca^{2+}$ influx, which acts at downstream of ET (Kwak and Lee 1997). $Ca^{2+}$ from intracellular pools, but not $Ca^{2+}$ from the apoplast, may interact with ET signal transduction (Petruzzelli et al. 2003). The CaM binding proteins (EICBPs) isolated from tomato, parsley, and *Arabidopsis* have been found to be induced by ET (Reddy et al. 2000). $Ca^{2+}$ may play an important role in ET signal transduction (Raz and Fluhr 1992; Kwak and Lee 1997; Gallardo et al. 1999; Reddy et al. 2000).

Several PAMPs are known to induce production of ethylene in plant cells. The oomycete PAMP cryptogein induced production of ethylene in tobacco cells (Milat et al. 1991). The bacterial PAMP flg22 up-regulated *ACS* genes encoding ACC synthase, which is the key enzyme involved in biosynthesis of ethylene (Denoux et al. 2008). Flg22 induces ET production through activation of ACS6, an ET biosynthetic enzyme (Liu and Zhang 2004). The PAMP Nep1 induced the expression of transcripts encoding ACC synthase and ethylene-responsive element binding factors (ERF), which are involved in ethylene signaling (Bae et al. 2006). The maize HAMP *Zm*Pep1-treated maize leaves emitted a five-fold increase in ethylene (ET). Expression of the gene encoding ACC oxidase also responded to ZmPep1 treatment (Huffaker et al. 2011).

Flg22 treatment up-regulated the expression of the ET-responsive transcription factor ERF1 (Ethylene-Responsive element-binding Factor 1) in *Arabidopsis* (Clay et al. 2009). ERF1 is a downstream component of ethylene signaling system (Berrocal-Lobo and Molina 2004). ET signaling is required for the full induction of ERF1 in response to flg22 (Clay et al. 2009). ERF5, another ethylene-responsive element-binding factor, is induced by the fungal PAMP chitin. It induces defense against *Pseudomonas syringae* pv. *tomato* in *Arabidopsis* (Son et al. 2012).

The transcription factor MYB51 is also induced by ET and MYB51 acts downstream of ET signaling for the callose response (Clay et al. 2009). MYB51 induced expression of all known indole glucosinolate (IGS) biosynthetic enzymes. MYB51 is involved in the transcriptional activation of IGS biosynthetic gene ASA1. ASA1 expression is ET-inducible (Clay et al. 2009). ASA1 gene expression was also induced by Flg22 treatment (Clay et al. 2009). The HAMP oligogalacturonide induced the ET biosynthesis enzymes ACC synthases (ACS7 and ACS8) (Denoux et al. 2008).

ET is required for the oxidative burst contributing to plant immunity (Mersmann et al. 2010). ROS production is triggered by flg22 in *Arabidopsis*. The ROS production was diminished in ethylene-insensitive mutants (Mersmann et al. 2010). Ethylene signaling also regulates accumulation of the FLS2 receptor (Mersmann et al. 2010). *FLS2* accumulation was reduced in *etr1 and ein2*, indicating a requirement of ethylene signaling for *FLS2* expression (Mersmann et al. 2010).

### 2.20.11 Abscisic Acid Signaling System

Abscisic acid (ABA) signaling system is another important component in plant immune system activated by PAMPs (Adie et al. 2007). PAMPs trigger increases in ABA concentrations, inducing disease resistance (Whenham et al. 1986) or susceptibility (Koga et al. 2004; Schmidt et al. 2008). Xanthoxin is the first cytoplasmic precursor for the biosynthesis of ABA. It is converted to abscisic aldehyde by the action of a dehydrogenase/reductase encoded by *ABA2*. Abscisic aldehyde is converted to ABA by the action of two enzymes, ABA3 and AAO3. *ABA3* encodes the molybdenum cofactor sulfurase that adds the sulfur atom to the molybdenum center. The *AAO3* encodes an aldehyde oxidase that requires molybdenum cofactor for its activity. Abscisic aldehyde is converted to abscisic alcohol, which in turn is converted to ABA through a shunt pathway (Wasilewska et al. 2008).

A regulator of G protein signaling (RGS) proteins, RGS1, has been identified in *Arabidopsis* (Chen et al. 2006). RGS1 significantly stimulated the expression of *NCED* encoding the 9-*cis*-epoxycarotenoid dioxygenase (NCED) enzyme, which cleaves 9-*cis* xanthophylls to xanthoxin. The NCED is the first committed step for ABA synthesis (Chen et al. 2006; Wasilewska et al. 2008). RGS1 also triggered increased expression of *ABA2* gene (Chen et al. 2006), which encodes dehydrogenase/reductase involved in conversion of xanthoxin to abscisic acid aldehyde (Wasilewska et al. 2008). These results suggest that RGS1 is involved in biosynthesis of ABA (Chen et al. 2006).

ABA regulates several signaling systems, including $Ca^{2+}$ signaling complex, ROS signaling pathway, and NO signaling system. ABA induces oscillations in $[Ca^{2+}]_{cyt}$ by inducing both $Ca^{2+}$ release from intracellular stores and $Ca^{2+}$ influx from the extracellular space (Hamilton et al. 2000; Pei et al. 2000; Klüsener et al. 2002). ROS production precedes activation of $Ca^{2+}$ influx, and $H_2O_2$ activates plasma membrane $Ca^{2+}$ channels (Coelho et al. 2002). Cytosolic $Ca^{2+}$ elevation induced by ABA activates slow (S-type) anion channels (Schroeder and Hagiwara 1990; Brault et al. 2004). ABA activates the plasma membrane anion channels and in several species, this response is associated with changes in the cytoplasmic $Ca^{2+}$ concentration (Marten et al. 2007). The expression of various calcium-dependent protein kinases (CDPKs) of tobacco was also upregulated by ABA (Ludwig et al. 2004). ABA response element binding factors (ABFs) have been shown to be activated by phosphorylation by protein kinases (Uno et al. 2000). ABA activates mitogen activated protein kinase (MAPK)-mediated signaling system (Gomi et al. 2005; Wang and Song 2008). ROS may act upstream of the MAPK cascade in the ABA signaling system in maize leaves (Zhang et al. 2006).

The PAMP flg22 triggers ABA synthesis in plants (Melotto et al. 2006). The HAMP oligogalacturonates (OG) induced the enzyme molybdenum cofactor sulfurase (ABA3), which is involved in the biosynthesis of ABA (Denoux et al. 2008). The PAMP-induced ABA signaling system has been reported to be involved in stomatal closure (Hubbard et al. 2010). Stomata serve as passive ports of bacterial entry during infection. They constitute one entry point for bacteria, which need to reach apoplastic spaces to multiply and cause disease (Nicaise et al. 2009). The stomata in the *Arabidopsis* leaf epidermis have been shown to act as innate immunity gates to actively prevent bacteria from entering the plant leaf (Melotto et al. 2006). The PAMP flg22 triggered closure of stomata which occurred within the first hour of contact with plant tissue (Melotto et al. 2006). Abscisic acid signaling system has been reported to be involved in stomatal closure (Hubbard et al. 2010). Flg22 triggered ABA synthesis, NO production, and OST1 (for OPEN STOMATA1) kinase, which are required for stomatal closure. ABA increase was the critical early event in stomatal closure induced by flg22 (Melotto et al. 2006).

### 2.20.12 PAMP-Induced Expression of Transcription Factors

Transcription factors are the master regulators of expression of genes involved in many cellular processes and they regulate the gene transcription processes by modulating the rate of transcription initiation of target genes (Du et al. 2009). They play important role in activating plant immune responses either positively (Liu et al. 2007; Qiu et al. 2007; Moreau et al. 2012) or negatively (Li et al. 2007; Nurmberg et al. 2007; Kim et al. 2008b; Sun et al. 2010; Moreau et al. 2012). In *Arabidopsis*, there are 72 expressed WRKY genes, and many of them are implicated in the regulation of the plant immune responses positively or negatively via modulation of the JA/SA signaling pathways (Eulgem and Somssich 2007). WRKY proteins are transcription factors that recognize the W-box elements that exist in the promoters of many *PR* genes (Ulker and Somssich 2004). WRKY transcription factors bind to W-box DNA elements (C/TTGACC/T) that are found in the promoters of many defense-related genes, including *PR-1* and *NPR1* (Maleck et al. 2000; Yu et al. 2001; Eulgem and Somssich 2007). Multiple W-box DNA elements were predicted in the promoter region of *PEPR1* and *PROPEP1-5* genes in *Arabidopsis* and therefore, the WRKY transcription factors may play an important role in the amplification of the HAMP Pep peptide signal (Yamaguchi et al. 2010). WRKY transcription factors also regulate the expression of their own genes and/or other WRKY genes (Eulgem and Somssich 2007).

After Flg22 treatment WRKY70 was consistently induced around 20-fold in *A. thaliana*. WRKY70 induces expression of SA-responsive PR genes (Li et al. 2004). WRKY70 has been shown to function downstream of ROS and SA (Knoth et al. 2007). The WRKY70 transcription factor acts as a node of convergence for jasmonate-mediated and salicylate-mediated signals in plant defense (Li et al. 2004). It modulates the selection of signaling pathways in plant defense (Li et al. 2006).

Flg22 induces activation of several other transcription factors in *Arabidopsis*. It activates WRKY33 and WRKY40 transcription factors (Denoux et al. 2008). These transcription factors function as activators of JA signaling system and repressors of SA signaling system (Zheng et al. 2006). WRKY33 positively regulates JA-mediated responses, while it retards the SA-mediated PR genes expression (Zheng et al. 2006). The transcription factors WRKY22 and WRKY29 are known to function downstream of the FLS2-mediated immune response in *Arabidopsis*. Overexpression of the AtWRKY29 constitutively activates the plant defense response against bacterial invasion (Asai et al. 2002). WRKY29, WRKY33, and WRKY53 are induced by the HAMP Pep1 and they are positive regulators of defense responses (Asai et al. 2002; Zheng et al. 2006; Murray et al. 2007).

Nine WRKY genes were induced by the PAMP Nep1 treatment, while WRKY65 was repressed in *A. thaliana* (Bae et al. 2006). Keates et al. (2003) detected the induction of WRKY18 within 15 min after treatment with the PAMP Nep1. These results suggest that PAMPs/HAMPs may trigger the expression of several transcription factors and trigger the plant innate immune responses.

### 2.20.13 Hierarchy of PAMP-Induced Signaling Systems

Several signaling systems induced by PAMPs have been described widely. Cross-talk between these signaling systems has also been reported. However, the exact sequence of these signaling events has not been well understood. The hierarchy of early signaling events induced by a PAMP in *Nicotiana benthamiana* has been reported (Segonzac et al. 2011). Two different calcium influx inhibitors suppressed the ROS burst, activation of MAPKs, and PAMP-induced gene expression. The calcium burst was unaffected in plants specifically silenced for components involved in ROS burst generation, or for MAPKs activated by PAMP treatment. ROS burst still occurred in plants silenced for the two MAPK genes, *NbSIPK* and *NtWIPK* or both genes simultaneously, demonstrating that these MAPKs are dispensable for ROS production. *NbSIPK* silencing is sufficient to prevent PAMP-induced gene expression but both the MAPKs are required for bacterial immunity against two strains of *P. syringae*. These results suggest that the PAMP-triggered calcium influx is upstream of separate signaling branches, one leading to MAPK activation and thence gene expression and the other to ROS production (Segonzac et al. 2011).

## 2.21 Different PAMPs and HAMPs May Induce Similar Early Signaling Systems

Each pathogen may contain or secrete several PAMPs. For example, bacterial pathogens may generally possess the PAMPs flg22, EF-Tu and lipopolysaccharides. Several HAMPs have also been identified in different plants. The multiple and highly variable PAMPs and HAMPs have been reported to induce almost similar early signaling systems. At 1 h, 2586 and 1672 genes had altered expression levels after flg22 or oligogalacturonides (OGs) treatment, respectively (Denoux et al. 2008). The transcriptome analysis revealed that both of them induced expression of almost same genes involved in $Ca^{2+}$-signaling, ROS signaling, NO signaling, MAPK signaling, SA signaling, JA signaling, ET signaling, and ABA signaling systems in *Arabidopsis* (Denoux et al. 2008).

The transcriptome changes observed in *Arabidopsis* 30 min after flg22 and 60 min after elf26 treatments were highly correlated (Zipfel et al. 2006). A clear overlap in the sets of genes with altered expression in response to flg22 and peptidoglycan (PGN) was also observed (Gust et al. 2007). A large number of genes (441 genes) were commonly upregulated in EF-Tu and chitin-treated cells (Wan et al. 2008b). Based on the comparative analysis of microarray data using *Arabidopsis* supplied with flg22, elf18, and chitin (Zipfel et al. 2004, 2006; Wan et al. 2008b) it was concluded that flg22, elf18, and chitin signaling share a conserved downstream signaling pathway leading to basal resistance.

Gene expression in tobacco cultured cells was monitored after application of two different PAMPs/MAMPs, PiE (the PAMP from the cell walls of *Phytophthora infestans*) and TvX (a xylanase MAMP from *Trichoderma viride*) (Suzuki et al. 2007). There was no substantial difference in the gene expression profiles between cells treated with the two different MAMPs, at least during the early phase of defense signaling system (Suzuki et al. 2007). Pep1 induced transcriptional induction of *MPK3*, *WRKY29*, *WRKY33*, and *WRKY53* in *Arabidopsis* (Yamaguchi et al. 2010). These have been reported to be induced by a fungal PAMP, chitin (Wan et al. 2004, 2008a, b), and bacterial PAMPs flg22 and elf18 (Zipfel et al. 2004, 2006). Collectively these results suggest that different PAMPs/HAMPs may induce almost same type of early signaling related genes. Several studies have revealed that the PAMPs/HAMPs activate conserved early basal defense responses (Garcia-Brugger et al. 2006; Jones and Dangl 2006; Qutob et al. 2006; Thilmony et al. 2006; Ferrari et al. 2007).

The very early responses induced within seconds or minutes by the PAMPs include protein phosphorylation followed by $Ca^{2+}$ influx, plasma membrane depolarization, anion and $K^+$ efflux, cytosol acidification, activation of MAP kinases, transient production of ROS and NO production (Boller and He 2009; Boller and Felix 2009). All these early events are induced by most of the PAMPs (Asai et al. 2002; Zhang et al. 2002a, b, c; Hu et al. 2004; Lecourieux et al. 2005; Garcia-Brugger et al. 2006; Moscatiello et al. 2006; Denoux et al. 2008). Most PAMPs and HAMPs (but not all) induce calcium ion influx (Lecourieux et al. 2002, 2006; Garcia-Brugger et al. 2006; Denoux et al. 2008; Erbs et al. 2008; Trouvelot et al. 2008; Aslam et al. 2009; Qi et al. 2010). MAPK signaling is activated by multiple PAMPs, including flagellin, EF-Tu, peptidoglycan, lipopolysaccharide, and bacterial HrpZ harpin, and fungal chitin in *Arabidopsis* (Wu et al. 2011). Production of ROS by various PAMPs and HAMPs has been widely reported (Huffaker et al. 2006; Denoux et al. 2008).

The PRRs FLS2 and EFR are induced also by bacterial LPS, fungal chitin, and the oomycete-derived NPP1 (Zipfel et al. 2006). Overall, these different PAMPs seem to trigger changes in a common set of genes, indicating that plants do not distinguish bacteria, fungi, and oomycetes on the basis of the signaling signature of their PAMPs. Rather, presence of one type of PAMP seems to serve as an indicator of injury or danger in general, resulting in plant innate immune systems (Zipfel et al. 2006).

## 2.22 Magnitude and Timing of Expression of Early Signaling Systems May Vary Depending on Specific PAMPs

Although various PAMPs/HAMPs may induce same set of early signaling events, such as the same $Ca^{2+}$ influx, activation of the same MAPK cascades, and similar production and accumulation of ROS and NO, the induction of these events may vary in magnitude and timing depending on the specific PAMPs/HAMPs. Lecourieux et al. (2005) showed that the PAMPs flg22, β-1,3-glucan, four different elicitins, and harpin, and the HAMP OGs induced changes in $Ca^{2+}$ concentration in tobacco cells, but these changes in $Ca^{2+}$ concentration varied in magnitude and timing, depending on the PAMP/HAMP. The proteinaceous PAMPs (flg22, elicitins, harpin) induced a pronounced and sustainable $[Ca^{2+}]_{nuc}$ elevation, relative to the small effects of the PAMP β-1,3-glucan and the HAMP oligogalacturonides in induction of calcium signatures (Lecourieux et al. 2005). Aslam et al. (2009) showed that each PAMP may elicit different calcium signature. A comparison of calcium influx patterns revealed that flg22 induced a rapid (about 2 min) response, usually with two or three minor, decreasing peaks, whereas elf18-induced calcium influx pattern was less defined and broader (Aslam et al. 2009).

The PAMPs xylanase and the HAMP OGs activate AtMPK6 with characteristic time courses in *Arabidopsis*. The OGs elicited a rapid and transient activation of AtMPK6, with a maximal activity around 3.5 min and a nearly complete return to basal activity after 15 min. The PAMP xylanase induced AtMPK6 activity slowly with a maximum at 20 min. The activity decreased slowly thereafter and did not reach the basal level within 40 min (Nühse et al. 2000). The PAMPs flg22 and chitin also behaved differently in activating AtMPK6 in different time courses (Nühse et al. 2000). OG induced AtMPK3 activity within 3 min of exposure to the HAMP, and the enzyme activity returned to basal levels by 10 min. In contrast, AtMPK3 activity induced by flg22, was strong at 3 min, and was still robust at 10 min following elicitation (Denoux et al. 2008).

Different elicitors may induce the same signaling system, but the intensity of gene expression may differ. The proteinaceous elicitors (elicitins and harpin) induced a pronounced and sustainable $[Ca^{2+}]_{nuc}$ elevation, relative to the small effects of oligosaccharide elicitors (oligogalacturonides and β-1,3-glucan) (Lecourieux et al. 2005). Both cryptogein and OGs, the two different elicitors, triggered $Ca^{2+}$ influx in *Nicotiana plumbaginifolia* cells, but the two $[Ca^{2+}]_{cyt}$ calcium signatures differed in both kinetics (lag time, peak time, and duration) and peak intensities (Lecourieux et al. 2002).

Various oligosaccharide elicitors (the PAMP β-1,3-glucan and the HAMPs oligogalacturonides [OGs] and cellodextrins [CDs]) induced transient increases in cytosolic calcium ion ($[Ca^{2+}]_{cyt}$) in grapevine cells almost immediately after treatment (Aziz et al. 2007). When the cells were treated with CD, a rapid and transient increase of $[Ca^{2+}]_{cyt}$ was observed within 1 min and peaked at 0.6 uM after 2.5 min and decreased to 0.32 uM after 8 min. Cytosolic $Ca^{2+}$ was then maintained at this level for about 4 min, and then decreased slowly to the background level. The $Ca^{2+}$ signature was different in peak time or intensity after treatment with the other two oligosaccharide elicitors, OGA and β-1,3-glucan (Aziz et al. 2007).

Both flg22 and chitosan induced callose deposition in *Arabidopsis*, but both responses showed differences in timing (Luna et al. 2011). Furthermore, flg22- and chitosan-induced callose differed in the requirement for the NADPH oxidase RBOHD, the glucosinolate regulatory enzymes VTC1 and PEN2, and the callose synthase PMR4 (Luna et al. 2011).

## 2.23 PAMPs May Differ in Eliciting Various Defense Responses

Peptidoglycan (PGN), flagellin, and chitin induced largely overlapping early signaling patterns. However, defense responses induced by them were not identical. PGN treatment resulted for instance in camalexin production whereas application of

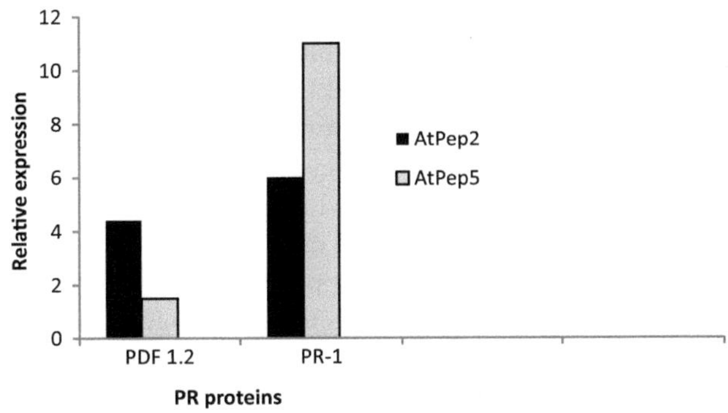


**Fig. 2.4** Differences in ability of AtPep2 and AtPep5 in triggering two different PR proteins in *Arabidopsis thaliana* (Adapted from Huffaker and Ryan 2007)

flg22 did not (Gust et al. 2007). Flg22 triggered the induction of *PR1* and *PDF1.2* genes encoding the pathogenesis-related proteins PR-1 and PR-10, respectively but did not induce *PAL1* encoding the enzyme phenylalanine ammonia-lyase involved in biosynthesis of phenolics, lignin, and phytoalexins. In contrast, OGA caused the upregulation of *PDF1.2* and *PAL1*, but not *PR-1* and elf18 induced all three *PR1*, *PDF1.2*, and *PAL1* genes (Denoux et al. 2008; Aslam et al. 2009).

Two HAMPs of *Arabidopsis*, *At*Pep2 and *At*Pep5, behaved differently in inducing two different PR proteins in *A. thaliana.* Pep2 induced very high expression of *PDF1.2*, while Pep5 was a poor inducer of *PDF1.2*. In contrast, Pep5 induced very high expression of *PR-1*, while Pep2 was poor inducer of *PR-1* (Fig. 2.4; Huffaker and Ryan 2007).

The pathogenesis-related protein-1 encoding gene *PR-1* was induced in *Arabidopsis* seedlings that were infiltrated with the PAMP flg22 solution. However, no induction of *PR1* was observed after addition of the HAMP OGs (Denoux et al. 2008). The PAMP NIP1 isolated from *Rhynchosporium secalis* induced accumulation of PR-1, PR-5, PR-9, and PR-10 proteins, but could not induce the pathogen-inducible germin-like protein (*OxOLP*), LOX gene (*LoxA*), and protease inhibitor gene (*SD10*) (Steiner-Lange et al. 2003). Harpin (HrpN) derived from the soft rot pathogen *Erwinia carotovora* subsp. *carotovora* induced *PR-1* and *PDF1.2* genes, while another *E. carotovora*-derived elicitor polygalacturonase (PehA) induced only *PDF1.2* (Kariola et al. 2003). The PAMP PGN (peptidoglycan) treatment induced camalexin but flg22 did not (Gust et al. 2007). While, the flg22 derived from *Pseudomonas aeruginosa* does not induce cell death, flagellin derived from *P. syringae* pv. *tabaci* 6605 did induce cell death (Naito et al. 2008). These results suggest that different PAMPs/HAMPs may induce distinctly different defense genes.

Although various PAMPs differ in inducing different defense genes, some PAMPs may behave similarly in inducing certain defense genes. ZmPep1 regulates the expression of various defense genes in maize. These include endochitinase,

*PR-4* gene, pathogenesis-related maize seed protein (*PRms*), peroxidase (PEX) genes, and *SerPIN* (Serine proteinase inhibitor) gene encoding a Bowman-Birk trypsin inhibitor in maize (Huffaker et al. 2011). *Rhizopus*-derived pectinase elicitor (PAMP) also induced increased transcript abundance of all above five defense genes to comparable levels as ZmPep1 (Huffaker et al. 2011). The HAMP Pep1 induced transcription of *PDF1.2*, and *PR-1* genes, which are also induced by Flg22, elf18, and chitin (Huffaker et al. 2006; Yamaguchi et al. 2010).

The PAMP-induced stomatal closure appears to be important component in induced plant immune responses against bacterial pathogens (Melotto et al. 2006; Gudesblat et al. 2009). The PAMPs, flg22 (Melotto et al. 2006; Zhang et al. 2008b; Zeng and He 2010), chitosan, a polymer of β-1,4-glucosamine residues derived from fungal chitin (Lee et al. 1999; Amborabé et al. 2008), and LPS (Melotto et al. 2006), and the HAMP oligogalacturonide (Lee et al. 1999) trigger stomatal closure. In contrast, the PAMP peptidoglycan has not been shown to trigger stomatal responses (Erbs et al. 2008; Zeng and He 2010).

## 2.24 Synergism and Antagonism in Induction of Plant Immune Responses by PAMPs/HAMPs

### *2.24.1 Multiple PAMPs May Be Required to Activate the Complex Defense Signaling Systems*

Each pathogen may contain or secrete several PAMPs. Most known PAMPs are essential components of a bacterial or fungal cell. Thus, probably every microbe bears several PAMPs (Zipfel et al. 2004). Several studies have identified the signaling events triggered by individual PAMPs. However, the PAMPs rather being released singly *in planta*, are likely to be presented as a cocktail, and the different PAMPs may interact synergistically or antagonistically in triggering different signaling systems (Aslam et al. 2009). A single PAMP may not be able to activate all the defense signaling-related genes and several PAMPs may be required to activate the complex signaling systems (Zipfel et al. 2004).

The defense signaling systems may be induced by several types of PAMPs and elicitors. Among the 126 genes that were up- or downregulated during incompatible rice-*Acididovorax avenae* interactions, expression of 45 genes was decreased when cultured rice cells were inoculated with a flagellin-deficient incompatible strain, indicating that approximately 37 % of the 126 genes were directly controlled by flagellin perception. The remaining 81 genes would have been activated by other PAMPs (Fujiwara et al. 2004). Of the 56 PAMP-induced kinases in *Arabidopsis thaliana*, only 31 were found to be induced by flg22 treatment (Thilmony et al. 2006). These results suggest that several PAMPs may be involved in activation of defense signaling system.

**Table 2.6** Synergistic action of PAMPs/elicitors in induction of different *PR* genes in *Arabidopsis*

| | HrpN | | PehA | | HrpN + PehA | |
|---|---|---|---|---|---|---|
| *PR* gene | 24 h | 48 h | 24 h | 48 h | 24 h | 48 h |
| *PR1* | + | ++ | – | – | ++ | +++ |
| *PDF 1.2* | – | – | – | ++ | +++ | +++ |

Adapted from Kariola et al. (2003)
–, not detected; +, ++, +++, increased intensity of expression of defense genes

### 2.24.2 Different PAMPs May Act Synergistically

The different PAMPs may act synergistically in inducing expression of defense response. N-acetylchitoheptaose and tetraglucosyl glucitol are the two oligosaccharide PAMPs derived from the rice blast pathogen *Magnaporthe grisea* cell walls. The two PAMPs were recognized by different receptors. However, they synergistically activated phytoalexin biosynthesis in cultured rice cells, suggesting that different PAMPs may act synergistically in inducing expression of defense response (Yamaguchi et al. 2002). There was a significant enhancement of PAMP-induced calcium ion influx when the following PAMPs were combined, compared with the effect of the individual PAMPs: flg22 + elf18; flg22 + LOS; flg22 + core oligosaccharides; and LOS + core oligosaccharides. These increases were additive, except for flg22 and LOS, where there appears to be synergy (Aslam et al. 2009).

Harpin (HrpN) and polygalacturonase (PehA) are the two *Erwinia carotovora* subsp. *carotovora*-derived elicitors. Both of them individually did not induce significant amount of production of superoxide in *Arabidopsis*. In contrast, they triggered increased production of superoxide when the two elicitors were applied simultaneously (Kariola et al. 2003). The harpin elicitor induced both SA and JA/ET signaling pathways, as indicated by induction of the SA-signaling system marker gene *PR-1* and the JA/ET-signaling system marker gene PDF1.2. PehA elicitor induced only JA/ET signaling system. However, when both the elicitors were applied simultaneously, they triggered faster and higher expression of both the SA and JA/ET signaling systems (Table 2.6; Kariola et al. 2003). The results suggest that some PAMPs/elicitors may act synergistically in induction of defense signaling systems. Some PAMPs have additive effect and the flagellin peptide (flg22) and elongation factor 18 (elf18) show additive effect in inducing defense signaling system (Aslam et al. 2009).

### 2.24.3 Some PAMPs May Show Antagonistic Effect in Activating Defense Responses

Some PAMPs show antagonistic effect between them. The flg22 and oligosaccharide elicitor (OGA) showed mutual interference between them. OGA suppresses flg22-induced defences in *Arabidopsis* (Aslam et al. 2009). Reduced calcium influx was observed when *Arabidopsis* was challenged with two PAMPs (flg22 + peptidoglycan [PGN], flg22 + oligogalacturonides [OGA] and elf18 + OGA) concurrently. Induction of ROS by flg22 was reduced by more than 90 %, whenflg22 was applied along with OGA (Aslam et al. 2009).

## 2.25 Amount of PAMP/HAMP Determines the Intensity of Expression of Defense Signaling Genes

The intensity of PAMP-induced expression of various defense signaling systems may depend on the amount of PAMP applied to activate the immune systems. The intensities of calcium signatures induced by both cryptogein and OGs were shown to depend on the PAMP/HAMP concentration (Lecourieux et al. 2002). Flg22 induced the transcription factor gene *WRKY40* in *Arabidopsis* and the induction of gene expression increased at increasing concentrations of the elicitor (Fig. 2.5; Denoux et al. 2008). The relationship between increase in concentrations of PAMPs and increased upregulation of expression of defense-related genes has been reported in barley (Fujita et al. 2004), rice (Schaffrath et al. 1995), and parsley cells (Davis and Hahlbrock 1987).

The HAMP ZmPep1 application induced increased expression of defense genes and the induced magnitude of change in transcript abundance was found to be dose-dependent. The leaves treated with *Zm*Pep1 displayed increased defense gene expression with increasing amounts of *Zm*Pep1 peptide application (Huffaker et al. 2011). These results suggest that the expression of defense signaling depends on the applied dosage of the PAMP.

Dosage required to trigger the plant immune response may differ among different PAMPs. At the same concentration (1 μM), flg22 is more potent than elf18 to induce the oxidative burst in *Arabidopsis* leaves (Zipfel et al. 2006) and elf26 peptides from *Agrobacterium tumefaciens* and *Erwinia amylovora* are 50 times more potent to induce alkalinization of *Arabidopsis* cell culture than elf26 from *P. syringae* pv. *tomato* DC3000 (Kunze et al. 2004).

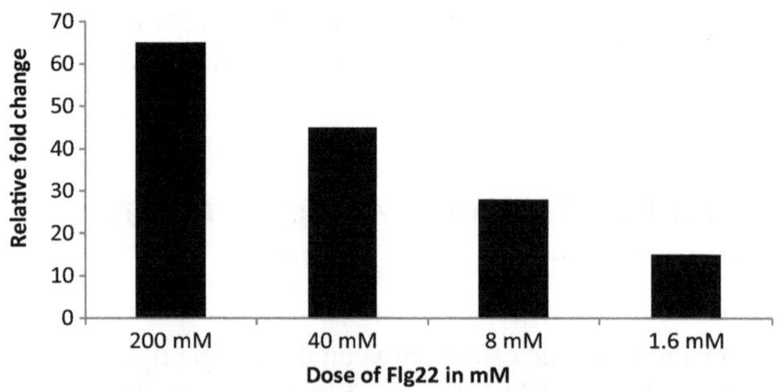


**Fig. 2.5** Effect of different concentrations of Flg22 in inducing expression of the transcription factor *WRKY40* in *Arabidopsis* (Adapted from Denoux et al. 2008)

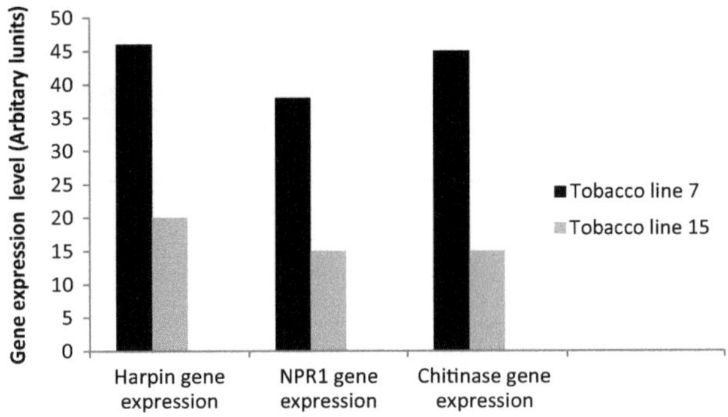


**Fig. 2.6** Relationship between intensity of expression of the PAMP harpin gene and plant defense genes (Adapted from Peng et al. 2004)

## 2.26 Amount of PAMP Available in the Infection Court May Determine the Level of Induction of Immune Responses

It has been suggested that the amount of PAMP available in the infection court may determine the intensity of induced defense gene expression (Lecourieux et al. 2002; Denoux et al. 2008). For example, transgenic tobacco plants expressing the *hpaG*$_{Xoo}$ gene encoding the PAMP harpin$_{Xoo}$ were obtained by Peng et al. (2004). The harpin gene expression varied in the transgenic lines. The line 7 showed highest expression of the harpin gene while the line 15 showed lowest expression of the gene. The line, which showed highest PAMP expression, showed greatest expression of the regulatory gene *NPR1* and the defense gene *Chia5* encoding chitinase. By contrast, the line 15, which showed lowest PAMP expression, showed lowest expression of the defense gene *Chia5* and the regulatory gene *NPR1* (Fig. 2.6; Peng et al. 2004). High and low expression of other defense genes *PR-1a* and *PR-1b* was also correlated with the high and low expression of the PAMP gene in the lines 7 and 15, respectively. The line 7 showed highest disease resistance, while the line 15 showed lowest disease resistance. These results suggest a link between higher level of PAMP in plants and higher level of induction of immune responses.

## 2.27 PAMPs May Trigger Different Signaling Systems

### *2.27.1 Some PAMPs May Activate Only Specific Signaling Systems*

The PAMPs may activate plant innate immune responses triggering different signaling pathways. Each PAMP/elicitor may regulate distinctly different signaling pathway(s). The harpin (HrpN) elicitor derived from the soft rot pathogen *Erwinia carotovora* subsp. *carotovora* induced both SA and JA/ET signaling pathways,

**Table 2.7** CBEL elicitor induces defense genes through different SA, JA, and ET signaling

| Plant defense genes | CBEL elicitor-induced expression of defense genes in wild-type and mutant *Arabidopsis* plants | | | |
|---|---|---|---|---|
| | Wild-type | SA-deficient mutant | JA-deficient mutant | ET-deficient mutant |
| *ASA1* (anthranilate synthase) | + | + | – | + |
| *PR-1* | + | – | + | + |
| *PDF1.2* (defensin) | + | + | – | – |

Adapted from Khatib et al. (2004)
+, expressed; –, not expressed

while *E. carotovora*-derived polygalacturonase (PehA) elicitor induced only the JA/ET signaling system in *Arabidopsis* (Kariola et al. 2003). The HAMPs AtPep1 and AtPep2 induced both JA/ET and SA signaling pathways in *A. thaliana.* These HAMPs triggered expression of *PR-1* (the gene induced by SA) and *PDF1.2* (the gene induced by JA/ET) genes in *Arabidopsis* (Huffaker and Ryan 2007). In contrast, ZmPep1, the ortholog of AtPep1, activated de novo synthesis of JA and ethylene, and not SA in maize (Huffaker et al. 2011). A β-1,3-glucan sulfate elicitor induced JA signaling system in grapevine (Trouvelot et al. 2008).

The PAMP $NLP_{Pp}$ from *Phytophthora parasitica* strongly induced ethylene biosynthesis enzymes ACC synthase (ACS) and ACC oxidase (ACO). However, none of the genes encoding JA biosynthetic enzymes, such as lipoxygenase, allene oxide synthase, allene oxide cyclase, 12-oxophytodienoate reductase, and jasmonate-O-methyl transferase were altered in potato plants (Qutob et al. 2006). By contrast, rapid accumulation of transcripts encoding SA biosynthetic enzymes such as isochorismate synthase 1 and phenylalanine ammonia-lyase was observed (Qutob et al. 2006). Collectively these studies indicate that $NLP_{Pp}$ may activate ET and SA signaling pathways, but JA signaling pathway may not be involved in NLPpp-mediated signaling system.

The PAMP CBEL induced different types of defense responses regulated by different signaling pathways (Table 2.7; Khatib et al. 2004). It induced the genes *PR-1* and *PDF1.2* encoding PR-1 and PR-12 defensin proteins, respectively and *ASA1* encoding anthranilate synthase involved in the biosynthesis of the phytoalexin camalexin in *Arabidopsis*. Different signaling pathways were involved in the induction of these defense genes. The induction of *PR-1* was abolished in *NahG* plants, which are deficient in SA signaling, while the induction of *PDF1.2* and *ASA1* genes was not altered in NahG plants. The results suggest that expression of PR-1 gene is dependent of SA signaling system, but the expression of *PDF1.2* and *ASA1* is independent of SA system. The CBEL-induced expression of *PDF1.2* was totally abolished in *coi1* (JA-deficient) and *ein2* (ET-deficient) mutants and the PAMP-induced expression of *ASA1* was abolished in *coi1* mutants. These results suggest that expression of *PDF1.2* is dependent on the PAMP-induced JA/ET signaling systems, whereas the expression of *ASA1* is dependent on JA signaling system in *Arabidopsis* (Khatib et al. 2004). Collectively, these results suggest that the PAMP CBEL activates different hormone signaling systems and triggers various defense responses.

### 2.27.2 Some PAMPs May Activate Multiple Hormone Signaling Systems

Some PAMPs may not have any specificity in activating different hormone signaling systems. The PAMP NPP1 from *Phytophthora sojae* triggers ethylene production and also SA signaling pathway (Fellbrich et al. 2002). Pep-13, a PAMP from *Phytophthora*, induces the accumulation of both SA and JA and also activates defense genes in potato (Halim et al. 2009). The PAMP flg22 induces both SA-dependent *PR-1* expression and JA/ET-dependent *PDF1.2* expression in *Arabidopsis* (Gómez-Gómez et al. 1999). Flg22 activates SA, JA and ET signaling pathways and OGs also activate these three pathways (Denoux et al. 2008). These studies demonstrate that different PAMPs may activate various signaling systems and some of them may be specific in activating specific signaling systems.

## 2.28 PAMPs May Function Differently in Different Plants

The PAMPs may behave differently in different plant system. They may trigger variable downstream signaling responses in different plants. Khatib et al. (2004) observed enhanced β-1,3-glucanase activities after infiltration of CBEL in leaves belonging to four different botanical families, i.e. tobacco (Solanaceae), *Arabidopsis* (Brassicaceae), French bean (Fabaceae) and *Zinnia* (Asteraceae). However, other species of the same families such as pea and sunflower did not respond to CBEL (Khatib et al. 2004).

The harpin ($HrpN_{ea}$) from *Erwinia amylovora* induces increases in anion current in apple cells, while it decreases the anion current in *Arabidopsis* cells (Reboutier et al. 2007). In apple cells, harpin did not trigger any significant $H_2O_2$ production, while in *Arabidopsis* cells, a rapid transitory increase of $H_2O_2$ production was observed within 30 min of the elicitor treatment (Reboutier et al. 2007). These results suggest that the PAMPs may trigger defense responses depending on the specific host plants.

## 2.29 Specificity of PAMPs in Triggering Immune Responses in Plants

The PAMPs may trigger immune responses in some specific plants. The bacterial PAMP CSP22 triggers defense responses specifically in Solanaceous plants (Felix and Boller 2003). The bacterial PAMP EF-Tu triggers immune responses only in members of the family Brassicaceae whereas responsiveness to Ax21 seems confined to rice (Kunze et al. 2004; Zipfel 2008; Segonzac and Zipfel 2011). Flagellin induces resistance in *Arabidopsis*, tobacco, tomato, and rice. The PAMP rhamnolipids were able to stimulate defense genes in tobacco, wheat, grapevine, and *Arabidopsis thaliana* (Vatsa et al. 2010).

The specificity of particular PAMP may be due to its specificity towards its PRR in plants. The PRR EFR for the PAMP EF-Tu has been detected in plants belonging to the family Brassicaceae. The transfer of EFR from Brassicaceae to Solanaceae species confers a broad-spectrum resistance to phytopathogenic bacteria in Solanaceous plants (Lacombe et al. 2010). The results suggest that EFR is important in triggering plant immune responses and its action is not specific to particular plants. Similarly the PRR *Xa21* gene is detected in rice plants and the PAMP Ax21 activates plant immune responses in rice. Transgenic *Citrus sinensis* plants expressing the rice *Xa21* gene were developed and these transgenic plants also conferred resistance against the citrus bacterial pathogen *Xanthomonas axonopodis* pv. *citri* (Mendes et al. 2010). The results suggest that the PRR from one plant species can be transferred to another plant species to extend the usefulness of PRRs in management of wide-spectrum of pathogens in a wide range of plants.

## 2.30 Role of PAMPs and Effectors in Activation of Plant Innate Immune Responses

Pathogen effectors are proteins and small molecules that alter host-cell structure and function. These alterations either facilitate infection or trigger defense responses or both (Hogenhout et al. 2009; Boureau et al. 2011). Effectors are double-edged swords that enhance virulence of pathogens in susceptible plants and trigger resistance in plants carrying cognate resistance (R) proteins (Zong et al. 2008). Effector-triggered immunity (ETI) and PAMP-triggered immunity (PTI) are two branches of the plant immune system. PTI uses transmembrane PRRs that respond to PAMPs, whereas ETI acts largely inside the cell, using polymorphic nucleotide binding-leucine-rich repeat (NB-LRR) protein products encoded by most *R*(resistance) genes (Jones and Dangl 2006). These NB-LRR proteins have been targeted by breeders for decades to elicit resistance to crop pathogens in the field (Tör et al. 2009). The ETI was formerly known as gene-for-gene resistance (Boller and He 2009).

Interactions between plants and pathogens can be classified as compatible, incompatible, and nonhost interactions. Incompatible interactions are cultivar-specific and determined by ETI. In contrast, compatible interactions are thought to lack ETI. Nonhost interactions refer to those between a plant species and non-adapted pathogens. Some effectors from nonadapted pathogens also trigger ETI and induce HR, typically seen in the incompatible interactions (Zhang et al. 2010b). Effector-triggered immunity results in amplification of PTI and it constitutes the second layer of defense (Day and He 2010).

Pathogen effectors recognized by NB-LRR proteins activate defense responses similar to those activated by PAMPs. However, ETI generally activates them in a more prolonged and robust fashion than PTI and usually includes the hypersensitive response (HR) (Tsuda and Katagiri 2010). The PAMP- and effector- activated signaling pathways include $Ca^{2+}$ fluxes, MAP kinase cascade, ROS production, hormone signaling network, and transcriptional reprogramming. The triggered plant immune responses include accumulation of pathogenesis-related proteins, deposition of lignin and callose in the cell wall, and production of anti-microbial compounds (Tsuda and Katagiri 2010; Gimenez-Ibanez and Rathjen 2010).

## 2.31 Effectors May Suppress PAMP-Triggered Immunity

### *2.31.1 Inhibition of PAMP-Triggered Immunity*

Several pathogens are capable of delivering effector proteins into the host cell to enhance virulence and these effectors often inhibit PAMP-triggered immunity (PTI) (Göhre and Robatzek 2008; Song and Yang 2010; Szczesny et al. 2010; Zhang et al. 2010b; Akimoto-Tomiyama et al. 2012). Some effectors may also suppress the immune response induced by another effector produced by the same pathogen (Szczesny et al. 2010). Some pathogens have acquired effectors to collectively overcome PTI and ETI in their host plants (Zhang et al. 2010b; Block and Alfano 2011). The bacterial pathogen *Pseudomonas syringae* suppresses both PTI and ETI by the injection of type III effector (T3E) proteins into host cells (Block and Alfano 2011). Pathogens secrete several effectors (virulence factors, toxins) into the host cell and suppress or disable PAMPs-induced signaling pathways. This results in effector-triggered susceptibility (ETS) (Hauck and Thilmony 2003; Li et al. 2005a; He et al. 2006; Jones and Dangl 2006; Nomura et al. 2006; Zhang et al. 2007a; Göhre et al. 2008; Cui et al. 2009; Hogenhout et al. 2009; Wu et al. 2011).

### *2.31.2 Effectors May Degrade PRRs*

Major function of effectors is to suppress PAMP-triggered defense responses. Some effectors have been shown to degrade the PRRs and inhibit PAMP-triggered immunity. Flagellin (flg22) is the PAMP recognized in several bacterial pathogens including *Pseudomonas syringae* pv. *tomato*. The PAMP flg22 is perceived by the PRR FLS2 in *Arabidopsis*. AvrPtoB *is* the effector secreted by the bacterial pathogen. The AvrPtoB associates with FLS2 through its N terminus and the interaction is enhanced by the PAMP flg22 activation. The effector AvrPtoB promotes degradation of the PRR FLS2. The AvrPtoB has been recognized as an E3 ubiquitin ligase and it structurally and functionally mimics E3 ubiquitin ligase (Janjusevic et al. 2006; Abramovitch et al. 2006). The C-terminal region (residues 400–550) of AvrPtoB encodes an ubiquitin E3-ligase domain and it ubiquitinates the PRR FLS2 to promote its degradation (Gimenez-Ibanez et al. 2009b). E3 ligases play a key role in the ubiquitin-proteasome-mediated degradation of protein. The effector catalyzes polyubiquitination of the kinase domain of FLS2, resulting in degradation of the PRR. The E3 ubiquitin ligase activity of AvrPtoB seems to be required for full enhancement of virulence of the bacterial pathogen and degradation of the receptor occurs during the bacterial infection (Göhre et al. 2008).

The effector AvrPtoB also inhibits PTI by targeting another PRR CERK1 (CHITIN ELICITOR RECEPTOR KINASE 1) for degradation (Gimenez-Ibanez et al. 2009a, b). To find out the role of the effector AvrPtoB in inhibiting the PAMP-triggered immunity by degrading the PRR CERK1, transgenic *Arabidopsis* plants expressing *avrPtoB* from an inducible promoter were developed. Treatment of these plants with chitin elicited generation of ROS, induction of defense gene expression and deposition of callose into cell walls. All of these responses were suppressed efficiently by prior induction of *avrPtoB* transgene with dexamethasone. The PAMP chitin treatment-induced activation of MAPKs was also abolished by AvrPtoB (Gimenez-Ibanez et al. 2009b). These results suggest that FLS2 and CERK1 are the targets of AvrPtoB, leading to their degradation (Göhre et al. 2008; Shan et al. 2008; Gimenez-Ibanez et al. 2009b; Hann et al. 2010).

### 2.31.3 Effectors May Bind the Receptor Kinase PRRs to Block PAMP-Triggered Immunity

Virulent pathogens may suppress the defense signal transduction-mediated by plant transmembrane receptor kinases, the PRRs (Xiang et al. 2008). *Pseudomonas syringae* effector protein AvrPto suppresses host defenses by directly targeting the transmembrane receptor kinases involved in bacterial perception (Zipfel and Rathjen 2008). *P. syringae* injects two sequence-distinct effectors, AvrPto and AvrPtoB, to intercept convergent defense responses stimulated by PAMPs. The AvrPto binds receptor kinases, including *Arabidopsis* FLS2 and EFR and tomato LeFLS2, to block plant defense responses in the plant cell (Xiang et al. 2008). Xiang et al. (2011) provided evidences to show that FLS2 is targeted by the *P. syringae* effector AvrPto in plants. AvrPto is a kinase inhibitor that inhibits the PRR receptor kinase activity (Xiang et al. 2008). The bacterial effector AvrPtoB targets the LysM receptor kinase CERK1 to promote bacterial virulence (Gimenez-Ibanez et al. 2009a). It directly targets the kinase domains of FLS2, CERK1, and Pto (Shan et al. 2008).

### 2.31.4 Effectors May Prevent Interaction of Co-receptor BAK1 with PAMPs

PRRs appear to interact with some transmembrane proteins that act as signaling adapters or amplifiers to achieve their full functionality (Zipfel 2009). BAK1 is an important transmembrane protein, which interacts with PRRs and acts downstream of PRRs. (Chinchilla et al. 2007a, b; Heese et al. 2007). BAK1 is a signal amplifier (Nicaise et al. 2009). BAK1has been shown to be required for full function of PAMP-PRR signaling complex to activate plant immune responses (Schulze et al. 2010). BAK1 is required for proper functionality of several PRRs including FLS2, EFR, CERK1, PEPR1, and PEPR2 (Postel et al. 2010; Zhang and Zhou 2010).

Some effectors block the action of BAK1. The effectors AvrPto and AvrPtoB bind to BAK1 and thereby blocking its interaction with the PRR FLS2 resulting in suppression of immunity (Shan et al. 2008). AvrPtoB binds BAK1 during infection and impede BAK1-dependent plant defense responses (Shan et al. 2008). AvrPtoB directly targets the kinase domains of BAK1 (Shan et al. 2008). Suppression of BAK1 by AvrPtoB may occur by inhibition of kinase activity and may have broad effects through the multiple BAK1-dependent PAMP-signaling pathways (Shan et al. 2008). AvrPtoB targets BAK1 to disrupt the complex (Hann et al. 2010). However, Xiang et al. (2011) provided evidences to show BAK1 is not targeted by the *P. syringae* effector AvrPto in plants.

### *2.31.5 Effectors May Target the Receptor-Like Cytoplasmic Kinases BIK1 and PBL1*

Downstream of PAMP-PRR-BAK1 signaling complex, several receptor-like cytoplasmic kinases (RLCK) play important role in regulation of the signaling pathways. The important RLCKs involved in the signaling complex in Arabidopsis include BIK1 (Botrytis-induced kinase 1), PBS1 (AvrPphB susceptible 1), and PBS1-like (PBL) proteins. The BIK1 plays an important role in mediating early flagellin signaling from the FLS2/BAK1 receptor complex (Lu et al. 2010). BIK1 forms a complex with unstimulated FLS2 in plants, and the PAMP flg22 induces a rapid phosphorylation of BIK1 in both an FLS2- and BAK-dependent manner (Lu et al. 2010; Zhang et al. 2010a; Wu et al. 2011). BIK1 links the PAMP-PRR signaling complex to downstream intracellular signaling (Lu et al. 2010). PBS1-like (PBL) cytoplasmic receptor-like kinases (RLCKs) act additively with BIK1 in plant innate immune system (Zhang et al. 2010a). PBS receptor-like kinases are required for signaling from multiple PAMPs and act downstream of FLS2, EFR, and CERK1 to trigger immune responses. It has been shown that BIK1 and PBS proteins integrate immune signaling from multiple PRRs (Zhang et al. 2010a).

Zhang et al. (2010a) presented evidence that several related Arabidopsis cytoplasmic receptor kinases, exemplified by BIK1 and PBS1 are cleaved by AvrPphB, an effector from the pathogenic bacterium *P. syringae* (Zhang et al. 2010a). In addition to cleavage of BIK1 and PBS1, AvrPphB targets about 10 PBL kinases, including PBL1 and PBL2 (Zhang et al. 2010a). The effector AvrPphB is a cysteine protease that cleaves PBS1 kinase (Ade et al. 2007). Plants lacking BIK1 and/or PBS1 were compromised in their PTI response to several PAMPs suggesting that at least these two targets of AvrPphB are important in PTI and are virulence targets of *P. syringae* (Zhang et al. 2010a; Lu et al. 2010).

### 2.31.6 *Effectors May Inhibit Autophosphorylation of PRRs*

Most of the PRRs identified are receptor kinases and the PAMPs induce autophosphorylation of these PRRs (Gómez-Gómez et al. 2001; Wang et al. 2001; Robatzek et al. 2006; Kanzaki et al. 2008; Xiang et al. 2008; Chen et al. 2010d; Kim et al. 2010; Petutschnig et al. 2010). PAMP-induced autophosphorylation of PRRs is required for activation of the PRRs (Chen et al. 2010d; Park et al. 2010a, b). The effector AvrPto produced by *P. syringae* pv. *tomato* interacts *in vivo* with the PRRs FLS2 and EFR and inhibits their autophosphorylation in the dose-dependent manner (Xiang et al. 2008). The effector thus prevents activation of the function of the PRRs. AvrPto inhibits all responses induced by several PAMPs (He et al. 2006; Hann and Rathjen 2007; Xiao et al. 2007).

The *P. syringae* effector HopF2 has been found to be a potent suppressor of early immune gene transcription signaling activated by multiple PAMPs, including bacterial flagellin, ef-Tu, peptidoglycan, lipopolysaccharide and HrpZ1 harpin, and fungal chitin (Wu et al. 2011). HopF2 is targeted to the plant plasma membrane through a putative myristoylation site and the membrane association appears to be required for its PAMP-suppression function (Wu et al. 2011). The plasma membrane-associated cytoplasmic kinase BIK1is phosphorylated within 1 min upon flagellin perception. Expression of HopF2 in plants potently diminished the flagellin-induced phosphorylation of BIK1 (Wu et al. 2011). These results suggest that HopF2 likely intercepts PAMP signaling at the plasma membrane immediately of signal perception by PRR by inhibiting the phosphorylation of the protein kinase BIK1 (Wu et al. 2011). These studies suggest that the effectors may interfere with the function of PRRs by inhibiting autophosphorylation of the PRRs.

## 2.32 PAMP-Induced Small RNA-Mediated RNA Silencing

### 2.32.1 *RNA Silencing Is an Immune System in Plants*

Small RNAs (sRNAs) are non-protein-coding RNAs of 20 to 30-nucleotide length (Ghildiyal and Zamore 2009; Havecker et al. 2010; Kulcheski et al. 2011; al. 2011). These small RNAs have been identified as regulatory RNAs that modulate gene expression at both the transcriptional and posttranscriptional levels (Zhao et al. 2012). They act as sequence-specific repressors of target gene expression, either at the transcriptional level through DNA and/or histone methylation or at the posttranscriptional level through transcriptional cleavage or translational inhibition (Ramachandran and Chen 2008a, b).

In general, small RNAs are grouped into two major classes: microRNAs (miRNAS) (Carthew and Sontheimer 2009; Katiyar-Agarwal and Jin 2010; Cuperus et al. 2010; Kulcheski et al. 2011), and small-interfering RNAs (siRNAs) (Llave et al. 2002; Song and Joshua-Tor 2006; Chellappan et al. 2010; Chen et al. 2010b; Dunoyer et al. 2010; Katiyar-Agarwal and Jin 2010). Small RNAs are classified into miRNAs and siRNAs based on their precursor structures and biogenesis pathways. The miRNAs are originated from hairpin-folded single-stranded RNAs transcribed from miRNA genes (Bartel 2004; Mallory and Vaucheret 2006), while siRNAs are produced usually from long double-stranded RNAs (dsRNAs) (Hamilton et al. 2002; Narry Kim 2005; Chapman and Carrington 2007).

Small RNAs are involved in a variety of phenomena that are essential for genome stability, development, and adaptive responses in biotic and abiotic stresses (Mallory and Vaucheret 2006; Vaucheret 2006; Chen 2009; Kulcheski et al. 2011). Small RNA molecules act as mobile signals that direct mRNA cleavage and DNA methylation in recipient cells (Ciomperlik et al. 2011; Molnar et al. 2011). They are key regulators of gene expression that guide both transcriptional and post-transcriptional silencing mechanisms in eukaryotes (Kulcheski et al. 2011). They function by guiding sequence-specific gene silencing at the transcriptional and/or post-transcriptional level (Vaucheret et al. 2006; Chellappan et al. 2010; Havecker et al. 2010). They are also big contributors to plant innate immunity (Chellappan et al. 2009; Dunoyer et al. 2010; Mosher et al. 2010; Molnar et al. 2011).

RNA silencing refers to a number of related cellular processes that employ the small RNAs to regulate the expression of genetic material in a sequence-specific manner (Qu et al. 2008; Jaubert et al. 2011; Zhao et al. 2012). RNA silencing is a conserved mechanism in plants that plays a role in various biological processes including regulation of gene expression. RNA silencing also plays a role in genome stability and protects plants against invading nucleic acids such as transgenes and viruses (Ellendorff et al. 2009). RNA silencing is a type of plant immune system conferring resistance against viruses and also against bacteria and fungi (Voinnet 2001; Katiyar-Agarwal and Jin 2010; Zhang et al. 2011). Plants use RNA silencing as a surveillance mechanism to protect against viral (Mlotshwa et al. 2002; Garcia-Ruiz et al. 2010), bacterial (Katiyar-Agarwal et al. 2006; Navarro et al. 2006, 2008; Agorio and Vera 2007; Jin 2008; Li et al. 2010; Zhang et al. 2011), and fungal pathogens (Lu et al. 2007; Ellendorff et al. 2009).

The important feature of RNA silencing is its ability to spread from cell to cell (Chitwood and Timmermans 2010). RNA silencing is a non-cell-autonomous process; it spreads both to neighboring cells and systemically over long distances (Dunoyer et al. 2010). A RNA silencing signal has been shown to move through plasmodesmata and the phloem (Molnar et al. 2010). Both exogenous and endogenous siRNAs, as opposed to their precursor molecules, act as mobile silencing signals between plant cells (Dunoyer et al. 2010). The silencing signal may involve siRNA duplexes, and not Argonaute1 (AGO1) – bound siRNA single strands. The small RNA signaling system has been shown to play important role in host defense responses against viral, bacterial, and fungal pathogens. As a counter-defense, pathogens encode specific proteins that function as suppressors of small RNA-directed RNA silencing (Alvarado and Scholthof 2009; Lewsey et al. 2010; Xie et al. 2010; Burgyán and Havelda 2011; Shimura and Pantaleo 2011).

Small RNAs are generated from dsRNA precursors by the ribonuclease III enzyme Dicer (Qu et al. 2008). Four paralogs of Dicer (Dicer-like, DCLs) have been detected in Arabidopsis. DCL1 excises miRNAs from intergenic stem-loop transcripts to promote cleavage of cellular transcripts carrying miRNA-complementary sequences (Bartel 2004). DCL2 produces viral-derived siRNAs (Xie et al. 2004) and siRNAs from antisense overlapping transcripts (Borsani et al. 2005). DCL3 generates DNA repeat-associated siRNAs (Xie et al. 2004), whereas DCL4 synthesizes trans-acting siRNAs and mediates RNA interference (Dunoyer et al. 2005; Xie et al. 2005; Howell et al. 2007).

The generated small RNAs are subsequently incorporated into RNA-induced silencing complexes (RISCs). The functions of RISCs are carried out in large part by the activity of RNase H-like Argonaute (AGO) proteins (Voinnnet 2009; Jaubert et al. 2011). Once produced, the miRNAs and siRNAs are recruited by the AGO proteins into RISCs to direct the cleavage or translational repression of homologous mRNAs (Baulcombe 2004; Dunoyer et al. 2010). These small RNAs bind to Argonaute nucleases and form base paired structures with their RNA targets. In many instances the target RNA is simply degraded, presumably by exonucleases. However, some of the targeted molecules, especially those that interact with two different small RNAs are degraded through a more complex mechanism. The targeted RNA is first copied into dsRNA by an independent RNA polymerase and is then cleaved into siRNAs by a Dicer nuclease. The secondary siRNAs are able to guide Argonaute nucleases in degradation of RNA targets (Fagard et al. 2000; Narry Kim 2005; Szittya et al. 2008).

RNA silencing operated through the production of small RNAs is an important antiviral plant immune system (Voinnet 2001; Szittya et al. 2008; Garcia-Ruiz et al. 2010). Certain endogenous small RNAs in plants, including miRNAs and siRNAs, are induced or repressed in response to bacterial and fungal pathogen attack and subsequently regulate the expression of genes involved in disease resistance and defense responses by mediating transcriptional or post-transcriptional gene silencing (Katiyar-Agarwal et al. 2006, 2007; Agorio and Vera 2007; Jin 2008; Navarro et al. 2008; Li et al. 2010). Small RNA signaling system has been shown to be involved in PAMP-triggered immune responses. In *Arabidopsis*, flg22 triggered rapid changes in transcript levels, including down-regulation of a gene subset, potentially by posttranscriptional mechanisms (Navarro et al. 2004). The posttranscriptional mechanism involves RNA silencing, a sequence-specific mRNA degradation process mediated by small RNAs. It suggests that the PAMP-triggered down-regulation of genes depends on small RNA-mediated RNA silencing system.

### 2.32.2 *Flg22 Triggers Accumulation of miRNAs, Which Cleave and Down-Regulate Auxin Signaling Genes*

Flg22 induced a two-fold increase in microRNA (miR393) accumulation in *Arabidopsis* seedlings. The up-regulation of miR393 by flg22 resulted from enhanced transcription of *At-miR393a* gene. Flg22 elicited a subset of mRNAs, including TIR1

(Transport Inhibitor Response 1) and two of its three functional paralogs, AFB2 and AFB3 (auxin signaling F-box proteins 2 and 3). The F-box proteins TIR1, AFB2 and AFB3 were specifically cleaved by miR393 in a DCL1-dependent manner. A two- to threefold reduction in the levels of TIR1, AFB1, AFB2, and AFB3 was observed 30 min after flg22 elicitation (Navarro et al. 2006). TIR 1 is part of the ubiquitin-ligase complex $SCF^{TIR1}$ that interacts with Aux/IAA proteins to promote their degradation. Aux/IAA proteins repress auxin signaling through heterodimerization with Auxin Response Factors (ARFs). These transcription factors ARFs bind to auxin-responsive elements (AuxREs) in promoters of primary auxin response genes and activate (or repress) transcription. The PAMP flg22 triggered events that contributed to rapid down-regulation of the primary auxin-response genes *GH3-like*, *BDL/IAA12*, and *AXR3/IAA17* (Navarro et al. 2006).

Transgenic *Arabidopsis* plants over expressing *miR393a* showed resistance to the pathogen (Navarro et al. 2006). Augmenting auxin signaling through over-expressing a TIR1 paralog that is partially refractory to miR393 enhanced susceptibility to the bacterial pathogen, and conversely, repressing auxin signaling through miR393 overexpression increased bacterial resistance. These results suggest that down-regulation of auxin signaling, resulting in ARF inactivation, is part of a PAMP-induced immune response (Navarro et al. 2006). It is known that auxin promotes susceptibility to bacterial diseases (O'Donnell et al. 2003a, b).

Li et al. (2010) showed flg22-triggered accumulation of another microRNA called miR160a besides miR393a already reported in *Arabidopsis*. Flg22 induces miR160a accumulation and represses its target genes *ARF16* and *ARF17* (Li et al. 2010). ARF proteins bind auxin-responsive elements to activate or repress transcription of primary auxin-response genes. Thus, multiple auxin pathway genes may be regulated by miRNAs during PAMP-triggered innate immunity (PTI) responses (Li et al. 2010).

### *2.32.3 Flg22 Suppresses Accumulation of Some miRNAs, Which Have a Negative Role in PAMP-Triggered Innate Immunity*

Some of the AGO1-bound miRNAs play a negative role in PTI resistance, although AGO1 overall positively regulates PTI resistance. Flg22 suppressed *miR398b* and *miR773* accumulation (Li et al. 2010). Consistent with this, flg22 treatment enhanced the expression of their target genes *COX5b.1*, *CSD2*, and *MET1*. COX5b.1 is a mitochondrial cytochrome c oxidase, CSD2 is a copper and zinc-containing superoxide dismutase enzyme that converts superoxide anion to hydrogen peroxide and MET1 is a DNA methyltransferase. The *miR398b* and *miR773* overexpression plants were compromised in PTI defenses exemplified by reduced callose deposition and supported greater *P. syringae* pv. *tomato* DC3000 and DC3000 *hrcC*- strains proliferation, indicating that *miR398b* and *miR773* negatively regulate plant disease resistance (Li et al. 2010).

### 2.32.4 Importance of miRNA-Directed RNA Silencing Pathway in PAMP-Triggered Immunity (PTI)

The Argonaute proteins AGO1 and AGO7 and the Dicer-like protein DCL1 have been shown to be involved in PTI responses induced by flg22 (Li et al. 2010). *ago1* and *dcl1* mutants are compromised in PTI responses and flg22-induced disease resistance, indicating that overall AGO1 and DCL1 positively regulate PTI (Li et al. 2010). The *ago1* and *dcl1* mutants showed defects in one or more of the late responses induced by PAMPs. However, these mutants displayed normal MAPK activation and transient oxidative burst, the events that occur less than 5 min after flg22 treatment (Li et al. 2010). It suggests that the PAMP-induced miRNA- mediated defense gene expression and callose deposition occur independent of MAPK activation and oxidative burst (Li et al. 2010).

The transgenic plants overexpressing *miR160a* exhibited enhanced callose deposition, suggesting that the PAMP triggers innate immune system by activating accumulation of specific miRNA (Li et al. 2010). The importance of miRNA pathway in PAMP-triggered immunity was assessed by employing miRNA-deficient *Arabidopsis* mutants (Navarro et al. 2008). The miRNA-deficient mutants sustained growth of non-pathogenic *Pseudomonas fluorescens* and *Escherichia coli* strains. These mutants also restored growth of a type III secretion-defective mutant of *Pseudomonas syringae*. Some *P. syringae* effectors suppress transcriptional activation of some PAMP-responsive miRNAs, miRNA biogenesis, stability or activity (Navarro et al. 2008). These results suggest that bacterial pathogens have evolved to suppress PAMP-triggered RNA silencing to cause disease.

### 2.32.5 Small RNAs May Also Be Involved in Effector-Triggered Immunity (ETI)

It has also been reported that inoculation of plants with incompatible strains *P. syringae* pv. *tomato* DC3000 (*avrRpm1*) and DC3000 (*avrRpt2*) but not the compatible strain DC3000 represses *miR398* levels (Jagadeeswaran et al. 2009). It suggests that *miR398*may also be involved in effector-triggered immunity (ETI). One of the long siRNAs (IsiRNAs), AtlsiRNA-1, is specifically induced by the bacterial pathogen *Pseudomonas syringae* pv. *tomato* carrying effector *avrRpt2* (Katiyar-Agarwal et al. 2007). *P. syringae* pv. *tomato* (avrRpt2)-mediated induction of AtlsiRNA-1 specifically targets the *AtRAP* gene, which encodes a RNA-binding protein containing a putative RNA-binding RAP domain (Katiyar-Agarwal et al. 2007; Katiyar-Agarwal and Jin 2010). Induction of AtlsiRNA-1 leads to down-regulation of its target, *AtRAP* mRNA. It silences *AtRAP*, which encodes a RAP-domain protein involved in disease resistance (Katiyar-Agarwal et al. 2007). AtRAP is a negative regulator of both PAMP-triggered immunity (PTI) and effector-triggered immunity (ETI) because the knockout mutant of this gene resulted in enhanced resistance to both avirulent *P. syringae* pv. *tomato* (*avrRpt2*) and a virulent strain *P. syringae* pv. *tomato* DC3000 (Katiyar-Agarwal and Jin 2010). These results suggest that small RNA-directed RNA silencing may play a role in both PTI and ETI.

# References


Abramovitch RB, Kim YJ, Chen SR, Dickman MB, Martin GB (2003) *Pseudomonas* type III effector AvrPtoB induces plant disease susceptibility by inhibition of host programmed cell death. EMBO J 22:60–69

Abramovitch RB, Janjusevic R, Stebbins CE, Martin GB (2006) Type III effector AvrPtoB requires intrinsic E3 ubiquitin ligase activity to suppress plant cell death and immunity. Proc Natl Acad Sci USA 103:2851–2856

AbuQamar S, Chen X, Dahwan R, BluhmB SJ, Lam S, Dietrich RA, Mengiste T (2006) Expression profiling and mutant analysis reveals complex regulatory networks involved in *Arabidopsis* response to *Botrytis* infection. Plant J 48:28–44

Acharya BR, Raina S, Maqbool SB, Jagadeeswaran G, Mosher SL, Appel HM, Schultz JC, Klessig DF, Raina R (2007) Overexpression of CRK13, an Arabidopsis cysteine-rich receptor-like kinase, results in enhanced resistance to *Pseudomonas syringae*. Plant J 50:488–499

Ade J, DeYoung BJ, Golstein C, Innes RW (2007) Indirect activation of a plant nucleotide binding site-leucine-rich repeat protein by a bacterial protease. Proc Natl Acad Sci USA 104:2531–2536

Adie BA, Perez-Perez J, Godoy M, Sanchez-Serrano JJ, Schmelz EA, Solano R (2007) ABA is an essential signal for plant resistance to pathogens affecting JA biosynthesis and activation of defenses in *Arabidopsis*. Plant Cell 19:1665–1681

Agorio A, Vera P (2007) ARGONAUTE4 is required for resistance to *Pseudomonas syringae* in *Arabidopsis*. Plant Cell 19:3778–3790

Ahmad S, Van Hulten M, Martin J, Pieterse CMJ, Van Wees SCM, Ton J (2011) Genetic dissection of basal defence responsiveness in accessions of *Arabidopsis thaliana*. Plant Cell Environ 34:1191–1208

Ahn I-P (2007) Disturbance of the Ca2+/calmodulin-dependent signaling pathway is responsible for the resistance of Arabidopsis *dnd1* against *Pectobacterium carotovorum* infection. Mol Plant Pathol 8:747–759

Ahn I-P (2008) Glufosinate ammonium-induced pathogen inhibition and defense responses culminate in disease protection in bar-transgenic rice. Plant Physiol 146:213–227

Ahn I-P, Kim S, Lee Y-H, Suh S-C (2007) Vitamin B1-induced priming is dependent on hydrogen peroxide and the *NPR1* gene in Arabidopsis. Plant Physiol 143:838–848

Akashi-Takamura S, Miyake K (2008) TLR accessory molecules. Curr Opin Immunol 20: 420–425

Aker J, de Vries SC (2008) Plasma membrane receptor complexes. Plant Physiol 147:1560–1564

Akimoto-Tomiyama C, Furutani A, Tsuge S, Washington EJ, Nishizawa Y, Minami E, Ochiai H (2012) XopR, a type III effector secreted by *Xanthomonas oryzae* pv. *oryzae*, suppresses microbe-associated molecular pattern-triggered immunity in *Arabidopsis thaliana*. Mol Plant Microbe Interact 25:505–514

Akira S, Uematsu S, Takeuchi O (2006) Pathogen recognition and innate immunity. Cells 124:783–801

Albert M, Jehle AK, Mueller K, Elsele C, Lipschis M, Felix G (2010) *Arabidopsis thaliana* pattern recognition receptors for bacterial elongation factor Tu and flagellin can be combined to form functional chimeric receptors. J Biol Chem 285:19035–19042

Alfano JR, Collmer A (2004) Type III secretion system effector proteins: double agents in bacterial disease and plant defense. Annu Rev Phytopathol 42:385–414

Ali R, Ma W, Lemtiri-Chlieh F, Tsaltas D, Leng Q, von Bodman S, Berkowitz GA (2007) Death don't have no mercy and neither does calcium: Arabidopsis CYCLIC NUCLEOTIDE GATED CHANNEL2 and innate immunity. Plant Cell 19:1081–1095

Alkan N, Fluhr R, Prusky D (2012) Ammonium secretion during *Colletotrichum coccodes* infection modulates salicylic and jasmonic acid pathways of ripe and unripe tomato fruit. Mol Plant Microbe Interact 25:85–96

Almasia NI, Bazzini AA, Hopp HE, Vazquez-Rovere C (2008) Overexpression of *snakin-1* gene enhances resistance to *Rhizoctonia solani* and *Erwinia carotovora* in transgenic potato plants. Mol Plant Pathol 9:329–338

Altenbach DD, Robatzek S (2007) Pattern recognition receptors: from the cell surface to intracellular dynamics. Mol Plant Microbe Interact 20:1031–1039

Alvarado VY, Scholthof HB (2009) Plant responses against invasive nucleic acids: RNA silencing and its suppression by plant viral pathogens. Semin Cell Dev Biol 20:1032–1040

Amborabé B-E, Bonmort J, Fleurat-Lessard P, Roblin G (2008) Early events induced by chitosan on plant cells. J Exp Bot 59:2317–2324

Anand A, Uppalapati SR, Ryu C-M, Allen SN, Kang L, Tang Y, Mysore KS (2008) Salicylic acid and systemic acquired resistance play a role in attenuating crown gall disease caused by *Agrobacterium tumefaciens*. Plant Physiol 146:703–715

Anderson JP, Badruzsaufari E, Schenk PM, Manners JM, Desmond OJ, Ehlert C, Maclean DJ, Ebert PR, Kazan K (2004) Antagonistic interaction between abscisic acid and jasmonate-ethylene signaling pathways modulates defense gene expression and disease resistance in *Arabidopsis*. Plant Cell 16:3460–3479

Andreasson E, Jenkins T, Brodersen P, Thorgrimsen S, Petersen NHT, Zhu S, Qiu J-L, Micheelsen P, Rocher A, Petersen M, Newman A, Nielsen HB, Hirt H, Somssich I, Mattsson O, Mundy J (2005) The MAP kinase substrate MKS1 is a regulator of plant defense responses. EMBO J 24:2579–2589

Anelli T, Sitia R (2008) Protein quality control in the early secretory pathway. EMBO J 27:507–517

Arenas YC, Kalkman ERIC, Schouten A, Dieho M, Vredenbregt P, Uwumukiza B, Ruiz MO, van Kan JAL (2010) Functional analysis and mode of action of phytotoxic Nep1-like proteins of *Botrytis cinerea*. Physiol Mol Plant Pathol 74:376–386

Asai S, Yoshioka H (2009) Nitric oxide as a partner of reactive oxygen species participates in disease resistance to necrotrophic pathogen *Botrytis cinerea* in *Nicotiana benthamiana*. Mol Plant Microbe Interact 22:619–629

Asai T, Stone JM, Heard JE, Kovtum Y, Yorgey P, Sheen J, Ausubel FM (2000) Fumonisin B1-induced cell death in *Arabidopsis* protoplasts requires jasmonate-, ethylene-, and salicylate-dependent signaling pathways. Plant Cell 12:1823–1835

Asai T, Tena G, Plotnikova J, Willmann MR, Chiu WL, Gómez-Gómez I, Boller T, Ausubel FM, Sheen J (2002) MAP kinase signaling cascade in *Arabidopsis* innate immunity. Nature 415:977–983

Asai S, Ohta K, Yoshioka H (2008) MAPK signaling regulates nitric oxide and NADPH oxidase-dependent oxidative bursts in *Nicotiana benthamiana*. Plant Cell 20:1390–1406

Asano T, Masuda D, Yasuda M, Nakashita H, Kudo T, Kimura M, Yamaguchi K, Nishiuchi T (2008) *AtNFXL1*, an Arabidopsis homologue of the human transcription factor *NF-X1*, functions as a negative regulator of the trichothecene phytotoxin-induced defense response. Plant J 53:450–464

Asensio JL, Canada FJ, Siebert H, Laynez J, Poveda A, Nieto PM, Soedjanaamadja UM, Gabius H, Jimenez-Barbero J (2000) Structural basis for chitin recognition by defense proteins: GlcNAc residues are bound in a multivalent fashion by extended binding sites in hevein domains. Chem Biol 7:529–543

Aslam SN, Newman MA, Erbs G, Morissey KL, Chinchilla D, Boller T, Jensen TT, De Castro C, Iearno T, Molinaro A, Jackson RW, Knight MR, Cooper RM (2008) Bacterial polysaccharides suppress induced innate immunity by calcium chelation. Curr Biol 18:1078–1083

Aslam SN, Erbs G, Morrissey KL, Newman M-A, Chinchilla D, Boller T, Molinaro A, Jackson RW, Cooper RM (2009) Microbe-associated molecular pattern (MAMP) signatures, synergy, size, and charge: influences on perception or mobility and host defense responses. Mol Plant Pathol 10:375–387

Asselbergh B, Höfte M (2007) Basal tomato defences to *Botrytis cinerea* include abscisic acid-dependent callose formation. Physiol Mol Plant Pathol 71:33–40

Asselbergh B, Achuo AE, Höfte M, Van Gijsegem F (2008) Absicisic acid deficiency leads to rapid activation of tomato defence responses upon infection with Erwinia chrysanthemi. Mol Plant Pathol 9:11–24

Atsumi G, Kagaya U, Kitazawa H, Nakahara KS, Uyeda I (2009) Activation of the salicylic acid signaling pathway enhances *Clover yellow vein virus* virulence in susceptible pea cultivars. Mol Plant Microbe Interact 22:166–175

Audenaert K, De Meyer GB, Höfte MM (2002) Abscisic acid determines basal susceptibility of tomato to *Botrytis cinerea* and suppresses salicylic acid-dependent signaling mechanisms. Plant Physiol 128:491–501

Avni A, Bailey BA, Mattoo AK, Anderson JD (1994) Induction of ethylene biosynthesis in *Nicotiana tabacum* by a *Trichoderma viride* xylanase is correlated to the accumulation of 1-aminocyclopropane-1-carboxylic acid (ACC) synthase and ACC oxidase transcripts. Plant Physiol 106:1049–1055

Aziz A, Poinssot B, Daire X, Adrian M, Bezier A, Lambert B, Joubert JM, Pugin A (2003) Laminarin elicits defense responses in grapevine and induces protection against *Botrytis cinerea* and *Plasmopara viticola*. Mol Plant Microbe Interact 16:1118–1128

Aziz A, Heyraud A, Lambert B (2004) Oligogalacturonide signal transduction, induction of defence-related responses and protection of grapevine against *Botrytis cinerea*. Planta 218:767–774

Aziz A, Gauthier A, Bezier A, Poinssot B, Joubert JM, Pugin A, Heyraud A, Baillieu F (2007) Elicitor and resistance-inducing activities of β-1,4 cellodextrins in grapevine, comparison with β-1,3 glucans and α-1,4 oligogalacturonides. J Exp Bot 58:1463–1472

Bae H, Kim MS, Sicher RC, Bae H-J, Bailey BA (2006) Necrosis- and ethylene-inducing peptide from *Fusarium oxysporum* induces a complex cascade of transcripts associated with signal transduction and cell death in *Arabidopsis*. Plant Physiol 141:1056–1067

Bai GH, Desjardins AEAF, Plattner RD (2002) Deoxynivalenol-nonproducing *Fusarium graminearum* causes initial infection, but does not cause disease spread in wheat spikes. Mycopathologia 153:91–98

Bailey BA (1995) Purification of a protein from culture filtrates of *Fusarium oxysporum* that induces ethylene and necrosis in leaves of *Erythroxylum coca*. Phytopathology 85:1250–1255

Bailey BA, Dean JFD, Anderson JD (1990) An ethylene biosynthesis-inducing endoxylanase elicits electrolyte leakage and necrosis in *Nicotiana tabacum* cv Xanthi leaves. Plant Physiol 94:1849–1854

Bailey BA, Jennings JC, Anderson JD (1997) The 24-kDa protein from *Fusarium oxysporum* f sp *erythroxyli*: occurrence in related fungi and the effect of growth medium on its production. Can J Microbiol 43:45–55

Baillieul F, de Ruffray P, Kauffmann S (2003) Molecular cloning and biological activity of alpha-, beta-, and gamma-megaspermin, three elicitins secreted by *Phytophthora megasperma* H20. Plant Physiol 131:155–166

Bakker EG, Traw MB, Toomajian C, Kreitman M, Bergelson J (2008) Low levels of polymorphism in genes that control the activation of defense response in *Arabidopsis thaliana*. Genetics 178:2031–2043

Balaji V, Mayrose M, Scherf O, Jacob-Hirsch J, Eichenlaub R, Iraki N, Manulis-Sasson S, Rechavi G, Barash I, Sessa G (2008) Tomato transcriptional changes in response to *Clavibacter michiganensis* reveal a role for ethylene in disease development. Plant Physiol 146:1797–1809

Balbi V, Devoto A (2008) Jasmonate signalling network in *Arabidopsis thaliana*: Crucial regulatory nodes and new physiological scenarios. New Phytol 177:301–318

Bar M, Avni A (2008) EHD2 inhibits ligand-induced endocytosis and signaling of the leucine-rich repeat receptor-like protein LeEix2. Plant J 59:600–611

Bar M, Aharon M, Benjamin S, Rothlat B, Horowitz M, Avni A (2008) AtEHDs, novel Arabidopsis EH-domain-containing proteins involved in endocytosis. Plant J 55:1025–1038

Bar M, Sharfman M, Schuster S, Avni A (2009) The coiled-coil domain of EHD2 mediates inhibition of LeEix2 endocytosis and signaling. PLoS One 4:e7963

Bar M, Sharfman M, Ron M, Avni A (2010) BAK1 is required for the attenuation of ethylene-inducing xylanase (Eix)-induced defense responses by the decoy receptor LeEix1. Plant J 63:791–800

Bar M, Sharfman M, Avni A (2011) LeEix1 functions as a decoy receptor to attenuate LeEix2 signaling. Plant Signal Behav 6:455–457

Barre A, Herve C, Lescure B, Rougé P (2002) Lectin receptor kinases in plants. Crit Rev Plant Sci 21:379–399

Bartel DP (2004) MicroRNAs: genomics, biogenesis, mechanism, and function. Cell 116: 281–297

Bartetzko V, Sonnewald S, Vogel F, Hartner K, Stadler R, Hammes UZ, Bornke F (2009) The *Xanthomonas campestris* pv. *vesicatoria* type III effector protein XopJ inhibits protein secretion evidence for interference with cell wall-associated defense responses. Mol Plant Microbe Interact 22:655–664

Bauer Z, Gómez-Gómez L, Boller T, Felix G (2001) Sensitivity of different ecotypes and mutants of *Arabidopsis thaliana* toward the bacterial elicitor flagellin correlates with the presence of receptor-binding sites. J Biol Chem 276:45669–45676

Baulcombe D (2004) RNA silencing in plants. Nature 431:356–363

Baureithel K, Felix G, Boller T (1994) Specific, high-affinity binding of chitin fragments to tomato cells and membranes. Competitive inhibition of binding by derivatives of chitooligosaccharides and a nod factor of *Rhizobium*. J Biol Chem 269:17931–17938

Baxter J, Moeder W, Urquhart W, Shahinas D, Chin K, Christendat D, Kang H-G, Angelova M, Kato N, Yoshioka K (2008) Identification of a functionally essential amino acid for Arabidopsis cyclic nucleotide gated ion channels using the chimeric *AtCNGC11/12* gene. Plant J 56:457–469

Baysal O, Zeller W (2004) Extract of *Hedera helix* induces resistance on apple rootstock M26 similar to Acibenzolar-S-methyl against fire blight (*Erwinia amylovora*). Physiol Mol Plant Pathol 65:305–315

Becker J, Nagel S, Tenhaken R (2000) Cloning, expression and characterization of protein elicitors from the soybean pathogenic fungus *Phytophthora sojae*. J Phytopathol 48:161–167

Bedini E, Castro C, Erbs G, Dow JM, Newman M-A, Parrilli M, Unverzagt C (2005) Structure-dependent modulation of a pathogen response in plants by synthetic O-antigen polysaccharides. J Am Chem Soc 127:2414–2416

Bednarek P, Osbourn A (2009) Plant-microbe interactions: chemical diversity in plant defense. Science 324:746–748

Belkhadir Y, Wang X, Chory J (2006) Arabidopsis brassinosteroid signaling pathway. Sci STKE 364:cm5

Bendahmane A, Köhn BA, Dedi C, Baulcombe DC (1995) The coat protein of potato virus X is a strain-specific elicitor of Rx1-mediated virus resistance in potato. Plant J 8:933–941

Bendahmane A, Kanyuka K, Baulcombe DC (1999) The *Rx* gene from potato controls separate virus resistance and cell death responses. Plant Cell 11:781–791

Benschop JJ, Mohammed S, O'Flaherty M, Heck AJR, Slijper M, Menke FLH (2007) Quantitative phosphoproteomics of early elicitor signaling in *Arabidopsis*. Mol Cell Proteomics 6:1198–1214

Bent AF, Mackey D (2007) Elicitors, effectors, R genes: the new paradigm and a lifetime supply of questions. Annu Rev Phytopathol 45:399–436

Berrocal-Lobo M, Molina A (2004) Ethylene response factor 1 mediates Arabidopsis resistance to the soilborne fungus *Fusarium oxysporum*. Mol Plant Microbe Interact 17:763–770

Berrocal-Lobo M, Molina A, Solano R (2002) Constitutive expression of ETHYLENE-RESPONSE-FACTOR1 in *Arabidopsis* confers resistance to several necrotrophic fungi. Plant J 29:23–32

Bethke G, Scheel DK, Lee J (2009a) Sometimes new results raise new questions: the question marks between mitogen-activated protein kinase and ethylene signaling. Plant Signal Behav 4:672–674

Bethke G, Unthan T, Uhrig JF, Poschl Y, Gust AA, Scheel DK, Lee J (2009b) Flg22 regulates the release of an ethylene response factor substrate from MAP kinase 6 in *Arabidopsis thaliana* via ethylene signaling. Proc Natl Acad Sci USA 106:8067–8072

Bethke G, Pecher P, Eschen-Lippold L, Tsuda K, Katagiri F, Glazebrook J, Scheel D, Lee J (2012) Activation of the *Arabidopsis thaliana* mitogen-activated protein kinase MPK11 by the flagellin-derived elicitor peptide, flg22. Mol Plant Microbe Interact 25:471–480

Bhuiyan NH, Selvaraj G, Wei Y, King J (2009) Gene expression and silencing reveal that monolignol biosynthesis plays a crucial role in penetration defence in wheat against powdery mildew invasion. J Exp Bot 60:509–521

Bi D, Cheng YT, Li X, Zhang Y (2010) Activation of plant immune responses by a gain-of-function mutation in an atypical receptor-like kinase. Plant Physiol 153:1771–1779

Bittel P, Robatzek D (2007) Microbe-associated molecular patterns (MAMPs) probe plant immunity. Curr Opin Plant Biol 10:335–341

Blee KA, Yang KY, Anderson AJ (2004) Activation of defense pathways: synergism between reactive oxygen species and salicylic acid and consideration of field applicability. Eur J Plant Pathol 110:203–212

Block A, Alfano JR (2011) Plant targets for *Pseudomonas syringae* type III effectors: virulence targets or quarded decoys? Curr Opin Microbiol 14:39–46

Block A, Li G, Fu ZQ, Alfano JR (2008) Phytopathogen type III effector weaponry and their plant targets. Curr Opin Plant Biol 11:396–403

Bohland C, Balkenhohl T, Loers G, Feussner I, Grambow HJ (1997) Differential induction of lipoxygenase isoforms in wheat upon treatment with rust fungus elicitor, chitin oligosaccharides, chitosan, and methyl jasmonate. Plant Physiol 114:679–685

Bohlmann H, Vignutelli A, Hilpert B, Miersch O, Wasternack C, Apel K (1998) Wounding and chemicals induce expression of the *Arabidopsis thaliana Thi21*, encoding a fungal defense thionin, via the octadecanoid pathway. FEBS Lett 43:281–286

Boller T, Felix G (2009) A renaissance of elicitors: perception of microbe-associated molecular patterns and danger signals by pattern-recognition receptors. Annu Rev Plant Biol 60:379–406

Boller T, He SY (2009) Innate immunity in plants: an arms race between pattern recognition receptors in plants and effectors in microbial pathogens. Science 324:742–744

Bolton MD (2009) Primary metabolism and plant defense – fuel for the fire. Mol Plant Microbe Interact 22:487–497

Bolton MD, van Esse HP, Vossen JH, de Jonge R, Sterigopoulos I, Stulemeijer IJ, van den Berg GC, Borrás-Hidalgo O, Dekker HL, de Koster CG, de Wit PJ, Joosten MH, Thomma BP (2008a) The novel *Cladosporium fulvum* lysin motif effector Ecp6 is a virulence factor with orthologues in other fungal species. Mol Microbiol 69:119–136

Bolton MD, van Esse HP, Vossen JH, de Jonge R, Sterigopoulos I, Stulemeijer IJ, van den Berg GC, Borrás-Hidalgo O, Dekker HL, de Koster CG, de Wit PJ, Joosten MH, Thomma BP (2008b) The novel *Cladosporium fulvum* lysin motif effector Ecp6 is a virulence factor with orthologues in other fungal species. Mol Microbiol 69:119–136

Bonnet P, Bourdon E, Ponchet M, Blein J-P, Ricci P (1996) Acquired resistance triggered by elicitins in tobacco and other plants. Eur J Plant Pathol 102:181–192

Borejsza-Wysocka E, Kader AA, Norelli JL, Bauer DW, Garr ER, Beer SV, Aldwinckle HS (2000) Effect of expressing *hrpN* in apple on resistance to *Erwinia amylovora*. In Vitro Cell Dev Biol Anim 34:1023

Borsani O, Zhu J, Verslues PE, Sunkar R, Zhu JK (2005) Endogenous siRNAs derived from a pair of natural *cis*-antisense transcripts regulate salt tolerance in *Arabidopsis*. Cell 123:1279–1291

Boudart G, Charpentier M, Lafitte C, Martinez Y, Jauneau A, Gaulin E, Esquerré-Tugayé M-T, Dumas B (2003) Elicitor activity of a fungal endopolygalacturonase in tobacco requires a functional catalytic site and cell wall localization. Plant Physiol 131:93–101

Boudsocq M, Willmann MR, McCormack M, Lee H, Shan L, He P, Bush J, ChengSH Sheen J (2010) Differential innate immune signalling via $Ca^{2+}$ sensor protein kinases. Nature 464: 418–422

Bouizgarne B, El-Maarouf-Bouteau H, Frankart C, Reboutier D, Madiona K, Pennarun AM, Monestiez M, Trouverie J, Amiar Z, Briand J, Brault M, Rona JP, Ouhdouch Y, El Hadrami I, Bouteau F (2006) Early physiological responses of *Arabidopsis thaliana* cells to fusaric acid: toxic and signalling effects. New Phytol 169:209–218

Boureau T, Siamer S, Perino C, Gaubert S, Patrit O, Degrave A, Fagard M, Chevreau E, Barny M-A (2011) The HrpN effector of *Erwinia amylovora*, which is involved in type III translocation, contributes directly or indirectly to callose elicitation on apple leaves. Mol Plant Microbe Interact 24:577–584

Bourque S, Binet MN, Ponchet M, Pugin A, Lebrun-Garcia A (1999) Characterization of the cryptogein binding sites on plant plasma membranes. J Biol Chem 274:34699–34705

Boutrot F, Segonzac C, Chang KN, Qiao H, Ecker JR, Zipfel JR, Rathjen JP (2010) Direct transcriptional control of the *Arabidopsis* immune receptor FLS2 by the ethylene-dependent transcription factor EIN3 and EIL1. Proc Natl Acad Sci U S A 107:14502–14507

Bretz JR, Mock NM, Charity JC, Zeyad S, Baker CJ, Hutcheson SW (2003) A translocated protein tyrosine phosphatase of *Pseudomonas syringae* pv. *tomato* DC3000 modulates plant defense response to infection. Mol Microbiol 49:389–400

Brummer M, Arend M, Fromm J, Schlenzig A, Oßwald WF (2002) Ultrastructural changes and immunocytochemical localization of the elicitin quercinin in *Quercus robur* L roots infected with *Phytophthora quercina*. Physiol Mol Plant Pathol 61:109–120

Brunner F, Rosahl S, Lee J, Rudd JJ, Geller C, Kauppinen S, Rasmussen G, Scheel D, Nürnberger T (2002) Pep-13, a plant defense-inducing pathogen-associated pattern from *Phytophthora* transglutaminases. EMBO J 21:6681–6688

Brutus A, Sicilia F, Macone A, Cervone F, De Lorenzo G (2010) A domain swap approach reveals a role of the plant wall-associated kinase 1 (WAK1) as a receptor of oligogalacturonides. Proc Natl Acad Sci USA 107:9452–9457

Buck TM, Wright CM, Brodsky JL (2007) The activities and function of molecular chaperones in the endoplasmic reticulum. Semin Cell Dev Biol 18:751–761

Buist G, Steen A, Kok J, Kuipers OP (2008) LysM, a widely distributed protein motif for binding to (peptido)glycans. Mol Microbiol 68:838–847

Burgyán J, Havelda Z (2011) Viral suppressors of RNA silencing. Trends Plant Sci 16:265–272

Cabral A, Oome S, Sander N, Küfner I, Nürnberger T, van den Ackerveken G (2012) Nontoxic Nep-1 like proteins of the downy mildew pathogen *Hyaloperonospora arabidopsidis*: repression of necrosis-inducing activity by a surface-exposed region. Mol Plant Microbe Interact 25:697–708

Cabrera JC, Boland A, Messiaen J, Cambier P, Cutsem PV (2008) Egg box conformation of oligogalacturonides: the time-dependent stabilization of the elicitor-active conformation increases its biological activity. Glycobiology 18:473–482

Caffali KH, Mohnen D (2009) The structure, function, and biosynthesis of plant cell wall pectic polysaccharides. Carbohydr Res 344:1879–1900

Campbell EJ, Schenk PM, Kazan K, Penninckx K, Anderson JP, Maclean DJ, Cammue BP, Ebert PR, Manners JM (2003) Pathogen-responsive expression of a putative ATP-binding cassette transporter gene conferring resistance to the diterpenoid sclareol is regulated by multiple defense signaling pathways in Arabidopsis. Plant Physiol 133:1272–1284

Cao Y, Wu Y, ZhengZ SF (2006) Overexpression of the rice EREBP-like gene *OsBIERF3* enhances disease resistance and salt tolerance in transgenic tobacco. Physiol Mol Plant Pathol 202–211

Caplan JL, Zhu X, Mamillapalli P, Marathe R, Anandalakshmi R, Dinesh-Kumar SP (2009) Induced ER chaperones regulate a receptor-like kinase to mediate antiviral innate immune response in plants. Cell Host Microbe 6:457–469

Caramelo JJ, Castro OA, de Prat GG, Parodi AJ (2003) The endoplasmic reticulum glucosyltransferase recognizes nearly native glycoprotein folding intermediates. J Biol Chem 279:46280–46285

Cavalcanti FR, Resende MLV, Lima J, Silveira JAG, Oliveira TA (2006) Activities of antioxidant enzymes and photosynthetic responses in tomato pre-treated by plant activators and inoculated by *Xanthomonas vesicatoria*. Physiol Mol Plant Pathol 68:198–208

Chapman FJ, Carrington JC (2007) Specialization and evolution of endogenous small RNA pathways. Nat Rev Genet 8:884–896

Chassot C, Nawrath C, Metraux J-P (2007) Cuticular defects lead to full immunity to a major plant pathogen. Plant J 49:972–980

Chassot C, Buchala A, Schoonbeek H-J, Metraux J-P, Lamotte O (2008) Wounding of Arabidopsis leaves causes a powerful but transient protection against *Botrytis* infection. Plant J 55:555–567

Che FS, Nakajima Y, Tanaka N, Iwano M, Yoshida T, Takayama S, Kadota I, Isogai A (2000) Flagellin from an incompatible strain of *Pseudomonas avenae* induces a resistance response in cultured rice cells. J Biol Chem 275:32347–32356

Chellappan P, Vanitharani R, Ogbe F, Fauquet CM (2005) Effect of temperature on *Geminivirus-induced* RNA silencing in plants. Plant Physiol 138:1828–1841

Chellappan P, Zhang X, Jin H (2009) Host small RNAs are big contributors to plant innate immunity. Curr Opin Plant Biol 12:465–472

Chellappan P, Xia J, Zhou X, Gao S, Zhang X, Coutino G, Vazquez F, Zhang W, Jin H (2010) siRNAs from miRNA sites mediate DNA methylation of target genes. Nucleic Acids Res 38:6883–6894

Chen JG (2008) Heterotrimeric G-proteins in plant development. Front Biosci 13:3321–3333

Chen X, Shang J, Chen D, Lei C, Zou Y, Zhai W, Liu G, Ling Z, Cao G, Ma B, Wang Y, Zhao X, Li S, Zhu L (2006) A B-lectin receptor kinase gene conferring rice blast resistance. Plant J 46:794–804

Chen YC, Siems WF, Pearce G, Ryan CA (2008) Six peptide wound signals derived from a single precursor protein in *Ipomoea batatas* leaves activate the expression of the defense gene sporamin. J Biol Chem 283:11469–11476

Chen F, Gao J, Miao S, Yuan Y- X, Wang M-Y, Li Q, Mao B-Z, Jiang L-W, He ZH (2010a) Plasma membrane localization and potential endocytosis of constitutively expressed XA21 proteins in transgenic rice. Mol Plant 3:917–926

Chen L, Hamada S, Fujiwara M, Zhu T, Thao NP, Wong HL, Krishna P, Ueda T, Kaku H, Shibuya N, Kawasaki T, Shimamoto K (2010b) The Hop/Sti-Hsp90 chaperone complex facilitates the maturation and transport of a PAMP receptor in rice innate immunity. Cell Host Microbe Interact 7:185–196

Chen X, Chern M, Canlas PE, Ruan D, Jiang C, Ronald PC (2010c) An ATPase promotes autophosphorylation of the pattern recognition receptor XA21 and inhibits XA21-mediated immunity. Proc Natl Acad Sci USA 107:8029–8034

Chen X, Chern M, Canlas PE, Jiang C, Ruan D, Cao P, Ronald PC (2010d) A conserved threonine residue in the juxtamembrane domain of the XA21 pattern recognition receptor is critical for kinase autophosphorylation and XA21-mediated immunity. J Biol Chem 285:10454–10463

Cheng B, Yu X, Ma Z, Dong S, Dou D, Wang Y, Zheng X (2012) *Phytophthora sojae* effector Avh331 suppresses the plant defence response by disturbing the MAPK signalling pathway. Physiol Mol Plant Pathol 77:1–9

Chern M, Fitzgerald HA, Canlas PE, Navarre DA, Ronald PC (2005) Overexpression of a rice NPR1 homolog leads to constitutive activation of defense response and hypersensitivity to light. Mol Plant Microbe Interact 18:511–520

Chern M, Canlas PE, Ronald PC (2008) Strong suppression of systemic acquired resistance in *Arabidopsis* by NRR is dependent on its ability to interact with NPR1 and its putative repressive domain. Mol Plant 1:552–559

Cheung M-Y, Zeng N-Y, Tong S-W, Li FW-Y, Zhao K-J, Zhang Q, Sun SS-M, Lam H-M (2007) Expression of a RING-HC protein from rice improves resistance to *Pseudomonas syringae* pv. *tomato* DC3000 in transgenic *Arabidopsis thaliana*. J Exp Bot 58:4147–4159

Chinchilla D, Bauer Z, Regenass M, Boller T, Felix G (2006) The *Arabidopsis* receptor kinase FLS2 binds flg22 and determines the specificity of flagellin perception. Plant Cell 18:465–476

Chinchilla D, Boller T, Robatzek S (2007a) Flagellin signaling in plant immunity. Adv Exp Med Biol 598:358–371

Chinchilla D, Zipfel C, Robatzek S, Kemmerling B, Nürnberger T, Jones JDG, Felix G, Boller T (2007b) A flagellin-induced complex of the receptor FLS2 and BAK1 initiates plant defence. Nature 448:497–500

Chinchilla D, Shan L, He P, de Vries S, Kemmerling B (2009) One for all: the receptor-associated kinase BAK1. Trends Plant Sci 14:535–541

Chisholm ST, Coaker G, Day B, Staskawicz BJ (2006) Host microbe interactions: shaping the evolution of the plant immune response. Cell 124:803–814

Choi DS, Hwang BK (2011) Proteomics and functional analyses of pepper *Abscisic acid-responsive 1* (*ABR1*), which is involved in cell death and defense signaling. Plant Cell 23:823–842

Choi HW, Kim YJ, Lee SC, Hong JK, Hwang BK (2007) Hydrogen peroxide generation by the pepper extracellular peroxidase CaPO2 activates local and systemic cell death and defense response to bacterial pathogens. Plant Physiol 145:890–904

Ciomperlik J, Omarov RT, Scholthof HB (2011) An antiviral RISC isolated from *Tobacco rattle virus*-infected plants. Virology 412:117–124

Cirillo L, Bedini E, Molinaro A, Parrilli M (2010) Synthesis of a β-GlcN-(1 → 4)-MurNAc building block en route to N-acetylated peptidoglycan fragments. Tetrahedron Lett 51:1117–1120

Clarke JD, Volko SM, Ledford H, Ausubel FM, Dong X (2000) Roles of salicylic acid, jasmonic acid, and ethylene in *cpr*-induced resistance in Arabidopsis. Plant Cell 12:2175–2190

Clay NK, Adio AM, Denoux C, Jander G, Ausubel FM (2009) Glucosinolate metabolites required for an *Arabidopsis* innate immune response. Science 323:95–101

Cloud-Hansen KA, Brook Petersen S, Stabb EV, Goldman WE, McFall-Ngai MJ, Handelsman J (2006) Breaching the great wall: peptidoglycan and microbial interactions. Nat Rev Microbiol 4:710–716

Coppinger P, Repetti PP, Day B, Dahlbeck D, Mehlert A, Staskawicz BJ (2004) Overexpression of the plasma membrane-localized NDR1 protein results in enhanced bacterial disease resistance in *Arabidopsis thaliana*. Plant J 40:225–237

Coram TE, Pang ECK (2007) Transcriptional profiling of chickpea genes differentially regulated by salicylic acid, methyl jasmonate and aminocyclopropane carboxylic acid to reveal pathways of defence-related gene regulation. Funct Plant Biol 34:52–64

Costet L, Fritig B, Kauffmann S (2002) Scopoletin expression in elicitor-treated and tobacco mosaic virus infected tobacco plants. Physiol Plant 115:228–235

Cox KD, Bryson PK, Schnabel G (2007) Instability of propiconazole resistance and fitness in *Monilinia fructicola*. Phytopathology 97:448–453

Crofts AJ, Leborgne Castel N, Pesca M, Vitale A, Dennecke J (1998) BiP and calreticulin form an abundant complex that is independent of endoplasmic reticulum stress. Plant Cell 10:813–823

Cui H, Xiang T, Zhou JM (2009) Plant immunity: a lesson from pathogenic bacterial effector proteins. Cell Microbiol 11:1453–1461

Cui H, Wang Y, Xue L, Chu J, Yan C, Fu J, Chen M, Innes RW, Zhou JM (2010) *Pseudomonas syringae* effector protein AvrB perturbs *Arabidopsis* hormone signaling by activating MAP kinase 4. Cell Host Microbe 7:164–175

Culver JN, Dawson WO (1991) Tobacco mosaic virus elicitor coat protein genes produce a hypersensitive phenotype in transgenic *Nicotiana sylvestris* plants. Mol Plant Microbe Interact 4:458–463

Culver JN, Stubbs G, Dawson WO (1994) Structure-function relationship between tobacco mosaic virus coat protein and hypersensitivity in *Nicotiana sylvestris*. J Mol Biol 242:130–138

Cunnac S, Lindeberg M, Collmer A (2009) *Pseudomonas syringae* type III secretion system effectors: repertoires in search of functions. Curr Opin Microbiol 12:53–60

Cuzick A, Lee S, Gezan S, Hammond-Kosack KE (2008) NPR1 and EDS11 contribute to host resistance against *Fusarium culmorum* in *Arabidopsis* buds and flowers. Mol Plant Pathol 9:697–704

D'Amelio R, Berta G, Gamalero E, Massa N, Avidano L, Cantamessa S, D'Agostino G, Bosco D, Marzachi C (2011) Increased plant tolerance against chrysanthemum yellows phytoplasma (*Candidatus Phytoplasma asteris*) following double inoculation with *Glomus mosseae* BEG12 and *Pseudomonas putida* SIPflRif. Plant Pathol 60:1014–1022

D'Ovidio R, Mattei B, Roberti S, Bellincampi D (2004a) Polygalacturonases, polygalacturonase-inhibiting proteins and pectic oligomers in plant-pathogen interactions. Biochim Biophys Acta 1696:237–244

D'Ovidio R, Mattei B, Roberti S, Bellincampi D (2004b) Polygalacturonases, polygalacturonase-inhibiting proteins and pectic oligomers in plant-pathogen interactions. Biochim Biophys Acta Proteins Proteomics 1696:237–244

Dardick C, Ronald P (2006) Plant and animal pathogen recognition receptors signal through non-RD kinases. PLoS Pathog 2:e2

Davis KR, Hahlbrock K (1987) Induction of defense responses in cultured parsley cells by plant cell wall fragments. Plant Physiol 85:1286–1290

Day B, He SY (2010) Battling immune kinases in plants. Cell Host Microbe 7:259–261

Day RB, Okada M, Ito Y, Tsukada K, Zaghouani H, Shibuya N, Stacey G (2001) Binding site for chitin oligosaccharides in the soybean plasma membrane. Plant Physiol 126:1162–1173

Dayakar BV, Lin HJ, Chen CH, Ger MJ, Lee BH, Pai CH, Chow D, Huang HE, Hwang SY, Chung MC, Feng TY (2003) Ferredoxin from sweet pepper (*Capsicum annuum* L) intensifying harpin(pss)-mediated hypersensitive response shows an enhanced production of active oxygen species (AOS). Plant Mol Biol 51:913–924

de Capdeville GD, Beer SV, Wilson CL, Aist JR (2008) Some cellular correlates of harpin-induced resistance to blue mold of apples. Trop Plant Pathol 33:103–113

de Freitas MB, Stadnik MJ (2012) Race-specific and ulvan-induced defense responses in bean (*Phaseolus vulgaris*) against *Colletotrichum lindemuthianum*. Physiol Mol Plant Pathol 78:8–13

de Jonge R, Thomma BPHJ (2009) Fungal LysM effectors: extinguishers of host immunity? Trends Microbiol 17:151–157

de Jonge R, van Esse HP, Kombrink A, Shinya T, Desaki Y, Bours R, van der Krol S, Shibuya N, Joosten MHAJ, Thomma BPHJ (2010) Conserved fungal LysM effector Ecp6 prevents chitin-triggered immunity in plants. Science 329:953–955

de Silva FG, Shen Y, Dardick G, Burdman S, Yadav RC, de Leon AL, Ronald RC (2004) Bacterial genes involved in type 1 secretion and sulfation are required to elicit the rice Xa21-mediated innate immune response. Mol Plant Microbe Interact 17:593–601

de Torres M, Mansfield JW, Grabov N, Brown IR, Ammouneh H, Tsiamis G, Forsyth A, Robatzek S, Grant M, Boch J (2006) *Pseudomonas syringae* effector AvrPtroB suppresses basal defence in Arabidopsis. Plant J 47:368–382

de Torres-Zabala M, Truman W, Bennett MH, Lafforgue G, Mansfield JW, Rodriguez Egea P, Bogre L, Grant M (2007) *Pseudomonas syringae* pv. *tomato* hijacks the *Arabidopsis* abscisic acid signalling pathway to cause disease. EMBO J 26:1434–1443

De Vos M, Van Oosten VR, Van Poecke RMP, Van Pelt JA, Pozo MJ, Mueller MJ, Buchala AJ, Métraux J-P, Van Loon LC, Dicke M, Pieterse MJC (2005) Signal signature and transcriptome changes of *Arabidopsis* during pathogen and insect attack. Mol Plant Microbe Interact 18:923–937

De Vos M, Van Zaanen W, Koornneef A, Korzelius JP, Dicke M, Van Loon LC, Pieterse CMJ (2006) Herbivore-induced resistance against microbial pathogens in *Arabidopsis*. Plant Physiol 142:352–363

De Vries RE, Trigiano RN, Windham MT, Windham AS, Sorocham JC, Rinehart TA, Vargas JM (2008) Genetic analysis of fungicide-resistant *Sclerotinia homeocarpa* isolates from Tennessee and Northern Mississippi. Plant Dis 92:83–90

DebRoy S, Thilmony R, Kwack Y-B, Nomura K, He S-Y (2004) A family of conserved bacterial effectors inhibits salicylic acid-mediated basal immunity and promotes disease necrosis in plants. Proc Natl Acad Sci USA 101:9927–9932

Decreux A, Messiaen J (2005) Wall-associated kinase WAK1 interacts with cell wall pectins in a calcium-induced conformation. Plant Cell Physiol 46:268–278

Decreux A, Thomas A, Spies B, Brasseur R, Van Cutsem P, Messiaen J (2006) In vitro characterization of the homogalacturonan-binding domain of the wall-associated kinase WAK1 using site-directed mutagenesis. Phytochemistry 67:1068–1079

Delker C, Zolman BK, Miersch O, Wasternack C (2007) Jasmonate biosynthesis in *Arabidopsis thaliana* requires peroxisomal β-oxidation enzymes – Additional proof by properties of *pex6* and *aim1*. Phytochemistry 68:1542–1650

Demidchik V, Shabala SN, Davies JM (2007) Spatial variation in $H_2O_2$ response of *Arabidopsis thaliana* root epidermal $Ca^{2+}$ flux and plasma membrane $Ca^{2+}$ channels. Plant J 49:377–386

Dempsey DA, Vlot AC, Wildermuth MC, Klessig DF (2011) Salicylic acid biosynthesis and metabolism. Arabidopsis Book 9:e0156. doi:101199/tab0156

Denny T (2000) *Ralstonia solanacearum* – a plant pathogen in touch with its host. Trends Microbiol 11:486–489

Denoux C, Galletti R, Mammarella N, Gopalan S, Werck D, De Lorenzo G, Ferrari S, Ausubel FM, Dewdney J (2008) Activation of defense response pathways by OGs and Flg22 elicitors in Arabidopsis seedlings. Mol Plant 1:423–445

Desaki Y, Miya A, Venkatesh B, Tsuyumu S, Yamane H, Kaku H, Minami E, Shibuya N (2006) Bacterial lipopolysaccharides induce defense responses associated with programmed cell death in rice cells. Plant Cell Physiol 47:1530–1540

Desender S, Klarzynski O, Potin P, Barzic MR, Andrivon D, Val F (2006) Lipopolysaccharides of *Pectobacterium atrosepticum* and *Pseudomonas corrugata* induce different defense response patterns in tobacco, tomato, and potato. Plant Biol (Stuttg) 8:636–645

Desmond OJ, Edgar CI, Manners JM, Maclean DJ, Schenk PM, Kazan K (2006) Methyl jasmonate induced gene expression in wheat delays symptom development by the crown rot pathogen *Fusarium pseudograminearum*. Physiol Mol Plant Pathol 67:171–179

Di Matteo A, Federici L, Mattei B, Salvi G, Johnson KA, Savino C, De Lorenzo G, Tsernoglou D, Cervone F (2003) The crystal structure of polygalacturonase-inhibiting protein (PGIP), a leucine-rich repeat protein involved in plant defense. Proc Natl Acad Sci USA 100:10124–10128

Di Matteo A, Bonivento D, Tsernoglou D, Federici L, Cervone F (2006) Polygalacturonase-inhibiting protein (PGIP) in plant defence: a structural view. Phytochemistry 67:526–533

Ding S-W (2010) RNA-based antiviral immunity. Nat Rev Immunol 10:632–644

Djamei A, Pitzschke A, Nakagami H, Rajh I, Hirt H (2007) Trojan horse strategy in agrobacterium transformation: abusing MAPK defense signaling. Science 318:453–456

Doares SH, Syrovets T, Weiler EW, Ryan CA (1995) Oligogalacturonides and chitosan activate plant defensive genes through the octadecanoid pathway. Proc Natl Acad Sci USA 92:4095–4098

Dodds PN, Rathjen JP (2010) Plant immunity: towards an integrated view of plant-pathogen interactions. Nat Rev Genet 11:539–548

Dodds PN, Lawrence GJ, Catanzariti AM, Teh T, Wang CIA, Ayliffe MA, Kobe B, Ellis JG (2006) Direct protein interaction underlies gene-for-gene specificity and coevolution of the flax resistance genes and flax rust avirulence genes. Proc Natl Acad Sci U S A 103:8888–8893

Domingo C, Andrés F, Tharreau D, Iglesias DJ, Talón M (2009) Constitutive expression of *OsGH31* reduces auxin content and enhances defense response and resistance to a fungal pathogen in rice. Mol Plant Microbe Interact 22:201–210

Dong H, Delaney TP, Bauer DW, Beer SV (1999) Harpin induces disease resistance in *Arabidopsis* through the systemic acquired resistance pathway mediated by salicylic acid and the *NIM1* gene. Plant J 20:207–215

Dong J, Chen C, Chen Z (2003) Expression profiles of the *Arabidopsis* WRKY gene superfamily during plant defense response. Plant Mol Biol 51:21–37

Dow JM, Osbourn AE, Wilson TJG, Daniels MJ (1995) A locus determining pathogenicity of Xanthomonas campestris is involved in lipopolysaccharide biosynthesis. Mol Plant Microbe Interact 8:768–777

Dow JM, Newman M-A, von Roepenack E (2000) The induction and modulation of plant defense responses by bacterial lipopolysaccharides. Annu Rev Phytopathol 38:241–261

Drutskaya MS, Belousov PV, Nedopasov SA (2011) Innate mechanisms of viral recognition. Mol Biol 45:5–15

Dubery IA, Sanabria NM, Huang JC (2012) Nonself perception in plant innate immunity. Adv Exp Med Biol 738:79–107

Duclos J, Fauconnier A, Coelho AC, Bollen A, Cravador A, Godfroid E (1998) Identification of an elicitin gene cluster in *Phytophthora cinnamomi*. DNA Seq 9:231–237

Dunning FM, Sun W, Jansen KL, Helft L, Bent AF (2007) Identification and mutational analysis of *Arabidopsis* FLS2 leucine-rich repeat domain residues that contribute to flagellin perception. Plant Cell 19:3297–3317

Dunoyer P, Voinnet O (2005) The complex interplay between plant viruses and host RNA-silencing pathways. Curr Opin Plant Biol 8:415–423

Dunoyer P, Himber C, Voinnet O (2005) DICER-LIKE4 is required for RNA interference and produces the 21-nucleotide small interfering RNA component of the plant cell-to-cell silencing signal. Nat Genet 37:1356–1360

Dunoyer P, Himber C, Voinnet O (2006) Induction, suppression and requirement of RNA-silencing pathways in virulent *Agrobacterium tumefaciens* infections. Nat Genet 38:258–263

Dunoyer P, Schott G, Himber C, Meyer D, Takeda A, Carrington JC, Voinnet O (2010) Small RNA duplexes function as mobile silencing signals between plant cells. Science 328:912–916

Ebel J, Mithofer A (1998) Early events in the elicitation of plant defence. Planta 206:335–348

Eckardt NA (2008) Chitin signaling in plants: insights into the perception of fungal pathogens and rhizobacterial strains. Plant Cell 20:241–243

Edgar CI, McGrath KC, Dombrecht B, Manners JM, Maclean DC, Schenk PM, Kazan K (2006) Salicylic acid mediates resistance to the vascular wilt pathogen *Fusarium oxysporum* in the model host *Arabidopsis thaliana*. Australas Plant Pathol 35:581–591

Elizabeth SV, Bender CL (2007) The phytotoxin coronatine from *Pseudomonas syringae* pv. *tomato* DC3000 functions as a virulence factor and influences defence pathways in edible brassicas. Mol Plant 8:83–92

Ellendorff U, Fradin EF, de Jonge R, Thomma BPHJ (2009) RNA silencing is required for *Arabidopsis* defence against *Verticillium* wilt disease. J Exp Bot 60:591–602

Engelhardt S, Lee J, Gäbler Y, Kemmerling B, Haapalainen M-L, Li C-M, Wei Z, Keller H, Joosten M, Taira S, Nürnberger T (2009) Separable roles of the *Pseudomonas syringae* pv. *phaseolicola* accessory protein HrpZ1 in ion-conducting pore formation and activation of plant immunity. Plant J 57:706–717

Epple P, Apel K, Bohlmann H (1997) Overexpression of an endogenous thionin enhances resistance of *Arabidopsis* against *Fusarium oxysporum*. Plant Cell 9:509–520

Erbs G, Newman M-A (2009) MAMPs/PAMPs – elicitors of innate immunity. In: Jackson RW (ed) Plant pathogenic bacteria: genomics and molecular biology. Academic/Norwich Caister Press, Norfolk, pp 227–240

Erbs G, Silipo A, Aslam S, De Castro C, Liparoti V, Flagiello A, Pucci P, Lanzetta R, Parrilli M, Molinaro A, Newman M-A, Cooper RM (2008) Peptidoglycan and muropeptides from pathogens *Agrobacterium* and *Xanthomonas* elicit plant innate immunity: structure and activity. Chem Biol 15:438–448

Erbs G, Molinaro A, Dow JM, Newman M-A (2010) Lipopolysaccharides and plant innate immunity. In: Wang X, Quino PJ (eds) Endotoxins: structure, function and recognition, Subcellular biochemistry. 53Part 2. Springer, Dordrecht, pp 387–403

Erickson IF, Holzberg S, Calderon-Urrea A, Handley V, Axtell M, Corr C, Baker B (1999) The helicase domain of the TMV replicase protein induces the N-mediated defense response in tobacco. Plant J 18:67–75

Espinosa A, Guo M, Tam VC, Fu ZQ, Alfano JR (2003) *Pseudomonas syringae* type III-secreted protein HopPtoD2 possesses protein tyrosine phosphatase activity and suppresses programmed cell death in plants. Mol Microbiol 49:377–387

Eulgem T, Somssich IE (2007) Networks of WRKY transcription factors in defense signaling. Curr Opin Plant Biol 10:366–371

Eyal Y, Sagee O, Fluhr R (1992) Dark induced accumulation of a basic pathogenesis-related (PR-1) transcript and a light requirement for its induction by ethylene. Plant Mol Biol 19:589–599

Fabro G, Di Rienzo JA, Voigt CA, Savchenko T, Dehesh K, Somerville S, Alvarez ME (2008) Genome-wide expression profiling Arabidopsis at the stage of *Golovinomyces cichoracearum* haustorium formation. Plant Physiol 146:1421–1439

Fagard M, Boutet S, Morel JB, Bellini C, Vaucheret H (2000) AGO1, QDE-2, RDE-1 are related proteins required for post-transcriptional gene silencing in plants, quelling in fungi, and RNA interference in animals. Proc Natl Acad Sci USA 97:11650–11654

Fagard M, Dellagi A, Roux C, Périno C, Rigault M, Boucher V, Shevchik VE, Expert D (2007) *Arabidopsis thaliana* expresses multiple lines of defense to counterattack *Erwinia chrysanthemi*. Mol Plant Microbe Interact 20:794–805

Felix G, Boller T (2003) Molecular sensing of bacteria in plants. The highly conserved RNA-binding motif RNP-1 of bacterial cold shock proteins is recognized as an elicitor signal in tobacco. J Biol Chem 278:6201–6208

Felix G, Duran JD, Volko S, Boller T (1999) Plants have a sensitive perception system for the most conserved domain of bacterial flagellin. Plant J 18:265–276

Fellbrich G, Romanski A, Varet A, Blume B, Brunner F, Stefan Engelhardt S, Felix G, Kemmerling B, Krzymowska M, Nürnberger T (2002) NPP1, a *Phytophthora*-associated trigger of plant defence in parsley and *Arabidopsis*. Plant J 32:375–390

Fellers JP, Tremblay D, Handest MF, Lommel SA (2002) The *Potato virusY* $M^SN^R$ Nlb-replicase is the elicitor of a veinal necrosis-hypersensitive response in root knot nematode resistant tobacco. Mol Plant Pathol 3:145–152

Fernández-Calvo P, Chini A, Fernández-Barbero G, Chico J-M, Gimenez-Ibanez S, Geerinck J, Eeckhout D, Schweizer F, Godoy M, Franco-Zorrilla JM, Pauwels L, Witters E, Puga MI, Paz-Ares J, Goossens A, Reymond P, De Jaeger G, Solano R (2011) The *Arabidopsis* bHLH transcription factors $MYC_3$ and $MYC_4$ are targets of JAZ repressors and act additively with $MYC_2$ in the activation of jasmonate responses. Plant Cell 23:701–715

Ferrari S, Galletti R, De Denoux C, Lorenzo G, Ausubel FM, Dewdney J (2007) Resistance to *Botrytis cinerea* induced in Arabidopsis by elicitors is independent of salicylic acid, ethylene, or jasmonate signaling but requires PHYTOALEXIN DEFICIENT3. Plant Physiol 144:367–379

Fischer U, Droge-Laser W (2004) Overexpression of NtERF5, a new member of the tobacco ethylene response transcription factor family enhances resistance to *Tobacco mosaic virus*. Mol Plant Microbe Interact 17:1162–1171

Fliegmann J, Mithöfer A, Wanner G, Ebel J (2004) An ancient enzyme domain hidden in the putative beta-glucan elicitor receptor of soybean may play an active part in the perception of pathogen-associated molecular patterns during broad host resistance. J Biol Chem 279:1132–1140

Fliegmann J, Montel E, Djulić A, Cottaz S, Driguez H, Ebel J (2005) Catalytic properties of the bifunctional soybean beta-glucan-binding protein, a member of family 81 glycoside hydrolases. FEBS Lett 579(29):6647–6652

Flors V, Ton J, van Doorn R, Garcia-Agustin P, Mauch-Mani B (2008) Interplay between JA, SA, and ABA signalling during basal and induced resistance against *Pseudomonas syringae* and *Alternaria brassicicola*. Plant J 54:81–92

Fry W (2008) Plant diseases that changed the world *Phytophthora infestans*: the plant (and *R* gene) destroyer. Mol Plant Pathol 9:385–402

Fu ZQ, Guo M, Alfano JR (2006) *Pseudomonas syringae* HrpJ is a type III secreted protein that is required for plant pathogenesis, injection of effectors, and secretion of the HrpZ1 harpin. J Bacteriol 188:6060–6069

Fu ZQ, Guo M, Jeong BR, Tian F, Elthon TE, Cerny RL, Staiger D, Alfano JR (2007) A type III effector ADP-ribosylates RNA-binding proteins and quells plant immunity. Nature 447:284–288

Fu D-Q, Ghabrial S, Kachroo A (2009) *GmRAR1* and *Gm SGT1* are required for basal, R gene-mediated and systemic acquired resistance in soybean. Mol Plant Microbe Interact 22:86–95

Fu L-J, Shi K, Gu M, Zhou Y-H, Dong D-K, Liang W-S, Song F-M, Yu J-Q (2010) Systemic induction and role of mitochondrial alternative oxidase and nitric oxide in a compatible tomato-*Tobacco mosaic virus* interaction. Mol Plant Microbe Interact 23:39–48

Fujita K, Suzuki T, Kunoh H, Carver TLW, Thomas BJ, Gurr S, Shiraishi T (2004) Induced inaccessibility in barley cells exposed to extracellular material released by non-pathogenic powdery mildew conidia. Physiol Mol Plant Pathol 64:169–178

Fujiwara S, Tanaka N, Kaneda T, Takayama S, Isogai A, Che FS (2004) Rice cDNA microarray-based gene expression profiling of the response to flagellin reception in cultured rice cells. Mol Plant Microbe Interact 17:986–998

Fujiwara M, Umemura K, Kawasaki T, Shimamoto K (2006) Proteomics of Rac GTPase signaling reveals its predominant role in elicitor-induced defense response of cultured rice cells. Plant Physiol 140:734–745

Fung RWM, Gonzalo M, Fekete C, Kovacs LG, He Y, Marsh E, McIntyre LM, Schchtman DP, Qiu W (2008) Powdery mildew induces defense-oriented reprogramming of the transcriptome in a susceptible but not in a resistant grapevine. Plant Physiol 146:236–249

Furman-Matarasso N, Cohen E, Du Q, Chejanovsky N, Hanania U, Avni A (1999) A point mutation in the ethylene-inducing xylanase elicitor inhibits the beta-1-4-endoxylanase activity but not the elicitation activity. Plant Physiol 121:345–352

Gaige AR, Ayella A, Shuai B (2010) Methyl jasmonate and ethylene induce partial resistance in *Medicago truncatula* against the charcoal rot pathogen *Macrophomina phaseolina*. Physiol Mol Plant Pathol 74:412–416

Galletti R, De Lorenzo G, Ferrari S (2009) Host-derived signals activate plant innate immunity. Plant Signal Behav 4:33–34

Gan Q, Bai H, Zhao X, Tao Y, Zeng H, Han Y, Song W, Zhu L, Liu G (2011) Transcriptional characteristics of *Xa21*-mediated defense responses in rice. J Integr Plant Biol 53:300–311

Gao M, Liu J, Zhang Z, Cheng F, Chen S, Zhang Y (2008a) MEKK1, MKK1/MKK2 and MPK4 function together in a mitogen-activated protein kinase cascade to regulate innate immunity in plants. Cell Res 18:1190–1198

Gao Y, Wang S, Asami T, Chen JG (2008b) Loss-of-function mutations in the *Arabidopsis* heterotrimeric G-protein alpha subunit enhance the developmental defects in brassinosteroid signaling and biosynthesis mutants. Plant Cell Physiol 49:1013–1024

Gao X, Brodhagen M, Isakeit T, Brown SH, Göbel C, Betran J, Feussner I, Keller NP, Kolomiets MV (2009a) Inactivation of the lipoxygenase *ZmLOX3* increases susceptibility of maize to *Aspergillus* spp. Mol Plant Microbe Interact 22:222–231

Gao M, Wang X, Wang D, Xu F, Ding X, Zhang Z, Bi D, Cheng YT, Chen S, Li X, Zhang Y (2009b) Regulation of cell death and innate immunity by two receptor-like kinases in *Arabidopsis*. Cell Host Microbe 23:34–44

Garcia-Brugger A, Lamotte O, Vandelle E, Bourque S, Lecourieux D, Poinssot B, Wendehenne D, Pugin A (2006) Early signaling events induced by elicitors of plant defenses. Mol Plant Microbe Interact 19:711–724

Garcia-Ruiz H, Takeda A, Chapman EJ, Sullivan CM, Fahlgren N, Brempelis KJ, Carrington JC (2010) *Arabidopsis* RNA dependent RNA polymerases and dicer-like proteins in antiviral defense and small interfering RNA biogenesis during *Turnip mosaic virus* infection. Plant Cell 22:481–496

Gassmann W, Bhattacharjee S (2012) Effector-triggered immunity signaling: from gene-for-gene pathways to protein-protein interaction networks. Mol Plant Microbe Interact 25:862–868

Gaulin E, Jauneau A, Villalba-Mateos F, Rickauer M, Esquerré-Tugayé M-T, Bottin A (2002) The CBEL glycoprotein of *Phytophthora parasitica* var *nicotianae* is involved in cell wall deposition and adhesion to cellulosic substrates. J Cell Sci 115:4565–4575

Gaulin E, Drame N, Lafitte C, Torto-Alalibo T, Martinez Y, Ameline-Torregrosa C, Khatib M, Mazarguil H, Villalba-Mateos F, Kamoun S, Mazars C, Dumas B, Bottin A, Esquerré-Tugayé M-T, Rickauer M (2006) Cellulose binding domains of a *Phytophthora* cell wall protein are novel pathogen-associated molecular patterns. Plant Cell 18:1766–1777

Geldner N, Robatzek S (2008) Plant receptors go endosomal: a moving view on signal transduction. Plant Physiol 147:1565–1574

Genger RK, Jurkowski GI, McDowell JM, Lu H, Jung HW, Greenberg JT, Bent AF (2008) Signaling pathways that regulate the enhanced disease resistance of *Arabidopsis* "Defense, No death" mutants. Mol Plant Microbe Interact 21:1285–1296

Geraats BPJ, Bakker PAHM, Lawrence CB, Achuo EA, Höfte M, van Loon LC (2003) Ethylene-insensitive tobacco shows differentially altered susceptibility to different pathogens. Phytopathology 93:813–821

Gerber IB, Zeidler D, Durner J, Dubery IA (2004) Early perception responses of *Nicotiana tabacum* cells in response to lipopolysaccharides from *Burkholderia cepacia*. Planta 218:647–657

Gerber IB, Laukens K, Witters E, Dubery IA (2006) Lipopolysaccharide-responsive phosphoproteins in *Nicotiana tabacum* cells. Plant Physiol Biochem 44:369–379

Gfeller A, Liechti R, Farmer EE (2010) Arabidopsis jasmonate signaling pathway. Sci Signal 3(109):cm4

Ghazala W, Varrelmann M (2007) *Tobacco rattle virus* 29K movement protein is the elicitor of extreme and hypersensitive-like resistance in two cultivars of *Solanum tuberosum*. Mol Plant Microbe Interact 20:1396–1405

Gijzen M, Nürnberger T (2006) Nep1-like proteins from plant pathogens: recruitment and diversification of the NPP1 domain across taxa. Phytochemistry 67:1800–1897

Gilardi P, Garcia-Luque I, Serra MT (2004) The coat protein of tobamovirus acts as elicitor of both *L2* and *L4* gene-mediated resistance in *Capsicum*. J Gen Virol 85:2077–2085

Gimenez-Ibanez S, Rathjen JP (2010) The case for the defense: plants versus Pseudomonas syringae. Microbes Infect 12:428–437

Gimenez-Ibanez S, Hann DR, Ntoukakis V, Petutschnig E, Lipka V, Rathjen JP (2009a) AvrPtoB targets the LysM receptor kinase CERK1 to promote bacterial virulence on plants. Curr Biol 19:423–429

Gimenez-Ibanez S, Ntoukakis V, Rathjen JP (2009b) The LysM receptor kinase CERK1 mediates bacterial perception in Arabidopsis. Plant Signal Behav 4:539–541

Godiard L, Sauviac L, Torii KU, Grenon O, Mangin B, Grimsley NH, Marco Y (2003) ERECTA, an LRR receptor-like kinase protein controlling development pleiotropically affects resistance to bacterial wilt. Plant J 36:353–365

Göhre V, Robatzek S (2008) Breaking the barriers: microbial effector molecules subvert plant immunity. Annu Rev Phytopathol 46:189–215

Göhre V, Spallek T, Häweker H, Mersmann S, Mentzel T, Boller T, de Torres M, Mansfield JW, Robatzek S (2008) Plant pattern-recognition receptor FLS2 is directed for degradation by the bacterial ubiquitin ligase AvrPtoB. Curr Biol 18:1824–1832

Gomés E, Sagot E, Gaillard C, Laquitaine L, Poinssot B, Sanejouand H, Delrot S, Coutos-Thévenot P (2003) Nonspecific lipid-transfer protein genes expression in grape (*Vitis* sp) cells in response to fungal elicitor treatments. Mol Plant Microbe Interact 16:456–464

Gómez-Gómez L, Boller T (2000) FLS2: an LRR- receptor-like kinase involved in the perception of bacterial elicitor flagellin in *Arabidopsis*. Mol Cell 5:1003–1011

Gómez-Gómez L, Boller T (2002) Flagellin perception: a paradigm for innate immunity. Trends Plant Sci 7:251–256

Gómez-Gómez L, Felix G, Boller T (1999) A single locus determines sensitivity to bacterial flagellin in *Arabidopsis thaliana*. Plant J 18:277–284

Gómez-Gómez L, Bauer Z, Boller T (2001) Both the extracellular leucine-rich repeat domain and the kinase activity of FLS2 are required for flagellin binding and signaling in *Arabidopsis*. Plant Cell 13:1155–1163

Grant MR, Jones JDG (2009) Hormone (Dis)harmony moulds plant health and disease. Science 324:750–752

Grant SR, Fisher EJ, Chang JH, Mole BM, Dangl JL (2006) Subterfuge and manipulation: type III effector proteins of phytopathogenic bacteria. Annu Rev Microbiol 60:425–449

Grennan AK (2007) Protein *S*-nitrosylation: protein targets and roles in signal transduction. Plant Physiol 144:1237–1239

Grennan AK (2008) Ethylene response factors in jasmonate signaling and defense response. Plant Physiol 146:1457–1458

Großkinsky DK, Koffler BE, Roitsch T, Maier R, Zechmann B (2012) Compartment-specific anti-oxidative defense in *Arabidopsis* against virulent and avirulent *Pseudomonas syringae*. Phytopathology 102:662–673

Groen AJ, de Vries SC, Lilley KS (2008) A proteomics approach to membrane trafficking. Plant Physiol 147:1584–1589

Gross A, KappD NT, Niehaus K (2005) Endocytosis of *Xanthomonas campestris* pathovar *campestris* lipopolysaccharides in non-host plant cells of *Nicotiana tabacum*. New Phytol 165:215–226

Guan R, Mariuzza RA (2007) Peptidoglycan recognition proteins of the innate immune system. Trends Microbiol 15:127–134

Gudesblat GE, Torres PS, Vojnov A (2009) *Xanthomonas campestris* overcomes Arabidopsis stomatal innate immunity through a DSF cell-to-cell signal-regulated virulence factor. Plant Physiol 149:1017–1027

Guimarães RL, Stotz HU (2004) Oxalate production by *Sclerotinia sclerotiorum* deregulates guard cells during infection. Plant Physiol 136:3703–3711

Gurr SJ, Ruston PJ (2005) Engineering plants with increased disease resistance: what are we going to express? Trends Biotechnol 23:275–282

Gust AA, Biswas R, Lenz HD, Rauhut T, Ranf S, Kemmerling B, Götz F, Glawischnig E, Lee J, Felix G, Nurnberger T (2007) Bacteria-derived peptidoglycans constitute pathogen-associated molecular patterns triggering innate immunity in *Arabidopsis*. J Biol Chem 282:32338–32348

Gust AA, Brunner F, Nürnberger T (2010) Biotechnological concepts for improving plant innate immunity. Curr Opin Biotechnol 21:204–210

Gutjahr C, Paszkowski U (2009) Weights in the balance: Jasmonic acid and salicylic acid signaling in root-biotroph interactions. Mol Plant-Microbe Interact 22:763–772

Haapalainen M, Engelhardt S, Küfner I, Li C-M, Nürnberger T, Lee J, Romantschuk M, Taira S (2011) Functional mapping of harpin HrpZ of *Pseudomonas syringae* reveals the sites responsible for protein oligomerization lipid interactions and plant defence induction. Mol Plant Pathol 12:151–166

Haas BJ, Kamoun S, Zody MC, Jiang RH, Cano LM, Grabherr M, Kodira CD, Raffaele S, Torto-Alalibo T, Bozkurt TO, Ah-Fong AM, Alvardo L, Anderson VL, Armstrong MR, Avrova A, Baxter L, Beynon J, Boevink PC, Bollmann SR, Bos JI, Burlone V, Cai G, Cakir C, Carrington JC, Chawner M, Conti L, Costanzo S, Ewan R, Fahlgren N, Fischback MA, Fugelstad J, Gilroy FM, Gnerre S, Green PJ, Greenville-Briggs LJ, Griffith J, Grünwald NJ, Horn K, Horner NR, Hu CH, Huitema E, Jeong DH, Jones AM, Jones JD, Jones RW, Karlsson EK, Kunjeti SG, Lamour K, Liu Z, Ma L, Maclean D, Chibucos MC, McDonald H, McWalters J, Meijer HJ, Morgan W, Morris PF, Munro CA, O'Neill K, Ospina-Giraldo M, Pinzón A, Pritchard L, Ramsahoye B, Ren Q, Restrepo S, Roy S, Sadanandom A, Savidor A, Schornack S, Schwartz DC, Schumarn UD, Schwessinger B, Seyer L, Sharpe T, Silvar C, Song J, Studholme DJ, Sykes S, Thines M, van de Vondervoort PJ, Phuntumart V, Wawra S, Weide R, Win J, Young C, Zhou S, Fry W, Meyers BC, van West P, Ristaino J, Govers F, Birch PR, Whisson SC, Judelson HS, Nusbaum C (2009) Genome sequence and comparative analysis of the Irish potato famine pathogen *Phytophthora infestans*. Nature 461:393–398

Hahlbrock K, Scheel D, Logemann E, Nürnberger T, Parniske M, Reinhold S, Sacks WR, Schmelzer E (1995) Oligopeptide elicitor-mediated defense gene activation in cultured parsley cells. Proc Natl Acad Sci USA 92:4150–4157

Hajimorad MR, Eggenberger AL, Hill JH (2005) Loss and gain of elicitor function of *Soybean mosaic virus* G7 provoking *Rsv1*-mediated lethal systemic hypersensitive response maps to P3. J Virol 79:1215–1222

Halim VA, Hunger A, Macioszek V, Landgraf P, Nürnberger T, Scheel D, Rosahl S (2004) The oligopeptide elicitor Pep-13 induces salicylic acid-dependent and – independent defense reactions in potato. Physiol Mol Plant Pathol 64:311–318

Halim VA, Altmann S, Ellinger D, Eschen-Lippold L, Miersch O, Scheel D, Rosahl S (2009) PAMP-induced defense responses in potato require both salicylic acid and jasmonic acid. Plant J 57:230–242

Hamel L-P, Beaudoin N (2010) Chitooligosaccharide sensing and downstream signaling contrasted outcomes in pathogenic and beneficial plant-microbe interactions. Planta 232:787–806

Hamilton A, Voinnet O, Chappell L, Baulcombe D (2002) Two classes of short interfering RNA in RNA silencing. EMBO J 21:4671–4679

Hanania U, Avni A (1997) High-affinity binding site for ethylene-inducing xylanase elicitor on *Nicotiana tabacum* membranes. Plant J 12:113–120

Hann DR, Rathjen JP (2007) Early events in the pathogenicity of *Pseudomonas syringae* on *Nicotiana benthamiana*. Plant J 49:607–618

Hann DR, Gimenez-Ibanez S, Rathjen JP (2010) Bacterial virulence effectors and their activities. Curr Opin Plant Biol 13:388–393

Hauck P, Thilmony R, He SY (2003) A *Pseudomonas syringae* type III effector suppresses cell wall-based extracellular defense in susceptible *Arabidopsis* plants. Proc Natl Acad Sci USA 100:8577–8582

Häweker H, Rips S, Koiwa H, Salomon S, Saijo Y, Chinchilla D, Robatzek S, von Schaewen A (2010) Pattern recognition receptors require *N*-glycosylation to mediate plant immunity. J Biol Chem 285:4629–4636

Hawes M, Ronald P (2009) Focus: plant interactions with bacterial pathogens. Plant Physiol 150:1621–1622

He SY, Huang HC, Collmer A (1993) *Pseudomonas syringae* pv. *syringae* harpin$_{Pss}$: a protein that is secreted via the Hrp pathway and elicits the hypersensitive response in plant cells. Cell 73:1255–1266

He Z-H, Cheeseman I, He D, Kohorn BD (1999) A cluster of live cell wall-associated receptor kinases Wak1-5 are expressed in specific organs of *Arabidopsis*. Plant Mol Biol 39:1189–1999

He P, Shan L, Lin NC, Martin GB, Kemmerling B, Nurnberger T, Sheen J (2006) Specific bacterial suppressors of MAMP signaling upstream of MAPKKK in *Arabidopsis* innate immunity. Cell 125:563–575

He P, Shan L, Sheen J (2007) Elicitation and suppression of microbe-associated molecular pattern-triggered immunity in plant-microbe interactions. Cell Microbiol 9:1385–1396

Heese A, Hann DR, Gimenez-Ibanez S, Jones AME, He K, Li J, Schroeder JI, Peck SC, Rathjen JP (2007) The receptor-like kinase SERK3/BAK1 is a central regulator of innate immunity in plants. Proc Natl Acad Sci USA 104:12217–12222

Heidel AJ, Dong X (2006) Fitness benefits of mutations affecting the systemic acquired resistance pathway in *Arabidopsis thaliana*. Genetics 173:1621–1628

Hernández-Blanco C, Feng DX, Hu J, Sanchez-Vallet A, Deslandes L, Llorente F, Berrocal-Lobo M, Keller H, Barlet X, Sánchez-Rodriguez C, Anderson LK, Somerville S, Marco Y, Molina A (2007) Impairment of cellulose synthase required for *Arabidopsis* secondary cell wall formation enhances disease resistance. Plant Cell 19:890–903

Hogenhout SA, van der Hoorn R, Terauchi R, Kamoun S (2009) Emerging concepts in effector biology of plant-associated organisms. Mol Plant Microbe Interact 22:115–122

Holst O, Molinaro A (2009) Core oligosaccharide and lipid A components of lipopolysaccharides. In: Moran A, Brennan P, Holst PO, von Itzstein M (eds) Microbial glycobiology: structures relevance and applications. Elsevier, San Diego, pp 29–56

Hondo D, Hase S, Kanayama Y, Yoshikawa N, Takenaka S, Takahashi H (2007) The *LeATL6*-associated ubiquitin/proteasome system may contribute to fungal elicitor-activated defense response via the jasmonic acid-dependent signaling pathway in tomato. Mol Plant Microbe Interact 20:72–81

Howell MD, Fahlgren N, Chapman EJ, Cumbie JS, Sullivan CM, Givan SA, Kasschau KD, Carrington JC (2007) Genome-wide analysis of the RNA-DEPENDENT RNA POLYMERASE6/DICER-LIKE4 pathway in *Arabidopsis* reveals dependency on miRNA- and tasiRNA-directed targeting. Plant Cell 19:926–942

Hubbard KE, Nishimura N, Hitoni K, Getzoff ED, Schroeder JI (2010) Early abscisic acid signal transduction mechanisms: newly discovered components and newly emerging questions. Genes Dev 24:1695–1708

Huffaker A, Ryan CA (2007) Endogenous peptide defense signals in Arabidopsis differentially amplify signaling for the innate immune responses. Proc Natl Acad Sci USA 104: 10732–10736

Huffaker A, Pearce G, Ryan CA (2006) An endogenous peptide signal in Arabidopsis activates components of the innate immune response. Proc Natl Acad Sci USA 103:10098–10103

Huffaker A, Dafoe NJ, Schmelz EA (2011) ZmPep1 an ortholog of *Arabidopsis* elicitor peptide 1 regulates maize innate immunity and enhances disease resistance. Plant Physiol 155:1325–1338

Hwang IS, Hwang BK (2010) The pepper 9-lipoxygenase gene *CaLOX1* functions in defense and cell death responses to microbial pathogens. Plant Physiol 152:948–967

Hwang IS, Hwang BK (2011) The pepper mannose-binding lectin gene *CaMBL1* is required to regulate cell death and defense responses to microbial pathogens. Plant Physiol 155:447–463

Ingle RA, Carstens M, Denby KJ (2006) PAMP recognition and plant-pathogen arms race. Bioessays 28:880–889

Inui H, Yamaguchi Y, Hirano S (1997) Elicitor actions of *N*-acetylchitooligosaccharides and laminooligosaccharides for chitinase and L-phenylalanine ammonia-lyase induction in rice suspension culture. Biosci Biotechnol Biochem 61:975–978

Irani NG, Russinova E (2009) Receptor endocytosis and signaling in plants. Curr Opin Plant Biol 12:653–659

Iriti M, Faoro F (2007) Review of innate and specific immunity in plants and animals. Mycopathologia 164:57–64

Iriti M, Faoro F (2009) Chitosan as a MAMP, searching for a PRR. Plant Signal Behav 4:66–68

Ishiga Y, Takeuchi K, Inagaki Y, Toyoda K, Shiraishi T, Ichinose Y (2005) Defense responses of *Arabidopsis thaliana* inoculated with *Pseudomonas syringae* pv. *tabaci* wild type and defective mutants for flagellin (Δ*fliC*) and flagellin-glycosylation (Δ*orf1*). J Gen Plant Pathol 71:302–307

Jacchieri SG, Torquato R, Brentani RR (2003) Structural study for binding flagellin by toll-like receptor 5. J Bacteriol 185:424–427

Jagadeeswaran G, Saini A, Sankar R (2009) Biotic and abiotic stress down-regulate miR398 expression in *Arabidopsis*. Planta 229:1009–1014

Jakob K, Kniskern JM, Bergelson J (2007) The role of pectate lyase and the jasmonic acid defense response in *Pseudomonas viridiflava* virulence. Mol Plant Microbe Interact 20:146–158

Janeway CAJ, Medzhitov R (2002) Innate immune recognition. Annu Rev Immunol 20:197–216

Jaubert M, Bhattacharjee S, Mello AFS, Perry KL, Moffett P (2011) ARGONAUTE2 mediates RNA-silencing antiviral defenses against *Potato virus X* in Arabidopsis. Plant Physiol 156:1556–1564

Jelenska J, Yao N, Vinatzer BA, Wright CM, Brodsky JL, Greenberg JT (2007) A J domain virulence effector of *Pseudomonas syringae* remodels host chloroplasts and suppresses defenses. Curr Biol 17:499–508

Jenner CE, Tomimura K, Ohshima K, Hughes SL, Walsh JA (2002) Mutations in *Turnip mosaic virus* P3 and cylindrical inclusion bodies are separately required to overcome two *Brassica napus* resistance genes. Virology 300:50–59

Jenner CE, Wang X, Tomimura K, Ohshima F, Ponz F, Walsh JA (2003) The dual role of the potyvirus P3 protein of *Turnip mosaic virus* as a symptom and avirulence determinant in brassicas. Mol Plant Microbe Interact 16:777–784

Jennings JC, Apel-Birkhold PC, Mock NM, Baker CJ, Anderson JD, Bailey BA (2001) Induction of defense responses in tobacco by the protein Nep1 from *Fusarium oxysporum*. Plant Sci 161:891–899

Jensen MK, Rung JH, Gregersen PL, Gjetting T, Fuglsang AT, Hansen M, Joehnk N, Lyngjaer MF, Collinge DB (2007) The HvNAC6 transcription factor: a positive regulator of penetration resistance in barley and *Arabidopsis*. Plant Mol Biol 65:137–150

Jeong BR, van Dijk K, Alfano JR (2009) *Pseudomonas syringae* type III-secreted proteins and their activities and effects on plant innate immunity. Annu Plant Rev 34:48–76

Jeworutzki E, Roelfsema MR, Anschütz U, Krol E, Elzenga JT, Felix G, Boller T, Hedrich R, Becker D (2010) Early signaling through the *Arabidopsis* pattern recognition receptors FLS2 and EFR involves $Ca^{2+}$-associated opening of plasma membrane anion channels. Plant J 62:367–378

Jia Y, McAdams SA, Bryan GT, Hershey HP, Valent B (2000) Direct interaction of resistance gene and avirulence gene products confers rice blast resistance. EMBO J 19:4004–4014

Jin H (2008) Endogenous small RNAs and antibacterial immunity in plants. FEBS Lett 582:2679–2684

Jin H, Yan Z, Nam KH, Li J (2007) Allele-specific suppression of a defective brassinsteroid receptor reveals a physiological role of UGCT in ER quality control. Mol Cell 26:821–830

Jin Y, Awad W, Petrova K, Hendershot LM (2008) Regulated release of ERdj3 from unfolded proteins by BiP. EMBO J 27:2873–2882

Jin Y, Zhuang M, Hendershot LM (2009) ERdj3 a luminal ER DnaJ homologue binds directly to unfolded proteins in the mammalian ER: identification of critical residues. Biochemistry 48:41–49

Johansson A, Staal J, Dixelius C (2006) Early responses in the *Arabidopsis-Verticillium longisporum* pathosystem are dependent on *NDR1* JA- and ET-associated signals via cytosolic NPR1 and *RFO1*. Mol Plant Microbe Interact 19:958–969

Johnson KL, Ingram GC (2005) Sending the right signals: regulating receptor kinase activity. Curr Opin Plant Biol 8:526–531

Jones JDG, Dangl JL (2006) The plant immune system. Nature 444:323–329

Jones-Rhoades MW, Bartel DP (2004) Computational identification of plant microRNAs and their targets including a stress-induced miRNA. Mol Cell 14:787–799

Joo JH, Wang S, Chen JG, Jones AM, Fedoroff NV (2005) Different signaling and cell death roles of heterotrimeric G protein α and β subunits in the *Arabidopsis* oxidative stress response to ozone. Plant Cell 17:957–970

Joo S, Liu Y, Leuth A, Zhang S (2008) MAPK phosphorylation-induced stabilization of ACS6 protein is mediated by the non-catalytic C-terminal domain, which also contains the *cis-determinant* for rapid degradation by the 26S proteasome pathway. Plant J 54:129–140

Jourdan E, Henry G, Duby F, Dommes J, Barthelemy JP, Thonart P, Ongena M (2009) Insights into the defense-related events occurring in plant cells following perception of surfactin-type lipopeptide from *Bacillus subtilis*. Mol Plant Microbe Interact 22:456–468

Jung HW, Tschaplinski TJ, Wang L, Glazebrook J, Greenberg JT (2009) Priming in systemic plant immunity. Science 324:89–91

Jung J, Won SY, Suh SC, Kim H, Wing R, Jeong Y, Hwang I, Kim M (2007) The barley ERF-type transcription factor HvRAF confers enhanced pathogen resistance and salt tolerance in *Arabidopsis*. Planta 225:575–588

Kachroo P, Shanklin J, Shah J, Whittle EJ, Klessig DF (2001) A fatty acid desaturase modulates the activation of defense signaling pathways in plants. Proc Natl Acad Sci USA 98:9448–9453

Kachroo P, Kachroo A, Lapchyk L, Hildebrand D, Klessig DF (2003a) Restoration of defective cross talk in *ssi2* mutants: role of salicylic acid, jasmonic acid, and fatty acids in *SSI2*-mediated signaling. Mol Plant Microbe Interact 16:1022–1029

Kachroo A, Lapchyk L, Fukushige H, Hildebrand D, Klessig D, Kachroo P (2003b) Plastidial fatty acid signaling modulates salicylic acid- and jasmonic acid-mediated defense pathways in the Arabidopsis *ssi2* mutant. Plant Cell 15:2952–2965

Kachroo A, Venugopal SC, Lapchyk L, Falcone D, Hildebrand D, Kachroo P (2004) Oleic acid levels regulated by glycerolipid metabolism modulate defense gene expression in *Arabidopsis*. Proc Natl Acad Sci USA 101:5152–5157

Kaku H, Nishizawa Y, Ishii-Minami N, Akimoto-Tomiyama C, Dohmae N, Takio K, Minami E, Shibuya N (2006) Plant cells recognize chitin fragments for defense signaling through a plasma membrane receptor. Proc Natl Acad Sci USA 103:11086–11091

Kalde M, Barth M, Somssich IE, Lippok B (2003) Members of the *Arabidopsis* WRKY group III transcription factors are part of different plant defense signaling pathways. Mol Plant Microbe Interact 16:295–305

Kaliff M, Staal J, Myrenas M, Dixelius C (2007) ABA is required for *Leptosphaeria maculans* resistance via ABI- and ABI4-dependent signaling. Mol Plant Microbe Interact 20:335–345

Kamoun S (2006) A catalogue of the effector secretome of plant pathogenic oomycetes. Annu Rev Phytopathol 44:41–60

Kamoun S, Klucher KM, Coffey MD, Tyler BM (1993a) A gene encoding a host-specific elicitor protein of *Phytophthora parasitica*. Mol Plant Microbe Interact 6:573–581

Kamoun S, Young M, Glascock CB, Tyler BM (1993b) Extracellular protein elicitors from *Phytophthora*: host-specificity and induction of resistance to bacterial and fungal phytopathogens. Mol Plant Microbe Interact 6:15–25

Kamoun S, Lindquist H, Govers F (1997a) A novel class of elicitin-like genes from *Phytophthora infestans*. Mol Plant Microbe Interact 10:1028–1030

Kamoun S, van West P, de Jong AJ, de Groot KE, Vleeshouwers VGAA, Govers F (1997b) A gene encoding a protein elicitor of *Phytophthora infestans* is down-regulated during infection of potato. Mol Plant Microbe Interact 10:13–20

Kamoun S, van West P, Vleeshouwers VGAA, de Groot KE, Govers F (1998) Resistance of *Nicotiana benthamiana* to *Phytophthora infestans* is mediated by the recognition of the elicitor protein INF1. Plant Cell 10:1413–1425

Kanetis L, Förster H, Adaskaveg JE (2008) Optimizing efficacy of new postharvest fungicides and evaluation of sanitizing agents for managing citrus green mold. Plant Dis 92:261–269

Kang S, Swiegard JA, Valent B (1995) The PWL host specificity gene family in the blast fungus *Magnaporthe grisea*. Mol Plant Microbe Interact 8:939–948

Kang L, Li J, Zhao T, Xiao F, Tang X, Thilmony R, He S, Zhou JM (2003) Interplay of the Arabidopsis nonhost resistance gene *NHO1* with bacterial virulence. Proc Natl Acad Sci USA 100:3519–3524

Kang S, Kim HB, Lee H, Choi JY, Heu S, Oh CJ, Kwon SI, An CS (2006) Overexpression in *Arabidopsis* of a plasma membrane-targeting glutamate receptor from small radish increases glutamate-mediated $Ca^{2+}$ influx and delays fungal infection. Mol Cells 21:418–427

Kanneganti T-D, Huitema E, Cakir C, Kamoun S (2006) Synergistic interactions of the plant cell death pathways induced by *Phytophthora infestans* Nep1-like protein PiNPP11 and INF1 elicitin. Mol Plant Microbe Interact 19:854–863

Kanzaki H, Saitoh H, Ito A, Fujisawa S, Kamoun S, Katou S, Yoshioka H, Terauchi R (2003) Cytosolic HSP90 and HSP70 are essential components of INF1-mediated hypersensitive response and non-host resistance to *Pseudomonas cichorii* in *Nicotiana benthamiana*. Mol Plant Pathol 4:383–391

Kanzaki H, Saitoh H, Takahashi Y, Berberich T, Ito A, Kamoun S, Terauchi R (2008) NbLRK1, a lectin-like receptor kinase protein of *Nicotiana benthamiana*, interacts with *Phytophthora infestans* INF1 elicitin and mediates INF1-induced cell death. Planta 228:977–987

Kariola T, Palomäki TA, Brader G, Palva ET (2003) *Erwinia carotovora* subsp *carotovora* and *Erwinia*-derived elicitors HrpN and PehA trigger distinct but interacting defense responses and cell death in *Arabidopsis*. Mol Plant Microbe Interact 16:179–187

Kariola T, Brader G, Li J, Palva ET (2005) Chlorophyllase 1, a damage control enzyme, affects the balance between defense pathways in plants. Plant Cell 17:282–294

Kasparovsky T, Milat M-L, Humbert C, Blein J-P, Havel L, Mikes V (2003) Elicitation of tobacco cells with ergosterol activates a signal pathway including mobilization of internal calcium. Plant Physiol Biochem 41:495–501

Katagiri F, Tsuda K (2010) Understanding the plant immune system. Mol Plant Microbe Interact 23:1531–1536

Katiyar-Agarwal S, Jin H (2007) Discovery of pathogen-regulated small RNAs in plants. Methods Enzymol 427:215–227

Katiyar-Agarwal S, Jin H (2010) Role of small RNAs in host-microbe interactions. Annu Rev Phytopathol 48:225–246

Katiyar-Agarwal S, Morgan R, Dahlbeck D, Borsani O, Villegas A, Zhu JK, Staskawicz BJ, Jin H (2006) A pathogen-inducible endogenous siRNA in plant immunity. Proc Natl Acad Sci USA 103:18002–18007

Katiyar-Agarwal S, Gao S, Vivian-Smith A, Jin H (2007) A novel class of bacteria-induced small RNAs in *Arabidopsis*. Genes Dev 21:3123–3134

Katsir L, Schilmiller A, Staswick P, He SY, Howe GA (2008) COI1 is a critical component of a receptor for jasmonate and the bacterial virulence factor coronatine. Proc Natl Acad Sci USA 105:7100–7105

Kawai T, Akira S (2009) The roles of TLRs, RLRs, and NLRs in pathogen recognition. Int Immunol 21:317–337

Kawamura Y, Hase S, Takenaka S, Kanayama Y, Yoshioka H, Kamoun S, Takahashi H (2009) INF1 elicitin activates jasmonic acid- and ethylene-mediated signaling pathways and induces resistance to bacterial wilt disease in tomato. J Phytopathol 157:287–297

Kazan K, Manners JM (2008) Jasmonate signaling: toward an integrated view. Plant Physiol 146:1459–1468

Kazemi-Pour N, Condemine G, Hugovieux-Cotte-Pattat N (2004) The secretome of the plant pathogenic bacterium *Erwinia chrysanthemi*. Proteomics 4:3177–3186

Keates SE, Kostman TA, Anderson JD, Bailey BA (2003) Altered gene expression in three plant species in response to treatment with Nep1, a fungal protein that causes necrosis. Plant Physiol 132:1610–1622

Keinath NF, Kierszniowska S, Lorek J, Bourdais G, Kessler SA, Asano H, Grossniklaus U, Schulze W, Robatzek S, Panstruga R (2010) PAMP-induced changes in plasma membrane compartmentalization reveal novel components of plant immunity. J Biol Chem 285:39140–39149

Kemmerling B, Schwedt A, Rodriguez P, Mazzotta S, Frank M, Qamar SA, Mengiste T, Betsuyaku S, Parker JE, Müssig C, Thomma BPHJ, Albrecht C, de Vries SC, Hirt H, Nürnberger T (2007) The BRI1-associated kinase 1 BAK1 has a brassinolide-independent role in plant cell-death control. Curr Biol 17:1116–1122

Keshavarzi M, Soylu S, Brown I, Bonas U, Nicole M, Rossiter J, Mansfield J (2003) Basal defenses induced in pepper by lipopolysaccharides are suppressed by *Xanthomonas campestris* pv. *vesicatoria*. Mol Plant Microbe Interact 17:805–815

Khang CH, Berruyer R, Giraldo MC, Kankanala P, Park SY, Czymmek K, Kang S, Valent B (2010) Translocation of *Magnaporthe oryzae* effectors into rice cells and their subsequent cell-to-cell movement. Plant Cell 22:1388–1403

Khatib M, Lafitte C, Esquerré-Tugayé M-T, Bottin A, Rickauer M (2004) The CBEL elicitor of *Phytophthora parasitica* var *nicotianae* activates defence in *Arabidopsis thaliana* via three different signalling pathways. New Phytol 162:501–510

Kim MG, da Cunha L, McFall AJ, Belkhadir Y, DebRoy S, Dangl JL, Mackey D (2005) Two *Pseudomonas syringae* type III effectors inhibit RIN4-regulated basal defense in *Arabidopsis*. Cell 121:749–759

Kim KC, Fan B, Chen Z (2006) Pathogen-induced Arabidopsis WRKY7 is a transcriptional repressor and enhances susceptibility to *Pseudomonas syringae*. Plant Physiol 142:1180–1192

Kim S-Y, Kim Y-C, Seong ES, Lee Y-H, Park M, Choi D (2007) The chili pepper CaATL1: an AT-hook motif-containing transcription factor implicated in defence responses against pathogens. Mol Plant Pathol 8:761–771

Kim B, Masuta C, Matsuura H, Takahashi H, Inukai T (2008a) Veinal necrosis induced by *Turnip mosaic virus* infection in *Arabidopsis* is a form of defense response accompanying HR-like cell death. Mol Plant Microbe Interact 21:260–268

Kim K-C, Lai Z, Fan B, Chen Z (2008b) *Arabidopsis* WRKY38 and WRKY62 transcription factors interact with histone deacetylase 19 in basal defense. Plant Cell 20:2357–2371

Kim Y-T, Oh J, Kim K-H, Uhm J-Y, Lee B-M (2010) Isolation and characterization of *NgRLK1* a receptor-like kinase of *Nicotiana glutinosa* that interacts with the elicitin of *Phytophthora capsici*. Mol Biol Rep 37:717–727

Kinoshita T, Caño-Delgado A, Seto H, Hiranuma S, Fujioka S, Yoshida S, Chory J (2005) Binding of brassinosteroids to the extracellular domain of plant receptor kinase BRI1. Nature 433:167–171

Klarzynski O, Plesse B, Joubert J-C, Yvin J-C, Kopp M, Kloareg B, Fritig B (2000) Linear β-13 glucans are elicitors of defense responses in tobacco. Plant Physiol 124:1027–1037

Klarzynski O, Descamps V, Plesse B, Yvin J-C, Kloareg B, Fritig B (2003) Sulfated fucan oligosaccharides elicit defense responses in tobacco and local and systemic resistance against *Tobacco mosaic virus*. Mol Plant Microbe Interact 16:115–122

Kleizen B, Braakman I (2004) Protein folding and quality control in the endoplasmic reticulum. Curr Opin Cell Biol 16:343–349

Knecht K, Seyffarth M, Desel C, Thurau T, Sherameti I, Lou B, Oelmüller R, Cai D (2010) Expression of *BvGLP-1* encoding a germin-like protein from sugar beet in *Arabidopsis thaliana* leads to resistance against phytopathogenic fungi. Mol Plant Microbe Interact 23:446–457

Knirel Y (2009) O-specific polysaccharides of Gram-negative bacteria. In: Moran A, Brennan P, Holst PO, von Itzstein M (eds) Microbial glycobiology: structures relevance and applications. Elsevier, San Diego, pp 29–56

Knoth C, Ringler J, Dangl JL, Eulgem T (2007) *Arabidopsis* WRKY70 is required for full *RPP4-mediated* disease resistance and basal defense against *Hyaloperonospora parasitica*. Mol Plant Microbe Interact 20:120–128

Kobayashi M, Ohura L, Kawakita K, Yokota N, Fujiwara M, Shimamoto K, Doke N, Yoshioka H (2007) Calcium-dependent protein kinases regulate the production of reactive oxygen species by potato NADPH oxidase. Plant Cell 19:1065–1080

Kobeasy MI, El-Beltagi HS, El-Shazly MA, Khattab EAH (2011) Induction of resistance in *Arachis hypogaea* L against *Peanut mottle virus* by nitric oxide and salicylic acid. Physiol Mol Plant Pathol 76:112–118

Koch PL, Grau CR, Jo Y-K, Jung G (2009) Thiophanate-methyl and propiconazole sensitivity in *Sclerotinia homoeocarpa* populations from golf courses in Wisconsin and Massachusetts. Plant Dis 93:100–105

Koga H, Dohi K, Mori M (2004) Abscisic acid and low temperatures suppress the whole plant-specific resistance reaction of rice plants to the infection of *Magnaporthe grisea*. Physiol Mol Plant Pathol 65:3–9

Kolade OO, Bamford VA, Ancillo Anton G, Jones JD, Vera P, Hammings AM (2006) In vitro characterization of the cysteine-rich capping domains in a plant leucine-rich repeat protein. Biochim Biophys Acta 1764:1043–1053

Konagaya K-I, Matsushita Y, Kasahara M, Nyunoya H (2004) Members of 14-3-3 protein isoforms interacting with the resistance gene product N and the elicitor of *Tobacco mosaic virus*. J Gen Plant Pathol 6:221–231

Koornneef A, Pieterse CMJ (2008) Cross talk in defense signaling. Plant Physiol 146:839–844

Korasick DA, McMichael C, Walker KA, Anderson JG, Bednarek SY, Heese A (2010) Novel functions of stomatal cytokinesis-defective 1 (SCD1) in innate immune responses against bacteria. J Biol Chem 285:23342–23350

Korolev N, David DR, Elad Y (2008) The role of phytohormones in basal resistance in *Trichoderma*-induced systemic resistance to *Botrytis cinerea* in *Arabidopsis thaliana*. Biocontrol 53:667–683

Kreuze JF, Klein IS, Lazaro MU, Cuellar Chuquiyuri WJ, Morgan GL, Cipriani Mejia PG, Ghislain M, Valkonen JPT (2008) RNA silencing-mediated resistance to a crinivirus (Closteroviridae) in cultivated sweetpotato (Ipomoea batatas L) and development of sweet potato virus disease following co-infection with a potyvirus. Mol Plant Pathol 9:589–598

Krol E, Mentzel T, Chinchilla D, Boller T, Felix G, Kemmerling B, Postel S, Arents M, Jeworutzki E, Al-Rasheid KAS, Becker D, Hedrich R (2010) Perception of the Arabidopsis danger signal peptide 1 involves the pattern recognition receptor AtPEPR1 and its close homologue AtPEPR2. J Biol Chem 285:13471–13479

Kromann P, Taipe A, Perez WG, Forbes GA (2009) Rainfall thresholds as support for timing fungicide applications in the control of potato late blight in Ecuador and Peru. Plant Dis 93:142–148

Kuchitsu K, Kikuyama M, Shibuya N (1993a) N-acetylchitooligosaccharides biotic elicitor for phytoalexin production induce transient membrane depolarization in suspension-cultured rice cells. Protoplasma 174:79–81

Kuchitsu K, Komatsu S, Hirano H, Shibuya N (1993b) Induction of protein phosphorylation by N-acetylchitooligosaccharide elicitor in suspension-cultured rice cells. Plant Cell Physiol 35:s90

Kuchitsu K, Kosaka T, Shiga T, Shibuya N (1995) EPR evidence for generation of hydroxyl radical triggered by N-acetylchitooligosaccharide elicitor and a protein phosphatase inhibitor in suspension-cultured rice cells. Protoplasma 188:138–142

Kuchitsu K, Yazaki Y, Sakano K, Shibuya N (1997) Transient cytoplasmic pH change and ion fluxes through the plasma membrane in suspension-cultured rice cells triggered by N-acetylchitooligosaccharide elicitor. Plant Cell Physiol 38:1012–1018

Kudo C, Suzuki T, Fukuoka S, Asai S, Suenaga H, Sasabe M, Takano Y, Okuno T, Toyoda K, Shiraishi T, Ichinose Y, Inagaki Y-S (2007) Suppression of Cdc27B expression induces plant defence responses. Mol Plant Pathol 8:365–373

Kufner I, Ottmann C, Oecking C, Nūrnberger T (2009) Cytolytic toxins as triggers of plant immune response. Plant Signal Behav 10:977–979

Kunze G, Zipfel C, Robatzek S, Niehaus K, Boller T, Felix G (2004) The N terminus of bacterial elongation factor Tu elicits innate immunity in *Arabidopsis* plants. Plant Cell 16:3496–3507

Kvitko BH, Ramos AR, Morello JE, Oh HS, Collmer A (2007) Identification of harpins in *Pseudomonas syringae* pv. *tomato* DC3000 which are functionally similar to HrpK1 in promoting translocation of type III secretion system effectors. J Bacteriol 189:8059–8072

Kvitko BH, Park DH, Velásquez AC, Wei CF, Russell AB, Martin GB, Schneider DJ, Collmer A (2009) Deletions in the repertoire of *Pseudomonas syringae* pv *tomato* DC3000 type III secretion effector genes reveal functional overlap among effectors. PLoS Pathog 5:e1000388

Kwon C (2010) Plant defense responses coming to shape. Plant Pathol J 26:115–120

Kwon SI, Kim SH, Battacharjee S, Noh J-J, Gassmann W (2008) *SRFR1* a suppressor of effector-triggered immunity encodes a conserved tetratricopeptide repeat protein with similarity to transcriptional repressors. Plant J 109–119

Lacombe S, Rougon-Cardoso A, Sherwood E, Peeters N, Dahlbeck D, Peter H, Smoker M, Rallapalli G, Thomma BPHJ, Staskawicz B, Jones JDG, Zipfel JDG (2010) Interfamily transfer of a plant pattern-recognition receptor confers broad-spectrum bacterial resistance. Nat Biotechnol 28:365–369

Lamotte O, Gould K, Lecourieux D, Sequeira-Legrand A, Lebrun-Garcia A, Durner J, Pugin A, Wendehenne D (2004) Analysis of nitric oxide signaling functions in tobacco cells challenged by the elicitor cryptogein. Plant Physiol 135:516–529

Lanteri ML, Laxalt AM, Lamattina L (2008) Nitric oxide triggers phosphatidic accumulation via phospholipase D during auxin-induced adventitious root formation in cucumber. Plant Physiol 147:188–198

Laquitaine L, Gomès E, Francois J, Marchive C, Pascal S, Hamdi S, Atanassova R, Delrot S, Coutos-Thévenot P (2006) Molecular basis of ergosterol-induced protection of grape against *Botrytis cinerea*: induction of type 1 LTP promoter activity WRKY and stilbene synthase gene expression. Mol Plant Microbe Interact 19:1103–1112

Laurie-Berry N, Joardar V, Street JH, Kunkel BN (2006) The *Arabidopsis thaliana JASMONATE INSENSITIVE 1* gene is required for suppression of salicylic acid-dependent defenses during infection by *Pseudomonas syringae*. Mol Plant Microbe Interact 19:789–800

Leborgne-Castel N, Lherminier J, Der C, Fromentin J, Houot V, Simon-Plas F (2008) The plant defense elicitor cryptogein stimulates clathrin-mediated endocytosis correlated with reactive oxygen species production in Bright Yellow-2 tobacco cells. Plant Physiol 146:1255–1266

Leclercq J, Fliegmann J, Tellström Nieber A, Cullimore JV, Niehaus K, Küster H, Ebel J, Mithöfer A (2008) Identification of a multigene family encoding putative β-glucan-binding proteins in *Medicago truncatula*. J Plant Physiol 165:766–776

Lecourieux – Ouaked D, Pugin A, Lebrun-Garcia A (2000) Phosphoproteins involved in the signal transduction of cryptogein an elicitor of defense reactions in tobacco. Mol Plant Microbe Interact 13:821–829

Lecourieux D, Mazars C, Pauly N, Ranjeva R, Pugin A (2002) Analysis and effects of cytosolic free calcium elevations in response to elicitors in *Nicotiana plumbaginifolia* cells. Plant Cell 14:2627–2641

Lecourieux D, Lamotte O, Bourque S, Wendehenne D, Mazars C, Ranjeva R, Pugin A (2005) Proteinaceous and oligosaccharidic elicitors induce different calcium signatures in the nucleus of tobacco cells. Cell Calcium 38:527–538

Lecourieux D, Ranjeva R, Pugin A (2006) Calcium in plant defence-signaling pathways. New Phytol 171:249–269

Lee S, Choi H, Suh SS, Doo IS, Oh KY, Choi E, Schroeder-Taylor AT, Low PS, Lee Y (1999) Oligogalacturonic acid and chitosan reduce stomatal aperture by inducing the evolution of reactive oxygen species from guard cells of tomato and *Commelina communis*. Plant Physiol 121:147–152

Lee J, Klessig DF, Nűrnberger T (2001a) A harpin binding site in tobacco plasma membranes mediates activation of the pathogenesis-related gene *HIN1* independent of extracellular calcium but dependent on mitogen-activated protein kinase activity. Plant Cell 13:1079–1093

Lee J, Klusener B, Tsiamis G, Stevens C, Neyt C, Tampakaki AP, Panopoulos NJ, Nöller J, Weiler EW, Cornelis GR, Mansfield JW, Nürnberger T (2001b) $HrpZ_{Psph}$ from the plant pathogen *Pseudomonas syringae* pv. *phaseolicola* binds to lipid bilayers and forms an ion-conducting pore in vitro. Proc Natl Acad Sci USA 98:289–294

Lee J, Bricker TM, Lefevre M, Pinson SRM, Oard JH (2006a) Proteomic and genetic approaches to identifying defence-related proteins in rice challenged with the fungal pathogen *Rhizoctonia solani*. Mol Plant Pathol 7:405–416

Lee SW, Han SW, Bartley LE, Ronald PC (2006b) Unique characteristics of *Xanthomonas oryzae* pv. *oryzae* AvrXa21 and implications for plant innate immunity. Proc Natl Acad Sci USA 103:18395–18400

Lee SW, Jeong KS, Han SW, Lee SE, Phee BK, Hahn TR, Ronald P (2008) The *Xanthomonas oryzae* pv. *oryzae* PhoPQ two-component system is required for AvrXa21 activity hrpG expression and virulence. J Bacteriol 190:2183–2197

Lee SW, Han S, Sririyanum M, Park C, Seo Y, Ronald PC (2009) A type 1-secreted sulfated peptide triggers XA21-mediated innate immunity. Science 326:850–853

Lehtonen NT, Akita M, Frank W, Reski R, Valkonen JPT (2012) Involvement of a class III peroxidase and the mitochondrial protein TSPO in oxidative burst upon treatment of moss plants with a fungal elicitor. Mol Plant Microbe Interact 25:363–371

Lenman M, Sörensson C, Andreasson E (2008) Enrichment of phosphoproteins and phosphopeptide derivatization identify universal stress proteins in elicitor-treated *Arabidopsis*. Mol Plant Microbe Interact 21:1275–1284

Leon-Reyes A, Du Y, Koornneer A, Proietti S, Körbes AP, Memelink J, Pieterse CMJ, Ritsema T (2010) Ethylene signaling renders the jasmonate response of *Arabidopsis* insensitive to future suppression by salicylic acid. Mol Plant Microbe Interact 23:187–197

Lewsey MG, Murphy AM, MacLean D, Dalchau N, Westwood JH, Macaulay K, Bennett M, Moulin M, Hanke DE, Powell G, Smith AG, Carr JP (2010) Disruption of two signaling pathways by a viral RNA silencing suppressor. Mol Plant Microbe Interact 23:835–845

Li J, Wen J, Lease KA, Doke JT, Tax FE, Walker JC (2002) BAK1 an Arabidospsis LRR receptor-like protein kinase interacts with BRI1 and modulates brassinosteroid signaling. Cell 110:213–222

Li M, Shao M, Lu X-Z, Wang J-S (2005a) Biological activities of purified $harpin_{Xoo}$ and $harpin_{Xoo}$ detection in transgenic plants using its polyclonal antibody. Acta Biochim Biophys Sin 37:713–718

Li XY, Lin H, Zhang W, Zou Y, Zhang J, Tang X, Zhou JM (2005b) Flagellin induces innate immunity in nonhost interactions that is suppressed by *Pseudomonas syringae* effectors. Proc Natl Acad Sci USA 102:12990–12995

Li H, Zhou SY, Zhao WS, Su SC, Peng YL (2009a) A novel wall-associated receptor-like protein kinase gene *OsWAK1* plays important roles in rice blast disease resistance. Plant Mol Biol 69:337–346

Li J, Chu ZH, Batoux M, Nekrasov V, Roux M, Chinchilla D, Zipfel C, Jones JD (2009b) Specific ER quality control components required for biogenesis of the plant innate immune receptor EFR. Proc Natl Acad Sci USA 106:15973–15978

Li Y, Zhang Q, Zhang J, Wu L, Qi Y, Zhou J-M (2010) Identification of microRNAs involved in pathogen-associated molecular pattern-triggered plant innate immunity. Plant Physiol 152: 2222–2231

Lin Z, Alexander L, Hackett R, Grierson D (2008) LeCTR2 a CTR1-like protein kinase from tomato plays a role in ethylene signalling development and defence. Plant J J54:1083–1093

Lindeberg M, Cartinhour S, Myers CR, Schechter LM, Schneider DJ, Collmer A (2006) Closing the circle on the discovery of genes encoding Hrp regulon members and type III secretion system effectors in the genomes of three model *Pseudomonas syringae* strains. Mol Plant Microbe Interact 19:1151–1158

Liners F, Letesson J-J, Didembourg C, Van Cutsem P (1989) Monoclonal antibodies against pectin: recognition of aq conformation induced by calcium. Plant Physiol 91:1419–1424

Liners F, Thibault J-F, Van Cutsem P (1992) Influence of the degree of polymerization of oligogalacturonates and of esterification pattern of pectin on their recognition by monoclonal antibodies. Plant Physiol 990:1099–1104

Liu H, Gale M Jr (2010) Hepatis virus evasion from RIG-I-dependent hepatic innate immunity. Gastroenterol Res Pract 2010. doi:101155/2010/548390

Liu Y, Zhang S (2004) Phosphorylation of 1-aminocyclopropane-1-carboxylic acid synthase by MPK6 a stress-responsive mitogen-activated protein kinase induces ethylene biosynthesis in *Arabidopsis*. Plant Cell 16:3386–3399

Liu X, Bai X, Wang X, Chu C (2007) OsWRKY71 a rice transcription factor is involved in rice defense response. J Plant Physiol 164:969–979

Liu Y, Kong X, Pan J, Li D (2010) VIP1: linking *Agrobacterium*-mediated transformation to plant immunity? Plant Cell Rep 29:805–812

Liu P-P, von Dahl CC, Klessig DF (2011a) The extent to which methyl salicylate is required for signaling systemic acquired resistance is dependent on exposure to light after infection. Plant Physiol 157:2216–2226

Liu P-P, von Dahl CC, Park S-W, Klessig DF (2011b) Interconnection between methyl salicylate and lipid-based long-distance signaling during the development of systemic acquired resistance in *Arabidopsis* and tobacco. Plant Physiol 155:1762–1768

Lizasa E, Mitsutomi M, Nagano Y (2010) Direct binding of a plant LysM receptor-like kinase LysM RLK1/CERK1 to chitin in vitro. J Biol Chem 285:2996–3004

Llorente F, Alonso-Blanco C, Sanchez-Rodriguez C, Jorda L, Molina A (2005) ERECTA receptor-like kinase and heterotrimeric G protein from *Arabidopsis* are required for resistance to the necrotrophic fungus *Plectosphaerella cucumerina*. Plant J 43:165–180

Lohmann GV, Shimoda Y, Nielsen W, Jørgensen FG, Grossmann C, SandalN SK, Thirup S, Madsen LH, Tabata S, Sato S, Stougaard J, Radutoiu S (2010) Evolution and regulation of the *Lotus japonica* LysM receptor gene family. Mol Plant Microbe Interact 23:510–521

Lorenzo O, Piqueras R, Sánchez-Serrano JJ, Solano R (2003) ETHYLENE RESPONSE FACTOR 1 integrates signals from ethylene and jasmonate pathways in plant defense. Plant Cell 15:165–178

Lotze MT, Zeh HJ, Rubartelli A, Sparvero LJ, Amoscato AA, Washburn NR, Devera ME, Liang X, Tor M, Billiar T (2007) The grateful dead: damage associated molecular pattern molecules and reduction/oxidation regulate immunity. Immunol Rev 220:60–81

Lovato FA, Inoue-Nagata AK, Nagata T, de Avita AC, Pereira LAR, Resende RO (2008) The N protein of *Tomato spotted wilt virus* (TSWV) is associated with the induction of programmed cell death (PCD) in *Capsicum chinense* plants a hypersensitive host to TSWV infection. Virus Res 137:245–252

Love AJ, Yun BW, Laval V, Loake GJ, Milner JL (2005) *Cauliflower mosaic virus* a compatible pathogen of *Arabidopsis* engages three distinct defense-signaling pathways and activates rapid systemic generation of reactive oxygen species. Plant Physiol 139:935–948

Love AJ, Laval V, Geri C, Laird J, Tomos AD, Hooks MA, Milner JL (2007) Components of *Arabidopsis* defense- and ethylene-signaling pathways regulate susceptibility to *Cauliflower mosaic virus* by restricting long-distance movement. Mol Plant Microbe Interact 20:659–670

Lu S, Sun Y-H, Amerson H, Chiang VL (2007) MicroRNAs in loblolly pine (*Pinus taeda* L) and their association with fusiform rust gall development. Plant J 51:1077–1098

Lu X, Tintor N, Mentzel T, Kombrink E, Boller T, Robatzek S, Schulze-Lefert P, Saijo Y (2009) Uncoupling of sustained MAMP receptor signaling from early outputs in an *Arabidopsis* endoplasmic reticulum glucosidase II allele. Proc Natl Acad Sci USA 106:22522–22527

Lu D, Wu S, Gao X, Zhang Y, Shan L, He P (2010) A receptor-like cytoplasmic kinase BIK1 associates with a flagellin receptor complex to initiate plant innate immunity. Proc Natl Acad Sci USA 107:495–501

Ludwig AA, Romeis T, Jones JDG (2004) CDPK-mediated signaling pathways: specificity and cross-talk. J Exp Bot 55:181–188

Luna E, Pastor V, Robert J, Flors V, Mauch-Mani B, Ton J (2011) Callose deposition: a multifaceted plant defense response. Mol Plant Microbe Interact 24:183–193

Luttrell LM (2006) Transmembrane signaling by G protein-coupled receptors. Methods Mol Biol 332:3–49

Ma W, Smigel A, Verma R, Berkowitz GA (2009) Cyclic nucleotide gated channels and related signaling components in plant innate immunity. Plant Signal Behav 4:277–282

Mackey D, McFall AJ (2006) MAMPs and MIMPs: proposed classification for inducers of innate immunity. Mol Microbiol 61:1365–1371

Mackey D, Holt BFIII, Wiig DJL (2002) RIN4 interacts with *Pseudomonas syringae* type III effector molecules and is required for RPM1-mediated resistance in *Arabidopsis*. Cell 108:743–754

MacNab RM (1996) Flagella and motility. In: Niedhard FC (ed) *Escherichia coli* and *Salmonella*: cellular and molecular biology. American Society for Microbiology Press, Washington, DC, pp 123–145

Makandar R, Essig JS, Schapaugh MA, Trick HJ, Shah J (2006) Genetically engineered resistance to Fusarium head blight in wheat by expression of *Arabidopsis NPR1*. Mol Plant Microbe Interact 19:123–129

Makandar R, Nalam V, Chaturvedi R, Jeannotte R, Sparks AA, Shah J (2010) Involvement of salicylate and jasmonate signaling pathways in *Arabidopsis* interaction with *Fusarium graminearum*. Mol Plant Microbe Interact 23:861–870

Makandar R, Nalam VJ, Lee H, Trick HN, Dong Y, Shah J (2012) Salicylic acid regulates basal resistance to Fusarium head blight in wheat. Mol Plant Microbe Interact 25:431–439

Mallory AC, Vaucheret H (2006) Functions of microRNAs and related small RNAs in plants. Nat Genet (Suppl) 38:S31–S36

Mallory AC, Bartel DP, Bartel B (2005) MicroRNA-directed regulation of *Arabidopsis* AUXIN RESPONSE FACTOR7 is essential for proper development and modulates expression of early auxin response genes. Plant Cell 17:1360–1375

Malock K, Levine A, Eulgem TM, Morgan A, Schmid J, Lawton KA, Dangl JL, Dietrich RA (2001) The transcriptome of *Arabidopsis thaliana* during systemic acquired resistance. Nat Genet 26:403–410

Mandal B, Mandal S, Csinos AS, Martinez N, Culbreath AK, Pappu HR (2008) Biological and molecular analyses of the acibenzolar-S-methyl-induced systemic acquired resistance in flue-cured tobacco against *Tomato spotted wilt virus*. Phytopathology 98:196–204

Manosalva PM, Davidson RM, Liu B, Zhu X, Hulbert SH, Leung H, Leach JE (2009) A germin-like protein gene family functions as a complex quantitative trait locus conferring broad-spectrum disease resistance in rice. Plant Physiol 149:286–296

Manosalva PM, Park S-W, Forouhar F, Tong L, Fry WE, Klessig DF (2010) *Methyl Esterase 1* (*StMES1*) is required for systemic acquired resistance in potato. Mol Plant Microbe Interact 23:1151–1163

Mao Y, Tyler BM (1996) Cloning and sequence analysis of elicitin genes of *Phytophthora sojae*. Fungal Genet Biol 20:169–172

Mao P, Duan M, Wei C, Li Y (2007) WRKY62 transcription factor acts downstream of cytosolic NPR1 and negatively regulates jasmonate responsive gene expression. Plant Cell Physiol 48:833–842

Martinez C, Blanc F, Claire E, Besnard O, Nicole M, Baccou JC (2001) Salicylic acid and ethylene pathways are differentially activated in melon cotyledons by active or heat-denatured cellulase from *Trichoderma longibrachiatium*. Plant Physiol 127:334–344

Matsumoto K, Sawada H, Matsumoto K, Hamada H, Yoshimoto E, Ito T, Takeuchi S, Tsuda S, Suzuki K, Kobayashi K, Kiba A, Okuno T, Hikichi Y (2008) The coat protein gene of tobamovirus $P_0$ pathotype is a determinant for activation of temperature-insensitive $L^{1a}$-gene-mediated resistance in *Capsicum* plants. Arch Virol 155:645–650

Mattinen L, Tshulkina M, Mae A, Pihonen M (2004) Identification and characterization of Nip necrosis-inducing virulence protein of *Erwinia carotovora* subsp *carotovora*. Mol Plant Microbe Interact 17:1366–1375

Mayer M, Reiinsteimn J, Buchner J (2003) Modulation of the ATPase cycle of BiP to peptides and proteins. J Mol Biol 330:137–144

McDonald G, Inohara N, Nujnez G (2005) Peptidoglycan signaling in innate immunity and inflammatory disease. J Biol Chem 280:20177–20180

McDowell JM, Simon SA (2008) Molecular diversity at the plant-pathogen interface. Dev Comp Immunol 32:736–744

McGrath KC, Dombrecht B, Manners JM, Schenk PM, Edgar CI, Maclean DJ, Scheible W-R, Udvardi MK, Kazan K (2005) Repressor- and activator-type ethylene response factors functioning in jasmonate signaling and disease resistance identified via a genome-wide screen of *Arabidopsis* transcription factor gene expression. Plant Physiol 139:949–959

McHale L, Tan X, Koehl P, Michelmore RW (2006) Plant NBS-LRR proteins: adaptable guards. Genome Biol 7:212

Medzhitov R (2001) Toll-like receptors and innate immunity. Nat Rev Immunol 1:135–145

Medzhitov R, Janeway CA Jr (1997) Innate immunity: the virtues of a monoclonal system of recognition. Cell 91:295–298

Mei C, Qi M, Sheng G, Yang Y (2006) Inducible overexpression of a rice allene oxide synthase gene increases the endogenous jasmonic acid level PR gene expression and host resistance to fungal infection. Mol Plant Microbe Interact 19:1127–1137

Meindl T, Boller T, Felix G (2000) The bacterial elicitor flagellin activates its receptor in tomato cells according to the address message concept. Plant Cell 12:1783–1794

Melotto M, Underwood W, Koczan J, Nomura K, He SY (2006) Plant stomata function in innate immunity against bacterial invasion. Cell 126:969–980

Melotto M, Underwood W, He SY (2008) Role of stomata in plant innate immunity and foliar bacterial diseases. Annu Rev Phytopathol 46:101–122

Ménard R, Albar S, de Ruffray P, Jamois F, Franz G, Fritig B, Yvin J-C, Kauffmann S (2004) β-1,3-glucan sulfate but not β-1,3-glucan induces the salicylic acid signaling pathway in tobacco and *Arabidopsis*. Plant Cell 16:3020–3032

Mendes BMJ, Cardoso SC, Boscarriol-Camargo RL, Cruz RB, Mourão Filho FAA, Bergamin Filho A (2010) Reduction in susceptibility to *Xanthomonas axonopodis* pv. *citri* in transgenic *Citrus sinensis* expressing the rice *Xa21* gene. Plant Pathol 59:68–75

Meng Y, Moscou MJ, Wise RP (2009) *Blufensin1* negatively impacts basal defense in response to barley powdery mildew. Plant Physiol 149:271–285

Menke FLH, van Pelt JA, Pieterse CMJ, Klessig DF (2004) Silencing of the mitogen-activated protein kinase MPK6 compromises disease resistance in Arabidopsis. Plant Cell 16:897–907

Merkouropoulos G, Andreasson F, Hess D, Boller T, Peck SC (2008) An *Arabidopsis* protein phosphorylated in response to microbial elicitation *AtPHOS32* is a substrate of MAP kinases 3 and 6. J Biol Chem 283:10493–10499

Mersmann S, Bourdais G, Rietz S, Robatzek S (2010) Ethylene signaling regulates accumulation of the FLS2 receptor and is required for the oxidative burst contributing to plant immunity. Plant Physiol 154:391–400

Messiaen J, Van Cutsem P (1999) Polyamines and pectins: II. Modulation of pectic signal transduction. Planta 208:247–256

Mészáros T, Helfer A, Hatzimasoura E, Magyar Z, Serazetdinova L, Rios G, Bardóczy V, Teige M, Koncz C, Peck S, Bögre L (2006) The *Arabidopsis* MAP kinase kinase MKK1 participates in defence responses to the bacterial elicitor flagellin. Plant J 48:485–498

Meusser B, Hirsch G, Jarosch E, Sommer T (2005) ERAD: the long road to destruction. Nat Cell Biol 7:766–772

Meziane H, Van Der Sluis I, Van Loon LC, Höfte M, Bakker PAHM (2005) Determinants of *Pseudomonas putida* WCS358 involved in inducing systemic resistance in plants. Mol Plant Pathol 6:177–185

Miao Y, Zentgraf U (2007) The antagonist function of *Arabidopsis* WRKY53 and ESR/ESP in leaf senescence is modulated by the jasmonic and salicylic acid equilibrium. Plant Cell 19:819–830

Michel K, Abderhalden O, Bruggmann R, Dudler R (2006) Transcriptional changes in powdery mildew infected wheat and Arabidopsis leaves undergoing syringolin-triggered hypersensitive cell death at infection sites. Plant Mol Biol 62:561–578

Minami E, Kuchitsu K, He DY, Kouchi H, Midoh N, Ohtsuki Y, Shibuya N (1996) Two novel genes rapidly and transiently activated in suspension-cultured rice cells by treatment with *N*-acetylchitoheptaose a biotic elicitor for phytoalexin production. Plant Cell Physiol 37:563–567

Mishina TE, Zeier J (2006) The Arabidopsis flavin-dependent monooxygenase FMO1 is an essential component of biologically induced systemic acquired resistance. Plant Physiol 141:1666–1675

Mishina TE, Zeier J (2007) Pathogen-associated molecular pattern recognition rather than development of tissue necrosis contributes to bacterial induction of systemic acquired resistance in *Arabidopsis*. Plant J 50:500–513

Mithöfer A, Boland W (2008) Recognition of herbivory-associated molecular patterns. Plant Physiol 146:825–831

Mithöfer A, Fliegmann J, Neuhaus-Url G, Schwarz H, Ebel J (2000) The hepta-β-glucoside elicitor-binding proteins from legumes represent a putative receptor family. Biol Chem 381:705–713

Mitsuhara I, Iwai T, Seo S, Yanagawa Y, Kawahigasi H, Hirose S, Ohkawa Y, Ohashi Y (2008) Characteristic expression of twelve *PR1* family genes in response to pathogen infection wounding and defense-related signal compounds. Mol Genet Genomics 279:415–427

Miya A, Albert P, Shinya T, Desaki Y, Ichimura K, Shirasu K, Narusaka Y, Kawakami N, Kaku H, Shibuya N (2007) CERK1 a LysM receptor kinase is essential for chitin elicitor signaling in *Arabidopsis*. Proc Natl Acad Sci USA 104:19613–19618

Mlotshwa S, Voinnet O, Florian Mette M, Matzke M, Vaucheret H, Ding SW, Pruss GT, Vance VB (2002) RNA silencing and the mobile silencing signal. Plant Cell 14:S289–S301

Mlotshwa S, Pruss GT, Vance V (2008) Small RNAs in viral infection and host defense. Trends Plant Sci 13:375–382

Mohamed N, Lherminier J, Farmer M-J, Fromentin J, Béno N, Houot V, Milat M-L, Blein J-P (2007) Defense responses in grapevine leaves against *Botrytis cinerea* induced by application of a *Pythium oligandrum* strain or its elicitin oligandrin to roots. Phytopathology 97:611–620

Mohr PG, Cahill D (2007) Suppression by ABA of salicylic acid and lignin accumulation and the expression of multiple genes in *Arabidopsis* infected with *Pseudomonas syringae* pv. *tomato*. Funct Integr Genomics 7:181–191

Molina L, Kahmann R (2007) A *Ustilago maydis* gene involved in $H_2O_2$ detoxification is required for virulence. Plant Cell 19:2293–2309

Molinari M, Helenins A (2000) Chaperone selection during glycoprotein translocation into the endoplasmic reticulum. Science 288:331–333

Molinaro A, Newman M-A, Lanzetta R, Parrilli M (2009) The structures of lipopolysaccharides from plant-associated gram-negative bacteria. Eur J Org Chem 34:5887–5896

Molloy S (2010) Plant immunity: helping plants to fight back. Nat Rev Microbiol 8:312–313

Montesano M, Koiv V, Mae A, Palva ET (2001) Novel receptor-like protein kinases induced by *Erwinia carotovora* and short oligouronides in potato. Mol Plant Pathol 2:339–346

Montesano M, Brader G, Palva ET (2003) Pathogen-derived elicitors: searching for receptors in plants. Mol Plant Pathol 4:73–78

Moreau M, Degrave A, Vedel R, Bitton F, Patrit O, Renou J-P, Barny M-A, Fagard M (2012) EDS1 contributes to nonhost resistance of *Arabidopsis thaliana* against *Erwinia amylovora*. Mol Plant Microbe Interact 25:421–430

Morris R, Walker JC (2003) Receptor-like protein kinases; the keys to response. Curr Opin Plant Biol 6:339–342

Moscatiello R, Mariani P, Sanders D, Maathuis FJ (2006) Transcriptional analysis of calcium-dependent and calcium-independent signalling pathways induced by ologogalacturonides. J Exp Bot 57:2847–2865

Mosher RA, Lewsey MG, Shivaprasad PV (2010) RNA silencing in plants: flash report. Silence 1:13

Motteram J, Kūfner I, Deller S, Brunner F, Hammond-Kosack KE, Nurnberger T, Rudd JJ (2009) Molecular characterization and functional analysis of *MgNLP* the sole NPP1 domain-containing protein from the fungal wheat leaf pathogen *Mycosphaerella graminicola*. Mol Plant Microbe Interact 22:790–799

Mukherjee M, Larrimore KE, Ahmed NJ, Bedick TS, Barghouthi NT, Traw MB, Barth C (2010) Ascorbic acid deficiency in *Arabidopsis* induces constitutive priming that is dependent on hydrogen peroxide, salicylic acid and the *NPR1* gene. Mol Plant Microbe Interact 23:340–351

Munford RS, Varley AW (2006) Shield as signal: lipopolysaccharides and the evolution of immunity to Gram-negative bacteria. PLoS Pathog 2:467–471

Mur LAJ, Bi YM, Darby RM, Firek S, Draper J (1997) Compromising early salicylic acid accumulation delays the hypersensitive response and increases viral dispersal during lesion development in TMV-infected tobacco. Plant J 12:1113–1126

Mur LAJ, Kenton P, Atzorn R, Miersch O, Wasternack C (2006) The outcomes of concentration-specific interactions between salicylate and jasmonate signaling include synergy antagonism and oxidative stress leading to cell death. Plant Physiol 140:249–262

Murray SL, Ingle RA, Petersen LN, Denby KJ (2007) Basal resistance against *Pseudomonas syringae* in *Arabidopsis* involves WRKY53 and a protein with homology to a nematode resistance protein. Mol Plant Microbe Interact 20:1431–1438

Naito K, Ishiga Y, Toyoda K, Shiraishi T, Ichinose Y (2007) N-terminal domain including conserved flg22 is required for flagellin-induced hypersensitive cell death in *Arabidopsis thaliana*. J Gen Plant Pathol 73:281–285

Naito K, Taguchi F, Suzuki T, Inagaki Y, Toyoda K, Shiraishi TT, Ichinose Y (2008) Amino acid sequence of bacterial microbe-associated molecular pattern flg22 is required for virulence. Mol Plant Microbe Interact 21:1165–1174

Nakagami H, Soukupová H, Schikora A, Zársk V, Hirt H (2006) A mitogen-activated protein kinase kinase kinase mediates reactive oxygen species homeostasis in *Arabidopsis*. J Biol Chem 281:38697–38704

Nakashima A, Chen L, Thao NP, Fujiwara M, Wong HL, Kuwano M, Umemura K, Shirasu K, Kawasaki T, Shimamoto K (2008) RACK1 functions in rice immunity by interacting with the Rac1 immune complex. Plant Cell 20:2265–2279

Nakashita H, Yoshioka K, Yasuda M, Nitta T, Arai Y, Yoshida S, Yamaguchi I (2002) Probenazole induces systemic acquired resistance in tobacco through salicylic acid accumulation. Physiol Plant Pathol 61:197–203

Nam KH, Li J (2002) BRI1/BAK1 a receptor kinase pair mediating brassinosteroid signaling. Cell 110:203–212

Nandi A, Kachroo P, Fukushige H, Hildebrand DF, Klessig DF, Shah J (2003a) Ethylene and jasmonic acid signaling pathways affect the NPR1-independent expression of defense genes without impacting resistance to *Pseudomonas syringae* and *Peronospora parasitica* in the *Arabidopsis ssi1* mutant. Mol Plant Microbe Interact 18:588–599

Nandi A, Krothapalli K, Buseman C, Li M, Welti R, Enyedi A, Shah J (2003b) The *Arabidopsis thalianasid* mutants affect plastidic lipid composition and suppress dwarfing cell death and the enhanced disease resistance phenotypes resulting from the deficiency of a fatty acid desaturase. Plant Cell 15:2383–2398

Nandi A, Moeder W, Kachroo P, Klessig DF, Shah J (2005) *Arabidopsis ssi-2* conferred susceptibility to *Botrytis cinerea* is dependent on *EDS5* and *PAD4*. Mol Plant Microbe Interact 18:363–370

Narry Kim V (2005) Small RNAs: classification biogenesis and function. Mol Cells 19:1–15

Navarre DA, Wolpert TJ (1999) Victorin induction of an apoptotic/senescence-like response in oats. Plant Cell 11:237–250

Navarro L, Zipfel C, Rowland O, Keller I, Robatzek S, Boller T, Jones JD (2004) The transcriptional innate immune response to flg22. Interplay and overlap with *Avr* gene-dependent defense responses and bacterial pathogenesis. Plant Physiol 135:1113–1128

Navarro L, Dunoyer P, Jay F, Arnold B, Dhamasiri N, Estelle M, Voinnet O, Jones JDG (2006) A plant miRNA contributes to antibacterial resistance by repressing auxin signaling. Science 312:436–439

Navarro L, Jay F, Nomura K, He SY, Voinnet O (2008) Suppression of the microRNA pathway by bacterial elicitor proteins. Science 312:436–439

Nekrasov V, Li J, Batoux M, Roux M, Chu ZH, Lacombe S, Rougon A, Bittel P, Kiss-Papp M, Chinchilla D, van Esse HP, Jorda L, Schwessinger B, Nicaise V, Thomma BP, Molina A, Jones JD, Zipfel C (2009) Control of the pattern-recognition receptor EFR by an ER protein complex in plant immunity. EMBO J 28:3428–3438

Newman M-A, Dow JM, Daniels MJ (2001) Bacterial lipopolysaccharides and plant-pathogen interactions. Eur J Plant Pathol 107:95–102

Newman M-A, Roepenack-Lahaye EV, Parr A, Daniels MJ, Dow JM (2002) Prior exposure to lipopolysaccharide potentiates expression of plant defenses in response to bacteria. Plant J 29:487–495

Newman M-A, Dow JM, Molinaro A, Parrilli M (2007) Priming induction and modulation of plant defense responses by bacterial lipopolysaccharides. Plant Immun 13:69–84

Nicaise V, Roux M, Zipfel C (2009) Recent advances in PAMP-triggered immunity against bacteria: pattern recognition receptors watch over and raise the alarm. Plant Physiol 150:1638–1647

Nie X (2006) Salicylic acid suppresses *Potato virus Y* isolate N: O-induced symptoms in tobacco plants. Phytopathology 96:255–263

Ning W, Chen F, Mao B, Li Q, Liu Z, Guo Z, He Z (2004) *N*-acetylchitooligosaccharides elicit rice defense responses including hypersensitive response-like cell death oxidative burst and defence gene expression. Physiol Mol Plant Pathol 64:263–271

Nishiuchi T, Masuda D, Nakashita H, Ichimura K, Shinozaki K, Yoshida S, Kimura M, Yamaguchi I, Yamaguchi K (2006) *Fusarium* phytotoxin trichothecenes have an elicitor-like activity in *Arabidopsis thaliana* but the activity differed significantly among their molecular species. Mol Plant Microbe Interact 19:512–520

Nojiri H, Sugimori M, Yamane H, Nishimura Y, Yamada A, Shibuya N, Kodama O, Murofushi N, Omori T (1996) Involvement of jasmonic acid in elicitor-induced phytoalexin production in suspension-cultured rice cells. Plant Physiol 110:138–142

Nomura K, DebRoy S, Lee YH, Pumplin N, Jones J, He SY (2006) A bacterial virulence protein suppresses host immunity to cause plant disease. Science 313:220–223

Nühse TS, Peck SC, Hirt H, Boller T (2000) Microbial elicitors induce activation and dual phosphorylation of the *Arabidopsis thaliana* MAPK 6. J Biol Chem 275:7521–7526

Nühse TS, Boller T, Peck SC (2003) A plasma membrane syntaxin is phosphorylated in response to the bacterial elicitor flagellin. J Biol Chem 278:45248–45254

Nühse TS, Stenballe A, Jensen ON, Peck SC (2004) Phosphoproteomics of the *Arabidopsis* plasma membrane and a new phosphorylation site database. Plant Cell 16:2394–2405

Nühse TS, Bottrill AR, Jones AM, Peck SC (2007) Quantitative phosphoproteomic analysis of plasma membrane proteins reveals regulatory mechanisms of plant innate immune responses. Plant J 51:931–940

Nürnberger T (1999) Signal perception in plant pathogen defense. Cell Mol Life Sci 55:167–182

Nürnberger T, Brunner F (2002) Innate immunity in plants and animals: emerging parallels between the recognition of general elicitors and pathogen-associated molecular patterns. Curr Opin Plant Biol 5:318–324

Nürnberger T, Kufner I (2011) The role of the plant plasma membrane in microbial sensing and innate immunity. Plant Cell Monogr 19:471–483

Nürnberger T, Lipka V (2005) Non-host resistance in plants – new insights into an old phenomenon. Mol Plant Pathol 6:335–345

Nürnberger T, Scheel D (2001) Signal transmission in the plant immune response. Trends Plant Sci 6:372–379

Nürnberger T, Nennstiel D, Jabs T, Sacks WR, Hahlbrock K, Scheel D (1994) High affinity binding of a fungal oligopeptide elicitor to parsley plasma membranes triggers multiple defense responses. Cell 78:449–460

Nürnberger T, Nennstiel D, Hahlbrock K, Scheel D (1995) Covalent cross-linking of the *Phytophthora megasperma* oligopeptide elicitor to its receptor in parsley membranes. Proc Natl Acad Sci USA 92:2338–2342

Nürnberger T, Brunner F, Kemmerling B, Piater L (2004) Innate immunity in plants and animals: striking similarities and obvious differences. Immunol Rev 198:249–266

O'Donnell PJ, Schmelz EA, Moussatche P, Lund ST, Jones JB, Klee HJ (2003a) Susceptible to intolerance – a range of hormonal actions in a susceptible Arabidopsis pathogen response. Plant J 33:245–257

O'Donnell PJ, Schmelz E, Block A, Miersch O, Wasternack C, Jones JB, Klee HJ (2003b) Multiple hormones act sequentially to mediate a susceptible tomato pathogen defense response. Plant Physiol 133:1181–1189

Ogasawara Y, Kaya H, Hiraoka G, Yumoto F, Kimura S, Kadota Y, Hishinuma H, Senzaki E, Yamagoe S, Nagata K, Nara M, Suzuki K, Tanokura M, Kuchitsu K (2008) Synergistic activation of the *Arabidopsis* NADPH oxidase AtrbohD by $Ca^{2+}$ and phosphorylation. J Biol Chem 283:8885–8892

Oh CS, Beer SV (2007) AtHIPM an ortholog of the apple HrpN-interacting protein is a negative regulator of plant growth and mediates the growth-enhancing effect of HrpN in *Arabidopsis*. Plant Physiol 145:426–436

Oñate-Sánchez L, Anderson JP, Young J, Singh KB (2007) AtERF14, a member of the ERF family of transcription factors plays a nonredundant role in plant defense. Plant Physiol 143:400–409

Ott PG, Varga GJ, Szatmári A, Bozsó Z, Klement É, Medzihradszky KF, Besenyei E, Czelleng A, Klement Z (2006) Novel extracellular chitinases rapidly and specifically induced by general bacterial elicitors and suppressed by virulent bacteria as a marker of early basal resistance in tobacco. Mol Plant Microbe Interact 19:161–172

Ottmann C, Luberacki B, Kūfner I, Koch W, Brunner F, Weyand M, Mattinen L, Pirhonen M, Anderluh G, Ulrich Seitz H, Nurnberger T, Oecking C (2009) A common toxin fold mediates microbial attack and plant defense. Proc Natl Acad Sci USA 106:10359–10364

Padgett HS, Watanabe Y, Beachey RN (1997) Identification of the TMV replicase sequence that activates the *N* gene-mediated hypersensitive response. Mol Plant Microbe Interact 10:709–715

Panabières F, Marais A, Le Berre J-Y, Penot I, Fournier D, Ricci P (1995) Characterization of a gene cluster of *Phytophthora cryptogea* which codes for elicitins proteins inducing a hypersensitive-like response in tobacco. Mol Plant Microbe Interact 8:996–1003

Panstruga R, Dodds PN (2009) Terrific protein traffic: the mystery of effector protein delivery by filamentous plant pathogens. Science 324:748–750

Papaefthimiou I, Hamilton A, Denti M, Baulcombe D, Tsagris M, Tabler M (2001) Replicating *Potato spindle tuber viroid* RNA is accompanied by short RNA fragments that are characteristic of post-transcriptional gene silencing. Nucleic Acids Res 29:2395–2400

Paradies I, Konze JR, Elstner EF, Paxton J (1980) Ethylene: indicator but not inducer of phytoalexin synthesis in soybean. Plant Physiol 76:1106–1109

Park SW, Kaimoyo E, Kumar D, Mosher S, Klessig DF (2007) Methyl salicylate is a critical mobile signal for plant systemic acquired resistance. Science 318:113–116

Park C-J, Peng Y, Chen X, Dardick C, Ruan D, BartR CPE, Ronald PC (2008) Rice XB15 a protein phosphatase 2C negatively regulates cell death and XA21-mediated innate immunity. PLoS Biol 6:e231

Park J-Y, Jin J, Lee Y-W, Kang S, Lee Y-H (2009) Rice blast fungus (*Magnaporthe oryzae*) infects *Arabidopsis* via a mechanism distinct from that required for the infection of rice. Plant Physiol 149:474–486

Park C-J, Bart R, Chern M, Canlas PE, Bai W, Ronald PC (2010a) Overexpression of the endoplasmic reticulum chaperone BiP3 regulates XA21-mediated innate immunity in rice. PLoS One 5(2):e9262

Park C-J, Han S-W, Chen X, Ronald PC (2010b) Elucidation of XA21-mediated innate immunity. Cell Microbiol 12:1017–1025

Pastor V, Luna E, Mauch-Mani B, Ton J, Flors V (2012) Primed plants do not forget. Environ Exp Bot (in press)

Pastuglia M, Roby D, Dumas C, Cock M (1997) Rapid induction by wounding and bacterial infection of a *S* gene family receptor-like kinase gene in *Brassica oleracea*. Plant Cell 9:49–60

Pattison RJ, Amtmann A (2009) N-glycan production in the endoplasmic reticulum of plants. Trends Plant Sci 14:92–99

Pavli OI, Kelaidi GI, Tampakaki AP, Skaracis GN (2011) The *hrpZ* gene of *Pseudomonas syringae* pv. *phaseolicola* enhances resistance to rhizomania disease in transgenic *Nicotiana benthamiana* and sugar beet. PLoS One 6(3):e17306

Pearce G, Yamaguchi Y, Munske G, Ryan CA (2008) Structure-activity studies of AtPep1, a plant peptide signal involved in the innate immune response. Peptides 29:2083–2089

Pearce G, Yamaguchi Y, Barona G, Ryan CA (2010) A subtilisin-like protein from soybean contains an embedded cryptic signal that activates defense-related genes. Proc Natl Acad Sci USA 107:14921–14925

Peck SC, Nühse TS, Hess D, Iglesias A, Meins F, Boller T (2001) Directed proteomics identifies a plant-specific protein rapidly phosphorylated in response to bacterial and fungal elicitors. Plant Cell 13:1467–1475

Pemberton CL, Salmond GPC (2004) The Nep1-like proteins – a growing family of microbial elicitors of plant necrosis. Mol Plant Pathol 5:353–359

Pemberton CL, Whitehead NA, Sebaihia M, Bell KS, Hyman IJ, Harris SJ, Martlin AJ, Robson ND, Birch PRJ, Cart JP, Toth IK, Salmond GPC (2005) Novel quorum-sensing-controlled genes in *Erwinia carotovora* subsp carotovora: identification of a fungal elicitor homologue in a soft-rotting bacterium. Mol Plant Microbe Interact 18:343–353

Peng J-L, Bao Z-L, Ren H-Y, Wang J-S, Dong HS (2004) Expression of harpin$_{Xoo}$ in transgenic tobacco induces pathogen defense in the absence of hypersensitive cell death. Phytopathology 94:1048–1055

Peng Y, Bartley LE, Chen X, Dardick C, Chern M, Ruan R, Canlas PE, Ronald PC (2008) OsWRKY62 is a negative regulator of basal and *XA21*-mediated defense against *Xanthomonas oryzae* pv. *oryzae* in rice. Mol Plant 1:446–458

Peng Y, Bartley LE, Canlas P, Ronald PC (2010) OsWRKY IIa transcription factors modulate rice innate immunity. Rice 3:36–42

Perchepied L, Balagué C, Riou C, Claudel-Renard C, Riviére N, Grezes-Besset B, Roby D (2010) Nitric oxide participates in the complex interplay of defense-related signaling pathways controlling disease resistance to *Sclerotinia sclerotiorum* in *Arabidopsis thaliana*. Mol Plant Microbe Interact 23:846–860

Pernollet JC, Sallankin M, Salle-Tourne M, Huet JC (1993) Elicitin isoforms from *Phytophthora* species: comparison of their physico-chemical properties and toxicity to tobacco and other plant species. Physiol Mol Plant Pathol 42:53–67

Petutschnig EK, Jones AME, Serazetdinova L, Lipka U, Lipka V (2010) The lysin motif receptor-like kinase (LysM-RLK) is a major chitin binding protein in *Arabidopsis thaliana* and subject to chitin-induced phosphorylation. J Biol Chem 285:28902–28911

Pfund C, Tans-Kersten J, Dunning FM, Alonso JM, Ecker JR, Allen C, Bent AF (2004) Flagellin is not a major defense elicitor in *Ralstonia solanacearum* cells or extracts applied to *Arabidopsis thaliana*. Mol Plant Microbe Interact 17:696–706

Pimpl P, Dennecke J (2000) ER retention of soluble proteins: retrieval retention or both? Plant Cell 12:1517–1521

Pitzschke A, Hirt H (2009) Disentangling the complexity of mitogen-activated protein kinases and reactive oxygen species signaling. Plant Physiol 149:606–615

Pitzschke A, Schikora A, Hirt H (2009a) MAPK cascade signaling networks in plant defence. Curr Opin Plant Biol 12:1–6

Pitzschke A, Djamei A, Bitton F, Hirt H (2009b) A major role of the MEKK1-MKK1/2-MPK4 pathway in ROS signaling. Mol Plant 2:120–137

Poinssot B, Vandelle E, Bentéjac M, Adrian M, Levis C, Brygoo Y, Garin J, Sicilia F, Coutos-Thévenot P, Pugin A (2003) The endopolygalacturonase I from *Botrytis cinerea* activates grapevine defense reactions unrelated to its enzymatic activity. Mol Plant Microbe Interact 16:553–564

Polverari A, Molesini B, Pezzotti M, Buonaurio R, Marte M, Delledonne M (2003) Nitric oxide-mediated transcriptional changes in *Arabidopsis thaliana*. Mol Plant Microbe Interact 16: 1094–1105

Popescu SC (2012) A model for the biosynthesis and transport of plasma membrane-associated signaling receptors to the cell surface. Front Plant Sci 3:71. doi:10.3389/fpls.2012.00071

Postel S, Kufner I, Beueter C, Mazzotta S, Schwedt A, Borlotti A, Halter T, Kemmerling B, Nürnberger T (2010) The multifunctional leucine-rich repeat receptor kinase BAK1 is implicated in Arabidopsis development and immunity. Eur J Cell Biol 89:169–174

Potlakayala SD, Reed DW, Covello PS, Fobert PR (2007) Systemic acquired resistance in canola is linked with pathogenesis-related gene expression and requires salicylic acid. Phytopathology 97:794–802

Pré M, Atallah M, Champion A, De Vos M, Pieterse CMJ, Memelink J (2008) The AP2/ERF domain transcription factor ORA59 integrates jasmonic acid and ethylene signals in plant defense. Plant Physiol 147:1347–1357

Prins M, Laimer PM, Noris E, Schubert J, Wassenegger M, Tepfer M (2008) Strategies for antiviral resistance in transgenic plants. Mol Plant Pathol 9:73–83

Prithiviraj B, Bais HP, Jha AK, Vivanco JM (2005) *Staphylococcus aureus* pathogenicity on *Arabidopsis thaliana* is mediated either by a direct effect of salicylic acid on the pathogen or by SA-dependent NPR1-independent host responses. Plant J 42:417–432

Pugin A, Frachisse J-M, Tavernier E, Bligny R, Gout E, Douce R, Guern J (1997) Early events induced by the elicitor cryptogein in tobacco cells: involvement of a plasma membrane NADPH oxidase and activation of glycolysis and the pentose phosphate pathway. Plant Cell 9:2077–2091

Pungartnik C, da Silva AC, de Melo SA, Gramacho KP, de Mattos Cascardo JC, Brendel M, Micheli F, da Silva GA (2009) High-affinity copper transport and Snq2 export permease of *Saccharomyces cerevisiae* modulate cytotoxicity of PR-10 from *Theobroma cacao*. Mol Plant Microbe Interact 22:39–51

Qi X, Bao FS, Xie Z (2009) Small RNA deep sequencing reveals role for *Arabidopsis thaliana* RNA-dependent RNA polymerases in viral siRNA biogenesis. PLoS One 4:e4971

Qi Z, Verma R, Gehring C, Yamaguchi Y, Zhao Y, Ryan CA, Berkowitz G (2010) $Ca^{2+}$ signaling by plant *Arabidopsis thaliana* Pep peptides depends on AtPepR1 a receptor with guanylyl cyclase activity and cGMP-activated $Ca^{2+}$ channels. Proc Natl Acad Sci USA 107:21193–21198

Qiu D, Xiao J, Ding X, Xiong M, Cai M, Cao Y, Li X, Xu C, Wang S (2007) OsWRKY13 mediates rice disease resistance by regulating defense-related genes in salicylate- and jasmonate-dependent signaling. Mol Plant Microbe Interact 20:492–499

Qiu J, Zhou L, Yun B, Nielsen H, Fiil BK, Petersen K, MacKinlay J, Loake GJ, Mundy J, Morris PC (2008a) *Arabidopsis* mitogen-activated protein kinase kinases MKK1 and MKK2 have overlapping functions in defense signaling mediated by MEKK1 MPK4 and MKS. Plant Physiol 148:212–222

Qiu JL, Fiil BK, Petersen K, Nielsen HB, Botanga CJ, Thorgrimsen S, Palma K, Suarez-Rodriguez MC, Sandbech-Clausen S, Lichota J, Brodersen P, Grasser KD, Mattson O, Glazebrook J, Mundy J, Petersen M (2008b) *Arabidopsis* MAP kinase 4 regulates gene expression through transcription factor release in the nucleus. EMBO J 27:2214–2221

Qu F, Morris TJ (2005) Suppressors of RNA silencing encoded by plant viruses and their role in viral infection. FEBS Lett 579:5958–5964

Qu F, Ye X, Morris TJ (2008) Arabidopsis DRB4, AGO1, AGO7, and RDR6 participate in a DCL4-initiated antiviral silencing pathway negatively regulated by DCL1. Proc Natl Acad Sci USA 105:14732–14737

Qutob D, Kamoun S, Gijzen M (2002) Expression of a *Phytophthora sojae* necrosis-inducing protein occurs during transition from biotrophy to necrotrophy. Plant J 32:361–373

Qutob D, Huitema E, Gijzen M, Kamoun S (2003) Variation in structure and activity among elicitins from *Phytophthora sojae*. Mol Plant Pathol 4:119–124

Qutob D, Kemmerling B, Brunner F, Küfner I, Engelhardt S, Gust AA, Luberacki B, Seitz HU, Stahl DO, Rauhut T, Glawischnig E, Schween G, Lacombe B, Watanabe N, Lam E, Schlichting R, Scheel D, Nau K, Dodt G, Hubert D, Gijzen M, Nürnberger T (2006) Phytotoxicity and innate immune responses induced by NEP1-like proteins. Plant Cell 18:3721–3744

Racapé J, Belbahri L, Engelhardt S, Lacombe B, Lee J, Lochman J, Marals A, Nicole M, Nurnberger T, Parlange F, Puverel S, Keller H (2005) $Ca^{2+}$-dependent lipid binding and membrane integration of PopA, a harpin-like elicitor of the hypersensitive response in tobacco. Mol Microbiol 58:1406–1420

Radman R, Saez T, Bucke C, Keshavarz T (2003) Elicitation of plants and microbial cell systems. Biotechnol Appl Biochem 37:91–102

Radutoiu S, Madsen LH, Madsen EB, Jurkiewicz A, Fukai E, Quistgaard EM, Albrektsen AS, James EK, Thirup S, Stougaard J (2007) LysM domains mediate lipochitin-oligosaccharide recognition and Nfr genes extend the symbiotic host range. EMBO J 26:3923–3935

Ralph SG, Hudgins JW, Jancsik S, Franceschi VR, Bohlmann J (2007) Aminocyclopropane carboxylic acid synthase is a regulated step in ethylene-dependent induced conifer defense full-length cDNA cloning of a multigene family differential constitutive and wound- and insect-induced expression and cellular and subcellular localization in spruce and Douglas Fir. Plant Physiol 143:410–424

Ramesh Sundar A, Velazhahan R, Viswanathan R, Padmanabhan P, Vidhyasekaran P (2001) Induction of systemic resistance to *Colletotrichum falcatum* in sugarcane by a synthetic signal molecule Acibenzolar-S-methyl (CGA-245704). Phytoparasitica 29:231–242

Ramonell K, Berrocal-Lobo M, Koh S, Wan J, Edwards H, Stacey G, Somerville S (2005) Loss-of-function in chitin responsive genes show increased susceptibility to the powdery mildew pathogen. Plant Physiol 138:1027–1036

Ramos HC, Rumbo M, Sirard JC (2004) Bacterial flagellins: mediators of pathogenicity and host immune responses in mucosa. Trends Microbiol 12:509–517

Reboutier D, Frankart C, Briand J, Biligui B, Laroche S, Rona J-P, Barny M-A, Bouteau F (2007) The $HrpN_{ea}$ harpin from *Erwinia amylovora* triggers differential responses on the nonhost *Arabidopsis thaliana* cells and on the host apple cells. Mol Plant Microbe Interact 20:94–100

Rentel MC, Leonelli L, Dahlbeck D, Zhao B, Staskawicz BJ (2008) Recognition of the *Hyaloperonospora parasitica* effector ATR13 triggers resistance against oomycete bacterial and viral pathogens. Proc Natl Acad Sci USA 105:1091–1096

Ridley BL, O'Neill MA, Mohnen D (2001) Pectins: structure biosynthesis and oligogalacturonide-related signaling. Phytochemistry 57:929–967

Rinaldi C, Kohler A, Frey P, Duchaussoy F, Ningre N, Couloux A, Wincker P, Le Thiec D, Fluch S, Martin F, Duplessis S (2007) Transcript profiling of poplar leaves upon infection with compatible and incompatible strains of the foliar rust *Melampsora larici-populina*. Plant Physiol 144:347–366

Robatzek S (2007) Vehicle trafficking in plant immune responses. Cell Microbiol 9:1–8

Robatzek S, Chinchilla D, Boller T (2006) Ligand-induced endocytosis of the pattern recognition receptor FLS2 in *Arabidopsis*. Genes Dev 20:537–542

Robatzek S, Bittel P, Chinchilla D, Köchner P, Felix G, Shiu SH, Boller T (2007) Molecular identification and characterization of the tomato flagellin receptor LeFLS2 an orthologue of *Arabidopsis* FLS2 exhibiting characteristically different perception specificities. Plant Mol Biol 64:539–547

Ron M, Avni A (2004) The receptor for the fungal elicitor ethylene-inducing xylanase is a member of a resistance-like gene family in tomato. Plant Cell 16:1604–1615

Ronald PC, Beutler B (2010) Plant and animal sensors of conserved microbial signatures. Science 308:1783–1786

Rossard S, Roblin G, Atanassova R (2010) Ergosterol triggers characteristic elicitation steps in *Beta vulgaris* leaf tissues. J Exp Bot 61:1807–1816

Rotblat B, Enshel-Seijffers D, Gershoni MJ, Schuster S, Avni A (2002) Identification of an essential component of the elicitation active site of the EIX protein elicitor. Plant J 32:1–7

Royet J, Reichhart J-M (2003) Detection of peptidoglycans by NOD proteins. Trends Cell Biol 13:610–614

Ruddock LW, Molinari M (2006) N-glycan processing in ER quality control. J Cell Sci 119:4373–4380

Ryals J, Neuenschwander UH, Willits MG, Molina A, Steiner HY, Hunt MD (1996) Systemic acquired resistance. Plant Cell 8:1809–1819

Ryan CA, Pearce G (2003) Systemins: a functionally defined family of peptide signals that regulate defensive genes in Solanaceae species. Proc Natl Acad Sci USA 100:14577–14580

Ryan CA, Huffaker A, Yamaguchi Y (2007) New insights into innate immunity in *Arabidopsis*. Cell Microbiol 9:1902–1908

Saijo Y (2010) ER quality control of immune receptors and regulators in plants. Cell Microbiol 12:716–724

Saijo Y, Tintor N, Lu X, Rauf P, Pajerowska-Mukhtar K, Häweker H, Dong X, Robatzek S, Schulze-Lefert P (2009) Receptor quality control in the endoplasmic reticulum for plant innate immunity. EMBO J 28:3439–3449

Sanabria NM, Dubery IA (2006) Differential display profiling of the *Nicotiana* response to LPS reveals elements of plant basal resistance. Biochem Biophys Res Commun 344:1001–1007

Sanabria NM, Goring D, Nürnberger T, Dubery IA (2008) Self/nonself perception and recognition mechanisms in plants a comparison of self incompatibility and innate immunity. New Phytol 178:503–513

Sänchez G, Gerhardt N, Siciliano F, Vojnov A, Malcuit I, Marono MR (2010) Salicylic acid is involved in the *Nb*-mediated defense responses to *Potato virus X* in *Solanum tuberosum*. Mol Plant Microbe Interact 23:394–405

Schaffrath U, Scheinpflug H, Reisner HJ (1995) An elicitor from *Pyricularia oryzae* induces resistance responses in rice: isolation characterization and physiological properties. Physiol Mol Plant Pathol 46:293–307

Scheer JM, Ryan CA (2002) The systemin receptor 160 from *Lycopersicon peruvianum* is a member of the LRR receptor kinase family. Proc Natl Acad Sci USA 99:9585–9590

Schenk PM, Kazan K, Wilson IW, Anderson JP, Richmond T, Somerville SC, Manners JM (2000) Coordinated plant defense responses in *Arabidopsis* revealed by microarray analysis. Proc Natl Acad Sci USA 97:11655–11660

Schlessinger J (2000) Cell signaling by receptor tyrosine kinases. Cell 103:211–225

Schmelz EA, Carroll MJ, LeClere S, Phipps SM, Meredith J, Chourey PS, Alborn HT, Teal PE (2006) Fragments of ATP synthase mediate plant perception of insect attack. Proc Natl Acad Sci USA 103:8894–8899

Schmidt K, Pflugmacher M, Klages S, Maser A, Mock A, Stahl DJ (2008) Accumulation of the hormone abscisic acid (ABA) at the infection site of the fungus *Cercospora beticola* supports the role of ABA as a repressor of plant defence in sugar beet. Mol Plant Pathol 9:661–673

Schornack S, Huitema E, Cano LM, Bozkurt TO, Oliva R, Van Damme M, Schwizer S, Raffaele S, Chaparro-Garcia A, Farrer R, Segretin ME, Bos J, Haas BJ, Zody MC, Nusbaum C, Win J, Thines M, Kamoun S (2009) Ten things to know about oomycete effectors. Mol Plant Pathol 10:795–801

Schouten A, van Baarlen P, van Kan JAL (2008) Phytotoxic Nep1-like proteins from the necrotrophic fungus *Botrytis cinerea* associated with membranes and the nucleus of plant cells. New Phytol 177:493–505

Schulze B, Mentzel T, Jehle A, Mueller K, Beeler S, Boller T, Felix G, Chinchilla D (2010) Rapid heteromerization and phosphorylation of ligand-activated plant transmembrane receptors and their associated kinase BAK1. J Biol Chem 285:9444–9451

Schulze-Lefert P (2004) Plant immunity: the origami of receptor activation. Curr Biol 14:R22–R24

Schuster SC, Khan S (1994) The bacterial flagellar motor. Annu Rev Biophys Biomol Struct 23:509–539

Schweizer P, Buchala A, Dudler RA, Metraux JP (1998) Induced systemic resistance in wounded rice plants. Plant J 14:475–481

Schweizer P, Kmecl A, Carpita N, Dudler RA (2000) A soluble carbohydrate elicitor from *Blumeria graminis* f sp *tritici* is recognized by a broad range of cereals. Physiol Mol Plant Pathol 56:157–167

Schweswinger B, Zipfel C (2008) News from the frontline: recent insights into PAMP-triggered immunity in plants. Curr Opin Plant Biol 11:389–395

Segonzac C, Zipfel C (2011) Activation of plant pattern-recognition receptors by bacteria. Curr Opin Microbiol 14:54–61

Segonzac C, Feike D, Gimenez-Ibanez S, Hann DR, Zipfel C, Rathjen JP (2011) Hierarchy and roles of PAMP-induced responses in *Nicotiana benthamiana*. Plant Physiol 156:687–699

Seifert GJ, Blaukopf C (2010) Irritable walls: the plant extracellular matrix and signaling. Plant Physiol 153:467–478

Séjalon N, Villaba F, Bottin A, Rickauer M, Dargent R, Esquerré-Tugayé MT (1995) Characterization of a cell surface antigen isolated from the plant pathogen *Phytophthora parasitica* var *nicotianae*. Can J Bot 73:1104–1108

Séjalon-Delmas N, Villalba J, Bottin A, Rickauer M, Dargent R, Esquerré-Tugayé MT (1997) Purification elicitor activity and cell wall localisation of a glycoprotein from *Phytophthora parasitica* var *nicotianae* a fungal pathogen of tobacco. Phytopathology 87:899–909

Sekine KT, Nandi A, Ishihara T, Hase S, Ikegami M, Shah J, Takahashi H (2004) Enhanced resistance to *Cucumber mosaic virus* in the *Arabidopsis thaliana ssi2* mutant is mediated via an SA-independent mechanism. Mol Plant Microbe Interact 17:623–632

Semenza K, Hardwick KG, Dean N, Pelham HR (1990) ERD2 a yeast gene required for the receptor mediated retrieval of luminal ER proteins from the secretory pathway. Cell 61:1349–1357

Seo HS, Song JT, Cheong J-J, Lee Y-H, Lee Y-W, Hwang I, Lee JS, Choi YD (2001) Jasmonic acid carboxyl methyltransferase: a key enzyme for jasmonate-regulated plant responses. Proc Natl Acad Sci U S A 98:4788–4793

Seo Y-S, Rojas MR, Lee J-Y, Lee S-W, Jeon J-S, Ronald P, Lucas WJ, Gilbertson RL (2006) A viral resistance gene from common bean functions across plant families and is up-regulated in a non-virus-specific manner. Proc Natl Acad Sci USA 103:11856–11861

Serrano M, Robatzek S, Torres M, Kombrink E, Somssich I, Robinson M, Schulze-Lefert P (2007) Chemical interference of pathogen-associated molecular pattern-triggered immune responses in *Arabidopsis* reveals a potential role for fatty-acid synthase type II complex-derived lipid signals. J Biol Chem 282:6803–6811

Shan L, He P, Sheen J (2007) Intercepting host MAPK signaling cascades by bacterial type III effectors. Cell Host Microbe 1:167–174

Shan L, He P, Li J, Heese A, Peck SC, Nümberger T, Martin GB, Sheen J (2008) Bacterial effectors target the common signaling partner BAK1 to disrupt multiple MAMP receptor-signaling complexes and impede plant immunity. Cell Host Microbe 4:17–27

Shao F, Golstein C, Ade J, Stoutemyer M, Dixon JE, Innes RW (2003) Cleavage of *Arabidopsis* PBS1 by a bacterial type III effector. Science 301:1230–1233

Shao M, Wang J, Dean RA, Lin Y, Gao X, Hu S (2008) Expression of a harpin-encoding gene in rice confers durable nonspecific resistance to *Magnaporthe grisea*. Plant Biotechnol J 6:73–81

Sharp JK, McNeil M, Albersheim P (1984) The primary structures of one elicitor active and seven elicitor inactive hexa-β-glucopyranosyl-D-glucitol isolated from the mycelial cell walls of *Phytophthora megasperma* f sp *glycinea*. J Biol Chem 259:11321–11336

Shen QH, Saijo Y, Mauch S, BiskupC BS, Keller B, Seki H, Ulker B, Somssich IE, Schulze-Lefert P (2007) Nuclear activity of MLA immune receptors links isolate-specific and basal disease resistance responses. Science 315:1098–1103

Shibuya N, Minami E (2001) Oligosaccharide signalling for defense responses in plant. Physiol Mol Plant Pathol 59:223–233
Shimizu T, Nakano T, Takamizawa D, Desaki Y, Ishi-Minami N, Nishizawa Y, Minami E, Okada K, Yamane H, Kaku H, Shibuya N (2010) Two LysM receptor molecules CEBiP and OsCERK1 cooperatively regulate chitin elicitor signaling in rice. Plant J 64:204–214
Shimono M, Sugano S, Nakayama A, Jiang C-J, Ono K, Toki S, Takatsuji H (2007) Rice WRKY plays a crucial role in benzothiadiazole-inducible blast resistance. Plant Cell 19:2064–2076
Shinya T, Galis I, Narisawa T, Sasaki M, Fukuda H, Matsuoka H, Saito M, Matsuoka K (2007) Comprehensive analysis of glucan elicitor-regulated expression in tobacco BY-2 cells reveals a novel MYB transcription factor involved in the regulation of phenylpropanoid metabolism. Plant Cell Physiol 48:1404–1413
Shinya T, Osada T, Desaki Y, Hatamoto M, Yamanaka Y, Hirano H, Takai R, Che F-S, Kaku H, Shibuya N (2010) Characterization of receptor proteins using affinity cross-linking with biotinylated ligands. Plant Cell Physiol 51:262–270
Shiu S-H, Bleecker AB (2001a) Receptor-like kinases from Arabidopsis form a monophyletic gene family related to animal receptor kinases. Proc Natl Acad Sci USA 98:10763–10768
Shiu S-H, Bleecker AB (2001b) Plant receptor-like kinase gene family: diversity function and signaling. Sci STKE 113:RE22
Shiu S-H, Bleecker AB (2003) Expansion of receptor-like kinase/Pelle gene family and receptor proteins in Arabidopsis. Plant Physiol 132:530–543
Shiu S-H, Karlowski WM, Pan R, Tzeng YH, Mayer KFX, Li W-H (2004) Comparative analysis of the receptor-like kinase family in *Arabidopsis* and rice. Plant Cell 16:1220–1234
Silipo A, Mollinaro A (2010) The diversity of the core oligosaccharide in lipopolysaccharides. In: Wang X, Quinn PJ (eds) Endotoxins: structure function and recognition, Subcellular biochemistry, vol 53, part 1. Springer Science Business Media B V, Dordrecht, pp 69–99
Silipo A, De Castro C, Lanzetta R, Molinaro A, Parrilli M (2004) Full structural characterization of the lipid A components from the *Agrobacterium tumefaciens* strain C58 lipopolysaccharide fraction. Glycobiology 14:805–815
Silipo A, Molinaro A, Sturiale L, Dow JM, Erbs JM, Lanzetta R, Newman M-A, Parrilli M (2005) The elicitation of plant innate immunity by lipooligosaccharide of *Xanthomonas campestris*. J Biol Chem 280:33660–33668
Silipo A, Sturiale L, Garozzo D, Erbs G, Jensen TT, Lanzetta R, Dow JM, Parrilli M, Newman M-A, Molinaro A (2008) The acylation and phosphorylation pattern of lipid A from *Xanthomonas campestris* strongly influence its ability to trigger the innate immune response in *Arabidopsis*. Chembiochem 9:896–904
Silipo A, Erbs G, Shinya T, Dow JM, Parrilli MT, Lanzetta Shibuya N, Newman M-A, Molinaro A (2010) Glyco-conjugates as elicitors or suppressors of plant innate immunity. Glycobiology 20:406–419
Simpson SD, Ashford DA, Harvey DJ, Bowles DJ (1998) Short chain oligogalacturonides induce ethylene production and expression of the gene encoding aminocyclopropane 1-carboxylic acid oxidase in tomato plants. Glycobiology 8:579–583
Sindhu A, Chintamanani S, Brandt AS, Zanis M, Scofield SR, Johal GS (2008) A guardian of grasses: specific origin and conservation of a unique disease-resistance gene in the grass lineage. Proc Natl Acad Sci USA 105:1762–1767
Singh S, Bockus WW, Sharma I, Bowden RL (2008) A novel source of resistance in wheat to *Pyrenophora tritici-repentis* race 1. Plant Dis 92:91–95
Smart CD, Myers KL, Restrepo S, Martin GB, Fry WE (2003) Partial resistance of tomato to *Phytophthora infestans* is not dependent upon ethylene jasmonic acid or salicylic acid signaling pathways. Mol Plant Microbe Interact 16:141–148
Smet ID, Voss U, Jürgens G, Beeckman T (2009) Receptor-like kinases shape the plant. Nat Cell Biol 11:1166–1173
Snedden WA, Fromm H (2001) Calmodulin as a versatile calcium signal transducer in plants. New Phytol 151:35–66

Sohn KH, Lei R, Nemri A, Jones JDG (2007) The downy mildew effectors ATR1 and ATR13 promote disease susceptibility in *Arabidopsis thaliana*. Plant Cell 19:4077–4090

Son GH, Wan J, Kim HJ, Nguyen XC, ChungW-S HJC, Stacey G (2012) Ethylene-responsive element-binding factor 5, ERF5, is involved in chitin-induced innate immunity response. Mol Plant Microbe Interact 25:48–60

Song C, Yang B (2010) Mutagenesis of 18 type II effectors reveals virulence function of $XopZ_{Pxo99}$ in *Xanthomonas oryzae* pv. *oryzae*. Mol Plant Microbe Interact 23:893–902

Song W-Y, Wang G-L, Chen L-L, Kim H-S, Pi L-Y, Holsten T, Gardner J, Wang B, Zhai W-X, Zhu L-H, Fauquet C, Ronald P (1995) A receptor kinase-like protein encoded by the rice disease resistance gene *Xa21*. Science 270:1804–1806

Song WY, Pi L-Y, Wang GL, Gardner J, Holsten TE, Ronald PC (1997) Evolution of the rice *Xa21* disease resistance gene family. Plant Cell 9:1279–1287

Song WY, Pi L-Y, Bureau TE, Ronald PC (1998) Identification and characterization of 14 transposon-like elements in the noncoding regions of members of the *Xa21* family of disease resistance genes in rice. Mol Gen Genet 258:449–456

Soto MJ, Dominguez-Ferreras A, Pérez-Mendoza D, Sanjuan J, Olivares J (2009) Mutualism versus pathogenesis: the give-and-take in plant-bacteria interactions. Cell Microbiol 11:381–388

Soulie MC, Perino C, Piffeteau A, Choquer M, Malfatti P, Cimerman A, Kunz O, Boccara M, Vidal-Cros A (2006) *Botrytis cinerea* virulence is drastically reduced after disruption of chitin synthesis class III gene (*Bcchs3a*). Cell Microbiol 8:1310–1321

Spoel SH, Dong X (2008) Making sense of hormone crosstalk during plant immune response. Cell Host Microbe 3:348–351

Spoel SH, Koornneef A, Claessens SMC, Korzelius JP, Van Pelt JA, Mueller MJ, Buchala AJ, Metraux J-P, Brown R, Kazan K, Van Loon LC, Dong X, Pieterse CMJ (2003) NPR1 modulates cross-talk between salicylate- and jasmonate-dependent defense pathways through a novel function in the cytosol. Plant Cell 15:760–770

Spoel SH, Johnson JS, Dong X (2007) Regulation of tradeoffs between plant defenses against pathogens with different lifestyles. Proc Natl Acad Sci USA 104:18842–18847

Staats M, Van Baarlen P, Schouten A, Van Kan JAL (2007) Functional analysis of NLP genes from *Botrytis elliptica*. Mol Plant Pathol 8:209–214

Steiner-Lange S, Fischer A, Boettcher A, Rouhara I, Liedgens H, Schmelzer E, Knogge W (2003) Differential defense reactions in leaf tissues of barley in response to infection by *Rhynchosporium secalis* and to treatment with a fungal avirulence gene product. Mol Plant Microbe Interact 16:893–902

Sterigopoulos I, van den Burg HA, Okmen B, Beenen HG, van Liere S, Kema GH, de Wit PJGM (2010) Tomato Cf resistance proteins mediate recognition of cognate homologous effectors from fungi pathogenic on dicots and monocots. Proc Natl Acad Sci U S A 107:7610–7615

Stone JM, Heard JE, Asai T, Ausubel FM (2000) Simulation of fungal-mediated cell death by fumonisin B1 and selection of fumonisin B1-resistant (fbr) *Arabidopsis* mutants. Plant Cell 12:1811–1822

Strauss E (2009) Phytoplasma research begins to bloom. Science 325:388–390

Suarez-Rodriguez MC, Adams-Phillips L, Liu Y, Wang H, Su S-H, Jester PJ, Zhang S, Bent AF, Krysan PJ (2007) MEKK1 is required for flg22-induced MPK4 activation in *Arabidopsis* plants. Plant Physiol 143:661–669

Sun W, Dunning FM, Pfund C, Weingarten R, Bent AF (2006) Within-species flagellin polymorphism in *Xanthomonas campestris* pv. *campestris* and its impact on elicitation of *Arabidopsis FLAGELLIN SENSING2*-dependent defenses. Plant Cell 18:764–779

Sun F, Liu P, Xu J, Dong H (2010) Mutation in RAP26L a transactivator of the ERF transcription factor family enhances *Arabidopsis* resistance to *Pseudomonas syringae*. Physiol Mol Plant Pathol 74:295–302

Sun W, Cao Y, Labby KJ, Bittel P, Boller T, Bent AF (2012) Probing the *Arabidopsis* flagellin receptor: FLS2-FLS2 association and contributions of specific domains to signaling function. Plant Cell 24:1096–1113

Sunkar R, Zhu J-K (2004) Novel and stress-regulated microRNAs and other small RNAs from *Arabidopsis*. Plant Cell 16:2001–2019

Suzuki K, Yano A, Nishiuchi T, Nakano T, Kodama H, Yamaguchi K, Shinshi H (2007) Comprehensive analysis of early response genes to different microbial elicitors in tobacco cells. Plant Sci 173:291–301

Suzuki F, Yamaguchi J, Koba A, Nakajima T, Arai M (2010) Changes in fungicide resistance frequency and population structure of Pyricularia oryzae after discontinuance of MBI-D fungicides. Plant Dis 94:329–334

Szczesny R, Buttner B, Escolar L, Schulze S, Seiferth A, Bonas U (2010) Suppression of the AvrBs1-specific hypersensitive response by the YopJ effector homolog AvrBsT from *Xanthomonas* depends on a SNF1-related kinase. New Phytol 187:1058–1074

Szittya G, Dalmay T, Burgyan J (2008) Plant antiviral defense: gene silencing pathway. In: Encyclopedia of virology, 3rd edn. Elsevier Ltd, Amsterdam, pp 141–149

Tada Y, Kusaka K, Betsuyaku S, Shinogi T, Sakamoto M, Ohura Y, Hata S, Mori T, Tosa Y, Mayama S (2005) Victorin triggers programmed cell death and the defense response via interaction with a cell surface mediator. Plant Cell Physiol 46:1787–1798

Taguchi F, Shimizu R, Inagaki Y, Toyoda K, Shiraishi T, Ichinose Y (2003a) Post-translational modification of flagellin determines the specificity of HR induction. Plant Cell Physiol 44:342–349

Taguchi F, Shimizu R, Nakagima R, Toyoda K, Shiraishi T, Ichinose Y (2003b) Differential effects of flagellins from *Pseudomonas syringae* pv. *tabaci, tomato* and *glycinea* on plant defense response. Plant Physiol Biochem 41:165–174

Taguchi F, Takeuchi K, Katoh E, Murata K, Suzuki T, Marutani M, Kawasaki T, Eguchi M, Katoh S, Kaku H, Kaku H, Yasuda C, Inagaki Y, Toyoda K, Shiraishi T, Ichinose Y (2006) Identification of glycosylation genes and glycosylated amino acids of flagellin in *Pseudomonas syringae* pv. *tabaci*. Cell Microbiol 8:923–938

Taguchi F, Yamamoto M, Ohnishi-Kameyama M, Iwaki M, Yoshida M, Ishii T, Ichinose Y (2010) Defects in flagellin glycosylation affect the virulence of *Pseudomonas syringae* pv. *tabaci* 6605. Microbiology 156(Pt 1):72–80

Takabatake R, Karita E, Seo S, Mitsuhara I, Kuchitsu K, Ohashi Y (2007) Pathogen-induced calmodulin isoforms in basal resistance against bacterial and fungal pathogens in tobacco. Plant Cell Physiol 48:414–423

Takahashi H, Kanayama Y, Zheng MS, Kusano T, Hase S, Ikegami M, Shah J (2004) Antagonistic interactions between the SA and JA signaling pathways in Arabidopsis modulate expression of defense genes and gene-for-gene resistance to *Cucumber mosaic virus*. Plant Cell Physiol 45:803–809

Takahashi Y, Kinoshita T, Shimazaki K-I (2007) Protein phosphorylation and binding of a 14-3-3 protein in *Vicia* guard cells in response to ABA. Plant Cell Physiol 48:1182–1191

Takai R, Isogai A, Takayama S, Che F-S (2008) Analysis of flagellin perception mediated by flg22 receptor OsFLS2 in rice. Mol Plant Microbe Interact 21:1635–1642

Takakura Y, Ishida Y, Inoue Y, Tsutsumi F, Kuwata S (2004) Induction of hypersensitive response-like reaction by powdery mildew in transgenic tobacco expressing $harpin_{pss}$. Physiol Mol Plant Pathol 64:83–89

Takakura Y, Che F-S, Ishida Y, Tsutsumi F, Kuotani K-I, Usami S, Isogai A, Imaseki H (2008) Expression of a bacterial flagellin gene triggers plant immune responses and confers disease resistance in transgenic rice plants. Mol Plant Pathol 9:525–529

Takeuchi K, Taguchi F, Inagaki Y, Toyoda K, Shiraishi T, Ichinose Y (2003) Flagellin glycosylation island in *Pseudomonas syringae* pv. *glycinea* and its role in host specificity. J Bacteriol 185:6658–6665

Takeuchi K, Ono H, Yoshida M, Ishi T, Katch E, Taguchi F, Miki R, Murata K, Kaku H, Ichinose Y (2007) Flagellin glycans from two pathovars of *Pseudomonas syringae* contain rhamnose in D and L configurations in different ratios and modified 4-amino-46-dideoxyglucose. J Bacteriol 189:6945–6956

Takken FLW, Tameling WIL (2009) To nibble at plant resistance proteins. Science 324:744–745

Tampakaki AP, Skandalis N, Gazi AD, Bastaki MN, Sarris PF, Charova SN, Kokkinidis M, Panapoulos NJ (2010) Playing the "Harp" evolution of our understanding of hrp/hrc genes. Annu Rev Phytopathol 48:347–370

Tanaka N, Che F-S, Watanabe N, Fujiwara S, Takayama S, Isogai A (2003) Flagellin from an incompatible strain of *Acidovorax avenae* mediates $H_2O_2$ generation accompanying hypersensitive cell death and expression of *PALCht-1* and *PBZ1* but not of *LOX* in rice. Mol Plant Microbe Interact 16:422–428

Tang W, Kim TW, Oses-Prieto JA, Sun Y, Deng Z, Zhu S, Wang R, Burlingame AL, Wang Z-Y (2008) BSKs mediate signal transduction from the receptor kinase BRI1 in *Arabidopsis*. Science 321:557–560

Taraporewala ZF, Culver JN (1997) Structural and functional conservation of the tobamovirus coat protein elicitor active site. Mol Plant Microbe Interact 10:597–604

Taylor KR, Gallo RL (2006) Glucosaminoglycans and their proteoglycans: host-associated molecular patterns for initiation and modulation of inflammation. FASEB J 20:9–22

Tellström V, Usadel B, Thimm O, Stitt M, Küster H, Niehaus K (2007) The lipopolysaccharide of *Sinorhizobium meliloti* suppresses defense-associated gene expression in cell cultures of the host plant *Medicago truncatula*. Plant Physiol 143:825–837

Temple BR, Jones AM (2007) The plant heterotrimeric G-protein complex. Annu Rev Plant Biol 58:249–266

Thaler JS, Bostock RM (2004) Interactions between abscisic acid-mediated responses and plant resistance to pathogens and insects. Ecology 85:48–58

Thaler JS, Fidantsef AL, Bostock RM (2002) Antagonism between jasmonate- and salicylate-mediated induced plant resistance: effects of concentration and timing of elicitation on defense-related proteins herbivore and pathogen performance in tomato. J Chem Ecol 28:1131–1159

Thaler JS, Owen B, Higgins VJ (2004) The role of the jasmonate response in plant susceptibility to diverse pathogens with a range of lifestyles. Plant Physiol 135:530–538

Thao NP, Chen L, Nakashima Hara S-I, Umemura K, Takahashi A, Shirasu K, Kawasaki T, Shimamoto K (2007) RAR1 and HSP90 form a complex with Rac/ROP GTPase and function in innate-immune responses in rice. Plant Cell 19:4035–4045

Thatcher LF, Manners JM, Kazan K (2009) *Fusarium oxysporum* hijacks COI1-mediated jasmonate signaling to promote disease development in *Arabidopsis*. Plant J 58:927–939

Thilmony R, Underwood W, He SY (2006) Genome-wide transcriptional analysis of the *Arabidopsis thaliana* interaction with the plant pathogen *Pseudomonas syringae* pv. *tomato* DC3000 and the human pathogen *Escherichia coli* 0157:H7. Plant J 46:34–53

Thomma BPHJ, Eggermont K, Penninckx IAMA, Mauch-Mani B, Vogelsang R, Cammue BPA, Broekaert WF (1998) Separate jasmonate-dependent and salicylate-dependent defense-response pathways in *Arabidopsis* are essential for resistance to distinct microbial pathogens. Proc Natl Acad Sci USA 95:15107–15111

Thomma B, Penninckx I, Broekaert WF, Cammue BPA (2001a) The complexity of disease signaling in *Arabidopsis*. Curr Opin Immunol 13:63–68

Thomma B, Tierens KFM, Penninckx I, Mauch-Mani B, Broekaert WF, Cammue BPA (2001b) Different microorganisms differentially induce Arabidopsis disease response pathways. Plant Physiol Biochem 39:673–680

Thomma BPHJ, Nürnberger T, Joosten MHAJ (2011) Of PAMPs and effectors: the blurred PTI-ETI dichotomy. Plant Cell 23:4–15

Thuerig B, Felix G, Binder A, Boller T, Tamm L (2006) An extract of *Penicillium chrysogenum* elicits early defense-related responses and induces resistance in *Arabidopsis thaliana* independently of known signaling pathways. Physiol Mol Plant Pathol 67:180–193

Ton J, Mauch-Mani B (2004) β-amino-butyric acid-induced resistance is based on ABA-dependent priming for callose. Plant J 38:119–130

Tör M, Lotze M, Holton N (2009) Receptor-mediated signaling in plants: molecular patterns and programmes. J Exp Bot 60:3645–3654

Torto TA, Li S, Styer S, Huitema E, Testa A, Gow NAR, van West P, Kamoun S (2003) EST mining and functional expression assays identify extracellular effector proteins from the plant pathogen *Phytophthora*. Genome Res 13:1675–1685

Trouvelot S, Varnier A-L, Allégre M, Mercier L, Baillieul C, Arnould C, Gianinazzi-Pearson V, Klarzynski O, Joubert J-M, Pugin A, Daire X (2008) A β-13-glucan sulfate induces resistance against *Plasmopara viticola* through priming of defense responses including HR-like cell death. Mol Plant Microbe Interact 21:232–243

Tsuda K, Katagiri P (2010) Comparing signaling mechanisms engaged in pattern-triggered and effector-triggered immunity. Curr Opin Plant Biol 13:459–465

Tsuda K, Sato M, Glazebrook J, Cohen JD, Katagiri F (2008) Interplay between MAMP-triggered and SA-mediated defense responses. Plant J 53:763–775

Tsuda K, Sato M, Stoddard T, Glazebrook J, Katagiri F (2009) Network properties of robust immunity in plants. PLoS Genet 5(12):e1000772, pp16

Tyler BM (2002) Molecular basis of recognition between *Phytophthora* pathogens and their hosts. Annu Rev Phytopathol 40:137–167

Tyler BM (2009) Entering and breaking: virulence effector proteins of oomycete plant pathogens. Cell Microbiol 11:13–20

Ueeda M, Kubota M, Nishi K (2006) Contribution of jasmonic acid to resistance against *Phytophthora capsici* in *Capsicum annuum* cv SCM334. Physiol Mol Plant Pathol 67:149–154

Ulker B, Somssich IE (2004) WRKY transcription factors: from DNA binding towards biological function. Curr Opin Plant Biol 7:491–498

Umemoto N, Kakitani M, Iwamatsu A, Yoshikawa M, Yamaoka N, Ishida I (1997) The structure and function of a soybean β-glucan-elicitor binding protein. Proc Natl Acad Sci USA 94:1029–1034

Umemura K, Tanino S, Nagatsuka T, Koga J, Iwata M, Nagashima K, Amemiya Y (2004) Cerebroside elicitor confers resistance to Fusarium disease in various plant species. Phytopathology 94:813–818

Underwood W, Zhang S, He SY (2007) The *Pseudomonas syringae* type III effector tyrosine phosphatase HopAO1 suppresses innate immunity in *Arabidopsis thaliana*. Plant J 52:658–672

Uppalapati SR, Toyoda K, Yasuhiro I, Ichinose Y, Shiraishi T (2004) Differential regulation of MBP kinases by a glycoprotein elicitor and a polypeptide suppressor from *Mycosphaerella pinodes* in pea. Physiol Mol Plant Pathol 64:17–25

Uppalapati SR, Ishiga Y, Wangdi T, Kunkel BN, Anand A, Mysore KS, Bender CL (2007) The phytotoxin coronatine contributes to pathogen fitness and is required for suppression of salicylic acid accumulation in tomato inoculated with *Pseudomonas syringae* pv. *tomato* DC3000. Mol Plant Microbe Interact 20:955–965

Uppalapati SR, Marek SM, Lee H-K, Nakashima J, Tang Y, Sledge MK, Dixon RA, Mysore KS (2009) Global gene expression profiling during *Medicago truncatula-Phymatotrichopsis omnivora* interaction reveals a role for jasmonic acid ethylene and the flavonoid pathway in disease development. Mol Plant Microbe Interact 22:7–17

Valent B, Khang CH (2010) Recent advances in rice blast effector research. Curr Opin Plant Biol 13:434–441

Van Breusegem F, Bailey-Serres J, Mittler R (2008) Unraveling the tapestry of networks involving reactive oxygen species in plants. Plant Physiol 147:978–984

Van Damme EJM, Peumans WJ, Barre A, Rouge P (1998) Plant lectins: a composite of several distinct families of structurally distinct families of structurally and evolutionary related proteins and diverse biological roles. Crit Rev Plant Sci 17:575–692

Van Damme M, Huibers RP, Elberse J, Van den Ackerveken G (2008) *Arabidopsis* DMR6 encodes a putative 2OG-Fe(II)oxygenase that is defense-associated but required for susceptibility to downy mildew. Plant J 54:785–793

Van der Ent S, Verhagen BWM, Van Doorn R, Bakker D, Verlaan MG, Pel MJC, Joosten RG, Proveniers MCG, Van Loon LC, Ton J, Pieterse CMJ (2008) *MYB72* is required in early signaling steps of rhizobacteria-induced systemic resistance in *Arabidopsis*. Plant Physiol 146:1293–1304

Van der Hoorn RAL, Wulff BB, Rivas S, Durrant MC, van der Ploeg A, de Wit PJ, Jones JD (2005) Structure-function analysis of Cf9 a receptor-like protein with extracytoplasmic leucine-rich repeats. Plant Cell 17:1000–1015

Van Verk MC, Pappaaioannou D, Neeleman L, Bol JF, Linthorst HJM (2008) A novel WRKY transcription factor is required for induction of *PR-1a* gene expression by salicylic acid and bacterial elicitors. Plant Physiol 140:1983–1995

van't Slot K, Knogge W (2002) A dual role for microbial pathogen-derived effector proteins in plant disease and resistance. Crit Rev Plant Sci 21:229–271

Varnier AL, Sanchez L, Valsa P, Boudesocque L, Garcia-Brugger A, Rabenoelina F, Sorokin A, Renault JH, Kauffmann S, Pugin A, Clement C, Baillieu F, Dorey S (2009) Bacterial rhamnolipids are novel MAMPs conferring resistance to *Botrytis cinerea* in grapevine. Plant Cell Environ 32:178–193

Vatsa P, Sanchez L, Clement C, Baillieul F, Dorey S (2010) Rhamnolipid biosurfactants as new players in animal and plant defense against microbes. Int J Mol Sci 11:5095–5108

Vaughn SF, Lulai EC (1992) Further evidence that lipoxygenase activity is required for arachidonic acid-elicited hypersensitivity in potato callus cultures. Plant Sci 84:91–98

Vaucheret H (2006) Post-transcriptional small RNA pathways in plants; mechanisms and regulations. Genes Dev 20:759–771

Vaucheret H, Mallory AC, Bartel DP (2006) AGO1 homeostasis entails coexpression of MIR168 and AGO1 and preferential stabilization of miR168 by AGO1. Mol Cell 22:129–136

Veit S, Worle JM, Nürnberger T, Koch W, Seitz HU (2001) A novel protein elicitor (PaNie) from *Pythium aphanidermatum* induces multiple defense responses in carrot, Arabidopsis, and tobacco. Plant Physiol 127:832–841

Vembar SS, Brodsky JL (2008) One step at a time: endoplasmic reticulum associated degradation. Nat Rev Mol Cell Biol 9:944–957

Veronese P, Nakagami H, Bluhm B, Abuqamar S, Chen X, Salmeron J, Dietrich RA, Hirt H, Mengiste T (2006) The membrane-anchored BOTRYTIS -INDUCED KINASE 1 plays distinct roles in Arabidopsis resistance to necrotrophic and biotrophic pathogens. Plant Cell 18:257–273

Véronési C, Delavault P, Simier P (2008) Acibenzolar-*S*-methyl induces resistance in oilseed rape (*Brassica napus* L) against branched broomrape (*Orobanche ramosa* L). Crop Prot 28:104–108

Vidal S, ErikksonARB MM, Dencke J, Palva T (1998) Cell wall-degrading enzymes from *Erwinia carotovora* cooperate in the salicylic acid-independent induction of a plant defence response. Mol Plant Microbe Interact 11:23–32

Vidhyasekaran P (2004) Concise encyclopedia of plant pathology. Haworth Press, Binghamton/New York, p 619

Vidhyasekaran P (2007a) Fungal pathogenesis in plants and crops: molecular biology and host defense mechanisms, 2nd edn. CRC Press/Taylor & Francis Group, Boca Raton, p 510

Vidhyasekaran P (2007b) Handbook of molecular technologies in crop disease management. Haworth Press, Binghamton, 462

Vijayan P, Shockey J, Levesque CA, Cook RJ, Browse J (1998) A role for jasmonate in pathogen defence of Arabidopsis. Proc Natl Acad Sci USA 95:7209–7214

Villalba Mateos F, Rickauer M, Esquerré-Tugayé MT (1997) Cloning and characterization of a cDNA encoding an elicitor of *Phytophthora parasitica* var *nicotianae* that shows cellulose-binding and lectin-like activities. Mol Plant Microbe Interact 10:1045–1053

Viterbo A, Wiest A, Brotman Y, Chet I, Kenerley C (2007) The 18mer peptaibols from *Trichoderma virens* elicit plant defense responses. Mol Plant Pathol 8:737–746

Vleeshouwers VGAA, Rietman H, Krenek P, Champouret N, Young C, Oh S-K, Wang M, Bouwmeester K, Vosman B, Visser RGF, Jacobsen E, Govers F, Kamoun S, Van der Vossen EAG (2008) Effector genomics accelerates discovery and functional profiling of potato disease resistance and *Phytophthora infestans* avirulence genes. PLoS One 3(8):e2875

Vogeli U, Vogeli-Lange R, Chappell J (1992) Inhibition of phytoalexins biosynthesis in elicitor-treated tobacco cell-suspension cultures by calcium/calmodulin antagonists. Plant Physiol 100:1369–1376

Voinnet O (2001) RNA silencing as a plant immune system against viruses. Trends Genet 17:449–459

Vollmer W (2008) Structural variation in the glycan strands of bacterial peptidoglycan. FEMS Microbiol Rev 32:287–306

Vollmer W, Born P (2009) Bacterial cell wall envelope peptidoglycan. In: Moran A, Brennan P, Holst O, von Itszstein M (eds) Microbial glycobiology: structures relevance and applications. Elsevier, San Diego, pp 15–28

Von Numers N, Survila M, Aalto M, Batoux M, Heino P, Palva ET, Li J (2010) Requirement of a homolog of glucosidase II β-subunit for EFR-mediated defense signaling in *Arabidopsis thaliana*. Mol Plant 3:740–750

Von Zastrow M, Sorkin A (2007) Signaling on the endocytic pathway. Curr Opin Cell Biol 19:436–445

Waller F, Müller A, Chung K-M, Yap Y-K, Nakamura K, Weiler E, Sano H (2006) Expression of a WIPK-activated transcription factor results in increase of endogenous salicylic acid and pathogen resistance in tobacco plants. Plant Cell Physiol 47:1169–1174

Walters DR, Cowley T, Weber H, Nashaat NI (2006) Changes in oxylipins in compatible and incompatible interactions between oilseed rape and the downy mildew pathogen *Peronospora parasitica*. Physiol Mol Plant Pathol 67:268–273

Wan J, Zhang S, Stacey G (2004) Activation of a mitogen-activated protein kinase pathway in *Arabidopsis* by chitin. Mol Plant Pathol 5:125–135

Wan J, Zhang X-C, Neece D, Ramonell KM, Clough S, Kim S-Y, Stacey MG, Stacey G (2008a) A LysM receptor-like kinase plays a critical role in chitin signaling and fungal resistance in *Arabidopsis*. Plant Cell 20:471–481

Wan J, Zhang X-C, Stacey G (2008b) Chitin signaling and plant disease resistance. Plant Signal Behav 3:831–833

Wang W, Jones C, Ciacci-Zanella J, Holt T, Gilchrist DG, Dickman MB (1996) Fumonisins and *Alternaria alternata lycopersici toxins*: sphinganine analog mycotoxins induce apoptosis in monkey kidney cells. Proc Natl Acad Sci USA 93:3461–3465

Wang ZY, Seto H, Fujioka S, Yoshida S, Chory J (2001) BRI1 is a critical component of plasma membrane receptor for plant steroids. Nature 410:380–383

Wang JY, Cai Y, Gou JY, Mao YB, Xu YH, Jiang WH, Chen XY (2004) VdNEP an elicitor from *Verticillium dahliae* induces cotton plant wilting. Appl Environ Microbiol 70:4989–4995

Wang X, Li X, Meisenhelder J, Hunter T, Yoshida S, Asami T, Chory J (2005) Autoregulation and homodimerization are involved in the activation of the plant steroid receptor, BRI1. Dev Cell 8:855–865

Wang Y-S, Pi L-Y, Chen X, Chakrabarty PK, Jiang J, De Leon AL, Liu G-Z, Li L, Benny U, Oard J, Ronald PC, Song W-Y (2006) Rice XA21 binding protein 3 is a ubiquitin ligase required for full *Xa21*-mediated disease resistance. Plant Cell 18:3635–3646

Wang X, Kota U, He K, Blackburn K, Li J, Goshe MB, Huber SC, Clouse SD (2008a) Sequential transphorylation of the BRI1/BAK1 receptor kinase complex impacts early events in brassinosteroid signaling. Dev Cell 15:220–235

Wang Y, Gao M, Li Q, Wang L, Jeon J-S, Qu N, Zhang Y, He Z (2008b) *OsRAR1* and *OsSGT1* physically interact and function in rice basal disease resistance. Mol Plant Microbe Interact 21:294–303

Wang L, Tsuda K, Sato M, Cohen JD, Katagiri F, Glazebrook J (2009a) Arabidopsis CaM binding protein CBP60g contributes to MAMP-induced SA accumulation and is involved in disease resistance against *Pseudomonas syringae*. PLoS Pathog 5(2):e1000301. doi:101371/journalppat1000301

Wang Z, Mao H, Dong C, Ji R, Cai L, Fu H, Liu S (2009b) Overexpression of *Brassica napusMPK4* enhances resistance to *Sclerotinia sclerotiorum* in oilseed rape. Mol Plant Microbe Interact 22:235–244

Wang G, Fiers M, Ellendorff U, Wang Z, deWit PJGM, Angenent GC, Thomma BPHJ (2010a) The diverse roles of extracellular leucine-rich repeat-containing receptor-like proteins in plants. Crit Rev Plant Sci 29:285–299

Wang X-B, Wu Q, Ito T, Cillo F, Li W-X, Chen X, Yu J-L, Ding S-W (2010b) RNAi-mediated viral immunity requires amplification of virus-derived siRNAs in *Arabidopsis thaliana*. Proc Natl Acad Sci USA 107:484–489

Wang Y, Li J, Hou S, Wang X, Li Y, Ren D, Chen S, Tang X, Zhou J-M (2010c) A *Pseudomonas syringae* ADP-ribosyltransferase inhibits *Arabidopsis* mitogen-activated protein kinase kinases. Plant Cell 22:2033–2044

Watt SA, Wike A, Patschkowski T, Niehaus K (2005) Comprehensive analysis of the extracellular proteins from *Xanthomonas campestris* pv. *campestris* B100. Proteomics 5:153–167

Watt SA, Tellstrom V, Patschkowski T, Niehaus K (2006) Identification of the bacterial superoxide dismutase (SodM) as plant-inducible elicitor of the oxidative burst reaction in tobacco cell suspension cultures. J Biotechnol 126:78–86

Wei ZM, Laby RJ, Zumoff CH, Bauer DW, He SY, Collmer A, Beer SV (1992) Harpin elicitor of the hypersensitive response produced by the plant pathogen *Erwinia amylovora*. Science 257:85–88

Wen W, Wang J (2001) Cloning and expressing a harpin gene from *Xanthomonas oryzae* pv. *oryzae*. Acta Phytopathol Sin 31:296–300

Williams DB (2006) Beyond lectins: the calnexin/calreticulin chaperone system of the endoplasmic reticulum. J Cell Sci 119:615–623

Willmann MR, Endres MW, Cook RT, Gregory BD (2011) The functions of RNA-dependent RNA polymerases in *Arabidopsis*. Arabidopsis Book 9:e0146. doi:10:1199/tab0146

Wilson DR, Beveridge TJ (1993) Bacterial flagellar filaments and their component flagellins. Can J Microbiol 39:451–472

Wilson ID, Neill SJ, Hancock JT (2008) Nitric oxide synthesis and signalling in plants. Plant Cell Environ 31:622–631

Wilton M, Subramaniam R, Elmore J, Felsensteiner C, Coaker G, Desveaux D (2010) The type III effector HopF2Pto targets Arabidopsis RIN4 protein to promote *Pseudomonas syringae* virulence. Proc Natl Acad Sci USA 106:8067–8072

Wroblewski T, Caldwell KS, Piskurewicz U, Cavanaugh KA, Xu H, Kozik A, Ochoa O, McHale LK, Lahre K, Jelenska J, Castillo JA, Blumenthal D, Vinatzer BA, Greenberg JT, Michelmore RW (2009) Comparative large-scale analysis of interactions between several crop species and the effector repertoires from multiple pathovars of *Pseudomonas* and *Ralstonia*. Plant Physiol 150:1733–1749

Wu T, Guo A, Zhao Y, Wang X, Wang Y, Zhao D, Li X, Ren H, Dong H (2010) Ectopic expression of the rice lumazine synthase gene contributes to defense responses in transgenic tobacco. Phytopathology 100:573–581

Wu S, Lu D, Kabbage M, Wei H-L, Swingle B, Records AR, Dickman M, He P, Shan L (2011) Bacterial effector HopF2 suppresses *Arabidopsis* innate immunity at the plasma membrane. Mol Plant Microbe Interact 24:585–593

Wypijewski K, Hornyik C, Shaw JA, Stephens J, Goraczniak R, Gunderson SI, Lacomme C (2009) Ectopic 5′ splice sites inhibit gene expression by engaging RNA surveillance and silencing pathways in plants. Plant Physiol 151:955–965

Xiang T, Zong N, Zou Y, Wu Y, Zhang J, Xing W, Li Y, Tang X, Zhu L, Chai J, Zhou J-M (2008) *Pseudomonas syringae* effector AvrPto blocks innate immunity by targeting receptor kinases. Curr Biol 18:74–80

Xiang T, Zong N, Zhang J, Chen J, Chen M, Zhou J-M (2011) BAK1 is not a target of the *Pseudomonas syringae* effector AvrPto. Mol Plant Microbe Interact 24:100–107

Xiao CL, Boal RJ (2009) Preharvest application of a boscalid and pyraclostrobin mixture to control post-harvest gray mold and blue mold in apples. Plant Dis 93:185–189

Xie Z, Johansen LK, Gustafson AM, Kasschau KD, Lellis AD, Ziberman D, Jacobsen SE, Carrington JC (2004) Genetic and functional diversification of small RNA pathways. PLoS Biol 2:E104

Xie Z, Allen E, Wilken A, Carrington JC (2005) DICER-LIKE 4 functions in trans-acting small interfering RNA biogenesis and vegetative phase change in *Arabidopsis thaliana*. Proc Natl Acad Sci USA 102:12984–12989

Xie Z, Khanna K, Ruan S (2010) Expression of microRNAs and its regulation in plants. Semin Cell Dev Biol 21:790–797

Xing D, Chen Z (2006) Effects of mutations and constitutive overexpression of *EDS1* and *PAD4* on plant resistance to different types of microbial pathogens. Plant Sci 171:251–262

Xing T, Jordan M (2000) Genetic engineering of plant signal transduction mechanisms. Plant Mol Biol Rep 18:309–318

Xing DH, Lai ZB, Zheng ZY, Vinod KM, Fan BF, Chen ZX (2008) Stress- and pathogen-induced *Arabidopsis* WRKY48 is a transcriptional activator that represses plant basal defense. Mol Plant 1:459–470

Xu WH, Wang YS, Liu GZ, Chen X, Tinjuangjun PK, Pi LY, Song WY (2006a) The autophosphorylated Ser686 Thr688 and Ser689 residues in the intracellular juxtamembrane domain of XA21 are implicated in stability control of rice receptor-like kinase. Plant J 45:740–751

Xu X, Chen C, Fan B, Chen Z (2006b) Physical and functional interactions between pathogen-induced *Arabidopsis* WRKY18, WRKY40 and WRKY60 transcription factors. Plant Cell 18:1310–1326

Yakushiji S, Ishiga Y, Inagaki Y, Toyoda K, Shiraishi T, Ichinose Y (2009) Bacterial DNA activates immunity in *Arabidopsis thaliana*. J Gen Plant Pathol 75:227–234

Yamaguchi T, Yamada A, Hong N, Ogawa T, Ishii T, Shibuya N (2000a) Differences in the recognition of glucan elicitor signals between rice and soybean: β-glucan fragments from the rice blast disease fungus *Pyricularia oryzae* that elicit phytoalexin biosynthesis in suspension-cultured rice cells. Plant Cell 12:817–826

Yamaguchi T, Ito Y, Shibuya N (2000b) Oligosaccharide elicitors and their receptors for plant defense responses. Trends Glyosci Glyotechnol 12:113–120

Yamaguchi T, Maehara Y, Kodama O, Matsumura M, Shibuya N (2002) Two purified oligosaccharide elicitors N-acetylchitoheptaose and tetraglucosyl glucitol derived from *Magnaporthe grisea* cell walls synergistically activate biosynthesis of phytoalexin in suspension-cultured rice cells. J Plant Physiol 159:1147–1149

Yamaguchi Y, Pearce G, Ryan CA (2006) The cell surface leucine-rich repeat receptor for AtPep1 an endogenous elicitor in *Arabidopsis* is functional in transgenic tobacco cells. Proc Natl Acad Sci USA 103:10104–10109

Yamaguchi Y, Huffaker A, Bryan AC, Tax FE, Ryan CA (2010) PEPR2 is a second receptor for the Pep1 and Pep2 peptides and contributes to defense responses in *Arabidopsis*. Plant Cell 22:508–522

Yamamizo C, Kuchimura K, Kobayashi A, Katou S, Kawakita K, Jones JDG, Doke N, Yoshioka H (2006) Rewiring mitogen-activated protein kinase cascade by positive feedback confers potato blight resistance. Plant Physiol 140:681–692

Yazawa K, Jiang C-J, Kojima M, Sakakibara H, Takatsuji H (2012) Reduction of abscisic acid signaling in rice during the early phase of *Magnaporthe oryzae* infection decreases its susceptibility to the fungus. Physiol Mol Plant Pathol 78:1–7

Yi SY, Kim JH, Joung YH, Lee S, Kim WT, Yu SH, Choi D (2004) The pepper transcription factor *CaPF1* confers pathogen and freezing tolerance in *Arabidopsis*. Plant Physiol 136:2862–2874

Yoo B-C, Kragler F, Varkonyi-Gasic E, Haywood V, Archer-Evans S, Lee YM, Lough TJ, Lucas WJ (2004) A systemic small RNA signaling system in plants. Plant Cell 16:1979–2000

Yoo SD, Cho Y-H, Tena G, Xiong Y, Sheen J (2008) Dual control of nuclear EIN3 by bifurcate MAPK cascades in $C_2H_4$ signalling. Nature 451:789–795

Yoshioka H, Numata N, Nakajima K, Katou S, Kawakita K, Rowland O, Jones JD, Doke N (2003) *Nicotiana benthamiana* gp91$^{phox}$ homologs *NbrbohA* and *NbrbohB* participate in $H_2O_2$ accumulation and resistance to *Phytophthora infestans*. Plant Cell 15:706–718

Yu D, Chen C, Chen Z (2001) Evidence for an important role of WRKY DNA binding proteins in the regulation of *NPR1* gene expression. Plant Cell 13:1527–1540

Yuan ZC, Edlind MP, Liu P, Saenkham P, Banta LM, Wise AA, Ronzone E, Binns AN, Kerr K, Nester EW (2007) The plant signal salicylic acid shuts down expression of the *vir* regulon and activates quormone-quenching genes in *Agrobacterium*. Proc Natl Acad Sci USA 104:11790–11795

Zago E, Morsa S, Dat JF, Alard P, Ferrarini A, Inzé D, Delledonne M, Van Breusegem F (2006) Nitric oxide- and hydrogen peroxide-responsive gene regulation during cell death induction in tobacco. Plant Physiol 141(2):404–411

Zamioudis C, Peterse CMJ (2012) Modulation of host immunity by beneficial microbes. Mol Plant Microbe Interact 25:139–150

Zeidler D, Zähringer U, Gerber I, Hartung T, Bors W, Hutzler P, Durner J (2004) Innate immunity in *Arabidopsis thaliana*: lipopolysaccharides activate nitric oxide synthase NOS and induce defense genes. Proc Natl Acad Sci USA 101:15811–15816

Zeidler D, Dubery IA, Schmitt-Kopplin P, Von Rad U, Durner J (2010) Lipopolysaccharide mobility in leaf tissue of *Arabidopsis thaliana*. Mol Plant Pathol 11:747–755

Zeng W, He S-Y (2010) A prominent role of the flagellin receptor FLAGELLIN-SENSING2 in mediating stomatal response to *Pseudomonas syringae* pv. *tomato* DC3000 in *Arabidopsis*. Plant Physiol 153:1188–1198

Zeng Q, Wang X, Running MP (2007) Dual lipid modification of Arabidopsis Gγ-subunits is required for efficient plasma membrane targeting. Plant Physiol 143:1119–1131

Zhang X, Mou Z (2009) Extracellular pyridine nucleotides induce PR gene expression and disease resistance in *Arabidopsis*. Plant J 57:302–312

Zhang J, Zhou J-M (2010) Plant immunity triggered by microbial molecular signatures. Mol Plant 3:783–793

Zhang B, Ramonell K, Somerville S, Stacey G (2002a) Characterization of early chitin-induced gene expression in *Arabidopsis*. Mol Plant Microbe Interact 15:963–970

Zhang DP, Wu ZY, Li XY, Zhao ZX (2002b) Purification and identification of a 42-kilodalton abscisic acid-specific-binding protein from epidermis of broad bean leaves. Plant Physiol 128:714–725

Zhang L, Liu BF, Liang S, Jones RL, Lu YT (2002c) Molecular and biochemical characterization of a calcium/calmodulin-binding protein kinase from rice. Biochem J 368:145–167

Zhang HK, Zhang X, Mao BZ, Li Q, He ZH (2004a) Alpha-picolinic acid a fungal toxin and mammal apoptosis-inducing agent elicits hypersensitive-like response and enhances disease resistance in rice. Cell Res 14:1–7

Zhang W, Qin C, Zhao J, Wang X (2004b) Phospholipase Dα1-derived phosphatidic acid interacts with ABI1 phosphatase 2C and regulates abscisic acid signaling. Proc Natl Acad Sci USA 101:9508–9513

Zhang Z, Henderson C, Gurr SJ (2004c) *Blumeria graminis* secretes an extracellular catalase during infection of barley: potential role in suppression of host defence. Mol Plant Pathol 5:537–547

Zhang Z-G, Wang C, Li J, Ji R, Shen G, Wang S-C, Zhou X, Zheng X-B (2004d) The role of SA in the hypersensitive response and systemic acquired resistance induced by elicitor PB90 from *Phytophthora boehmeriae*. Physiol Mol Plant Pathol 65:31–38

Zhang Y, Guenzi AC, Anderson MP, Taliaferro CM, Gonzales RA (2006) Enrichment of bermudagrass genes associated with tolerance to the spring dead spot fungus *Ophiosphaerella herpotricha*. Physiol Mol Plant Pathol 68:105–118

Zhang J, Shao F, Li Y, Cui H, Chen L, Li H, Zou Y, Long C, Lan L, Chai J, Chen S, Tang X, Zhou J-M (2007a) A *Pseudomonas syringae* effector inactivates MAPKs to suppress PAMP-induced immunity in plants. Cell Host Microbe 1:175–185

Zhang X-C, Wu X, Findley S, Wan J, Libault M, Nguyen HT, Cannon SB, Stacey G (2007b) Molecular evolution of LysM type receptor-like kinases in plants. Plant Physiol 144:623–636

Zhang Z, Li Q, Staswick PE, Wang M, Zhu Y, He Z (2007c) Dual regulation role of *GH35* in salicylic acid and auxin signaling Arabidopsis-*Pseudomonas syringae* interaction. Plant Physiol 145:450–464

Zhang Z, Yao W, Dong N, Liang H, Liu H, Huang R (2007d) A novel ERF transcription activator in wheat and its induction kinetics after pathogen and hormone treatments. J Exp Bot 58:2993–3003

Zhang W, Chen J, Zhang H, Song F (2008a) Overexpression of a rice diacylglycerol kinase gene *OsBIDK1* enhances disease resistance in transgenic tobacco. Mol Cells 26:258–264

Zhang W, He SY, Assmann SM (2008b) The plant innate immunity response in stomatal guard cells invokes G-protein-dependent ion channel regulation. Plant J 56:984–996

Zhang Z, Lenk A, Andersson MX, Gjetting T, Pedersen C, Nielsen ME, Newmann M-A, Hou B-H, Somerville SC, Thordal-Christensen H (2008c) A lesion-mimic double mutant in *Arabidopsis* reveals novel complexity of pathogen defense signaling. Mol Plant 1:510–527

Zhang X, Cannon SB, Stacey G (2009a) Evolutionary genomics of LysM genes in land plants. BMC Evol Biol 9:183

Zhang YJ, Fan PS, Zhang X, Chen CJ, Zhou MG (2009b) Quantification of *Fusarium graminearum* in harvested grain by real-time polymerase chain reaction to assess efficacies of fungicides on Fusarium head blight deoxynivalenol contamination and yield of winter wheat. Phytopathology 99:95–100

Zhang J, Li W, Xiang J, Liu Z, Laluk K, Ding X, Zou Y, Gao M, Zhang X, Chen S, Mengiste T, Zhang Y, Zhou J-M (2010a) Receptor-like cytoplasmic kinases integrate signaling from multiple plant immune receptors and are targeted by a *Pseudomonas syringae* effector. Cell Host Microbe 7:290–301

Zhang J, Lu H, Li X, Li Y, Cui H, Wen C-K, Tang X, Su Z, Zhou J-M (2010b) Effector-triggered and PAMP-triggered immunity differentially contribute to basal resistance to *Pseudomonas syringae*. Mol Plant Microbe Interact 23:940–948

Zhang X, Francis MI, Dawson WO, Graham JH, Orbovic V, Triplett EW, Mou Z (2010c) Overexpression of the *ArabidopsisNPR1* gene in citrus increases resistance to citrus canker. Eur J Plant Pathol 128:91–100

Zhang W, Gao S, Zhou X, Chellappan P, Chen Z, Zhou X, Zhang X, Fromuth N, Coutino G, Coffey M, Jin N (2011) Bacteria-responsive microRNAs regulate plant innate immunity by modulating plant hormone networks. Plant Mol Biol 175:93–105

Zhao S, Qi X (2008) Signaling in plant disease resistance and symbiosis. J Integr Plant Biol 50:799–807

Zheng Z, Qamar SA, Chen Z, Mengiste T (2006) *Arabidopsis* WRKY33 transcription factor is required for resistance to necrotrophic fungal pathogens. Plant J 48:596–605

Zheng Z, Mosher SL, Fan B, Klessig DF, Chen Z (2007) Functional analysis of *Arabidopsis* WRKY25 transcription factor in plant defense against *Pseudomonas syringae*. BMC Plant Biol 7:2

Zhou J-M, Chai JJ (2008) Plant pathogenic bacterial type III effectors subdue host responses. Curr Opin Microbiol 11:179–185

Zhou C, Zhang L, Duan J, Miki B, Wu K (2005) Histone deacetylase19 is involved in jasmonic acid and ethylene signaling of pathogen response in *Arabidopsis*. Plant Cell 17:1196–1204

Zhou F, Mosher S, Tian M, Sassi G, Parker J, Klessig DF (2008) The *Arabidopsis* gain-of-function mutant *ssi4* requires *RAR1* and *SGT1b* differentially for defense activation and morphological alterations. Mol Plant Microbe Interact 21:40–49

Zhu H, Li GJ, Ding L, Cui X, Berg H, Assmann SM, Xia Y (2009) Arabidopsis extra large G-protein 2 (XLG2) interacts with the Gβ subunit of heterotrimeric G protein and functions in disease resistance. Mol Plant 2:513–525

Zhu L, Li Y, Li L, Yang J, Zhang M (2011) Ethylene is involved in leafy mustard systemic resistance to *Turnip mosaic virus* infection through the mitochondrial alternative oxidase pathway. Physiol Mol Plant Pathol 76:166–172

Zimmerli L, Metraux J-P, Mauch-Mani B (2001) β-Aminobutyric acid-induced protection of *Arabidopsis* against the necrotrophic fungus *Botrytis cinerea*. Plant Physiol 126:517–523

Zimmerli L, Stein M, Lipka V, Schulze-Lefert P, Somerville S (2004) Host and non-host pathogens elicit different jasmonate/ethylene responses in *Arabidopsis*. Plant J 40:633–646

Zimmermann P, Hirsch-Hoffmann M, Hennig L, Gruissem W (2004) GENEVESTIGATOR: Arabidopsis microarray database and analysis toolbox. Plant Physiol 136:2621–2632

Zipfel C (2008) Pattern-recognition receptors in plant innate immunity. Curr Opin Immunol 20:10–16

Zipfel C (2009) Early molecular events in PAMP-triggered immunity. Curr Opin Plant Biol 12:414–420

Zipfel C, Felix G (2005) Plants and animals: a different taste for microbes? Curr Opin Plant Biol 8:353–360

Zipfel C, Rathjen P (2008) Plant immunity: AvrPto targets the frontline. Curr Biol 18:R218–R220

Zipfel C, Robatzek S, Navarro L, Oakeley EJ, Jones JDG, Felix G, Boller T (2004) Bacterial disease resistance in *Arabidopsis* through flagellin perception. Nature 428:764–767

Zipfel C, Kunze G, Chinchilla D, Caniard A, Jones JD, Boller T, Felix G (2006) Perception of the bacterial PAMP EF-Tu by the receptor EFR restricts *Agrobacterium*-mediated transformation. Cell 125:749–760

Zong N, Xiang T, Zou Y, Chai J, Zhou JM (2008) Blocking and triggering of plant immunity by *Pseudomonas syringae* effector AvrPto. Plant Signal Behav 3:583–585

Zuo K-J, Qin J, Zhao J-Y, Ling H, Zhang L-D, Cao Y-F, Tang K-X (2007) Over-expression of *GbERF2* transcription factor in tobacco enhances brown spots disease resistance by activating expression of downstream genes. Gene 391:80–90

# Chapter 3
# G-Proteins as Molecular Switches in Signal Transduction

**Abstract** Guanosine triphosphate (GTP)-binding proteins (G-proteins) are the regulatory GTPases that have the ability to bind GTP and hydrolyze it to guanosine diphosphate (GDP). GDP locks G proteins into their inactive state, while GTP locks G-proteins into their activated state. Active or inactive states of G-proteins depend on the binding of GTP or GDP, respectively. G-proteins have been found to be key players in plant innate immunity. The GTPases act as molecular switches controlling the transmission of extracellular signals like pathogen-associated molecular patterns (PAMPs) to intracellular signaling pathways. The PAMPs have been shown to activate GTP binding to G-protein. The GTPase is normally inactive. The PAMP stimulates exchange of GTP for GDP and thus converts the G-proteins from their inactive state to their active state. Upon stimulation by an upstream PAMP signal, a guanine nucleotide exchange factor (GEF) converts the GDP-bound inactive form into the GTP-bound active form through GDP/GTP replacement. Through its effector domain, the GTP form interacts with specific downstream effector proteins. The GTP form exhibits a weak intrinsic GTPase activity for GTP hydrolysis, requiring a GTPase-activating protein (GAP) for efficient deactivation. Most small GTPases cycle between membrane-bound and cytosolic forms. Only membrane-associated GTPases can be activated by GEF and their removal by a cytosolic factor called guanine nucleotide dissociation inhibitor (GDI) negatively regulates these GTPases.

G-proteins include two major subfamilies: heterotrimeric G-proteins and small G-proteins (also called small GTPases). The heterotrimeric G-proteins contain Gα-, Gβ-, and Gγ- subunits. The small G-proteins are monomeric G-proteins and they appear to be similar to α-subunits, operating without the β-, and γ-subunits. Both classes of G-proteins use the GTP/GDP cycle as a molecular switch for signal transduction. Both heteromeric and monomeric small G-proteins trigger immune responses by activating several immune signaling systems. These include $Ca^{2+}$ channel activation, $K^+$ channel regulation, generation of reactive oxygen species through activation of NADPH oxidase, regulation of redox signaling, activation of nitric oxide (NO) signaling system, activation of mitogen-activated protein kinase (MAPK) signaling cascade, activation of phospholipases, efflux of vacuolar $H^+$, biosynthesis

P. Vidhyasekaran, *PAMP Signals in Plant Innate Immunity: Signal Perception and Transduction*, Signaling and Communication in Plants 21,
DOI 10.1007/978-94-007-7426-1_3, © Springer Science+Business Media Dordrecht 2014

of polyamines, biosynthesis of phosphatidic acid and programmed cell death. G-proteins also activate various plant hormone signaling systems including salicylic acid-, jasmonic acid-, ethylene-, abscisic acid-, auxin-, brassinosteroid-, and gibberellic acid- mediated signaling systems. The different subunits in heterotrimeric G-proteins and the monomeric small G-proteins may behave differently in activating defense responses against various pathogens. Ability of G-proteins to trigger immune responses also varies depending upon the type of invading pathogen.

**Keywords** GTPases • Molecular switches • PAMP signal • G-proteins • GTPase activating protein • GDI • Heterotrimeric G-proteins • Small G-proteins

## 3.1 G-Proteins Switch on Plant Innate Immunity Signaling Systems

Guanosine triphosphate (GTP)-binding proteins (G-proteins) are the regulatory GTPases, which act as molecular switches in signal transduction system (Gilman 1987; Cabrera-Vera et al. 2003; Nibau et al. 2006; Zeng et al. 2007; Fujiwara et al. 2009; Yalovsky et al. 2010; Zhang et al. 2011, 2012). Plant cells contain many G-proteins (Xing et al. 1997; Roos et al. 1999; Suharsono et al. 2002; Morel et al. 2004; Zeng et al. 2007; Yong et al. 2010; Zhang et al. 2012). Two classes of signaling G-proteins have been reported. These include heterotrimeric G-proteins and small monomeric G-proteins (Ras/Ras-like small GTPases) (Gu et al. 2004; Perfus-Barboch et al. 2004). In the Ras superfamily of small GTPases, only the Ras and Rho families have been shown to transmit extracellular signals (Gu et al. 2004). Ras superfamily is named the Ras superfamily because the founding members are encoded by human Ras genes initially discovered as cellular homologs of the viral *ras* oncogene. Plants do not possess a true Ras GTPase such as those that are pivotal signaling in animals. Instead, they have a unique subfamily of Rho-family GTPases, called ROPs (Rho-related GTPase of plants). ROP is the sole subfamily of Rho GTPase in plants. ROPs are also referred to as RAC (for Ras [rat sarcoma oncogene product] related C3 botulinum toxin substrate) proteins (Gu et al. 2004; Kiirika et al. 2012). RAC/ROP small GTPases share a common ancestor with Rho, cdc42 and Rac and they are the only Rho-like GTPases in plants (Gu et al. 2004).

Many studies using inhibitors and agonists of G-proteins in different plant species have suggested that G-proteins are involved in defense signaling initiated by pathogen-associated molecular patterns (PAMPs)/elicitors (Legendre et al. 1992; Beffa et al. 1995; Gelli et al. 1997; Ono et al. 2001; Park et al. 2000; Zhang et al. 2011, 2012). Transgenic tobacco plants that expressed an antisense construct derived from a *Medicago sativa* cDNA encoding a G-protein failed to show disease resistance-related hypersensitive reaction when infiltrated with an elicitor (Schiene et al. 2000). It suggests that the G-protein is involved in defense signaling.

Both heteromeric G-proteins and small G-proteins (RAC/ROP small GTPases) play an important role in activating various signal transduction systems initiated by PAMPs (Fig. 3.1; Gao et al. 2010a, b; Kiirika et al. 2012). $Ca^{2+}$ is a master regulator

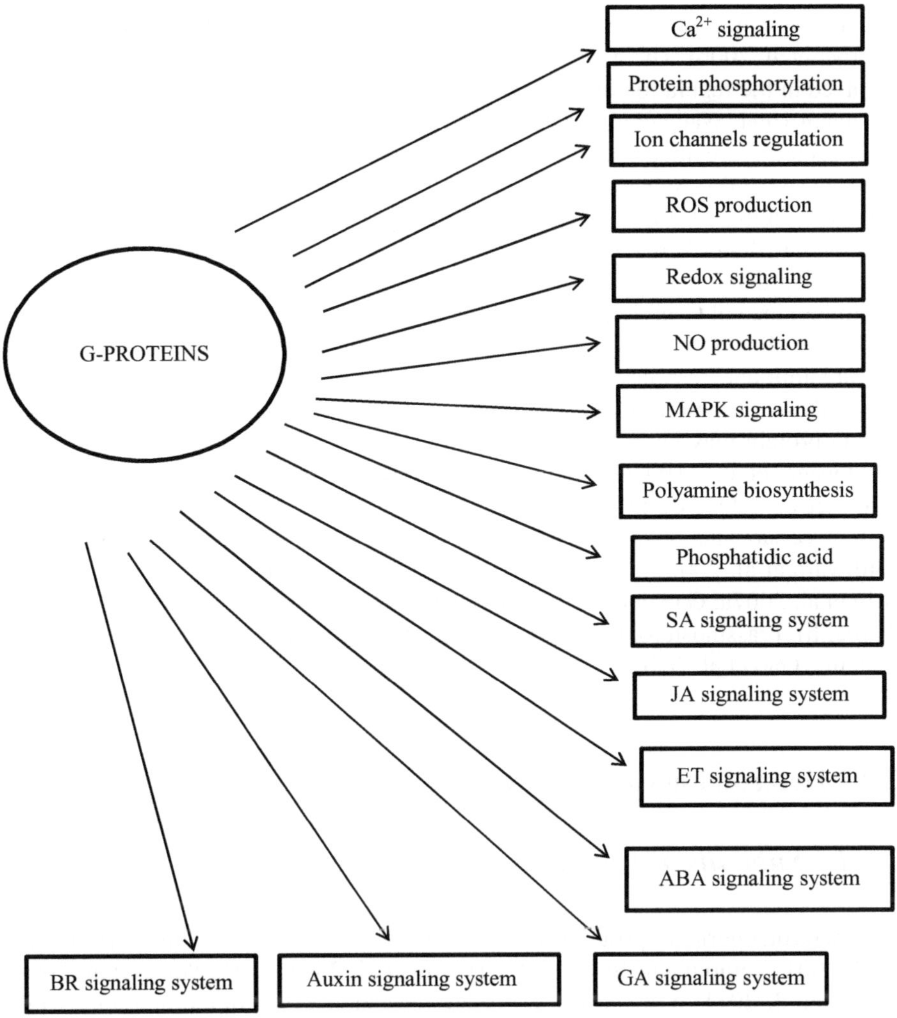


**Fig. 3.1** G-proteins-triggered signaling systems

of gene expression in plants (Galon et al. 2010) and it acts as intracellular second messenger that is used by plants to encode information and deliver it downstream to proteins which decode/interpret signals and initiate defense responses (Abdul Kadar and Lindsberg 2010; DeFalco et al. 2010; Dodd et al. 2010; Stael et al. 2012). G-proteins trigger changes in cytosolic $Ca^{2+}$ concentrations (Schultheiss et al. 2003). The G-proteins induce $Ca^{2+}$ channel opening in plants through the action of PAMPs (Gelli et al. 1997). Protein phosphorylation precedes $Ca^{2+}$ influx in tobacco cells treated with a PAMP isolated from the oomycete pathogen *Phytophthora cryptogea* (Tavernier et al. 1995). The G-proteins modulate the phosphorylation/dephosphorylation system in the plasma membrane of tomato cells and transduce the signal (Vera-Estrella et al. 1994a). Phosphorylation of proteins involved in G-protein coupled signaling has been reported in tobacco cells treated with a bacterial PAMP (Gerber et al. 2006). The *Arabidopsis* G-protein GPA1 has been demonstrated to be involved in the regulation of inward $K^+$ channels and slow anion channels (Wu and Assmann 1994; Wang et al. 2001; Zhang et al. 2008).

G-proteins are involved in PAMP-activated ROS-mediated signaling system (Park et al. 2000; Suharsono et al. 2002). The PAMP flg22 induces G-protein-activated ROS signaling systems. The gene *AGB1*, encoding the β-subunit of G-protein in *Arabidopsis*, is highly induced after flg22 treatment (Zipfel et al. 2004). The *agb1* mutants are impaired in the oxidative burst triggered by flg22, suggesting the importance of G-proteins in ROS signaling system (Ishikawa 2009). G-proteins have been shown to be involved in generation of NO which is involved in stomatal closure immune responses (Li et al. 2009; He et al. 2013). G-protein induces biosynthesis of the important second messenger polyamine (Fujiwara et al. 2006). The G-protein may be involved in generation of phospholipid second messengers (Viehweger et al. 2006). G-proteins are also involved in salicylate signaling system (Sano et al. 1994; Beffa et al. 1995; Fujiwara et al. 2006), jasmonate signaling system (Zhao and Sakai 2003; Trusov et al. 2006), ethylene signaling system (Fujiwara et al. 2006; Steffens and Sauter 2010), abscisic acid signaling system (Liu et al. 2007a; Gao et al. 2010a, b), gibberellic acid signaling system (Gao et al. 2010a, b), brassinosteroid signaling system (Oki et al. 2009), and auxin signaling systems (Gao et al. 2010b).

## 3.2 Heterotrimeric G-Protein Signaling

### *3.2.1 Subunits of Heterotrimeric G-Proteins*

The heterotrimeric G-proteins contain Gα-, Gβ-, and Gγ- subunits (Fujisawa et al. 2001; Temple and Jones 2007; Trusov et al. 2008, 2010; Wang et al. 2008; Zhang et al. 2011). Gβ and Gγ are tightly associated as a functional unit, while Gα can signal independently or through Gβγ (Zeng et al. 2007). In *Arabidopsis* genome only one gene is present for the Gα and Gβ subunits (Mason and Botella 2001; Assmann 2002; Zeng et al. 2007), while three genes have been identified for Gγ subunits (Chakravorty et al. 2011, Thung et al. 2012). In addition to a single prototypical Gα protein (GPA1), *Arabidopsis* has three unique Gα-like proteins, known as Extra Large G-protein 1 (XLG1), XLG2, and XLG3 (Ding et al. 2008; Zhu et al. 2009). The rice genome harbors one gene for each of Gα (*RGA1*) and Gβ (*RGB1*) and two genes for Gγ (RGG1 and RGG2) (Kato et al. 2004). *TaGA1* and *TaGA2* genes encoding G-protein α subunits have been cloned from wheat (Hossain et al. 2003). Four Gα, four Gβ, and two Gγ proteins have been detected in soybean (Bisht et al. 2011). Choudhury et al. (2011) identified 10 Gγ proteins and these can be grouped into three distinct families based on sequence features: the archetypal Gγ proteins, the prenylation-less Gγ proteins and the cysteine-rich Gγ proteins.

### 3.2.2 G-Protein-Coupled Receptor

G-proteins physically couple the recognition of extracellular signals like pathogen-associated molecular patterns (PAMPs) with specific cell-surface receptors called G-protein coupled receptors (GPCRs) (Colucci et al. 2002; Apone et al. 2003; Chen et al. 2004; Gookin et al. 2008; Hu et al. 2010; Tesmer 2010). The interaction of GPCRs with heterotrimeric G proteins is an important biological process in activating various defense responses (Hu et al. 2010). The most promising GPCR, *GCR1*, has been cloned from *Arabidopsis thaliana* (Josefsson and Rask 1997). *Arabidopsis GCR1* encodes a protein with predicted seven-transmembrane-spanning domain (Chen et al. 2004). It has been shown that GCR1 physically interacts with GPA1, the G protein α-subunit (Pandey and Assmann 2004).

### 3.2.3 Gγ Protein Triggers Plasma Membrane Targeting of Gβγ to Trigger Immune Responses

The Gγ subunit is an essential part of the heterotrimer, binding tightly to Gβ and anchoring the Gβγ dimer to the plasma membrane (Anderson and Botella 2007; Marrari et al. 2007). The G-protein γ subunits are responsible for providing functional selectivity in Gβγ dimer (Trusov et al. 2007, 2012). The Gγ protein of the G-protein heterotrimer is crucial for its proper targeting at the plasma membrane and correct functioning (Choudhury et al. 2011). Most of the Gγ subunits are small proteins of about 8–11 kDa (Trusov et al. 2012) and contain a conserved prenylation signal at their C-termini, which is a target for posttranslational prenylation (McIntyre 2009). The heterotrimer formation, together with isoprenylation, is required for plasma membrane targeting of Gβγ (Takida and Wedegaertner 2003).

Variations among the Gγ proteins in their size and presence of prenylation signal in the C-terminus have also been reported. The size of the *Arabidopsis* Gγ subunit AGG3 is approximately of 25 kDa, and the protein contains a large cysteine-rich C-terminus (Chakravorty et al. 2011). The rice Gγ subunit RGG2 docs not contain a C-terminal prenylation signal (Kato et al. 2004).

All known Gγ proteins contain a signature DPLL/l motif which together with few additional conserved amino acids in the middle coiled-coil region is required for interaction with the Gβ proteins. Most of the known Gγ proteins also contain a CAXX motif at C termini which is isoprenylated, resulting in the targeting of the proteins to the plasma membrane (Fukada 1995; Clapham and Neer 1997).

The *Arabidopsis* Gγ proteins AGG1 and AGG2 are involved in regulation of defense responses of plants (Mason and Botella 2000, 2001; Trusov et al. 2007). The rice Gγ protein RGG1 and soybean GmGγ1 and GmGγ2 proteins are highly homologous to the AGG1 protein and contain all the conserved features and motifs of Gγ proteins. The rice RGG2 protein has an extra 57 amino acid extension at its N terminus (compared to RGG1) and does not contain the signature prenylation motif.

The two pea Gγ proteins PGG1 and PGG2 do not contain the highly conserved DPLL/1 motif even though a possible prenylation motif is present at its C termini (Misra et al. 2007).

### 3.2.4 *Activation of G-Protein Heterotrimer in Elementary G-Protein Signaling*

Heterotrimeric G-proteins are composed of α, β, γ subunits, which exist as associated heterotrimers in their inactive state. The heterotrimeric G-protein signaling begins with ligand (PAMP signal) binding, which results in a conformational change in a G-protein-coupled receptor. Once activated by the GPCR, the Gα protein, which possesses a GDP/GTP-nucleotide-binding site and GTP-hydrolase activity, changes its form to a structure that allows exchange of GDP for GTP (Pandey et al. 2010). The GPCR works as a guanine exchange factor (GEF) for Gα and facilitates the Gα subunit to exchange GDP for GTP and become active (Oki et al. 2009). GTP binding is accompanied by structural rearrangements that disengage the Gβγ interaction and result in heterotrimer dissociation. The free subunits then relay signals by interacting with downstream proteins called effectors. The GTP-bound Gα separates from the associated Gβγ dimer and the freed Gα and Gβγ proteins can then interact with downstream effector molecules, alone or in combination, to transduce the signal (Pandey et al. 2010). Gα and Gβγ independently interact with multiple downstream effectors mediating specific signal transduction pathways. Subsequent to signal propagation, the intrinsic GTPase activity of Gα eventually results in hydrolysis of bound GTP to GDP, which inactivates Gα and allows its re-association with the Gβγ dimer to reform the inactive G-protein complex (Oldham and Hamm 2008; Oki et al. 2009; Liu et al. 2010; Pandey et al. 2010; Trusov et al. 2012).

## 3.3 Small G-Proteins Signaling

Small G-proteins (small GTPases) are monomeric guanine nucleotide binding proteins related to the α subunit of heterotrimeric G proteins (Yang 2002). All small G-proteins belong to Ras superfamily. The small G-proteins constitute the sole group of Rho family of small GTPases, called ROPs in plants (Berken 2006; Yang and Fu 2007; Nakashima et al. 2008). Rops are also referred to as RAC (Gu et al. 2004; Kiirika et al. 2012). The common features of this super family include four guanine nucleotide binding domains and an effector binding domain (Yang 2002). *Arabidopsis* contains 11 *Rac* genes (Yang 2002), while rice contains seven *Rac* genes (Miki et al. 2005). Six small G-proteins have been detected in barley (Schultheiss et al. 2003).

The family of Rho GTPases termed either ROPs or RACs is known to act as the major molecular switches in multitude of signal transduction pathways in plants

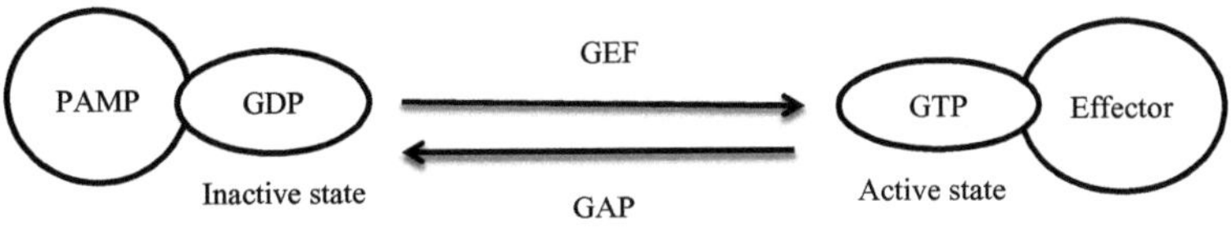


**Fig. 3.2** Small GTPases-mediated elementary signal transduction pathway

(Yalovsky et al. 2010; Wu et al. 2011). These GTPases act as a simple binary switch (the 'off' GDP-bound and the 'on' GTP-bound states). Shuttling between GDP-bound and GTP-bound states is controlled by two major regulators, guanine nucleotide exchange factors (GEFs) and GTPase activating proteins (GAPs). GEFs catalyze GDP release, which is exchanged with GTP, while GAPs enhance GTP hydrolysis, thereby accelerating RAC/ROP inactivation (Mucha et al. 2011; Wu et al. 2011).

Upon stimulation by an upstream PAMP signal, GEF, which is also known as a G-protein-coupling receptor (GPCR) (Pandey et al. 2010), converts the GDP-bound inactive form of the small GTPase into the GTP-bound active form through GDP/GTP replacement (Yang 2002). Subsequent to signal propagation, the intrinsic GTPase activity results in hydrolysis of bound GTP to GDP. The GTP form exhibits a weak intrinsic GTPase activity for GTP hydrolysis, requiring specific GTPases activating proteins (GAPs) with unique domain composition for efficient deactivation (Fig. 3.2; Yang 2002; Mucha et al. 2011).

Most small G-proteins cycle between membrane-bound and cytosolic forms. Only membrane-associated GTPases can be activated by GEF and their removal by a cytosolic factor called guanine nucleotide dissociation inhibitor (GDI) negatively regulates these GTPases (Yang 2002). Activated RAC/ROPs are capable of receiving a wide variety of inputs and accordingly generating a multitude of specific inputs (Yang and Fu 2007; Liu et al. 2010). They interact with immediate cellular effectors that interact with cellular components, relaying the signal to the ultimate target systems to effect the corresponding signal-induced responses (Wu et al. 2011). Rho is known to orchestrate a great number of signaling networks through a large number of interacting partners in plants (Yang and Fu 2007).

## 3.4 Heterotrimeric G-Protein Gα May Act Upstream of Small G-Protein in Immune Signaling

Both the heterotrimeric G-protein and small G-protein may function together and trigger immune responses in plants (Suharsono et al. 2002). In rice cells it was shown that the PAMP signal was recognized by unknown receptor in the plasma membrane and transmitted to the heterotrimeric G-protein Gα subunit. *Gα* mRNA accumulation was induced by the signals from the receptor. The activated Gα triggered accumulation of the small G-protein OsRac1 mRNA, which in turn strongly induced transcription of the defense-related PR10 protein gene *PBZ1* (Fig. 3.3; Suharsono et al. 2002).

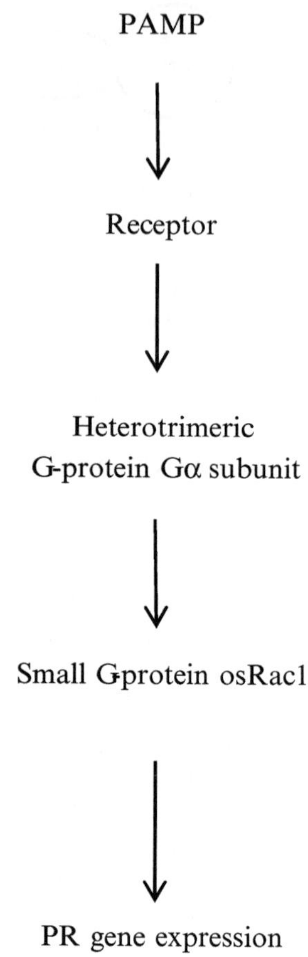


**Fig. 3.3** Heterotrimeric G-protein acting upstream of small G-protein in immune signaling system in rice cells (Adapted from Suharsono et al. 2002)

These results suggest the involvement of both the heterotrimeric G-protein Gα subunit and small G-protein OsRac1 in defense signaling and the heterotrimeric G-protein Gα acts upstream of small G protein OsRac1 in rice cells.

## 3.5 Different G-Protein Subunits in Heterotrimeric G-Proteins Play Distinct Roles in Plant Innate Immunity

Heterotrimeric G-proteins contain α, β, and γ subunits and these subunits may behave differently in inducing defense responses. For example, the β and γ subunits were involved in triggering immune responses against necrotrophic fungal pathogens, while the α subunit conferred susceptibility against these pathogens in *Arabidopsis*. The *Arabidopsis* α subunit *gpa1* mutant exhibited enhanced resistance to several necrotrophic fungal pathogens, including *Plectosphaerella cucumerina*,

*Alternaria brassicicola*, and *Fusarium oxysporum* in *A. thaliana* (Llorente et al. 2005; Trusov et al. 2006). In contrast, mutations in the *Arabidopsis* β subunit AGB1 and the Arabidopsis γ subunit AGG1 enhanced plant susceptibility to necrotrophic fungal pathogens (Llorente et al. 2005; Trusov et al. 2006, 2007). These results suggest that the Gα subunit is a negative regulator, while Gβ and Gγ subunits are positive regulators of disease resistance.

Besides the three subunits of the heterotrimeric G-proteins, *Arabidopsis* has three unique Gα-like proteins, known as Extra Large G-protein 1 (XLG1), XLG2, and XLG3 (Zhu et al. 2009). *XLG2* and *XLG 3* were rapidly induced by infection with the bacterial pathogen *Pseudomonas syringae*, whereas the *XLG1* transcript level was not affected by the pathogen infection. The *xlg2* loss-of-function mutation caused enhanced susceptibility to *P. syringae* (Zhu et al. 2009). The *xlg2* mutation affected pathogen-triggered induction of a small set of defense-related genes (Zhu et al. 2009). Constitutive overexpression of *XLG2* leads to the accumulation of transcripts from multiple defense-related genes. These results suggest that XLG2 is involved in triggering immune responses. In contrast, *xlg1* and *xlg3* mutants showed no difference from wild-type in resistance to *P. syringae*, suggesting that XLG1 and XLG3 are not involved in triggering defense responses against the pathogen (Zhu et al. 2009).

AGB1, the sole Gβ subunit in *Arabidopsis* has been found to be a positive regulator in resistance against necrotrophic fungal pathogens. The *agb1* mutant impaired in the Gβ subunit shows enhanced susceptibility to these pathogens (Llorente et al. 2005; Trusov et al. 2006, 2009). The Gβ subunit forms an obligate dimer with either one of the *Arabidopsis* Gγ subunits (γ1/AGG1 and γ2/AGG2). The *agg1 agg2* double mutant is as susceptible as *agb1*plants to *Plectosphaerella cucumerina* (Delgado-Cerezo et al. 2012). This heteromeric G-protein-mediated resistance was found to be independent of SA-, JA-, ethylene-, and abscisic acid-mediated signaling pathways. However, this G-protein-mediated resistance was found to be modulated by cell wall defense responses. The xylose content was lower in *agb1* and *agg1 agg2* mutants than in wild-type plants, suggesting that cell wall modifications may be the immune response triggered by Gβ and Gγ subunits of the heteromeric G-proteins (Delgado-Cerezo et al. 2012).

The rice *dl* mutant, which is deficient in the Gα subunit, exhibited a reduced hypersensitive response after infection by the rice blast pathogen *Magnaporthe oryzae* (formerly known as *M. grisea*) (Suharsono et al. 2002, Iwata et al. 2002) and bacterial blight pathogen *Xanthomonas oryzae* pv. *oryzae* (Komatsu et al. 2004). These results suggest that the heteromeric Gα subunit may play an important role in both fungal and bacterial disease resistance (Izawa et al. 2010).

## 3.6 Small G-Proteins Activate Plant Innate Immunity

The small G-proteins have been reported to play important role in activating immune responses against fungal, oomycete, bacterial, and viral diseases in different plants (Sano and Ohashi 1995; Ono et al. 2001; Suharsono et al. 2002; Shirasu and

Schulze-Lefert 2003; Wong et al. 2004; Moeder et al. 2005; Nakashima et al. 2008). The small G-protein OsRac1 is involved in basal and R protein-mediated resistance to the rice blast fungal pathogen *M. oryzae* and the bacterial blight pathogen *X. oryzae* pv. *oryzae* (Ono et al. 2001; Suharsono et al. 2002). The constitutively active *OsRac1* greatly reduced blast and bacterial blight disease development (Ono et al. 2001). The role of *Medicago truncatula* small GTPase *MtROP9*, orthologous to *Medicago sativa Rac1*, in the *Aphanomyces* root rot development in *M. truncatula* was studied by silencing the *MtROP9* gene using the RNA interference (RNAi) vector (Kiirika et al. 2012). MtROP9 knockdown promoted the root rot disease development, suggesting the role of the small G-protein in conferring resistance against the oomycete pathogen *Aphanomyces euteiches* (Kiirika et al. 2012). Transgenic tobacco plants that expressed a dominant negative form of the small G-protein OsRac1 from rice showed reduced resistance against *Tobacco mosaic virus* (TMV) compared to the wild-type plants, suggesting involvement of OsRac1 in triggering resistance against the virus disease in tobacco (Moeder et al. 2005).

## 3.7 Small G-Proteins May Be Involved in Susceptible Interactions

Although small monomeric G-proteins/small GTPases are involved in activation of immune responses against a wide range of pathogens (Ono et al. 2001; Suharsono et al. 2002; Shirasu and Schulze-Lefert 2003; Nakashima et al. 2008), they may also be involved in disease development process in some plant-pathogen interactions (Schultheiss et al. 2003). Barley RAC/ROP G-protein family members have been shown to be involved in susceptibility to the powdery mildew pathogen *Blumeria graminis* f. sp. *hordei*. Five Rac/Rop genes were constitutively expressed in the barley leaf epidermis and none of these genes showed enhancement of mRNA abundance after inoculation with *B. graminis* f. sp. *hordei*. The small GTPases HvRACB, HvRAC3 and HvROP6 proteins were found to be potentially involved in the establishment of susceptibility to the barley powdery mildew fungus *B. graminis* f. sp. *hordei*. These small G-proteins may be involved in processes supporting parasitic entry into epidermal host cells (Schultheiss et al. 2003). Interaction of plant RAC homologs with the NADPH oxidase complex appears to regulate activity of NADPH oxidase that produces $O^{-}_{2}$ in response to pathogen attack (Ono et al. 2001). In barley, $O^{-}_{2}$ production takes place during attack by *B. graminis* f. sp. *hordei* at sites of successful penetration of epidermal cells, but not at sites where fungal penetration is prevented (Hückelhoven and Kogel 1998). In contrast, $H_2O_2$ accumulates in barley at sites where penetration by *B. graminis* f. sp. *hordei* is successfully prevented (Thordal-Christensen et al. 1997; Hückelhoven et al. 2000). These results suggest that RAC small GTP-binding protein might have activated NADPH oxidase-dependent $O^{-}_{2}$ production that would have facilitated the fungal penetration. ROS has been reported to be involved in both activation and suppression of immune responses depending on spatial and quantitative differences in the occurrence of ROS

(Levine et al. 1994; Jabs et al. 1996; Schultheiss et al. 2002). Probably, the RAC small GTP-binding protein may be required for successful fungal haustorium establishment in barley-powdery mildew pathogen interactions (Schultheiss et al. 2002).

Hoefle et al. (2011) have reported that the barley small G-protein RACB is required for full susceptibility of the leaf epidermis to invasion by *B. graminis* f. sp. *hordei*. Stable transgenic knockdown of RACB reduced the ability of barley to accommodate haustoria of *B. graminis* in intact epidermal leaf cells. A ROP-GTPase activating protein interacting with RACB, MICROTUBULE-ASSOCIATED ROP-GTPase ACTIVATING PROTEIN (MAGAP1), has been identified. Under fungal attack, MAGPAP1-labeled microtubules built a polarized network at sites of successful defense. By contrast, microtubules loosened where the fungus succeeded in penetration (Hoefle et al. 2011; Huesmann et al. 2012). The results suggest that RACB and MAGAP1 may play antagonistic roles in cytoskeleton organization for fungal entry.

## 3.8 RAR1-SGT1-HSP90-HSP70 Molecular Chaperone Complex: A Core Modulator of Small G-Protein-Triggered Plant Innate Immunity

Molecular chaperones have been reported to participate in the small G-protein-mediated signal transduction, besides taking part in the folding of newly synthesized proteins (Seo et al. 2008). HEAT SHOCK PROTEIN 90 (HSP90), HSP70, REQUIRED FOR MLA12 RESISTANCE1 (RAR1), and SUPPRESSOR OF THE G2 ALLELE OF SKP1 (SGT1) are components of molecular chaperone complexes that are conserved in the plant kingdom. They form a complex with the small G-protein Rac1 in rice cells. HSP90, SGT1, and RAR1 functionally co-operate as a molecular chaperone complex to transduce plant immune responses (Seo et al. 2008). SGT1 is a co-chaperone of HSP90 and functions in plant immunity. HSP70 is also a target of SGT1 and facilitates its transfer to HSP90 (Seo et al. 2008).

Receptor for Activated C-Kinase1 (RACK1) has been identified as an interactor with Rac1 in rice. Rice contains two *RACK1* genes, *RACK1A* and *RACK1B* and RACK1A protein interacts with the GTP form of Rac1 (Nakashima et al. 2008). RACK1 homologs have been isolated from several plant species (Shirasu and Schulze-Lefert 2003). RACK1 shares significant homology to the β subunit of G-proteins (Adams et al. 2011). RACK1 constitutes a component of the Rac1 immune complex consisting of Rac1, RAR1, SGT1, HSP90, and HSP70 and it functions as a scaffolding protein for the immune complex (Fig. 3.4; Thao et al. 2007; Nakashima et al. 2008). The abundance of each of the chaperones/cochaperones (HSP90, HSP70, SGT1 and RAR1) present in the immune complex may need to be finely regulated to ensure a rapid and stable immune response (Nakashima et al. 2008).

OsRac1, RAR1, and HSP90 have been found to be functionally related in rice cells. OsRAR1-RNA interference (RNAi) rice plants had impaired basal

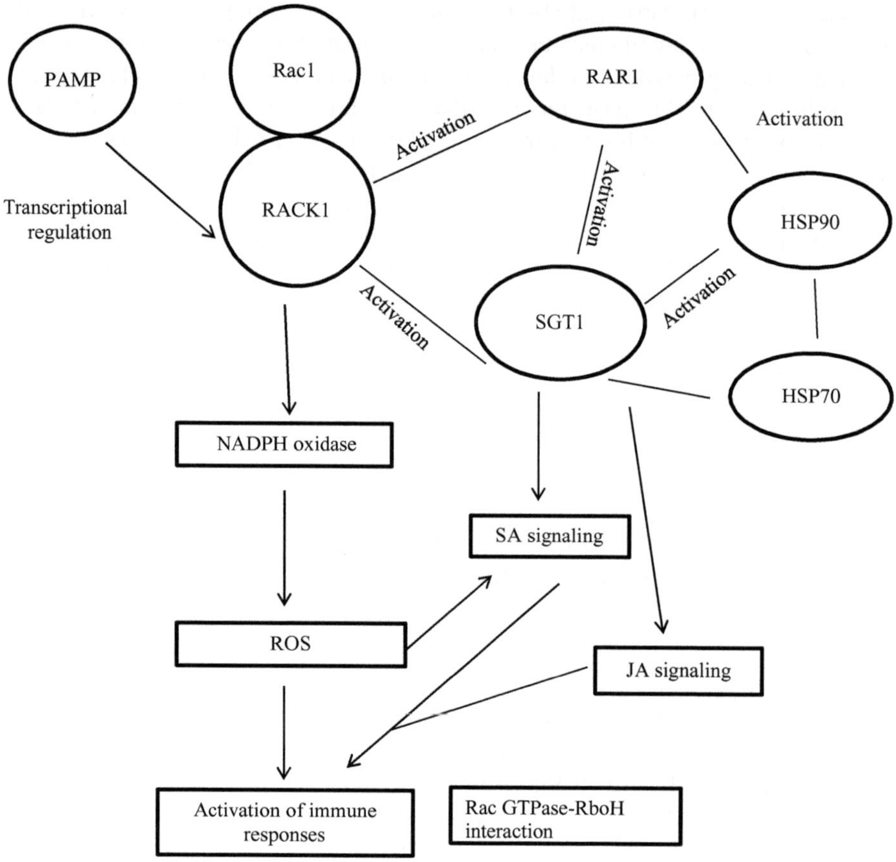


**Fig. 3.4** Function of RAR1-SGT1-HSP90-HSP70 molecular chaperone complex in small G-protein-triggered immunity in rice (Adapted from Thao et al. 2007; Nakashima et al. 2009; Shirasu 2009)

resistance to the rice blast pathogen *M. oryzae* and the bacterial blight pathogen *X. oryzae* pv. *oryzae* (Thao et al. 2007). Constitutively active *OsRac1* complemented the loss of resistance, suggesting that OsRac1 and RAR1 are functionally linked. Studies with OsRAR1-RNAi and treatment with geldanamycin, an HSP90-specific inhibitor, showed that RAR1 and HSP90 are essential for the Rac1-mediated enhancement of PAMP-triggered immune responses in rice cell cultures. It was also shown that the function of HSP 90, but not RAR1, may be essential for their association with the Rac1 complex. OsRac1 also regulates RAR1 expression at both the mRNA and protein levels. It was found that Rac1, RAR1, HSP90, and HSP70 form a complex in rice cells and these proteins play important roles in plant innate immunity (Thao et al. 2007).

Silencing of the co-chaperone genes *RAR1* and *SGT1-2* involved in basal and *R* gene-mediated defense resulted in susceptibility to *Pseudomonas syringae* in soybean

(Fu et al. 2009), suggesting the role of these genes in disease development. The *RAR1* and *SGT1* genes are required for SA accumulation in *Arabidopsis* and both are required in a genetically additive manner for induction of disease resistance (Zhou et al. 2008). Overexpression of *OsRAR1* and *OsSGT1* in rice significantly induced basal resistance to both the bacterial pathogen *X. oryzae* pv. *oryzae* and the fungal pathogen *M. oryzae* (Wang et al. 2008). RAR1 and SGT1 together contribute to basal resistance in *Arabidopsis*. Both RAR1 and SGT1 are required for inducing disease resistance, SA accumulation, and lesion formation after pathogen infection. RAR1 and SGT1 trigger expression of various SA-regulated defense-related genes including *PR-1*, *PR-2*, *PR-5*, *RPW8.1*, *RPW8.2*, *WRKY6*, *WRKY29*, and *EDS1* (Zhou et al. 2008).

RAR1 and SGT1 have also been shown to be involved in JA-mediated signaling system. Up-regulation of JA-inducible *PDF1.2* and *JR2* expression was found to be compromised in *rar1* and *sgt1* mutants, suggesting that both RAR1 and SGT1 play important role in JA signaling system (Kawamura et al. 2009). SGT1 is also required for the activation of the $SCF^{COI1}$-mediated JA response (Gray et al. 2003; Lorenzo and Solano 2005). The Arabidopsis *COI1* gene is required for the JA-mediated defense response against pathogens (Xie et al. 1998). COI1 protein contains an F-box motif and associates physically with AtCUL1, AtRbx1 and the Skp1-like proteins ASK1 and ASK2 to assemble $SCF^{COI1}$ ubiquitin-ligase complexes (Skp1-cdc53-F-box protein) (Xu et al. 2002). $SCF^{COI1}$ targets key regulators of JA signaling pathway for ubiquitination and subsequent degradation by the 26S proteasome (Kawamura et al. 2009).

SGT1 and HSP90 are highly expressed in plants infected with pathogens (Azevedo et al. 2006; Takahashi et al. 2003a). HSP90 and HSP70 have a tight functional link (Thao et al. 2007). SGT1, which associates with HSP70, is required for its nuclear localization (Shirasu 2009). Small monomeric G-protein Rac1 forms a complex with RAR1, SGT1, HSP90 and HSP70 in rice cells and these proteins play important roles in plant innate immunity.

## 3.9 PAMP Signal May Convert the G-Proteins from Their Inactive State to Their Active State to Trigger Immune Responses

G-proteins have the ability to bind guanosine-5′-triphosphate (GTP) and hydrolyze it to guanosine diphosphate (GDP). GDP locks G proteins into their inactive state, while GTP locks G- proteins into their activated state (Gelli et al. 1997; Oki et al. 2009; Pandey et al. 2010). Active or inactive states of G-proteins depend on the binding of GTP or GDP, respectively (Xing et al. 1997; Cabrera-Vera et al. 2003). Both classes of G-proteins, heterotrimeric G-proteins and small G-proteins, use the GTP/GDP cycle as a molecular switch for signal transduction (Xing et al. 1997; Pandey et al. 2010).

Heterotrimeric G-proteins act as the specific reaction partners of G-protein-coupled receptors. The GTPase is normally inactive. In the basic state, the Gα

–GDP- Gβγ complex and the receptor that can activate it are separately associated with the membrane. On receptor activation, the receptor becomes highly affine for the G-protein complex. On binding with the complex, GDP dissociates from the complex and the free complex has a high affinity for GTP. Upon GTP binding, both Gα –GTP and Gβγ separate from both the receptor and from each other. Both Gα –GTP and Gβγ may activate separate effector molecules sending the signal further down in the signal transduction chain. The GTPase activated by the intracellular receptor domain activates other molecules of the signal transduction chain, either via the α unit or the βγ complex (Yang 2002; Agrawal et al. 2003).

G-proteins are located in the cytoplasmic face of the cell plasma membrane (Casey 1995). Posttranslational lipid modifications of *Arabidopsis* Gγ-subunits have been shown to be required for plasma membrane targeting (Zeng et al. 2007). Many G-protein coupled receptors have been identified in plasma membrane (Kaziro et al. 1991). The PAMPs have been shown to activate GTP binding to G-protein (Gelli et al. 1997; Zhao and Sakai 2003). Binding of the PAMP with a pattern-recognition receptor (PRR) activates the G-protein (Kaziro et al. 1991; Tsukada et al. 2002). The PAMP stimulates conformational changes in the G- proteins. It stimulates exchange of GTP for GDP and thus converts the G-proteins from their inactive state to their active state (Gelli et al. 1997; Pandey and Assmann 2004).

## 3.10 PAMP-Activated G-Proteins Switch on Calcium Ion-Mediated Immune Signaling System

### *3.10.1 G-Proteins Activate InsP3-Gated Channels*

Perception of PAMP signals by pattern recognition receptors activates G-proteins. One of the earliest events in the immune signaling system appears to be G-proteins-triggered transient changes in permeability of the plasma membrane to $Ca^{2+}$ and influx of extracellular $Ca^{2+}$ through the membrane (Garcia-Brugger et al. 2006; Laohavisit et al. 2009, 2010; Vadassery and Oelmūller 2009). G-protein activates $Ca^{2+}$ channels and enhances $Ca^{2+}$ influx through $Ca^{2+}$-permeable channels (Wang et al. 2001; Zhang et al. 2011). G-proteins have been shown to be involved in triggering changes in cytosolic $Ca^{2+}$ concentrations (Blumwald et al. 1998; Schultheiss et al. 2003). Massive influx of $Ca^{2+}$ was observed within 15–30 min after PAMP treatment in tobacco-cultured cells (Lecourieux-Ouaked et al. 2000). The induced calcium ($[Ca^{2+}]_{cyt}$) elevations predominantly result from a continuous $Ca^{2+}$ influx through the plasma membrane (Hu et al. 2004; Vandelle et al. 2006).

RACK1 is an interactor with the G-protein Rac1 in rice. RACK1 homologs have been isolated from several plant species (Shirasu and Schulze-Lefert 2003). RACK1 protein interacts with the GTP form of the G-protein Rac1 (Nakashima et al. 2008). RACK1 binds to Gβ and occurs in a complex with the Gα and Gγ units (Patterson et al. 2004). RACK1 binds inositol 1,4,5-trisphosphate (InsP3) receptors and regulates $Ca^{2+}$ release by enhancing InsP3 receptor binding affinity for InsP3. Overexpression of RACK1 markedly augments $Ca^{2+}$ release, while depletion of RACK1 by interference RNA diminishes $Ca^{2+}$ release (Patterson et al. 2004).

InsP3-activated $Ca^{2+}$ channel is the important $Ca^{2+}$ release channel (Alexandre and Lassales 1992). InsP3-gated channels release $Ca^{2+}$ from the vacuole and endoplasmic reticulum (ER) (Berridge 1993). The calcium released through this channel induces calcium waves and oscillations in the cytosol (Berridge 1993). Calcium ion acts as a signal carrier (Kudla et al. 2010) and calcium signaling is modulated by specific "calcium signatures". The specific changes in calcium transients, oscillations, or repeated spikes/waves are called calcium signatures (Lecourieux et al. 2006). Collectively the results suggest that G-protein triggers the InsP3-activated $Ca^{2+}$ channel and modulates $Ca^{2+}$ signature – mediated immune signaling system.

### 3.10.2 G-Proteins Stimulate $H^+$-ATPase and Regulate $Ca^{2+}$ Channel

PAMP-PRR signaling activates G-proteins and the activated G-proteins stimulate the plasma membrane $H^+$-ATPase (Vera-Estrella et al. 1994a, b; Xing et al. 1997; Blumwald et al. 1998). The plasma membrane $H^+$-ATPases generate an $H^+$-gradient across the plant plasma membrane. The concomitant hyperpolarization of the membrane potential induces the opening of the $Ca^{2+}$ channel. The proton gradient creates an electrical potential, which drives $Ca^{2+}$ uptake through ion channels (Palmgren and Harper 1999). The results suggest that the G-proteins may also modulate the expression of $H^+$-ATPase and activate $Ca^{2+}$ signaling.

### 3.10.3 G-Proteins Activate $Ca^{2+}$ Signaling System Through Modulation of Phosphorylation/Dephosphorylation System

G-proteins may be involved in $Ca^{2+}$ channel opening (Gelli et al. 1997). Protein phosphorylation precedes $Ca^{2+}$ influx in tobacco cells treated with a PAMP isolated from the oomycete pathogen *Phytophthora cryptogea* (Tavernier et al. 1995). The G-proteins modulate the phosphorylation/dephosphorylation system in the plasma membrane of tomato cells and transduce the signal (Vera-Estrella et al. 1994a). Phosphorylation of proteins involved in G-protein coupled signaling has been reported in tobacco cells treated with a bacterial PAMP (Gerber et al. 2006).

The activation of the $Ca^{2+}$ channel by PAMPs was modulated by a heterotrimeric G-protein–dependent phosphorylation of the channel protein in tomato, probably by activating protein kinase, and inhibiting protein phosphatase (Gelli et al. 1997). The activated G-protein transduced the signal by activating phosphorylation/dephosphorylation system in the plasma membrane in tomato cells (Vera-Estrella et al. 1994b).

An elicitor-induced increase in ATPase activity was shown to be activated by the G-proteins in tomato (Vera-Estrella et al. 1994a). G-protein has been shown to activate a membrane-bound phosphatase that mediates the dephosphorylation of the host plasma membrane $H^+$-ATPase in tomato (Xing et al. 1997). The dephosphorylation of $H^+$-ATPase was followed by rephosphorylation by protein kinase and $Ca^{2+}$-dependent kinase (Xing et al. 1996).

## 3.11 G-Proteins May Trigger Efflux of Vacuolar Protons into Cytoplasm to Activate pH-Dependent Signaling Pathway

Transient shifts of intracellular and apoplastic pH appear to be essential steps in signal transduction processes (Viehweger et al. 2002). Cytoplasmic acidification induced by biotic stress is considered as a plant-specific trigger for the synthesis of defense-related compounds including phytoalexins and other secondary metabolites (Sakano 2001). Acidification of the cytoplasm increases the expression of mRNAs encoding several defense-related enzymes (Lapous et al. 1998; He et al. 1998). The G-protein may activate phospholipase and pH-dependent signal path (Viehweger et al. 2006). The function of a Gα protein in defense signal transduction system was studied by developing transgenic cell cultures of California poppy (*Eschscholzia californica*) in which the Gα content was decreased via antisense transformation or by the expression of recombinant anti-Gα single-chain antibodies. All transgenic cell types were deficient in two elicitor-triggered early signal events: activation of phospholipase $A_2$ ($PLA_2$), and efflux of vacuolar protons. The lacking $H^+$ efflux could be restored by adding lysophosphatidylcholine (LPC), a product of $PLA_2$ activity, to vacuoles in situ. The results suggest that Gα mediates the stimulation of $PLA_2$ by the elicitor and the resulting peak of LPC initiates a transient efflux of vacuolar protons into the cytoplasm. This results in acidic peak of the cytoplasmic pH which may induce the expression of enzymes of phytoalexin production and induce resistance against pathogens (Viehweger et al. 2006).

## 3.12 G-Proteins Switch on ROS Signaling System

### *3.12.1 G-Proteins Trigger Generation of ROS to Induce Immune Responses*

G-proteins activate primarily reactive oxygen species (ROS) production and ROS-mediated signaling systems (Park et al. 2000; Bokoch and Diebold 2002; Suharsono et al. 2002; Wong et al. 2007) to trigger innate immune responses.

Both small G-proteins (Yang 2002; Schultheiss et al. 2003; Morel et al. 2004; Wong et al. 2007; Kiirika et al. 2012) and heterotrimeric G-proteins (Zhu et al. 2009; Zhao et al. 2010; Zhang et al. 2011) are known to trigger generation of ROS.

Small monomeric G-proteins (small GTPases) are involved in the regulation of ROS generation in the innate immune responses via the activation of NADPH oxidase homologs of plants termed respiratory burst oxidase homolog (RBOH) (Agrawal et al. 2003; Kiirika et al. 2012). A small G-protein from rice, OsRac1, has been shown to induce ROS production in rice cells (Kawasaki et al. 1999; Ono et al. 2001; Wong et al. 2007). OsRac1 was found to be a positive regulator of disease resistance. It activates RBOH-mediated ROS signaling through direct binding of Rac1 to the catalytic subunits of the RBOH protein's N-terminal extension, which is specific for the plant RBOH proteins (Kawasaki et al. 2006; Wong et al. 2007; Nakashima et al. 2008).

The small GTPase MtROP9 triggered the expression of *MtRBOH* gene involved in ROS generation and conferred resistance against root rot of *Medicago truncatula* caused by *Aphanomyces euteiches* (Kiirika et al. 2012). Transgenic tobacco plants that expressed a dominant negative form of the small G-protein OsRac1 from rice showed reduced resistance against *Tobacco mosaic virus* (TMV) compared to the wild-type plants. The dominant-negative *OsRac1* gene in tobacco suppressed ROS accumulation (Moeder et al. 2005), suggesting that OsRac1 is involved in ROS production. Collectively these results suggest that the small G-protein is involved in activating immune responses by activating ROS signaling.

Heterotrimeric G-proteins are also involved in ROS production. The *Arabidopsis* Gα subunit, GPA1, is involved in ROS production (Zhao et al. 2010). The *Arabidopsis gpa1* mutants have been shown to be disrupted in production of ROS (Zhang et al. 2011). It has been reported that different subunits of a heterotrimeric G-protein may signal ROS production (oxidative burst) in different manner. The *Arabidopsis* heterotrimeric G-protein is known to contain three subunits, Gα, Gβ, and Gγ. The Gα and Gβ proteins were shown to be necessary component of the biphasic oxidative burst, while only the Gα protein was found to be required for the late phase of the oxidative burst in *Arabidopsis thaliana* induced by an external signal. The early endogenous ROS production induced by the elicitor signal was mediated by both the Gα and Gβ proteins, while the extracellular ROS signal induced by the elicitor was generated by membrane-bound NADPH oxidases mediated by only the Gα protein (Joo et al. 2005). In addition to a single prototypical Gα protein (GPA1), *Arabidopsis* has three unique Gα-like proteins, known as Extra Large G-proteins (XLG). The transgenic *Arabidopsis* plants overexpressing one of the genes encoding XLG proteins, *XLG2*, showed constitutive accumulation of transcripts from *RbohC* and the NADPH oxidase RbohC is known to be involved in ROS production (Zhu et al. 2009). These studies suggest that both the groups of G-proteins, small G-proteins and heterotrimeric G-proteins, trigger immune responses by activating ROS production.

### *3.12.2 G-Proteins Switch on $Ca^{2+}$ Influx – RBOH-Mediated ROS Signaling Pathway*

Numerous Rboh (for respiratory burst oxidase homolog) genes have been isolated in plants. All *rboh* genes identified to date possess a conserved N-terminal extension that contains two $Ca^{2+}$ binding EF-hand motifs. A substantial part of the N-terminal region of Rboh, including the two EF-hand motifs, is required for the Rac GTPases and the N-terminal extension interaction. Cytosolic $Ca^{2+}$ concentration has been shown to regulate Rac-Rboh interaction. Transient coexpression of *OsRac1* and *rbohB* enhanced ROS production in *Nicotiana benthamiana*, suggesting that direct Rac-Rboh interaction may activate NADPH oxidase activity in plants (Fig. 3.5; Wong et al. 2007; Zhao et al. 2010). Collectively these results suggest that cytosolic $Ca^{2+}$ concentration may modulate NADPH oxidase activity by regulating the interaction between Rac GTPase and Rboh.

### *3.12.3 Interplay Between ROP, RBOH, CDPK, $Ca^{2+}_{[cyt]}$, and ROS in G-Protein-Mediated ROS Signaling*

An interplay between the monomeric small G-protein Rho-like GTPase of plants (ROP), the plant orthologs of the respiratory burst NADPH oxidases (RBOH), calcium-dependent protein kinase (CDPK), cytosolic calcium transients ($Ca^{2+}_{[cyt]}$), and reactive oxygen species (ROS) production has been described (Van Breusegem et al. 2008). It has been reported that activation of the plasma membrane-localized RBOHs involves phosphorylation of two N-terminal Ser by a calcium-dependent protein kinase as well as interaction with ROP. RBOH phosphorylation as well as binding to calcium synergizes its activation, raising the possibility that it may function as a calcium sensor (Ogasawara et al. 2008; Takeda et al. 2008). The RBOH/ROP-GTP interaction is regulated by the binding of calcium to two EF-hand motifs at the N terminus of the NADPH oxidase (Wong et al. 2007). The RBOH activation results in production of ROS (Van Breusegem et al. 2008). The activation of RBOH promotes calcium channel activation and calcium influx, thereby stimulating RBOH activity and amplification of the initial signal (Takeda et al. 2008).

### *3.12.4 Phosphatidic Acid Activates G-Protein-Mediated Pathway of ROS Generation*

Phosphatidic acid (PA), a second messenger generated in a phospholipid signaling pathway, has been shown to play an important role in Rho-related small G protein GTPase-mediated pathway of ROS generation. G proteins activate phospholipase C (PLC) and phospholipase D (PLD) (Munnik et al. 1995; Ritchie and Gilroy 2000)

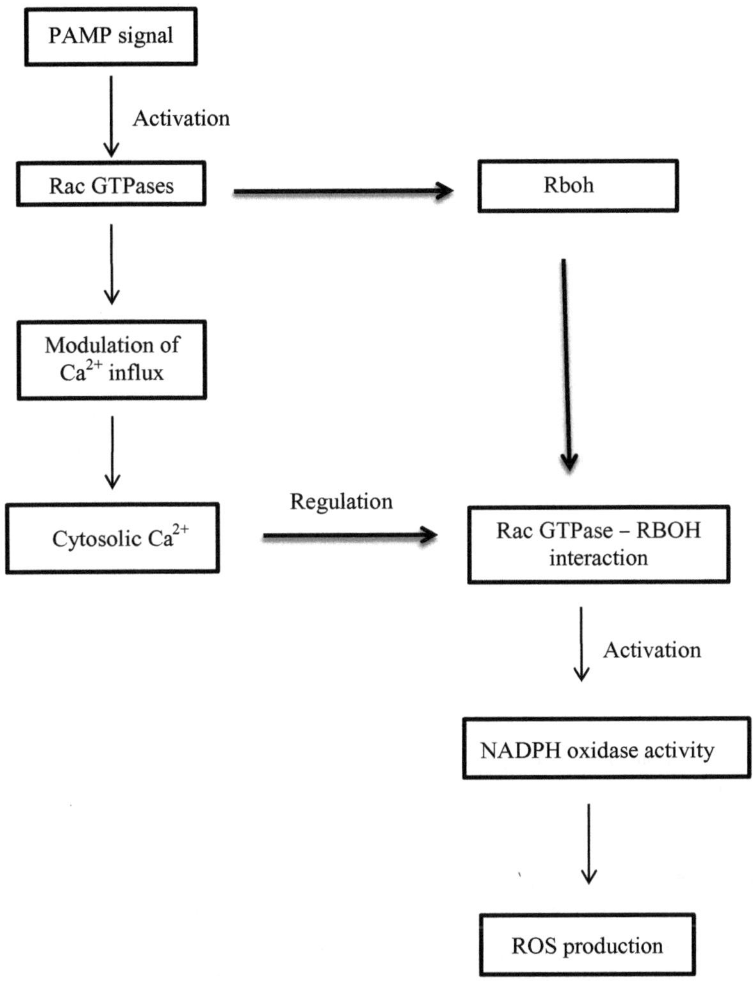


**Fig. 3.5** Interplay between Rac GTPases – RBOHs – cytosolic calcium transient in ROS generation

and the phospholipases are involved in ROS generation (Laxalt et al. 2007; Guo et al. 2012). Heterotrimeric G protein α-subunit regulates PLD through a motif analogous to the DRY motif in G-protein-coupled receptors in *Arabidopsis* (Zhao and Wang 2004). The activated PLD hydrolyzes phospholipids to produce the lipid second messenger PA (Zhao and Wang 2004).

PA has been shown to be able to trigger an oxidative burst (Sang et al. 2001; Park et al. 2004). PA is involved in the activation of NADPH oxidase and ROS is generated through the action of NADPH oxidase (Laxalt et al. 2007). PA directly activates NADPH oxidase by interacting with the enzyme components (Palicz et al. 2001). It promotes superoxide-generating activity in plants through the activation of

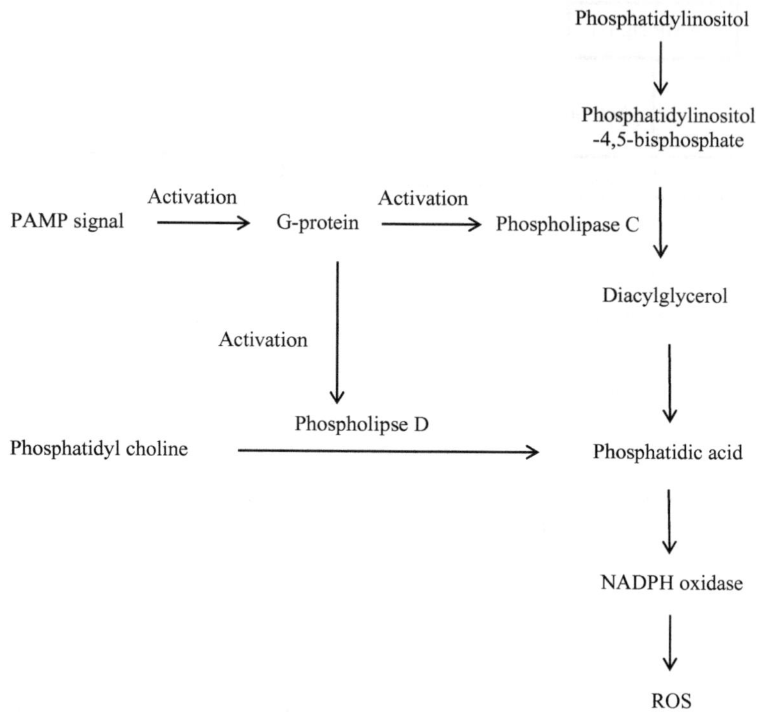


**Fig. 3.6** Generation of ROS by phosphatidic acid through the action of G-protein-activated phospholipases

NADPH oxidase (Sang et al. 2001). PA has been shown to induce ROS in tomato cells. Inhibition of the PLC diminished elicitor-induced ROS production (Laxalt et al. 2007). These results suggest that PA is generated through the action of G-proteins-activated phospholipases and the generated PA activates NADPH oxidase involved in generation of ROS (Fig. 3.6).

### 3.12.5 *Small G-Proteins May Trigger Accumulation of ROS by Suppressing the Action of ROS Scavengers*

Small G-proteins may act as a suppressor of ROS scavenging, resulting in accumulation of ROS (Wong et al. 2004). Metallothioneins are small, ubiquitous Cys-rich proteins involved in ROS scavenging. The expression of the metallothionein gene (*OsMT2b*) was synergistically down-regulated by OsRac1 and PAMPs from *Magnaporthe oryzae*, the rice blast pathogen. *OsMT2b*-overexpressing cells showed reduced PAMP-induced $H_2O_2$ production. In contrast, *OsMT2b*-RNAi-silenced transgenic cells showed significantly higher PAMP-induced OsRac1-dependent

$H_2O_2$ production than the wild-type cells. Transgenic plants overexpressing *OsMT2b* showed increased susceptibility to the rice bacterial blight pathogen *X. oryzae* pv. *oryzae* and the blast pathogen *M. oryzae* (Wong et al. 2004). The results suggest that OsRac1 plays a role as a suppressor of ROS scavenging and triggers immune responses against bacterial and fungal pathogens.

### *3.12.6 G-Proteins Act as Redox Regulators in ROS Signaling*

The G-protein appears to be a key redox regulator in defense signaling (Ono et al. 2001; Baxter-Burrell et al. 2002). Fujiwara et al. (2006) detected activation of several redox regulators in cultured rice cells transformed with OsRac1. The induced redox regulators included glyceraldehyde-3-P dehydrogenase, NADPH-thioredoxin reductase, ferredoxin-NADPH reductase, NADPH dependent oxidoreductase, quinine oxidoreductase, and glutathione-S-transferase (GST1) (Fujiwara et al. 2006). Redox signaling is known to play an important role in innate immune system (Desikan et al. 2005; Fedoroff 2006). Cytosolic glyceraldehyde-3-phosphate dehydrogenases interact with phospholipase D (PLDδ) to transduce the ROS hydrogen peroxide signal in *Arabidopsis thaliana* (Guo et al. 2012).

## 3.13 G-Proteins Activate Nitric Oxide Signaling System

Nitric oxide (NO) plays a key role in immune signaling transduction system (Lindermayr et al. 2010; Perchepied et al. 2010; Wang et al. 2010). Heterotrimeric G-proteins have been shown to be involved in generation of NO (Li et al. 2009; He et al. 2013). The calcium ion sensor protein calmodulin activates GPA1, the Gα-subunit of heterotrimeric protein in *Arabidopsis*. The activated heterotrimeric G-protein in turn activates NADPH oxidases (Li et al. 2009). GPA1 has been shown to function upstream of the NADPH oxidases AtrbohD and AtrbohF (Zhang et al. 2011). These NADPH oxidases are involved in the production of $H_2O_2$ (Li et al. 2009). The *gpa1* mutants have been shown to be disrupted in production of ROS (Zhang et al. 2011). The modulation of NO production by Gα protein has been shown to require NADPH oxidase-dependent $H_2O_2$ generation (Fig. 3.7; Li et al. 2009).

Stress-induced $H_2O_2$ and NO generation were found to be regulated by GPA1, the Gα-subunit of heterotrimeric G-protein (He et al. 2013). The $H_2O_2$ and NO accumulation were nullified in *gpa1* knockout mutants but enhanced by overexpression of a constitutively active form of GPA1 (He et al. 2013). The results suggest that the heterotrimeric protein is involved in NO production. Further it has been demonstrated that Gα activation of NO production depends on $H_2O_2$ (He et al. 2013). It suggests that the signaling pathway involves G-protein-dependent activation of $H_2O_2$ production and subsequent NO accumulation.

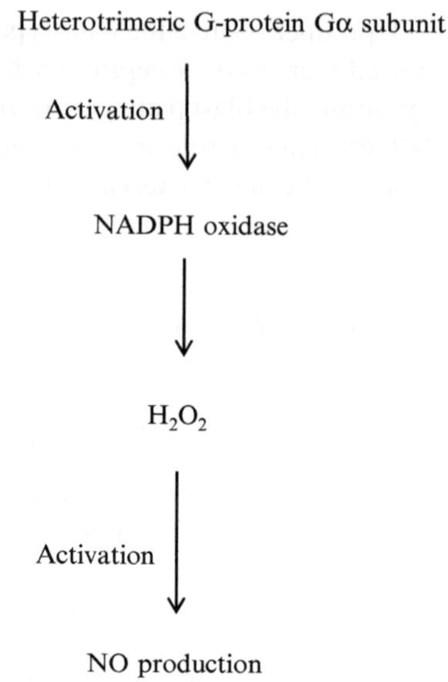


**Fig. 3.7** Heterotrimeric G protein-mediated nitric oxide (NO) production in *Arabidopsis* (Adapted from Li et al. 2009)

$H_2O_2$-induced NO production depended mainly on NITRIC OXIDE ASSOCIATED 1 (NOA1) (also called NITRIC OXIDE SYNTHASE1 [NOS1]) (Li et al. 2009). However Bright et al. (2006) reported that it is NITRATE REDUCTASE (NR), not NOA, that is responsible for NO generation in *Arabidopsis* in response to $H_2O_2$.

## 3.14 Close Relationship Between G-Proteins and MAPKs in Signal Transduction

Both hetrotrimeric and small monomer G-proteins may be required for defense signaling (Zhao et al. 2010; Zhang et al. 2011, 2012; Kiirika et al. 2012). Mitogen-activated protein kinases (MAPKs) are involved in defense signaling system (Liu et al. 2011; Mao et al. 2011). A close relationship between MAPKs and G-proteins in the transduction of external signals into intracellular responses has been reported (Lieberherr et al. 2005).

To study the importance of these two types of G-proteins in defense signaling, transgenic rice cell lines in which the small G-protein *OsRac1* gene was specifically silenced by RNA interference and loss-of-function mutant rice cells containing heterotrimeric Gα mutation were obtained (Lieberherr et al. 2005). There was a strong reduction of OsRac1 protein level in the transgenic plants and these transgenic plants also showed strong reduction in MAPK protein level. In the *OsRac1*-RNAi line, the MAPK activity was not induced by an elicitor, suggesting a requirement for OsRac1 for the kinase activity response. In rice cell cultures containing the Gα mutation, the MAPK protein level was very much reduced, indicating that Gα is also required for the kinase activity response (Lieberherr et al. 2005).

The Arabidopsis extra-large heterotrimeric G-protein *XLG2* overexpression lines showed constitutive accumulation of transcripts from *AtMPK3* (Zhu et al. 2009). MPK3 is involved in MAPK signaling cascade (Mészáros et al. 2006). A protein-protein interaction of β-subunit of heterotrimeric G-proteins (PsGβ) with a MAPK (PsMPK3) has been reported in *Pisum sativum* (Bhardwaj et al. 2011). The transcription of these two genes also showed co-regulation under abscisic acid (ABA) and methyl jasmonate treatments. β-subunit of G-proteins from rice also showed interaction with PsMPK3, suggesting that the β-subunit from a heterologous system also shows interaction with MPK3. MPK3 may function as an effector molecule for Gβ subunit of heterotrimeric G-proteins from *Pisum sativum* (Bhardwaj et al. 2011).

## 3.15 G-Proteins Induce Biosynthesis of Polyamines Which Act as Second Messengers Triggering Early Signaling Events

Polyamines are polycationic, ubiquitous aliphatic amines that occur in all plant cells (Tun et al. 2006; Nambeesan et al. 2010, 2012). The diamine putrescine and the polyamines spermidine and spermine are involved in activation of immune signaling. They play important role as second messengers in immune response signaling (Walters 2000; Takahashi et al. 2003b, 2004; Walters 2003; Tun et al. 2006; Ozawa et al. 2009, 2010; Szepesi et al. 2011). The polyamine biosynthesis is activated during pathogenesis (Marini et al. 2001; Gardiner et al. 2010). *Tobacco mosaic virus* (TMV) infection resulted in increased concentration of the putrescine and spermidine in tobacco leaves (Torrigiani et al. 1997). The polyamines also may accumulate during the defense response (Mo and Pus 2002; Walters et al. 2002; Fujiwara et al. 2006). Sugar beet plants treated with methyl jasmonate showed increased resistance against *Beet mosaic virus* (BtMV) and the increased resistance was associated with increased accumulation of polyamines (Haggag et al. 2010). Increased levels of putrescine, spermidine and spermine were observed in barley leaves inoculated with the powdery mildew fungus *Blumeria graminis* f. sp. *hordei* (Cowley and Walters 2002a, b). Marini et al. (2001) showed that hypersensitive response to TMV infection was accompanied by increase in activities of polyamine biosynthetic enzymes.

The polyamines have also been reported to trigger defense responses (Yamakawa et al. 1998; Walters 2003). Polyamines have been shown to be involved in resistance against *Ascochyta rabiei* in chickpea (Angelini et al. 1993). Tobacco leaves infected with TMV showed accumulation of spermine and the accumulated spermine induced both acidic PR proteins and resistance to TMV via a salicylic acid-independent pathway (Yamakawa et al. 1998; Hiraga et al. 2000). Polyamines may also be involved in

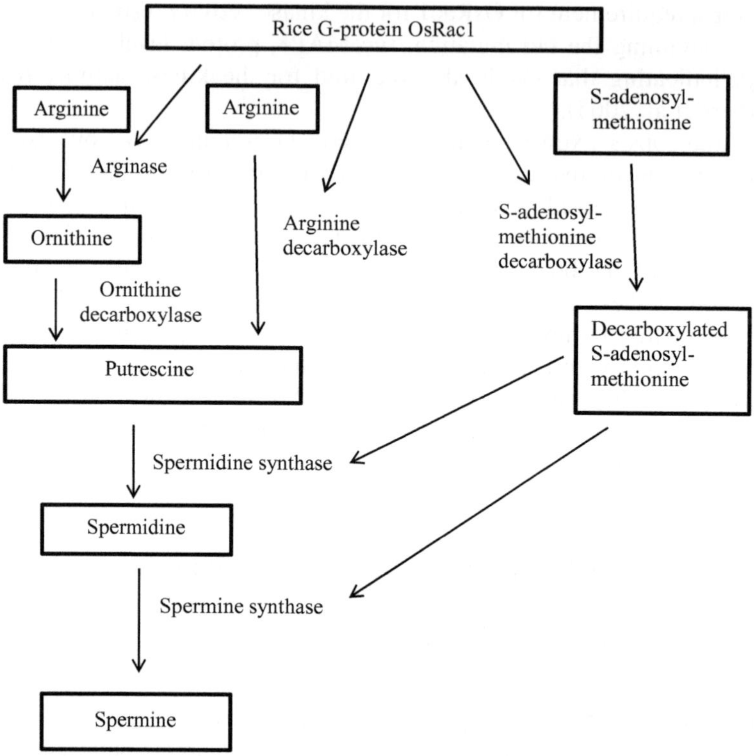


**Fig. 3.8** Small G-protein OsRac1-mediated polyamine biosynthesis in rice cells (Adapted from Fujiwara et al. 2006)

conferring susceptibility to some pathogens. Polyamines induced susceptibility to the necrotrophic pathogen *Botrytis cinerea* in tomato (Nambeesan et al. 2012).

The starting point for polyamine biosynthesis is the basic amino acids ornithine and arginine, which are decarboxylated by ornithine decarboxylase (ODC) and arginine decarboxylase (ADC), respectively (Walters 2003). Decarboxylation of ornithine by ODC or arginine by ADC leads to the synthesis of putrescine, which is converted to spermidine by spermidine synthase. Spermidine, in turn, is then converted to spermine by spermine synthase (Nambeesan et al. 2012). Fujiwara et al. (2006) showed the induction of the two key enzymes in biosynthesis of polyamines, arginase and spermidine synthase, in rice cells by OsRac1, besides induction of S-adenosylmethionine decarboxylase. Arginase produces ornithine which is transformed into putrescine via ornithine decarboxylase, while spermidine synthase converts putrescine to spermidine. S-adenosylmethionine decarboxylase is involved in decarboxylation of S-adenosyl methionine. In these reactions, both sperimidine synthase and spermine enzymes use aminopropyl residues derived from decarboxylated S-adenosyl-methionine (Fig. 3.8; Fujiwara et al. 2006; Kresge et al. 2007; Nambeesan et al. 2012).

The G-protein-triggered synthesis of polyamines is involved in activation or suppression of various signaling systems. Spermine treatment elicited biosynthesis of jasmonic acid in lima bean leaves (Ozawa et al. 2009, 2010). JA treatment enhanced polyamine biosynthesis in sugarbeet (Haggag et al. 2010). Spermine treatment induced calcium influx and ROS production (Ozawa et al. 2010). Polyamines have been found to be a common source of hydrogen peroxide in host- and nonhost hypersensitive response during pathogen infection (Yoda et al. 2009). Polyamine oxidases are $H_2O_2$ producing enzymes (Angelini and Federico 1989; Angelini et al. 2008). Polyamines induce rapid biosynthesis of nitric oxide (NO) in *Arabidopsis thaliana* (Tun et al. 2006). Polyamines also induced the activities of various enzymes involved in redox signaling system and the activated enzymes included ascorbate peroxidase, glutathione peroxidase, glutathione reductase, superoxide dismutase and catalase (Ozawa et al. 2010). Interaction between polyamines and SA signaling system in tomato has been reported (Szepesi et al. 2011). Polyamines induced susceptibility to the necrotrophic pathogen *Botrytis cinerea* in tomato. The polyamine-mediated susceptibility to *B. cinerea* was shown to be linked to interference with the functions of ethylene in plant defense (Nambeesan et al. 2012).

## 3.16 G-Proteins Modulate Salicylic Acid Signaling Pathway

G-proteins may trigger salicylic acid signaling system (Beffa et al. 1995). Cholera toxin from *Vibrio cholerae* is a multimeric protein consisting of A1, A2, and five B subunits. The A1 subunit catalyses the ADP-ribosylation of Gα, which irreversibly blocks the GTPase activity of G-proteins leading to the sustained activation of the downstream signaling pathway (Beffa et al. 1995). Cholera toxin does not activate G-proteins directly; it acts to maintain the active state of G-proteins with bound GTP (Beffa et al. 1995). Transgenic tobacco plants expressing A1 subunit of cholera toxin were developed and tissues of these transgenic plants showed accumulation of high levels of salicylic acid (Beffa et al. 1995). Sano et al. (1994) reported that expression of a small G- protein in transgenic tobacco abnormally induced salicylic acid in response to an external stimulus. Transgenic tobacco plants expressing a rice gene encoding small GTPase, *rgp1*, showed high accumulation of salicylic acid (Yoda and Sano 2003; Sano et al. 1994). These studies reveal that G-proteins are involved in SA biosynthesis.

Tobacco plants transformed with the *rgp1*gene showed increase in the mRNA levels of genes encoding acidic pathogenesis-related proteins, which are inducible by SA (Sano et al. 1994). Cultured rice cells were transformed with the rice *OsRac1* gene encoding a small G-protein. A salicylic acid induced protein, glucosyltransferase IS5a, was found to accumulate in these transformed cells (Fujiwara et al. 2006). Engineering the *rgp1* gene in tobacco has been shown to increase resistance to *Tobacco mosaic virus* infection by inducing salicylic acid (Sano et al. 1994). Collectively these studies suggest that G-proteins play an important role in SA biosynthesis and signal transduction in plant innate immune responses.

## 3.17 G-Proteins Trigger Ethylene Signaling Pathway

G-protein may trigger ethylene-dependent signaling pathway. G-proteins have been recognized as crucial signal transducers in ethylene-mediated signaling (Steffens and Sauter 2010). The rice small G-protein OsRac1 induced methionine synthase and S-adenosyl-methionine synthase in rice cells (Fujiwara et al. 2006). Methionine synthase is involved in conversion of homocysteine to methionine and S-adenosyl-methionine synthase catalyses the formation of S-adenosyl-methionine. S-adenosylmethionine is a precursor of ethylene and it is converted into 1-amino-cyclopropane-1-carboxylic acid (ACC) by the action of ACC synthase. ACC is oxidized by ACC oxidase resulting in the formation of ethylene (Fig. 3.9; Wang et al. 2002; Vidhyasekaran 2007).

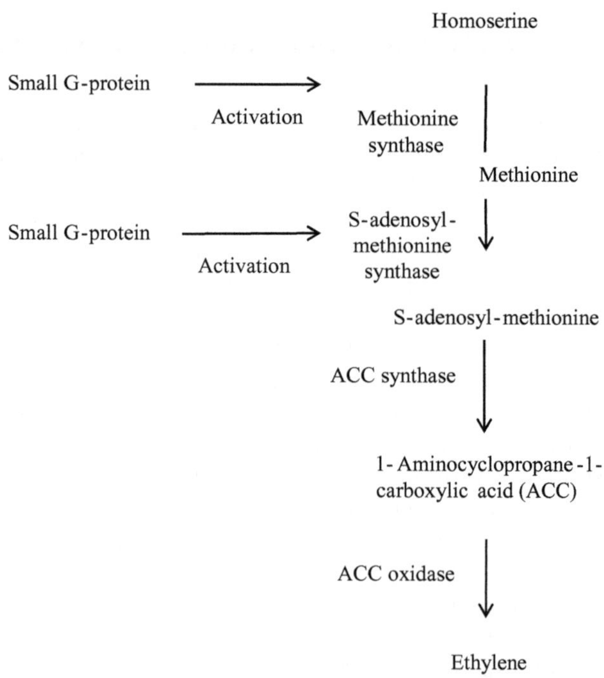


**Fig. 3.9** G-protein – induced ethylene biosynthesis

## 3.18 G-Proteins Switch on Jasmonate Signaling System

G-proteins may switch on jasmonate signaling pathway. Inhibitors of G-proteins signaling pathway suppressed elicitor-induced increases in lipoxygenase activity, whereas activators of G-proteins signaling pathway increased lipoxygenase activity in Mexican cypress (*Cupressus lusitanica*) cell cultures (Zhao and Sakai 2003). Lipoxygenase is an important enzyme in the octadecanoid pathway leading to the biosynthesis of jasmonate (Schaller 2001; Zhao et al. 2004). Calcium ion influx and $H_2O_2$ production appear to precede increases in lipoxygenase activity (Zhao and Sakai 2003; Zhao et al. 2004). Another important enzyme involved in jasmonate biosynthesis is 12-oxo-10,15(Z)-phytodienoic acid (OPDA)-reductase, which converts OPDA to jasmonate (Schaller 2001). Gβ mutants defective in Gβ activity showed lower induction of *OPR1* gene encoding *OPDA* in Arabidopsis. It suggests that Gβ is involved in biosynthesis of jasmonate (Trusov et al. 2006). The extracellular signals activate the receptor-coupled G-proteins and then the active G-proteins further switch on $Ca^{2+}$ channel. $Ca^{2+}$ influx and the subsequent $Ca^{2+}$ wave may initiate calmodulin-dependent protein kinase cascade, ROS production, and eventually the jasmonate biosynthesis (Fig. 3.10; Zhao and Sakai 2003; Zhao et al. 2004; Trusov et al. 2006; He et al. 2013).

Defense against the necrotrophic pathogens *Alternaria brassicicola* and *Fusarium oxysporum* has been shown to be impaired in *Arabidopsis thaliana* mutants lacking functional Gβ (Trusov et al. 2006). The induction of a number of defense-related genes in Gβ-deficient mutants was severely reduced in response to *A. brassicicola* infection. In addition, Gβ-deficient mutants exhibit decreased sensitivity to a number of methyl jasmonate-induced responses such as induction of the plant defensin gene *PDF1.2* (Trusov et al. 2006). These results suggest that JA signaling is influenced by Gβ functional subunit of heterotrimeric G-protein. In contrast, Gα-deficient mutants did not show any significant reduction in defense responses against *A. brassicicola*, suggesting JA signaling was not influenced by Gα (Trusov et al. 2006).

Heterotrimeric G-proteins are involved in activating defense responses against necrotrophic fungal pathogens such as *A. brassicicola* and *F. oxysporum* in *A. thaliana*. In contrast, these G-proteins did not activate defense responses against the hemibiotrophic bacterial pathogen *Pseudomonas syringae* pv. *tomato* (Trusov et al. 2006). It is known that JA signaling is involved in conferring disease resistance only against necrotrophic fungal pathogens, and not against hemibiotrophic bacterial pathogens (Spoel et al. 2007; Vidhyasekaran 2007; Zheng et al. 2006). These results provide additional evidences to show that heterotrimeric G-proteins participate in JA signaling system.

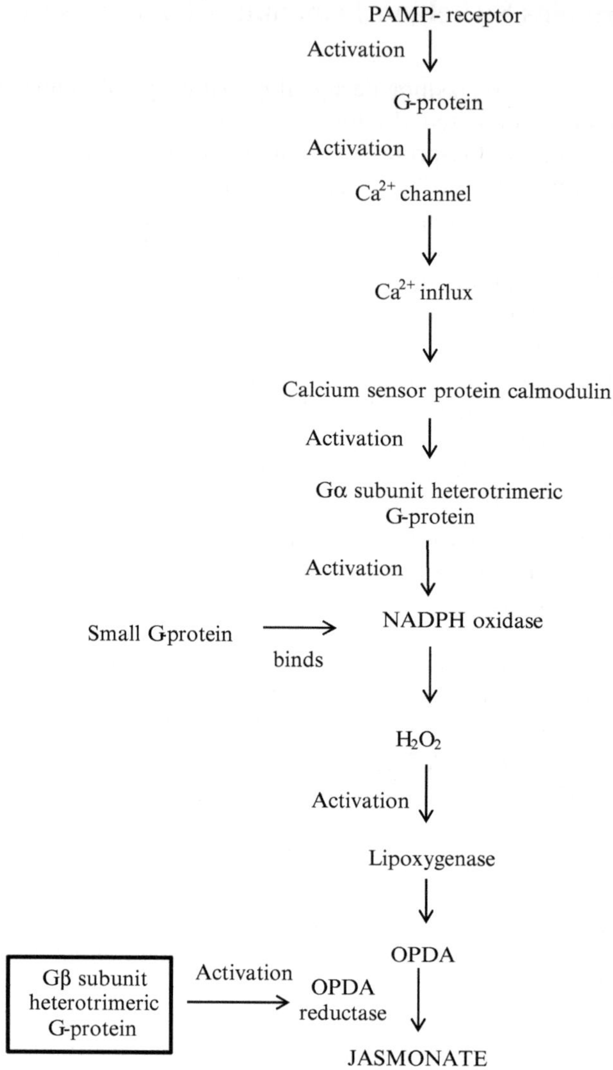


**Fig. 3.10** Small and heterotrimeric G-proteins- induced jasmonate biosynthesis

## 3.19 G-Proteins Switch on Abscisic Acid Signaling System

### *3.19.1 G-Proteins May Be Involved in Abscisic Acid Biosynthesis*

G-proteins may regulate ABA signaling system (Pandey and Assmann 2004; Pandey et al. 2006; Liu et al. 2007a). A regulator of G protein signaling (RGS) proteins, RGS1, has been identified in *Arabidopsis* (Chen et al. 2006). Transgenic *Arabidopsis* plants overexpressing RGS1 were developed and RGS1 overexpression significantly stimulated the expression of *NCED* encoding the 9-*cis*-epoxycarotenoid dioxygenase (NCED) enzyme, which cleaves 9-*cis* xanthophylls to xanthoxin. The NCED is the first committed step for ABA synthesis (Chen et al. 2006; Wasilewska et al. 2008). The transgenic plants overexpressing RGS1 also showed increased expression of *ABA2* gene (Chen et al. 2006), which encodes dehydrogenase/reductase involved in conversion of xanthoxin to abscisic acid aldehyde (Wasilewska et al. 2008). Abscisic aldehyde oxidase (AAO) catalyses the final step of ABA biosynthesis, which converts ABA aldehyde to ABA (Iuchi et al. 2001). These results indicate that RGS1 is involved in biosynthesis of ABA (Chen et al. 2006).

### 3.19.2 G-Proteins May Act as Abscisic Acid Receptors

The receptor for G-protein called G protein-coupled receptor (GPCR) has also been identified as a plasma membrane receptor for the plant hormone abscisic acid (Liu et al. 2007a, b). The G-protein receptor, GCR2, has been identified as a plasma membrane-localized receptor for ABA in *Arabidopsis* (Liu et al. 2007a). GCR2 is a membrane protein with seven transmembrane helices. The GCR2 genetically and physically interacts with the *Arabidopsis* G protein α subunit GPA1 to mediate all known ABA responses in *Arabidopsis*. This receptor binds ABA with high affinity at physiological concentration. The GCR2 interacts with the Gαβγ complex. Binding of ABA to GCR2 results in the release of Gα and Gβγ dimer to activate downstream ABA signaling events (Liu et al. 2007a).

Another novel class of GPCR-type G proteins, GTGs has been suggested as ABA receptors (Pandey et al. 2009). GTG proteins contain GTP binding/GTPase activity. GTGs interact with the heterotrimeric G-protein α-subunit, GPA1 and the GTP-bound form of GPA1 inhibits the GTPase activity of GTG proteins. The GDP-bound GTGs bind ABA stronger than GTP-bound form, thus it might be the active form for perceiving and transducing ABA signal (Guo et al. 2011).

### 3.19.3 G-Proteins May Regulate Inward $K^+$ Channels and Slow Anion Channels Activated by ABA

In the G protein signaling system, the Gα subunit GPA1, the Gβ subunit (AGB1), and the G-protein coupled receptor (GCR1) regulate the downstream events in ABA signaling (Pandey et al. 2006). GCR1 may interact with GPA1 and the ligand binding to GCR1 may regulate heterotrimeric G protein signaling via GPA1 and the βγ dimer (Eckardt 2004). GPA1 acts upstream of the small GTPase Rac in inducing disease resistance response in rice (Suharsono et al. 2002).

The *Arabidopsis* GPA1 has been demonstrated to be involved in the regulation of inward $K^+$ channels and slow anion channels activated by ABA (Wang et al. 2001). $Ca^{2+}$-permeable channels, like inward $K^+$ channels and S-type anion channels are regulated by Gα-dependent pathways. These channels are activated by ABA and this activation is lost in *gpa1* mutant lines (Zhang et al. 2011). The results suggest that ABA activation of these channels is dependent on G-proteins.

### 3.19.4 ABA Increases Expression of Genes Encoding Heterotrimeric G-Proteins

ABA treatment triggered an increase in expression of *BnGβ1* encoding a putative G-protein β subunit (Gβ) in *Brassica napus*. The expression of *BnGβ1* was induced by low ABA concentrations, but high ABA concentrations inhibited the expression, suggesting that there might be an ABA-mediated feedback expression of *BnGβ1* expression (Gao et al. 2010b). In contrast, the expression of *BnGA1* encoding Gα G-protein subunit in *B. napus* was significantly induced by the high concentrations of ABA (Gao et al. 2010a). These results suggest that the heteromeric G-protein may participate in ABA signaling system.

### 3.19.5 G-Protein May Play a Role in ABA Signaling Between ABA Reception and ROS Production

The participation of G-proteins in ABA signaling system was demonstrated using *Arabidopsis* mutants deficient in GPA1, the sole canonical *Arabidopsis* Gα subunit expression (Zhang et al. 2011). *Arabidopsis* mutants deficient in GPA1 were found to be impaired in ABA inhibition of $K^+$ influx channels, and in pH-independent activation of anion efflux channels. The *gpa1* mutants were disrupted both in ABA-induced $Ca^{2+}$-channel activation, and in production of reactive oxygen species (ROS) in response to ABA. These studies further suggested that absence of GPA1 interrupts ABA signaling between ABA reception and ROS production, with a consequent impairment in $Ca^{2+}$-channel activation (Zhang et al. 2011). These results strongly suggest that the heterotrimeric G-protein is involved in ABA signaling system.

### 3.19.6 G-Proteins May Be Involved in Modulation of ABA-Induced Stomatal Closure Immune Response

G-proteins may be involved in modulation of ABA-induced stomatal closure immune response. ABA is known to be involved in stomatal pore closure which is a key component in plant immune responses against bacterial pathogens (Takahashi et al. 2007; Hettenhausen et al. 2012; Uraji et al. 2012). The

*Arabidopsis* small GTPase protein AtRac1 has been identified as a central component in the ABA-mediated stomatal closure (Lemichez et al. 2001). ABA treatment induced inactivation of AtRac GTPases. In contrast, ABA-treatment-induced AtRac inactivation was not observed in the *Arabidopsis* ABA-insensitive mutant *abi1-1*, which is impaired in stomatal closure. Expression of a dominant-positive mutant of *AtRac1* blocked the ABA-mediated stomatal closure in wild-type plants, whereas expression of a dominant-negative AtRac1 mutant recapitulated the ABA effects in the absence of the hormone. It was also observed that the dominant-negative form of AtRac1 could restore stomatal closure in the *abi-1* mutant (Lemichez et al. 2001). These results demonstrate that a small monomeric G-protein plays a key role in the stomatal closure immune response induced by ABA.

## 3.20 G-Proteins May Participate in Gibberellic Acid Signaling

Heteromeric G-protein has been reported to play a role in GA signaling system (Iwasaki et al. 2003). GA treatment triggered an increase in expression of *BnGβ1*encoding a putative G-protein β subunit (Gβ) in *Brassica napus*. *BnGβ1* was more prominently induced by high concentrations of GA (Gao et al. 2010b). The expression of *BnGA1* encoding a G-protein α subunit of *B. napus* was induced by low gibberellin 3 (GA3) concentrations and higher GA3 concentrations inhibit the expression of *BnGA1* (Gao et al. 2010a). These results suggest that heterotrimeric G-proteins may be involved in signaling pathways modulated by GA.

## 3.21 G-Proteins Participate in Brassinosteroid Signaling

G-proteins also take part in activation of brassinosteroid (BR) signaling system. The α subunit of plant heterotrimeric G proteins (Gα) has been shown to participate in BR signaling responses in *Arabidopsis* and rice (Ullah et al. 2002; Wang et al. 2006; Gao et al. 2008; Oki et al. 2009). The expression of the *Brassica napus* heterotrimeric G protein α subunit (*BnGA1*) was induced by exogenous application of brassinosteroid (Gao et al. 2010a).

Wang et al. (2006) showed that the rice *d1* mutant was less responsive to 24-epibrassinolide compared to wild-type plants, suggesting that the rice Gα was involved in brassinosteroid signaling. Rice Gα has been shown to affect BR signaling cascade, but Gα may not be a signaling molecule in BRI1-mediated perception/transduction (Oki et al. 2009). These studies suggest that G-proteins may participate in BR signaling system.

## 3.22 Interplay Between G-Proteins and Auxin Signaling Systems

The auxin signaling system plays an important role in plant innate immunity (Fu and Wang 2011). The auxin indole-3-acetic acid activates small monomeric G-proteins (Rac-like GTPases) and they in turn stimulate auxin-responsive gene expression (Tao et al. 2002; Xu et al. 2010; Wu et al. 2011). Addition of exogenous auxin to tobacco seedlings stimulates activation of endogenous Rac/Rops (Tao et al. 2002). It has been demonstrated that overexpressing a wild-type tobacco Rac/Rop GTPase, NtRac1, and its constitutively active mutant form activates auxin-responsive gene expression (Tao et al. 2002). On the other hand, overexpressing dominant-negative NtRac1and Rac-negative regulators, or reducing the endogenous NtRac1 level, suppresses auxin –induced gene expression (Tao et al. 2002).

GH3 and DR5 are natural and synthetic auxin-inducible promoters, respectively. *GH3-GUS* (β-glucuronidase) and *DR5-GUS* were found to be induced by naphthalene acetic acid (NAA) in transfected protoplasts (Tao et al. 2002). Coexpression of NtRac1 activated expression of the auxin-responsive genes. The Rac-GTPase regulators Rac-GAP (GTP-ase activating protein) and Rac-GD1 (guanine nucleotide dissociation inhibitor) downregulate the activity of the Rac GTPases by maintaining them predominantly in the GDP-bound inactive state. Expression of these negative regulators in transfected protoplasts counteracted considerably the ability of coexpressed NtRac1 to activate auxin –responsive promoters. These results suggest that active NtRac1 upregulates a subset of auxin-responsive promoters in transfected protoplasts (Tao et al. 2002).

It has been reported that the tobacco RAC-like GTPase Ntrac1 activates auxin-responsive gene expression in the absence of auxin (Tao et al. 2002). It suggests a signaling role for auxin-regulated gene expression for this small GTPase. The dominant-negative mutant forms of NtRac1(NtRac1[DN], with either a D121A or a T20N conversion rendering them constitutively in the inactive GDP-bound state) and the negative Rac-GTPase regulators Rac GAP and Rac-GDI block auxin-induced gene expression in the presence of exogenous auxin, indicating that endogenous active NtRacs are important for mediating the auxin signal to responsive genes. Further it has been shown that auxin activates NtRacs (Tao et al. 2002). Collectively these results suggest the existence of a signaling pathway whereby auxin-activated NtRacs stimulate downstream responsive gene expression.

## 3.23 G-Proteins Activate Defense-Related Enzymes

G-proteins may regulate expression of several defense-related enzymes. OsRac1, the rice small G-protein, induced several enzymes associated with the phenylpropanoid pathway, including caffeic acid-*O*-methyltransferase (the enzyme involved in lignin biosynthesis) and isoflavone reductase (the enzyme involved in phytoalexin biosynthesis) (Fujiwara et al. 2006). Cinnamoyl-CoA reductase, a key enzyme in lignin biosynthesis, is activated by Rac1 in rice (Kawasaki et al. 2006). The G-protein would have activated the phenylpropanoid pathway, probably by triggering MAPK signaling system (Lieberherr et al. 2005; Fujiwara et al. 2006). Silencing of the rice MAPK, OsMAPK6, reduced the level of phenylalanine ammonia-lyase (*PAL*) mRNA in rice (Lieberherr et al. 2005). The *PAL* gene is an important defense gene and the PAL enzyme represents the entry point of phenylpropanoid pathway for the synthesis of phenolics, phytoalexins, lignins (Vidhyasekaran 2007).

## References


Abdul Kadar M, Lindsberg S (2010) Cytosolic calcium and pH signaling in plants under salinity stress. Plant Signal Behav 5:233–238

Adams DR, Ron D, Kiely PA (2011) RACK1, a multifaceted scaffolding protein: structure and function. Cell Commun Signal 9:22

Agrawal GK, Iwahashi H, Rakwal R (2003) Small GTPase 'Rop': molecular switch for plant defense responses. FEBS Lett 546:173–180

Alexandre J, Lassales JP (1992) Intracellular $Ca^{2+}$ release by InsP3 in plants and effects of buffers on $Ca^{2+}$ diffusion. Philos Trans Biol Sci 338:53–61

Anderson DJ, Botella JR (2007) Expression analysis and subcellular localization of the *Arabidopsis thaliana* G-protein β subunit AGB1. Plant Cell Rep 26:1469–1480

Angelini R, Federico R (1989) Histochemical evidence of polyamine oxidation and generation of hydrogen peroxide in the cell wall. J Plant Physiol 135:212–217

Angelini R, Bragaloni M, Federico R, Infantino A, Porta-Puglia A (1993) Involvement of polyamines, diamine oxidase and peroxidase in resistance of chick-pea to *Ascochyta rabiei*. J Plant Physiol 142:704–709

Angelini R, Tisi A, Rea G, Chen MM, Botta M, Federico R, Cona A (2008) Involvement of polyamine oxidase in wound healing. Plant Physiol 146:162–177

Apone F, Alyeshmerni N, Wiens K, Chalmers D, Chrispeels MJ, Colucci G (2003) The G-protein-coupled receptor GCR1 regulates DNA synthesis through activation of phosphatidylinositol-specific phospholipase C. Plant Physiol 133:571–579

Assmann SM (2002) Heterotrimeric and unconventional GTP binding proteins in plant cell signaling. Plant Cell 14S:S355–S373

Azevedo C, Betsuyaku S, Peart J, Takahashi A, Noel L, Sadanandom A, Casals C, Parker J, Shirasu K (2006) Role of SGT1 in resistance protein accumulation in plant immunity. EMBO J 25:2007–2016

Baxter-Burrell A, Yang Z, Springer PS, Bailey-Serres J (2002) RopGAP4-dependent Rop GTPase rheostat control of *Arabidopsis* oxygen deprivation tolerance. Science 296:2026–2028

Beffa R, Szell M, Meuwly P, Pay A, Vogeli-Lange R, Metraux JP, Neuhaus G, Meins F, Ferenc N (1995) Cholera toxin elevates pathogen resistance and induces pathogenesis-related gene expression in tobacco. EMBO J 14:5753–5761

Berken A (2006) ROPs in the spotlight of plant signal transduction. Cell Mol Life Sci 63:2446–2459

Berridge MJ (1993) Inositol trisphosphate and calcium signaling. Nature 361:315–325

Bhardwaj D, Sheikh AH, Sinha AK, Tuteja N (2011) Stress induced β subunit of heterotrimeric G-proteins from *Pisum sativum* interacts with mitogen activated protein kinase. Plant Signal Behav 6:287–292

Bisht NC, Jez JM, Pandey S (2011) An elaborate heterotrimeric G-protein family from soybean expands the diversity of plant G-protein networks. New Phytol 190:35–48

Blumwald E, Aharon GS, Lam BC-H (1998) Early signal transduction pathways in plant-pathogen interactions. Trends Plant Sci 3:342–346

Bokoch GM, Diebold BA (2002) Current molecular models for NADPH oxidase regulation by RacGTPase. Blood 100:2692–2696

Bright J, Desikan R, Hancock JT, Weir IS, Neill SJ (2006) ABA-induced NO generation and stomatal closure in *Arabidopsis* are dependent on $H_2O_2$ synthesis. Plant J 45:113–122

Cabrera-Vera TM, Vanhauwe J, Thomas TO, Medkova M, Preininger A, Mazzoni MR, Hamm HE (2003) Insights into G protein structure, function, and regulation. Endocr Rev 24:765–781

Casey PJ (1995) Protein lipidation in cell signaling. Science 268:221–225

Chakravorty D, Trusov Y, Zhang W, Sheahan MB, Acharya BW, McCurdy DW, Assmann SM, Botella JR (2011) A highly atypical heterotrimeric G protein γ subunit is involved in guard cell $K^+$ channel regulation and morphological development in *Arabidopsis thaliana*. Plant J 67:840–851

Chen J-G, Pandey S, Huang J, Alonso JM, Ecker JR, Assmann SM, Jones AM (2004) GCR1 can act independently of heterotrimeric G-protein in response to brassinosteroids and gibberellins in Arabidopsis and germination. Plant Physiol 135:907–915

Chen Y, Fangfang JI, Xie H, Liang J (2006) Overexpression of the regulator of G-protein signalling protein enhances ABA-mediated inhibition of root elongation and drought tolerance in *Arabidopsis*. J Exp Bot 57:2101–2110

Choudhury SR, Bisht NC, Thompson R, Todorov O, Pandey S (2011) Conventional and novel Gγ protein families constitute the heterotrimeric G-protein signaling network in soybean. PLoS One 6:e23361

Clapham DE, Neer EJ (1997) G protein βγ subunits. Annu Rev Pharmacol Toxicol 37:167–203

Colucci G, Apone E, Alyeshmerni N, Chalmers D, Chrispeels MJ (2002) GCR1, the putative Arabidopsis G protein-coupled receptor gene is cell cycle-regulated, and its overexpression abolishes seed dormancy and shortens time to flowering. Proc Natl Acad Sci U S A 99:4736–4741

Cowley T, Walters DR (2002a) Polyamine metabolism in barley reacting hypersensitively to the powdery mildew fungus *Blumeria graminis* f. sp. *hordei*. Plant Cell Environ 25:461–468

Cowley T, Walters DR (2002b) Polyamine metabolism in an incompatible interaction between barley and the powdery mildew fungus, *Blumeria graminis f. sp hordei*. J Phytopathol 150:1–7

DeFalco TA, Bender KW, Snedden WA (2010) Breaking the code: $Ca^{2+}$ sensors in plant signaling. Biochem J 426:27–40

Delgado-Cerezo M, Sánchez-Rodriguez C, Escudero V, Miedes E, Fernández PV, Jordá L, Hernández-Blanco C, Sánchez-Vallet A, Bednarek P, Schulze-Lefert P, Somerville S, Estevez JM, Persson S, Molina A (2012) *Arabidopsis* heteromeric G-protein regulates cell wall defense and resistance to necrotrophic fungi. Mol Plant 6:98–114

Desikan R, Hancock JT, Bright J, Harrison J, Weir I, Hooley R, Neill SJ (2005) A role for ETR1 in hydrogen peroxide signaling in stomatal guard cells. Plant Physiol 137:831–834

Ding L, Pandey S, Assmann SM (2008) Arabidopsis extra-large G proteins (XLGs) regulate root morphogenesis. Plant J 53:248–263

Dodd AN, Kudla J, Sanders D (2010) The language of calcium signaling. Annu Rev Plant Biol 61:593–620

Eckardt NA (2004) Abscisic acid signal transduction: function of G protein-coupled receptor 1 in Arabidopsis. Plant Cell 16:1353–1354

Fedoroff N (2006) Redox regulatory mechanisms in cellular stress responses. Ann Bot (Lond) 98:289–300

Fu J, Wang S (2011) Insights into auxin signaling in plant-pathogen interactions. Front Plant Sci 2:1–7. doi:10.3389/fpls.2011.00074, article 74

Fu DQ, Ghabrial S, Kachroo A (2009) *GmRAR1* and *Gm SGT1* are required for basal, *R* gene-mediated and systemic acquired resistance in soybean. Mol Plant Microbe Interact 22:86–95

Fujisawa Y, Kato H, Iwasaki Y (2001) Structure and function of heterotrimeric G proteins in plants. Plant Cell Physiol 42:789–794

Fujiwara M, Umemura K, Kawasaki T, Shimamoto K (2006) Proteomics of Rac GTPase signaling reveals its predominant role in elicitor-induced defense response of cultured rice cells. Plant Physiol 140:734–745

Fujiwara M, Hamada S, Hiratsuka M, Fukao Y, Kawasaki T, Shimamoto K (2009) Proteome analysis of detergent-resistant membranes (DRMs) associated with OsRac1-mediated innate immunity in rice. Plant Cell Physiol 50:1191–1200

Fukada Y (1995) Prenylation and carboxylmethylation of G-protein γ subunit. Methods Enzymol 250:91–105

Galon Y, Finkler A, Fromm H (2010) Calcium-regulated transcription in plants. Mol Plant 3:653–669

Gao Y, Wang S, Asami T, Chen JG (2008) Loss-of-function mutations in the Arabidopsis heterotrimeric G-protein α subunit enhance the developmental defects of brassinosteroid signaling and biosynthesis mutants. Plant Cell Physiol 49:1013–1024

Gao Y, Li T, Liu Y, Ren C, Zhao Y, Wang M (2010a) Isolation and characterization of gene encoding G protein α subunit protein responsive to plant hormones and abiotic stresses in *Brassica napus*. Mol Biol Rep 37:3957–3965

Gao Y, Li T, Zhao Y, Ren C, Liu Y, Wang M (2010b) Cloning and characterization of a G protein β subunit gene responsive to plant hormones and abiotic stresses in *Brassica napus*. Plant Mol Biol Rep 28:450–459

Garcia-Brugger A, Lamotte O, Vandelle E, Bourque S, Lecourieux D, Poinssot B, Wendehenne D, Pugin A (2006) Early signaling events induced by elicitors of plant defenses. Mol Plant Microbe Interact 19:711–724

Gardiner DM, Kazan K, Praud S, Torney FJ, Rusu A, Manners JM (2010) Early activation of wheat polyamine biosynthesis during *Fusarium* head blight implicates putrescine as an inducer of trichothecene mycotoxin production. BMC Plant Biol 10:289

Gelli A, Higgins VJ, Blumwald E (1997) Activation of plant plasma membrane $Ca^{2+}$-permeable channels by race-specific fungal elicitors. Plant Physiol 113:269–279

Gerber IB, Laukens K, Witters E, Dubery IA (2006) Lipopolysaccharide-responsive phosphoproteins in *Nicotiana tabacum* cells. Plant Physiol Biochem 44:369–379

Gilman AG (1987) G proteins: transducers of receptor-generated signals. Annu Rev Biochem 56:615–649

Gookin T, Kim J, Assmann S (2008) Whole proteome identification of plant candidate G-protein coupled receptors in *Arabidopsis*, rice, and poplar: computational prediction and in-vivo protein coupling. Genome Biol 9:R120

Gray WM, Muskett PR, Chuang H-W, Parker JE (2003) *Arabidopsis* SGT1b is required for $SCF^{TIR1}$-mediated auxin response. Plant Cell 15:1310–1319

Gu Y, Wang Z, Yang Z (2004) ROP/RAC GTPase: an old new master regulator for plant signaling. Curr Opin Plant Biol 7:527–536

Guo J, Yang X, Weston DJ, Chen J-G (2011) Abscisic acid receptors: past, present and future. J Integr Plant Biol 53:469–479

Guo L, Devaiah SP, Narasimhan R, Pan X, Zhang Y, Zhang W, Wang X (2012) Cytosolic glyceraldehyde-3-phosphate dehydrogenases interact with phospholipase Dδ to transduce hydrogen peroxide signals in the *Arabidopsis* response to stress. Plant Cell 24:2200–2212

Haggag WM, Mahmoud YS, Farag EM (2010) Signaling necessities and function of polyamines/jasmonate-dependent induced resistance in sugar beet against *Beet mosaic virus* (BtMV) infection. New York Sci J 3:95–103

He DY, Yazaki Y, Nishizawa Y, Takai R, Yamada K, Sakano K, Shibuya N, Minami E (1998) Gene activation by cytoplasmic acidification in suspension-cultured rice cells in response to the potent elicitor, N-acetylchitoheptaose. Mol Plant Microbe Interact 11:1167–1174

He JM, Ma XG, Zhang Y, Sun TF, Xu FF, Chen YP, Liu X, Yue M (2013) Role and interrelationship of Gα protein, hydrogen peroxide and nitric oxide in UV-B-induced stomatal closure in *Arabidopsis* leaves. Plant Physiol 161(3):1570–1583

Hettenhausen C, Baldwin IT, Wu J (2012) Silencing *MPK4* in *Nicotiana attenuata* enhances photosynthesis and seed production but compromises abscisic acid-induced stomatal closure and guard cell-mediated resistance to *Pseudomonas syringae* pv. *tomato* DC3000. Plant Physiol 158:759–776

Hiraga S, Ito H, Yamakawa H, Obtsubo N, Seo S, Mitshuhara I, Matsui H, Honma M, Ohashi Y (2000) An HR-induced tobacco peroxidase gene is responsive to spermine, but not to salicylate, methyl jasmonate, and ethephon. Mol Plant Microbe Interact 13:210–216

Hoefle C, Huesmann C, Schultheiss H, Börnke F, Hensel G, Kumlehn J, Huckelhoven R (2011) A barley ROP GTPase ACTIVATING PROTEIN associates with microtubules and regulates entry of the barley powdery mildew fungus into leaf epidermal cells. Plant Cell 23:2422–2439

Hossain MS, Koba T, Harada K (2003) Cloning and characterization of two full-length cDNAs, *TaGA1* and *TaGA2*, encoding G-protein α subunits expressed differentially in wheat genome. Genes Genet Syst 78:127–138

Hu X-Y, Neill SJ, Cai WM, Tang Z-C (2004) Induction of defence gene expression by oligogalacturonic acid requires increases in both cytosolic calcium and hydrogen peroxide in *Arabidopsis thaliana*. Cell Res 14:234–240

Hu J, Wang Y, Zhang X, Lloyd JR, Li JH, Karpiak J, Costanzi S, Wess J (2010) Structural basis of G protein-coupled receptor – G protein interactions. Nat Chem Biol 6:541–548

Hückelhoven R, Kogel K-H (1998) Tissue-specific superoxide generation at interaction sites in resistant and susceptible near-isogenic barley lines attacked by the powdery mildew fungus. Mol Plant Pathol 1:287–292

Hückelhoven R, Trujillo M, Kogel KH (2000) Mutations in *Ror1* and *Ror2* genes cause modification of hydrogen peroxide accumulation in *mlo*-barley under attack from the powdery mildew fungus. Mol Plant Pathol 1:287–292

Huesmann C, Reiner T, Hoefle C, Preuss J, Jurca ME, Domoki M, Fehér A, Hückelhoven R (2012) Barley ROP binding kinase1 is involved in microtubule organization and in basal penetration resistance to the barley powdery mildew fungus. Plant Physiol 159:311–320

Ishikawa A (2009) The Arabidopsis G-protein beta-subunit is required for defense response against *Agrobacterium tumefaciens*. Biosci Biotechnol Biochem 73:47–52

Iuchi S, Kobayashi M, Taji T, Nakamoto M, Seki M, Kato T, Tabata S, Kakubari Y, Yamaguchi-Shinozaki K, Shinozaki K (2001) Regulation of drought tolerance by gene manipulation of 9-cis-epoxycarotenoid dioxygenase, a key enzyme in abscisic acid biosynthesis in *Arabidopsis*. Plant J 27:325–333

Iwasaki Y, Fujisawa Y, Kato H (2003) Function of heterotrimeric G protein in gibberellin signaling. J Plant Growth Regul 22:126–133

Iwata M, Umemura K, Teraoka T, Usami H, Fujisawa Y, Iwasaki Y (2002) Role of the α subunit of heterotrimeric G-protein in probenazole-inducing defense signaling in rice. J Gen Plant Pathol 63:83–86

Izawa Y, Minami M, Ohki S, Iwasaki Y (2010) Expression profile of the α subunit of the heterotrimeric G protein in rice. Plant Signal Behav 5:845–847

Jabs T, Dietrich RA, Dangl JL (1996) Initiation of runaway cell death in an *Arabidopsis* mutant by extracellular superoxide. Science 273:1853–1856

Joo JH, Wang S, Chen JG, Jones AM, Fedoroff NV (2005) Different signaling and cell death roles of heterotrimeric G protein α and β subunits in the Arabidopsis oxidative stress response to ozone. Plant Cell 17:957–970

Josefsson LG, Rask L (1997) Cloning of a putative G-protein-coupled receptor from *Arabidopsis thaliana*. Eur J Biochem 249:415–420

Kato C, Mizutani T, Tamaki H, Kumagai H, Kamiya T, Hinbe A, Fujisawa Y, Kato H, Iwasaki Y (2004) Characterization of heterotrimeric G protein complexes in rice plasma membrane. Plant J 38:320–331

Kawamura Y, Takenaka S, Hase S, Kubota M, Kanayama Y, Nakaho K, Klessig DF, Takahashi H (2009) Enhanced defense responses in *Arabidopsis* induced by the cell wall protein fractions from *Pythium oligandrum* require *SGT1*, *RAR1*, *NPR1* and *JAR1*. Plant Cell Physiol 50:924–934

Kawasaki T, Henmi K, Ono E, Hatakeyama S, Iwano M, Satoh H, Shimamoto K (1999) The small GTP-binding protein rac is a regulator of cell death in plants. Proc Natl Acad Sci U S A 96:10922–10926

Kawasaki T, Koita H, Nakatsubo T, Hasegawa K, Wakabayashi K, Takahashi H, Umemura K, Umezawa T, Shimamoto K (2006) Cinnamoyl-CoA reductase, a key enzyme in lignin biosynthesis, is an effector of small GTPase Rac in defense signaling in rice. Proc Natl Acad Sci U S A 103:230–235

Kaziro Y, Itoh H, Kozasa T, Nakafuku M, Satoh T (1991) Structure and function of signal-transducing GTP- binding properties. Annu Rev Biochem 60:349–400

Kiirika LM, Bergmann HF, Schikowsky C, Wimmer D, Korte J, Schmitz U, Niehaus K, Colditz F (2012) Silencing of the Rac1 GTPase *MtROP9* in *Medicago truncatula* stimulates early mycorrhizal and oomycete root colonizations but negatively affects rhizobial infection. Plant Physiol 159:501–516

Komatsu S, Yang G, Hayashi N, Kaku H, Umemura K, Iwasaki Y (2004) Alterations by a defect in a rice protein α subunit in probenazole and pathogen-induced responses. Plant Cell Environ 27:947–957

Kresge N, Simoni RD, Hill RL (2007) The biosynthesis of polyamines: the work of Herbert Tabor and Celia White Tabor. J Biol Chem 282:e26–e28

Kudla J, Batistić O, Hashimoto K (2010) Calcium signals: the lead currency of plant information processing. Plant Cell 22:541–563

Laohavisit A, Mortimer JC, Demidchik V, Coxon KM, Stancombe MA, Macpherson N, Brownlee C, Hofmann A, Webb AAR, Miedema H, Battey NH, Davies JM (2009) *Zea mays* annexins modulate cytosolic free $Ca^{2+}$ and generate a $Ca^{2+}$-permeable conductance. Plant Cell 21:479–493

Laohavisit A, Brown AT, Cicuta P, Davies JM (2010) Annexins: components of the calcium and reactive oxygen signaling network. Plant Physiol 152:1824–1829

Lapous D, Mathieu Y, Guern J, Lauriere C (1998) Increase of defense gene transcripts by cytoplasmic acidification in tobacco cell suspensions. Planta 205:452–458

Laxalt AM, Raho N, ten Have A, Lamattina L (2007) Nitric oxide is critical for inducing phosphatidic acid accumulation in xylanase-elicited cells. J Biol Chem 282:21160–21168

Lecourieux – Ouaked D, Pugin A, Lebrun-Garcia A (2000) Phosphoproteins involved in the signal transduction of cryptogein, an elicitor of defense reactions in tobacco. Mol Plant Microbe Interact 13:821–829

Lecourieux D, Ranjeva R, Pugin A (2006) Calcium in plant defence-signaling pathways. New Phytol 171:249–269

Legendre L, Heinstein PF, Low PS (1992) Evidence for participation of GTP-binding proteins in elicitation of the rapid oxidative burst in cultured soybean cells. J Biol Chem 267:20140–20147

Lemichez E, Wu Y, Sanchez J-P, Mettouchi A, Mathur J, Chua N-H (2001) Inactivation of AtRac1 by abscisic acid is essential for stomatal closure. Genes Dev 15:1808–1816

Levine A, Tenhaken R, Dixon R, Lamb C (1994) $H_2O_2$ from the oxidative burst orchestrates the plant hypersensitive disease resistance response. Cell 79:583–593

Li J-H, Liu Y-Q, Pin L, Lin H-F, Bai Y, Wang X-C, Chen Y-L (2009) A signaling pathway linking nitric oxide production to heterotrimeric G protein and hydrogen peroxide regulates extracellular calmodulin induction of stomatal closure in Arabidopsis. Plant Physiol 150:114–124

Lieberherr D, Thao NP, Nakashima A, Umemura K, Kawasaki T, Shimamoto K (2005) A sphingolipid elicitor-inducible mitogen-activated protein kinase is regulated by the small GTPase OsRac1 and heterotrimeric G-protein in rice. Plant Physiol 138:1644–1652

Lindermayr C, Sell S, Müller B, Leister D, Dumer J (2010) Redox regulation of NPR1-TGA1 system of Arabidopsis thaliana by nitric oxide. Plant Cell 22:2894–2907

Liu X, Yue Y, Li W, Nie Y, Li W, Wu W-H, Ma L (2007a) A G protein-coupled receptor is a plasma membrane receptor for the plant hormone abscisic acid. Science 315:1712–1716

Liu X, Yue Y, Li W, Ma L (2007b) Response to comment on "A G protein-coupled receptor is a plasma membrane receptor for the plant hormone abscisic acid". Science 318:914d

Liu W, Chen A-M, Luo L, Sun J, Cao L-P, Yu G-O, Zhu J-B, Wang Y-Z (2010) Characterization and expression analysis of *Medicago truncatula* ROP GTPase family during the early stage of symbiosis. J Integr Plant Biol 52:639–652

Liu J-Z, Horstman HD, Braun E, Graham MA, Zhang C, Navarre D, Qiu W-L, Lee Y, Nettleton D, Hill JH, Whitham SA (2011) Soybean homologs of MPK4 negatively regulate defense responses and positively regulate growth and development. Plant Physiol 157:1363–1378

Llorente F, Alonso-Blanco C, Sanchez-Rodriquez C, Jorda L, Molina A (2005) ERECTA receptor-like kinase and heterotrimeric G protein from Arabidopsis are required for resistance to the necrotrophic fungus *Plectosphaerella cucumerina*. Plant J 43:165–180

Lorenzo O, Solano R (2005) Molecular players regulating the jasmonate signaling network. Curr Opin Plant Biol 8:532–540

Mao G, Meng X, Liu Y, Zheng Z, Chen Z, Zhang S (2011) Phosphorylation of a WRKY transcription factor by two pathogen-responsive MAPKs drives phytoalexin biosynthesis in *Arabidopsis*. Plant Cell 23:1639–1653

Marini F, Betti L, Scaramagli S, Biondi S, Torrigiani P (2001) Polyamine metabolism is upregulated in response to tobacco mosaic virus in hypersensitive, but not in susceptible, tobacco. New Phytol 149:301–309

Marrari Y, Crouthamel M, Irannejad R, Wedegaertner PB (2007) Assembly and trafficking of heterotrimeric G proteins. Biochemistry 46:7665–7677

Mason MG, Botella JR (2000) Completing the heterotrimer: isolation and characterization of an *Arabidopsis thaliana* G-protein γ-subunit cDNA. Proc Natl Acad Sci U S A 97:14784–14788

Mason MG, Botella JR (2001) Isolation of a novel G-protein γ subunit from *Arabidopsis thaliana* and its interaction with Gβ. Biochim Biophys Acta 1520:147–153

McIntyre WE (2009) Structural determinants involved in the formation and activation of G protein βγ dimers. Neurosignals 17:82–99

Mészáros T, Helfer A, Hatzimasoura E, Magyar Z, Serazetdinova L, Rios G, Bardóczy V, Teige M, Koncz C, Peck S, Bögre L (2006) The Arabidopsis MAP kinase kinase MKK1 participates in defence responses to the bacterial elicitor flagellin. Plant J 48:485–495

Miki D, Itoh R, Shimamoto K (2005) RNA silencing of single and multiple members in a gene family of rice. Plant Physiol 138:1903–1913

Misra S, Wu Y, Venkataraman G, Sopory SK, Tuteja N (2007) Heterotrimeric G-protein complex and G-protein-coupled receptor from a legume *Pisum sativum*: role in salinity and heat stress and cross-with phospholipase C. Plant J 51:656–669

Mo H, Pus EC (2002) Up-regulation of arginine decarboxylase gene expression and accumulation of polyamines in mustard (*Brassica juncea*) in response to stress. Physiol Plant 114:439–449

Moeder W, Yoshioka K, Klessig DF (2005) Involvement of the small GTPase Rac in the defense responses of tobacco to pathogens. Mol Plant Microbe Interact 18:116–124

Morel J, Fromentin J, Blein J-P, Simon-Plas F, Elmayan T (2004) RAC regulation of NtrbohD, the oxidase responsible for the oxidative burst in elicited tobacco cell. Plant J 37:282–293

Mucha E, Fricke I, Schaefer A, Wittinghofer A, Berken A (2011) Rho proteins of plants-functional cycle and regulation of cytoskeletal dynamics. Eur J Cell Biol 90:934–943

Munnik T, Arisz SA, de Vrije T, Musgrave A (1995) G protein activation stimulates phospholipase D signaling in plants. Plant Cell 7:2197–2210

Nakashima A, Chen L, Thao NP, Fujiwara M, Wong HL, Kuwano M, Umemura K, Shirasu K, Kawasaki T, Shimamoto K (2008) RACK1 functions in rice immunity by interacting with the Rac1 immune complex. Plant Cell 20:2265–2279

Nakashima K, Fujita Y, Kanamori N, Katagiri T, Umezawa T, Kidokoro S, Maruyama K, Yoshida T, Ishiyama K, Kobayashi M, Shinozaki K, Yamaguchi-Shinozaki K (2009) Three Arabidopsis SnRK2 protein kinases, SRK2D/SnRK2.2, SRK2E/SnRK2.6/OST1 and SRK2I/SnRK2.3 involved in ABA signaling are essential for the control of seed development and dormancy. Plant Cell Physiol 50:1345–1363

Nambeesan S, Datsenka T, Ferruzzi MG, Malladi A, Mattoo AK, Handa AK (2010) Overexpression of yeast spermidine synthase impacts ripening, senescence and decay symptoms in tomato. Plant J 63:836–847

Nambeesan S, AbuQamar S, Laluk K, Mattoo AK, Mickelbart MV, Ferruzzi MG, Mengiste T, Handa AK (2012) Polyamines attenuate ethylene-mediated defense responses to abrogate resistance to *Botrytis cinerea* in tomato. Plant Physiol 158:1034–1045

Nibau C, Wu HM, Cheung AY (2006) RAC/ROP GTPases: 'hubs' for signal integration and diversification in plants. Trends Plant Sci 11:309–315

Ogasawara Y, Kaya H, Hiraoka G, Yumoto F, Kimura S, Kadota Y, Hishinuma H, Senzaki E, Yamagoe S, Nagata K, Nara M, Suzuki K, Tanokura M, Kuchitsu K (2008) Synergistic activation of the *Arabidopsis* NADPH oxidase AtrbohD by $Ca^{2+}$ and phosphorylation. J Biol Chem 283:8885–8892

Oki K, Inaba N, Kitagawa K, Fujioka S, Kitano H, Fujisawa Y, Kato H, Iwasaki Y (2009) Function of the α subunit of rice heterotrimeric G protein in brassinosteroid signaling. Plant Cell Physiol 50:161–172

Oldham WM, Hamm HE (2008) Heterotrimeric G protein activation by G-protein-coupled receptors. Nat Rev Mol Cell Biol 9:60–71

Ono E, Wong HL, Kawasaki T, Hasegawa M, Kodama O, Shimamoto K (2001) Essential role of the small GTPase Rac in disease resistance of rice. Proc Natl Acad Sci U S A 98:759–764

Ozawa R, Bertea CM, Foti M, Narayana R, Arimura G-I, Muroi A, Horiuchi JI, Nishioka T, Maffei ME, Takabayashi J (2009) Exogenous polyamines elicit herbivore-induced volatiles in lima bean leaves: involvement of calcium, $H_2O_2$ and jasmonic acid. Plant Cell Physiol 50:2183–2199

Ozawa R, Bertea CM, Foti M, Narayana R, Arimura G-I, Muroi A, Maffei ME, Takabayashi J (2010) Polyamines and jasmonic acid induce plasma membrane potential variations in Lima bean. Plant Signal Behav 5:308–310

Palicz A, Foubert TR, Jesaitis AJ, Marodi L, McPhail LC (2001) Phosphatidic acid and diacylglycerol directly activate NADPH oxidase by interacting with enzyme components. J Biol Chem 276:3090–3097

Palmgren MG, Harper JF (1999) Pumping with plant P-type ATPases. J Exp Bot 50:883–893

Pandey S, Assmann SM (2004) The Arabidopsis putative G-protein-coupled receptor GCR1 interacts with the G protein α subunit GPA1 and regulates abscisic acid signaling. Plant Cell 16:1616–1632

Pandey S, Chen J-G, Jones AM, Assmann SM (2006) G-protein complex mutants are hypersensitive to abscisic acid regulation of germination and postgermination development. Plant Physiol 141:243–256

Pandey S, Nelson DC, Assmann SM (2009) Two novel GPCR-type G proteins are abscisic acid receptors in *Arabidopsis*. Cell 136:136–148

Pandey S, Wang R-S, Wilson L, Li S, Zhao Z, Gookin TE, Assmann SM, Albert R (2010) Boolean modeling of transcriptome data reveals novel modes of heterotrimeric G-protein action. Mol Syst Biol 6:372

Park J, Choi HJ, Lee S, Lee T, Yang Z, Lee Y (2000) Rac-related GTP-binding protein in elicitor-induced reactive oxygen generation by suspension-cultured soybean cells. Plant Physiol 124:725–732

Park J, Gu Y, Lee Y, Yang Z, Lee Y (2004) Phosphatidic acid induces leaf cell death in *Arabidopsis* by activating the Rho-related small G protein GTPase-mediated pathway of reactive oxygen species generation. Plant Physiol 134:129–136

Patterson RL, van Rossum DB, Barrow RK, Snyder SH (2004) RACK1 binds to inositol 1,4,5-trisphosphate receptors and mediates $Ca^{2+}$ release. Proc Natl Acad Sci U S A 101:2328–2332

Perchepied L, Balagué C, Riou C, Claudel-Renard C, Rivière N, Grezes-Besset B, Roby D (2010) Nitric oxide participates in the complex interplay of defense-related signaling pathways controlling disease resistance to *Sclerotinia sclerotiorum* in *Arabidopsis thaliana*. Mol Plant Microbe Interact 23:846–860

Perfus-Barboch I, Jones AM, Assmann SM (2004) Plant heterotrimeric G protein function: insights from *Arabidopsis* and rice mutants. Curr Opin Plant Biol 7:719–731

Ritchie S, Gilroy S (2000) Abscisic acid stimulation of phospholipase D in the barley aleurone is G-protein-mediated and localized to the plasma membrane. Plant Physiol 124:693–702

Roos W, Dordschbal B, Steighardt J, Heike M, Weiss D, Saalbach G (1999) A redox-dependent, G-protein-coupled phospholipase A of the plasma membrane is involved in the elicitation of alkaloid biosynthesis in Eschcholtzia californica. Biochim Biophys Acta Mol Cell Res 1448:390–402

Sakano K (2001) Metabolic regulation of pH in plant cells: role of cytoplasmic pH in defense reaction and secondary metabolism. Int Rev Cytol 206:1–44

Sang Y, Cui D, Wang X (2001) Phospholipase D and phosphatidic acid-mediated generation of superoxide in Arabidopsis. Plant Physiol 126:1449–1458

Sano H, Ohashi Y (1995) Involvement of small GTP-binding proteins in defense signal-transduction pathways of higher plants. Proc Natl Acad Sci U S A 92:4138–4144

Sano H, Seo S, Orudgev S, Youssefian K, Ishizuka K, Ohashi Y (1994) Expression of the gene for a small GTP binding protein in transgenic tobacco elevates endogenous cytokinin levels, abnormally induces salicylic acid in response to wounding, and increases resistance to tobacco mosaic virus infection. Proc Natl Acad Sci U S A 91:10556–10560

Schaller F (2001) Enzymes of the biosynthesis of octadecanoid-derived signaling molecules. J Exp Bot 52:11–23

Schiene K, Puhler A, Niehaus K (2000) Transgenic tobacco plants that express an antisense construct derived from a *Medicago sativa* cDNA encoding a Rac-related small GTP-binding protein fail to develop necrotic lesions upon elicitor infiltration. Mol Gen Genet 263:761–770

Schultheiss H, Dechert C, Kogel K-H, Hückelhoven R (2002) A small GTP-binding host protein is required for entry of powdery mildew fungus into epidermal cells of barley. Plant Physiol 128:1447–1454

Schultheiss H, Dechert C, Kogel K-H, Hückelhoven R (2003) Functional analysis of barley RAC/ROP G-protein family members in susceptibility to the powdery mildew fungus. Plant J 36:589–601

Seo W-S, Lee S-K, Song M-Y, Suh J-P, Hahn T-R, Ronald P, Jeon J-S (2008) The HSP90-SGT1-RAR1 molecular chaperone complex: a core modulator in plant immunity. J Plant Biol 51:1–10

Shirasu K (2009) The HSP90-SGT1 chaperone complex for NLR immune sensors. Annu Rev Plant Biol 60:139–164

Shirasu K, Schulze-Lefert P (2003) Complex formation, promiscuity, and multi-functionality: protein interactions in disease-resistance pathways. Trends Plant Sci 8:252–258

Spoel SH, Johnson JS, Dong X (2007) Regulation of tradeoffs between plant defenses against pathogens with different lifestyles. Proc Natl Acad Sci U S A 104:18842–18847

Stael S, Wurzinger B, Mair A, Mehlmer N, Vothknecht UC, Teige M (2012) Plant organellar calcium signaling: an emerging field. J Exp Bot 63:1525–1542

Steffens B, Sauter M (2010) G proteins as regulators in ethylene-mediated hypoxia signaling. Plant Signal Behav 5:375–378

Suharsono U, Fujisawa Y, Kawasaki T, Iwasaki Y, Satoh H, Shimamoto K (2002) The heterotrimeric G protein subunit acts upstream of the small GTPase Rac in disease resistance of rice. Proc Natl Acad Sci U S A 99:13307–13312

Szepesi A, Gémes K, Orosz G, Petö A, Takács Z, Vorák M, Tari I (2011) Interaction between salicylic acid and polyamines and their possible roles in tomato hardening processes. Acta Biol Szeged 55:165–166

Takahashi A, Casais C, Ichimura K, Shirasu K (2003a) HSP90 interacts with RAR1 and SGT1 and is essential for RPS2-mediated disease resistance in *Arabidopsis*. Proc Natl Acad Sci U S A 100:11777–11782

Takahashi Y, Berberich T, Miyazaki A, Seo S, Ohashi Y, Kusano T (2003b) Spermine signaling in tobacco: activation of mitogen-activated protein kinases by spermine is mediated through mitochondrial dysfunction. Plant J 36:820–829

Takahashi Y, Uehara Y, Berberich T, Ito A, Saitoh H, Miyazaki A, Terauchi R, Kusano T (2004) A subset of hypersensitive response marker genes, including HSR203J, is the downstream target of a spermine signal transduction pathway in tobacco. Plant J 40:586–595

Takahashi F, Yoshida R, Ichimura K, Mizoguchi T, Seo S, Yonezawa M, Maruyama K, Yamaguchi-Shinozaki K, Shinozaki K (2007) The mitogen-activated protein kinase cascade MKK3-MPK6 is an important part of the jasmonate signal transduction pathway in Arabidopsis. Plant Cell 19:805–818

Takeda S, Gapper C, Kaya H, Bell E, Kuchitsu K, Dolan L (2008) Local positive feedback regulation determines cell shape in root hair cells. Science 319:1241–1244

Takida S, Wedegaertner PB (2003) Heterotrimer formation, together with isoprenylation, is required for plasma membrane targeting of Gβγ. J Biol Chem 278:17284–17290

Tao LZ, Cheung AY, Wu HM (2002) Plant Rac-like GTPases are activated by auxin and mediate auxin-responsive gene expression. Plant Cell 14:2745–2760

Tavernier E, Wendehenne J-P, Blein J-P, Pugin A (1995) Involvement of free calcium in action of cryptogein, a proteinaceous elicitor of a hypersensitive reaction in tobacco cells. Plant Physiol 109:1025–1031

Temple BR, Jones AM (2007) The plant heterotrimeric G-protein complex. Annu Rev Plant Biol 58:249–266

Tesmer JJG (2010) The quest to understand heterotrimeric G protein signaling. Nat Struct Mol Biol 17:650–652

Thao NP, Chen L, Nakashima HS-I, Umemura K, Takahashi A, Shirasu K, Kawasaki T, Shimamoto K (2007) RAR1 and HSP90 form a complex with Rac/ROP GTPase and function in innate-immune responses in rice. Plant Cell 19:4035–4045

Thordal-Christensen H, Zhang Z, Wei Y, Collinge DB (1997) Subcellular localization of $H_2O_2$ in plants: $H_2O_2$ accumulation in papillae and hypersensitive response during the barley-powdery mildew interaction. Plant J 11:1187–1194

Thung L, Trusov Y, Chakravorty D, Botella JR (2012) Gγ1 + Gγ2 + Gγ3 = Gβ: the search for heterotrimeric G-protein γ subunits in arabidopsis is over. J Plant Physiol 169:542–545

Torrigiani P, Rabiti AL, Bortolotti G, Betti I, Marani F, Canova A, Bagni N (1997) Polyamine synthesis and accumulation in the hypersensitive response to TMV in *Nicotiana tabacum*. New Phytol 135:467–473

Trusov Y, Rookes JE, Chakravorty D, Armour D, Schenk PM, Botella JR (2006) Heterotrimeric G proteins facilitate *Arabidopsis* resistance to necrotrophic pathogens and are involved in jasmonate signaling. Plant Physiol 140:210–220

Trusov Y, Rookes JE, Tilbrook K, Chakravorty D, Mason MG, Anderson D, Chen J-G, Jones AM, Botella JR (2007) Heterotrimeric G protein γ subunits provide functional selectivity in Gβγ dimer signaling in Arabidopsis. Plant Cell 19:1235–1250

Trusov Y, Zhang W, Assmann SM, Botella JR (2008) Gγ1 + Gγ2 ≠ GB. Heterotrimeric G protein Gγ-deficient mutants do not re-capitulate all phenotypes of Gβ-deficient mutants. Plant Physiol 147:636–640

Trusov Y, Sewelam N, Rookes JE, Kunkel M, Nowak E, Schenk PM, Botella JR (2009) Heterotrimeric G proteins-mediated resistance to necrotrophic pathogens includes mechanisms independent of salicylic acid-, jasmonic acid/ethylene- and abscisic acid-mediated defense signaling. Plant J 58:69–81

Trusov Y, Jorda I, Molina A, Botella JP (2010) G proteins and plant innate immunity. Springer, Heidelberg

Trusov Y, Chakravorty D, Botella JR (2012) Diversity of heterotrimeric G-protein γ subunits in plants. BMC Res Notes 5:608

Tsukada K, Ishizaka M, Fujisawa Y, Iwasaki Y, Yamaguchi T, Minami E, Shibuya N (2002) Rice receptor for chitin oligosaccharide elicitor does not couple to heterotrimeric G-protein: elicitor responses of suspension cultured rice cells from Daikoku dwarf (*d1*) mutants lacking a functional G-protein α-subunit. Physiol Plant 116:373–382

Tun NN, Santa-Catarina C, Begum T, Silveira V, Handro W, Floh EIS, Scherer GFE (2006) Polyamines induce rapid biosynthesis of nitric oxide (NO) in *Arabidopsis thaliana* seedlings. Plant Cell Physiol 47:346–354

Ullah H, Chen JG, Wang S, Jones AM (2002) Role of a heterotrimeric G protein in regulation of *Arabidopsis* seed germination. Plant Physiol 129:897–907

Uraji M, Katagiri T, Okuma E, Te W, Hossain M-A, Masuda C, Miura A, Nakamura Y, Mori I-C, Shinozaki K, Murata Y (2012) Cooperative function of PLDδ and PLDα1in abscisic acid-induced stomatal closure in Arabidopsis. Plant Physiol 159:450–460

Vadassery J, Oelmũller R (2009) Calcium signaling in pathogenic and beneficial plant microbe interactions What can we learn from the interaction between *Piriformospora indica* and *Arabidopsis thaliana*. Plant Signal Behav 411:1024–1027

Van Breusegem F, Bailey-Serres J, Mittler R (2008) Unraveling the tapestry of networks involving reactive oxygen species in plants. Plant Physiol 147:978–984

Vandelle E, Poinssot B, Wendehenne D, Bentéjac M, Pugin A (2006) Integrated signaling network involving calcium, nitric oxide, active oxygen species but not mitogen-activated protein kinases in BcPG-1 elicited grapevine defenses. Mol Plant Microbe Interact 19:429–440

Vera-Estrella R, Barkla BJ, Higgins VJ, Blumwald E (1994a) Plant defense response to fungal pathogens. Activation of host-plasma membrane $H^+$-ATPase by elicitor-induced enzyme dephosphorylation. Plant Physiol 104:209–215

Vera-Estrella R, Higgins VJ, Blumwald E (1994b) Plant defense responses to fungal pathogens. II. G protein mediated changes in host plasma membrane redox reactions. Plant Physiol 106:97–103

Vidhyasekaran P (2007) Fungal pathogenesis in plants and crops: molecular biology and host defense mechanisms, IIth edn. CRC Press/Taylor & Francis Group, Boca Raton

Viehweger K, Dordschbal B, Roos W (2002) Elicitor-activated phospholipase $A_2$ generates lysophosphatidylcholines that mobilize the vacuolar $H^+$ pool for pH signaling via the activation of $Na^+$-dependent proton fluxes. Plant Cell 14:1509–1525

Viehweger K, Schwartze W, Schumann B, Lein W, Roos W (2006) The Gα protein controls a pH-dependent signal path to the induction of phytoalexin biosynthesis in *Eschscholzia californica*. Plant Cell 18:1510–1523

Walters DR (2000) Polyamines in plant-microbe interactions. Physiol Mol Plant Pathol 57:137–146

Walters D (2003) Resistance to plant pathogens: possible roles for free polyamines and polyamine catabolism. New Phytol 159:109–115

Walters D, Cowley T, Mitchell A (2002) Methyl jasmonate alters polyamine metabolism and induces systemic protection against powdery mildew infection in barley seedlings. J Exp Bot 53:747–756

Wang XQ, Ullah H, Jones AM, Assmann SM (2001) G protein regulation of ion channels and abscisic acid signaling in *Arabidopsis* guard cells. Science 292:2070–2072

Wang KLC, Li H, Ecker IP (2002) Ethylene biosynthesis and signaling networks. Plant Cell 14:S131–S155

Wang L, Xu YY, Ma QB, Li D, Xu ZH, Chong K (2006) Heterotrimeric G protein α subunit is involved in rice brassinosteroid response. Cell Res 16:916–922

Wang S, Assmann SM, Fedoroff NV (2008) Characterization of the *Arabidopsis* heterotrimeric G protein. J Biol Chem 283:13913–13922

Wang W-H, Yi X-Q, Han A-D, Liu T-W, Chen J, Wu F-H, Dong X-J, He J-X, Pei Z-X, Pei Z-M, Zheng H-L (2010) Calcium-sensing receptor regulates stomatal closure through hydrogen peroxide and nitric oxide in response to extracellular calcium in *Arabidopsis*. J Exp Bot 63:177–190

Wasilewska A, Vlad F, Sirichandra C, Redko Y, Jammes F, Valon C, Frei dit Frey F, Leung J (2008) An update on abscisic acid signaling in plants and more. Mol Plant 1:198–217

Wong HL, Sakamoto T, Kawasaki T, Umemura K, Shimamoto K (2004) Down-regulation of metallothionein, a reactive oxygen scavenger, by the small GTPase OsRac1 in rice. Plant Physiol 135:1447–1456

Wong H, Pinontoan R, Hayashi K, Tabata R, Yaeno T, Hasegawa K, Kojima C, Yoshioka H, Iba K, Kawasaki T, Shimamoto K (2007) Regulation of rice NADPH oxidase by binding of Rac GTPase to its N-terminal extension. Plant Cell 19:4022–4034

Wu W-H, Assmann SM (1994) A membrane-delimited pathway of G-protein regulation of the guard-cell inward $K^+$ channel. Proc Natl Acad Sci U S A 91:6310–6314

Wu H-M, Hazak O, Cheung AY, Yalovsky S (2011) RAC/ROP GTPases and auxin signaling. Plant Cell 23:1208–1218

Xie DX, Feys BF, James S, Nieto-Rostro M, Turner JG (1998) *COI1*: an *Arabidopsis* gene required for jasmonate-regulated defense and fertility. Science 280:1091–1094

Xing T, Higgins VJ, Blumwald E (1996) Regulation of plant defense response to fungal pathogens: two types of protein kinases in the reversible phosphorylation of the host plasma membrane $H^+$-ATPase. Plant Cell 8:555–564

Xing T, Higgins VJ, Blumwald E (1997) Identification of G proteins mediating fungal elicitor-induced dephosphorylation of host plasma membrane $H^+$-ATPase. J Exp Biol 48:229–237

Xu L, Liu F, Lechner E, Genschik P, Crosby WL, Ma H, Peng W, Huang D, Xie D (2002) The SCF(COI1) ubiquitin-ligase complexes are required for jasmonate response in *Arabidopsis*. Plant Cell 14:1919–1935

Xu T, Wen M, Nagawa S, Fu Y, Chen JG, Wu MJ, Perrot-Rechnmann C, Friml J, Jones AM, Yang Z (2010) Cell surface- and rho GTPase-based auxin signaling controls cellular interdigitation in *Arabidopsis*. Cell 143:99–110

Yalovsky S, Baluska F, Jones A (2010) Integrated G proteins signaling in plants. Springer, Heidelberg

Yamakawa H, Kamada H, Satoh M, Ohashi Y (1998) Spermine is a salicylate-independent endogenous inducer for both tobacco acidic pathogenesis-related proteins and resistance against tobacco acidic pathogenesis-related proteins and resistance against tobacco mosaic virus infection. Plant Physiol 118:1213–1222

Yang Z (2002) Small GTPases: versatile signaling switches in plants. Plant Cell 14(Suppl): S375–S388

Yang Z, Fu Y (2007) ROP/RAC GTPase signaling. Curr Opin Plant Biol 10:490–494

Yoda H, Sano H (2003) Activation of hypersensitive response genes in the absence of pathogens in transgenic tobacco plants expressing a small GTPase. Planta 217:993–997

Yoda H, Fujimura K, Takahashi H, Munemura I, Uchimiya H, Sano H (2009) Polyamines as a common source of hydrogen peroxide in host- and nonhost hypersensitive response during pathogen infection. Plant Mol Biol 70:103–112

Yong G, Tingting L, Yang L, Cabda R, Yun Z, Maolin W (2010) Isolation and characterization of gene encoding G protein α subunit protein responsive to plant hormones and abiotic stresses in *Brassicus napus*. Mol Biol Rep 37:3957–3965

Zeng Q, Wang X, Running MP (2007) Dual lipid modification of Arabidopsis Gγ-subunits is required for efficient plasma membrane targeting. Plant Physiol 143:1119–1131

Zhang W, He SY, Assmann SM (2008) The plant innate immunity response in stomatal guard cells invokes G-protein-dependent ion channel regulation. Plant J 56:984–996

Zhang W, Jeon BW, Assmann SM (2011) Heterotrimeric G-protein regulation of ROS signalling and calcium currents in *Arabidopsis* guard cells. J Exp Bot 62:2371–2379

Zhang H, Gao Z, Zheng X, Zhang Z (2012) The role of G-proteins in plant immunity. Plant Signal Behav 7:1284–1288

Zhao J, Sakai K (2003) Multiple signalling pathways mediate fungal elicitor-induced β-thujalplicin biosynthesis in *Cupressus lusitanica* cell cultures. J Exp Bot 54:647–656

Zhao J, Wang X (2004) *Arabidopsis* phospholipase Dα1 interacts with the heterotrimeric G-protein α-subunit through a motif analogous to the DRY motif in G-protein-coupled receptors. J Biol Chem 279:1794–1800

Zhao J, Zheng S-H, Fujita K, Sakai K (2004) Jasmonate and ethylene signalling and their interaction are integral parts of the elicitor signalling pathway leading to β-thujaplicin biosynthesis in *Cupressus lusitanica* cell cultures. J Exp Bot 55:1003–1012

Zhao Z, Stanley BA, Zhang W, Assmann SM (2010) ABA-regulated G protein signaling in *Arabidopsis* guard cells: a proteomic perspective. J Proteome Res 9:1637–1647

Zheng Z, Qamar SA, Chen Z, Mengiste T (2006) Arabidopsis WRKY33 transcription factor is required for resistance to necrotrophic fungal pathogens. Plant J 48:596–605

Zhou F, Mosher S, Tian M, Sassi G, Parker J, Klessig DF (2008) The *Arabidopsis* gain-of-function mutant *ssi4* requires *RAR1* and *SGT1b* differentially for defense activation and morphological alterations. Mol Plant Microbe Interact 21:40–49

Zhu H, Li G-J, Ding L, Cui X, Berg H, Assmann SM, Xia Y (2009) *Arabidopsis* extra-large G-protein 2 (XLG2) interacts with the Gβ subunit of heterotrimeric G protein and functions in disease resistance. Mol Plant 2:513–525

Zipfel C, Robatzek S, Navarro L, Oakeley EJ, Jones JDG, Felix G, Boller T (2004) Bacterial disease resistance in *Arabidopsis* through flagellin perception. Nature 428:764–767

# Chapter 4
# Calcium Ion Signaling System: Calcium Signatures and Sensors

**Abstract** Plant innate immune system is a potential surveillance system against possible attack by pathogens. The invading pathogen's signature (pathogen associated molecular pattern [PAMP]) is perceived by pattern recognition receptors (PRRs) of host plants. The informations generated by the PAMP alarm signals activate the plant innate immune system. The plant immune system uses several second messengers to encode information generated by the PAMPs and deliver the information downstream of PRRs to proteins which decode and interpret the signals and initiate defense gene expression. Calcium ion is an important intracellular second messenger and carries the PAMP signal downstream to initiate immune responses. Transient changes in permeability of the plasma membrane to $Ca^{2+}$ and influx of extracellular $Ca^{2+}$ through various plasma membrane-resident ion channels are the earliest events in defense signaling system. Different PAMPs trigger $Ca^{2+}$ influx through activation/'opening' of specifically different ion channels. Massive influx of $Ca^{2+}$ occurs within a few minutes after PAMP treatment. The PAMP-activated calcium signaling is modulated by calcium signatures. $Ca^{2+}$ signatures (single calcium transients, oscillations, or waves) are generated in the cytosol, and in noncytosolic locations including the nucleus and chloroplast through the coordinated action of $Ca^{2+}$ influx and efflux pathways. PAMP signals induce an oscillation in the cytosolic free $Ca^{2+}$ concentration and the information encoded in the induced transient $Ca^{2+}$ changes is decoded by an array of $Ca^{2+}$-binding proteins that serve as $Ca^{2+}$ sensors. The $Ca^{2+}$ sensors identified in plants include calmodulins (CaMs), CaM-like and CaM-related proteins, calcineurin B-like (CBL) proteins, $Ca^{2+}$-dependent protein kinases (CDPKs), and $Ca^{2+}$-binding proteins without EF hands. Specific calcium signatures are recognized by different calcium sensors to transduce calcium-mediated signals into downstream events. The information encoded by specific calcium signatures is decoded by an array of sensors and the extracellular signals are transmitted to cellular calcium-dependent effectors. Thus the extracellular PAMP signals are transmitted to the calcium-dependent effectors through calcium sensors resulting in activation of the complex immune signaling network.

P. Vidhyasekaran, *PAMP Signals in Plant Innate Immunity: Signal Perception and Transduction*, Signaling and Communication in Plants 21,
DOI 10.1007/978-94-007-7426-1_4, © Springer Science+Business Media Dordrecht 2014

**Keywords** $Ca^{2+}$ influx • $Ca^{2+}$ signature • Calmodulin • $Ca^{2+}$ sensors • Calcium-dependent effectors • Calcium-dependent protein kinases • $Ca^{2+}$-mediated signals

## 4.1 Calcium Signature in Plant Immune Signal Transduction System

Pathogen-associated molecular patterns (PAMPs) of invading pathogens is recognized by pattern recognition receptors (PRRs) of host plants (Zipfel 2008; Boutrot et al. 2010; Keinath et al. 2010; Shimizu et al. 2010; Nürnberger and Küfner 2011; Segonzac and Zipfel 2011). Binding of PAMPs to PRRs and subsequent receptor kinase activation are key steps in plant innate immunity (Kwaaitaal et al. 2011). One of the earliest detectable signaling events after PAMP perception is a rapid and transient rise in cytosolic $Ca^{2+}$ levels through the function of plasma membrane-resident $Ca^{2+}$ channels (Blume et al. 2000; Kwaaitaal et al. 2011; Ranf et al. 2011). In plant cells, the calcium ion is a ubiquitous intracellular second messenger involved in numerous signaling pathways (McAinsh et al. 1995; Himmelbach et al. 2003; Lecourieux et al. 2006; Zhu et al. 2007; Luan 2009; McAinsh and Pittman 2009; Abdul Kadar and Lindsberg 2010; DeFalco et al. 2010; Dodd et al. 2010; Hamada et al. 2012; Stael et al. 2012). Second messengers are molecules that are used by plants to encode information and deliver it downstream to proteins which decode/interpret signals and initiate cellular responses (e.g. changes in enzyme activity, gene expression, and cytoskeletal rearrangement) (Snedden and Fromm 2001). $Ca^{2+}$ is a master regulator of gene expression in plants (Galon et al. 2010a, b).

Calcium ion acts as a signal carrier (Allen et al. 2000). Calcium signaling is modulated by specific calcium signatures. Spatial and temporal changes in cytosolic calcium ($[Ca^{2+}]_{cyt}$) are referred to as "calcium signature". These changes may proceed as single calcium transients, oscillations, or repeated spikes/waves (Lecourieux et al. 2006). A combination of changes in all $Ca^{2+}$ parameters produced by a particular signal is called as a $Ca^{2+}$ signature (Luan et al. 2002). $Ca^{2+}$ signatures are generated in the cytosol, and in noncytosolic locations including the nucleus and chloroplast through the coordinated action of $Ca^{2+}$ influx and efflux pathways (McAinsh and Pittman 2009). Specific calcium signatures are recognized by different calcium sensors to transduce calcium-mediated signals into downstream events (Harmon et al. 2000; Rudd and Franklin-Tong 2001; Sanders et al. 2002; Reddy and Reddy 2004; Reddy et al. 2003, 2011a, b; Wang et al. 2012; Hashimoto et al. 2012).

The $Ca^{2+}$ signature controls diverse cellular processes via $Ca^{2+}$ sensors (DeFalco et al. 2010). Plant cells employ an array of $Ca^{2+}$-binding proteins that serve as $Ca^{2+}$ sensors. The $Ca^{2+}$ binding proteins that function as sensors undergo conformational changes upon $Ca^{2+}$ binding that allow them to interact with downstream effectors (Clapham 2007; DeFalco et al. 2010; Hashimoto et al. 2012). The $Ca^{2+}$ sensors identified in plants include calmodulins (CaM) (Snedden and Fromm 2001; Reddy and Reddy 2004; Kang et al. 2006a, b; Takabatake et al. 2007), CaM-like and CaM-related proteins (Jakobek et al. 1999; Rodriguez-Concepcion et al. 1999; Reddy 2001; Snedden and Fromm 2001; Luan et al. 2002; Lecourieux et al. 2006; DeFalco et al. 2010), calcineurin B-like (CBL) proteins (Luan et al. 2002; Hashimoto et al. 2012), $Ca^{2+}$-dependent protein kinases (Harmon et al. 2001; Harmon 2003; Ludwig et al. 2004: Kobayashi et al. 2007; Batistić and Kudla 2009; Luan 2009; Hashimoto et al. 2012), and $Ca^{2+}$-binding proteins without EF hands (Clark and Roux 1995; Reddy 2001; Tomsig and Cruetz 2002; Tomsig et al. 2003).

The calcium sensor proteins fall into two main classes, referred to as sensor relays and sensor responders (Sanders et al. 2002; Reddy and Reddy 2004). Sensor relays include calmodulin (CaM), CaM-related proteins, and calcineurin B-like (CBL) proteins. They function through bimolecular interactions. They undergo a conformational change induced by $Ca^{2+}$ before interacting with and changing the activity or structure of the target protein (Lecourieux et al. 2006). Sensor responders are $Ca^{2+}$-dependent protein kinases (CDPKs) (Cheng et al. 2002). They function at first through intramolecular interactions and undergo a $Ca^{2+}$-induced conformational change that alters the protein's own activity or structure (Harmon et al. 2000; Harper et al. 2004). These two groups of proteins are involved in decoding calcium signals (Lecourieux et al. 2006; Kudla et al. 2010).

Changes and oscillations in cytosolic free $Ca^{2+}$ concentration are associated with transduction of signals in plant cells (Sanders et al. 1999, 2002). Many biotic and abiotic signals elicit transient increases in cytoplasmic free $Ca^{2+}$ ($[Ca^{2+}]_{cyt}$) concentration in plants (Luan et al. 2002). Increases in $Ca^{2+}]_{cyt}$ may be mostly due to influx of external $Ca^{2+}$. However, internal $Ca^{2+}$ release may also contribute for increases in $[Ca^{2+}]_{cyt}$ (Staxen et al. 1999; Hwang et al. 2000a, b; Karita et al. 2004). Cellular $Ca^{2+}$ levels are tightly regulated in plant cells and hence small changes in intracellular $Ca^{2+}$ can provide information for the modification of enzyme activity and gene expression (Gong et al. 2004). Signals may induce an oscillation in the cytosolic free $Ca^{2+}$ concentration. The information encoded in transient $Ca^{2+}$ changes is decoded by an array of $Ca^{2+}$ binding proteins giving rise to a cascade of downstream effects, including altered phosphorylation and gene expression patterns (Sanders et al. 2002; Hashimoto et al. 2012). Thus the extracellular signals are transmitted to cellular calcium-dependent effectors to activate the transcription of immune response-related genes (Luan et al. 2002; Sanders et al. 2002; Gong et al. 2004; Lecourieux et al. 2006; Ma and Berkowitz 2007; Dodd et al. 2010; Reddy et al. 2011a).

## 4.2 Upstream Events Leading to Activation of $Ca^{2+}$ – Permeable Channels

### *4.2.1 PAMP-Triggered $Ca^{2+}$ Influx and Elevations in Cytosolic Free Calcium*

Transient changes in permeability of the plasma membrane to $Ca^{2+}$ and influx of extracellular $Ca^{2+}$ through the membrane appear to be one of the earliest events in defense signaling system (Atkinson et al. 1996; Wendehenne et al. 2002; Garcia-Brugger et al. 2006; Laohavisit et al. 2009; Vadassery and Oelmüller 2009;

Kwaaitaal et al. 2011; Vincill et al. 2012). Massive influx of $Ca^{2+}$ was observed within 15–30 min after PAMP/elicitor treatment in tobacco-cultured cells (Lecourieux-Ouaked et al. 2000). The initiation of innate immune responses upon the PAMPs (flg22, elf18 and chitin) recognition by their PRRs (FLS2, EFR and CEK1, respectively) has been shown to be due to apoplastic $Ca^{2+}$ influx via ion channels (Kwaaitaal et al. 2011).

PAMPs/elicitors induce elevations in the cytosolic free calcium ($[Ca^{2+}]_{cyt}$) (Navazio et al. 2007). The induced calcium ($[Ca^{2+}]_{cyt}$) elevations predominantly result from a continuous $Ca^{2+}$ influx through the plasma membrane (Mithofer et al. 1999; Blume et al. 2000; Lecourieux et al. 2002; Hu et al. 2004; Vandelle et al. 2006). The transient increase of cytosolic $Ca^{2+}$ ($[Ca^{2+}]_{cyt}$) results from passive $Ca^{2+}$ influx from the apoplast and/or from intracellular $Ca^{2+}$ stores through dedicated channels (Bush 1993, 1995). High accumulation of $Ca^{2+}$ in the cytosol may result in cytotoxicity and hence low $Ca^{2+}$ concentrations are maintained in the cytosol by the action of $H^+/Ca^{2+}$ antiporters (carriers) and $Ca^{2+}$-ATPases (ATP-dependent $Ca^{2+}$ pumps) (Sanders et al. 1999, 2002; Zhu et al. 2010). These pumps and antiporters expel $Ca^{2+}$ in an energy-dependent manner from the cytosol to the apoplast and into vacuole and endoplasmic reticulum (ER) (Kwaaitaal et al. 2011). As a result, $Ca^{2+}$ in the apoplast and intracellular stores is up to 20,000 times higher than in the cytosol. The combination of this concentration gradient and a proton-based electrochemical gradient over cellular membranes may be the driving force for the PAMP-dependent influx of $Ca^{2+}$ into the cytosol (Bush 1995; Kwaaitaal et al. 2011).

### *4.2.2 PAMP – Activated G-Proteins May Initiate $Ca^{2+}$ Influx*

G-proteins may be involved in $Ca^{2+}$ channel opening (Gelli and Blumwald 1997). Protein phosphorylation precedes $Ca^{2+}$ influx in tobacco cells treated with a PAMP isolated from the oomycete pathogen *Phytophthora cryptogea* (Tavernier et al. 1995). The G-proteins modulate the protein phosphorylation system in the plasma membrane of tomato cells and transduce the signal (Vera-Estrella et al. 1994). Phosphorylation of proteins involved in G-protein coupled signaling has been reported in tobacco cells treated with a bacterial elicitor (Gerber et al. 2006).

An interplay between the monomeric GTPase Rho-like GTPase of plants (ROP), the plant orthologs of the respiratory burst NADPH oxidases (RBOH), ROS production, and cytosolic calcium transients has been reported (Fig. 4.1; Van Breusegem et al. 2008). GTPase Rac modulates activity of NADPH oxidases (RBOH) (Petry et al. 2010). The NADPH oxidase – GTPase interaction is regulated by the binding of calcium to two EF-hand motifs at the N terminus of the NADPH oxidase (Wong et al. 2007). The NADPH oxidase activation results in production of ROS (Van Breusegem et al. 2008; Petry et al. 2010). Activation of NADPH oxidase promotes calcium channel activation and calcium influx (Takeda et al. 2008). $H_2O_2$ and OH° may serve as distinct signals in the regulation of calcium influx, due to the existence of calcium channels that are distinctively sensitive to the generated $H_2O_2$ (Demidchik et al. 2007).

**Fig. 4.1** Role of G protein-ROS-$Ca2^{+}$-phosphorylation in activation of $Ca^{2+}$ channels

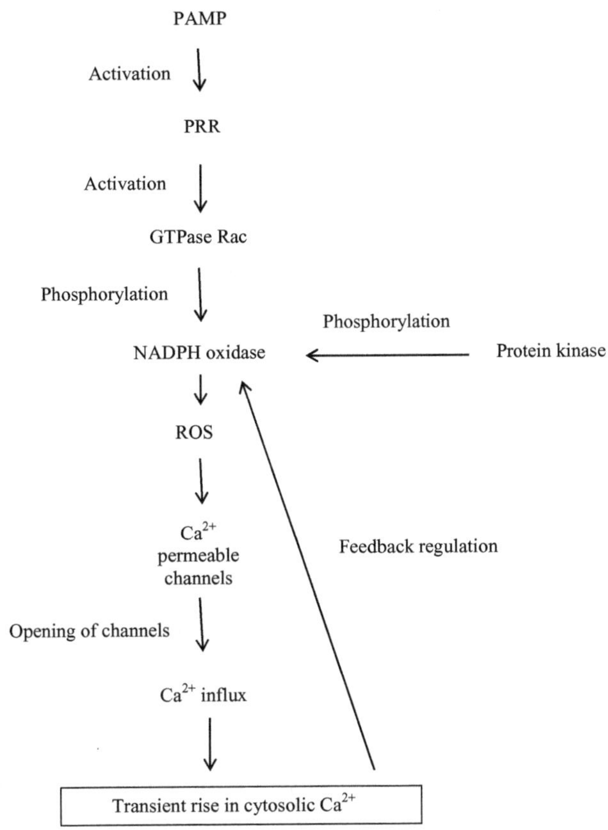


### 4.2.3 *Reactive Oxygen Species (ROS) Regulate $Ca^{2+}$ Influx*

Reactive oxygen species (ROS) may regulate $Ca^{2+}$ influx through plasma membrane transport proteins (Mori and Schroeder 2004; Laohavisit et al. 2012). Elicitor-induced $H_2O_2$ participates in $[Ca^{2+}]_{cyt}$ increase, probably through the activation of $H_2O_2$ – sensitive $Ca^{2+}$ channels located in the plasma membrane (Lecourieux et al. 2002). $H_2O_2$ triggers calcium influx in tobacco (Takahashi et al. 1998; Kawano and Muto 2000). An external stimulus, such as PAMP, increases activity of NADPH oxidase which is involved in ROS production. Activation of the plasma membrane-localized NADPH oxidase involves phosphorylation of two N-terminal Ser by a calcium-dependent protein kinase as well as interaction with GTPase. NADPH phosphorylation as well as binding to calcium synergizes NADPH activation (Ogasawara et al. 2008; Takeda et al. 2008; Petry et al. 2010). The increased activity of NADPH oxidases induces ROS production. The ROS activates hyperpolarization- activated $Ca^{2+}$ influx current (Pei et al. 2000; Foreman et al. 2003). $H_2O_2$ may also be produced indirectly by producing more NADPH by means of $Ca_{2+}$/

CaM-regulated NAD kinase (Harding et al. 1997). The indirectly induced $H_2O_2$ production would have activated the $Ca^{2+}$ influx.

The NADPH oxidase-derived ROS stimulates a $Ca^{2+}$ influx into the cytoplasm. The rise in $Ca^{2+}$ level in turn activates NADPH oxidase to produce ROS (Takeda et al. 2008), suggesting a positive feedback regulation of $Ca^{2+}$ influx – ROS signaling system. Protein phosphorylation has been shown to be a prerequisite for the $Ca^{2+}$-dependent activation of *Arabidopsis* NADPH oxidases (Kimura et al. 2012). In *Arabidopsis*, the respiratory oxidase homologue F (*AtrbohF*) encodes NADPH oxidase. AtRbohF exhibited ROS-producing activity that was synergistically activated by protein phosphorylation and $Ca^{2+}$. The two EF-hand motifs of AtRbohF in the terminal cytosolic region were found to be crucial for its $Ca^{2+}$-dependent activation. A protein kinase inhibitor inhibited the $Ca^{2+}$-dependent activation of AtRbohD in a dose-dependent manner, suggesting that protein phosphorylation is a prerequisite for the $Ca^{2+}$-dependent activation of RbohF (Kimura et al. 2012). These results suggest a positive feedback regulation of $Ca^{2+}$ and ROS in the $Ca^{2+}$ influx signaling system (Fig. 4.1).

## 4.3 $Ca^{2+}$ Influx Channels in Plant Cell Plasma Membrane

### *4.3.1 Voltage-Gated $Ca^{2+}$-Permeable Channels*

Several $Ca^{2+}$-permeable channels have been found in plant plasma membranes and they have been implicated in immune response signaling systems (Gaxiola et al. 2007; Hamilton et al. 2000; White and Broadley 2003; Ma et al. 2009b; Qi et al. 2010; Kwaaitaal et al. 2011; Michard et al. 2011; Vatsa et al. 2011). Calcium ion channels are integral membrane proteins that are involved in transport of solutes across the cell membrane in plants (Maathuis et al. 1997). Diffusion of ions through the channel is mostly due to membrane voltage (Miedema et al. 2008). Ion channels remain in "closed" conformational state and the ion channels "open" in response to ligands or to a change in membrane voltage (Maathuis et al. 1997). Some channels open in hyperpolarizing conditions (i.e. at rather negative membrane voltages, inward current) (Schroeder et al. 1994). Another class of channels open in depolarizing conditions (at relatively positive voltage, outward current) (Tester 1990). Channels are increasingly activated at more negative and more positive membrane voltage (Maathuis et al. 1997). Changes in membrane potential are associated with the initiation of a number of signal transduction pathways (Ward et al. 1995). Most of these $Ca^{2+}$ channels are not strictly selective for $Ca^{2+}$, and they also facilitate the transport of other cations (Very and Sentenac 2002).

Depolarization-activated $Ca^{2+}$-permeable channels are common in plasma membrane of plant cells. They activate significantly at voltages more positive than about −150 to −100 mV. These voltage-gated $Ca^{2+}$ channels mediate $Ca^{2+}$ influx across the plasma membrane of cells (Huang et al. 1994). $Ca^{2+}$ channel opens upon depolarization of the membrane electrical potential (Huang et al. 1994; White 2000, 2004). Membrane depolarization may be due to the activation of anion channels (Sanders et al. 2002; Jeworutzki et al. 2010). Anion efflux results in plasma membrane depolarization, which in turn triggers the activation of voltage-dependent $Ca^{2+}$ channels that mediate $Ca^{2+}$ influx (Ward et al. 1995).

Application of the PAMPs flg22 and elf18 induced rapid dose-dependent membrane potential depolarizations and pronounced anion currents in *Arabidopsis thaliana*. The depolarization was superimposed by an increase in cytosolic calcium that was indispensable for depolarization. The early immune signaling events induced by the PAMPs flg22 and elf18 were found to involve $Ca^{2+}$-associated opening of plasma membrane anion channels (Jeworutzki et al. 2010). The elicitor-induced $Ca^{2+}$ influx is inhibited by different anion-channel blockers (Ward et al. 1995; Ebel and Mithofer 1998). It suggests that anion flux precedes and controls $Ca^{2+}$ influx. The anion channels may initiate and amplify plasma membrane depolarization, which in turn may activate $Ca^{2+}$ voltage-dependent channels.

It has also been reported that $Ca^{2+}$ influx is a prerequisite for the activation of plasma membrane anion channels in several systems (Ward et al. 1995; Jabs et al. 1997; Wendehenne et al. 2002). In cryptogein-treated tobacco cells, the major calcium influx did not result from plasma membrane depolarization. Instead, the $Ca^{2+}$ influx occurred upstream and it triggered anion efflux and plasma membrane depolarization, which in turn may mobilize some $Ca^{2+}$ voltage-dependent channels (Pugin et al. 1997; Wendehenne et al. 2002). Collectively these results suggest that initial $Ca^{2+}$ -influx may be through some voltage-dependent channels, independent of anion efflux and subsequent activation of $Ca^{2+}$ influx may be through the anion efflux-activated voltage-dependent channels.

The outward-rectifying $K^+$ channels found in the plasma membrane of plant cells are also $Ca^{2+}$-permeable depolarization-activated channels. These $Ca^{2+}$-permeable outward rectifying $K^+$ channels activate significantly at voltages more positive than about −50 mV under most physiological conditions and catalyze a large $K^+$ efflux simultaneously with a small $Ca^{2+}$ influx (White and Broadley 2003). These channels may play a role in the initial $Ca^{2+}$ influx into the cytosol.

The rice two-pore channel1 (OsTPC1) is a putative voltage-gated $Ca^{2+}$-permeable channel (Kurusu et al. 2005; Hamada et al. 2012). Overexpression of *OsTPC1* induced several defense-related signaling systems, resulting in induction of oxidative burst and activation of a mitogen-activated protein kinase and hypersensitive cell death. Retrotransposon-insertional mutagenesis of *OsTPC1* resulted in suppression of activation of the MAP kinase and hypersensitive-related cell death (Kurusu et al. 2005). The OsTPC1 has been shown to play a critical role in hypersensitive cell death induced by a fungal xylanase protein (TvX) elicitor in suspension-cultured rice cells (Hamada et al. 2012). TvX induced a prolonged increase in cytosolic $Ca^{2+}$, mainly due to a $Ca^{2+}$ influx through the plasma membrane. TvX induced production of major diterpenoid phytoalexins and the expression of diterpene cyclase genes involved in phytoalexin biosynthesis (Hamada et al. 2012). These results suggest that the $Ca^{2+}$-permeable voltage-gated channels may act as key regulators of defense signaling system.

Although membrane depolarization-activated $Ca^{2+}$-permeable channels are common in plants, the presence of hyperpolarization-activated $Ca^{2+}$-permeable channels has also been reported in plasma membrane of certain plant cells including tomato (Gelli and Blumwald 1997) and *Arabidopsis* (Pei et al. 2000) cells. Hyperpolarization-activated cation channels are activated at voltages more negative than about −100 to −150 mV (White and Broadley 2003). Reactive oxygen species increase the activity of these channels through the activity of NADPH oxidase (Pei et al. 2000; Murata et al. 2001; Foreman et al. 2003). Abscisic acid (ABA) activates a hyperpolarization-dependent $Ca^{2+}$-permeable channel in the plasma membrane of *Arabidopsis* guard cells, leading to $Ca^{2+}$ influx and to an increase in $[Ca^{2+}]_{cyt}$ (Hamilton et al. 2000; Pei et al. 2000). ABA has been shown to increase the level of ROS in an NADPH-dependent manner and ROS is known to stimulate hyperpolarization-activated $Ca^{2+}$ influx current in the plasma membrane termed $I_{Ca}$ (Pei et al. 2000). Reactive oxygen species (ROS) has been shown to stimulate hyperpolarization-activated $Ca^{2+}$-permeable channels and cause a transient increase in cytoplasmic calcium ($[Ca^{2+}]_{cyt}$) content in *Arabidopsis* (Pei et al. 2000).

### *4.3.2 Cyclic Nucleotide-Gated Ion Channels (CNGCs)*

Cyclic nucleotide-gated ion channels (CNGCs) have been found in plant cell plasma membrane (Leng et al. 1999; Maathuis and Sanders 2001; Sanders et al. 2002; Kaplan et al. 2007; Baxter et al. 2008; Ma and Berkowitz 2011; Qi et al. 2010; Abdel-Hamid et al. 2011; Moeder et al. 2011). Twenty genes encoding putative CNGCs have been detected in *Arabidopsis* (Talke et al. 2003). The *HLM1* gene encodes a cyclic nucleotide-gated channel, CNGC4 in *Arabidopsis* (Balagué et al. 2003). *AtCNGC11* and *AtCNGC12* genes encode two CNG ion channels involved in host defense response (Yoshioka et al. 2006). *DND1* gene encodes AtCNGC2 that allows passage of $Ca^{2+}$, $K^+$, and other cations (Clough et al. 2000). A CNG ion channel has been identified in barley and it was homologous to the *Arabidopsis* HLM1. Ten other members of the barley CNG channel gene family have been identified (Rostoks et al. 2006).

CNG ion channels are heterotetrameric in structure. Six membrane-spanning domains flanked by hydrophilic amino and carboxy terminus have been detected in these proteins. A putative cyclic nucleotide binding domain was located in the carboxy terminus (Zhong et al. 2003). AtCNGC subunits share many similarities with voltage-gated outward rectifying $K^+$-selective ion channel proteins, including a cytoplasmic N terminus, six membrane-spanning regions, a pore domain, and a cytoplasmic C terminus. AtCNGCs are gated primarily by binding of cAMP (cyclic adenosine monophosphate) or cGMP (cyclic guanosine-3′,5′-cyclic monophosphate) rather than by voltage (Yoshioka et al. 2006). AtCNGC4 is activated by both cGMP and cAMP (Balagué et al. 2003). Cyclic nucleotides cAMP and cGMP have been shown to be linked with $Ca^{2+}$ signaling (Moutinho et al. 2001).

CNGCs are involved in $Ca^{2+}$-dependent signaling pathways (Talke et al. 2003; Yoshioka et al. 2006). The CNGC detected in *Arabidopsis*, AtCNGC2, is a plasma membrane protein permeable to $Ca^{2+}$ (Yu et al. 1998). Leng et al. (1999, 2002) and Balague et al. (2003) showed cyclic nucleotide-dependent conductance of $Ca^{2+}$ in *Arabidopsis*. The *atcngc2* knock-out plants showed an altered $Ca^{2+}$ signature in response to the bacterial PAMP lipopolysaccharide (LPS) (Ma et al. 2009b). The unique position of CNGCs as ligand-gated $Ca^{2+}$-permeable channels suggests that they function at key sites where cyclic nucleotide and $Ca^{2+}$ signaling pathways interact (Talke et al. 2003).

The CNGCs contain a C-terminal cyclic nucleotide-binding domain with overlapping calmodulin-binding domain. Calmodulin-binding domains are common in CNG ion channels and calmodulin binding has been demonstrated for the *Arabidopsis* CNGC proteins AtCNGC1 and AtCNGC2 (Kohler et al. 1999; Kohler and Neuhaus 2000). A tobacco plasma membrane calmodulin-binding channel protein (designated NtCBP4) had a putative cyclic nucleotide-binding domain. The NtCBP4 calmodulin-binding domain was found to perfectly coincide with an alpha-C-helix motif of its putative cyclic nucleotide-binding domain. The coinciding calmodulin- and cyclic nucleotide-binding domains may serve as a point of communication between calcium and cyclic nucleotide signal transduction pathways in plants (Arazi et al. 2000). CNGCs that are regulated by calmodulin play essential roles in signal transduction (Borsics et al. 2007).

CNGCs play important role in plant innate immunity signaling (Baxter et al. 2008; Abdel-Hamid et al. 2011; Moeder et al. 2011). In *Arabidopsis*, the CNGCs AtCNGC2 and AtCNGC4 have been shown to trigger hypersensitive reaction (HR), an important component of defense signaling pathway. *Arabidopsis* mutants, *dnd1* (without functional CNGC2) and *dnd2/HLM1* (without functional AtCNGC4), fail to produce HR (Balague et al. 2003; Jurkowski et al. 2004). AtCNGC11 and AtCNGC12 are positive regulators of defense signaling in *Arabidopsis*. These two AtCNGCs play a role in SA-signaling system dependent on EDS1 and PAD4 (Yoshioka et al. 2006). The chimeric gene *AtCNGC11/12* induced resistance associated HR-related cell death. The vacuolar processing enzyme (VPE), a caspase-like protein, was involved in this process. In VPE-silenced plants, development of cell death was much slower and weaker compared to control plants, suggesting the involvement of VPE as a caspase in AtCNGC11/12-induced HR-related cell death (Urquhart et al. 2011). *DND1* negatively regulates disease resistance to bacterial pathogens in *Arabidopsis* (Ahn 2007). The resistance of *dnd1* mutant against *Pectobacterium carotovorum* was dependent on calmodulin and inhibition of $Ca^{2+}$ increment (Ahn 2007).

The signaling cascade initiated by the host-associated molecular pattern (HAMP)/damage-associated molecular pattern (DAMP) AtPep1 leads to expression of defense genes in a $Ca^{2+}$-dependent manner (Qi et al. 2010). The endogenous elicitor AtPep1 after binding with its pattern recognition receptor (PRR) AtPepR1 activates plant membrane inwardly conducting $Ca^{2+}$ permeable channels in mesophyll cells, resulting in cytosolic $Ca^{2+}$ elevation. This activity is dependent on the PRR AtPepR1 as well as a cyclic nucleotide-gated channel (CNGC2). The PRR

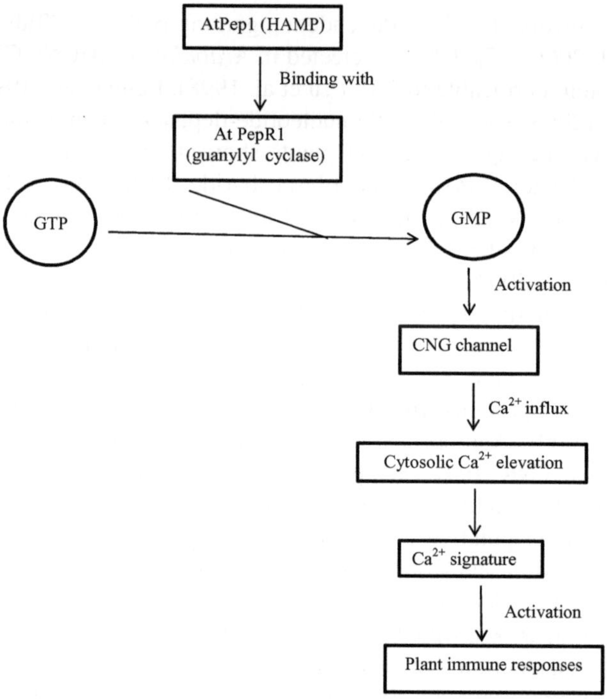


**Fig. 4.2** Activation of CNG channel-mediated $Ca^{2+}$ influx by the HAMP AtPep1 (Adapted from Qi et al. 2010)

AtPepR1 has guanylyl cyclase activity and this activity generates cGMP from GTP. The cGMP activates CNGC2-dependent cytosolic $Ca^{2+}$ elevation (Qi et al. 2010). AtPep-dependent expression of defense genes such as PDF1.2, MPK3, and WRKY33, is mediated by the $Ca^{2+}$ signaling pathway associated with AtPep peptides and their receptor (Fig. 4.2; Qi et al. 2010). These studies indicate that downstream from AtPep and AtPepR1 in a signaling cascade, the cGMP-activated channel CNGC2 is involved in AtPep- and AtPepR1-dependent inward $Ca^{2+}$ conductance and resulting cytosolic $Ca^{2+}$ elevation.

### *4.3.3 Glutamate Receptor-Like Ion Channels*

Glutamate receptors (GLRs) comprise another class of ion channel that might provide a calcium-permeable pathway across the plasma membrane (Sanders et al. 2002; Vatsa et al. 2011; Price et al. 2012; Vincill et al. 2012). The glutamate receptors are ligand-gated non-selective cation channels permeable to calcium (Demidchik et al. 2002, 2004; Kang et al. 2006b; Tapken and Hollmann 2008; Tikhonov and

Magazanik 2009; Vatsa et al. 2011). The ionotropic glutamate receptor consists of a ligand binding domain and a channel-forming domain (Price et al. 2012). The ionotropic glutamate receptor-like channels possess a large N-terminal extracellular domain, three transmembrane regions, a hydrophobic loop defining the pore region and a cytosolic C-terminal domain (Tapken and Hollmann 2008; Kwaaitaal et al. 2011). Twenty glutamate receptor-like channels have been detected in *A. thaliana* genome (Chiu et al. 1999, 2002; Wheeler and Brownlee 2008). Glutamate receptor-like genes have been shown to form $Ca^{2+}$ channels (Michard et al. 2011). $Ca^{2+}$ permeability of a glutamate receptor channel in *Arabidopsis* has been demonstrated (Vincill et al. 2012). Exogenous glutamate application triggers a large transient elevation in $[Ca^{2+}]_{cyt}$ and a membrane depolarization in *Arabidopsis* (Dennison and Spalding 2000).

In *A. thaliana* overexpressing a full-length cDNA clone (*RsGluR*) encoding a putative glutamate receptor from small radish, glutamate treatment triggered greater $Ca^{2+}$ influx in the root cells of transgenic plants than in those of the wild type. Jasmonate-responsive genes including defensins and JA-biosynthetic genes were upregulated in the transgenic plants. *RsGluR* overexpression also inhibited growth of the fungal pathogen *Botrytis cinerea*. The RsGluR is a glutamate –gated $Ca^{2+}$ channel located in the plasma membrane of higher plants and plays a direct or indirect role in defense against pathogen infection by triggering JA biosynthesis (Kang et al. 2006a, b).

The glutamate receptors were found to be activated by the PAMP cryptogein and they were involved in triggering $Ca^{2+}$ influx (Vatsa et al. 2011). The downstream event in the glutamate receptor-mediated signaling pathway included NO production (Vatsa et al. 2011). Kwaaitaal et al. (2011) showed that the initiation of innate immune responses upon the PAMPs flg22, elf18 and chitin recognition involves apoplastic $Ca^{2+}$ influx via glutamate receptor-like channels in *A. thaliana*. The downstream events in the glutamate receptor channels-mediated immune response signaling pathway included mitogen-activated protein kinase cascade, activation of calcium-dependent protein kinase (CDPK) and accumulation of defense gene transcripts (Kwaaitaal et al. 2011).

### *4.3.4 Annexins as Calcium Transporters*

Elevation of cytoplasmic free calcium ($[Ca^{2+}]_{cyt}$) is a regulatory step in plant innate immunity and it relies spatiotemporal control of $Ca^{2+}$-permeable channel activity at endomembranes and the plasma membrane (McAinsh and Pittman 2009). The plant genomes encode multiple potential $Ca^{2+}$-permeable channel subunits that could contribute to $[Ca^{2+}]_{cyt}$ elevation (McAinsh and Pittman 2009; Ward et al. 2009). Plant annexins appear capable of mediating passive, channel-like $Ca^{2+}$ transport (Mortimer et al. 2008; Laohavisit et al. 2009, 2010; Laohavisit and Davies 2011). Annexins are membrane binding proteins that can form $Ca^{2+}$-permeable conductances *in vitro* (Laohavisit et al. 2012). Pepper annexin mediates $Ca^{2+}$ influx into artificial vesicles (Hofmann et al. 2000). A maize annexin preparation was found to promote $Ca^{2+}$ influx into root epidermal protoplasts (Laohavisit et al. 2009). It also formed a $Ca^{2+}$-permeable conductance in planar lipid bilayers that resembled plant plasma membrane $Ca^{2+}$-permeable nonselective cation channel (Laohavisit et al. 2009).

Laohavisit et al. (2010) reported that annexins form a ROS-stimulated passive $Ca^{2+}$ transport pathway in maize. *Arabidopsis* loss-of-function mutant for annexin1 (*Atann1*) was found to lack epidermal ROS-activated $Ca^{2+}$ conductance (Laohavisit et al. 2012). An ROS-activated $Ca^{2+}$ conductance was reconstituted by recombinant annexin1 (ANN1) in planar lipid bilayers (Laohavisit et al. 2012). The results suggest that annexin1 presents a novel $Ca^{2+}$-permeable transporter providing a molecular link between ROS and cytosolic $Ca^{2+}$ in plants.

## 4.4 $Ca^{2+}$ Release Channels Involved in Releasing Stored $Ca^{2+}$ in Vacuole and Endoplasmic Reticulum into Cytosol

### *4.4.1 Inositol 1,4,5-Trisphosphate(InsP3)-Activated $Ca^{2+}$ Channel*

Besides the $Ca^{2+}$ influx channels, $Ca^{2+}$ efflux channels located in the membrane of intracellular organelles have also been shown to be involved in signal transduction (Schroeder and Thuleau 1991). The plasma membrane $Ca^{2+}$ channels allow $Ca^{2+}$ influx from the cell wall space into the cytosol, while $Ca^{2+}$ release channels allow release of stored $Ca^{2+}$ in vacuole and endoplasmic reticulum into the cytosol (Sanders et al. 1999). A number of $Ca^{2+}$ release channels have been found in the vacuolar membranes. Inositol 1,4,5-trisphosphate (InsP3)-activated $Ca^{2+}$ channel is the important $Ca^{2+}$ release channel (Alexandre et al. 1990). InsP3-gated channels release $Ca^{2+}$ from the vacuole (Alexandre and Lassales 1992; Berridge 1993) and endoplasmic reticulum (ER) (Martinec et al. 2000). InsP3 is a second messenger that controls many cellular processes by generating internal calcium signals (Berridge 1984, 1993; Berridge and Irvine 1989). InsP3 releases calcium through an intact intracellular plant membrane by activating a $Ca^{2+}$ channel (Alexandre et al. 1990). It operates through receptors which resemble ryanodine receptors of human muscle (Berridge 1993). This calcium channel is voltage-dependent and opened only on depolarization of the vacuoles (Alexandre et al. 1990). The calcium released through this channel induces calcium waves and oscillations (calcium signature) in the cytosol (Berridge 1993).

RACK1 (for Receptor for Activated C Kinase1) is an interactor which binds with phosphorylated active forms of protein kinase C (PKC). It acts as a scaffold protein, bringing activated PKC into contact with its various substrates (Shirasu and Schulze-Lefert 2003; Patterson et al. 2004; Nakashima et al. 2008). RACK1 binds InsP3 receptors and regulates $Ca^{2+}$ release by enhancing InsP3 receptor binding affinity for InsP3. Overexpression of RACK1 markedly augments $Ca^{2+}$ release, while depletion of RACK1 by interference RNA diminishes $Ca^{2+}$ release (Patterson et al. 2004).

### *4.4.2 Cyclic Adenosine 5′-Diphospho-Ribose (cADPR) Gated Channels*

Cyclic adenosine 5′-diphosphoribose (cADPR) gated channels are another family of channels, which release $Ca^{2+}$ from the vacuole and endoplasmic reticulum (Navazio et al. 2001). The cADPR-gated pathway has been shown to be voltage-dependent and it does not spontaneously desensitize. It is colocalized with an InsP3-gated calcium release pathway in individual vacuoles (Allen et al. 1995). cADPR elevates cytosolic free calcium in plants (Navazio et al. 2001). cADPR-mediated induction of abscisic acid-responsive gene expression has been shown to be exerted by means of mobilization of internal $Ca^{2+}$ stores (Wu et al. 1997).

### *4.4.3 Slowly Acting Vacuolar (SV) Channel*

Another class of $Ca^{2+}$ release channel type residing in the vacuolar membrane is gated by voltage and acts by membrane depolarization (Sanders et al. 2002). This channel is known as the Slow Vacuolar (SV) channel in reference to its voltage-activation kinetics (Hedrich and Neher 1987; Hedrich and Marten 2011). The SV channel is activated by rises in $[Ca^{2+}]_{cyt}$ and this response potentially endows the channel with the capacity to catalyze $Ca^{2+}$-induced $Ca^{2+}$ release (Bewell et al. 1999). The channel is regulated by $[Ca^{2+}]_{cyt}$ and by phosphorylation (Allen and Sanders 1995). This channel is downregulated by 14-3-3 proteins (van den Wijngaard et al. 2001).

### *4.4.4 NAADP-Activated $Ca^{2+}$ Efflux Channel*

A distinct $Ca^{2+}$ release channel activated by the nicotinamide adenosine dinucleotide phosphate (NADP) metabolite nicotinic acid adenine dinucleotide phosphate (NAADP) has been detected in ER of higher plants (Navazio et al. 2000; Patel 2004). The NAADP potently mobilizes $Ca^{2+}$ from ER in red beet (*Beta vulgaris*) and cauliflower (Navazio et al. 2000). The exclusively endoplasmic reticulum location of the NAADP-sensitive $Ca^{2+}$ pathway distinguishes it from the InsP3- and cADPR-gated pathways. The NAADP-gated $Ca^{2+}$ release pathway was found to be independent of cytosolic free $Ca^{2+}$ and therefore incapable of operating $Ca^{2+}$ –induced $Ca^{2+}$ release (Navazio et al. 2000).

## 4.5 $Ca^{2+}$ Efflux from Cytosol to Vacuole and Endoplasmic Reticulum (ER)

### 4.5.1 $H^+/Ca^{2+}$ Antiport System

$Ca^{2+}$ efflux from cytosol to vacuole and ER through $H^+/Ca^{2+}$ antiporters (carriers) and calcium ATPases (pumps) may also be involved in signal transduction (Hirschi et al. 1996; Sanders et al. 1999, 2002; Zhu et al. 2010). $Ca^{2+}$ is sequestered into vacuoles through the $H^+/Ca^{2+}$ antiport system that is driven by the proton-motive force of the tonoplast $H^+$-translocating ATPase (Schumaker and Sze 1986). $Ca^{2+}$ uptake is tightly coupled to $H^+$ loss. At least one $Ca^{2+}$ is exchanged for each $H^+$ (Schumaker and Sze 1986). The $H^+/Ca^{2+}$ exchange in tonoplast vesicles is electrogenic, generating a membrane potential, interior positive. This exchange depends on the pH gradient between the vacuole and cytoplasm. The transmembrane movement of $H^+$ is catalyzed by $H^+$-conducting proteins (Schumaker and Sze 1986). A $H^+/Ca^{2+}$ antiporter has been cloned from *Arabidopsis*, and the protein has been designated as CAX1 (calcium exchanger 1; Hirschi 2001). CAX1 appears to transport cytosol $Ca^{2+}$ into vacuole and ER and the $Ca^{2+}$ entering into these organelles may regulate calcium release channels in them (Hirschi 2001). Cell-specific vacuolar calcium storage mediated by CAX1 has been shown to regulate apoplastic calcium concentration (Conn et al. 2011). $Ca^{2+}$ efflux systems may also participate in signal transduction (Sanders et al. 2002).

### 4.5.2 Calcium Ion Pumps

Calcium ion pumps may also exist in plant cell membrane. These include P-type ATPases, which are ATP-fuelled pumps sharing a common enzymatic mechanism involving a phosphorylated reaction cycle intermediate, hence P type (Palmgren and Harper 1999). The pumps become phosphorylated at an aspartate residue in the sequence DKTGT. The P-type ATPases directly use ATP to drive ion translocation (Sanders et al. 2002). More than 40 P-type ATPases have been reported in *A. thaliana*. Two P-type ATPases, $Ca^{2+}$-ATPases and $H^+$-ATPases, are involved in $Ca^{2+}$ signaling.

Plant cells contain two distinct types of $Ca^{2+}$ pumps, types IIA and IIB, based on protein sequence identities (Axelsen and Palmgren 1998). The $Ca^{2+}$-ATPases are regulated by binding with calmodulin (Sze et al. 2000). ACA2, a type IIB pump detected in ER of *Arabidopsis* cells is normally kept in an autoinhibited conformation and is activated when $Ca^{2+}$ induces calmodulin to bind to the N-terminal domain of the pump (Harper et al. 1998; Hwang et al. 2000a). This binding of calmodulin might have disrupted an inhibitory interaction within the pump, thereby resulting in a "release of inhibition" (Hwang et al. 2000a). It has been demonstrated that a CDPK can inhibit the basal and calmodulin-stimulated activities of the $Ca^{2+}$ pump

ACA2 through phosphorylation at position $Ser^{45}$ in the N-terminal regulatory domain. Thus there may be a crosstalk between two different $Ca^{2+}$ signaling pathways (calmodulin and CDPKs), providing a mechanism to control $Ca^{2+}$ efflux through opposing inhibitory and stimulatory activities. Factors that shift this balance may alter the rate of $Ca^{2+}$ efflux, and thereby alter the magnitude or duration of a $Ca^{2+}$ signal (Hwang et al. 2000b).

The *Arabidopsis* $Ca^{2+}$-ATPases ACS8 and ACS10 have been shown to form a complex with the PRR FLS2 in planta. The mutant *aca8* and *aca10* plants showed decreases in the PAMP flg22-induced $Ca^{2+}$ transient increase in cytosol (Frey et al. 2012), suggesting that $Ca^{2+}$-ATPases modulate the $Ca^{2+}$ signaling.

$Ca^{2+}$-ATPases have been shown to be involved in plant innate immune responses. Programmed cell death (PCD) initiated at the infection sites is a defense response against pathogens (Zhu et al. 2010). The ER-localized type IIb $Ca^{2+}$-ATPase of *Nicotiana benthamiana* (*NbCA1*) has been shown to function as a regulator of PCD. Silencing of *NbCA1* accelerated the PAMP cryptogein-induced cell death. Downregulation of *NbCA1* resulted in the modulation of intracellular calcium signaling in response to the PAMP cryptogein treatment (Zhu et al. 2010). The results indicate that ER-$Ca^{2+}$-ATPase is a component of the calcium efflux pathway that controls PCD in an innate immune response. Disruption of the vacuolar calcium-ATPases in Arabidopsis results in the activation of salicylic acid signaling pathway involved in innate immune responses, suggesting the role of $Ca^{2+}$-ATPases in innate immunity (Boursiac et al. 2010).

## 4.6 Plasma Membrane $H^+$-ATPases in $Ca^{2+}$ Signaling

$H^+$-ATPases belong to the large P-type ATPases superfamily. The plasma membrane $H^+$-ATPases generate an $H^+$-gradient across the plant plasma membrane. The proton gradient energizes many important transport systems in plants. The proton gradient also creates an electrical potential, which is used to drive cation uptake through ion channels. The plasma membrane $H^+$-ATPase is an $H^+$ pump and is electrogenic, i.e. it can establish a membrane potential: positive on the outside, negative on the inside. It establishes the pH gradient (acidic on the outside) and creates the plasma membrane potential (typically −100 to −200 mV; negative on the inside) (Palmgren 1998; Palmgren and Harper 1999; Elmore and Coaker 2011). $H^+$-ATPases use energy from ATP hydrolysis to pump protons from the cytosol to the extracellular space (Sondergaard et al. 2004). Activation or inhibition of the $H^+$-ATPase modulates membrane potential (Ward et al. 2009). Changes in membrane potential alter the activities of voltage-gated channels and control ion flux at the plasma membrane (Haruta et al. 2010; Elmore and Coaker 2011).

Plant $H^+$-ATPases become activated as the C-terminal domain is degraded by controlled proteolysis. C-terminus of the ATPase of these pumps may be a regulatory domain inhibiting pump activity (Palmgren et al. 1991). It has been suggested that the C-terminus of the ATPase might block the active site, and that following its displacement the ATPase gets activated (Palmgren and Harper 1999). The activation of $H^+$-ATPase appears to be modulated by $Ca^{2+}$ influx – induced calcium dependent protein kinase (CDPK) (Camoni et al. 1998b; Schaller and Oecking 1999). The CDPK phosphorylates $H^+$-ATPase and the phosphorylation site is located at the C-terminal domain of $H^+$-ATPase (Camoni et al. 1998b). The $H^+$-ATPase was found to be phosphorylated at serine and threonine residues (Schaller and Sussman 1988). Phosphorylation of $H^+$-ATPase is stimulated by the addition of $Ca^{2+}$ and by a decrease in pH, from 7.2 to 6.2, suggesting that changes in the cytoplasmic $Ca^{2+}$ and pH are potentially important elements in modulating the kinase-mediated phosphorylation (Schaller and Sussman 1988). It is suggested that the $H^+$-ATPase might actually become activated following a dephosphorylation reaction preceded by phosphorylation reaction by CDPK (Palmgren and Harper 1999). The elicitor-induced stimulation of the plasma membrane $H^+$-ATPase was inhibited by okadaic acid, a phosphatase inhibitor, but not by staurosporine, a protein kinase inhibitor in tomato, suggesting that protein dephosphorylation was required for increased $H^+$-ATPase activity (Vera-Estrella et al. 1994). G proteins may be involved in elicitor-receptor binding, which in turn may stimulate the $H^+$-ATPase by dephosphorylation (Vera-Estrella et al. 1994; Xing et al. 1997).

Regulation of $H^+$-ATPases appears to depend on the presence or absence of 14-3-3 proteins (Chung et al. 1999; Fuglsang et al. 1999; Kanczewska et al. 2005; Ottmann et al. 2007; Duby and Boutry 2009). An in vitro interaction between a phosphorylated CDPK and 14-3-3 isoforms from *Arabidopsis* has been reported (Camoni et al. 1998a). There may be a functional link among phosphorylated CDPK, $H^+$-ATPase and 14-3-3 protein in defense signaling (Romeis et al. 2000; Duby et al. 2009). Binding of 14-3-3 proteins to the plasma membrane $H^+$-ATPase involves the three C-terminal residues Tyr-Thr-Val and requires phosphorylation of Thr (Fuglsang et al. 1999; Duby and Boutry 2009). The penultimate threonine residue and accompanying mode III motif is widely conserved across $H^+$ – ATPases throughout the plant kingdom, suggesting that this mechanism of activation is highly conserved (Duby and Boutry 2009). Additional phosphorylated residues within the C-terminal domain have been reported to affect the enzyme activity (Speth et al. 2010). 14-3-3 proteins recognize phosphate-bearing amino acid and regulate the $H^+$-ATPase enzyme activity (Romeis et al. 2000). Phosphorylation of the penultimate Thr residue in the C-terminal, autoregulatory domain of the $H^+$-ATPase results in 14-3-3 protein-dependent activation of the pump (Fuglsang et al. 1999, 2007; Svennelid et al. 1999; Maudoux et al. 2000). Phosphorylation at a second or additional unidentified sites inhibits the $H^+$-ATPase and this reaction appears to be regulated by CDPK (Vera Estrella et al. 1994; Xing et al. 1996; De Nishi et al. 1999; Rutschmann et al. 2002).

PAMPs have been shown to cause either $H^+$-ATPase activation concomitant with extracellular acidification and membrane hyperpolarization, or $H^+$-ATPase inactivation resulting in the depolarization of the plasma membrane (Wevelsiep et al. 1993; Vera-Estrella et al. 1994; Hammond-Kosack et al. 1996; Xing et al. 1996). Depolarization/hyperpolarization of cell membrane may modulate $Ca^{2+}$ influx in plant cells (White and Broadley 2003). Rutschmann et al. (2002) observed the apparent colocalization of the plasma membrane $H^+$-ATPase and the tomato CDPK in vivo, suggesting a potential role in the regulation of $H^+$-ATPase pump activity by $Ca^{2+}$-induced CDPK.

Plasma membrane $H^+$ – ATPase has been reported to play important role in plant innate immune responses. The early responses to the PAMP flg22 treatment include dynamic regulation of the plasma membrane $H^+$ – ATPases (Nuhse et al. 2007; Keinath et al. 2010). Perception of flg22 by the PRR FLS2 results in rapid membrane depolarization and alkalinization of the apoplast, probably induced by inhibition of $H^+$ – ATPases and activation of anion channels (Jeworutzki et al. 2010). Calcium influx at the plasma membrane occurs very rapidly after PAMP treatment and may contribute to plasma membrane $H^+$ – ATPase regulation in plant cells during PAMP-triggered immunity (Boller and Felix 2009; Kim et al. 2010). Stimulation of plasma membrane ATPase activity induces the accumulation of salicylic acid (SA) and the transcription of pathogenesis-related (PR) genes (Schaller and Oecking 1999). Stomatal closure is an immune response against bacterial pathogens. Bacterial PAMPs induce stomatal closure, which is dependent on both SA- and abscisic acid (ABA)-biosynthesis and associated signaling components (Melotto et al. 2006; Zhang et al. 2008; Zeng and He 2010). It has been reported that down-regulation of plasma membrane $H^+$ – ATPase activity via ABA signaling system contributes to PAMP-induced stomatal closure (Melotto et al. 2006; Merlot et al. 2007; Liu et al. 2009; Elmore and Coaker 2011).

## 4.7 Anion Channels in $Ca^{2+}$ Influx and Increase in $[Ca^{2+}]_{cyt}$

Anion channels which mediate $Cl^-$ and $NO_3^-$ efflux (Barbier-Brygoo et al. 2000; Vidhyasekaran 2007) have been shown to be involved in defense signaling system. Perception of flg22 by the PRR FLS2 results in activation of anion channels (Jeworutzki et al. 2010). Activation of $NO_3^-$ efflux has been shown to be dependent on protein phosphorylation (Binet et al. 2001; Wendehenne et al. 2002). Protein kinases act as positive regulators, while phosphatases negatively control the chain of events leading to anion channel activity. Because of the outward-directed anion gradients across the plasma membrane, the anion channels drive passive effluxes from the cytoplasm into the extracellular space (Wendehenne et al. 2002). Anion efflux results in plasma membrane depolarization (Sanders et al. 2002). Membrane depolarization induces $Ca^{2+}$ influx across the plasma membrane (Thuleau et al. 1994). It has also been shown that $Ca^{2+}$ influx may activate anion channels and increases in cytoplasmic $Ca^{2+}$ activate anion channels (Ward et al. 1995; Marten et al. 2007; Suh et al. 2007). Cytoplasmic $Ca^{2+}$ elevation has been shown to result in activation of S-type anion channels via phosphorylation (Schmidt et al. 1995; Allen et al. 1999; Geiger et al. 2010). The activated anion channels further intensify $Ca^{2+}$ influx which will pave the way for triggering $Ca^{2+}$-mediated signaling system (Fig. 4.3).

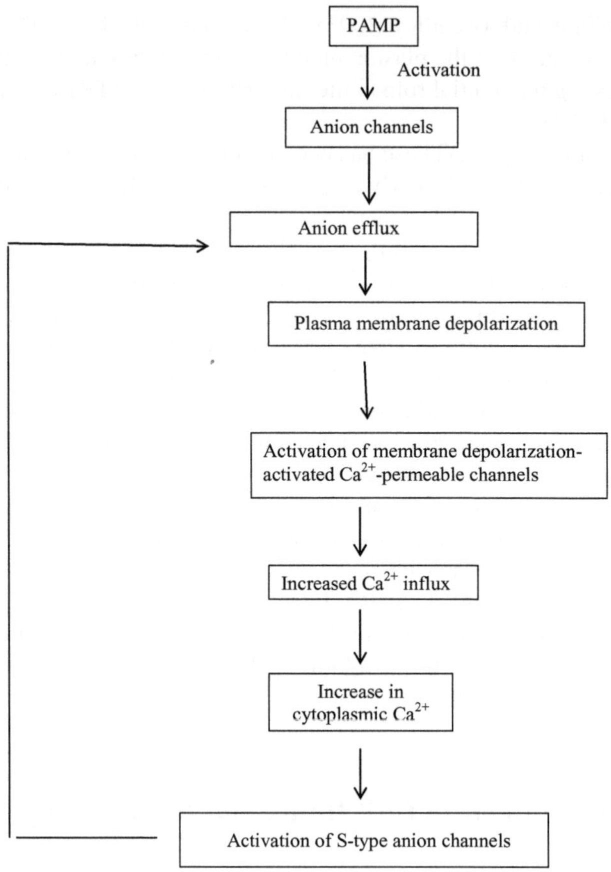


**Fig. 4.3** Role of anion channels in induction of $Ca^{2+}$ influx by PAMPs

## 4.8 $K^+$ Channels in $Ca^{2+}$ Influx

The major function of $K^+$ channels is in regulation of membrane voltage control (Maathuis et al. 1997). The $K^+$ channels play a role in regulating both the influx and efflux of $K^+$ from cells (Maathuis et al. 1997). Two types of $K^+$ channels have been identified, each with a characteristic voltage dependence, one opens in hyperpolarizing (inward rectifying $K^+$ channel, $K^+_{in}$) and another opens at depolarizing (outward-rectifying $K^+$ channel, $K^+_{out}$) conditions. Stimulation of $H^+$-ATPase will hyperpolarize the membrane and direct the $K^+$ gradient inward, while anion channel activation will depolarize the membrane and enhance the activity of $K^+_{out}$ channel activity (Maathuis et al. 1997).

Membrane depolarization (White 2000) or hyperpolarization (Pei et al. 2000) triggers $Ca^{2+}$ channel activation as part of the $Ca^{2+}$ signaling system. $K^+$ channels regulate both depolarization and hyperpolarization of cell membrane (Maathuis et al. 1997) and hence $K^+$ channels play important role in modulation of $Ca^{2+}$ signaling system. Excessive hyperpolarization or depolarization may result in membrane damage and affect signal transduction (Maathuis et al. 1997). When excessive hyperpolarization occurs, $K^+_{in}$ channels exhibit a strongly increasing tendency to open and $K^+$ uptake through these channels prevents the membrane voltage from becoming too negative. When signals induce rapid and excess membrane depolarization, $K^+_{out}$ channels will open, and the resulting efflux of $K^+$ will tend to limit the extent of the depolarization. In this way, $K^+$ channels will modulate the membrane potential and thereby activate $Ca^{2+}$ signaling system.

## 4.9 $K^+/H^+$ Exchange Response in $Ca^{2+}$ Signaling System

$K^+/H^+$ exchange response is an important component in the defense signaling system (Orlandi et al. 1992). The elicitors have been shown to trigger apoplastic alkalinization combined with cytosolic acidification in plant cells (Sakano 2001; Felle et al. 2004). Transient shifts of intracellular and apoplastic pH have been reported to be essential steps in several signal transduction processes (Felle et al. 2004). NAD kinase is an important enzyme involved in $Ca^{2+}$ signaling and in ROS signaling systems. NAD kinase is activated over a pH range of 7.1–6.8. The cytosolic pH in unstimulated tobacco cells was 7.5, which changed to lower level after being stimulated. At the acidic pH, NAD kinase was activated resulting in HR-related defense responses (Karita et al. 2004).

The enzyme $H^+/K^+$-ATPase is a proton pump which is responsible for the acidification of cytoplasm. The enzyme is a member of the P-type ATPase superfamily, a large family of related proteins that transport ions across cell membranes. As an ion pump, the $H^+/K^+$-ATPase is able to transport ions against a concentration gradient using energy derived from the hydrolysis of ATP. A phosphate group is transferred from adenosine triphosphate (ATP) to the $H^+/K^+$-ATPase during the transport cycle. This phosphate transfer powers a conformational change in the enzyme that helps drive ion transport. The enzyme transports one $H^+$ in exchange of one $K^+$ (Kuhlbrandt 2004). The $K^+/H^+$ exchange response may be mediated by $Ca^{2+}$ influx (Atkinson et al. 1990). The $Ca^{2+}$-dependent protein kinase, activated by increased cytosolic $Ca^{2+}$, may evoke ion fluxes that lead to extracellular alkalinization and depolarization of the plasma membrane (Schaller and Oecking 1999).

## 4.10 PAMPs and DAMPs May Trigger Calcium Ion Influx/Efflux Through Different $Ca^{2+}$ Channels

It has been shown that DND1 (a cyclic nucleotide-gated ion channel, CNGC) is important for cytosolic $Ca^{2+}$ elevation in response to bacterial PAMP lipopolysaccharides (LPS) and the damage-associated molecular pattern (DAMP)/

host-associated molecular pattern (HAMP)/endogenous elicitor *Arabidopsis* Pep peptides (Ma et al. 2009a, b; Qi et al. 2010). However, DND1 is not required for flg22 and elf18 activation of $Ca^{2+}$ signaling (Jeworutzki et al. 2010). Early signaling through the *Arabidopsis* PRRs FLS2 and EFR has been shown to involve calcium ion-associated opening of plasma membrane anion channels (Jeworutzki et al. 2010). Glu-receptor-like-type $Ca^{2+}$ channels are involved in cryptogein- and flg22-triggered immune responses (Kwaaitaal et al. 2011; Michard et al. 2011; Vatsa et al. 2011). The flg22/FLS2 signaling show greater requirement for intracellular $Ca^{2+}$ stores and inositol phosphate signaling, whereas Pep/PEPR signaling requires extracellular $Ca^{2+}$ to activate the $Ca^{2+}$ signaling system (Ma et al. 2012). $Ca^{2+}$ signaling by the *Arabidopsis thaliana* Pep peptides depends on CGMP-activated $Ca^{2+}$ channels (Qi et al. 2010). These results show the requirement of different $Ca^{2+}$ channels for the different PAMPs to activate $Ca^{2+}$ signaling system and innate immune responses.

## 4.11 Induction of Increases in Concentration, Oscillations and Waves in Cytoplasmic Calcium Ion ($[Ca^{2+}]_{cyt}$)

Various elicitors have been shown to induce biphasic $[Ca^{2+}]_{cyt}$ perturbations in plant cells These elicitors elicit an immediate transient increase in $[Ca^{2+}]_{cyt}$ in plant cells, which is followed by a more prolonged elevation of $[Ca^{2+}]_{cyt}$ lasting many minutes or hours (Lecourieux et al. 2002). The sustained increase in $[Ca^{2+}]_{cyt}$ alone is correlated with the induction of defense responses (Cessna and Low 2001; Rudd and Franklin-Tong 2001; Lecourieux et al. 2002). The elicitor-induced $[Ca^{2+}]_{cyt}$ elevations predominantly result from a continuous $Ca^{2+}$ influx through the plasma membrane (Hu et al. 2004; Vandelle et al. 2006).

The cytoplasmic $Ca^{2+}$ spikes (oscillations and waves) result from two opposing reactions, $Ca^{2+}$ influx through channels and $Ca^{2+}$ efflux through pumps and transport systems (Hwang et al. 2000a, b). Different messages can be encoded by changing a $Ca^{2+}$ spike's magnitude, duration, location, or frequency (Sanders et al. 1999; McAinsh and Pittman 2009). $Ca^{2+}$ signal is presented by the concentration of $Ca^{2+}$ (Trewavas and Malho 1998; Trewavas 1999). PAMPs/elicitors may activate $Ca^{2+}$ influx and the different signals may induce different $Ca^{2+}$ concentrations in the cytosol. The different concentrations of $Ca^{2+}$ may activate different $Ca^{2+}$-induced proteins (Karita et al. 2004).

The changes in $[Ca^{2+}]_{cyt}$ concentrations are monitored by the $Ca^{2+}$ sensors, such as calmodulins. Varying concentration of $[Ca^{2+}]_{cyt}$ may differentially regulate the calmodulin (CaM)-stimulated expression of CaM-binding proteins. Three types of tobacco CaM isoforms have been reported in tobacco. Type I CaM induced NAD kinase at 1–5 μM, which is the increased $Ca^{2+}$ concentration in stimulated cells. Type II CaM activated NAD kinase at lower $Ca^{2+}$ concentration of around 0.1 μM, which is the cytosolic concentration in unstimulated cells. The type II CaM is expressed constitutively and remained unchanged after external stimuli application.

Type III CaM did not induce NAD kinase at any $Ca^{2+}$ level (Karita et al. 2004). $Ca^{2+}$ spike frequency optimizes gene expression (Li et al. 1998). Calcium oscillations increase the efficiency and specificity of gene expression (Dolmetsch et al. 1998). A combination of changes in all $Ca^{2+}$ parameters produced by a particular signal determines $Ca^{2+}$ signature (Luan et al. 2002).

## 4.12 $Ca^{2+}$ Sensors in $Ca^{2+}$ Signal Transduction

The calcium signature is perceived by different $Ca^{2+}$-binding proteins (Kudla et al. 2010). These intracellular $Ca^{2+}$-binding proteins are also known as $Ca^{2+}$ sensors. The changes in $[Ca^{2+}]_{cyt}$ concentrations are monitored by the $Ca^{2+}$ sensors and the $Ca^{2+}$ signals are subsequently decoded and propagated downstream to activate plant defense responses. $Ca^{2+}$ signaling pathways are composed of molecular relays; the first runner after $Ca^{2+}$ is $Ca^{2+}$ "sensor", which monitors temporal and spatial changes in $Ca^{2+}$ concentrations. Several $Ca^{2+}$ sensors have been identified in plants (Fig. 4.4).

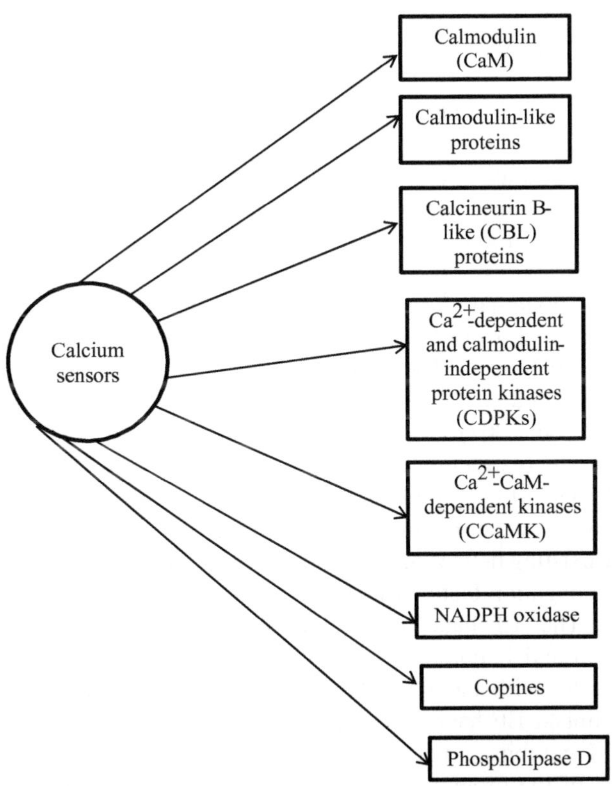


**Fig. 4.4** Calcium-binding proteins as calcium sensors

These include calmodulin (CaM) and CaM-related proteins, which typically contain four elongation factor (EF)-hand domains for $Ca^{2+}$ binding (Snedden and Fromm 2001). Calcineurin B-like (CBL) proteins detected in *Arabidopsis* are another family of $Ca^{2+}$ sensors (Luan et al. 2002). Another class of calcium binding proteins is $Ca^{2+}$-dependent and calmodulin-independent protein kinases (CDPKs). $Ca^{2+}$-dependent protein kinases have protein kinase and calmodulin-like $Ca^{2+}$-binding domains in a single protein, which allows their direct activation by $Ca^{2+}$ (Hrabak et al. 2003). The CDPK proteins function both as $Ca^{2+}$ sensors and as effectors of their $Ca^{2+}$-sensing activity (Luan et al. 2002; Geiger et al. 2010). CDPKs contain a C-terminal calmodulin-like regulatory domain that functions to couple the calcium sensor (calmodulin-like domain) directly to its responder (kinase) (Luan et al. 2002).

## 4.13 Calmodulins as $Ca^{2+}$ Sensors

Calmodulin (CaM) is one of the best characterized $Ca^{2+}$ sensors. CaM has no catalytic activity of its own, but upon binding $Ca^{2+}$, it activates numerous target proteins involved in a variety of cellular processes (Snedden and Fromm 2001; Reddy et al. 2011a). CaMs contain an autoinhibitory domain that occludes the active site in the resting state. $Ca^{2+}$ binds to a site near or overlapping the autoinhibitory domain, thereby releasing it from the active site and activating the protein (Luan et al. 2002).

Calmodulin genes are activated during pathogenesis and transcription of these genes occurs within a few minutes of pathogen invasion. Transcription of calmodulin isoform 4 (*GmCaM4*) is rapidly induced within 30 min after pathogen (*Pseudomonas syringae* pv. *glycinea*) stimulation in soybean (Park et al. 2007). Two zing finger homeodomain transcription factors, GmZF-HD1 and GmZF-HD2 proteins have been shown to activate the *GmCaM4* gene expression in response to the bacterial pathogen. The pathogen induced binding of GmZF-HD1 and GmZF-HD2 to repeats of ATTA homeodomain binding site in the *GMCaM4* promoter (Park et al. 2007).

CaM contains two structurally similar domains connected by a flexible central linker. Each domain of the protein binds two calcium ions with positive cooperativity. The binding of $Ca^{2+}$ transforms the protein into its active form through a reorientation of the existing helices of the protein (Zhang and Yuan 1998). CaM typically contains four elongation factor (EF)-hand domains for $Ca^{2+}$ binding (Snedden and Fromm 2001). The 'EF hand' is a helix-loop-helix structure. CaM is an acidic EF-hand protein and is composed of 148 amino acids arranged in two globular domains connected with a long flexible helix. Each globular domain contains a pair of intimately linked EF hand (Snedden and Fromm 2001; Rainaldi et al. 2007). $Ca^{2+}$-free CaM exhibits a flat, hydrophilic molecular surface, while the $Ca^{2+}$-saturated form of the protein contains a Met-rich cavity containing hydrophobic surface in each domain. These hydrophobic surfaces are largely responsible for the binding of CaM to its targets. The unique flexibility and high polarizability of the Met residues located at the entrance of each hydrophobic pocket together with other hydrophobic amino acid residues create adjustable, sticky interaction surface areas that can accommodate CaM targets, which have various sizes and shapes (Zhang and Yuan 1998). The binding of calcium to calmodulin induces a conformational change that exposes hydrophobic binding sites that interact with target proteins, altering the activity of those proteins (Harmon 2003).

Plant cells contain multiple CaM isoforms with varying degrees of sequence homology to the single CaM reported in mammals. Several *CaM* genes have been isolated from plants (Lee et al. 1995, 1999; Heo et al. 1999; Kim et al. 2009). Thirteen CaM genes have been detected in tobacco (Takabatake et al. 2007). In soybean five CaM isoforms with varying degrees of sequence homology to the single mammalian CaM have been identified (Lee et al. 1995). Not all, but specific CaM genes are involved in defense signaling. The tobacco CaM gene, *NtCaM13*, was found to induce resistance against the bacterial pathogen *Ralstonia solanacearum*, the fungal pathogen *Rhizoctonia solani*, the oomycete pathogen *Pythium aphanidermatum*, and not against *Tobacco mosaic virus* in tobacco, while *NtCaM1* did not have any role in inducing resistance against the pathogens (Takabatake et al. 2007). The CaM isoforms have different expression patterns in various plant tissue types, suggesting that they play unique roles in the many different $Ca^{2+}$ signaling pathways of plants (Lee et al. 1995, 1999; Cho et al. 1998; Takabatake et al. 2007).

Elicitors activate CaM isoforms which participate in $Ca^{2+}$-mediated induction of defense response (Heo et al. 1999). Upon increase of $Ca^{2+}$ to submicromolar levels, all CaM molecules are activated. Full activation of the CaM occurs in a narrow region of calcium concentration during a signaling event (Luan et al. 2002). Induction of CaM genes, *SCaM-4* and *SCaM-5* genes in soybean depended on the increase of intracellular $Ca^{2+}$ level (Heo et al. 1999). The constitutive expression of these soybeans genes in transgenic tobacco plants constitutively expressed genes encoding PR-1a, PR-1b, PR2, PR3, PR4, PR5, class III acidic chitinase and class III basic chitinase (Heo et al. 1999). The expression of tobacco NtCaM13, which is closely related to SCaM4 and SCaM5, was elevated both at the RNA and protein level in TMV-infected leaves (Yamakawa et al. 2001).

$Ca^{2+}$-CaM binds and regulates the activity of a wide range of proteins. Three types of tobacco calmodulin (CaM) isoforms originated from 13 genes. These CaMs differentially activate target enzymes. Plant NAD kinase was activated most effectively by type II (NtCaM3 – NtCaM12), moderately by type I (NtCaM1 and NtCaM2), and weakly by type III (NtCaM13) CaMs. By contrast, NO synthase was activated most effectively by type III, moderately by I, and weakly by type II CaMs (Karita et al. 2004). In soybean, SCaM-4 activates cyclic nucleotide phosphodiesterase while it is unable to activate the CaM-dependent NAD kinase (Lee et al. 1995). By contrast, SCaM-1 activates NAD kinase (Lee et al. 1997). SCaM-1 activates the protein phosphatase calcineurin, while SCaM-4 antagonizes its activation (Cho et al. 1998).

## 4.14 Calmodulin-Binding Proteins

### *4.14.1 Function of Calmodulin-Binding Proteins in Different Signaling Systems*

Calmodulin (CaM) is a small (~150 residues), acidic protein comprised of a flexible central helical region which joins two globular domains, each with two $Ca^{2+}$-binding EF-hand motifs (Gilford et al. 2007). Upon binding to $Ca^{2+}$, the hydrophobic surfaces in each globular domain are exposed which then interact with the characteristic amphiphilic structure, called calmodulin-binding domain (CBD), present in calmodulin-binding proteins (CBPs) (Reddy 2001; Snedden and Fromm 2001; Reddy et al. 2002a, b). This interaction leads to conformational changes in CBPs and modulation of their activity. The specificity of a response evoked due to characteristic $Ca^{2+}$ signature may depend upon the expression kinetics of CBPs (Ali et al. 2003).

CaM binds to a variety of proteins involved in various signaling systems (Fig. 4.5). These include proteins involved in pathogen-associated molecular pattern (PAMP) – receptor mediated signaling pathway (receptor-like kinases), $Ca^{2+}$ signaling system (CNG ion channels; $Ca^{2+}$-ATPases, $Ca^{2+}$-CaM-dependent kinases), reactive oxygen species (ROS) signaling system (NAD kinase, *BAG* gene), redox signaling

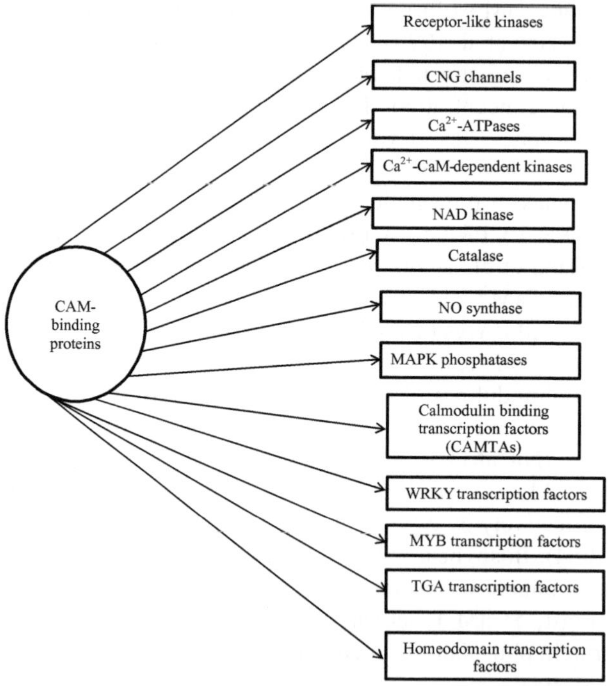


**Fig. 4.5** Calmodulin-binding proteins

system (catalase), nitric oxide (NO) signaling system (NO synthase [NOS]), mitogen-activated protein kinase (MAPK) cascade (MAPK phosphatases), transcription factors (OsCBTs, WRKY7, MYB2, TGA3, CAMTAs, ethylene-responsive genes-encoded proteins), glucosinolate metabolism (nuclear factors), and MLO proteins (receptor-like protein/modulator of defense response).

### *4.14.2 CaM Binds to CNG Channel Protein*

Cyclic nucleotide-gated ion channels (CNGCs) are involved in $Ca^{2+}$-dependent signaling pathways (Talke et al. 2003; Yoshioka et al. 2006). These channels have been found to be CaM-binding proteins. CNGCs have a CaM-binding domain near the C-terminus suggesting a role for CaM in modulating the activity of these channels (Leng et al. 1999). Calmodulin binding has been demonstrated for the *Arabidopsis* CNGC proteins AtCNGC1, AtCNGC2, and AtCNGC10 (Kohler et al. 1999; Kohler and Neuhaus 2000; Borsics et al. 2007). *Arabidopsis DND1* codes for a cyclic nucleotide-gated channel 2 (AtCNGC2) involved in plant defense responses (Clough et al. 2000). In bean, three CNGC isoforms (PvCNGC-A, PvCNGC-B, and PvCNGC-C) have been shown to be involved in defense responses. These proteins bound CaM in a $Ca^{2+}$-dependent manner. Expression of an isoform of CNGCs, PvCNGC was induced, whereas the expression of two other isoforms PvCNGC –B and PvCNGC-C was repressed in response to incompatible pathogens in bean. It suggests that there may be functionally distinct role for each CNGC in plants (Ali et al. 2003). CNGCs constitute a link between cyclic nucleotide and $Ca^{2+}$ signals (Talke et al. 2003). High-affinity CaM-binding site in tobacco plasma-membrane channel protein coincides with cyclic nucleotide-binding domains (Arazi et al. 2000).

### *4.14.3 $Ca^{2+}$-ATPases as CaM-Binding Proteins*

$Ca^{2+}$-ATPases are localized in the endomembranes or plasma membrane. They play an important role in removing $Ca^{2+}$ from the cytoplasm to terminate a signaling event (Sze et al. 2000; McAinsh and Pittman 2009). Among the $Ca^{2+}$-ATPases in plants, type IIB $Ca^{2+}$-ATPases are involved in $Ca^{2+}$ signaling. These ATPases bind with calmodulin activated by $Ca^{2+}$. $Ca^{2+}$-CaM interacts with type IIB ATPases to activate the pump by releasing an autoinhibitory domain from the active site (Luan et al. 2002).

### *4.14.4 Protein Kinases as CAM-Binding Proteins*

Some protein kinases are regulated by CaMs. A chimeric plant $Ca^{2+}$-CaM-dependent protein kinase, CCaMK, with a visinin-like $Ca^{2+}$ binding domain and a CaM binding domain in one molecule has been identified in lily (*Lilium longiflorum*) plants (Patil et al. 1995). The predicted structure of CCaMK contains a catalytic domain followed by two regulatory domains, a calmodulin-binding domain and a visinin-like $Ca^{2+}$-binding domain. The calmodulin-binding region contains three $Ca^{2+}$-binding EF-hand motifs and a biotin-binding site (Patil et al. 1995). Although $Ca^{2+}$ can regulate the kinase activity via a visinin-like domain, $Ca^{2+}$-CaM enhances the kinase activity toward a substrate and inhibits its autophosphorylation activity, suggesting that $Ca^{2+}$-CaM may regulate substrate specificity in vivo (Takezawa et al. 1996). CCaMK homologs have been detected in tobacco, apple, maize, and *Arabidopsis* (Patil et al. 1995; Watillon et al. 1995; Lu and Feldman 1997).

### 4.14.5 Receptor-Like Kinases as CaM-Binding Proteins

Some receptor-like kinases (RLKs) have been identified as CaM-binding proteins (Charpenteau et al. 2004). A CaM-binding protein, AtCaMRLK, has been identified as a RLK in *A. thaliana*. AtCaMRLK polypeptide shows a sequence characteristic of receptor kinases: an amino terminal signal sequence, a domain containing seven leucine-rich repeats, a single putative membrane-spanning segment and a protein kinase domain. A region of 23 amino acids, located near the kinase domain binds CaM in a calcium-dependent manner. The CaM-binding motif of AtCaMRLK was found to be conserved in several other members of the plant RLK family, suggesting a role for $Ca^{2+}$/CaM in the regulation of RLK-mediated pathways (Charpenteau et al. 2004). Several RLKs are known to be involved in plant defense responses (Navarro et al. 2004; Zipfel et al. 2004; Benschop et al. 2007).

### 4.14.6 NAD Kinase as CaM-Binding Protein

NAD kinase is the enzyme involved in elevation of NADPH levels in plant cells and it has been found to be a calmodulin-binding protein (Harding et al. 1997). NADPH oxidase triggers ROS production. The calcium-binding protein calmodulin is involved in generation of ROS. Transgenic tobacco plants expressing a foreign calmodulin gene showed enhanced NADPH oxidase – dependent production of ROS (Harding et al. 1997; Harding and Roberts 1998). NADPH levels were elevated rapidly through the activation of NAD kinase in the stimulated tobacco cells. Elicitor treatment also induced burst of ROS in transgenic tobacco cell cultures. Higher levels of NADPH in transgenic calmodulin cells led to a more rapid and intense burst of ROS, suggesting the involvement of an NADPH oxidase in the CaM-induced ROS production (Harding et al. 1997). These studies suggest that the calmodulin-binding protein NAD kinase is involved in triggering ROS generation which plays an important role in triggering plant immune responses (Fig. 4.6).

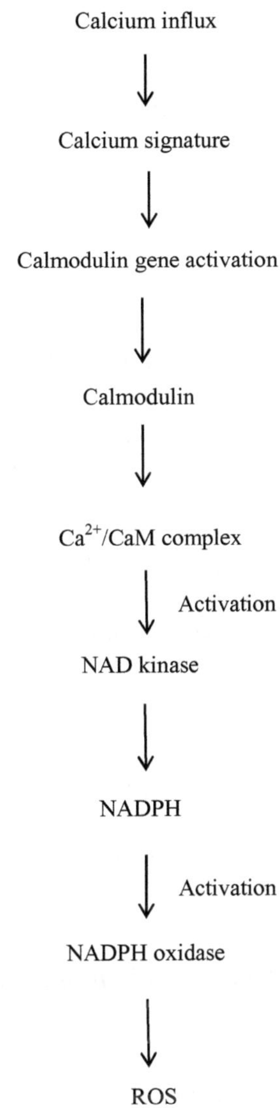


**Fig. 4.6** Role of calmodulin in $Ca^{2+}$-triggered ROS generation

## *4.14.7 Catalases as CaM-Binding Proteins*

CaM binds to and activates some plant catalases in the presence of calcium. $Ca^{2+}$/CaM may down-regulate $H_2O_2$ levels in plants by stimulating the catalytic activity of plant catalase. The results suggest that calcium has dual functions in regulating $H_2O_2$ homeostasis, which in turn influences redox signaling in plants (Yang and Poovaiah 2002b). Redox signaling system has been reported to be involved in plant defense response against pathogens (Önnerud et al. 2002; Cumming et al. 2004; Fobert and Després 2005; Fedoroff 2006).

### 4.14.8 NO Synthase as CaM-Binding Protein

In *Arabidopsis*, a NO synthase (NOS) protein contains CaM-binding motifs and full activation of the enzyme needs both $Ca^{2+}$ and CaM (Guo et al. 2003; Zeidler et al. 2004). NO synthesis is tightly regulated by a signaling cascade involving $Ca^{2+}$ influx in elicitor-treated tobacco cells (Lamotte et al. 2004). NOS is involved in NO production, which is stimulated by CaM and changes in $[Ca^{2+}]$cyt (Guo et al. 2003).

### 4.14.9 MAPK Phosphatase as a CaM-Binding Protein

Mitogen-activated protein kinase (MAPK) cascade is an important signaling cascade involved in defense response (Mészáros et al. 2006; Dóczi et al. 2007; Hall et al. 2007). The activity of MAPKs is strictly regulated via phosphorylation of the conserved TXY motif, which is accomplished by a corresponding MAPK kinase. After activation, the dephosphorylation and inactivation of MAPK is performed by MAPK phosphatases (Theodosiou and Ashworth 2002; Yamakawa et al. 2003; Katou et al. 2005). A putative MAPK phosphatase in *Nicotiana tabacum* (NtMKP1) has been identified as a CaM-binding protein (Yamakawa et al. 2003). In the presence of free $Ca^{2+}$, NtMKP1 binds to CaMs. The CaM-binding domain was identified as a 52-amino acid sequence between the conserved gelsolin and Serrich domain in the middle of the NtMKP1 protein (Rainaldi et al. 2007). A rice MAPK phosphatase, OsMKP1, has been shown to bind calmodulin (Katou et al. 2007). These results suggest an interaction between $Ca^{2+}$-CaM and a component of the MAPK signaling cascade in plants.

### 4.14.10 Calmodulin Binding Transcription Activators (CAMTAs)

One family of CaM-binding proteins, designated as the calmodulin-binding transcription activator (CAMTA) family resembles a group of putative transcription activators identified in the human genome (Bouché et al. 2002; Galon et al. 2010a, b). The C-terminal CaM binding domain of CAMTAs mediates interactions with calmodulin (Kudla et al. 2010). The CAMTA family of proteins contains a transcription activation domain and two types of DNA-binding domains designated the CG1 domain and the transcription factor immunoglobulin domain, ankyrin repeats, and a varying number of IQ CaM-binding motifs (Bouché et al. 2002). *Arabidopsis thaliana* contains six *CAMTA* genes (*AtCAMTA1* – *AtCAMTA6*). CAMTAs comprise a conserved family of transcription factors (Bouché et al. 2002). AtCAMTA homologue is found in rapeseed (BnCAMTA) (Zegzouti et al. 1999; Bouché et al. 2002).

Finkler et al. (2007) found that the $Ca^{2+}$/CaM-responsive CAMTAs bind to the ABA-responsive ABRE *cis* elements in *Arabidopsis*, suggesting a link between $Ca^{2+}$-responsive transcription factors and ABA-responsive *cis*-elements. CAMTA3 directly interacts with the promoter of the *EDS1* gene, a regulator of salicylic acid levels, and represses its expression (Du et al. 2009). $Ca^{2+}$/calmodulin binding to CAMTA3 is required for the suppression of plant defense, indicating a direct role of $Ca^{2+}$/calmodulin in regulating the function of CAMTA3 (Du et al. 2009).

### *4.14.11 Calmodulin-Binding OsCBT, NtER1, and AtSR1 Transcription Factors*

Some transcription factors of the basic helix-loop-helix family were shown to bind calmodulin (Onions et al. 2000). A CaM-binding transcription factor, OsCBT (*Oryza sativa* CaM-binding transcription factor), has been isolated from rice (Choi et al. 2005). It contains a CG-1 homology DNA binding domain, three ankyrin repeats, a putative transcriptional activation domain, and five putative CaM-binding motifs (Choi et al. 2005). OsCBT has two different types of functional CaM-binding domains, an IQ motif, and a $Ca^{2+}$-dependent motif (Choi et al. 2005).

Some ethylene-responsive genes-encoded proteins in tobacco, tomato, rapeseed, *Arabidopsis* have structural similarity to the rice CaM-binding transcription factor, OsCBT (Choi et al. 2005). The tobacco early ethylene-responsive gene *NtER1* encodes a CaM-binding protein (Yang and Poovaiah 2000). One *NtER1* homolog (*AtSR1*) and five related genes (*AtSR2-6*) have been identified in *Arabidopsis* and they are designated as *AtSR* (for *Arabidopsis thaliana Signal Responsive*) genes (Yang and Poovaiah 2002a). These six genes exhibit rapid and differential response to signal elicitors such as ethylene, jasmonate (JA), salicylic acid, ABA, and $H_2O_2$. $Ca^{2+}$/CaM binds to a 23-mer peptide in all AtSRs that corresponds to the CaM-binding region of NtER1. Each AtSR has a conserved structural feature with a DNA-binding domain in the N terminus and a CaM-binding domain in the C terminus. AtSR1 targets the nucleus and specifically recognizes a novel 6-bp CGCG box (A/C/G) CGCG (G/T/C). The multiple CGCG *cis*-elements are found in promoters of various genes involved in ET- and ABA signaling systems. The results suggest that AtSR may interact with ET and ABA signaling systems (Yang and Poovaiah 2002a).

### *4.14.12 Calmodulin-Binding WRKY Transcription Factors*

WRKY proteins constitute a large class of transcription factors. More than 70 WRKY genes have been reported in *Arabidopsis* (Eulgem et al. 2000; Robatzek and Somssich 2001; Dong et al. 2003; Bhattarai et al. 2010). About 109 WRKY genes have been identified in rice (Qu and Zhu 2006; Liu et al. 2005, 2007; Qiu et al. 2007). The WRKY transcription factor family is defined by a domain of 60 amino acids, which contains the amino acid sequence WRKY (tryptophan-arginine-lysine-tyrosine) at its amino-terminal end and a putative zinc finger motif at its carboxy-terminal end. Most of the WRKY proteins contain one WRKY domain, while some of the WRKY proteins have two WRKY domains (Eulgem et al. 2000; Maeo et al. 2001; Zheng et al. 2006). WRKYs are subdivided into three subgroups (Eulgem et al. 2000). Members of group I have two WRKY domains, whereas members of groups II and III have one WRKY domain. Group III domains contain a $Cx_7CX_{23}HXC$ pattern of zinc ligands which is distinct from the $Cx_{4-5}CX_{22-23}HXH$ zinc finger pattern of group I and group II WRKY domains (Maeo et al. 2001; Knoth et al. 2007). Members of the *Arabidopsis* WRKY group III transcription factors are part of different plant defense signaling pathways (Kalde et al. 2003)

WRKY is localized to the nucleus of plant cells and recognizes DNA molecules containing the *TTGACC* W-box sequence (Zheng et al. 2006). The WRKY domain binds specifically to various W box elements containing a (C/T)TGAC(C/T) core sequence. The promoters of a large number of defense-related genes contain W-box sequences that are recognized by WRKY proteins and the WRKY transcription factors have been shown to be necessary for the inducible expression of these defense genes (Eulgem et al. 1999, 2000; Yu et al. 2001; Shimono et al. 2007; van Verk et al. 2008). Several WRKY transcription factors have been shown to be involved in activation of SA biosynthesis genes in plant immune responses (van Verk et al. 2011).

CaM binds to the $Ca^{2+}$-dependent CaM-binding domain of several WRKY transcription factors. CaM binds specifically to the $Ca^{2+}$-dependent CaM-binding domain of AtWRKY7 transcription factor (Park et al. 2005). WRKY7 is a member of the WRKYIId subfamily, and all members of this subfamily including WRKY11, WRKY15, WRKY17, WRKY21, WRKY39, and WRKY74 were found to interact with $Ca^{2+}$/CaM (Park et al. 2005). Other WRKYs, WRKY43, WRKY45, WRKY50, and WRKY 53, have been shown to interact with different isoforms of CaM in a $Ca^{2+}$-dependent manner (Popescu et al. 2007).

The defense signal salicylic acid induces expression of *WRKY7* gene, suggesting that the *WRKY* gene is involved in plant's defense response. The transcription factor WRKY7 negatively regulates the expression of defense-related genes in *Arabidopsis* (Kim et al. 2006). The CaM binding transcription factors WRKY11 and WRKY17 act as negative regulators of basal resistance in *Arabidopsis thaliana* (Journot-Catalino et al. 2006). Another transcription factor interacting with CAM is WRKY45, which is involved in triggering plant defense responses. Transgenic rice plants overexpressing the transcription factor *WRKY45* showed enhanced resistance to the blast pathogen *Magnaporthe oryzae* (Shimono et al. 2007). All these WRKYs are CaM binding proteins.

### *4.14.13 Calmodulin-Binding MYB Transcription Factors*

Several *MYB* transcription factor genes are found in plants. They are characterized by the presence of a highly conserved MYB domain at their N-termini (Du et al. 2009).

MYB transcription factors contain one or more MYB domains (Stracke et al. 2001). MYB proteins are classified into subfamilies depending on the number of conserved repeats of the MYB domain they contain (Mengiste et al. 2003). A common feature of MYB proteins is the presence of a functional DNA binding domain that typically consists of one to three repeats of the MYB domain. The three MYB repeats are referred to as R1, R2, and R3. Each repeat is about 50–53 amino acids long and encodes three α-helices, with the second and third helices forming a helix-turn-helix structure which intercalates in the major groove of DNA when bound to it. MYB repeats typically contain regularly spaced tryptophan residues, which build a central tryptophan cluster in the three-dimensional helix-turn-helix fold (Du et al. 2009). MYB transcription factors bind to the *cis*-regulatory element such as MYB boxes (Laquitaine et al. 2006).

The R2R3-MYB subfamily of transcription factors is the most common in plants (Stracke et al. 2001). A soybean CaM, Gm-Cam4, has been reported to mediate $Ca^{2+}$ signaling response by activating an R2R3-MYB2 transcription factor (Yoo et al. 2005). A grapevine R2R3-MYB transcription factor gene, *VvMYB5a*, induces the expression of genes controlling the biosynthesis of defense-related phenylpropanoids (Deluc et al. 2006). The Arabidopsis *BOS1* (*BOTRYTIS SUSCEPTIBLE 1*) gene, which encodes a R2R3-MYB transcription factor, induces disease resistance against *P. syringae* and disruption of the gene enhances disease symptom development after infection by *P. syringae* (Mengiste et al. 2003). A R2R3-MYB-like transcription factor, MYB72, has been shown to be involved in triggering defense responses against broad-spectrum of pathogens in *Arabidopsis thaliana* (Van der Ent et al. 2008). *Arabidopsis* MYB2 has been reported to function as transcriptional activator in abscisic acid signaling (Abe et al. 2003). Several members of the MYB class of transcription factors have been shown to bind $Ca^{2+}$/CaM (Popescu et al. 2007). Direct interaction of a CaM isoform with the transcription factor MYB2 has been reported in *Arabidopsis* (Yoo et al. 2005).

### *4.14.14 Calmodulin-Binding TGA Transcription Factors*

The basic leucine zipper (bZIP) proteins belong to a large family of transcription factors. The bZIP family transcription factors contain a basic region for binding DNA and a leucine zipper dimerization domain (Jakoby et al. 2002). These proteins have a conserved region rich in basic amino acid residues that binds to the target DNA and contains nuclear localization signals (NLSs) and dimerization or multimerization domains. A leucine zipper region, which consists of several heptad repeats of hydrophobic residues, is found close to the basic region. The leucine zipper region is alpha-helical and prone to dimer formation via a coiled-coil arrangement (Kuhlmann et al. 2003; Meng et al. 2005).

The TGA class of transcription factors is the important group of bZIP transcription factors involved in defense signaling. Members of the TGA family of transcription factors bind tandem repeats of a *cis*-element within the promoters called activation sequence-1 (*as-1*), which contains a TGACG motif (Lebel et al. 1998; Després et al. 2000). Several TGA transcription factors have been shown to regulate expression of defense-related genes (Kesarwani et al. 2007). This family of transcription factors recognizes the TGACG/*as-1* elements found in the promoters of a variety of plant genes, including those regulating the expression of *Arabidopsis* and tobacco *PR-1* and the *Cauliflower mosaic virus* 35S promoter (Lebel et al. 1998; Kim and Delaney 2002). The *as-1* elements are responsible for SA responsiveness of these promoters. SA treatment increases the TGACG/*as-1* binding activity. Thus, the TGA transcription factors may play an important role in SA signaling system (Zhou et al. 2000). Several SA-responsive genes are regulated by bZIP transcription factors of TGA family (Ndamukong et al. 2007). Some of the bZIP transcription factors, such as TGA2 and TGA5 in *Arabidopsis*, interact with NPR1 and recognize the *as-1 cis* element found within the promoter of several *PR* genes (Kim and Delaney 2002).

Transgenic plants overexpressing different TGA transcription factor genes have been generated to develop disease resistant plants (Kim and Delaney 2002; Fitzgerald et al. 2005). Transgenic *Arabidopsis* plants containing sense or antisense *TGA5* gene constructs were developed by Kim and Delaney (2002). None of the *TGA5* sense lines showed an apparent increase in *TGA5* transcript levels compared to wild-type plants, whereas the *TGA5*-antisense lines showed a large increase in *TGA5* transcript accumulation. Increased *TGA5* accumulation in antisense lines may be due to negative autoregulation of the *TGA5* gene (Kim and Delaney 2002). The transgenic *TGA5*- antisense lines showed reduced induction of SA-mediated expression of *PR-1* gene by the oomycete pathogen *Hyaloperonospora parasitica*. The transgenic antisense lines showed enhanced resistance to *H. parasitica* (Kim and Delaney 2002). The induced resistance by *TGA5* to the pathogen has been suggested to act independent of SA signaling system.

The rice TGA factor, rTGA2.1, has been shown to bind to defense gene promoters (Chern et al. 2001). It binds to oligonucleotides containing the *as-1* like elements from the *PR-1* gene promoter and to the promoter of the rice chitinase gene, *RCH10* (Chern et al. 2001). It appears that *rTGA2.1* negatively regulates a subset of rice defense genes (Fitzgerald et al. 2005). Transgenic rice plants that have the endogenous rTGA2.1 transcripts silenced via dsRNA-mediated silencing (Sl) were also generated. The loss of *rTGA2.1* activity in the Sl lines resulted in reduced disease symptom development (Fitzgerald et al. 2005).

Several TGA proteins have been shown as CaM-binding proteins. TGA3, a member of a family of basic leucine zipper (bZIP) transcription factors, has been identified as a CaM binding protein that binds the promoter of CaM3 (Jakoby et al. 2002). Eighteen bZIP family members have been identified as CaM binding proteins in *Arabidopsis* (Popescu et al. 2007). An abscisic acid (ABA) – responsive bZIP transcription factor, ABF2, has been shown to bind CaM (Popescu et al. 2007).

### 4.14.15 Calmodulin-Binding Homeodomain Transcription Factors

Homeodomain proteins in the homoeobox gene family play important roles as transcription factors in plants (Williams 1998). Some of the homeodomain transcription factors have been shown to be involved in modulating plant immune responses (Park et al. 2007). An *Arabidopsis* homeodomain transcription factor, OVEREXPRESSOR OF CATIONIC PEROXIDASE 3, mediates resistance to infection by necrotrophic pathogens (Coego et al. 2005). The homeodomain transcription factors have also been reported as calmodulin-binding proteins. In soybean, transcription of calmodulin isoform 4 (*GmCaM4*) is rapidly induced within 30 min after pathogen stimulation. The GmCaM4 promoter contains two repeats of conserved homeodomain binding site, ATTA (Park et al. 2007). Two proteins, GmZF-HD1 and GmZF-HD2, belonging to the zinc finger homeodomain (ZF-HD) transcription factor family have been detected in soybean. These transcription factors bind to the two repeats of ATTA homeodomain binding site in the calmodulin GmCaM4. This binding was induced in response to the pathogen. Regulation of GmCaM4 gene by the GMZF-HD transcription factors may be a significant component of the plant defense-signaling pathway (Park et al. 2007).

### 4.14.16 CaM-Binding Protein Involved in Glucosinolate Metabolism

A nuclear protein, IQ-DOMAIN1 (IQD1), is a CaM-binding protein and integrates intracellular $Ca^{2+}$ signals towards stimulation of glucosinolate accumulation and plant defense (Levy et al. 2005). *IQD1* encodes a novel protein that contains putative nuclear localization signals and several motifs known to mediate calmodulin binding, which are arranged in a plant-specific segment of 67 amino acids, called the IQ67 domain. IQD1 protein is targeted to the cell nucleus and binds to calmodulin in a $Ca^{2+}$-dependent fashion. IQD1 affects expression of multiple genes with roles in glucosinolate metabolism. It is suggested that IQD1 is a nuclear factor that integrates intracellular $Ca^{2+}$ signals to fine-tune glucosinolate accumulation in response to pathogens (Levy et al. 2005).

### 4.14.17 CaM Binds with MLO Protein to Regulate Defense Response

CaM binds with MLO protein, which exhibits a dual role as docking molecule and defense modulator for the powdery mildew pathogen (Panstruga and Schulze-Lefert 2005). MLO has been detected in *Arabidopsis*, barley, and rice (Kim et al. 2002a, b;

Opalski et al. 2005; Panstruga 2005). MLO protein resides in the plasma membrane and has seven transmembrane domains. MLO interacts constitutively with the cytoplasmic calcium sensor calmodulin (Panstruga 2005). A 20-amino acid CaM-binding domain has been located in the rice OsMLO C-terminal cytoplasmic tail (Kim et al. 2002a). Loss of calmodulin binding halves the ability of MLO to negatively regulate defense against powdery mildew (Kim et al. 2002b). The result suggests that MLO is a CAM-binding protein involved in the modulation of defense reactions.

### *4.14.18 CaM-Binding Protein Involved in HR-Associated Cell Death*

A calmodulin-binding protein, AtBAG2, has been isolated from *Arabidopsis* (Kang et al. 2006a, b). The CaM-binding protein contains a central BCL-2-associated athanogene (BAC) shown to be involved in programmed cell death involved in host defense response. Agents generating ROS induced the *AtBAG6* transcript, indicating relationship between CaM and ROS signaling (Kang et al. 2006a).

### *4.14.19 CBP60 Family of Calmodulin-Binding Proteins*

A plant-specific family of CaM binding proteins called CaM60s has been detected in several plant species. CBP60s have been identified in *Arabidopsis* (Reddy et al. 2002b; Wang et al. 2009; Zhang et al. 2010), maize (Reddy et al. 1993), bean (Ali et al. 2003) and tobacco (Lu and Harrington 1994). Some of these CBPs have their CaM binding domain (CBD) at very close to the C-terminal ends (Reddy et al. 1993, 2002a; Lu and Harrington 1994) and others at the N terminus (Wang et al. 2009; Zhang et al. 2010). Other than the presence of CBD, CBP60s do not show sequence similarity to any other known domains (Ali et al. 2003).

The two CaM-binding proteins detected in bean (*Phaseolus vulgaris*), PvCBP60-C and PvCBP60-D, were found to be associated with host defense responses (Ali et al. 2003). A calmodulin binding protein, CBP60g, has been shown to be involved in activating SA biosynthesis (Wang et al. 2009; Zhang et al. 2010). *Arabidopsis* CBP60g positively affects the expression of *SID2*, which encodes an isochorismate synthase (ICS) that is involved in biosynthesis of SA (Wang et al. 2011). CBP60g shows DNA binding activity, and it preferentially binds to a DNA sequence that contains AATTTT, which is present in the promoter of ICS1. Mutants that abolish CaM binding activity of CBP60g did not complement the mutant phenotype, suggesting that binding of CaM to CBP60g is essential for its function (Wang et al. 2009).

## 4.15 Calmodulin-Like Proteins as $Ca^{2+}$ Sensors

Some CaM-like proteins are also involved in calcium signaling system. Typical CaMs are highly similar to animal CaM, whereas CaM-like proteins share 50–75 % identity to typical CaM and have CaM activity. A calmodulin-like protein, Hra32, involved in defense response has been detected in bean. The predicted Hra32 product contains four putative EF-hand calcium-binding domains that are separated by a spacing of nine amino acids (Jakobek et al. 1999). *Arabidopsis* CaM8 is a CaM-like protein. This protein can function as a CaM in $Ca^{2+}$ binding, but it appears to interact with a more limited set of target proteins compared with typical CaM isoforms (Zielinski 2002).

A tomato gene (*APR34*) encoding a CaM –like protein has been characterized. *APR134*-like genes (*CML42* and *CML43*) have been isolated from Arabidopsis. The *CML43* gene was rapidly induced in disease-resistant *Arabidopsis* leaves. Overexpression of *CML43* in *Arabidopsis* accelerated the hypersensitive response (Chiasson et al. 2005), suggesting the role of the CAM-like protein in plant immune response.

Centrins are CaM-like proteins with four $Ca^{2+}$-binding EF-hand motifs (Lecourieux et al. 2006). Centrins have been found to be associated with cytoskeleton and a role in microtubule severing and cytoskeleton reorganization has been shown (Lecourieux et al. 2006). Arrangement of microtubules and microfilaments was found to play an important role in the expression of nonhost resistance in barley (Kobayashi et al. 1997). A rapid and $Ca^{2+}$-dependent disruption of microtubular cytoskeleton is associated with disease resistance-associated cell death in a fungal elicitor-treated tobacco cells (Binet et al. 2001). *CCD-1* mRNA accumulates rapidly in elicitor-treated wheat cells (Takezawa 2000). *CCD-1* encodes a 14-kDa $Ca^{2+}$-binding protein that shares homology with the C-terminal half domain of centrin (Takezawa 2000). These studies showed that centrins are involved in disease resistance responses and in $Ca^{2+}$ signaling.

Some CaM proteins have additional non-CaM domains and they are called CaM-related proteins (Luan et al. 2002). Petunia CaM53 is a CaM-related protein. It has CaM activity but it contains a polybasic C-terminal domain that is not found in typical CaMs. This extra domain in CaM53 regulates the cellular localization (Rodriguez-Concepcion et al. 1999).

## 4.16 Calcineurin B-Like Proteins as $Ca^{2+}$ Sensors

Another family of $Ca^{2+}$ sensors consists of proteins similar to both the regulatory B-subunit of calcineurin and the neuronal $Ca^{2+}$ sensor in animals and these $Ca^{2+}$ sensors are called calcineurin B-like (CBL) proteins (Luan et al. 2002; Batistić et al. 2008; Hashimoto et al. 2012). Calcineurin is a $Ca^{2+}$/calmodulin-dependent protein phosphatase belonging to PP2B type (Kudla et al. 1999; Trewavas 1999).

CBLs in *Arabidopsis* are encoded by a multigene family of at least 10 members that have similar structural domains with small variations in the length of the coding regions (Kim et al. 2000; Albrecht et al. 2001; Luan et al. 2002). CBLs have a helix-loop-helix structural motif (the EF hands) that acts as the $Ca^{2+}$ binding site. CBLs contain three EF hands, whereas CaMs contain four EF hands. The EF-hand sequence consists of a 12-amino acid loop that uses amino acids at positions 1, 3, 5, 7, and 12 for interaction with $Ca^{2+}$. The Asp at position 1, Gly at position 6, and Glu at position 12 are the most highly conserved amino acids in the loop. Both CaMs and CBLs have similar amino acid sequences in their EF-hand motifs. However, they do not show significant similarity in their primary amino acid sequences (Luan et al. 2002).

Several CBLs have a conserved myristoylation site in their N-terminal regions. The myristoylation site is required for its location to membrane (Batistić et al. 2008). In the $Ca^{2+}$-free state, the myristoyl moiety in recoverin is inaccessible to membranes. The $Ca^{2+}$-induced conformational change exposes the myristoyl group and facilitates the association of recoverin with the membrane (Ishitani et al. 2000; Kim et al. 2000; Batistić et al. 2008).

Unlike CaMs, which interact with a large number of target proteins, CBLs are known to interact only with the family of SNF1-like protein kinases. These kinases are called as CBL-interacting kinases (CIPKs) and CBL interacts with CIPKs through the C-terminal nonkinase domain that contains a conserved region among different CIPK members (Batistić et al. 2008; Luan 2009). Micromolar levels of $Ca^{2+}$ are required for the interaction of CBL and CIPK (Halfter et al. 2000). With at least 10 CBLs and 25 CIPKs reported in *Arabidopsis*, many functional CBL-CIPK pairs can be formed that potentially function in a large array of signaling processes involving $Ca^{2+}$ signaling (Luan et al. 2002; Lecourieux et al. 2006; Batistić et al. 2008).

## 4.17 NADPH Oxidase as Calcium-Binding Protein

NADPH oxidase, resembling the human neutrophil respiratory burst NADPH oxidase has been detected in plasma membrane fractions in plant cells (Torres et al. 2002; Yoshioka et al. 2003). The respiratory burst oxidase homolog (Rboh) in plants has EF-hand calcium-binding motifs in its N-terminal extension. Rboh is an intrinsic plasma membrane protein and the extended N-terminal domain of Rboh projects into the cytosol (Keller et al. 1998). All plant *rboh* genes carry EF-hands that bind $Ca^{2+}$ and plant Rboh proteins were shown to be stimulated directly by $Ca^{2+}$ (Sagi and Fluhr 2001). $Ca^{2+}$-stimulated Rboh enzymes may be positioned close to $Ca^{2+}$ channels localized on the plasma membrane (Lecourieux et al. 2006). The activated NADPH oxidase generates superoxide radicals in plant cells, which are converted to $H_2O_2$ (Sagi and Fluhr 2006). These results suggest that $Ca^{2+}$ influx may activate ROS signaling system.

## 4.18 $Ca^{2+}$-Binding Proteins Without EF-Hands

Some $Ca^{2+}$-binding proteins do not contain EF-hand structural motifs. These proteins contain other $Ca^{2+}$-binding domains such as the C2 domain (Reddy 2001). The C2 domain is a $Ca^{2+}$-phospholipid-binding site, and $Ca^{2+}$ binding is coordinated by four to five amino acid residues provided by bipartite loops (Rizo and Sudhof 1998). These domains often mediate $Ca^{2+}$-dependent phospholipid binding.

The copines are a family of $Ca^{2+}$-dependent, phospholipids-binding proteins (Tomsig and Cruetz 2002). Copine proteins (in French, copine means "friend", these proteins are named copine because of their tight association with lipid membranes) contain two protein kinase C conserved 2 (C2) domains in the terminal region and a von Willebrand A (VWA) domain in the C-terminal region (Jambunathan et al. 2001; Jambunathan and McNellis 2003; Laxal and Munnik 2002). Copines bind membrane phospholipids due to the presence of two C2 domains in the N-terminal portion that are activated by calcium. C2 domain-containing proteins include protein kinase C and phospholipase C. The C-terminal half of the copine molecule, called the VWA domain, may be involved in targeted protein-protein interactions (Lecourieux et al. 2006). Copines have a calcium-dependent phospholipid-binding activity (Tomsig and Cruetz 2000).

In *Arabidopsis CPN1* (*copine1*) is a negative regulator of plant defense-related hypersensitive response (HR) (Jambunathan et al. 2001; Jambunathan and McNellis 2003). Mutation of the *CPN1* (*COPINE 1*) gene in *Arabidopsis* results in increased resistance to bacterial (*Pseudomonas syringae* pv. *tomato*) and oomyceteous (*Peronospora parasitica*) pathogens. The mutant showed constitutive expression of *PR* genes (Jambunathan and McNellis 2003). The results suggest that *CPN1* plays a role in disease resistance responses, possibly as a suppressor of defense responses. Copines may play a role in membrane trafficking (Hua et al. 2001) and they may represent a universal transduction pathway (Tomsig et al. 2003).

Copines have been suggested to represent a universal transduction pathway for calcium signaling because the copines are capable of interacting with a wide variety of target proteins including a MAP kinase kinase (MEK1), a protein phosphatase, a Cdc-42-binding kinase, the transcription factor Myc binding protein, ubiquitin conjugating enzyme, and an enzyme involved in exocytosis (homolog of SNIP [SNAP-25 Interacting protein]) (Tomsig et al. 2003). The copine binds to a domain of the target protein that is predicted to form a characteristic coiled coil. The interaction with copines may result in recruitment of target proteins to membrane surfaces and regulation of the enzymatic activities of target proteins (Tomsig et al. 2003).

C2 domain has been detected in all characterized plant phospholipase D enzymes (PLDs) (Wang 1999, 2001). $Ca^{2+}$ may associate directly with PLD through the C2 domain. A positive correlation between increased $[Ca^{2+}]_{cyt}$ and PLD activity has been reported (De Vrije and Munnik 1997). It suggests that PLDs may be involved in increased $[Ca^{2+}]_{cyt}$.

## 4.19 Calcium-Dependent Protein Kinases as $Ca^{2+}$ Sensors

### 4.19.1 *Structure of CDPKs*

Calcium-dependent protein kinases (CDPKs) are involved in $Ca^{2+}$ signaling (Boudsocq et al. 2010; Bush et al. 2010). They belong to the CDPK/SnRK family of protein kinases (Hrabak et al. 2003). The protein kinase activity of CDPK is stimulated by binding of calcium or its intrinsic CaM-like domain. The CDPKs bind up to four moles of $Ca^{2+}$ per mole of enzyme (Lee et al. 1998). CDPKs contain five domains: amino-terminal variable domain, catalytic domain, autoinhibitory domain, CaM-like domain, and a short domain. The amino-terminal variable domain varies in length from 40 to 180 amino acids. The catalytic domain is typical of serine/threonine protein kinases. Adjacent to the catalytic domain is an autoinhibitory domain that contains a pseudosubstrate sequence that can interact with the active site and inhibit activity. Also in the inhibitory domain is a site that can bind to the calmodulin-like domain in the presence of calcium. Binding of the CaM-like domain to this site is proposed to contribute to the stimulation of kinase activity in the presence of calcium (Yoo and Harmon 1996; Harmon 2003). The calmodulin-like domain is adjacent to the autoinhibitory domain, and contains four $Ca^{2+}$ binding helix-turn-helix structures known as EF-hands (Lee et al. 1998; Reddy et al. 2011a, b). Following the calmodulin-like domain is a short domain of variable length (Harmon 2003; Fig. 4.7).

### 4.19.2 *PAMP/Elicitor Triggers Activation of CDPK*

The PAMP/elicitor treatment induces $Ca^{2+}$ influx by activating the ion channels in the plasma membrane (Kwaaitaal et al. 2011; Ranf et al. 2011). The increases in $Ca^{2+}$ influx result in elevation in cytoplasmic $Ca^{2+}$ concentration (Luan et al. 2002). Several biotic stimuli trigger an increase in the concentration of cytoplasmic free $Ca^{2+}$, which then acts as a second messenger mediating a variety of cellular responses. The cytoplasmic free $Ca^{2+}$ concentration under resting conditions is maintained at very low levels (10–200 nM), ensuing low CDPK activity. An increase in the cytoplasmic calcium results in CDPK activation. CDPKs function as sensors of fluctuations in cytosolic $Ca^{2+}$ and initiate downstream signaling events (Harmon et al. 2000; Hrabak 2000; Kobayashi et al. 2007; Ito et al. 2010). CDPKs require $Ca^{2+}$ for their activation. CDPKs contain a kinase domain, an autoinhibitory domain and a CaM-like domain. The inhibitory domain contains a pseudosubstrate sequence. CDPK is autoinhibited by an interaction of the pseudosubstrate site within its junction domain that blocks the active site of the kinase domain. Binding of $Ca^{2+}$ to the CaM-like domain of the CDPK causes conformational change that extends to the adjacent junction domain and finally disengages the autoinhibitor of the active site (Huang et al. 1996). The release of the pseudosubstrate domain from the active site

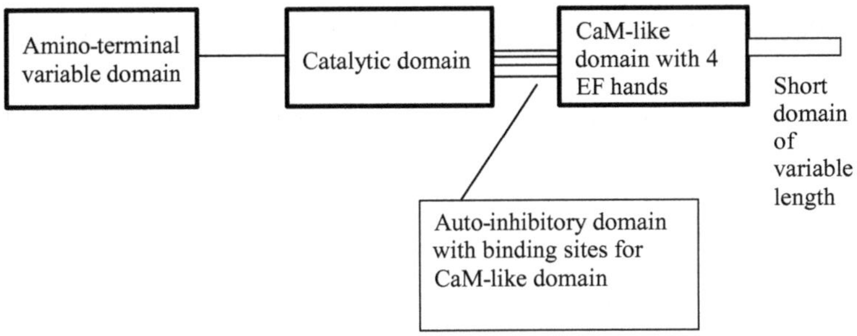


**Fig. 4.7** Structure of calcium-dependent protein kinase

results in kinase activation (Harmon et al. 2000). These results suggest $Ca^{2+}$ binding to CDPK is essential for the activation of the enzyme.

PAMP/elicitor may be involved in enhanced transcription of CDPK genes, probably through the action of calcium signature. The tobacco CDPK gene *NtCDPK1* was found to be induced by fungal elicitors (Yoon et al. 1999). Two other CDPK genes from tobacco, *NtCDPK2* and *NtCDPK3*, showed mRNA up-regulation after the application of the Avr9 race-specific elicitor (Romeis et al. 2001). Fungal elicitors triggered the transcription of the tomato *LeCDPK1* gene (Chico et al. 2002). A CDPK gene in maize, *ZmCPK10*, was induced both during a fungal infection and after treatment with fungal elicitors (Murillo et al. 2001). Activation of the *ZmCPK10* gene was very rapid. *ZmCPK10* transcripts could be detected 5 min after elicitation and reached maximum levels at 30 min after treatment (Murillo et al. 2001).

Several CDPKs have been reported in plants. *Arabidopsis thaliana* has 34 genes that encode CDPKs (Hrabak et al. 2003). Several *CDPK* genes have been detected in soybean, rice, tomato, maize, and *Arabidopsis* (Harmon et al. 2001). Individual isoforms of CDPKs may have different functions and participate in multiple distinct signaling pathways (Harmon et al. 2001). Depending on the calcium signatures (variations in $[Ca^{2+}]_{cyt}$ concentration, oscillations, and waves), specific CDPK isoforms are activated (Ludwig et al. 2004).

The *Nicotiana benthamiana* CDPK, NtCDPK2, induced enhanced ROS production and activation of defense-related genes in *N. benthamiana*. By contrast, a homologous isoform, NtCDPK3, did not induce the HR-associated cell death. The results suggest that NtCDPK2 kinase, but not the closely related NtCDPK3 protein, is specifically involved in the defense response (Ludwig et al. 2004). Dependent on the incoming signals, NtCDPK2 enzyme activation varied in strength and duration (Romeis et al. 2001). A short and weak NtCDPK2 activation results in the induction of the wound signaling pathway, whereas a much stronger and sustained elicitation leads to defense response signaling system (Romeis et al. 2001). Several CDPK genes have been detected in maize. However, only one specific CDPK gene, *ZmCPK10* gene is transcriptionally activated in response to both fungal infection and treatment with fungal elicitors/PAMPs. This gene was found to be involved in activation of defense signaling pathways, leading to the induction of PR genes (Murillo et al. 2001).

### 4.19.3 Stimulation of CDPK Activity by 14-3-3 Proteins

14-3-3 proteins have been shown to modulate the activity of CPDK signal transduction pathways. The 14-3-3 proteins are given this nomenclature based on their chromatography and electroprofiles. The CDPK isoform CPK1 from *Arabidopsis thaliana* is stimulated almost twofold by three different 14-3-3 proteins (Camoni et al. 1998b), suggesting that 14-3-3 proteins may modulate the activity of CDPK signal transduction pathways in plants. Protein kinases and phosphatases are regulated by 14-3-3 proteins (Ferl 2004). 14-3-3 proteins specifically bind and activate the *Arabidopsis thaliana* enzyme AtCPK1 in vitro in the presence of $Ca^{2+}$ (Camoni et al. 1998b). The 14-3-3 proteins play a role in the completion of signal transduction events. Phosphorylation may tag the proteins for association with 14-3-3 and the subsequent binding of 14-3-3s may complete the signal-induced changes in the protein activity (Ferl 2004). 14-3-3 proteins occur as homo- and heterodimers in vitro and in vivo and these dimers may mediate interaction between pairs of associated proteins (Jones et al. 1995). It has been shown that 14-3-3 proteins bind to phosphorylated Ser residues present within one of a small number of consensus sequences found in many of the proteins with which they interact (Yaffe et al. 1997).

### 4.19.4 Enhancement of CDPK Activity by Phospholipids

Specific phospholipids enhance in vitro substrate phosphorylation by CDPKs (Farmer and Choi 1999; Szczegielniak et al. 2000). Certain phospholipids, including phosphatidic acid, phosphatidylserine, and phosphatidylinositol, act as second messengers and enhanced the CDPK activity (Farmer and Choi 1999). In the presence of $Ca^{2+}$, specific phospholipids enhance phosphorylation by CDPKs by 2–30 times above that observed with $Ca^{2+}$ alone (Harper et al. 1993; Farmer and Choi 1999). Both phosphatidylinositol and lyso-phosphatidylcholine increase substrate phosphorylation by CPK1 from *Arabidopsis*, while only phosphatidylinositol enhances the CDPK autophosphorylation (Binder et al. 1994). A binding site for phosphatidylinositol has been detected in the N terminus of CPK1 (Binder et al. 1994).

### 4.19.5 CDPKs Target Proteins Involved in Immune Signaling System

CDPKs target several proteins involved in immune signaling systems (Fig. 4.8). CDPKs are involved in activation of various $Ca^{2+}$-permeable channels in plant cell plasma membrane. Arabidopsis CDPKs CPK6 and CPK3 have been shown

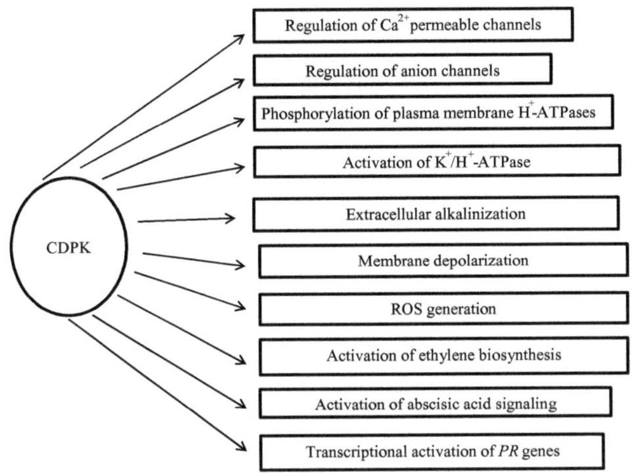


**Fig. 4.8** Immune signaling systems regulated by CDPKs

to be involved in regulation of anion- and $Ca^{2+}$-permeable channels (Mori et al. 2006). CDPKs regulate plasma membrane $H^+$-ATPases by phosphorylation (Schaller and Oecking 1999). Anion channels and $H^+$-ATPases have been shown to be involved in $Ca^{2+}$ influx (Sanders et al. 2002; White and Broadley 2003). The enzyme $H^+/K^+$-ATPase is a proton pump which is responsible for the acidification of cytoplasm. The activation of $K^+/H^+$-ATPase is mediated by $Ca^{2+}$ influx and activated by CDPK (Atkinson et al. 1990). The CDPK may evoke ion fluxes that lead to extracellular alkalinization and depolarization of the plasma membrane (Schaller and Oecking 1999). The activation of the $Ca^{2+}$ channel by elicitors was modulated by a heterotrimeric G-protein dependent phosphorylation of the channel protein in tomato, probably by activating a CDPK and inhibiting a protein phosphatase (Gelli and Blumwald 1997). These results suggest that CDPK targets several proteins involved in Ca2+ signaling system for phosphorylation and activation.

CDPKs may be involved in activation of enzymes/proteins involved in various defense signaling pathways (Fig. 4.8). Plasma membrane-associated NADPH oxidase is a substrate for CDPK (Xing et al. 2001). NADPH oxidase is known to stimulate ROS production (Kobayashi et al. 2007). Overexpression of the *Arabidopsis* CDPK gene *AtCPK1* in a heterologous tomato protoplast system resulted in an enhanced NADPH oxidase activity and increased production of ROS (Xing et al. 2001). Two CDPKs in potato, StCDPK4 and StCDPK5, phosphorylated Ser-82 and Ser-97 in the terminus of StRBOHB, a potato NADPH oxidase, in a calcium-dependent manner. Ectopic expression of the constitutively active mutant of StCDPK5, StCDPK5VK, provoked ROS production in *Nicotiana benthamiana* leaves. The heterologous expression of StCDPK5VK phosphorylated Ser-82 of StRBOHB in *N. benthamiana* (Kobayashi et al. 2007). These results suggest that the CDPK functions as a calcium sensor and phosphorylates the plasma membrane bound NADPH oxidase, which is the key enzyme in ROS signaling system.

CDPK activates amino-1-cyclopropane carboxylate synthase, the enzyme involved in ethylene biosynthesis (Sebastiá et al. 2004). *Arabidopsis* CDPK AtCPK32 interacts with ABF4, a transcriptional regulator of ABA-responsive gene expression, and modulates its activity (Choi et al. 2005). Expression of a grape CDPK, ACPK1, in *Arabidopsis thaliana* activates abscisic acid (ABA) signaling (Yu et al. 2007). Phenylalanine ammonia-lyase (PAL), the key enzyme in phenylpropanoid pathway involved in biosynthesis of phytoalexins, has been shown to be a substrate of a specific constitutively active *Arabidopsis* CDPK expressed in corn protoplasts (Cheng et al. 2001). The maize CDPK, ZmCPK10, triggers a rapid transcriptional activation of *PRms* encoding PR proteins in maize (Murillo et al. 2001). Collectively these results suggest that CDPKs may target various enzymes involved in plant defense systems.

## 4.20 Nuclear Free Calcium Ion ($[Ca^{2+}]_{nuc}$) in $Ca^{2+}$ Signaling

Increases in nuclear free calcium concentration ($[Ca^{2+}]_{nuc}$) have been reported to occur in plants in response to some external stimuli. The bacterial (harpin) and oomycete (elicitins) elicitors induced a pronounced and sustainable $[Ca^{2+}]_{nuc}$ elevation (Lecourieux et al. 2006; Mazars et al. 2009, 2010). The $[Ca^{2+}]_{nuc}$ rise depends on free cytosolic calcium ($[Ca^{2+}]$cyt), 1,4,5-trisphosphate (IP3) and reactive oxygen species (ROS) (Lecourieux et al. 2006).

Levy et al. (2005) identified an *Arabidopsis* gene, *IQDI* (*IQ-DOMAIN I*), which encodes a calmodulin-binding nuclear protein. *IQD1* integrates intracellular $Ca^{2+}$ signals towards stimulation of plant defenses, including accumulation of glucosinolates, the secondary metabolites involved in plant defense. CaM, CaM-binding proteins (Bouche et al. 2005), CDPK (Damman et al. 2003), and $Ca^{2+}$-CaM-regulated protein phosphatase (Andreeva and Kutuzov 2001) have been detected in plant nucleus. $Ca^{2+}$-ATPase (Downie et al. 1998) and some components of the phosphoinositide signaling pathway (Dröbak and Heras 2002) have been found to be localized to the plant nucleus. The nuclei isolated from tobacco were capable of producing $H_2O_2$ in a calcium-dependent manner (Astamker et al. 2007). The $[Ca^{2+}]_{nuc}$ rise depends on free cytosolic calcium, 1,4,5-trisphosphate (IP3), and ROS (Lecourieux et al. 2005). The interplay between nuclear and cytosolic calcium elaborates a global calcium signature and elicits biological responses (Pauly et al. 2001). Nuclear calcium may be responsible for the activation of $Ca^{2+}$-dependent proteins in the nucleus, and may be involved in the regulation of nuclear activities such as gene expression (White and Broadley 2003; Mazars et al. 2010). It has also been reported that the nucleus exhibits a $Ca^{2+}$ signature independently of the cytosol in response to stresses (Mazars et al. 2010).

# 4.21 Downstream Events in $Ca^{2+}$ Signaling System

## *4.21.1 ROS Generation*

### 4.21.1.1 $Ca^{2+}$ and CDPK–Mediated ROS Generation

Several signaling systems are activated by calcium influx and transient increase in cytosolic calcium levels. Calcium influx activates ROS production (Jabs et al. 1997). NADPH oxidase, resembling the human neutrophil respiratory burst NADPH oxidase has been detected in plasma membrane fractions in plant cells (Torres et al. 2002; Yoshioka et al. 2003). The respiratory burst oxidase homolog (Rboh) in plants has EF-hand calcium-binding motifs in its N-terminal extension. Rboh is an intrinsic plasma membrane protein and the extended N-terminal domain of Rboh projects into the cytosol (Keller et al. 1998). All plant *rboh* genes carry EF-hands that bind $Ca^{2+}$ and plant Rboh proteins were shown to be stimulated directly by $Ca^{2+}$ (Sagi and Fluhr 2001). $Ca^{2+}$-stimulated Rboh enzymes may be positioned close to $Ca^{2+}$ channels localized on the plasma membrane (Lecourieux et al. 2006). The activated NADPH oxidase generates superoxide radicals in plant cells, which are converted to $H_2O_2$ (Sagi and Fluhr 2006). Calcium-dependent protein kinases (CDPKs) may also play important role in ROS generation. The activation of the plasma membrane-located NADPH oxidases involves phosphorylation of two N-terminal serine residues by a CDPK (Kobayashi et al. 2007). $Ca^{2+}$ binding on EF-hand domains and phosphorylation by CDPK activate the NADPH oxidase enzyme in a synergistic manner to generate ROS (Fig. 4.9; Ogasawara et al. 2008).

### 4.21.1.2 Calmodulin-Mediated ROS Generation

The $Ca^{2+}$ sensor protein calmodulin (CaM) also has been reported to be involved in $Ca^{2+}$-triggered ROS generation. The pepper CaM gene *CaCaM1* has been shown to be involved in ROS generation (Choi et al. 2009). Treatment with calcium channel blocker suppressed ROS burst that was triggered by *CACAM1* expression in pepper and *Arabidopsis*, suggesting that calcium influx is required for the activation of *CaCaM1*-mediated ROS generation in plants (Choi et al. 2009). An increase in $Ca^{2+}$ amount in cytosol triggered by PAMP elicitor stimuli is perceived by the calmodulin. The calmodulin gene is strongly induced by the $Ca^{2+}$ influx in *Arabidopsis* (Desikan et al. 2001). The complex $Ca^{2+}$/CaM has been shown to regulate NAD kinase, which generates NADPH for NADPH oxidase activity (Harding et al. 1997). Increase in NADPH oxidase activity results in generation of ROS (Desikan et al. 1997; Harding et al. 1997; Pitzschke et al. 2009a; Mazars et al. 2010).

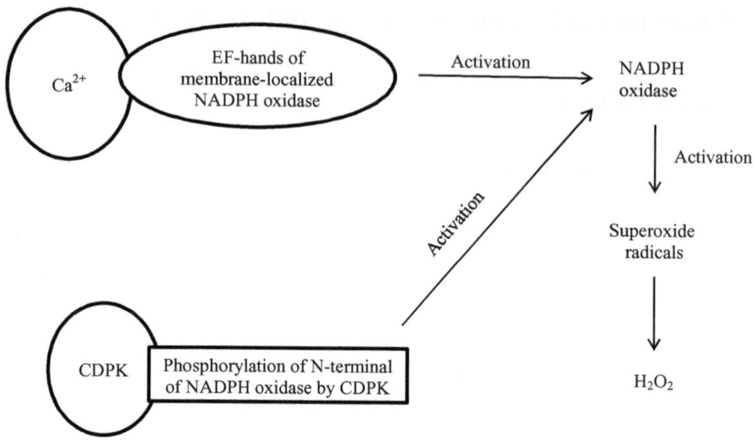


**Fig. 4.9** Synergistic action of $Ca^{2+}$ binding on EF-hand domains of NADPH oxidase and phosphorylation of NADPH oxidase by CDPK in ROS generation

### 4.21.2 NO Generation

Nitric oxide (NO) is a key mediator for rapid induction of plant immune responses (Bellin et al. 2013). Rapid NO production has been shown to be dependent on $Ca^{2+}$ signaling system (Lamotte et al. 2004; Ali et al. 2007; Courtois et al. 2008; Choi et al. 2009; Ma et al. 2008, 2012; Vatsa et al. 2011). It has been demonstrated that different calcium channels involved in calcium influx are also involved in NO production. The bacterial PAMP lipopolysaccharide (LPS) activates the $Ca^{2+}$ influx through $Ca^{2+}$ channels and this $Ca^{2+}$ current leads to downstream NO production (Ali et al. 2007). Plants without functional CNGC2 lack this cell membrane $Ca^{2+}$ current and do not display immune responses (hypersensitive response, HR). The impaired HR phenotype to an avirulent pathogen in *cngc2* mutant plants can be complemented by the addition of an NO donor (Ali et al. 2007). The results suggest the importance of the cyclic nucleotide gated channel in induction of NO in the immune signaling system (Ali et al. 2007). Another type of ion channels involved in $Ca^{2+}$influx is glutamate receptor (GLR)-like channels Ma et al. 2012). The oomycete PAMP elicitor signal cryptogein activates GLR calcium channels triggering NO production (Vatsa et al. 2011). The addition of the $Ca^{2+}$ channel blocker $Gd^{3+}$ or the $Ca^{2+}$ chelator EGTA abolished LPS-induced NO synthesis (Ali et al. 2007). The results indicate that NO synthesis occurs downstream of cytosolic $Ca^{2+}$ elevation.

NO synthase (NOS) is the key enzyme involved in NO production. NOS has been reported to be a CaM-binding protein. NOS contains CaM-binding motifs and full activation of the enzyme needs both $Ca^{2+}$ and CaM (Guo et al. 2003; Lamotte et al. 2004; Zeidler et al. 2004). CaM dependent NOS production has been reported in many plants (Delledonne et al. 1998; Courtois et al. 2008; Ma and Berkowitz 2011). The pathogen-induced $Ca^{2+}$ signals lead to CaM activation of NOS (Ma et al. 2008). The use of a CaM antagonist prevents NO generation and induction of immune responses. Application of a CaM antagonist does not prevent pathogen-induced cytosolic $Ca^{2+}$ elevation, suggesting that CaM does not act upstream from $Ca^{2+}$. The CaM antagonist and $Ca^{2+}$ chelation abolish NO generation. It suggests that plant NOS activity is $Ca^{2+}$/CaM dependent. Ma et al. (2008) suggested that the initial pathogen recognition signal of $Ca^{2+}$ influx into the cytosol activates CaM, which then acts to induce downstream NO synthesis, leading to innate immune responses.

NOS activity has been reported to be induced by the bacterial PAMP LPS (Delledonne 2005). LPS-induced NO production is suggested to be dependent on NOS enzyme activity and this process is regulated by CaM (Zeidler et al. 2004; Ali et al. 2007). The pepper CaM gene *CaCaM1* has been shown to be involved in NO generation (Choi et al. 2009). Upon treatment with the CaM antagonist, virulent *Pseudomonas syringae* pv. *tomato*–induced NO generation was also compromised in *CaCaM1* overexpressing plants (Choi et al. 2009).

It has also been reported that the generated NO can induce cytosolic $Ca^{2+}$ increase through activation of plasma membrane- and intracellular membrane-localized $Ca^{2+}$ channels during pathogen induced signaling cascades (Ali et al. 2007). The bacterial PAMP LPS could elicit NO generation in leaf guard cells and facilitate $Ca^{2+}$ influx into the cytosol (Ali et al. 2007). NO synthesis occurring during the plant-pathogen interactions causes elevation of cytosolic $Ca^{2+}$ level (Lamotte et al. 2004, 2006; Vandelle et al. 2006; Besson-Bard et al. 2008a, b). It is suggested that the NO generated downstream $Ca^{2+}$ influx may diffuse to neighboring cells and activate new $Ca^{2+}$ signals, which may amplify the NO generation process (Fig. 4.10; Ma et al. 2007; Ma and Berkowitz 2011).

### 4.21.3 MAPK Signaling System

Mitogen-activated protein kinase (MAPK) cascades are major pathways downstream of sensors/receptors that transduce extracellular stimuli into intracellular responses in plants (Liu et al. 2003; Pedley and Martin 2005). A typical MAPK signaling module consists of three protein kinases: a MAP kinase kinase kinase (MAPKKK or MEKK [for MAPK/Extracellular signal-regulated kinase Kinase Kinase]), a MAP kinase kinase (MAPKK or MKK), and a MAP kinase (MAPK or MPK) (Mészáros et al. 2006). MAP kinase cascade involves sequence of phosphorylation events (Hirt 2000). Among the 20 Arabidopsis MAP kinases, MPK3, MPK4, and MPK6 are implicated in plant immune responses (Petersen et al. 2000; Innes 2001; Asai et al. 2002; Menke et al. 2005; Takahashi et al. 2007; Gao et al. 2008; Ren et al. 2008; Pitzschke et al. 2009b; Liu et al. 2011). MPK3, MPK6, MKK4, and MKK5 form a cascade that positively regulates plant defenses (Pitzschke et al. 2009b). MPK3 has been shown to be required for camalexin accumulation upon *Botrytis cinerea* infection (Ren et al. 2008). Inactivation of MPK3

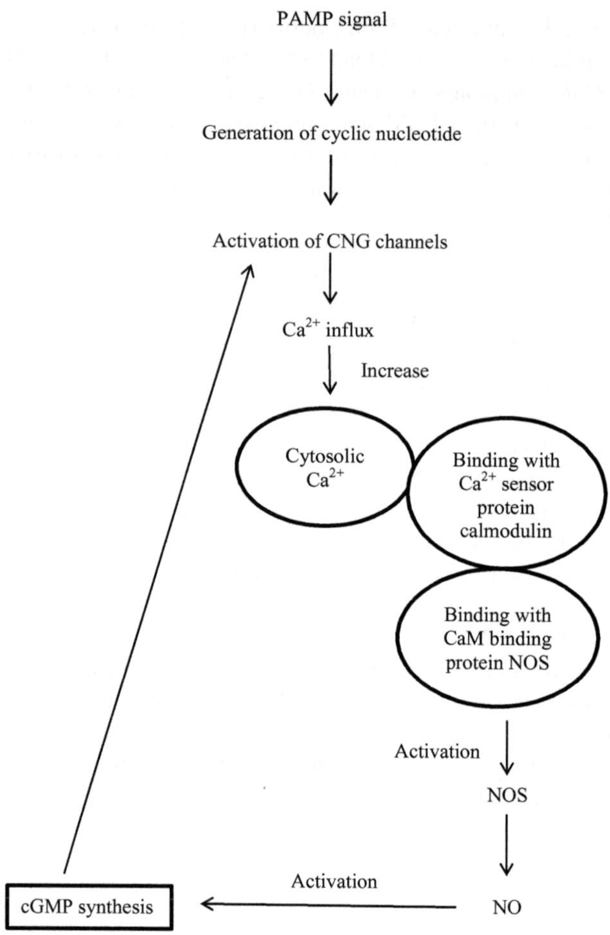


**Fig. 4.10** Amplification of PAMP signal through $Ca^{2+}$-triggered NO synthesis

and MPK6 by the *P. syringae* effector HopA/1 and inactivation of MKKs by the *P. syringae* effector HopF2 severely impair PAMP-induced defenses and render plants highly susceptible to nonpathogenic *P. syringae* bacteria (Zhang et al. 2007; Wang et al. 2010). MPK3 and MPK6 have been shown to be required for priming of defense responses during induced resistance (Beckers et al. 2009). The PAMPs are known to activate the expression of MPK3, MPK4, and MPK6 (Asai et al. 2002; Bethke et al. 2012).

A typical array of early defense responses induced by PAMPs includes $Ca^{2+}$-influx and the generation of ROS, nitric oxide, and ethylene. Much of this follows the mitogen-activated protein kinase cascades, leading to transcriptional changes of many defense-related genes (Zhang and Klessig 2001; Asai et al. 2002; Aslam et al.

2008, 2009; Boller and He 2009; Boller and Felix 2009). The tobacco MAP kinase SIPK is activated by the oomycete PAMP, β-megaspermin (Hall et al. 2007). The SIPK activation induced by the PAMP required external calcium influx, suggesting that SIPK activation occurs downstream of $Ca^{2+}$ influx (Hall et al. 2007). The signaling cascade initiated by the endogenous elicitor AtPep1 leads to expression of *MPK3* gene involved in MAPK signaling system in a $Ca^{2+}$-dependent manner (Qi et al. 2010). The MAPK, MPK6 has been shown to act downstream of cytosolic calcium signature (Yue et al. 2012).

## *4.21.4 Salicylate Signaling System*

### 4.21.4.1 $Ca^{2+}$ Signature May Modulate SA Biosynthesis and Accumulation Pathway

Salicylate (SA) signaling system has been identified as an important downstream event of $Ca^{2+}$ influx (Wang et al. 2010; Boursiac et al. 2010; Chen et al. 2011; Truman and Glazebrook 2012). $Ca^{2+}$ signaling has been reported to play an important role in modulation of salicylate signaling system (Du et al. 2009; Wang et al. 2011; Wan et al. 2012). $Ca^{2+}$ influx plays an important role in triggering SA signaling system (Garcia-Brugger et al. 2006; Ahn et al. 2007).

Cytosolic $Ca^{2+}$ signals result from a combined action of $Ca^{2+}$ influx through channels and $Ca^{2+}$ efflux through pumps and cotransporters (McAinsh and Pittman 2009; Ward et al. 2009; Boursiac et al. 2010). Influx occurs down the electrochemical gradient through various ion channels, such as voltage-gated channels or $Ca^{2+}$-permeable cyclic nucleotide-gated channels (CNGCs) or glutamate-gated ion channels (Qi et al. 2006, 2010; Moeder et al. 2011; Vatsa et al. 2011; Price et al. 2012; Vincill et al. 2012). Efflux requires energy-dependent $Ca^{2+}$ pumps (autoinhibited $Ca^{2+}$-ATPases (ACAs) and ER-type $Ca^{2+}$-ATPases) (McAinsh and Pittman 2009; Boursiac et al. 2010). The cytoplasmic $Ca^{2+}$ signal is shaped by the balance of activity between $Ca^{2+}$ influx and efflux (Boursiac et al. 2010).

Disruption of the vacuolar calcium-ATPases in *Arabidopsis* results in the activation of salicylic acid signaling pathway, probably by generating specific $Ca^{2+}$ signature in the cytosol (Boursiac et al. 2010). A double knockout mutation of the vacuolar $Ca^{2+}$ pumps ACA1 and ACA11 in *Arabidopsis thaliana* resulted in the activation of SA signaling system triggering programmed cell death. Initiation and spread of hypersensitive response that protects plants from pathogens could also be suppressed by disrupting the production of SA in *Arabidopsis* mutants with combined *aca4/11* mutations and a *sid2* (for *salicylic acid induction-deficient2*) mutation. SID2 is an isochorismate synthase that is involved in biosynthesis of SA (Wildermuth et al. 2001). These studies suggest that disruption of the vacuolar calcium-ATPases may result in the activation of *SID2* -mediated SA signaling pathway (Boursiac et al. 2010).

#### 4.21.4.2 Arabidopsis Calmodulin Binding Protein CBP60g Is Involved in SA Biosynthesis

A calmodulin binding protein, CBP60g, has been shown to be involved in activating SA biosynthesis (Wang et al. 2009). Overexpression of CBP60g in *Arabidopsis* caused elevated SA accumulation, increased expression of the defense genes, and enhanced defense responses, and enhanced resistance to *Pseudomonas syringae* (Wan et al. 2012). CBP60g has been shown to participate in SA signaling biosynthesis and accumulation (Wang et al. 2009). It has been suggested that the signal coming from CBP60g may act upstream from SA synthesis, as SA levels are reduced in *cbp60g* mutants (Wang et al. 2009). The effect of *cbp60g* mutant in SA biosynthesis was most similar to that of *pad4* mutant, suggesting that CBP60 may act upstream of PAD4 (Wang et al. 2009). PAD4, a key regulator of SA signaling system, contributes to SA levels. The *pad4* mutant plants showed reduced accumulation of SA after PAMP treatment (Tsuda et al. 2008). PAD4 is a key regulator acting at upstream of SA (Lippok et al. 2007). *Arabidopsis* plants carrying *pad4* mutations have a defect in accumulation of SA upon pathogen infection (Zhou et al. 1998). PAD4 is required for amplification of weak signals to a level sufficient for activation of SA signaling (Jirage et al. 1999). The PAD4 protein sequence displays similarity to triacyl glycerol lipases and other esterases (Jirage et al. 1999). It was also observed that the effect of *cbp60g* mutant in SA biosynthesis was almost similar to that of *sid2* mutant (Wang et al. 2009). It suggests that CBP60g may also act upstream of SID2, an isochorismate synthase that is involved in biosynthesis of SA (Wang et al. 2009, 2011). Isochorismate synthase encoded by *SID2* is essential for the biosynthesis of salicylic acid in response to pathogen challenge (Garcion et al. 2008; Truman and Glazebrook 2012). Both the calmodulin binding protein *CBP60g* and its closest homolog, the non-calmodulin binding *SARD1* (for SYSTEMIC ACQUIRED RESISTANCE DEFICIENT1), have been shown to bind to the promoter region of *SID2* (Zhang et al. 2010). CBP60g is strongly induced in response to PAMPs treatment (Wang et al. 2009). Plants carrying *cbp60g* null mutations were compromised in the induction of *SID2* and accumulation of SA (Wang et al. 2009). A central domain of CBP60g was found to bind to an oligomer with the sequence GAAATTTTGG selected from the *SID2* promoter (Zhang et al. 2010). PAMPs triggered signaling is greatly affected by the loss of *CBP60g* (Wang et al. 2011). Loss of CBP60g severely impacts the plants ability to produce SA in response to bacterial inoculation and renders the plant susceptible to infection. CBP60 was shown to bind specifically to a 10mer oligonucleotide with the sequence GAAATTTTGG (Truman and Glazebrook 2012). These results suggest that the calmodulin binding protein CBP60g binds with *SID2* gene and promotes SA biosynthesis through activation of SID2 (Fig. 4.11).

### *4.21.5 Jasmonate Signaling System*

Calcium signaling has been shown to act upstream of jasmonate (JA) biosynthesis pathway. PAMP elicitor signals activate a receptor-coupled G-protein and the activated G-proteins further switch on calcium ion channels (Zhao and Sakai 2003).

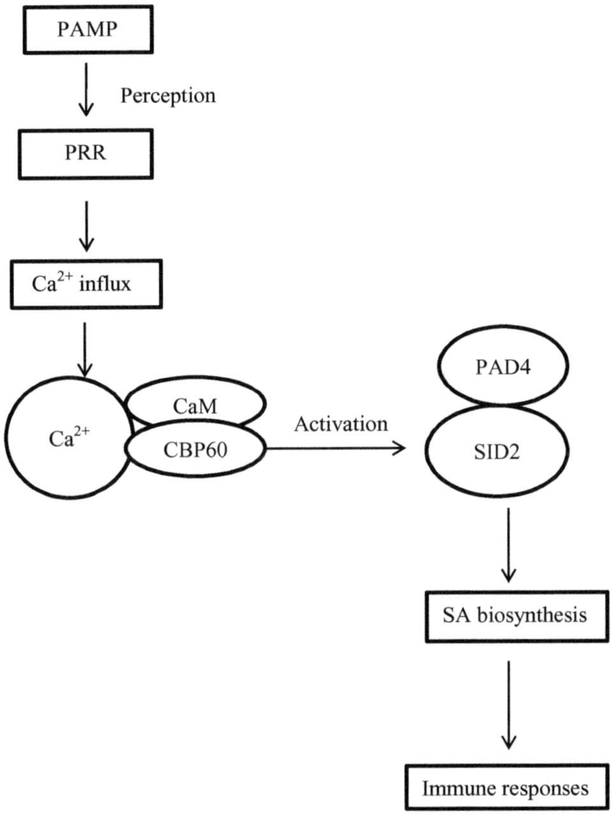


**Fig. 4.11** Role of calmodulin binding protein CBP60 in induction of salicylic acid biosynthesis (Adapted from Wang et al. 2009, 2011; Wan et al. 2012)

$Ca^{2+}$ influx and subsequent $Ca^{2+}$ wave (calcium signature) may activate NADPH oxidase and $H_2O_2$ production (Zhao and Sakai 2003). Calcium ion influx – induced $H_2O_2$ production triggers increases in lipoxygenase activity (Zhao and Sakai 2003). Lipoxygenase is an important enzyme in the octadecanoid pathway leading to the biosynthesis of jasmonate (Fig. 4.12; Schaller 2001; Vidhyasekaran 2007). Exogenous application of $H_2O_2$ has been found to induce the biosynthesis of endogenous jasmonate and initiate the JA pathway triggering downstream defense responses (Zhao and Sakai 2003). These studies suggest that $Ca^{2+}$ influx may be involved in activation of JA signaling system.

Calcium-dependent protein kinases (CDPKs) are unique enzymes found in plants and they are characterized as $[Ca^{2+}]_{cyt}$ sensors in plants. The *Arabidopsis* CDPK CPK6, has been shown to function as $[Ca^{2+}]_{cyt}$ sensor in the methyl jasmonate (MeJA) signaling (Munemasa et al. 2011). CPK6 has been found to be essential for the regulation of plasma membrane $Ca^{2+}$ – permeable channel activity (Munemasa et al. 2011). MeJA signaling is involved in stomatal closure (Suhita et al. 2003), a plant immune response against bacterial pathogens and CPK6 functions as a positive regulator of MeJA signaling (Munemasa et al. 2011).

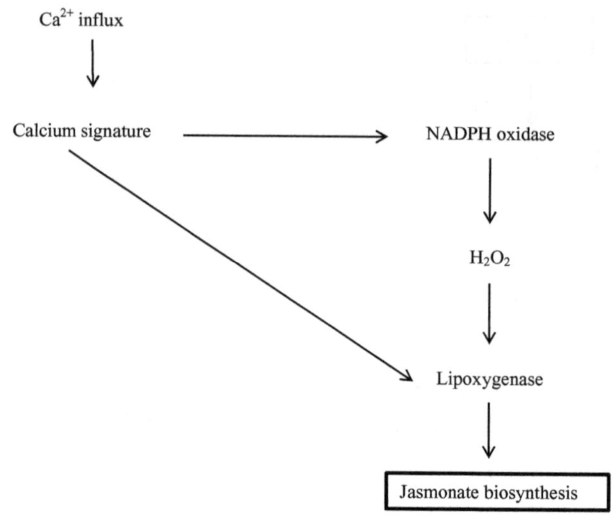


**Fig. 4.12** $Ca^{2+}$ influx/ROS/lipoxygenase-mediated jasmonate biosynthesis

### *4.21.6 Ethylene Signaling System*

Ethylene production is an important downstream event in $Ca^{2+}$ signaling system. $Ca^{2+}$ influx through $Ca^{2+}$ channels results in $Ca^{2+}$ waves (calcium signature) in cytosol. The calcium waves-activated calcium-dependent protein kinase (CDPK) activates 1-aminocyclopropane-1-carboxylic acid (ACC) synthase, the key enzyme involved in biosynthesis of ACC, a precursor for biosynthesis of ethylene (Sebastiá et al. 2004). The $Ca^{2+}$ influx is involved in activation of ACC oxidase, which is a key enzyme in biosynthesis of ethylene (Fig. 4.13; Gallardo et al. 1999). ACC oxidase activities as well as ethylene production from chickpea seeds were strongly inhibited by EGTA (ethylene glycol-bis (γβ-aminoethyl ether) N,N,N′,N′-tetraacetic acid), a selective extracellular calcium ion chelator, indicating that the influx of $Ca^{2+}$ is important for the ACC oxidase activity. The EGTA inhibition was restored by exogenous calcium ion treatment. Treatment of embryonic axes with either verapamil or LaCl3 (both $Ca^{2+}$ channel blockers) or TMB8 (an intracellular $Ca^{2+}$ antagonist) provoked an inhibition of both ACC oxidase activity and ethylene production (Gallardo et al. 1999). These results suggest the involvement of calcium ion fluxes and intracellular calcium levels in the activity of the last step of the ethylene biosynthetic pathway.

$Ca^{2+}$ influx may play an important role in ethylene signaling system (Raz and Fluhr 1992; Kwak and Lee 1997; Gallardo et al. 1999; Reddy et al. 2000). Blocking of $Ca^{2+}$ fluxes inhibits ethylene-induced gene expression and artificial elevation of $[Ca^{2+}]_{cyt}$ induces the expression of ethylene regulated genes in the absence of ethylene (Kwak and Lee 1997; Gallardo et al. 1999). Expression of several ethylene-inducible defense genes was found to be modulated by $Ca^{2+}$. The induction of acidic

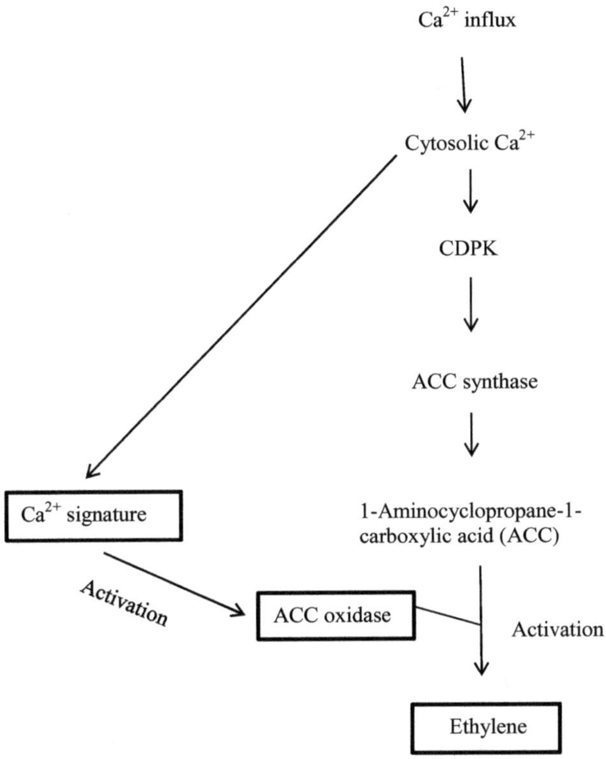


**Fig. 4.13** $Ca^{2+}$ influx and calcium-dependent protein kinase (CDPK) in downstream biosynthesis of ethylene

and basic PR-1 protein families and chitinase accumulation were found to be inhibited by EGTA, the $Ca^{2+}$ chelator. Addition of calcium or application of $Ca^{2+}$ ionophore induced the ethylene-inducible defense-related proteins (Raz and Fluhr 1992). These results suggest that downstream ethylene signaling system is triggered by $Ca^{2+}$ signature.

### *4.21.7 Abscisic Acid Signaling System*

*Arabidopsis* CDPK AtCPK32 interacts with ABF4, a transcriptional regulator of ABA-responsive gene expression, and modulates its activity (Choi et al. 2005). Expression of a grape CDPK, ACPK1, in *Arabidopsis thaliana* activates abscisic acid (ABA) signaling (Yu et al. 2006, 2007). Analysis of transcriptome changes in *Arabidopsis* revealed 230 calcium-responsive genes, of which 162 were upregulated and 68 were downregulated. Analysis of upstreams revealed, exclusively in the upregulated genes, a highly significant occurrence of a consensus sequence comprising two abscisic acid (ABA)-specific *cis* elements, the ABA-responsive element (ABRE, CACGTG [T/C/G]) and its coupling element ([C/A] ACGCG[T/C/A]). A tetramer of the ABRE *cis* element was sufficient to confer transcriptional activation in response to cytosolic $Ca^{2+}$ transients (Yu et al. 2007). Thus, for some specific $Ca^{2+}$ transients and motif combinations, ABREs function as $Ca^{2+}$-responsive *cis* elements (Kaplan et al. 2006; Yu et al. 2007). The link between $Ca^{2+}$-responsive transcription factors with ABA-responsive *cis*-elements suggests a link between $Ca^{2+}$ signaling and ABA signaling systems.

Interaction between $Ca^{2+}$ signaling and ABA signaling systems was further demonstrated by showing that calcium-dependent protein kinases (CDPKs) regulate ABA signal transduction in *Arabidopsis* (Zhu et al. 2007; Geiger et al. 2010). ABA stimulated two homologous CDPKs (CPK4 and CPK11) in *A. thaliana*. Loss of function mutations of *CPK4* and *CPK11* resulted in ABA insensitive phenotypes. The CPK4 and CPK11 kinases both phosphorylated two ABA-responsive transcription factors, ABF1 and ABF4, in vitro, suggesting that the two kinases may regulate ABA signaling through these transcription factors (Zhu et al. 2007). These results suggest that CDPK/calcium signaling pathway may regulate ABA signaling pathways.

### 4.21.8 *Phytoalexin Biosynthesis*

Elicitor-induced $[Ca^{2+}]_{cyt}$ has been shown to be required for the induction of phytoalexin production (Stab and Ebel 1987; Tavernier et al. 1995). Suppression of the sustained $[Ca^{2+}]_{cyt}$ increase in elicitor-treated tobacco cells suppressed the accumulation of transcripts of *PAL*, the gene encoding phenylalanine ammonialyase, the first committed enzyme in the phenylpropanoid pathway leading to biosynthesis of phytoalexins (Lecourieux et al. 2002). A prolonged $[Ca^{2+}]_{cyt}$ elevation is correlated with phytoalexin accumulation in elicitor-stimulated soybean or parsley cells (Mithöfer et al. 1999; Blume et al. 2000).

## 4.22 Importance of Calcium Signaling System in Activation of Plant Innate Immunity

Calcium signaling system involves activation of several ion channels, pumps and transporters resulting in generation of specific calcium signatures. The signals carried by the calcium signatures are transduced to different sensor proteins. Specific calcium signatures are recognized by different calcium sensors to transduce calcium-mediated signals into downstream events. Plant cells employ an array of $Ca^{2+}$-binding proteins that serve as $Ca^{2+}$ sensors. The $Ca^{2+}$ binding proteins that function as sensors undergo conformational changes upon $Ca^{2+}$ binding that allow them to interact with downstream effectors. The calcium sensor proteins fall into two main classes, referred to as sensor relays and sensor responders. The information encoded in transient $Ca^{2+}$ changes is decoded by an array of $Ca^{2+}$ binding proteins giving rise to a cascade of downstream effects. Thus the extracellular signals are transmitted to cellular calcium-dependent effectors to activate the transcription of immune response-related genes. The activation of calcium signaling system enhances host defense responses against a wide range of plant pathogens (Lecourieux et al. 2006; McAinsh and Pittman 2009; Abdul Kadar and Lindsberg 2010; DeFalco et al. 2010; Dodd et al. 2010; Hamada et al. 2012; Hashimoto et al. 2012).

Thus, $Ca^{2+}$ signaling system involves several ion channels, pumps, transporters, $Ca^{2+}$ sensor proteins, calmodulin-binding proteins, and calcium-dependent protein kinases. Activation of any one key component in the calcium signaling pathway triggers defense responses against pathogens, suggesting interplay of the different components in the signaling pathway. CNGC is an important $Ca^{2+}$ channel involved in $Ca^{2+}$ influx into cytosol. Application of cAMP results in CNGC2-dependent elevation of cytosolic $Ca^{2+}$ in *Arabidopsis* leaves (Ma et al. 2009a). Activation of the $Ca^{2+}$ ion channel by cAMP leads to NO generation, ROS generation, and enhanced defense response gene expression (Ma et al. 2009a). Glutamate receptor gated channel is another $Ca^{2+}$ channel detected in plant cell plasma membrane. In *Arabidopsis thaliana* overexpressing a glutamate receptor gene (*RsGluR*) from small radish, glutamate treatment triggered greater $Ca^{2+}$ influx in the root cells of transgenic plants (Kang et al. 2006a, b). The increased $Ca^{2+}$influx through the glutamate receptor-gated channel triggered several defense-related genes including JA-biosynthetic genes (Kang et al. 2006a, b). The rice two-pore channel1 (OsTPC1) is a putative voltage-gated $Ca^{2+}$-permeable channel (Kurusu et al. 2005; Hamada et al. 2012). Overexpression of *OsTPC1* induced several defense-related signaling systems, resulting in induction of oxidative burst and activation of a mitogen-activated protein kinase and hypersensitive cell death (Kurusu et al. 2005). Plant annexins appear capable of mediating passive, channel-like $Ca^{2+}$ transport (Mortimer et al. 2008; Laohavisit et al. 2009, 2010; Laohavisit and Davies 2011). Annexins are $Ca^{2+}$ transporter involved in $Ca^{2+}$ influx (Laohavisit and Davies 2011). Transgenic tobacco plants expressing the annexin gene (*AnnBj1*) isolated from *Brassica juncea* showed enhanced defense responses against *Phytophthora parasitica* var. *nicotianae*. The transgenic plants showed increased expression of several defense-related proteins (Jami et al. 2008).

Calmodulin (CaM) and calmodulin-like (CML) proteins are the potential $Ca^{2+}$ signal sensor proteins. Activation of the expression of these proteins triggers several immune responses and confers resistance against several pathogens. Transgenic overexpression of pepper calmodulin gene *CaCaM1* activates ROS and NO generation, and triggers defense responses against *Xanthomonas campestris* pv. *vesicatoria* in pepper leaves (Choi et al. 2009). Overexpression of the same pepper *CaCaM1* gene in *Arabidopsis* enhanced ROS and NO generation and conferred resistance against *Pseudomonas syringae* and *Hyaloperonospora parasitica* (Choi et al. 2009). Overexpression of the pathogen-inducible tobacco calmodulin gene *NtCaM13*, which is a component in the $Ca^{2+}$ signaling pathway, triggered defense responses against the oomycete pathogen *Pythium aphanidermatum*, the fungal pathogen

*Rhizoctonia solani*, and the bacterial pathogen *Ralstonia solanacearum* in tobacco (Takabatake et al. 2007). Overexpression of the soybean calmodulins GMCaM-4/5 induces constitutive *PR* gene expression and activates *trans*-acting elements that bind to *cis*-acting elements in the *Arabidopsis PR-1* promoter. The up-regulation of *PR* genes by these GmCaM isoforms was found to be dependent on NIM1 (Nonimmunity1) and unknown transcription factors (Park et al. 2004). Constitutive expression of soybean CaMs SCaM-4 and SCaM-5 in transgenic tobacco plants induced an array of defense-related genes. The transgenic plants showed defense responses against the oomycete *Phytophthora parasitica* var. nicotianae, the bacterial pathogen *Pseudomonas syringae* pv. *tabaci*, and the viral pathogen *Tobacco mosaic virus* (TMV) (Heo et al. 1999).

Calmodulin-like (CML) proteins also have been shown to trigger $Ca^{2+}$ signaling system. Overexpression of a tomato CML, *APR134*, in *Arabidopsis* accelerated hypersensitive immune response (Chiasson et al. 2005). Calmodulin-binding proteins (CBPs), another important components in calcium signaling system also trigger a series of defense responses. Overexpression of CBP60g in *Arabidopsis* caused elevated SA accumulation, increased expression of the defense genes, and enhanced defense responses, and the transgenic plants showed enhanced resistance to *Pseudomonas syringae* (Wan et al. 2012). Transgenic potato plants carrying a calcium-dependent protein kinase, which induces ROS, show high defense responses against the oomycete *P. infestans* (Kobayashi et al. 2007). Collectively these studies suggest that manipulation of even one component in the $Ca^{2+}$ signaling system may be able to trigger the entire gamut of immune response signaling systems to confer resistance against a wide-spectrum of pathogens.

## References

Abdel-Hamid H, Chin K, Moeder W, Yoshioka K (2011) High throughput chemical screening supports the involvement of $Ca^{2+}$ in cyclic nucleotide-gated ion channel-mediated cell death in *Arabidopsis*. Plant Signal Behav 6:1817–1819

Abdul Kadar M, Lindsberg S (2010) Cytosolic calcium and pH signaling in plants under salinity stress. Plant Signal Behav 5:233–238

Abe H, Urao T, Seki M, Shinozaki K, Yamaguchi-Shinozaki Y (2003) *Arabidopsis* AtMYC2 (bHLH) and AtMYB2 (MYB) function as transcriptional activators in abscisic acid signaling. Plant Cell 15:63–78

Ahn I-P (2007) Disturbance of the $Ca^{2+}$/calmodulin-dependent signaling pathway is responsible for the resistance of *Arabidopsis dnd1* against *Pectobacterium carotovorum* infection. Mol Plant Pathol 8:747–759

Ahn I-P, Kim S, Lee Y-H, Suh S-C (2007) Vitamin $B_1$-induced priming is dependent on hydrogen peroxide and the *NPR1* gene in *Arabidopsis*. Plant Physiol 143:838–848

Albrecht V, Ritz O, Linder S, Harter K, Kudla J (2001) The NAF domain defines a novel protein-protein interaction module conserved in $Ca^{2+}$-regulated kinases. EMBO J 20:1051–1053

Alexandre J, Lassales JP (1992) Intracellular $Ca^{2+}$ release by Ins*P3* in plants and effects of buffers on $Ca^{2+}$ diffusion. Philos Trans Biol Sci 338:53–61

Alexandre J, Lassales JP, Kado RT (1990) Opening of $Ca^{2+}$ channels in isolated red beet vacuole membrane by inositol-1,4,5-trisphosphate. Nature 343:567–570

Ali GS, Reddy VS, Lindgren PB, Jakobek JL, Reddy AS (2003) Differential expression of genes encoding calmodulin-binding proteins in response to bacterial pathogen and inducers of defense responses. Plant Mol Biol 51:803–815

Ali R, Ma W, Lemtiri-Chlieh F, Tsaltas D, Leng Q, von Bodman S, Berkowitz GA (2007) Death don't have no mercy and neither does calcium: *Arabidopsis* CYCLIC NUCLEOTIDE GATED CHANNEL2 and innate immunity. Plant Cell 19:1081–1095

Allen GJ, Sanders D (1995) Calcineurin, a type 2B protein phosphatase, modulates the $Ca^{2+}$-permeable slow vacuolar ion channel of stomatal guard cells. Plant Cell 7:1473–1483

Allen GJ, Muir SR, Sanders D (1995) Release of $Ca^{2+}$ from individual plant vacuoles by both InsP3 and cyclic ADP-ribose. Science 268:735–737

Allen GJ, Kuchitsu K, Chu SP, Murata Y, Schroeder JI (1999) *Arabidopsis abi1-1* and *abi2-1* phosphatase mutations reduce abscisic acid-induced cytoplasmic calcium rises in guard cells. Plant Cell 11:1785–1798

Allen GJ, Chu SP, Schumacher K, Shimazaki CT, Vafeados D, Kemper A, Hawke SD, Tallman G, Tsien RY, Harper JF, Chory J, Schroeder JI (2000) Alteration of stimulus-specific guard cell calcium oscillations and stomatal closing in *Arabidopsis det3* mutant. Science 289:2338–2342

Andreeva AV, Kutuzov MA (2001) PPP family of protein Ser/Thr phosphatases: two distinct branches? Mol Biol Evol 18:448–452

Arazi T, Kaplan B, Fromm H (2000) A high affinity calmodulin-binding site in a tobacco plasma-membrane channel protein coincides with a characteristic element of cyclic nucleotide-binding domains. Plant Mol Biol 42:591–601

Asai T, Tena G, Plotnikova J, Willmann MR, Chiu W-L, Gómez-Gómez L, Boller T, Ausubel FM, Sheen J (2002) MAP kinase signalling cascade in *Arabidopsis* innate immunity. Nature 415:977–983

Aslam SN, Newman MA, Erbs G, Morissey KL, Chinchilla D, Boller T, Jensen TT, De Castro C, Ierano T, Molinaro A, Jackson RW, Knight MR, Cooper RM (2008) Bacterial polysaccharides suppress induced innate immunity by calcium chelation. Curr Biol 18:1078–1083

Aslam SN, Erbs G, Morrissey K, Newman M-A, Chinchilla D, Boller T, Molinaro A, Jackson RW, Cooper RM (2009) Microbe-associated molecular pattern (MAMP) signatures, synergy, size, and charge: influences on perception or mobility and host defense responses. Mol Plant Pathol 10:375–387

Astamker C, Kiss V, Sagi M, Davydov O, Fluhr R (2007) Diverse subcellular locations of cryptogein-induced reactive oxygen species production in tobacco Bright Yellow-2 cells. Plant Physiol 143:1817–1826

Atkinson MM, Keppler LD, Orlandi EW, Baker CJ, Mischke CF (1990) Involvement of plasma membrane calcium influx in bacterial induction of the $K^+/H^+$ and hypersensitive responses in tobacco. Plant Physiol 92:215–221

Atkinson MM, Midland SL, Sims JJ, Keen NT (1996) Syringolide I triggers $Ca^{2+}$ infux, $K^+$ efflux, and extracellular alkalinization in soybean cells carrying the disease resistance gene *Rpg4*. Plant Physiol 112:297–302

Axelsen KB, Palmgren MG (1998) Evolution of substrate specificities in the P-type ATPase super-family. J Mol Evol 46:84–101

Balagué C, Lin B, Alcon C, Flottes G, Malmstrom S, Kohler C, Neuhaus G, Pelletier G, Gaymard F, Roby D (2003) HML1, an essential signaling component in the hypersensitive response, is a member of the cyclic nucleotide-gated ion channel family. Plant Cell 15:365–379

Barbier-Brygoo H, Vinauger M, Colcombet J, Ephritikhine G, Frachisee J-M, Maurel C (2000) Anion channels in higher plants: functional characterization, molecular structure and physio-logical role. Biochim Biophys Acta 1465:199–218

Batistič O, Kudla J (2009) Plant calcineurin B-like proteins and their interacting kinases. Biochim Biophys Acta 1793:985–992

Batistič O, Sorek N, Schültke S, Yalovsky S, Kudla J (2008) Dual fatty acyl modification determines the localization and plasma membrane targeting of CBL/CIPK $Ca^{2+}$ signaling complexes in *Arabidopsis*. Plant Cell 20:1346–1362

Baxter J, Moeder W, Urquhart W, Shahinas D, Chin K, Christendat D, Kang H-G, Angelova M, Kato N, Yoshioka K (2008) Identification of a functionally essential amino acid for Arabidopsis cyclic nucleotide gated ion channels using the chimeric *AtCNGC11/12* gene. Plant J 56:457–469

Beckers GJM, Jaskiewicz M, Liu Y, Underwood WR, He SY, Zhang S, Conrath U (2009) Mitogen-activated protein kinases 3 and 6 are required for full priming of stress responses in *Arabidopsis thaliana*. Plant Cell 21:944–953

Bellin D, Asai S, Delledonne M, Yoshioka H (2013) Nitric oxide as a mediator for defense responses. Mol Plant Microbe Interact 26:271–277

Benschop JJ, Mohammed S, O'Flaherty M, Heck AJR, Slijper M, Menke FLH (2007) Quantitative phosphoproteomics of early elicitor signaling in *Arabidopsis*. Mol Cell Proteomics 6:1198–1214

Berridge MJ (1984) Inositol trisphosphate and diacylglycerol as second messenger. Biochem J 220:345–360

Berridge MJ (1993) Inositol trisphosphate and calcium signaling. Nature 361:315–325

Berridge MJ, Irvine RF (1989) Inositol phosphates and cell signalling. Nature 341:197–205

Besson-Bard A, Courtois C, Gauthier A, Dahan J, Dobrowolska G, Jeandroz S, Pugin A, Wendehenne D (2008a) Nitric oxide in plants: production and cross-talk with $Ca^{2+}$ signaling. Mol Plant 1:218–228

Besson-Bard A, Pugin A, Wendehenne D (2008b) New insights into nitric oxide signaling in plants. Annu Rev Plant Biol 99:21–39

Bethke G, Pecher P, Eschen-Lippold L, Tsuda K, Katagiri F, Glazebrook J, Scheel D, Lee J (2012) Activation of the *Arabidopsis thaliana* mitogen-activated protein kinase MPK11 by the flagellin-derived elicitor peptide, flg22. Mol Plant Microbe Interact 25:471–480

Bewell MA, Maathuis FJ, Allen GJ, Sanders D (1999) Calcium-induced calcium release mediated by a voltage-activated cation channel in vacuolar vesicles from red beet. FEBS Lett 458:41–44

Bhattarai KK, Atamian HS, Kaloshian I, Eulgem T (2010) WRKY72 transcription factors contribute to basal immunity in tomato and Arabidopsis as well as gene-for-gene resistance mediated by the tomato *R* gene *Mi-1*. Plant J 63:229–240

Binder BM, Harper JF, Sussman MR (1994) Characterization of an *Arabidopsis* calmodulin-like domain protein kinase purified from *Escherichia coli* using an affinity sandwich technique. Biochemistry 33:2033–2044

Binet M, Humbert C, Lecourieux D, Vantard M, Pugin A (2001) Disruption of microtubular cytoskeleton induced by cryptogein, an elicitor of hypersensitive response in tobacco cells. Plant Physiol 125:564–572

Blume B, Nürnberger T, Nass N, Scheel D (2000) Receptor-mediated increase in cytoplasmic free calcium required for activation of pathogen defense in parsley. Plant Cell 12:1425–1440

Boller T, Felix G (2009) A renaissance of elicitors: perception of microbe-associated molecular patterns and danger signals by pattern-recognition receptors. Annu Rev Plant Biol 60:379–406

Boller T, He SY (2009) Innate immunity in plants: an arms race between pattern recognition receptors in plants and effectors in microbial pathogens. Science 324:742–744

Borsics T, Webb D, Andeme-Ondzihi C, Staehelin A, Christopher DA (2007) The cyclic nucleotide-gated calmodulin-binding channel AtCNGC10 localizes to the plasma membrane and influences numerous growth responses and starch accumulation in *Arabidopsis thaliana*. Planta 225:563–573

Bouché N, Scharlat A, Snedden WA, Bouchez D, Fromm H (2002) A novel family of calmodulin-binding transcription activators in multicellular organisms. J Biol Chem 277:21851–21861

Bouché N, Yellin A, Snedden WA, Fromm H (2005) Plant-specific calmodulin-binding proteins. Annu Rev Plant Biol 56:435–466

Boudsocq M, Willmann MR, McCormack M, Lee H, Shan L, He P, Bush J, Cheng SH, Sheen J (2010) Differential innate immune signalling via $Ca^{2+}$ sensor protein kinases. Nature 464:418–422

Boursiac Y, Lee SM, Romanowsky S, Blank R, Stadek C, Chung WS, Harper JF (2010) Disruption of the vacuolar calcium-ATPase in Arabidopsis results in the activation of a salicylic acid-dependent programmed cell death pathway. Plant Physiol 154:1158–1171

Boutrot F, Segonzac C, Chang KN, Qiao H, Ecker JR, Zipfel C (2010) Direct transcriptional control of the *Arabidopsis* immune receptor FLS2 by the ethylene-dependent transcription factors EIN3 and EIL1. Proc Natl Acad Sci U S A 107:14502–14507

Bush DS (1993) Regulation of cytosolic calcium in plants. Plant Physiol 103:7–13

Bush DS (1995) Calcium regulation in plant cells and its role in signaling. Annu Rev Plant Physiol Plant Mol Biol 46:95–122

Bush J, Cheng SH, Sheen J (2010) Differential innate immune signalling via $Ca^{2+}$ sensor protein kinases. Nature 464:418–422

Camoni L, Fullone MR, Marra M, Aducci P (1998a) The plasma membrane $H^+$-ATPase from maize roots is phosphorylated in the C-terminal domain by a calcium-dependent protein kinase. Physiol Plantarum 104:549–555

Camoni L, Harper JF, Palmgren MG (1998b) 14-3-3 proteins activate a plant calcium-dependent protein kinase (CDPK). FEBS Lett 430:381–384

Cessna SG, Low PS (2001) Activation of the oxidative burst in aequorin-transformed *Nicotiana tabacum* cells is mediated by protein kinase- and anion channel-dependent release of $Ca^{2+}$ from internal stores. Planta 214:126–134

Charpenteau M, Jaworski K, Ramirez BC, Tretyn A, Ranjeva R, Ranty B (2004) A receptor-like kinase from *Arabidopsis thaliana* is a calmodulin-binding protein. Biochem J 379:841–848

Chen J, Zhang Y, Wang C, Lu W, Jin JB, Hua X (2011) Proline induces calcium-mediated oxidative burst and salicylic acid signaling. Amino Acids 40:1473–1484

Cheng S-H, Sheen J, Gerrish C, Bolwell GP (2001) Molecular identification of phenylalanine ammonia-lyase as a substrate of a specific constitutively active *Arabidopsis* CDPK expressed in maize protoplasts. FEBS Lett 503:185–188

Cheng S-H, Willmann MR, Chen H-C, Sheen J (2002) Calcium signaling through protein kinases: the *Arabidopsis* calcium-dependent protein kinase gene family. Plant Physiol 129:469–485

Chern M-S, Fitzgerald H, Yadav RC, Canlas PE, Dong X, Ronald PC (2001) Evidence for a disease-resistance pathway in rice similar to the *NPR1*-mediated signaling pathway in Arabidopsis. Plant J 27:101–113

Chiasson D, Ekengren SK, Martin GB, Dobney SL, Snedden WA (2005) Calmodulin-like proteins from *Arabidopsis* and tomato are involved in host defense against *pseudomonas syringae* pv. *tomato*. Plant Mol Biol 58:887–897

Chico JM, Raíces M, Téllez-Iñón MT, Ulloa RM (2002) A calcium-dependent protein kinase is systemically induced upon wounding in tomato plants. Plant Physiol 128:256–270

Chiu J, DeSalle R, Lam HM, Meisel L, Coruzzi G (1999) Molecular evolution of glutamate receptors: a primitive signaling mechanism that existed before plants and animals diverged. Mol Biol Evol 16:826–838

Chiu JC, Brenner ED, DeSalle R, Nitabach MN, Holmes TC, Coruzzi GM (2002) Phylogenetic and expression analysis of the glutamate-receptor-like gene family in *Arabidopsis thaliana*. Mol Biol Evol 19:1066–1082

Cho MJ, Vaghy PI, Kondo R, Lee SH, Davis JP, Rehl R, Heo WD, Johnson JD (1998) Reciprocal regulation of mammalian nitric oxide synthase and calcineurin by plant calmodulin isoforms. Biochemistry 37:15593–15597

Choi MS, Kim MC, Yoo JH, Moon BC, Koo SC, Park BO, Lee JH, Koo YD, Han HJ, Lee SY, Chung WS, Lim CO, Cho MJ (2005) Isolation of a calmodulin-binding transcription factor from rice (*Oryza sativa* L.). J Biol Chem 280:40820–40831

Choi HW, Lee DH, Hwang BK (2009) The pepper calmodulin gene CaCaM1 is involved in reactive oxygen species and nitric oxide generation required for cell death and the defense response. Mol Plant Microbe Interact 22:1389–1400

Chung H-J, Sehnke PC, Fert RJ (1999) The 14-3-3 proteins: cellular regulation of plant metabolism. Trends Plant Sci 4:367–371

Clapham DE (2007) Calcium signaling. Cell 131:1047–1058

Clark GB, Roux SJ (1995) Annexins of plant cells. Plant Physiol 109:1133–1139

Clough SJ, Fengler KA, Yu IC, Lippok B, Smith RK Jr, Bent AF (2000) The *Arabidopsis dnd1* "defense, no death" gene encodes a mutated cyclic nucleotide-gated ion channels. Proc Natl Acad Sci U S A 97:9323

Coego A, Ramirez V, Jose Gil M, Flors V, Mauch-Mani B, Vera P (2005) An *Arabidopsis* homeodomain transcription factor, OVEREXPRESSOR OF CATIONIC PEROXIDASE 3, mediates resistance to infection by necrotrophic pathogens. Plant Cell 17:2123–2137

Conn J, Gilliham M, Athman A, Schreiber AW, Baumann U, Moller I, Cheng N-H, Stancombe MA, Hirschi KD, Webb AAR, Burton R, Kaiser BN, Tyerman SD, Leigh RA (2011) Cell-specific vacuolar calcium storage by *CAX1* regulates apoplastic calcium concentration, gas exchange, and plant productivity in *Arabidopsis*. Plant Cell 23:240–257

Courtois C, Besson A, Dahan J, Bourque S, Dobrowolska G, Pugin A, Wendehenne D (2008) Nitric oxide signaling in plants: interplays with $Ca^{2+}$ and protein kinases. J Exp Bot 59:155–163

Cumming RC, Andon NL, Haynes PA, Park M, Fischer WH, Schubert D (2004) Protein disulfide bond formation in the cytoplasm during oxidative stress. J Biol Chem 279:21749–21758

Damman C, Ichida A, Hong B, Romanowsky SM, Hrabak EM, Harmon AC, Pickard BG, Harper JF (2003) Subcellular targeting of nine calcium-dependent protein kinase isoforms from *Arabidopsis*. Plant Physiol 132:1840–1848

DeFalco TA, Bender KW, Snedden WA (2010) Breaking the code: $Ca^{2+}$ sensors in plant signalling. Biochem J 425:27–40

Deluc L, Barrieu F, Marchive C, Lauvergeat V, Decendit A, Richard T, Carde JP, Merillon JM, Hamdi S (2006) Characterization of a grapevine R2R3-MYB transcription factor that regulates the phenylpropanoid pathway. Plant Physiol 140:499–511

Delledonne M (2005) NO news is good news for plants. Curr Opin Plant Biol 8:390–396

Delledonne M, Xia Y, Dixon RA, Lamb C (1998) Nitric oxide functions as a signal in plant disease. Nature 394:585–588

De Nishi P, Dell'Orto M, Pirovano I, Zocchi G (1999) Calcium-dependent phosphorylation regulates the plasma-membrane $H^+$-ATPase activity of maize (*Zea mays* L.) roots. Planta 209:187–194

De Vrije T, Munnik T (1997) Activation of phospholipase D by calmodulin antagonists and mastoparan in carnation petal tissue. J Exp Bot 48:1631–1637

Demidchik V, Bowen HC, Maathuis FJ, Shabala SN, Tester MA, White PJ, Davies JM (2002) *Arabidopsis thaliana* root non-selective cation channels mediate calcium uptake and are involved in growth. Plant J 32:799–808

Demidchik V, Essah PA, Tester M (2004) Glutamate activates cation currents in the plasma membrane of *Arabidopsis* root cells. Planta 219:167–175

Demidchik V, Shabala SN, Davies JM (2007) Spatial variation in $H_2O_2$ response of *Arabidopsis thaliana* root epidermal $Ca^{2+}$ flux and plasma membrane $Ca^{2+}$ channels. Plant J 49:377–386

Dennison KL, Spalding EP (2000) Glutamate-gated calcium fluxes in *Arabidopsis*. Plant Physiol 124:1511–1514

Desikan R, Neill SJ, Hancock JT (1997) Generation of active oxygen in *Arabidopsis thaliana*. Free Radic Biol Med 28:773–778

Desikan R, Mackerness S, Hancock JT, Neill SJ (2001) Regulation of the *Arabidopsis* transcriptome by oxidative stress. Plant Physiol 127:159–172

Després C, DeLong C, Glaze S, Liu E, Fobert PR (2000) The Arabidopsis NPR1/NIM1 protein enhances the DNA binding activity of a subgroup of the TGA family of bZIP transcription factors. Plant Cell 12:279–290

Dodd AN, Kudla J, Sanders D (2010) The language of calcium signaling. Annu Rev Plant Biol 61:593–620

Dolmetsch RE, Xu K, Lewis RS (1998) Calcium oscillations increase the efficiency and specificity of gene expression. Nature 392:933–941

Dong J, Chen C, Chen Z (2003) Expression profiles of the *Arabidopsis WRKY* gene superfamily during plant defense response. Plant Mol Biol 51:21–37

Downie L, Priddle J, Hawes C, Evans DE (1998) A calcium pump at the higher plant nuclear envelope? FEBS Lett 429:44–48

Dröbak BK, Heras B (2002) Nuclear phosphoinositides and bring FYVE alive. Trends Plant Sci 7:132–138

Du L, Ali GS, Simons KA, Hou J, Yang T, Reddy AS, Poovaiah BW (2009) $Ca^{2+}$/calmodulin regulates salicylic acid-mediated immunity. Nature 457:1154–1158

Duby G, Boutry M (2009) The plant plasma membrane proton pump ATPase: a highly regulated P-type ATPase with multiple physiological roles. Pflugers Arch 457:645–655

Duby G, Poreba W, Piotrowiak D, Bobik K, Derua R, Waelkens E, Boutry M (2009) Activation of plant plasma membrane $H^+$-ATPase by 14-3-3 proteins is negatively controlled by two phosphorylation sites within the $H^+$-ATPase C-terminal region. J Biol Chem 284:4213–4221

Ebel J, Mithofer A (1998) Early events in the elicitation of plant defence. Planta 206:335–348

Elmore J, Coaker G (2011) The role of the plasma membrane $H^+$-ATPase in plant-microbe interactions. Mol Plant 4:416–427

Eulgem T, Rushton PJ, Schmelzer E, Hahlbrock K, Somssich IE (1999) Early nuclear events in plant defence signaling: rapid gene activation by WRKY transcription factors. EMBO J 18:4689–4699

Eulgem T, Rushton PJ, Robatzek S, Somssich IE (2000) The WRKY superfamily of plant transcription factors. Trends Plant Sci 5:199–206

Farmer PK, Choi JH (1999) Calcium and phospholipids activation of a recombinant calcium-dependent protein kinase (DcCPK1) from carrot (*Daucus carota* L). Biochim Biophys Acta 1434:6–17

Fedoroff N (2006) Redox regulatory mechanisms in cellular stress responses. Ann Bot (Lond) 98:289–300

Felle HH, Hermann A, Hanstein S, Hückelhoven R, Kogel K-H (2004) Apoplasic pH signaling in barley leaves attacked by the powdery mildew fungus *Blumeria graminis* f. sp. *hordei*. Mol Plant Microbe Interact 17:118–123

Ferl RJ (2004) 14-3-3 proteins: regulation of signal-induced events. Physiol Plantarum 120:173–178

Finkler A, Ashery-Padan R, Fromm H (2007) CAMTAs: calmodulin-binding transcription activators from plants to human. FEBS Lett 581:3893–3898

Fitzgerald HA, Canlas PE, Chern M-S, Ronald PC (2005) Alteration of TGA factor activity in rice results in enhanced tolerance to *Xanthomonas oryzae* pv. *oryzae*. Plant J 43:335–347

Fobert PR, Després C (2005) Redox control of systemic acquired resistance. Curr Opin Plant Biol 8:378–382

Foreman J, Demidchik V, Bothwell JH, Mylona P, Miedema H, Torres MA, Linstead P, Costa S, Brownlee C, Jones JD, Davies JM, Dolan I (2003) Reactive oxygen species produced by NADPH oxidase regulate plant cell growth. Nature 422:442–446

Frey F, Mbengue M, Kwaaitaal M, Nitsch L, Altenbach D, Häweker H, Lozano-Duran R, Njo MF, Beeckman T, Huettel B, Borst JW, Panstruga R, Robatzek S (2012) Plasma membrane calcium ATPases are important components of receptor-mediated signaling in plant immune responses and development. Plant Physiol 159:798–809

Fuglsang AT, Visconti S, Drumm K, Jahn T, Stensballe A, Mattei B, Jensen ON, Aducci P, Palmgren MG (1999) Binding of 14-3-3 protein to the plasma membrane $H^+$-ATPase AHA2 involves the three C-terminal residues Tyr (946)-Thr-Val and requires phosphorylation of Thr (947). J Biol Chem 274:36774–36780

Fuglsang AT, Guo Y, Cuin TA, Qiu Q, Song C, Kristiansen KA, Bych K, Schulz A, Shabala S, Schumaker KS, Palmgren MG, Zhu JK (2007) *Arabidopsis* protein kinase PK55 inhibits the plasma membrane $H^+$-ATPase by preventing interaction with 14-3-3 protein. Plant Cell 19:1617–1634

Gallardo M, Gomez-Jimenez MC, Matilla A (1999) Involvement of calcium in ACC-oxidase activity from *Cicer arietinum* seed embryonic axes. Phytochemistry 50:373–376

Galon Y, Nave R, Boyce JM, Nachmias D, Knight MR, Fromm H (2008) Calmodulin-binding transcription activator (CAMTA) 3 mediates biotic defense responses in *Arabidopsis*. FEBS Lett 582:943–948

Galon Y, Aloni R, Nachmias D, Snir O, Feldmesser E, Scrase-Field S, Boyce JM, Bouché N, Knight MR, Fromm H (2010a) Calmodulin-binding transcription activator 1 mediates auxin signaling and responds to stresses in *Arabidopsis*. Planta 232:165–178

Galon Y, Finkler A, Fromm H (2010b) Calcium-regulated transcription in plants. Mol Plant 3:653–669

Gao M, Liu J, Zhang Z, Cheng F, Chen S, Zhang Y (2008) MEKK1, MKK1/MKK2 and MPK4 function together in a mitogen-activated protein kinase cascade to regulate innate immunity in plants. Cell Res 18:1190–1198

Garcia-Brugger A, Lamotte O, Vandelle E, Bourque S, Lecourieux D, Poinssot B, Wendehenne D, Pugin A (2006) Early signaling events induced by elicitors of plant defenses. Mol Plant Microbe Interact 19:711–724

Garcion C, Lohmann A, Lamodiere E, Catinot J, Buchala A, Doermann P, Métraux J-P (2008) Characterization and biological function of the *ISOCHORISMATE SYNTHASE2* gene of the *Arabidopsis*. Plant Physiol 147:1279–1287

Gaxiola RA, Palmgren MG, Schumacher K (2007) Plant proton pumps. FEBS Lett 581:2204–2214

Geiger D, Scherzer S, Mumm P, Marten I, Ache P, Matschi S, Liese A, Wellmann C, Al-Rasheid KA, Grill E, Romeis T, Hedrich R (2010) Guard cell anion channel SLAC1 is regulated by CDPK protein kinases with distinct $Ca^{2+}$ affinities. Proc Natl Acad Sci U S A 107:8023–8028

Gelli A, Blumwald E (1997) Hyperpolarization-activated $Ca^{2+}$-permeable channels in the plasma membrane of tomato cells. J Membr Biol 155:35–45

Gerber IB, Laukens K, Witters E, Dubery IA (2006) Lipopolysaccharide-responsive phosphoproteins in *Nicotiana tabacum* cells. Plant Physiol Biochem 44:369–379

Gilford JL, Walsh M, Vogel H (2007) Structure and metal-ion-binding properties of the $Ca^{2+}$-binding helix-loop-helix EF-hand motifs. Biochem J 405:199–221

Gong D, Guo Y, Schumaker KS, Zhu J-K (2004) The SOS3 family of calcium sensors and SOS2 family of protein kinases in *Arabidopsis*. Plant Physiol 134:919–926

Guo F-Q, Okamoto M, Crawford NM (2003) Identification of a plant nitric oxide synthase gene involved in hormonal signaling. Science 302:100–103

Halfter U, Ishitani M, Zhu JK (2000) The *Arabidopsis* SOS2 protein kinase physically interacts with and is activated by the calcium-binding protein SOS3. Proc Natl Acad Sci U S A 97:3735–3740

Hall HC, Samuel MA, Ellis BE (2007) SIPK conditions transcriptional responses unique to either bacterial or oomycete elicitation in tobacco. Mol Plant Pathol 8:581–594

Hamada H, Kurusu T, Okuma E, Nokajima H, Kiyoduka M, Koyano T, Sugiyama Y, Okada K, Koga J, Saji H, Miyao A, Hirochika H, Yamane H, Murata Y, Kuchitsu K (2012) Regulation of a proteinaceous elicitor-induced $Ca^{2+}$ influx and production of phytoalexins by a putative voltage-gated cation channel, OsTPC1, in cultured rice cells. J Biol Chem 287:9931–9939

Hamilton DWA, Hills A, Köhler B, Blatt MR (2000) $Ca^{2+}$ channels at the plasma membrane of stomatal guard cells are activated by hyperpolarization and abscisic acid. Proc Natl Acad Sci U S A 97:4967–4972

Hammond-Kosack KE, Silverman P, Raskin I, Jones JDG (1996) Race-specific elicitors of *Cladosporium fulvum* induce changes in cell morphology and the synthesis of ethylene and salicylic acid in tomato plants carrying the corresponding *Cf* disease resistance gene. Plant Physiol 110:1381–1394

Harding SA, Roberts DM (1998) Incompatible pathogen infection results in enhanced reactive oxygen and cell death responses in transgenic tobacco expressing a hyperactive mutant calmodulin. Planta 206:253–258

Harding SA, Oh SH, Roberts DM (1997) Transgenic tobacco expressing a foreign calmodulin gene shows an enhanced production of active oxygen species. EMBO J 16:1137–1144

Harmon AC (2003) Calcium-regulated protein kinases of plants. Gravit Space Biol Bull 16(2):83–90

Harmon AC, Gribskov M, Harper JF (2000) CDPKs – a kinase for every $Ca^{2+}$ signal? Trends Plant Sci 5:154–159

Harmon AC, Gribskov M, Gubrium E, Harper JF (2001) The CDPK superfamily of protein kinases. New Phytol 151:175–183
Harper JF, Binder BM, Sussman MR (1993) Calcium and lipid regulation of an *Arabidopsis* protein kinase expressed in *Escherichia coli*. Biochemistry 32:3282–3290
Harper JF, Hong B, Hwang I, Guo HQ, Stoddard R, Huang JF, Palmgren MG, Sze H (1998) A novel calmodulin-regulated $Ca^{2+}$-ATPase (ACA2) from *Arabidopsis* with an N-terminal autoinhibitory domain. J Biol Chem 273:1099–1106
Harper JF, Bretyon G, Harmon A (2004) Decoding $Ca^{2+}$ signals through plant protein kinases. Annu Rev Plant Biol 55:263–288
Haruta M, Burch HL, Nelson RB, Barrett-Wilt G, Kline KG, Mohsin SB, Young JC, Otegui MS, Sussman MR (2010) Molecular characterization of mutant *Arabidopsis* plants with reduced plasma membrane proton pump activity. J Biol Chem 285:17918–17929
Hashimoto K, Eckert C, Anschūtz U, Scholz M, Held K, Waadt R, Reyer A, Hippler M, Becker D, Kudla J (2012) Phosphorylation of calcineurin B-like (CBL) calcium sensor proteins by their CBL-interacting protein kinases (CIPKs) is required for full activity of CBL-CIPK complexes toward their target proteins. J Biol Chem 287:7956–7968
Hedrich R, Marten I (2011) TPC1 – SV channels gain shape. Mol Plant 4:428–441
Hedrich R, Neher E (1987) Cytoplasmic calcium regulates voltage-dependent ion channels in plant vacuoles. Nature 329:833–836
Heo WD, Lee SH, Kim MC, Kim JC, Chung WS, Chun HJ, Lee KJ, Park CY, Park HC, Choi JY, Cho MJ (1999) Involvement of specific calmodulin isoforms in salicylic acid-independent activation of plant disease resistance responses. Proc Natl Acad Sci U S A 96:766–771
Himmelbach A, Yang Y, Grill E (2003) Relay and control of abscisic acid signaling. Curr Opin Plant Biol 6:470–479
Hirschi KD (2001) Vacuolar $H^+/Ca^{2+}$ transport: who's directing the traffic? Trends Plant Sci 6:100–104
Hirschi KD, Zhen R-G, Cunningham KW, Rea PA, Fink GR (1996) CAX1, an $H^+/Ca^{2+}$ antiporter from *Arabidopsis*. Proc Natl Acad Sci U S A 93:8782–8786
Hirt H (2000) Connecting oxidative stress, auxin, and cell cycle regulation through a plant mitogen-activated protein kinase pathway. Proc Natl Acad Sci U S A 97:2405–2407
Hofmann A, Proust J, Dorowski A, Schantz R, Huber R (2000) Annexin24 from *Capsicum annuum*: X-ray structure and biochemical characterization. J Biol Chem 275:8072–8082
Hrabak EM (2000) Calcium-dependent protein kinases and their relatives. Adv Bot Res 32:185–223
Hrabak EM, Chan CW, Gribskov M, Harper JF, Choi JH, Halford N, Kudla J, Luan S, Nimmo HG, Sussman MR, Thomas M, Walker-Simmons K, Zhu J-K, Harmon AC (2003) The Arabidopsis CDPK-SnRK superfamily of protein kinases. Plant Physiol 132:666–680
Hu X-Y, Neill SJ, Cai WM, Tang Z-C (2004) Induction of defence gene expression by oligogalacturonic acid requires increases in both cytosolic calcium and hydrogen peroxide in *Arabidopsis thaliana*. Cell Res 14:234–240
Hua J, Grisafi P, Cheng SH, Fink GR (2001) Plant growth homeostasis is controlled by the Arabidopsis *BON1* and *BAP1* genes. Genes Dev 15:2263–2272
Huang JW, Grunes DL, Kochian LV (1994) Voltage-dependent $Ca^{2+}$ influx into right-side-out plasma membrane vesicles isolated from wheat roots. Proc Natl Acad Sci U S A 91: 3473–3477
Huang J-F, Teyton L, Harper JF (1996) Activation of a $Ca^{2+}$-dependent protein kinase involves intramolecular binding of a calmodulin-like regulatory domain. Biochemistry 35: 13222–13230
Hwang I, Harper JF, Sze H (2000a) Calmodulin activation of an endoplasmic reticulum-located calcium pump involves an interaction with the N-terminal autoinhibitory domain. Plant Physiol 122:157–167
Hwang I, Sze H, Harper JF (2000b) A calcium-dependent protein kinase can inhibit a calmodulin-stimulated $Ca^{2+}$ pump (ACA2) located in the endoplasmic reticulum of *Arabidopsis*. Proc Natl Acad Sci U S A 97:6224–6229

Innes RW (2001) Mapping out the roles of MAP kinases in plant defense. Trends Plant Sci 6:392–394

Ishitani M, Liu J, Halfter U, Kim CS, Shi W, Zhu JK (2000) SOS3 function in plant salt tolerance requires N-myristoylation and calcium binding. Plant Cell 12:1667–1678

Ito T, Nakata M, Fukazawa J, Ishida S, Takahashi Y (2010) Alteration of substrate specificity: the variable N-terminal domain of tobacco $Ca^{2+}$-dependent protein kinase is important for substrate recognition. Plant Cell 22:1592–1604

Jabs T, Tschöpe M, Colling C, Hahlbrock K, Scheel D (1997) Elicitor-stimulated ion fluxes and $O_2^-$ from the oxidative burst are essential components in triggering defense gene activation and phytoalexin synthesis in parsley. Proc Natl Acad Sci U S A 94:4800–4805

Jakobek J, Smith-Becker JA, Lindgren PB (1999) A bean cDNA expressed during a hypersensitive reaction encodes a putative calcium-binding protein. Mol Plant Microbe Interact 12:712–719

Jakoby M, Weisshaar B, Droge-Laser W, Vicente-Carbajosa J, Tiedemann J, Kroj T, Parcy F (2002) bZIP transcription factors in *Arabidopsis*. Trends Plant Sci 7:106–111

Jambunathan N, McNellis TW (2003) Regulation of Arabidopsis *COPINE1* gene expression in response to pathogens and abiotic stimuli. Plant Physiol 132:1370–1381

Jambunathan N, Siani JM, McNellis TW (2001) A humidity-sensitive Arabidopsis copine mutant exhibits precocious cell death and increased disease resistance. Plant Cell 13:2225–2240

Jami SK, Clark GB, Turlapati SA, Handley C, Roux SJ, Kirti PB (2008) Ectopic expression of an annexin from *Brassica juncea* confers tolerance to abiotic and biotic stresses in transgenic tobacco. Plant Physiol Biochem 46:1019–1030

Jeworutzki E, Roelfsema MR, Anschütz U, Krol E, Elzenga JT, Felix G, Boller T, Hedrich R, Becker D (2010) Early signaling through the *Arabidopsis* pattern recognition receptors FLS2 and EFR involves $Ca^{2+}$-associated opening of plasma membrane anion channels. Plant J 62:367–378

Jirage D, Tootle TL, Reuber TL, Frost LN, Feys BJ, Parker JE (1999) *Arabidopsis thaliana* PAD4 encodes a lipase-like gene that is important for salicylic acid signaling. Proc Natl Acad Sci U S A 96:13583–13588

Jones DH, Ley S, Aitkan A (1995) Isoforms of 14-3-3 protein can form homo-and heterodimers in vivo and in vitro: implications for function as adapter proteins. FEBS Lett 368:55–58

Journot-Catalino N, Somssich IE, Roby D, Kroj T (2006) The transcription factors WRKY11 and WRKY17 act as negative regulators of basal resistance in *Arabidopsis thaliana*. Plant Cell 18:3289–3302

Jurkowski GI, Smith RK, Yu I-C, Ham JH, Sharma SB, Klessig DF, Fengler KA, Bent AF (2004) *Arabidopsis DND2*, a second cyclic nucleotide-gated ion channel gene for which mutation causes the "defense, no death" phenotype. Mol Plant Microbe Interact 17:511–520

Kalde M, Barth M, Somssich IE, Lippok B (2003) Members of the *Arabidopsis* WRKY group III transcription factors are part of different plant defense signaling pathways. Mol Plant Microbe Interact 16:295–305

Kanczewska J, Marco S, Vandermeeren C, Maudoux O, Rigaud J-L, Boutry M (2005) Activation of the plant plasma membrane $H^+$-ATPase by phosphorylation and binding of 14-3-3 proteins converts a dimer into a hexamer. Proc Natl Acad Sci U S A 102:11675–11680

Kang CH, Jung WY, Kang YH, Kim JY, Kim DG, Jeong JC, Baek DW, Jin JB, Lee JY, Kim MO, Chung WS, Mengiste T, Koiwa H, Kwak SS, Bahk JD, Lee SY, Nam JS, Yun DJ, Cho MJ (2006a) AtBAG6, a novel calmodulin-binding protein, induces programmed cell death in yeast and plants. Cell Death Differ 13:84–95

Kang S, Kim HB, Lee H, Choi JY, Heu S, Oh CJ, Kwon SI, An CS (2006b) Overexpression in *Arabidopsis* of a plasma membrane-targeting glutamate receptor from small radish increases glutamate-mediated $Ca^{2+}$ influx and delays fungal infection. Mol Cells 21:418–427

Kaplan B, Davydov O, Knight H, Galon Y, Knight MR, Fluhr R, Fromm H (2006) Rapid transcriptome changes induced by cytosolic $Ca^{2+}$ transients reveal ABRE-related sequences as $Ca^{2+}$-responsive *cis* elements in *Arabidopsis*. Plant Cell 10:2733–2748

Kaplan B, Sherman T, Fromm H (2007) Cyclic nucleotide-gated channels in plants. FEBS Lett 581:2237–2246

Karita E, Yamakawa H, Mitsuhara I, Kuchitsu K, Ohashi Y (2004) Three types of tobacco calmodulins characteristically activate plant NAD kinase at different $Ca^{2+}$ concentrations and pHs. Plant Cell Physiol 45:1371–1379

Katou S, Karita E, Yamakawa H, Seo S, Mitsuhara I, Kuchitsu K, Ohashi Y (2005) Catalytic activation of the plant MAPK phosphatase NtMKP1 by its physiological substrate salicylic acid-induced protein kinase but not by calmodulins. J Biol Chem 280:39569–39581

Katou S, Kuroda K, Seo S, Yanagawa Y, Tsuge T, Yamazaki M, Miyao A, Hirochika H, Ohashi Y (2007) A calmodulin-binding mitogen-activated protein kinase phosphatase is induced by wounding and regulates the activities of stress-related mitogen-activated protein kinases in rice. Plant Cell Physiol 48:332–344

Kawano T, Muto S (2000) Mechanism of peroxidase actions for salicylic acid-induced generation of active oxygen species and an increase in cytosolic calcium in tobacco cell suspension culture. J Exp Bot 51:685–693

Keinath NF, Kierszniowska S, Lorek J, Bourdais G, Kessler SA, Asano H, Grossniklaus U, Schulze W, Robatzek S, Panstruga R (2010) PAMP-induced changes in plasma membrane compartmentalization reveal novel components of plant immunity. J Biol Chem 285: 39140–39149

Keller T, Damude HG, Werner D, Doerner P, Dixon RA, Lamb CA (1998) A plant homolog of the neutrophil NADPH oxidase gp91phox subunit gene encodes a plasma membrane protein with $Ca^{2+}$ binding motifs. Plant Cell 10:255–266

Kesarwani M, Joo J, Dong X (2007) Genetic interactions of TGA transcription factors in the regulation of pathogenesis-related genes and disease resistance in *Arabidopsis*. Plant Physiol 144:336–346

Kim HS, Delaney TP (2002) Overexpression of *TGA5*, which encodes a bZIP transcription factor that interacts with NIM1/NPR1, confers SAR-independent resistance in *Arabidopsis thaliana* to *Peronospora parasitica*. Plant J 32:151–163

Kim KN, Cheong YH, Gupta R, Luan S (2000) Interaction specificity of *Arabidopsis* calcineurin B-like calcium sensors and their target kinases. Plant Physiol 124:1844–1853

Kim MC, Lee SH, Kim JK, Chun HJ, Choi MS, Chung WS, Moon BC, Kang CH, Park CY, Yoo JH, Kang YH, Koo SC, Koo YD, Jung JC, Kim ST, Schulze-Lefert P, Lee SY, Cho MJ (2002a) Mlo, a modulator of plant defense and cell death, is a novel calmodulin-binding protein. Isolation and characterization of a rice Mlo homologue. J Biol Chem 277:19304–19314

Kim MC, Panstruga R, Elliott C, Muller J, Devoto A, Yoon HW, Park HC, Schulze-Lefert P (2002b) Calmodulin interacts with MLO protein to regulate defence against mildew in barley. Nature 416:447–451

Kim KC, Fan B, Chen Z (2006) Pathogen-induced *Arabidopsis* WRKY7 is transcriptional repressor and enhances plant susceptibility to *Pseudomonas syringae*. Plant Physiol 142:1180–1192

Kim MC, Chung WS, Yun DJ, Cho MJ (2009) Calcium and calmodulin-mediated regulation of gene expression in plants. Mol Plant 2:13–21

Kim TH, Bohmer M, Hu H, Nishimura N, Schroeder JI (2010) Guard cell signal transduction network: advances in understanding abscisic acid, $CO_2$, and $Ca^{2+}$ signaling. Annu Rev Plant Biol 61:561–591

Kimura S, Kaya H, Kawarazaki T, Hiraoka G, Senzaki E, Michikawa M, Kuchitsu K (2012) Protein phosphorylation is a prerequisite for the $Ca^{2+}$-dependent activation of *Arabidopsis* NADPH oxidases and may function as a trigger for the positive feedback regulation of $Ca^{2+}$ and reactive oxygen species. Biochim Biophys Acta 1823:398–405

Knoth C, Ringler J, Dangl JL, Eulgem T (2007) *Arabidopsis* WRKY70 is required for full *RPP4-mediated* disease resistance and basal defense against *Hyaloperonospora parasitica*. Mol Plant Microbe Interact 20:120–128

Kobayashi Y, Kobayashi I, Funaki Y, Fujimoto S, Takemoto T, Kunoh H (1997) Dynamic reorganization of microfilaments and microtubules is necessary for the expression of non-host resistance in barley coleoptile cells. Plant J 11:525–537

Kobayashi M, Ohura L, Kawakita K, Yokota N, Fujiwara M, Shimamoto K, Doke N, Yoshioka H (2007) Calcium-dependent protein kinases regulate the production of reactive oxygen species by potato NADPH oxidase. Plant Cell 19:1065–1080

Köhler C, Neuhaus G (2000) Characterization of calmodulin binding to cyclic nucleotide-gated ion channels from *Arabidopsis thaliana*. FEBS Lett 14:133–136

Köhler C, Merkle T, Neuhaus G (1999) Characterisation of a novel gene family of a putative cyclic nucleotide-ion channel in *Arabidopsis thaliana*. Plant J 18:97–104

Kudla J, Xu G, Harter K, Gruissem W, Luan S (1999) Genes for calcineurin B-like proteins in *Arabidopsis* are differentially regulated by stress signals. Proc Natl Acad Sci U S A 96:4718–4723

Kudla J, Batistić O, Hashimoto K (2010) Calcium signals: the lead currency of plant information processing. Plant Cell 22:541–563

Kuhlbrandt W (2004) Biology, structure, and mechanism of P-type ATPases. Nat Rev Mol Cell Biol 5:282–295

Kuhlmann M, Horvay K, Strathmann A, Heinekamp T, Fischer U, Böttner S, Dröge-Laser W (2003) The α-helical D1 domain of the tobacco bZIP transcription factor BZ1-1 interacts with the ankyrin-repeat protein ANK1 and is important for BZ1-1 function, both in auxin signaling and pathogen response. J Biol Chem 278:8786–8794

Kurusu T, Yagala T, Miyao A, Hirochika H, Kuchitsu K (2005) Identification of a putative voltage-gated $Ca^{2+}$ channel as a key regulator of elicitor-induced hypersensitive cell death and mitogen-activated protein kinase activation of elicitor-induced hypersensitive cell death and mitogen-activated protein kinase activation in rice. Plant J 42:798–809

Kwaaitaal M, Huisman R, Maintz J, Reinstädler A, Panstruga R (2011) Ionotropic glutamate receptor (iGluR)-like channels mediate MAMP-induced calcium influx in *Arabidopsis thaliana*. Biochem J 440:355–365

Kwak SH, Lee SH (1997) The requirements for $Ca^{2+}$, protein phosphorylation, and dephosphorylation for ethylene signal transduction in *Pisum sativum* L. Plant Cell Physiol 38:1142–1149

Lamotte O, Gould K, Lecourieux D, Sequeira-Legrand A, Lebrun-Garcia A, Durner J, Pugin A, Wendehenne D (2004) Analysis of nitric oxide signaling functions in tobacco cells challenged by the elicitor cryptogein. Plant Physiol 135:516–529

Lamotte O, Courtois C, Dobrowolska G, Besson A, Pugin A, Wendehenne D (2006) Mechanisms of nitric oxide-induced increase of free cytosolic $Ca^{2+}$ concentration in *Nicotiana plumbaginifolia* cells. Free Radic Biol Med 40:1369–1376

Laohavisit A, Davies JM (2011) Annexins. New Phytol 189:40–53

Laohavisit A, Mortimer JC, Demidchik V, Coxon KM, Stancombe MA, Macpherson N, Brownlee C, Hofmann A, Webb AAR, Miedema H, Battey NH, Davies JM (2009) *Zea mays* annexins modulate cytosolic free $Ca^{2+}$ and generate a $Ca^{2+}$-permeable conductance. Plant Cell 21:479–493

Laohavisit A, Brown AT, Cicuta P, Davies JM (2010) Annexins components of the calcium and reactive oxygen signaling network. Plant Physiol 152:1824–1829

Laohavisit A, Shang Z, Rubio L, Cuin TA, Véry AA, Wang A, Mortimer JC, Macpherson N, Coxon KM, Battey NH, Brownlee C, Park OK, Sentenac H, Shabala S, Webb AA, Davies JM (2012) Arabidopsis annexin1 mediates the radical-activated plasma membrane $Ca^{2+}$- and $K^+$-permeable conductance in root cells. Plant Cell 24:1522–1533

Laquitaine L, Gomès E, Francois J, Marchive C, Pascal S, Hamdi S, Atanassova R, Delrot S, Coutos-Thévenot P (2006) Molecular basis of ergosterol-induced protection of grape against *Botrytis cinerea*: induction of type 1 LTP promoter activity, WRKY, and stilbene synthase gene expression. Mol Plant Microbe Interact 19:1103–1112

Laxal AM, Munnik T (2002) Phospholipid signaling in plant defence. Curr Opin Plant Biol 5:332–338

Lebel E, Heifetz P, Thorne L, Uknes S, Ryals J, Ward E (1998) Functional analysis of regulatory sequences controlling *PR-1* gene expression in *Arabidopsis*. Plant J 16:223–233

Lecourieux D, Mazars C, Pauly N, Ranjeva R, Pugin A (2002) Analysis and effects of cytosolic free calcium elevations in response to elicitors in *Nicotiana plumbaginifolia* cells. Plant Cell 14:2627–2641

Lecourieux D, Lamotte O, Bourque S, Wendehenne D, Mazars C, Ranjeva R, Pugin A (2005) Proteinaceous and oligosaccharidic elicitors induce different calcium signatures in the nucleus of tobacco cells. Cell Calcium 38:527–538

Lecourieux D, Ranjeva R, Pugin A (2006) Calcium in plant defence-signaling pathways. New Phytol 171:249–269

Lecourieux-Ouaked D, Pugin A, Lebrun-Garcia A (2000) Phosphoproteins involved in the signal transduction of cryptogein, an elicitor of defense reactions in tobacco. Mol Plant Microbe Interact 13:821–829

Lee SH, Kim JC, Lee MS, Heo WD, Seo HY, Yoon HW, Hong JC, Lee SY, Bahk JD, Hwang I, Cho MJ (1995) Identification of a novel divergent calmodulin isoform from soybean which has differential ability to activate calmodulin-dependent enzymes. J Biol Chem 270:21806–21812

Lee SH, Seo HY, Kim JC, Heo WD, Chung W-S, Lee KJ, Kim MC, Cheong YH, Choi JY, Lim CO, Cho MJ (1997) Differential activation of NAD kinase by plant calmodulin isoforms. The critical role of domain I. J Biol Chem 272:9252–9259

Lee JY, Yoo BC, Harmon AC (1998) Kinetic and calcium-binding properties of three calcium-dependent protein kinase isoenzymes from soybean. Biochemistry 37:6801–6809

Lee SH, Kim MC, Heo WD, Kim JC, Chung WS, Park CY, Park HC, Cheong YH, Kim CY, Lee S-H, Lee KJ, Bahk JD, Lee SY, Cho MJ (1999) Competitive binding of calmodulin-binding proteins: implications for the function of calmodulin isoforms in plants. Biochim Biophys Acta 1433:56–67

Leng Q, Mercier RW, Yao W, Berkowitz GA (1999) Cloning and first functional characterization of a plant cyclic nucleotide-gated cation channel. Plant Physiol 121:753–761

Leng Q, Mercier RW, Hua B-G, Fromm H, Berkowitz GA (2002) Electrophysiological analysis of cloned cyclic nucleotide-gated ion channels. Plant Physiol 128:400–410

Levy M, Wang Q, Kaspi R, Parrella MP, Abel S (2005) Arabidopsis IQD1, a novel calmodulin-binding nuclear protein, stimulates glucosinolate accumulation and plant defense. Plant J 43:79–96

Li W, Llopis J, Whitney M, Zlokarnik G, Tslen RY (1998) Cell-permeant caged InsP3 ester shows that $Ca^{2+}$ spike frequency can optimize gene expression. Nature 392:936–941

Lippok B, Birkenbihl RP, Rivory G, Brummer J, Schmelzer E, Logemann E, Somssich IE (2007) Expression of *AtWRKY33* encoding a pathogen- or PAMP-responsive WRKY transcription factor is regulated by a composite DNA motif containing W box elements. Mol Plant Microbe Interact 20:420–429

Liu Y, Jin H, Yang K, Kim C, Baker B, Zhang S (2003) Interaction between two mitogen-activated protein kinases during tobacco defense signaling. Plant J 34:149–160

Liu XQ, Bai XQ, Qian Q, Wang XJ, Chen MS, Chu CC (2005) OsWRKY03, a rice transcriptional activator that functions in defense signaling pathway upstream of OsNPR1. Cell Res 15:593–603

Liu X, Bai X, Wang X, Chu C (2007) OsWRKY71, a rice transcription factor, is involved in rice defense response. J Plant Physiol 164:969–979

Liu J, Elmore JM, Fuglsang AT, Palmgren M, Staskawicz B, Coaker G (2009) RIN4 functions with the plasma membrane $H^+$-ATPase to regulate stomatal apertures during pathogen attack. PLoS Biol 7:e1000139

Liu J-Z, Horstman HD, Braun E, Graham MA, Zhang C, Navarre D, Qiu W-L, Lee Y, Nettleton D, Hill JH, Whitham SA (2011) Soybean homologs of MPK4 negatively regulate defense responses and positively regulate growth and development. Plant Physiol 157:1363–1378

Lu YT, Feldman LJ (1997) Light-regulated root gravitropism: a role for, and characterization of, a calcium/calmodulin-dependent protein kinase homolog. Planta 203(suppl):S91–S97

Lu Y, Harrington HM (1994) Isolation of tobacco cDNA clones encoding calmodulin-binding proteins and characterization of a known calmodulin-binding domain. Plant Physiol Biochem 32:413–422

Luan S (2009) The CBL-CIPK network in plant calcium signaling. Trends Plant Sci 14:37–42

Luan S, Kudla J, Rodriguez-Concepcion M, Yalovsky S, Gruissem W (2002) Calmodulins and calcineurin B-like proteins: calcium sensors for specific signal response coupling in plants. Plant Cell 14:S389–S400

Ludwig AA, Romeis T, Jones JDG (2004) CDPK-mediated signaling pathways: specificity and cross-talk. J Exp Bot 55:181–188

Ma W, Berkowitz GA (2007) The grateful dead: calcium and cell death in plant innate immunity. Cell Microbiol 9:2571–2585

Ma W, Berkowitz GA (2011) $Ca^{2+}$ conduction by plant cyclic nucleotide gated channels and associated signaling components in pathogen defense signal transduction cascades. New Phytol 190:566–572

Ma W, Yoshioka K, Berkowitz GA (2007) Cyclic nucleotide gated channels and $Ca^{2+}$-mediated signal transduction during plant innate immune response to pathogens. Plant Signal Behav 2:548–550

Ma W, Smigel A, Tsai YC, Braam J, Berkowitz GA (2008) Innate immunity signaling: cytosolic $Ca^{2+}$ elevation is linked to downstream nitric oxide generation through the action of calmodulin or a calmodulin-like protein. Plant Physiol 148:818–828

Ma W, Qi Z, Smigel A, Walker RK, Verma R, Berkowitz GA (2009a) $Ca^{2+}$, cAMP, and transduction of non-self perception during plant immune responses. Proc Natl Acad Sci U S A 106:20995–21000

Ma W, Smigel A, Verma R, Berkowitz GA (2009b) Cyclic nucleotide gated channels and related signaling components in plant innate immunity. Plant Signal Behav 4:277–282

Ma Y, Walker RK, Zhao Y, Berkowitz GA (2012) Linking ligand perception by PEPR pattern recognition receptors to cytosolic $Ca^{2+}$ elevation and downstream immune signaling in plants. Proc Natl Acad Sci U S A 109:19852–19857

Maathuis FJM, Sanders D (2001) Sodium uptake in Arabidopsis roots is regulated by cyclic nucleotides. Plant Physiol 127:1617–1625

Maathuis FJM, Ichida AM, Sanders D, Schroeder JI (1997) Roles of higher $K^+$ channels. Plant Physiol 114:1141–1149

Maeo K, Hayashi S, Kojima-Suzuki H, Morikami A, Nakamura D (2001) Role of conserved residues of the WRKY domain in the DNA-binding of tobacco WRKY family proteins. Biosci Biotechnol Biochem 65:2428–2436

Marten H, Konrad KR, Dietrich P, Roelfsema RG, Hedrich R (2007) $Ca^{2+}$-dependent and -independent abscisic acid activation of plasma membrane anion channels in guard cells of *Nicotiana tabacum*. Plant Physiol 143:28–37

Martinec J, Feltl T, Scanlon CH, Lumsden PJ, Macháčková I (2000) Subcellular localization of a high affinity binding site for D-*myo*-inositol 1,4,5-trisphosphate from *Chenopodium rubrum*. Plant Physiol 124:475–483

Maudoux O, Batoko H, Oecking C, Gevaert K, Vandekerckhove J, Boutry M, Morsomme P (2000) A plant plasma membrane $H^+$-ATPase expressed in yeast is activated by phosphorylation at its penultimate residue and binding of 14-3-3 regulatory proteins in the absence of fusicoccin. J Biol Chem 275:17762–17770

Mazars C, Bourque S, Mithofer A, Pugin A, Ranjeva R (2009) Calcium homeostasis in plant cell nuclei. New Phytol 181:275–294

Mazars C, Thuleau P, Lamotte O, Bourque S (2010) Cross-talk between ROS and calcium in regulation of nuclear activities. Mol Plant 3:706–718

McAinsh MR, Pittman JK (2009) Shaping the calcium signature. New Phytol 181:275–294

McAinsh MR, Webb ARR, Taylor JE, Hetherington AM (1995) Stimulus-induced oscillations in guard cell cytosolic free calcium. Plant Cell 7:1207–1219

Melotto M, Underwood W, Koczan J, Nomura K, He SY (2006) Plant stomata function in innate immunity against bacterial invasion. Cell 126:969–980

Meng X-B, Zhao W-S, Lin R-M, Wang M, Peng Y-L (2005) Identification of a novel rice bZIP-type transcription factor gene, *OsbZIP1*, involved in response to infection of *Magnaporthe grisea*. Plant Mol Biol Rep 23:301–302

Mengiste T, Chen X, Salmeron J, Dietrich R (2003) The *Botrytis Susceptible 1* gene encodes an R2R3 MYB transcription factor protein that is required for biotic and abiotic stress responses in *Arabidopsis*. Plant Cell 15:2551–2565

Menke FLH, Kang H-G, Chen Z, Mee Park J, Kumar D, Klessig DF (2005) Tobacco transcription factor WRKY1 is phosphorylated by the MAP kinase SIPK and mediates HR-like cell death in tobacco. Mol Plant Microbe Interact 10:1027–1034

Merlot S, Leonhardt N, Fenzi F, Valon C, Costa M, Piette L, Vavasseur A, Genty B, Bolvin K, Müller A, Giraudat J, Leung J (2007) Constitutive activation of a plasma membrane $H^+$-ATPase prevents abscisic acid-mediated stomatal closure. EMBO J 26:3216–3226

Mészáros T, Helfer A, Hatzimasoura E, Magyar Z, Serazetdinova L, Rios G, Bardóczy V, Teige M, Koncz C, Peck S, Bögre L (2006) The Arabidopsis MAP kinase kinase MKK1 participates in defence responses to the bacterial elicitor flagellin. Plant J 48:485–495

Michard E, Lima PT, Borges F, Silva AC, Portes MT, Carvalho JE, Gilliham M, Liu LH, Obermeyer G, Feijó JA (2011) Glutamate receptor-like genes form $Ca^{2+}$ channels in pollen tubes and are regulated by pistil D-serine. Science 332:434–437

Miedema H, Demidchik V, Very AA, Bothwell JH, Brownlee C, Davies JM (2008) Two voltage-dependent calcium channels co-exist in the apical plasma membrane of *Arabidopsis thaliana* root hairs. New Phytol 179:378–385

Mithöfer A, Ebel J, Bhagwat AA, Boller T, Neuhaus-Url G (1999) Transgenic aequorin monitors cytosolic calcium transients in soybean cells challenged with β-glucan or chitin elicitors. Planta 207:566–574

Moeder W, Urguhart W, Ung H, Yoshioka K (2011) The role of cyclic nucleotide-gated ion channels in plant immunity. Mol Plant 4:442–452

Morel J, Fromentin J, Blein J-P, Simon-Plas F, Elmayan T (2004) RAC regulation of NtrbohD, the oxidase responsible for the oxidative burst in elicited tobacco cell. Plant J 37:282–293

Mori IC, Schroeder JI (2004) Reactive oxygen species activation of plant $Ca^{2+}$ channels. A signaling mechanism in polar growth, hormone transduction, stress signaling, and hypothetically mechanotransduction. Plant Physiol 135:702–708

Mori IC, Murata Y, Yang Y, Munemasa S, Wang Y-F, Andreoli S, Tiriac H, Alonso JM, Harper JF, Ecker JR, Kwak JM, Schroeder JI (2006) CDPKs CPK6 and CPK3 function in ABA regulation of guard cell S-type anion- and $Ca^{2+}$-permeable channels and stomatal closure. PLoS Biol 4:1749–1762

Mortimer JC, Laohavisit A, Macpherson N, Webb A, Brownlee C, Battey NH, Davies JM (2008) Annexins: multifunctional components of growth and adaptation. J Exp Biol 59: 533–544

Moutinho A, Hussey PJ, Trewavas AJ, Malho R (2001) cAMP acts as a second messenger in pollen tube growth and reorientation. Proc Natl Acad Sci U S A 98:10481–10486

Munemasa S, Hossain MA, Nakamura Y, Mori IC, Murata Y (2011) The Arabidopsis calcium-dependent protein kinase, CPK6, functions as a positive regulator of methyl jasmonate signaling in guard cells. Plant Physiol 155:553–561

Murata Y, Pei ZM, Mori IC, Schroeder J (2001) Abscisic acid activation of plasma membrane $Ca^{2+}$ channels in guard cells requires cytosolic NAD(P)H and is differentially disrupted upstream and downstream of reactive oxygen species production in *abi1-1* and *abi2-1* protein phosphatase 2C mutants. Plant Cell 13:2513–2523

Murillo I, Jaeck E, Cordero MJ, Secondo BS (2001) Transcriptional activation of a maize calcium-dependent protein kinase gene in response to fungal elicitors and infection. Plant Mol Biol 45:145–158

Nakashima A, Chen L, Thao NP, Fujiwara M, Wong HL, Kuwano M, Umemura K, Shirasu K, Kawasaki T, Shimamoto K (2008) RACK1 functions in rice immunity by interacting with the Rac1 immune complex. Plant Cell 20:2265–2279

Navarro L, Zipfel C, Rowland O, Keller I, Robatzek S, Boller T, Jones JD (2004) The transcriptional innate immune response to flg22. Interplay and overlap with Avr gene-dependent defense responses and bacterial pathogenesis. Plant Physiol 135:1113–1128

Navazio L, Bewell MA, Siddiqua A, Dickinson GD, Galione A, Sanders D (2000) Calcium release from the endoplasmic reticulum of higher plants by NADP metabolite nicotinic acid adenine dinucleotide phosphate. Proc Natl Acad Sci U S A 97:8693–8698

Navazio L, Mariani P, Sanders D (2001) Mobilization of $Ca^{2+}$ by cyclic ADP-ribose from the endoplasmic reticulum of cauliflower florets. Plant Physiol 125:2329–2338

Navazio L, Moscatiello R, Genre A, Novero M, Baldan B, Bonfante P, Mariani P (2007) A diffusible signal from arbuscular mycorrhizal fungi elicits a transient cytosolic calcium elevation in host plant cells. Plant Physiol 144:673–681

Ndamukong I, Abdallat AA, Thurow C, Fode B, Zander M, Weigel R, Gatz C (2007) SA-inducible *Arabidopsis* glutaredoxin interacts with TGA factors and suppresses JA-responsive *PDF1.2* transcription. Plant J 50:128–139

Nühse TS, Bottrill AR, Jones AM, Peck SC (2007) Quantitative phosphoproteomic analysis of plasma membrane proteins reveals regulatory mechanisms of plant innate immune responses. Plant J 51:931–940

Nürnberger T, Küfner I (2011) The role of the plant plasma membrane in microbial sensing and innate immunity: the plant plasma membrane. Plant Cell Monogr 19:471–483

Ogasawara Y, Kaya H, Hiraoka G, Yumoto F, Kimura S, Kadota Y, Hishinuma H, Senzaki E, Yamagoe S, Nagata K, Nara M, Suzuki K, Tanokura M, Kuchitsu K (2008) Synergistic activation of the *Arabidopsis* NADPH oxidase AtrbohD by $Ca^{2+}$ and phosphorylation. J Biol Chem 283:8885–8892

Onions J, Hermann S, Grundstrom T (2000) A novel type of calmodulin interaction in the inhibition of basic helix-loop-helix transcription factors. Biochemistry 39:4366–4374

Önnerud H, Zhang L, Gellerstedt G, Henriksson G (2002) Polymerization of monolignols by redox shuttle-mediated enzymatic oxidation: a new model in lignin biosynthesis. Plant Cell 14:1953–1962

Opalski KS, Schultheiss H, Kogel KH, Hückelhoven R (2005) The receptor-like MLO protein and the RAC/ROP family G-protein RACB modulate actin reorganization in barley attacked by the biotrophic powdery mildew fungus *Blumeria graminis* f. sp. *hordei*. Plant J 41:291–303

Orlandi WE, Hutcheson SW, Baker CJ (1992) Early physiological responses associated with race-specific recognition in soybean leaf tissue and cell suspensions treated with *Pseudomonas syringae* pv. *glycinea*. Physiol Mol Plant Pathol 40:173–180

Ottmann C, Marco S, Jaspert N, Marcon C, Schauer N, Weyand M, Vandermeeren C, Duby G, Boutry M, Wittinghofer A, Rigaud JL, Oecking C (2007) Structure of a 14-3-3 coordinated hexamer of the plant plasma membrane $H^+$-ATPase by combining X-ray crystallography and electron cryomicroscopy. Mol Cell 25:427–440

Palmgren MG (1998) Proton gradients and plant growth: role of the plasma membrane $H^+$-ATPase. Adv Bot Res 28:2–70

Palmgren MG, Harper JF (1999) Pumping with plant P-type ATPases. J Exp Bot 50:883–893

Palmgren MG, Soummarin M, Serrano R, Larsson C (1991) Identification of an autoinhibitory domain in the C-terminal region of the plant plasma membrane $H^+$-ATPase. J Biol Chem 266:20740–20745

Panstruga R (2005) Serpentine plant MLO proteins as entry portals for powdery mildew fungi. Biochem Soc Trans 33(Pt 2):389–392

Panstruga R, Schulze-Lefert P (2005) Corruption of host seven-transmembrane proteins by pathogenic microbes: a common theme in animals and plants? Microbes Infect 5:429–437

Park CY, Heo WD, Yoo JH, Lee JH, Kim MC, Chun HJ, Moon BC, Kim IH, Park HC, Choi MS, Ok HM, Cheong MS, Lee SM, Kim HS, Lee KH, Lim CO, Chung WS, Cho MJ (2004) Pathogenesis-related gene expression by specific calmodulin isoforms is dependent on NIM1, a key regulator of systemic acquired resistance. Mol Cells 18:207–213

Park CY, Lee JH, Yoo JH, Moon BC, Choi MS, Kang YH, Lee SM, Kim HS, Kang KY, Chung WS, Lim C, Cho MJ (2005) WRKY group IId transcription factors interact with calmodulin. FEBS Lett 579:1545–1550

Park HC, Kim ML, Lee SM, Bahk JD, Yun D-J, Lim CO, Hong JC, Lee SY, Cho MJ, Chung WS (2007) Pathogen-induced binding of the soybean zinc finger homoeodomain proteins GmZF-HD1 and GmZF-HD2 to two repeats of ATTA homeodomain binding site to calmodulin isoform 4 (*GmCaM4*) promoter. Nucleic Acids Res 35:3612–3623

Patel S (2004) NAADP-induced $Ca^{2+}$ release – a new signalling pathway. Biol Cell 96:19–28

Patil S, Takezawa D, Poovaiah BW (1995) Chimeric plant calcium/calmodulin-dependent protein kinase gene with a neural visinin-like calcium-binding domain. Proc Natl Acad Sci U S A 92:4897–4901

Patterson RL, van Rossum DB, Barrow RK, Snyder SH (2004) RACK1 binds to inositol 1,4,5-trisphosphate receptors and mediates $Ca^{2+}$ release. Proc Natl Acad Sci U S A 101:2328–2332

Pauly N, Knight MR, Thuleau P, Graziana A, Muto S, Ranjeva R, Mazars C (2001) The nucleus together with the cytosol generates patterns of specific cellular calcium signatures in tobacco suspension culture cells. Cell Calcium 30:413–421

Pedley KF, Martin GB (2005) Role of mitogen-activated protein kinases in plant immunity. Curr Opin Plant Biol 8:541–547

Pei Z-M, Murata Y, Benning G, Thomine S, Klüsener B, Allen GJ, Grill E, Schroeder JI (2000) Calcium channels activated by hydrogen peroxide mediate abscisic acid signaling in guard cells. Nature 406:731–734

Petersen M, Brodersen P, Naested H, Andreasson E, Lindhart U, Johansen B, Nielsen HB, Lacy M, Austin MJ, Parker JE, Sharma SB, Klessig DF, Martienssen R, Mattsson O, Jensen AB, Mundy J (2000) *Arabidopsis* MAP kinase 4 negatively regulates systemic acquired resistance. Cell 103:1111–1120

Petry A, Weitnauer M, Gorlach A (2010) Receptor activation of NADPH oxidases. Antioxid Redox Signal 13:437–487

Pitzschke A, Djamei A, Bitton F, Hirt H (2009a) A major role of the MEKK1-MKK1/2-MPK4 pathway in ROS signalling. Mol Plant 2:120–137

Pitzschke A, Schikora A, Hirt H (2009b) MAPK cascade signaling networks in plant defence. Curr Opin Plant Biol 12:1–6

Popescu SC, Popescu GV, Bachan S, Zhang Z, Seay M, Gerstein M, Snyder M, Dinesh-Kumar SP (2007) Differential binding of calmodulin-related proteins to their targets revealed through high-density Arabidopsis protein microarrays. Proc Natl Acad Sci U S A 104:4730–4735

Price MB, Jelesko J, Okumoto S (2012) Glutamine receptor homologs in plants: functions and evolutionary origins. Front Plant Sci 3:Article 235. doi:10:3389/fpls.2012.00235

Pugin A, Frachisse J-M, Tavernier E, Bligny R, Gout E, Douce R, Guern J (1997) Early events induced by the elicitor cryptogein in tobacco cells: involvement of a plasma membrane NADPH oxidase and activation of glycolysis and the pentose phosphate pathway. Plant Cell 9:2077–2091

Qi Z, Stephens NR, Spadling EP (2006) Calcium entry mediated by GLR3.3 in *Arabidopsis* glutamate receptor with a broad agonist profile. Plant Physiol 142:963–971

Qi Z, Verma R, Gehring C, Yamaguchi Y, Zhao Y, Ryan CA, Berkowitz GA (2010) $Ca^{2+}$ signaling by plant *Arabidopsis thaliana* Pep peptides depends on AtPepR1, a receptor with guanylyl cyclase activity, and cGMP-activated $Ca^{2+}$ channels. Proc Natl Acad Sci U S A 107:21193–21198

Qiu D, Xiao J, Ding X, Xiong M, Cai M, Cao Y, Li X, Xu C, Wang S (2007) OsWRKY13 mediates rice disease resistance by regulating defense-related genes in salicylate- and jasmonate-dependent signaling. Mol Plant Microbe Interact 20:492–499

Qu LJ, Zhu YX (2006) Transcription factor families in *Arabidopsis*: major progress and outstanding issues for future research. Curr Opin Plant Biol 9:544–549

Rainaldi M, Yamniuk AP, Murase T, Vogel HJ (2007) Calcium-dependent and -independent binding domain of tobacco MAPK phosphatase-1. J Biol Chem 282:6031–6042

Ranf S, Eschen-Lippold L, Pecher P, Lee J, Scheel D (2011) Interplay between calcium signaling and early signaling elements during defense responses to microbe-or damage-associated molecular patterns. Plant J 68:100–113

Raz V, Fluhr A (1992) Calcium requirement for ethylene-dependent responses. Plant Cell 4:1123–1130

Reddy AS (2001) Calcium: silver bullet in signaling. Plant Sci 160:381–404

Reddy VS, Reddy AS (2004) Proteomics of calcium-signaling components in plants. Phytochemistry 65:1745–1776

Reddy ASN, Takezawa D, Fromm H, Poovaiah BW (1993) Isolation and characterization of two cDNAs that encode for calmodulin-binding proteins from corn root tips. Plant Sci 94:109–117

Reddy AS, Reddy VS, Golovkin M (2000) A calmodulin binding protein from *Arabidopsis* is induced by ethylene and contains a DNA-binding motif. Biochem Biophys Res Commun 279:762–769

Reddy AS, Day IS, Narasimhulu SB, Safadi F, Reddy VS, Golovkin M, Harnly MJ (2002a) Isolation and characterization of a novel calmodulin-binding protein from potato. J Biol Chem 277:4206–4214

Reddy VS, Ali GS, Reddy ASN (2002b) Genes encoding calmodulin-binding proteins in the *Arabidopsis* genome. J Biol Chem 277:9840–9852

Reddy VS, Ali GS, Reddy AS (2003) Characterization of a pathogen-induced calmodulin-binding protein: mapping of four $Ca^{2+}$-dependent calmodulin-binding domains. Plant Mol Biol 52:143–159

Reddy ASN, Ali GS, Celesnik H, Day IS (2011a) Coping with stresses: roles of calcium- and calcium/calmodulin-regulated gene expression. Plant Cell 23:2010–2032

Reddy ASN, Ben-Hur A, Day IS (2011b) Experimental and computational approaches for the study of calmodulin interactions. Phytochemistry 72:1007–1019

Ren D, Liu Y, Yang K-Y, Han L, Mao G, Glazebrook J, Zhang S (2008) A fungal-responsive MAPK cascade regulates phytoalexin biosynthesis in *Arabidopsis*. Proc Natl Acad Sci U S A 105:5638–5643

Rizo J, Südhof TC (1998) C2-domains, structure and function of a universal $Ca^{2+}$-binding domain. J Biol Chem 273:15879–15882

Robatzek S, Somssich IE (2001) A new member of the *Arabidopsis* WRKY transcription factor family, *AtWRKY6*, is associated with both senescence- and defense-related processes. Plant J 28:123–133

Rodriguez-Concepcion M, Yalovsky S, Zik M, Fromm H, Gruissem W (1999) The prenylation status of a novel plant calmodulin directs plasma membrane or nuclear localization of the protein. EMBO J 18:1996–2007

Romeis T, Piedras P, Jones JD (2000) Resistance gene-dependent activation of a calcium-dependent protein kinase in the plant defense response. Plant Cell 12:803–815

Romeis T, Ludwig AA, Martin R, Jones JD (2001) Calcium-dependent protein kinases play an essential role in a plant defence response. EMBO J 20:5556–5567

Rostoks N, Schmierer D, Mudie S, Drader T, Brueggeman R, Caldwell DG, Waugh R, Kleinhofs A (2006) Barley necrotic locus *nec1* encodes the cyclic nucleotide-gated ion channel 4 homologous to the *Arabidopsis* HLM1. Mol Genet Genomics 275:159–168

Rudd JJ, Franklin-Tong VE (2001) Unravelling response-specificity in $Ca^{2+}$ signaling in plant cells. New Phytol 151:7–33

Rutschmann F, Stalder U, Piotrowski M, Oecking C, Schaller A (2002) *Le*CPK1, a calcium-dependent protein kinase from tomato: plasma membrane targeting and biochemical characterization. Plant Physiol 129:156–168

Sagi M, Fluhr R (2001) Superoxide production by plant homologues of the gp91(phox)NADPH oxidase: modulation of activity by calcium and by tobacco mosaic virus infection. Plant Physiol 126:1281–1290

Sagi M, Fluhr R (2006) Production of reactive oxygen species by plant NADPH oxidases. Plant Physiol 336–340

Sakano K (2001) Metabolic regulation of pH in plant cells: role of cytoplasmic pH in defense reaction and secondary metabolism. Int Rev Cytol 206:1–44

Sanders D, Brownlee C, Harper JF (1999) Communicating with calcium. Plant Cell 11:691–706

Sanders D, Pelloux J, Brownlee C, Harper JF (2002) Calcium at the cross-roads of signaling. Plant Cell 14:S401–S417

Schaller F (2001) Enzymes of the biosynthesis of octadecanoid-derived signaling molecules. J Exp Bot 52:11–23

Schaller A, Oecking C (1999) Modulation of plasma membrane $H^+$-ATPase activity differentially activates wound and pathogen defense responses in tomato plants. Plant Cell 11:263–272

Schaller GE, Sussman MR (1988) Phosphorylation of the plasma-membrane $H^+$-ATPase of oat roots by a calcium-stimulated protein kinase. Planta 173:509–518

Schmidt C, Schelle I, Liao YJ, Schroeder JI (1995) Strong regulation of slow anion channels and abscisic acid signaling in guard cells by phosphorylation and dephosphorylation events. Proc Natl Acad Sci U S A 92:9535–9539

Schroeder JI, Thuleau P (1991) $Ca^{2+}$ channels in higher plant cells. Plant Cell 3:555–559

Schroeder JI, Ward JM, Gassmann W (1994) Perspectives on the physiology and structure of inward rectifying $K^+$ channels in higher plants: biophysical implications of $K^+$ uptake. Annu Rev Biophys Biomol Struct 23:441–471

Schumaker KS, Sze H (1986) Calcium transport into the vacuole of oat roots characterization of $H^+/Ca^{2+}$ exchange activity. J Biol Chem 261:12172–12178

Sebastiá CH, Hardin SC, Clouse SD, Kieber JJ, Huber SC (2004) Identification of a new motif for CDPK phosphorylation in vitro that suggests ACC synthase may be a CDPK substrate. Arch Biochem Biophys 428:81–91

Segonzac C, Zipfel C (2011) Activation of plant pattern-recognition receptors by bacteria. Curr Opin Microbiol 14:54–61

Shimizu T, Nakano T, Takamizawa D, Desaki Y, Ishi-Minami N, Nishizawa Y, Minami E, Okada K, Yamane H, Kaku H, Shibuya N (2010) Two LysM receptor molecules, CEBiP and OsCERK1, cooperatively regulate chitin elicitor signaling in rice. Plant J 64:204–214

Shimono M, Sugano S, Nakayama A, Jiang C-J, Ono K, Toki S, Takatsuji H (2007) Rice WRKY45 plays a crucial role in benzothiadiazole-inducible blast resistance. Plant Cell 19:2064–2076

Shirasu K, Schulze-Lefert P (2003) Complex formation, promiscuity, and multi-functionality: protein interactions in disease-resistance pathways. Trends Plant Sci 8:252–258

Snedden WA, Fromm H (2001) Calmodulin as a versatile calcium signal transducer in plants. New Phytol 151:35–66

Sondergaard TE, Schulz A, Palmgren MG (2004) Energization of transport processes in plants: roles of the plasma membrane $H^+$-ATPase. Plant Physiol 136:2475–2482

Speth C, Jaspert N, Marcon C, Oecking C (2010) Regulation of the plant plasma membrane $H^+$-ATPase by its C-terminal domain: what do we know for sure? Eur J Cell Biol 89:145–151

Stáb MR, Ebel J (1987) Effects of $Ca^{2+}$ on phytoalexin induction by fungal elicitor in soybean cells. Arch Biochem Biophys 257:416–423

Stael S, Wurzinger B, Mair A, Mehlmer N, Vothknecht UC, Telge M (2012) Plant organellar calcium signaling: an emerging field. J Exp Bot 63:1525–1542

Staxén I, Pical C, Montgomery LT, Gray JE, Hetherington AM, McAinsh MR (1999) Abscisic acid induces oscillations in guard-cell cytosolic free calcium that involve phosphoinositide-specific phospholipase C. Proc Natl Acad Sci U S A 96:1779–1784

Stracke R, Werber M, Weisshaar B (2001) The *R2R3-MYB* gene family in *Arabidopsis thaliana*. Curr Opin Plant Biol 4:447–456

Suh SJ, Wang Y-F, Frelet A, Leonhardt A, Klein M, Forestier C, Mueller-Roeber B, Cho MH, Martinoia E, Schroeder JI (2007) The ATP binding cassette transporter AtMRP5 modulates anion and calcium channel activities in *Arabidopsis* guard cells. J Biol Chem 282:1916–1924

Suhita D, Kolla VA, Vavasseur A, Raghavendra AS (2003) Different signaling pathways during the suppression of stomatal opening by methyl jasmonate or abscisic acid. Plant Sci 164: 481–488

Svennelid F, Olsson A, Piotrowski M, Rosenquist M, Ottman C, Larsson C, Oecking C, Sommarin M (1999) Phosphorylation of Thr-948 at the C terminus of the plasma membrane $H^+$-ATPase creates a binding site for the regulatory 14-3-3 protein. Plant Cell 11:2379–2391

Szczegielniak J, Liwosz A, Jurkowski I, Loog M, Dobrowolska G, Ek P, Harmon ASC, Muszyriska G (2000) Calcium-dependent protein kinase from maize seedlings activated by phospholipids. Eur J Biochem 267:3818–3827

Sze H, Liang F, Hwang I, Curran AC, Harper JF (2000) Diversity and regulation of plant $Ca^{2+}$ pumps: insights from expression in yeast. Annu Rev Plant Physiol Plant Mol Biol 51: 433–462

Takabatake R, Karita E, Seo S, Mitsuhara I, Kuchitsu K, Ohashi Y (2007) Pathogen-induced calmodulin isoforms in basal resistance against bacterial and fungal pathogens in tobacco. Plant Cell Physiol 48:414–423

Takahashi K, Isobe M, Muto S (1998) Mastoparan induces an increase in cytosolic calcium ion concentration and subsequent activation of protein kinases in tobacco suspension culture cells. Biochim Biophys Acta 1401:339–346

Takahashi F, Yoshida R, Ichimura K, Mizoguchi T, Seo S, Yonezawa M, Maruyama K, Yamaguchi-Shinozaki K, Shinozaki K (2007) The mitogen-activated protein kinase cascade MKK3-MPK6 is an important part of the jasmonate signal transduction pathway in Arabidopsis. Plant Cell 19:805–818

Takeda S, Gapper C, Kaya H, Bell E, Kuchitsu K, Dolan L (2008) Local positive feedback regulation determines cell shape in root hair cells. Science 319:1241–1244

Takezawa D (2000) A rapid induction by elicitors of the mRNA encoding CCD-1, a 14 kDa $Ca^{2+}$-binding protein in wheat cultured cells. Plant Mol Biol 42:807–817

Takezawa D, Ramachandiran S, Paranjape V, Poovaiah BW (1996) Dual regulation of a chimeric plant serine/threonine kinase by calcium and calcium/calmodulin. J Biol Chem 271:8126–8132

Talke IN, Blaudez D, Maathuis FJM, Sanders D (2003) CNGCs: prime targets of plant cyclic nucleotide signaling? Trend Plant Sci 8:286–293

Tapken D, Hollmann M (2008) *Arabidopsis thaliana* glutamate receptor ion channel function demonstrated by ion pore transplantation. J Mol Biol 383:36–48

Tavernier E, Wendehenne J-P, Blein J-P, Pugin A (1995) Involvement of free calcium in action of cryptogein, a proteinaceous elicitor of a hypersensitive reaction in tobacco cells. Plant Physiol 109:1025–1031

Tester M (1990) Plant ion channels: whole-cell and single-channel studies. New Phytol 114:305–340

Theodosiou A, Ashworth A (2002) MAP kinase phosphatases. Genome Biol 3:1–10

Thuleau P, Ward JM, Ranjeva R, Schroeder JI (1994) Voltage-dependent calcium-permeable channels in the plasma membrane of a higher plant cell. EMBO J 13:2970–2975

Tikhonov DB, Magazanik LG (2009) Origin and molecular evolution of ionotropic glutamate receptors. Neurosci Behav Physiol 39:763–773

Tomsig JL, Cruetz CE (2000) Biochemical characterization of copine: a ubiquitous $Ca^{2+}$-dependent, phospholipid-binding protein. Biochemistry 39:16163–16175

Tomsig JL, Cruetz CE (2002) Copines: a ubiquitous family of $Ca^{2+}$-dependent phospholipid-binding proteins. Cell Mol Life Sci 59:1467–1477

Tomsig JL, Snyder SL, Cruetz CE (2003) Identification of targets for calcium signaling through the copine family of proteins: characterization of a coiled-coil copine-binding motif. J Biol Chem 278:10048–10054

Torres MA, Dangl JL, Jones JDG (2002) Arabidopsis gp91phox homologues *AtrbohD* and *AtrbohF* are required for accumulation of reactive oxygen intermediates in plant defense response. Proc Natl Acad Sci U S A 99:517–522

Trewavas AJ (1999) How plants learn. Proc Natl Acad Sci U S A 96:4216–4218

Trewavas AJ, Malho R (1998) $Ca^{2+}$ signaling in plant cells: the big network! Curr Opin Plant Biol 1:428–433

Truman G, Glazebrook J (2012) Coexpression analysis identifies putative targets for CBP60g and SARD1 regulation. BMC Plant Biol 12:216

Tsuda K, Sato M, Glazebrook J, Cohen JD, Katagiri F (2008) Interplay between MAMP-triggered and SA-mediated defense responses. Plant J 53:763–775

Urquhart W, Chin K, Ung H, Moeder W, Yoshioka K (2011) The cyclic nucleotide-gated channels AtCNGC11 and 12 are involved in multiple $Ca^{2+}$-dependent physiological responses and act in a synergistic manner. J Exp Bot 62:3671–3682

Vadassery J, Oelmüller R (2009) Calcium signaling in pathogenic and beneficial plant microbe interactions. What can we learn from the interaction between *Piriformospora indica* and *Arabidopsis thaliana*. Plant Signal Behav 411:1024–1027

Van Breusegem F, Bailey-Serres J, Mittler R (2008) Unraveling the tapestry of networks involving reactive oxygen species in plants. Plant Physiol 147:978–984

van den Wijngaard PW, Bunney TD, Roobeek I, Schönknecht G, de Boer AH (2001) Slow vacuolar channels from barley mesophyll cells are regulated by 14-3-3 proteins. FEBS Lett 488:100–104

Van der Ent S, Verhagen BWM, Van Doorn R, Bakker D, Verlaan MG, Pel MJC, Joosten RG, Proveniers MCG, Van Loon LC, Ton J, Pieterse CMJ (2008) *MYB72* is required in early signaling steps of rhizobacteria-induced systemic resistance in *Arabidopsis*. Plant Physiol 146:1293–1304

Van Verk MC, Pappaaioannou D, Neeleman L, Bol JF, Linthorst HJM (2008) A novel WRKY transcription factor is required for induction of *PR-1a* gene expression by salicylic acid and bacterial elicitors. Plant Physiol 140:1983–1995

van Verk MC, Bol JF, Linthorst HJM (2011) WRKY transcription factors involved in activation of SA biosynthesis genes. BMC Plant Biol 11:89. doi:10.1186/1471-2229-11-89

Vandelle E, Poinssot B, Wendehenne D, Bentéjac M, Pugin A (2006) Integrated signaling network involving calcium, nitric oxide, active oxygen species but not mitogen-activated protein kinases in BcPG-1 elicited grapevine defenses. Mol Plant Microbe Interact 19:429–440

Vatsa P, Chiltz A, Bourque S, Wendehenne D, Garcia-Brugger A, Pugin A (2011) Involvement of putative glutamate receptors in plant defence signaling and NO production. Biochimie 93:2095–2101

Vera-Estrella R, Barkla BJ, Higgins VJ, Blumwald E (1994) Plant defense response to fungal pathogens. Activation of host-plasma membrane $H^+$-ATPase by elicitor-induced enzyme dephosphorylation. Plant Physiol 104:209–215

Véry AA, Sentenac H (2002) Cation channels in the *Arabidopsis* plasma membrane. Trends Plant Sci 7:168–175

Vidhyasekaran P (2007) Fungal pathogenesis in plants and crops: molecular biology and host defense mechanisms, IIth edn. Taylor Francis Group, CRC Press, Boca Raton, p 510

Vincill ED, Bieck AM, Spalding EP (2012) $Ca^{2+}$ conduction by an amino acid-gated ion channel related to glutamate receptors. Plant Physiol 159:40–46

Wan D, Li R, Zou B, Zhang X, Cong J, Wang R, Xia Y, Li G (2012) Calmodulin-binding protein CBP60g is a positive regulator of both disease resistance and drought tolerance in *Arabidopsis*. Plant Cell Rep 31:1269–1281

Wang X (1999) The role of phospholipase D in signaling cascades. Plant Physiol 120:645–651

Wang X (2001) Plant phospholipases. Annu Rev Plant Physiol Plant Mol Biol 52:211–231

Wang L, Tsuda K, Sato M, Cohen JD, Katagiri F, Glazebrook J (2009) Arabidopsis CaM binding protein CBP60g contributes to MAMP-induced SA accumulation and is involved in disease resistance against *Pseudomonas syringae*. PLoS Pathog 5:e1000301

Wang Y, Li J, Hou S, Wang X, Li Y, Ren D, Chen S, Tang X, Zhou J-M (2010) A *Pseudomonas syringae* ADP-ribosyltransferase inhibits *Arabidopsis* mitogen-activated protein kinase kinases. Plant Cell 22:2033–2044

Wang L, Tsuda K, Truman W, Sato M, Le Nguyen V, Katagiri F, Glazebrook J (2011) CBP60g and SARD1 play partially redundant critical roles in salicylic acid signaling. Plant J 67:1029–1041

Wang W-H, Yi X-Q, Han A-D, Liu T-W, Chen J, Wu F-H, Dong X-J, He J-X, Pei Z-M, Pei Z-M, Zheng H-L (2012) Calcium-sensing receptor regulates stomatal closure through hydrogen peroxide and nitric oxide in response to extracellular calcium in *Arabidopsis*. J Exp Bot 63:177–190

Ward JM, Pei Z-M, Schroeder JI (1995) Roles of ion channels in initiation of signal transduction in higher plants. Plant Cell 7:833–844

Ward JM, Maser P, Schroeder JI (2009) Plant ion channels: gene families, physiology and functional genomic analyses. Annu Rev Physiol 71:59–82

Watillon B, Kettmann R, Boxus P, Burny A (1995) Structure of a calmodulin-binding protein kinase gene from apple. Plant Physiol 108:847–848

Wendehenne D, Lamotte O, Frachisse J-M, Barbier-Brygoo H, Pugin A (2002) Nitrate efflux is an essential component of the cryptogein signaling pathway leading to defense responses and hypersensitive cell death in tobacco. Plant Cell 14:1937–1951

Wevelsiep L, Rüpping E, Knogge W (1993) Stimulation of barley plasmalemma $H^+$-ATPase by phytotoxic peptides from the fungal pathogen *Rhynchosporium secalis*. Plant Physiol 101:297–301

Wheeler GL, Brownlee C (2008) $Ca^{2+}$ signaling in plants and green algae: changing channels. Trends Plant Sci 13:506–514

White PJ (2000) Calcium channels in higher plants. Biochim Biophys Acta 1465:171–189

White PJ (2004) Calcium signals in root cells: the roles of plasma membrane calcium channels. Biologia 59(S13):77–83

White PJ, Broadley MR (2003) Calcium in plants. Ann Bot (Lond) 92:487–511

Wildermuth MC, Dewdney J, Wu G, Ausubel FM (2001) Isochrorismate synthase is required to synthesize salicylic acid for plant defence. Nature 414:562–566

Williams RW (1998) Plant homeobox genes: many functions stem from a common motif. Bioessays 20:280–282

Wong HL, Pinontoan R, Hayashi K, Tabata R, Yaeno T, Hasegawa K, Kojima C, Yoshioka H, Iba K, Kawasaki T, Shimamoto K (2007) Regulation of rice NADPH oxidase by binding of Rac GTPase to its N-terminal extension. Plant Cell 19:4022–4034

Wu Y, Kuzma J, Maréchal E, Graeff R, Lee HC, Foster R, Chua NH (1997) Abscisic acid signaling through cyclic ADP-ribose in plants. Science 278:2126–2130

Xing T, Higgins VJ, Blumwald E (1996) Regulation of plant defense response to fungal pathogens: two types of protein kinases in the reversible phosphorylation of the host plasma membrane $H^+$-ATPase. Plant Cell 8:555–564

Xing T, Higgins VJ, Blumwald E (1997) Identification of G proteins mediating fungal elicitor-induced dephosphorylation of host plasma membrane $H^+$-ATPase. J Exp Biol 48:229–237

Xing T, Wang XJ, Malik K, Miki B (2001) Ectopic expression of an *Arabidopsis* calmodulin-like domain protein kinase-enhanced NADPH oxidase activity and oxidative burst in tomato protoplasts. Mol Plant Microbe Interact 14:1261–1264

Yaffe MB, Rittinger K, Volinia S, Caron PR, Aitken A, Leffers H, Gamblin SJ, Smerdon SJ, Cantley LC (1997) The structural basis for 14-3-3 phosphopeptide binding specificity. Cell 91:961–971

Yamakawa H, Katou S, Seo S, Mitsuhara I, Kamada H, Ohashi Y (2003) Plant mitogen-activated protein kinase phosphatase interacts with calmodulin. J Biol Chem 279:928–936

Yamakawa H, Mitsuhara I, Ito N, Seo S, Kamada H, Ohashi Y (2001) Transcriptionally and post-transcriptionally regulated response of 13 calmodulin genes to tobacco mosaic virus-induced cell death and wounding in tobacco plant. Eur J Biochem 268:3916–3929

Yang T, Poovaiah BW (2000) An early ethylene up-regulated gene encoding a calmodulin-binding protein involved in plant senescence and death. J Biol Chem 275:38467–38473

Yang T, Poovaiah BW (2002a) A calmodulin-binding/CGCG box DNA-binding protein family involved in multiple signaling pathways in plants. J Biol Chem 277:45049–45058

Yang T, Poovaiah BW (2002b) Hydrogen peroxide homeostasis: activation of plant catalase by calcium/calmodulin. Proc Natl Acad Sci U S A 99:4097–4102

Yoo BC, Harmon AC (1996) Intramolecular binding contributes to the activation of CDPK, a protein kinase with a calmodulin-like domain. Biochemistry 35:12029–12037

Yoo JH, Park CY, Kim JC, Heo WD, Cheong MS, Park HC, Kim MC, Moon BC, Choi MS, Kang YA, Lee JH, Kim HS, Lee SM, Yoon H-W, Lim CO, Yun DJ, Lee SY, Chung WS, Cho MJ (2005) Direct interaction of a divergent CaM isoform and the transcription factor, MYB2, enhances salt tolerance in *Arabidopsis*. J Biol Chem 280:3697–3706

Yoon GM, Cho HS, Ha HJ, Liu JR, Lee H-P (1999) Characterization of NtCDPK1, a calcium-dependent protein kinase gene in *Nicotiana tabacum*, and the activity of its encoded protein. Plant Mol Biol 39:991–1001

Yoshioka H, Numata N, Nakajima K, Katou S, Kawakita K, Rowland O, Jones JD, Doke N (2003) *Nicotiana benthamiana* $gp91^{phox}$ homologs *NbrbohA* and *NbrbohB* participate in $H_2O_2$ accumulation and resistance to *Phytophthora infestans*. Plant Cell 15:706–718

Yoshioka K, Moeder W, Kang H-G, Kachroo P, Masmoudi K, Berkowitz G, Klessig D-F (2006) The chimeric *Arabidopsis* CYCLIC NUCLEOTIDE-GATED ION CHANNEL11/12 activates multiple pathogen responses. Plant Cell 18:747–763

Yu IC, Parker J, Bent AF (1998) Gene-for-gene disease resistance without the hypersensitive response in *Arabidopsis dnd1* mutant. Proc Natl Acad Sci U S A 95:7819–7824

Yu D, Chen C, Chen Z (2001) Evidence for an important role of WRKY DNA binding proteins in the regulation of *NPR1* gene expression. Plant Cell 13:1527–1539

Yu X-C, Li M-J, Gao G-F, Feng H-Z, Geng X-Q, Peng C-C, Zhu S-Y, Wang X-J, Shen Y-Y, Zhang D-P (2006) Abscisic acid stimulates a calcium-dependent protein kinase in grape berry. Plant Physiol 140:558–579

Yu X-C, Zhu S-Y, Gao G-F, Wang X-J, Zhao R, Zou K-Q, Wang X-F, Zhang X-Y, Wu F-Q, Peng C-C, Zhang D-P (2007) Expression of a grape calcium-dependent protein kinase ACPK1 in *Arabidopsis thaliana* promotes plant growth and confers abscisic acid-hypersensitivity in germination, post germination growth, and stomatal movement. Plant Mol Biol 64:531–538

Yue H, Nie S, Xing D (2012) Over-expression of *Arabidopsis* Bax inhibitor-1 delays methyl jasmonate-induced leaf senescence by suppressing the activation of MAP kinase 6. J Exp Bot 63(12):4463–4474. doi:1093/jxb/ers 122

Zegzouti H, Jones B, Frasse P, Marty C, Maitre B, Latch A, Pech JC, Bouzayen M (1999) Ethylene-regulated gene expression in tomato fruit: characterization of novel ethylene-responsive and ripening-related genes isolated by differential display. Plant J 589–600

Zeidler D, Zähringer U, Gerber I, Hartung T, Bors W, Hutzler P, Durner J (2004) Innate immunity in *Arabidopsis thaliana*: lipopolysaccharides activate nitric oxide synthase NOS and induce defense genes. Proc Natl Acad Sci U S A 101:15811–15816

Zeidler D, Dubery IA, Schmitt-Kopplin P, Von Rad U, Durner J (2010) Lipopolysaccharide mobility in leaf tissue of *Arabidopsis thaliana*. Mol Plant Pathol 11:747–755

Zeng W, He S-Y (2010) A prominent role of the flagellin receptor FLAGELLIN-SENSING2 in mediating stomatal response to *Pseudomonas syringae* pv. *tomato* DC3000 in *Arabidopsis*. Plant Physiol 153:1188–1198

Zhang SQ, Klessig DF (2001) MAP cascades in plant defense signaling. Trends Plant Sci 6:5320–5327

Zhang M, Yuan T (1998) Molecular mechanisms of calmodulin's functional versatility. Biochem Cell Biol 76:313–323

Zhang J, Shao F, Li Y, Cui H, Chen L, Li H, Zou Y, Long C, Lan L, Chai J, Chen S, Tang X, Zhou JM (2007) A *Pseudomonas syringae* effector inactivates MAPKs to suppress PAMP-induced immunity in plants. Cell Host Microbe 1:175–188

Zhang W, He SY, Assmann SM (2008) The plant innate immunity response in stomatal guard cells invokes G-protein-dependent ion channel regulation. Plant J 56:984–996

Zhang Y, Xu S, Ding P, Wang D, Cheng YT, He J, Geo M, Xu F, Li Y, Zhu Z, Li X, Zhang Y (2010) Control of salicylic acid synthesis and systemic acquired resistance by two members of a plant-specific family of transcription factors. Proc Natl Acad Sci U S A 107:18220–18225

Zhao J, Sakai K (2003) Multiple signalling pathways mediate fungal elicitor-induced β-thujalplicin biosynthesis in *Cupressus lusitanica* cell cultures. J Exp Bot 54:647–656

Zhao S, Qi X (2008) Signaling in plant disease resistance and symbiosis. J Integr Plant Biol 50:799–807

Zheng Z, Qamar SA, Chen Z, Mengiste T (2006) *Arabidopsis* WRKY33 transcription factor is required for resistance to necrotrophic fungal pathogens. Plant J 48:596–605

Zhong H, Lai J, Yau K-W (2003) Selective heteromeric assembly of cyclic nucleotide-gated channels. Proc Natl Acad Sci U S A 100:5509–5513

Zhou N, Tootle T, Tsui F, Klessig D, Glazebrook J (1998) *PAD4* functions upstream from salicylic acid to control defense responses to *Arabidopsis*. Plant Cell 10:1021–1030

Zhou JM, Trifa Y, Silva H, Pontier D, Lam E, Shah J, Klessig DF (2000) NPR1 differentially interacts with members of the TGA/OBF family of transcription factors that bind an element of the *PR-1* gene required for induction by salicylic acid. Mol Plant Microbe Interact 13:191–202

Zhu S-Y, Yu X-C, Wang X-J, Zhao R, Li Y, Fan R-C, Shang Y, Du S-Y, Wang X-F, Wu F-Q, Xu Y-H, Zhang X-Y, Zhang D-P (2007) Two calcium-dependent protein kinases, CPK4 and CPK11, regulate abscisic acid signal transduction in *Arabidopsis*. Plant Cell 19: 3019–3036

Zhu X, Caplan J, Mamillapalli P, Czymmek K, Dinesh-Kumar SP (2010) Function of endoplasmic reticulum calcium ATPase in innate immunity-mediated programmed cell death. EMBO J 29:1007–1018

Zielinski RE (2002) Characterization of three new members of the *Arabidopsis thaliana* calmodulin gene family: conserved and highly diverged members of the gene family functionally complement a yeast calmodulin null. Planta 214:446–456

Zipfel C (2008) Pattern-recognition receptors in plant innate immunity. Curr Opin Immunol 20:10–16

Zipfel C, Robatzek S, Navarro L, Oakeley EJ, Jones JDG, Felix G, Boller T (2004) Bacterial disease resistance in *Arabidopsis* through flagellin perception. Nature 428:764–767

# Chapter 5
# Reactive Oxygen Species and Cognate Redox Signaling System in Plant Innate Immunity

**Abstract** PAMP elicitor signals trigger a rapid and transient production of reactive oxygen species (ROS) including hydrogen peroxide ($H_2O_2$), superoxide ($O_2^-$), singlet oxygen ($^1O_2$), and hydroxyl radical (OH°). The oxidative burst is often a very rapid response induced by elicitor, occurring within seconds to few minutes, suggesting that the process may not require *de novo* protein synthesis but involves the activation of pre-existing enzymes. Several enzymes including NADPH oxidases, peroxidases, xanthine oxidase, amine oxidases, oxalate oxidase, glycollate oxidase, urate oxidase and lipoxygenase have been implicated in the PAMP-induced ROS production. $H_2O_2$ is generated in apoplasts but accumulates to a greater extent in the cytoplasm than in the apoplast. It has been suggested that the apoplastic $H_2O_2$ is translocated to cytoplasm for participation in the pathogen defense. ROS plays a central role in launching the defense response against invading pathogens. ROS induces $Ca^{2+}$ signaling system, reversible phosphorylation, ubiquitin-proteasome signaling pathway, NO signaling system, salicylic acid signaling system, ethylene-mediated signaling system, jasmonic acid signaling system, and abscisic acid-mediated signaling system. The ROS signal functions are manifested as a consequence of their ability to act as mobile carriers of an unpaired electron and are involved in redox signaling system. Redox signaling occurs when at least one step in a signaling event involves one of its components being specifically modified by a reactive oxygen species. Signaling through the redox active molecule $H_2O_2$ has been shown to be important in inducing plant defense responses. Expression of several transcription factors has been shown to be regulated by $H_2O_2$ and these transcription factors may be direct targets for redox modification by $H_2O_2$. ROS signaling system and cognate redox signaling have been shown to be involved in activation of several defense genes. Some pathogens could cause disease mostly by interfering with the $H_2O_2$ signaling pathway.

**Keywords** Reactive oxygen species (ROS) • PAMP-induced ROS production • ROS-induced signaling systems • Redox signaling • $H_2O_2$-regulated transcription factors

P. Vidhyasekaran, *PAMP Signals in Plant Innate Immunity: Signal Perception and Transduction*, Signaling and Communication in Plants 21,
DOI 10.1007/978-94-007-7426-1_5, 

## 5.1 Reactive Oxygen Intermediates Involved in Oxidative Burst

The rapid and transient production of reactive oxygen species (ROS), also called oxidative burst or respiratory burst, is one of the most rapid defense responses observed in plants due to pathogen-associated molecular pattern (PAMP)/elicitor treatment (Yang et al. 1997; Grant and Loake 2000; Faize et al. 2004; Asada 2006; Sagi and Fluhr 2006; Vidhyasekaran, 2007; Lehtonen et al. 2012). The ROS, which are also called as reactive oxygen intermediates (ROI; Grant and Loake 2000; Pieterse and Van Loon 2004), include hydrogen peroxide ($H_2O_2$), superoxide ($O_2^-$), singlet oxygen ($^1O_2$), and hydroxyl radical ($OH^\circ$) (Grant and Loake 2000; Vidhyasekaran 2007).

$O_2^-$ is the first ROS induced by elicitor treatment (Haga et al. 1995; Faize et al. 2004). The superoxide is only short lived and is thought to be produced in the outer surface of the cell within a few minutes of elicitor treatment (Sagi and Fluhr 2006). The half life of $O_2^-$ is less than a second and is usually rapidly dismutated either nonenzymatically or via superoxide dismutase (SOD) to $H_2O_2$, which is relatively stable (Grant and Loake 2000). The negatively charged $O_2^-$ could traverse the plasma membrane as the neutral hydroperoxyl ($HO_2$) or being converted to the membrane-permeable $H_2O_2$ (Sagi and Fluhr 2006). Protonation of $O_2^-$ can produce the hydroperoxyl radical $HO_2^-$, which can convert fatty acids to toxic lipid peroxides. Moreover, in the presence of divalent metal ions such as $Fe^{2+}$, $H_2O_2$ can undergo Fenton reaction, producing the hydroxyl radical ($OH^\circ$) (Grant and Loake 2000). Singlet oxygen ($^1O_2$) is an excited state of molecular oxygen that can be generated in a number of ways including the spontaneous dismutation of two $O_2^-$ radicals (Elstner 1982; Scandalios 1993). Among the different ROS, $H_2O_2$ is the most attractive candidate for defense signaling because of its relatively long life and high permeability across membranes (Allan and Fluhr 1997).

## 5.2 Upstream Events in ROS Signaling System

### *5.2.1 Enzymes Involved in ROS Generation*

The oxidative burst is often a very rapid response induced by elicitor, occurring within seconds in some systems, such as cultured cells of French bean (*Phaseolus vulgaris*) and soybean (Bolwell et al. 1995). In other systems, such as rose (*Rosa damascena*) cultured cells (Arnott and Murphy 1991), it may be delayed for few minutes or hours. These observations suggest that the oxidative burst may not require *de novo* protein synthesis but involves the activation of pre-existing enzymes.

Several enzymes have been implicated in the PAMP/elicitor-induced apoplastic ROS production (Fig. 5.1). NADPH oxidases, which are inhibited by

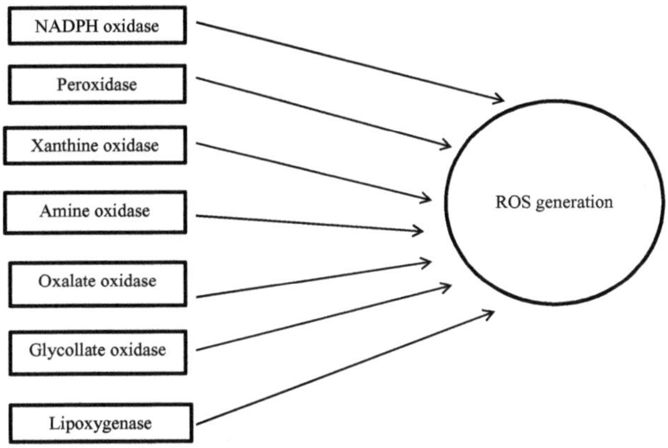


**Fig. 5.1** Enzymes involved in generation and accumulation of ROS in plant cells

diphenyleneiodonium (DPI) but not by cyanide or azide, and cell wall peroxidases, which are inhibited by cyanide or azide but not by DPI (Grant et al. 2000b; Bolwell et al. 2002) are the two important groups of enzymes involved in ROS production (Suzuki et al. 2011; Daudi et al. 2012; Lehtonen et al. 2012; O'Brein et al. 2012). The NADPH oxidases or cell wall peroxidases have been implicated in the ROS production in different plant systems. In rose cells $H_2O_2$ is produced by a plasma membrane NADPH oxidase, whereas in bean cells $H_2O_2$ is derived directly from cell wall peroxidases (Bolwell et al. 1998).

An additional source of ROS may emanate intracellularly from xanthine oxidase activity (Allan and Fluhr 1997). Xanthine oxidase is a reductase supplying electrons to $NAD^+$ to produce NADH (Halliwell and Gutteridge 1989) and produces $O_2^-$ (Montalbini 1992). Amine oxidases can induce ROS production by acting on amines as substrates for the enzymes. These are a ubiquitous group of plant enzymes and catalyze the oxidation of a variety of monoamines, diamines, and polyamines to the corresponding aldehyde and release $H_2O_2$ (Tipping and McPherson 1995). Oxalate oxidase is also a $H_2O_2$-generating enzyme (Zhou et al. 1998). Some enzymes, such as glycolate oxidase (Rojas and Mysore 2012; Rojas et al. 2012) and urate oxidase (Halliwell and Gutteridge 1989), can produce $H_2O_2$ (Halliwell and Gutteridge 1989). Lipoxygenase catalyzes the direct oxygenation of polyunsaturated fatty acids and produces $O_2^-$ (Thompson et al. 1987).

Different elicitors may induce ROS production by different types of enzymes. The cryptogein-induced ROS burst was insensitive to the NO synthase inhibitor L-NMMA, whereas L-arginine-induced ROS was sensitive to this inhibitor. Cryptogein-induced ROS was significantly inhibited by DPI, whereas L-arginine-induced ROS burst remained unaffected (Allan and Fluhr 1997). These studies indicated that cryptogein-induced ROS burst is due to the action of NADPH oxidase-type enzymes or xanthine oxidase. The L-arginine-induced ROS may be due to the action of peroxidase- or amine oxidase-type enzymes (Allan and Fluhr 1997).

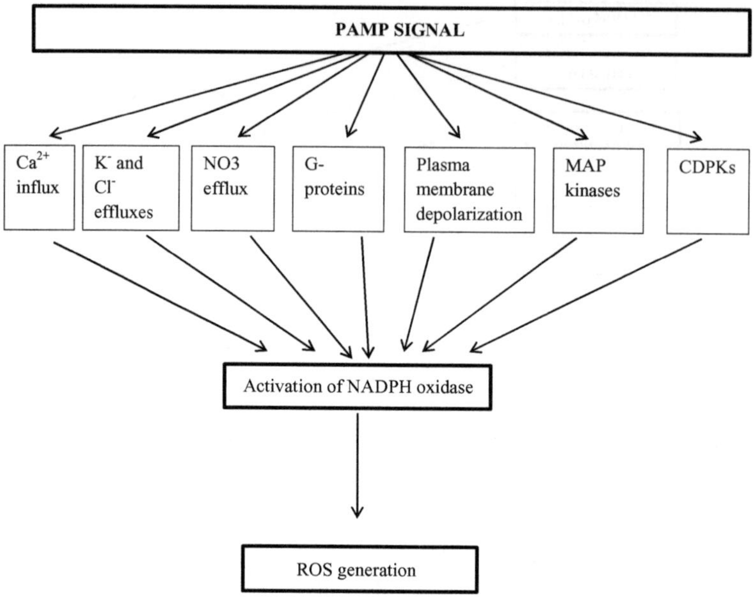


**Fig. 5.2** PAMP-triggered early signaling events activating NADPH oxidase to produce ROS

### 5.2.2 *Early PAMP-Induced Events Leading to Activation of NADPH Oxidase to Generate ROS*

The early PAMP elicitor-induced events include $Ca^{2+}$ influx, anion effluxes, cytosolic acidification, plasma membrane depolarization, MAPK signaling cascade activation, calcium-dependent protein kinase (CDPK) activation, and protein phosphorylation/dephosphorylation. These events have been shown to activate NADPH oxidase homologs of plants termed respiratory burst oxidase homolog (Rboh) which is responsible for ROS production (Fig. 5.2; Nürnberger and Scheel 2001; Simon-Plas et al. 2002; Wendehenne et al. 2002; Apel and Hirt 2004; Kadota et al. 2004; Torres and Dangl 2005; Wong et al. 2007; Asai et al. 2008; Zhu et al. 2009; Zhao et al. 2010; Zhang et al. 2011; Kiirika et al. 2012). The NADPH oxidases are localized to plasma membrane fractions (Keller et al. 1998) and are stimulated directly by $Ca^{2+}$ (Sagi and Fluhr 2001). All plant *rboh* genes carry EF-hands that bind $Ca^{2+}$ and plant Rboh proteins were shown to be stimulated directly by $Ca^{2+}$ (Jabs et al. 1997; Sagi and Fluhr 2001; Torres and Dangl 2005; Van Breusegem et al. 2008). $NO_3^-$ efflux activation seems to be essential to induce NADPH oxidase (Wendehenne et al. 2002). Anion efflux has been shown to be necessary for the induction of ROS production in parsley (Jabs et al. 1997) and soybean cells (Ebel et al. 1995). Phosphorylation events also occur both upstream and downstream of ROS production (Nürnberger and Scheel 2001; Apel and Hirt 2004).

The early signaling events induced by PAMPs include the increased expression of genes encoding G-proteins. The gene *AGB1*, encoding the β-subunit of G-protein in *Arabidopsis*, is highly induced after flg22 treatment (Zipfel et al. 2004). The PAMP-activated small G-proteins (Morel et al. 2004; Wong et al. 2007; Kiirika et al. 2012) and heterotrimeric G-proteins (Zhu et al. 2009; Zhao et al. 2010; Zhang et al. 2011) have been shown to trigger generation of ROS. The G-proteins are involved in the regulation of ROS generation via the activation of NADPH oxidase (RBOH) (Fig. 5.1; Agrawal et al. 2003; Kiirika et al. 2012). The small GTP-binding protein Rac2 homologs (called Rho-like proteins) regulate the production of ROS by the NADPH oxidase (Kawasaki et al. 1999; Moeder et al. 2005; Wong et al. 2007). Different plant Rac proteins appear to act as either positive or negative regulators of ROS production. *Osrac1* is a positive regulator of ROS production in rice (Ono et al. 2001), whereas *Ntrac5* acts as negative regulator of ROS production in tobacco (Morel et al. 2004). In Arabidopsis, heterotrimeric G protein signaling mediates the oxidative burst (Joo et al. 2005). The *Arabidopsis agb1* mutants are impaired in the oxidative burst triggered by flg22, suggesting the importance of G-proteins in ROS production (Ishikawa 2009).

MAP kinases may be involved in generation of ROS by activating the NADPH oxidase (Asai et al. 2008). Two different MAPK cascades have been shown to be involved in induction of ROS in *Nicotiana benthamiana*. The MAPK cascades NPK1-MEK2-SIPK/NTF4 and NPK1-MEK1-and NTF6 are involved in activation of NADPH oxidase which is involved in production of ROS (Asai et al. 2008). The MAP kinases may induce the NADPH oxidase at the gene transcriptional level and also by post-translational level (Yoshioka et al. 2003). *N. benthamiana* MAPK kinase induced the NADPH oxidase gene *NbrbohB* at the transcriptional level. At the post-translational level, the NADPH oxidase-induced oxidative burst is controlled through phosphorylation activation by its upstream MAP kinase and dephosphorylation inactivation by its negative regulator, phosphatase (Yoshioka et al. 2003). Calcium-dependent protein kinase (CDPK) has also been shown to phosphorylate NADPH oxidase (Xing et al. 1997; Blumwald et al. 1998). Accumulation of ROS requires both $Ca^{2+}$ influx and protein kinase activity (Romeis et al. 1999).

### *5.2.3 Cell Wall Peroxidases Are Involved in ROS Production in Some Plant Systems*

Cell wall peroxidases have been shown to be involved in ROS production in some plant systems (Daudi et al. 2012; O'Brein et al. 2012). In horseradish (*Armoracia lapathifolia*), the accumulation of $H_2O_2$ has been suggested to be due to the action of peroxidase producing phenolic and NAD radicals which reduce $O_2$ to superoxide (Halliwell 1978). In this model, the source of electrons in the apoplast is said to be malate, exported across the plasma membrane by a malate/oxalacetate carrier and used to reduce $NAD^+$ by apoplastic malate dehydrogenase. $H_2O_2$ is formed by the dismutation of superoxide (Bolwell et al. 1998).

The oxidative burst induced in French bean cultured cells by a fungal elicitor, involves an apoplastic peroxidase (Bolwell et al. 1995). The $O_2$-heme complex of peroxidase is reduced to compound III by reductants exported from the cell. Under elevated pH conditions, the complex is effectively hydrolyzed to release $H_2O_2$. In this model the source of electrons has not been identified, but the release of a reductant from elicited cells has been observed (Bolwell et al. 1995). Bolwell et al. (1998) showed that in bean cells treated with a fungal elicitor, $H_2O_2$ was derived directly from cell wall peroxidases following extracellular alkalinization and the appearance of a reductant.

Production of apoplastic ROS by chitin treatment in *Physcomitrella patens* has been shown to require peroxidase. The fungal elicitor chitin caused an immediate oxidative burst in wild-type *P. patens* but not in the Δ*Prx34* mutants lacking the chitin-responsive secreted class III peroxidase (Prx34), suggesting the requirement of peroxidase for the production of ROS (Lehtonen et al. 2012).

## 5.3 ROS-Scavenging Systems May Be Involved in Fine-Tuning Accumulation of ROS

Various ROS-scavenging systems, including catalases, ascorbate peroxidases, glutathione, superoxide dismutases are involved in increases in ROS in the plant cell (Mittler et al. 2004) and in activation of plant defense responses (Mittler et al. 1999; Klessig et al. 2000). It is widely reported that inhibition of catalase leads to accumulation of $H_2O_2$ (Takahashi et al. 1997). Salicylic acid (SA), which inhibits catalase, increases accumulation of $H_2O_2$ in elicited cells (Delaney et al. 1994; Willekens et al. 1994). Xanthine oxidase and peroxidase also reduce the level of catalase and hence increase the production of $H_2O_2$ (Milosevic and Slusarenko 1996). Compartmentalization of both ROS production and activation of ROS-scavenging systems contribute to fine-tuning of ROS levels and their signaling properties (Torres et al. 2006).

## 5.4 Site of Production of ROS

Production of ROS induced by various signals has been detected in the apoplast and also within cells. Plasma membranes and organelles, such as mitochondria, peroxisomes, and chloroplasts have been shown to act as ROS generators (Grant and Loake 2000; Asada 2006; Torres et al. 2006; Astamker et al. 2007). It has been shown that isolated nuclei can also generate $H_2O_2$ in response to calcium addition (Astamker et al. 2007). ROS accumulated within the tobacco cells more rapidly than the response outside the cell (Astamker et al. 2007). Sang et al. (2012) showed that the PAMP harpin activated NADP oxidase located in the plasma membrane and induced generation of ROS in the apoplast. $H_2O_2$ was generated in apoplasts in a

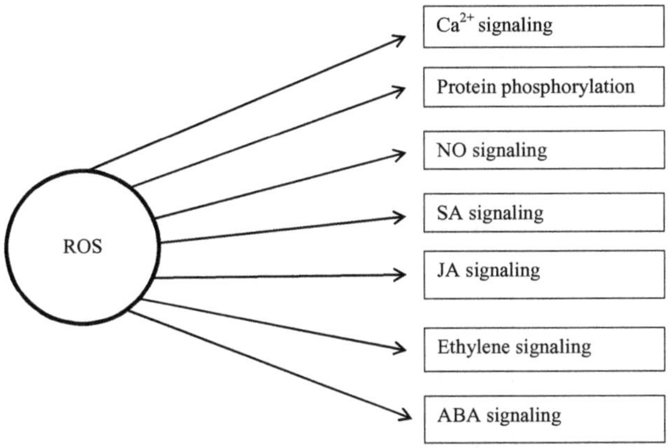


**Fig. 5.3** ROS-triggered immune response signaling pathways

NADPH oxidase-dependent manner but accumulated to a greater extent in the cytoplasm than in the apoplast. Inhibiting apoplastic $H_2O_2$ generation abrogated both cytoplasmic $H_2O_2$ accumulation and plant resistance to bacterial pathogens (Sang et al. 2012). These results suggest that the apoplastic $H_2O_2$ is translocated to cytoplasm for participation in the pathogen defense.

## 5.5 Biphasic ROS Production

Two phases of ROS induction by PAMPs/elicitors have been reported in plant cell suspension cultures. Very rapid responses (within minutes) have been termed phase 1 (Baker and Orlandi 1995) and have been shown to be specifically inhibited by DPI, calcium influx inhibitors, and kinase inhibitors (Baker and Orlandi 1995; Hammond-Kosac and Jones 1996). The phase I responses are not always correlated with defense responses. Later ROS production (many hours) is termed phase II and it appears to take part in defense signaling system (Allan and Fluhr 1997).

## 5.6 ROS Plays a Central Role in Triggering Immune Responses

ROS appears to interact with various defense signaling systems (Fig. 5.3). It plays a central role in launching the defense response (Vandenabeele et al. 2003). ROS induces $Ca^{2+}$ signaling system, reversible phosphorylation system, ubiquitin-proteasome signaling pathway, NO signaling system, salicylic acid signaling system, ethylene-mediated signaling system, and jasmonic acid – dependent signaling system (Fig. 5.3; Desikan et al. 2001; Vranová et al. 2002; Gupta and Luan 2003; Vandenabeele et al. 2003; Desikan et al. 2005; Fedoroff 2006; Hancock et al. 2006; Torres et al. 2006).

## 5.7 Interplay Between ROS and $Ca^{2+}$ Signaling System

Increase in $Ca^{2+}$ influx is known to trigger ROS generation (Desikan et al. 2001; Choi et al. 2009). The $Ca^{2+}$ sensor protein calmodulin (CaM) has been suggested to be involved in $Ca^{2+}$-triggered ROS generation. An increase in $Ca^{2+}$ amount in cytosol triggered by PAMP elicitor stimuli is perceived by the calmodulin. The calmodulin gene is strongly induced by the $Ca^{2+}$ influx in *Arabidopsis* (Desikan et al. 2001). The $Ca^{2+}$/CaM complex regulates NAD kinase, which generates NADPH for NADPH oxidase activity (Harding et al. 1997). Increase in NADPH oxidase activity results in generation of ROS (Choi et al. 2009; Pitzschke and Hirt 2006, 2009; Pitzschke et al. 2009a, b; Mazars et al. 2010).

While $Ca^{2+}$ influx –dependent activation of calmodulin gene is required for ROS production, ROS also triggers $Ca^{2+}$ influx (Levine et al. 1996). Intracellular $Ca^{2+}$ concentrations increase in response to oxidative burst (Price et al. 1994). A calmodulin gene was strongly induced by $H_2O_2$ in *Arabidopsis* (Desikan et al. 2001), suggesting that ROS is involved in triggering $Ca^{2+}$ signaling system. $H_2O_2$ has been shown to trigger calcium influx in tobacco (Kawano and Muto 2000).

It has been suggested that ROS might have regulated $Ca^{2+}$ influx through plasma membrane transport proteins (Laohavisit et al. 2010, 2012). The PAMP-induced $H_2O_2$ may trigger $[Ca^{2+}]_{cyt}$ increase, probably through the activation of $H_2O_2$ – sensitive $Ca^{2+}$ channels located in the plasma membrane (Lecourieux et al. 2002). The ROS activates hyperpolarization- activated $Ca^{2+}$ influx current (Pei et al. 2000; Foreman et al. 2003).

The NADPH oxidase-derived ROS stimulates a $Ca^{2+}$ influx into the cytoplasm (Takeda et al. 2008). $H_2O_2$ and $OH^{\circ}$ may serve as distinct signals in the regulation of calcium influx, due to the existence of calcium channels that are distinctively sensitive to the generated $H_2O_2$ (Demidchik et al. 2007). The rise in $Ca^{2+}$ level in turn activates NADPH oxidase to produce ROS (Takeda et al. 2008), suggesting a positive feedback regulation of $Ca^{2+}$ influx – ROS signaling system. These results suggest that a significant amount of cross-talk occurs between ROS and calcium signaling systems.

ROS may also act as a second messenger activating $Ca^{2+}$ signaling (Kwak et al. 2003). Abscisic acid (ABA) promotes ROS production, activates plasma membrane $Ca^{2+}$-permeable channels, and triggers cytosolic $Ca^{2+}$ increases in *Arabidopsis thaliana*. Disruption of NADPH oxidase catalytic subunits genes, *AtrbohD* and *AtrbohF*, has been shown to impair ABA signaling, ABA promotion of ROS production, ABA-induced cytosolic $Ca^{2+}$ increases and ABA-activation of plasma membrane $Ca^{2+}$-permeable channels in guard cells (Kwak et al. 2003). Application of $H_2O_2$ rescued $Ca^{2+}$ channel activation in the *atrbohD/F* mutants, suggesting that the ROS acts as a second messenger activating $Ca^{2+}$ signaling (Kwak et al. 2003).

## 5.8 Interplay Between ROS and NO Signaling Systems

$H_2O_2$ production has been shown to be required for NO synthesis. NO synthesis is severely reduced in the NADPH oxidase *Arabidopsis* double mutant *atrbohD atrbohF*, suggesting that endogenous $H_2O_2$ production is required for NO synthesis (Bright et al. 2006). Nitric oxide is often produced at the same time and in the same locations in plants as ROS (Neill et al. 2003). NO may also react with thiol groups on proteins in a process known as S-nitrosylation, to produce a –S-NO group (Hancock et al. 2006). NO is known to modify the same type of proteins which are modified by $H_2O_2$ (Lindermayr et al. 2005). These results suggest that there may be a competition between $H_2O_2$ and NO at the level of thiol modification which may determine the exact signaling processes that ensue (Hancock et al. 2006). It has also been reported that ROS activates the mitogen-activated protein kinase MPK6. The ROS-activated MAPK modulates nitric oxide biosynthesis in *Arabidopsis* (Wang et al. 2010).

## 5.9 Interplay Between ROS and MAPK Signaling Systems

$H_2O_2$ activates MAP kinases that modulate gene expression and transduce cellular responses to extracellular stimuli (Desikan et al. 1999; Kovtun et al. 2000; Samuel et al. 2000; Pitzschke and Hirt 2006, 2009; Pitzschke et al. 2009a, b). Activation of MAP kinases by ROS has been reported in *Arabidopsis* (Grant et al. 2000a; Kovtun et al. 2000; Desikan et al. 2001), tobacco (Samuel and Ellis 2002), and maize (Zhang et al. 2006). The MAPKKKs ANP1 and OMTK1 have been shown to be activated by $H_2O_2$ (Kovtun et al. 2000; Nakagami et al. 2004).

## 5.10 Interplay Between ROS and Salicylic Acid Signaling Systems

$H_2O_2$ stimulates salicylic acid (SA) biosynthesis in tobacco (León et al. 1995). SA is synthesized in tobacco leaves from benzoic acid (BA) after elicitation (Yalpani et al. 1993). Elicitors trigger the oxidative burst, which results in production of $H_2O_2$. $H_2O_2$ causes an intracellular accumulation of BA. The conversion of BA to SA is catalyzed by benzoic acid 2-hydroxylase (BA2H), an inducible enzyme that is synthesized de novo in response to increased BA level (León et al. 1993). BA2H is a soluble Cyt P-450 monoxygenase that uses molecular oxygen for the 2-hydroxylation of benzoic acid (León et al. 1995). The increased 2-hydroxylation activity may be due to the additional oxygen arising from the $H_2O_2$ degraded by catalase (León et al. 1995). The catalase-mediated release of molecular oxygen from peroxide may lead to the activation of BA2H, resulting in enhanced accumulation of SA (Fig. 5.4; León et al. 1995).

An UDP-glucose:SA:glucosyltransferase that converts SA to SA β-glucoside in tobacco is up-regulated after 30 min of exposure to external stimulus and it suggests the accumulation of SA in elicited tobacco plants at a transcriptional level

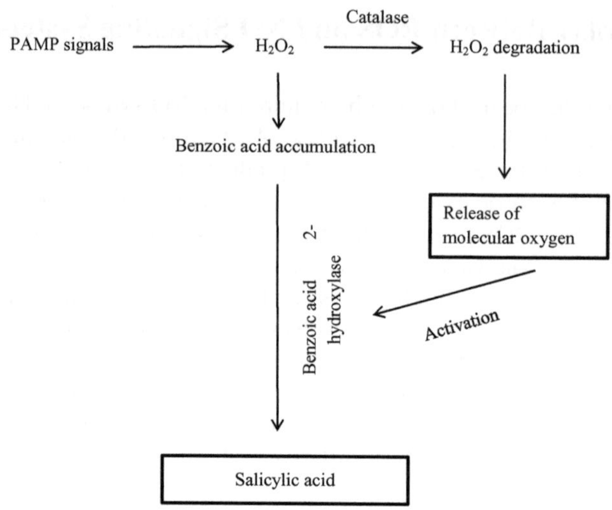


**Fig. 5.4** ROS-mediated salicylic acid biosynthesis

(Vandenabeele et al. 2003). SA is known to bind catalase (Chen et al. 1993; Conrath et al. 1995). This binding inhibits the activity of the enzyme in vitro, and hence it is suggested that the inactivation of catalase by SA may increase the levels of $H_2O_2$ (León et al. 1995; Torres et al. 2006). Several effector proteins involved in SA-induced defense responses have been identified. The first protein identified as a salicylic acid-binding protein is the cytosolic (peroxisomal) tobacco catalase (CAT) that reversibly binds SA (Chen et al. 1993; Conrath et al. 1995). SA inhibits CAT's $H_2O_2$-degrading activity (Durner and Klessig, 1996; Wendehenne et al. 1998). SA-mediated inhibition of CAT may generate $H_2O_2$, which may activate the ROS signaling system inducing expression of defense genes (Chen et al. 1993). ROS signaling system may act both in upstream and downstream of SA signaling (Slaymaker et al. 2002).

Thus $H_2O_2$ may increase accumulation of SA and SA may increase accumulation of $H_2O_2$. ROS metabolism can affect the function of NPR1 by controlling NPR1 redox state (Mou et al. 2003). NPR1 is an important regulator of responses downstream of SA (Mou et al. 2003; Zhang et al. 2003). It has also been demonstrated that calcium-dependent generation of ROS subsequently induces the production of salicylic acid and PR genes expression (Chen et al. 2011).

## 5.11 Interplay Between ROS and Ethylene Signaling Systems

ROS may function in ethylene signaling system. An increase in $H_2O_2$ was able to trigger transcriptional changes of genes involved in the biosynthesis of ethylene. S-adenosyl-L-methionine synthetase is the first enzyme in the ethylene biosynthesis pathway and 1-aminocyclopropane-1-carboxylate (ACC) oxidase is the final stage enzyme involved in conversion of ACC to ethylene. Transcript tags coding for both the enzymes were up-regulated within 2 h of increase in $H_2O_2$, followed by an increase in ethylene-responsive proteins, such as EREBP/AP2 domain proteins (Vandenabeele et al. 2003).

Ethylene receptor ETR1 (for ethylene response 1) can function as an ROS sensor (Desikan et al. 2005). The ethylene receptor ETR1 involved in ethylene perception and signaling has been shown to mediate $H_2O_2$ signaling in *Arabidopsis* (Desikan et al. 2005). Mutation of a Cys residue in the N-terminal region of ETR1 disrupts $H_2O_2$ signaling in plants (Desikan et al. 2005). Four ethylene-responsive element-binding proteins (EREBPs), the ethylene-responsive transcription factor (ERF1), and a CEO1-like protein, which is a potential cofactor of EREBP transcription factors, were found to be up-regulated by $H_2O_2$ in tobacco (Vandenabeele et al. 2003).

Ethylene receptor ETR1 has been shown to mediate $H_2O_2$ signaling in *Arabidopsis* (Desikan et al. 2005). ETR1 functions as an ROS sensor. Mutation of a Cys residue in the N-terminal region of ETR1 disrupts $H_2O_2$ signaling in plants (Desikan et al. 2005). The transcription factors induced by $H_2O_2$ include ethylene-responsive element binding protein (EREBP) in *Arabidopsis* (Desikan et al. 2001). $H_2O_2$ up-regulates four ethylene-responsive element-binding proteins (EREBPs), the ethylene-responsive transcription factor (ERF1), and a CEO1-like protein, which is a potential cofactor of EREBP transcription factors in tobacco (Vandenabeele et al. 2003).

## 5.12 Interplay Between ROS and Jasmonate Signaling Systems

A very rapid and sustained up-regulation of transcript tags that are involved in the production of JA signals was observed in ROS-elicited tobacco plants (Vandenabeele et al. 2003). Lipoxygenase is a key enzyme in JA biosynthesis. The gene encoding lipoxygenase (*Lox1*) is induced by ROS (Vranová et al. 2002). The genes encoding lipase, lipoxygenase, 12-oxophytodienoate reductase (12-ODPR), and divinyl ether synthase are activated in tobacco leaves accumulating ROS (Vandenabeele et al. 2003). These results suggest that ROS is involved in induction of JA biosynthesis.

## 5.13 Interplay Between ROS and Abscisic Acid (ABA) Signaling Systems

ABA has been shown to activate ROS signaling system (Hung et al. 2008; Wang and Song 2008; Zhang et al. 2011). ABA is shown to induce the production of ROS (Pei et al. 2000; Zhang et al. 2001; Sakamoto et al. 2008). $H_2O_2$ is synthesized in response to exogenous ABA (Guan et al. 2000; Pei et al. 2000). NADPH oxidase *AtrbohD* and *AtrbohF* genes function in ROS-dependent ABA signaling in *Arabidopsis*

(Kwak et al. 2003; Wang et al. 2006). Calcium-permeable channels are activated by ROS and ABA signaling network is involved in the ROS-induced activation of calcium-permeable channels (Pei et al. 2000; Murata et al. 2001; Trouverie et al. 2008; Zhang et al. 2011). Activation of these channels requires the presence of NAD(P)H in the cytosol. Type 2C protein phosphatases (PP2Cs) also play a role in the ABA-induced ROS-dependent activation of calcium channels (Murata et al. 2001). Anion channel activation and proton pumping inhibition involved in the plasma membrane depolarization induced by ABA in *A. thaliana* suspension cells were shown to be ROS dependent (Trouverie et al. 2008). $H_2O_2$ mediates, at least in part, ABA responses including stomatal closure and gene expression (Guan et al. 2000; Pei et al. 2000).

## 5.14 ROS Activates Phosphorylation/Dephosphorylation Systems

Protein phosphorylation plays a central role in plant immune response signaling (Peck et al. 2001; Nürnberger et al. 2004; Zipfel et al. 2004; Benschop et al. 2007; Tena et al. 2011). Protein kinases and protein phosphatases and their corresponding protein substrates play key roles in signal transduction. The proteins that make up the signal transduction pathway are present in the cell prior to the perception of elicitor (Benschop et al. 2007). These proteins are activated by post-translational modifications and conformational changes. The most widely recognized post-translational modification involved in signal transduction is protein phosphorylation (Benschop et al. 2007). Post-translational protein modification by phosphorylation/dephosphorylation is an important component in defense signaling system (Lecourieux-Ouaked et al. 2000; Peck et al. 2001; Romeis et al. 2001; Lecourieux et al. 2002, 2006).

ROS plays important role in triggering phosphorylation of proteins involved in defense signaling system. The genes encoding serine/threonine/tyrosine kinases involved in protein phosphorylation are induced by $H_2O_2$ in *Arabidopsis* (Desikan et al. 2001) and *Nicotiana tabacum* (Vranová et al. 2002; Vandenabeele et al. 2003). Calcium-dependent protein kinase is induced by ROS in *Arabidopsis* (Desikan et al. 2001). Phosphorylation induced by MAP kinases has been shown to be involved in ROS signaling (Ren et al. 2002; Yoshioka et al. 2003; Nakagami et al. 2006; Zhang et al. 2006; Xing et al. 2007, 2008). MAP kinase cascades have been shown to be involved in phosphorylation of transcription factors (Asai et al. 2002; Nakagami et al. 2006). Another kinase involved in $H_2O_2$ signaling is oxidative signal-inducible 1 (OXI1) (Rentel et al. 2004). One of the genes activated by $H_2O_2$ in *Arabidopsis* is that encoding a protein Tyr phosphatase (Desikan et al. 2001). Protein Tyr phosphatases are important signaling enzymes that regulate protein phosphorylation events (Fauman and Saper 1996). The *Arabidopsis* protein phosphatase 2C enzymes ABI1 and ABI2 and the protein tyrosine phosphatase AtPTP1 are inactivated by ROS (Meinhard and Grill 2001; Gupta and Luan 2003). These studies suggest that ROS is involved in modulating the phosphorylation process in innate immune responses.

## 5.15 Function of ROS in Ubiquitin-Proteasome System

Proteasomes are large protein complexes located in the nucleus and the cytoplasm (Peters et al. 1994). The main function of the proteasome is to degrade unneeded or damaged proteins by proteolysis. Proteasomes regulate the concentration of particular proteins and degrade misfolded proteins. Proteins are tagged for degradation by a small protein called ubiquitin (Pickart 2001). Ubiquitin- and proteasome-mediated degradation of proteins plays an important role in plant defense signaling system (Dreher and Callis 2007; Goritsching et al. 2007; Yao and Ndoja 2012). $H_2O_2$ triggers the up-regulation of ubiquitin precursor proteins, ubiquitin-conjugating enzymes (E2), and ubiquitin-protein ligases (E3) in tobacco (Vandenabeele et al. 2003). An Arabidopsis MAPKKK, MEKK1 has been shown to act downstream of $H_2O_2$. MEKK1 kinase activity and protein stability was regulated by $H_2O_2$ in a proteasome-dependent manner (Nakagami et al. 2006).

## 5.16 ROS May Regulate Expression of Transcription Factors

Expression of several transcription factors has been shown to be regulated by $H_2O_2$ (Desikan et al. 2001; Vandenabeele et al. 2003). These transcription factors may be direct targets for redox modification by $H_2O_2$ (Hancock et al. 2006). The transcription factors induced by $H_2O_2$ include ethylene-responsive element binding protein (EREBP), MYB-related transcription factor, and zinc finger transcription factor in *Arabidopsis* (Desikan et al. 2001) and WRKY 11 in tobacco (Vranová et al. 2002). Vandenabeele et al. (2003) identified four main classes of transcription factors which are regulating downstream gene expression in ROS signaling system in tobacco: MYB family, WRKY, AP2, and SCARECROW. The identified MYB transcription factor was homologous with the N-terminal MYB domain of a JA-dependent transcription factor. The induced WRKY transcription factors contained a redox-sensitive zinc-finger DNA-binding domain. Four EREBP/AP2 and ethylene-responsive transcription factor (ERF1) were also up-regulated together with a CEO1-like protein. CEO1 is a potential cofactor of EREBP transcription factors (Vandenabeele et al. 2003). Activation of these transcription factors may result in activation of transcription of several defense genes.

## 5.17 Redox Signaling System

The ROS signal functions are manifested as a consequence of their ability to act as mobile carriers of an unpaired electron (Forman et al. 2004). In redox signaling, the reaction of the ROS with the target protein is reminiscent of on-off signaling associated with phosphorylation/dephosphorylation events. Redox signaling occurs when at least one step in a signaling event involves one of its components being specifically modified by a reactive oxygen species through a reaction that is chemically reversible under physiological conditions and/or enzymatically catalyzed (Forman et al. 2004). Signaling through the redox active molecule $H_2O_2$ is important in inducing plant defense responses (Desikan et al. 2005).

For proteins to perceive the presence of ROS and to act as signal intermediaries there should be a direct chemical interaction which leads to signal propagation (Hancock et al. 2006). It is suggested that the ROS sensors must have some specific characteristic that enables them to propagate this signal (Hancock et al. 2006). Modification of thiol residues in the proteins may be the key component in the ROS signaling system (Cooper et al. 2002; Vranová et al. 2002; Foyer and Noctor 2005). If there are two cysteine residues in the target protein involved in the reaction, a disulphide bridge (S-S) may be formed. However, if there is only one cysteine, the –SH group may be oxidized to varying degrees. The thiol group may be oxidized to sulphenic acid (–SOH), and this may be further oxidized to sulphinic acid ($-SO_2H$) or sulphonic acid ($-SO_3H$). Since, any oxidation of the thiol is dependent on its mid-point redox potential and its availability to the oxidant, only a low proportion of the –SH groups within any protein will be able to modified in these ways. Thiols in different proteins have different mid-point potentials, and hence the proteins will be differentially controlled by fluctuations in the intracellular redox state. Some proteins may be regulated earlier, or later than others as the redox state becomes more oxidized (Hancock et al. 2006).

These oxidation states of the –SH group within cysteine may be restored by re-reduction. Thioredoxins (Schürmann and Jacquot 2000) and glutaredoxins (Lemaire 2004) may act as protein disulphide reductases as well as re-oxidizing –SOH groups (Collin et al. 2004). Sulphinic acid groups may be reduced back to the sulphenic group by sulphiredoxins. The sulphenic acid group created may then be reduced further by thioredoxins or glutaredoxins to regenerate the thiol, –SH (Hancock et al. 2006). These observations suggest that there are redox groups within proteins that can potentially toggle between oxidation and reduction states in a rapid and ROS dose-dependent manner, and in doing so the structures of the proteins will be altered and such proteins may partake in $H_2O_2$-mediated signaling (Hancock et al. 2006).

It has been suggested that It has been suggested that $H_2O_2$ signaling can activate responses such as gene expression and reversible protein phosphorylation through oxidative modification of reactive Cys residues within proteins (Danon 2002). It has been suggested that $H_2O_2$ signaling can activate responses such as gene expression and reversible protein phosphorylation through oxidative modification of reactive Cys residues within proteins (Danon 2002). It has been suggested that $H_2O_2$ signaling can activate responses such as gene expression and reversible protein phosphorylation through oxidative modification of reactive Cys residues within proteins (Danon 2002). Phosphatases contain readily oxidizable active site cysteine residues (Stone 2004). Since phosphatases are involved in regulation of protein kinases, redox regulation of phosphatase activity can, in turn, regulate the activity of its target protein kinases (Tonks 2005). Some protein kinases are directly redox regulated by thioredoxins and peroxiredoxins (Veal et al. 2004; Fedoroff 2006).

NPR1 (for Nonexpressor of Pathogenesis Related genes 1), the transcriptional regulatory cofactor, is activated by redox signaling in plants (Fedoroff 2006). In unstressed cells, NPR1 is maintained in the cytoplasm in a large complex comprising disulfide-bonded intermolecular oligomers (Mou et al. 2003). NPR1 is activated when SA accumulates in cells in response to stress signals. Upon activation, the intermolecular disulfide bonds are reduced, releasing monomeric NPR1. The released NPR1 then moves into the nucleus to interact with TGA transcription factors and activate defense gene expression (Fedoroff 2006). Mutations of either Cys82 or Cys216 render the protein both constitutively monomeric and nuclear and constitutively activate expression of defense genes (Mou et al. 2003). It has also been shown that an intramolecular disulfide bond between Cys260 and Cys266 in TGA1 prevents interaction with NPR1 (Després et al. 2003). Reduction of the disulfide bond permits TGA1 to interact with NPR1, which in turn stimulates its DNA-binding activity (Fedoroff 2006). SA promotes the reduction of NPR1 and TGA1 (Fobert and Després 2005), probably by producing $H_2O_2$ (Torres et al. 2006). Along with increase in $H_2O_2$, the transcript levels of genes encoding antioxidant proteins, such as peroxidase and glutathione-S-transferases have increased (Joo et al. 2005). The build-up of antioxidants that enhance the general cellular reducing capacity would have reduced the NPR1 complex (Cumming et al. 2004). Thus redox signaling may be involved in activation of NPR1.

The activities of NPR1 and of the TGA factors TGA1 and TGA4 have been shown to be modulated by SA-induced oxidoreduction modifications of key cysteine residues (Fobert and Després 2005). Reduction of two conserved cysteines in NPR1 leads to its monomerization and nuclear localization, which is required for the activation of pathogenesis-related (PR) genes. Reduction of conserved cysteines in TGA1 and TGA4 enables their interaction with NPR1, which acts as a redox-sensitive cofactor in stimulating TGA1 DNA-binding activity (Fobert and Després 2005).

Metallothioneins are small cysteine-rich proteins involved in ROS scavenging and metallothionein is down-regulated by OsRac1, a G-protein in rice (Wong et al. 2004). Down regulation of ROS scavengers may play an important role in redox-mediated defense signaling (Wong et al. 2004). The induced redox regulators detected in rice cells transformed with *OsRac1* included glyceraldehyde-3-P dehydrogenase, NADPH-thioredoxin reductase, ferredoxin-NADPH reductase, NADPH dependent oxidoreductase, quinine oxidoreductase, and glutathione-S-transferase (GST1) (Fujiwara et al. 2006). These results suggest that G-proteins may be involved in redox signaling system.

Thioredoxins are ubiquitous disulfide reductases that regulate the redox status of target proteins. They may act as regulators of scavenging mechanisms and as components of signaling pathways in the plant antioxidant network (Dos Santos and Rey 2006). A thioredoxin has been shown to interact with the disease resistance protein Cf-9 and modulate Cf-9 dependent signaling (Fobert and Després 2005).

Redox enzymes have been shown to be involved in lignin biosynthesis (Önnerud et al. 2002) and lignification is one of the key defense responses in plants (Vidhyasekaran 2007). Lignin may be formed by a radical polymerization initiated by redox enzymes. In this system, manganese oxalate works as a diffusible redox shuttle, first being oxidized from Mn(II) to Mn(III) by a peroxidase and then being reduced to Mn(II) by a simultaneous oxidation of the lignin monomers to radicals that form covalent linkages of the lignin. It suggests involvement of a redox shuttle/peroxidase system in lignin biosynthesis through activation of polymerization of monolignols (Önnerud et al. 2002).

## 5.18 ROS Signaling System May Activate Transcription of Defense Genes

ROS signaling system and cognate redox signaling have been shown to be involved in activation of several defense genes. Redox control of activation of PR genes has been reported in plants (Fobert and Després 2005). ROS stress induced the PR protein PRB1-b, and the antimicrobial protein chitinase class 4 in tobacco (Vranová et al. 2002). Thaumatin-like protein (PR-5 protein) is induced by $H_2O_2$ (Desikan et al. 2001). $H_2O_2$ induces PR-1 protein accumulation in tobacco (Neill et al. 2002). The ROS stress induced various enzymes involved in phytoalexins synthesis, including 5-epi-aristoiochene synthase and vestispiradiene synthase in tobacco (Vranová et al. 2002). $H_2O_2$ are capable of inducing expression of phenylalanine ammonia-lyase (Neill et al. 2002). Lignin biosynthesis is also activated by ROS (Vranová et al. 2002) and redox signaling system (Önnerud et al. 2002).

## 5.19 Pathogens May Cause Disease by Interfering with ROS Signaling System in Host Plants

ROS signaling system is involved in host plant defense. In the genome of the maize smut pathogen *Ustilago maydis*, an ortholog of *YAP1* (for *Yeast AP-1-like*) from *Saccharomyces cerevisae* has been identified (Molina and Kahmann 2007). The gene is involved in degradation of $H_2O_2$. The Yap1-regulated genes include peroxidase genes and peroxidases are known to degrade $H_2O_2$. *Yap1* gene was found to be essential for virulence of *U. maydis* and deletion mutants of this gene were attenuated in virulence. These results suggest that the biotrophic pathogen *U. maydis* causes the disease by interfering with the $H_2O_2$ signaling pathway (Molina and Kahmann 2007).

Another biotrophic fungal pathogen, *Blumeria graminis* f. sp. *hordei*, elicits a burst of $H_2O_2$ in its host barley at sites of germ tube invasion. The fungus produces catalase during the infection process. The fungal *catB* gene encoding catalase has been characterized. Enhanced numbers of *catB* transcripts were detected at mature primary germ tube and appressorium germ tube stages in a susceptible host. Areas of $H_2O_2$ clearing were observed at sites of fungal invasion (Zhang et al. 2004).

Collectively, these results suggest that the catalase activity of *B. graminis* f. sp. *hordei* may contribute to the fungal pathogenicity, probably by interfering with ROS signaling system (Zhang et al. 2004).

## References


Agrawal GK, Iwahashi H, Rakwal R (2003) Small GTPase 'Rop': molecular switch for plant defense responses. FEBS Lett 546:173–180

Allan AG, Fluhr R (1997) Two distinct sources of elicited reactive oxygen species in tobacco epidermal cells. Plant Cell 9:1559–1572

Apel K, Hirt H (2004) Reactive oxygen species, metabolism, oxidative stress, and signal transduction. Annu Rev Plant Biol 55:373–399

Arnott T, Murphy TM (1991) A comparison of the effects of a fungal elicitor and ultraviolet radiation on ion transport and hydrogen peroxide synthesis by rose cells. Environ Exp Bot 31:209–216

Asada K (2006) Production and scavenging of reactive oxygen species in chloroplasts and their functions. Plant Physiol 141:391–396

Asai T, Tena G, Plotnikova J, Willmann MR, Chiu WL, Gomez-Gomez I, Boller T, Ausubel FM, Sheen J (2002) MAP kinase signaling cascade in *Arabidopsis* innate immunity. Nature 415:977–983

Asai S, Ohta K, Yoshioka H (2008) MAPK signaling regulates nitric oxide and NADPH oxidase-dependent oxidative bursts in *Nicotiana benthamiana*. Plant Cell 20:1390–1406

Astamker C, Kiss V, Sagi M, Davydov O, Fluhr R (2007) Diverse subcellular locations of cryptogein-induced reactive oxygen species production in tobacco Bright Yellow-2 cells. Plant Physiol 143:1817–1826

Baker CJ, Orlandi EW (1995) Active oxygen in plant pathogenesis. Annu Rev Phytopathol 33:299–321

Benschop JJ, Mohammed S, O'Flaherty M, Heck AJR, Slijper M, Menke FLH (2007) Quantitative phosphoproteomics of early elicitor signaling in *Arabidopsis*. Mol Cell Proteomics 6:1198–1214

Blumwald E, Aharon GS, Lam BC-H (1998) Early signal transduction pathways in plant-pathogen interactions. Trends Plant Sci 3:342–346

Bolwell GP, Butt VS, Davies DR, Zimmerlin A (1995) The origin of the oxidative burst in plants. Free Radic Res 23:517–532

Bolwell GP, Davies DR, Gerrish C, Auh C-K, Murphy TM (1998) Comparative biochemistry of the oxidative burst produced by rose and French bean cells reveals two distinct mechanisms. Plant Physiol 116:1379–1385

Bolwell GP, Bindschedler LV, Blee KA, Butt VS, Davies DR, Gardner SL, Gerrish C, Minibayeva F (2002) The apoplastic oxidative burst in response to biotic stress in plants: a tree component system. J Exp Bot 53:1367–1376

Bright J, Desikan R, Hancock JT, Weir IS, Neill SJ (2006) ABA-induced NO generation and stomatal closure in *Arabidopsis* are dependent on $H_2O_2$ synthesis. Plant J 45:113–122

Chen Z, Silva H, Klessig DF (1993) Active oxygen species in the induction of plant systemic acquired resistance by salicylic acid. Science 262:1883–1886

Chen J, Zhang Y, Wang C, Lu W, Jin JB, Hua X (2011) Proline induces calcium-mediated oxidative burst and salicylic acid signaling. Amino Acids 40:1473–1484

Choi HW, Lee DH, Hwang BK (2009) The pepper calmodulin gene CaCaM1 is involved in reactive oxygen species and nitric oxide generation required for cell death and the defense response. Mol Plant Microbe Interact 22:1389–1400

Collin V, Lamkemeyer P, Miginiac-Maslow M, Hirasawa M, Knaff DB, Dietz K-J, Issakidis-Bourguet E (2004) Characterization of plasticidal thioredoxins from *Arabidopsis* belonging to the new γ-type. Plant Physiol 136:4088–4095

Conrath U, Chen Z, Ricigliano JR, Klessig DF (1995) Two inducers of plant defense responses, 2,6-dichloroisonicotinic acid and salicylic acid, inhibit catalase activity in tobacco. Proc Natl Acad Sci U S A 92:7143–7147

Cooper CE, Patel RP, Brookes PS, Darley-Usmar VM (2002) Nanotransducers in cellular redox signaling: modification of thiols by reactive oxygen and nitrogen species. Trends Biochem Sci 27:489–492

Cumming RC, Andon NL, Haynes PA, Park M, Fischer WH, Schubert D (2004) Protein disulfide bond formation in the cytoplasm during oxidative stress. J Biol Chem 279:21749–21758

Danon A (2002) Redox reactions of regulatory proteins: do kinetics promote specificity? Trends Biochem Sci 27:197–203

Daudi A, Cheng Z, O'Brein JA, Mammarella N, Khan S, Ausubel FM, Bolwell GP (2012) The apoplastic oxidative burst peroxidase in *Arabidopsis* is a major component of pattern-triggered immunity. Plant Cell 24:275–287

Delaney TP, Uknes S, Vernooij B, Friedrich L, Weymann K, Negrotto D, Gaffney T, Gut-Rella M, Kessmann H, Ward E, Ryals J (1994) A central role of salicylic acid in plant disease resistance. Science 266:1247–1250

Demidchik V, Shabala SN, Davies JM (2007) Spatial variation in $H_2O_2$ response of *Arabidopsis thaliana* root epidermal $Ca^{2+}$ flux and plasma membrane $Ca^{2+}$ channels. Plant J 49:377–386

Desikan R, Clarke A, Hancock JT, Neill SJ (1999) $H_2O_2$ activates a MAP kinase-like enzyme in *Arabidopsis thaliana* suspension cultures. J Exp Bot 50:1863–1866

Desikan R, Hancock JT, Bright J, Harrison J, Weir I, Hooley R, Neill SJ (2005) A role for ETR1 in hydrogen peroxide signaling in stomatal guard cells. Plant Physiol 137:831–834

Desikan R, Mackerness S, Hancock JT, Neill SJ (2001) Regulation of the *Arabidopsis* transcriptome by oxidative stress. Plant Physiol 127:159–172

Després C, Chubak C, Rochon A, Clark R, Bethune T, Desveaux D, Fobert PR (2003) The Arabidopsis *NPR1* disease resistance protein is a novel cofactor that confers redox regulation of DNA binding activity to the basic domain/leucine zipper transcription factor TGA1. Plant Cell 15:2181–2191

Dos Santos CV, Rey P (2006) Plant thioredoxins are key actors in the oxidative stress response. Trends Plant Sci 11:329–334

Dreher K, Callis J (2007) Ubiquitin, hormones and biotic stress in plants. Ann Bot 99:787–822

Durner J, Klessig DF (1996) Salicylic acid is a modulator of tobacco and mammalian catalases. J Biol Chem 271:28492–28501

Ebel J, Bhagwat AA, Cosio EG, Feger M, Kissel U, Mithöfer A, Waldmüller T (1995) Elicitor-binding proteins and signal transduction in the activation of a phytoalexin defense response. Can J Bot 73:506–510

Elstner EF (1982) Oxygen activation and oxygen toxicity. Annu Rev Plant Physiol 33:73–96

Faize M, Faize L, Ishizaka M, Ishii H (2004) Expression of potential responses of Asian and European pears to infection with *Venturia nashicola*. Physiol Mol Plant Pathol 64:319–330

Fauman EB, Saper MA (1996) Structure and function of the protein tyrosine phosphatases. Trends Biochem Sci 21:413–417

Fedoroff N (2006) Redox regulatory mechanisms in cellular stress responses. Ann Bot (Lond) 98:289–300

Fobert PR, Després C (2005) Redox control of systemic acquired resistance. Curr Opin Plant Biol 8:378–382

Foreman J, Demidchik V, Bothwell JH, Mylona P, Miedema H, Torres MA, Linstead P, Costa S, Brownlee C, Jones JD, Davies JM, Dolan I (2003) Reactive oxygen species produced by NADPH oxidase regulate plant cell growth. Nature 422:442–446

Forman HJ, Fukuto JM, Torres M (2004) Redox signaling: thiol chemistry defines which reactive oxygen species can act as second messengers. Am J Physiol Cell Physiol 287:C246–C256

Foyer CH, Noctor G (2005) Redox homeostasis and antioxidant signaling: a metabolic interface between stress perception and physiological responses. Plant Cell 17:1866–1875

Fujiwara M, Umemura K, Kawasaki T, Shimamoto K (2006) Proteomics of Rac GTPase signaling reveals its predominant role in elicitor-induced defense response of cultured rice cells. Plant Physiol 140:734–745

Goritsching S, Zhang Y, Li X (2007) The ubiquitin pathway is required for innate immunity in *Arabidopsis*. Plant J 49:540–551

Grant JJ, Loake GJ (2000) Role of reactive oxygen intermediates and cognate redox signaling in disease resistance. Plant Physiol 8:629–643

Grant JJ, Yun B-W, Loake GJ (2000a) Oxidative burst and cognate redox signaling reported by luciferase imaging: identification of a signal network that functions independently of ethylene, SA, and Me-JA but is dependent on MAPKK activity. Plant J 24:569–582

Grant M, Brown I, Adams S, Knight M, Ainslie A, Mansfield J (2000b) The *RPM1* plant disease resistance gene facilitates a rapid and sustained increase in cytosolic calcium that is necessary for the oxidative burst and hypersensitive cell death. Plant J 23:441–450

Guan LM, Zhao J, Scandalios JG (2000) *Cis*-elements and *trans*-factors that regulate expression of the maize *Cat1* antioxidant gene response to ABA and osmotic stress: $H_2O_2$ is the likely intermediary signaling molecule for the response. Plant J 22:87–95

Gupta R, Luan S (2003) Redox control of protein tyrosine phosphatases and mitogen-activated protein kinases in plants. Plant Physiol 132:1149–1152

Haga M, Kohno Y, Iwata M, Sekizawa Y (1995) Superoxide anion generation in rice blade protoplasts with the blast fungus proteoglucomannan elicitor as determined by CLA-phenyl luminescence and in suppression by treating the elicitor with α-D-mannosidase. Biosci Biotechnol Biochem 59:969–973

Halliwell B (1978) Superoxide-dependent formation of hydroxyl radicals in the presence of iron chelates. Is it a mechanism for hydroxyl radical production in biochemical systems? FEBS Lett 92:321–326

Halliwell B, Gutteridge JMC (1989) Free radicals in biology and medicine, 2nd edn. Clarendon Press, Oxford, p 346

Hammond-Kosac KE, Jones JDG (1996) Resistance gene-dependent plant defense responses. Plant Cell 8:1773–1791

Hancock J, Desikan R, Harrison J, Bright J, Hooley R, Neill S (2006) Doing the unexpected: proteins involved in hydrogen peroxide perception. J Exp Bot 57:1711–1718

Harding SH, Oh SH, Roberts DM (1997) Transgenic tobacco expressing a foreign calmodulin gene shows an enhanced production of active oxygen species. EMBO J 16:1137–1144

Hung KT, Cheng DG, Hsu YT, Kao CH (2008) Abscisic acid-induced hydrogen peroxide is required for anthocyanin accumulation in leaves of rice seedlings. J Plant Physiol 165:1280–1287

Ishikawa A (2009) The Arabidopsis G-protein beta-subunit is required for defense response against *Agrobacterium tumefaciens*. Biosci Biotechnol Biochem 73:47–52

Jabs T, Tschöpe M, Colling C, Hahlbrock K, Scheel D (1997) Elicitor-stimulated ion fluxes and $O_2^-$ from the oxidative burst are essential components in triggering defense gene activation and phytoalexin synthesis in parsley. Proc Natl Acad Sci U S A 94:4800–4805

Joo JH, Wang S, Chen JG, Jones AM, Fedoroff NV (2005) Different signaling and cell death roles of heterotrimeric G protein α and β subunits in the *Arabidopsis* oxidative stress response to ozone. Plant Cell 17:957–970

Kadota Y, Furuichi T, Ogasawara Y, Goh T, Higashi K, Muto S, Kuchitsu K (2004) Identification of putative voltage-dependent $Ca^{2+}$-permeable channels involved in cryptogein-induced $Ca^{2+}$ transients and defense responses in tobacco BY-2 cells. Biochem Biophys Res Commun 317:823–830

Kawano T, Muto S (2000) Mechanism of peroxidase actions for salicylic acid-induced generation of active oxygen species and an increase in cytosolic calcium in tobacco cell suspension culture. J Exp Bot 51:685–693

Kawasaki T, Henmi K, Ono E, Hatakeyama S, Iwano M, Satoh H, Shimamoto K (1999) The small GTP-binding protein rac is a regulator of cell death in plants. Proc Natl Acad Sci U S A 96:10922–10926

Keller T, Damude HG, Werner D, Doerner P, Dixon RA, Lamb CA (1998) A plant homolog of the neutrophil NADPH oxidase gp91phox subunit gene encodes a plasma membrane protein with $Ca^{2+}$ binding motifs. Plant Cell 10:255–266
Kiirika LM, Bergmann HF, Schikowsky C, Wimmer D, Korte J, Schmitz U, Niehaus K, Colditz F (2012) Silencing of the Rac1 GTPase *MtROP9* in *Medicago truncatula* stimulates early mycorrhizal and oomycete root colonizations but negatively affects Rhizobial infection. Plant Physiol 159:501–516
Klessig DF, Durner J, Noad R, Navarre DA, Wendehenne D, Kumar D, Zhou JM, Shah J, Zhang S, Kachroo P, Trifa Y, Pontier D, Lam E, Silva H (2000) Nitric oxide and salicylic acid signalling in plant defense. Proc Natl Acad Sci U S A 97:8849–8855
Kovtun Y, Chiu W-L, Tena G, Sheen J (2000) Functional analysis of oxidative stress-activated mitogen activated protein kinase cascade in plants. Proc Natl Acad Sci U S A 97:2940–2945
Kwak JM, Mori IC, Pei ZM, Leonhardt N, Torres MA, Dangl JL, Bloom RE, Bodde S, Jones JD, Schroeder JI (2003) NADPH oxidase *AtrbohD* and *AtrbohF* genes function in ROS-dependent ABA signaling in *Arabidopsis*. EMBO J 22:2623–2633
Laohavisit A, Brown AT, Cicuta P, Davies JM (2010) Annexins: components of the calcium and reactive oxygen signaling network. Plant Physiol 152:1824–1829
Laohavisit A, Shang Z, Rubio L, Cuin TA, Véry AA, Wang A, Mortimer JC, Macpherson N, Coxon KM, Battey NH, Brownlee C, Park OK, Sentenac H, Shabala S, Webb AA, Davies JM (2012) Arabidopsis annexin1 mediates the radical-activated plasma membrane $Ca^{2+}$- and $K^{+}$-permeable conductance in root cells. Plant Cell 24:1522–1533
Lecourieux – Ouaked D, Pugin A, Lebrun-Garcia A (2000) Phosphoproteins involved in the signal transduction of cryptogein, an elicitor of defense reactions in tobacco. Mol Plant Microbe Interact 13:821–829
Lecourieux D, Mazars C, Pauly N, Ranjeva R, Pugin A (2002) Analysis and effects of cytosolic free calcium elevations in response to elicitors in *Nicotiana plumbaginifolia* cells. Plant Cell 14:2627–2641
Lecourieux D, Ranjeva R, Pugin A (2006) Calcium in plant defence-signalling pathways. New Phytol 171:249–269
Lehtonen MT, Akita M, Frank W, Reski R, Valkonen JPT (2012) Involvement of a class III peroxidase and the mitochondrial protein TSPO in oxidative burst upon treatment of moss plants with a fungal elicitor. Mol Plant Microbe Interact 25:363–371
Lemaire SD (2004) The glutaredoxin family in oxygenic photosynthetic organisms. Photosynth Res 79:305–318
León J, Yalpani N, Raskin I, Lawton MA (1993) Induction of benzoic acid 2-hydroxylase in virus-inoculated tobacco. Plant Physiol 103:323–328
León J, Shulav V, Yalpani N, Lawton MA, Raskin J (1995) Benzoic acid 2-hydroxylase, a soluble oxygenase from tobacco, catalyzes salicylic acid biosynthesis. Proc Natl Acad Sci U S A 92:10413–10417
Levine A, Pennell RI, Alvarez ME, Palmer R, Lamb C (1996) Calcium-mediated apoptosis in a plant hypersensitive disease resistance response. Curr Biol 6:427–437
Lindermayr C, Saalbach G, Durner J (2005) Proteomic identification of *S*-nitrosylated proteins in *Arabidopsis*. Plant Physiol 137:921–930
Mazars C, Thuleau P, Lamotte O, Bourque S (2010) Cross-talk between ROS and calcium in regulation of nuclear activities. Mol Plant 3:706–718
Meinhard M, Grill E (2001) Hydrogen peroxide is a regulator of ABI1, a protein phosphatase 2C from *Arabidopsis*. FEBS Lett 508:443–444
Milosevic N, Slusarenko AJ (1996) Active oxygen metabolism and lignification in the hypersensitive response in bean. Physiol Mol Plant Pathol 49:143–158
Mittler R, Herr EH, Orvar BL, van Camp W, Willekens H, Inzé D, Ellis BE (1999) Transgenic tobacco plants with reduced capability to detoxify reactive oxygen intermediates are hyperresponsive to pathogen infection. Proc Natl Acad Sci U S A 96:14165–14170
Mittler R, Vanderauwera S, Gollery M, Van Breusegem F (2004) Reactive oxygen network of plants. Trends Plant Sci 9:490–498

Moeder W, Yoshioka K, Klessig DF (2005) Involvement of the small GTPase Rac in the defense responses of tobacco to pathogens. Mol Plant Microbe Interact 18:116–124

Molina L, Kahmann R (2007) A *Ustilago maydis* gene involved in $H_2O_2$ detoxification is required for virulence. Plant Cell 19:2293–2309

Montalbini P (1992) Changes in xanthine oxidase activity in bean leaves induced by *Uromyces phaseoli* infection. J Phytopathol 134:218–228

Morel J, Fromentin J, Blein J-P, Simon-Plas F, Elmayan T (2004) RAC regulation of NtrbohD, the oxidase responsible for the oxidative burst in elicited tobacco cell. Plant J 37:282–293

Mou Z, Fan W, Dong X (2003) Inducers of plant systemic acquired resistance regulate NPR1 function through redox changes. Cell 113:935–944

Murata Y, Pei ZM, Mori IC, Schroeder J (2001) Abscisic acid activation of plasma membrane $Ca^{2+}$ channels in guard cells requires cytosolic NAD(P)H and is differentially disrupted upstream and downstream of reactive oxygen species production in *abi1-1* and *abi2-1* protein phosphatase 2C mutants. Plant Cell 13:2513–2523

Nakagami H, Kiegerl S, Hirt H (2004) OMTK1, a novel MAPKKK, channels oxidative stress signaling through direct MAPK activation. J Biol Chem 279:26959–26966

Nakagami H, Soukupová H, Schikora A, Zársk V, Hirt H (2006) A mitogen-activated protein kinase kinase kinase mediates reactive oxygen species homeostasis in *Arabidopsis*. J Biol Chem 281:38697–38704

Neill SJ, Desikan R, Hancock JT (2002) Hydrogen peroxide signalling. Curr Opin Plant Biol 5:388–395

Neill SJ, Desikan R, Hancock JT (2003) Nitric oxide signalling in plants. New Phytol 159:11–35

Nürnberger T, Scheel D (2001) Signal transmission in the plant immune response. Trends Plant Sci 6:372–379

Nürnberger T, Brunner F, Kemmerling B, Piater L (2004) Innate immunity in plants and animals: striking similarities and obvious differences. Immunol Rev 198:249–266

O'Brein JA, Daudi A, Finch P, Butt VS, Whitelegge JP, Souda P, Ausubel FM, Bolwell GP (2012) A peroxidase-dependent apoplastic oxidative burst in cultured *Arabidopsis* cells functions in MAMP-elicited defense. Plant Physiol 158:2013–2027

Önnerud H, Zhang L, Gellerstedt G, Henriksson G (2002) Polymerization of monolignols by redox shuttle-mediated enzymatic oxidation: a new model in lignin biosynthesis. Plant Cell 14:1953–1962

Ono E, Wong HL, Kawasaki T, Hasegawa M, Kodama O, Shimamoto K (2001) Essential role of the small GTPase Rac in disease resistance of rice. Proc Natl Acad Sci U S A 98:759–764

Peck SC, Nühse TS, Hess D, Iglesias A, Meins F, Boller T (2001) Directed proteomics identifies a plant-specific protein rapidly phosphorylated in response to bacterial and fungal elicitors. Plant Cell 13:1467–1475

Pei Z-M, Murata Y, Benning G, Thomine S, Klüsener B, Allen GJ, Grill E, Schroeder JI (2000) Calcium channels activated by hydrogen peroxide mediate abscisic acid signaling in guard cells. Nature 406:731–734

Peters JM, Franke WW, Kleinschmidt JA (1994) Distinct 19S and 20S subcomplexes of the 26S proteasome and their distribution in the nucleus and the cytoplasm. J Biol Chem 269:7709–7718

Pickart CM (2001) Mechanisms underlying ubiquitination. Annu Rev Biochem 70:503–533

Pieterse CMJ, Van Loon LC (2004) NPR1: the spider in the web of induced resistance signaling pathways. Curr Opin Plant Biol 7:456–464

Pitzschke A, Hirt H (2006) Mitogen-activated protein kinases and reactive oxygen species signaling in plants. Plant Physiol 141:351–356

Pitzschke A, Hirt H (2009) Disentangling the complexity of mitogen-activated protein kinases and reactive oxygen species signaling. Plant Physiol 149:606–615

Pitzschke A, Djamei A, Bitton F, Hirt H (2009a) A major role of the MEKK1-MKK1/2-MPK4 pathway in ROS signaling. Mol Plant 2:120–137

Pitzschke A, Schikora A, Hirt H (2009b) MAPK cascade signaling networks in plant defence. Curr Opin Plant Biol 12:1–6

Price AH, Taylor A, Ripley SJ, Griffiths A, Trewavas AJ, Knight MR (1994) Oxidative signals in tobacco increase cytosolic calcium. Plant Cell 6:1301–1310
Ren D, Yang H, Zhang S (2002) Cell death mediated by MAPK is associated with hydrogen peroxide production in *Arabidopsis*. J Biol Chem 277:559–565
Rentel MC, Lecourieux D, Ouaked F, Usher SL, Petersen L, Okamoto H, Knight H, Peck SC, Grierson CS, Hirt H, Knight MR (2004) OXI1 kinase is necessary for oxidative burst-mediated signaling in *Arabidopsis*. Nature 427:858–861
Rojas C, Mysore KS (2012) Glycolate oxidase is an alternative source for $H_2O_2$ production during plant defense responses and functions independently from NADPH oxidase. Plant Signal Behav 7:752–755
Rojas CM, Senthil-Kumar M, Wang K, Ryu CM, Kaundal A, Mysore KS (2012) Glycolate oxidase modulates reactive oxygen species-mediated signal transduction during nonhost resistance in *Nicotiana benthamiana* and *Arabidopsis*. Plant Cell 24:336–352
Romeis T, Piedras P, Zhang S, Klessig DF, Hirt H, Jones JD (1999) Rapid Avr9- and Cf-9 – dependent activation of MAP kinases in tobacco cell cultures and leaves: convergence of resistance gene, elicitor, wound, and salicylate responses. Plant Cell 11:273–287
Romeis T, Ludwig AA, Martin R, Jones JD (2001) Calcium-dependent protein kinases play an essential role in a plant defence response. EMBO J 20:5556–5567
Sagi M, Fluhr R (2001) Superoxide production by plant homologues of the gp91(phox)NADPH oxidase: modulation of activity by calcium and by tobacco mosaic virus infection. Plant Physiol 126:1281–1290
Sagi M, Fluhr R (2006) Production of reactive oxygen species by plant NADPH oxidases. Plant Physiol 141:336–340
Sakamoto H, Matsuda O, Iba K (2008) *ITN1*, a novel gene encoding an ankyrin-repeat protein that affects the ABA-mediated production of reactive oxygen species and is involved in salt-stress tolerance in *Arabidopsis thaliana*. Plant J 56:411–422
Samuel MA, Ellis BE (2002) Double jeopardy: both overexpression and suppression of a redox-activated protein kinase render tobacco plants ozone sensitive. Plant Cell 14:2059–2069
Samuel MA, Miles GP, Ellis BE (2000) Ozone treatment rapidly activates MAP kinase signalling in plants. Plant J 22:367–376
Sang S, Li X, Gao R, You Z, Lü B, Liu P, Ma Q, Dong H (2012) Apoplastic and cytoplasmic location of harpin protein $Hpa1_{Xoo}$ plays different roles in $H_2O_2$ generation and pathogen resistance in *Arabidopsis*. Plant Mol Biol 79:375–391
Scandalios JG (1993) Oxygen stress and superoxide dismutases. Plant Physiol 101:7–12
Schürmann P, Jacquot JP (2000) Plant thioredoxin systems revisited. Annu Rev Plant Physiol Plant Mol Biol 51:371–400
Simon-Plas F, Elmayan T, Blein J-P (2002) The plasma membrane oxidase NtrbohD is responsible for AOS production in elicited tobacco cells. Plant J 31:137–148
Slaymaker DH, Navarre DA, Clark D, del Pozo O, Martin GB, Klessig DF (2002) The tobacco salicylic acid-binding protein-3 (SABP3) is the chloroplast carbonic anhydrase, which exhibits antioxidant activity and plays a role in the hypersensitive defense response. Proc Natl Acad Sci U S A 99:11640–11645
Stone JR (2004) An assessment of proposed mechanisms for sensing hydrogen peroxide in mammalian systems. Arch Biochem Biophys 422:119–124
Suzuki N, Miller G, Morales J, Shulaev V, Torres MA, Mittler R (2011) Respiratory burst oxidases: the engines of ROS signaling. Curr Opin Plant Biol 14:691–699
Takahashi H, Chen Z, Du H, Liu Y, Klessig D (1997) Development of necrosis and activation of disease resistance in transgenic tobacco plants with severely reduced catalase levels. Plant J 11:993–1005
Takeda S, Gapper C, Kaya H, Bell E, Kuchitsu K, Dolan L (2008) Local positive feedback regulation determines cell shape in root hair cells. Science 319:1241–1244
Tena G, Boudsocq M, Sheen J (2011) Protein kinase signalling networks in plant innate immunity. Curr Opin Plant Biol 14:519–529
Thompson JE, Legge RL, Barber RF (1987) The role of free radicals in senescence and wounding. New Phytol 105:317–344

Tipping AJ, McPherson MJ (1995) Cloning and molecular analysis of the pea seedling copper amine oxidase. J Biol Chem 270:16939–16946

Tonks NK (2005) Redox redux: revisiting PTPs and the control of cell signaling. Cell 121:667–670

Torres MA, Dangl JL (2005) Functions of the respiratory burst oxidase in biotic interactions, abiotic stress and development. Curr Opin Plant Biol 8:397–403

Torres MA, Jones JDG, Dangl JL (2006) Reactive oxygen species signaling in response to pathogens. Plant Physiol 141:373–378

Trouverie J, Vidal G, Zhang Z, Sirichandra C, Madiona K, Amiar Z, Prioul J-L, Jeannette E, Rona J-P, Brault M (2008) Anion channel activation and proton pumping inhibition involved in the plasma membrane depolarization induced by ABA in *Arabidopsis thaliana* suspension cells are both ROS dependent. Plant Cell Physiol 49:1495–1507

Van Breusegem F, Bailey-Serres J, Mittler R (2008) Unraveling the tapestry of networks involving reactive oxygen species in plants. Plant Physiol 147:978–984

Vandenabeele S, Van Der Kelen K, Dat J, Gadjev I, Boonefaes T, Morsa S, Rottiers P, Slooten L, Van Montagu M, Zabeau M, Inzé D, Van Breusegem F (2003) A comprehensive analysis of hydrogen peroxide-induced gene expression in tobacco. Proc Natl Acad Sci U S A 100:16113–16118

Veal EA, Findlay VJ, Day AM, Bozonet SM, Evans JM, Quinn J, Morgan BA (2004) A Cys peroxiredoxin regulates peroxide-induced oxidation and activation of a stress-activated MAP kinase. Mol Cell 15:129–139

Vidhyasekaran P (2007) Fungal pathogenesis in plants and crops: molecular biology and host defense mechanisms, IIth edn. CRC Press/Taylor & Francis Group, Boca Raton, 510

Vranová E, Inzé D, Van Brueusegem F (2002) Signal transduction during oxidative stress. J Exp Bot 53:1227–1236

Wang P, Song C-P (2008) Guard-cell signaling for hydrogen peroxide and abscisic acid. New Phytol 178:703–718

Wang W, Esch JJ, Shiu S-H, Agula H, Binder BM, Chang C, Patterson S, Bleecker AB (2006) Identification of important regions for ethylene binding and signaling in the transmembrane domain of the ER1 ethylene receptor of *Arabidopsis*. Plant Cell 18:3429–3442

Wang P, Du Y, Li Y, Ren D, Song C-P (2010) Hydrogen peroxide-mediated activation of MAP kinase 6 modulates nitric oxide biosynthesis and signal transduction in *Arabidopsis*. Plant Cell 22:2981–2998

Wendehenne D, Durner J, Chen Z, Klessig DF (1998) Benzothiadiazole, an inducer of plant defenses, inhibits catalase and ascorbate peroxidases. Phytochemistry 47:651–657

Wendehenne D, Lamotte O, Frachisse J-M, Barbier-Brygoo H, Pugin A (2002) Nitrate efflux is an essential component of the cryptogein signaling pathway leading to defense responses and hypersensitive cell death in tobacco. Plant Cell 14:1937–1951

Willekens H, Langebartels C, Tiré C, Van Montagu M, Inze D, Van Camp W (1994) Differential expression of catalase genes in *Nicotiana plumbaginifolia* (L.). Proc Natl Acad Sci U S A 91:10450–10454

Wong HL, Sakamoto T, Kawasaki T, Umemura K, Shimamoto K (2004) Down-regulation of metallothionein, a reactive oxygen scavenger, by the small GTPase OsRac1 in rice. Plant Physiol 135:1447–1456

Wong HL, Pinontoan R, Hayashi K, Tabata R, Yaeno T, Hasegawa K, Kojima C, Yoshioka H, Iba K, Kawasaki T, Shimamoto K (2007) Regulation of rice NADPH oxidase by binding of Rac GTPase to its N-terminal extension. Plant Cell 19:4022–4034

Xing T, Higgins VJ, Blumwald E (1997) Race-specific elicitors of *Cladosporium fulvum* promote translocation of cytosolic components of the NADPH-oxidase to the plasma membrane of tomato cells. Plant Cell 9:249–259

Xing Y, Jia W, Zhang J (2007) AtMEK1 mediates stress-induced gene expression of CAT1 catalase by triggering $H_20_2$ production in *Arabidopsis*. J Exp Bot 58:2969–2981

Xing Y, Jia W, Zhang J (2008) AtMKK1 mediates ABA-induced *CAT1* expression and $H_2O_2$ production via AtMPK6-coupled signaling in *Arabidopsis*. Plant J 54:440–451

Yalpani N, Altier DJ, Barbour E, Cigan AL, Scelonge CJ (1993) Production of 6-methylsalicylic acid by expression of a fungal polyketide synthase activates disease resistance in tobacco. Plant Cell 13:1401–1409

Yang Y, Shah J, Klessig DF (1997) Signal perception and transduction in plant defense responses. Genes Dev 11:1621–1639

Yao T, Ndoja A (2012) Regulation of gene expression by the ubiquitin-proteasome system. Semin Cell Dev Biol 23:523–529

Yoshioka H, Numata N, Nakajima K, Katou S, Kawakita K, Rowland O, Jones JD, Doke N (2003) *Nicotiana benthamiana* gp91$^{phox}$ homologs *NbrbohA* and *NbrbohB* participate in $H_2O_2$ accumulation and resistance to *Phytophthora infestans*. Plant Cell 15:706–718

Zhang X, Miao YC, An GY, Zhou Y, ShangGuan ZP, Gao JF, Song C-P (2001) $K^+$ channels inhibited by hydrogen peroxide mediate abscisic acid signaling in *Vicia faba* guard cells. Cell Res 11:195–202

Zhang C, Czymmek KJ, Shapiro AD (2003) Nitric oxide does not trigger early programmed cell death events but may contribute to cell-to-cell signaling governing progression of the *Arabidopsis* hypersensitive response. Mol Plant Microbe Interact 16:962–972

Zhang AY, Jiang MY, Zhang JH, Tan M, Hu XL (2006) Mitogen-activated protein kinase is involved in abscisic acid-induced antioxidant defense and acts downstream of reactive oxygen species production in leaves of maize plants. Plant Physiol 141:475–487

Zhang W, Qin C, Zhao J, Wang X (2004) Phospholipase Dα1-derived phosphatidic acid interacts with ABI1 phosphatase 2C and regulates abscisic acid signaling. Proc Natl Acad Sci U S A 101:9508–9513

Zhang L, Xi D, Luo L, Meng F, Li Y, Wu CA, Guo X (2011) Cotton GhMPK2 is involved in multiple signaling pathways and mediates defense responses to pathogen infection and oxidative stress. FEBS J 278:1267–1378

Zhao Z, Stanley BA, Zhang W, Assmann SM (2010) ABA-regulated G protein signaling in *Arabidopsis* guard cells: a proteomic perspective. J Proteome Res 9:1637–1647

Zhou F, Zhang Z, Gregersen PL, Mikkelsen JD, de Neergaard E, Collinge DB, Thordal-Christensen H (1998) Molecular characterization of the oxalate oxidase involved in the response of barley to the powdery mildew fungus. Plant Physiol 117:33–41

Zhu H, Li G-J, Ding L, Cui X, Berg H, Assmann SM, Xia Y (2009) *Arabidopsis* extra- large G-protein 2 (XLG2) interacts with the Gβ subunit of heterotrimeric G protein and functions in disease resistance. Mol Plant 2:513–525

Zipfel C, Robatzek S, Navarro L, Oakeley EJ, Jones JDG, Felix G, Boller T (2004) Bacterial disease resistance in *Arabidopsis* through flagellin perception. Nature 428:764–767

# Chapter 6
# Nitric Oxide Signaling System in Plant Innate Immunity

**Abstract** Nitric oxide (NO) is a diffusible molecular messenger that plays an important role in plant immune response signal transduction. The pathogen-associated molecular pattern (PAMP) signal molecules trigger a very rapid NO burst in plant cells. NO is synthesized predominantly by the enzyme nitric oxide synthase (NOS). NOS contains calmodulin (CaM)-binding motifs and full activation of the enzyme needs both $Ca^{2+}$ and CaM. NO is involved in influx of $Ca^{2+}$ into the cytosol. It modulates the activity of plasma membrane as well as intracellular $Ca^{2+}$ -permeable channels. NO acts as an important key redox-active signal for the activation of various defense responses. Both an oxidative and a NO burst have been reported to occur prior to activation of the signal cascade that eventually activates the transcription of defense genes. NO and reactive oxygen species (ROS) work in strong partnership during induction of the defense genes. A tight interrelationship between NO and SA in plant defense has been reported. NO is required for the full function of SA as a systemic acquired resistance (SAR) inducer. NO induces the key enzymes of the JA and ethylene biosynthesis pathways. NO acts substantially in cellular signal transduction through stimulus-coupled S-nitrosylation of cysteine residues. NO rapidly induces reversible *S*-nitrosylation of proteins involved in signal transduction. This redox-based post-translational modification is a key regulator of protein function in plant immunity. NO reacts rapidly with glutathione (GSH) to yield S-nitrosoglutathione (GSNO). GSNO is a bioactive, stable, and mobile reservoir of NO. It acts synergistically with SA in SAR. GSNO reductase (GSNOR) is the main enzyme responsible for the in vivo control of intracellular levels of GSNO and also the levels of *S*-nitrosylated proteins. NO bioactivity is controlled by NO synthesis by the different routes and by NO degradation, which is mainly performed by the GSNOR. S-nitrosothiols (SNOs) also play important role in SAR. GSNOR controls SNO in vivo levels and the SNO content positively regulates plant defense responses.

**Keywords** Nitric oxide (NO) burst • Nitric oxide synthase (NOS) • $Ca^{2+}$ influx • Redox-active signal • Systemic acquired resistance (SAR) • S-nitrosylation • S-nitrosoglutathione (GSNO) • GSNO reductase (GSNOR)

P. Vidhyasekaran, *PAMP Signals in Plant Innate Immunity: Signal Perception and Transduction*, Signaling and Communication in Plants 21,
DOI 10.1007/978-94-007-7426-1_6, © Springer Science+Business Media Dordrecht 2014

## 6.1 Nitric Oxide as a Component of the Repertoire of Signals Involved in Plant Immune Signaling System

Nitric oxide (NO) is a gaseous readily diffusible free radical identified as a signaling molecule in animals and plants (Besson-Bard et al. 2008a, b). Nitric oxide (NO) is a key mediator for rapid induction of plant immune responses (Bellin et al. 2013). It is now well-accepted that NO is a component of the repertoire of signals that a plant uses to both thrive and survive (Wilson et al. 2008). On perception of pathogen associated molecular patterns (PAMPs), plant pattern recognition receptors (PRRs) modulate signaling networks for defense responses that rely on rapid production of reactive nitrogen species (RNS) and reactive oxygen species (ROS) (Bellin et al. 2013). NO is a diffusible molecular messenger that plays an important role in the plant immune response signal transduction system (Grennan 2007). Involvement of NO in defense signaling has been well demonstrated (Durner et al. 1998; Delledonne et al. 2001; Polverari et al. 2003; Xu et al. 2004; Perchepied et al. 2010; Chun et al. 2012). NO biosynthetic genes *NIA1* and *NIA2* have been shown to induce NO synthesis and defense responses in *Arabidopsis*. Expression of the defense-related genes was either abolished or delayed in the double mutant. The double mutant *nia1 nia2* plants were highly susceptible to *Sclerotinia sclerotiorum* (Perchepied et al. 2010). The transgenic tobacco plants constitutively expressing a mammalian neuronal nitric oxide synthase (NOS) involved in NO production exhibited enhanced resistance to a spectrum of pathogens, including bacteria, fungi, and viruses (Chun et al. 2012). Collectively these results demonstrate the role of NO in defense signaling.

An oomycete PAMP/elicitor triggers a NO burst within minutes in tobacco cells (Foissner et al. 2000; Lamotte et al. 2004). A transient burst of NO has been observed in roots of *Arabidopsis thaliana* as an early response after contact with *Verticillium longisporum* (Tischner et al. 2010). NO acts substantially in cellular signal transduction through stimulus-coupled S-nitrosylation of cysteine residues (Benhar et al. 2008). It serves as a key redox-active signal for the activation of various defense responses (Klessig et al. 2000). NPR1 and TGA1 are key redox-controlled regulators of systemic acquired resistance (SAR) in plants. The translocation of NPR1 into the nucleus has been shown to be promoted by NO, suggesting that NO is a redox regulator of the NPR1/TGA1 system involved in SAR (Lindermayr et al. 2010). Collectively these studies suggest that NO plays an important role in plant innate immunity signaling system.

## 6.2 PAMP-Induced Biosynthesis of NO in Plants

NO may be synthesized through different pathways (Fig. 6.1; Planchet et al. 2006; Yamasaki and Cohen 2006; Zhao et al. 2007; Zottini et al. 2007). NO is synthesized predominantly by the enzyme NOS in mammals (Bethke et al. 2004). NOS catalyses NO production from the substrate arginine and requires $Ca^{2+}$/CaM activation (Crawford et al. 2006).

1. L-Arginine $\xrightarrow{\text{NO synthase}}$ L-Citrulline + **NO**

2. Arginine $\xrightarrow{\text{Ornithine decarboxylase}}$ Polyamine $\xrightarrow{\text{Unknown enzyme}}$ **NO**

3. Nitrate $\xrightarrow{\text{Nitrate reductase}}$ Nitrite $\longrightarrow$ **NO**

4. $2NO_2 + 2H^+$ (Nitrogen dioxide) $\longleftrightarrow$ $2HNO_2$ (Nitrous acid) $\longleftrightarrow$ **NO** $+ NO_2 + H_2O$

**Fig. 6.1** Four different pathways in NO synthesis in plants

Although studies concerning NO synthesis and signaling in animals are well-advanced, in plants there are still fundamental questions concerning how NO is produced and used. There is a range of potential NO-generating enzymes in plants, but no obvious plant nitric oxide synthase homolog has yet been identified (Neill et al. 2008; Wilson et al. 2008). The PAMP/elicitor-induced NO in tobacco cells has been shown to be reduced by NOS inhibitors, suggesting the occurrence of a NOS-like enzyme in plants (Lamotte et al. 2004). In *Arabidopsis thaliana*, a NOS-like enzyme, AtNOS1, has been identified and NO levels were found to be lower in the *Atnos1* mutants impaired in AtNOS1 expression (Guo et al. 2003). Inhibitors of mammalian NOS have been found to suppress NO production in plants subjected to biotic and abiotic stresses (Zhang et al. 2003; Lamotte et al. 2004; Zeidler et al. 2004; Mur et al. 2005; Zhao et al. 2007; Zottini et al. 2007; Mur et al. 2013). These observations suggest that NOS may be involved in NO generation. However, a clear homologue of animal NOS has not yet been identified in plants (Guo et al. 2003; Crawford et al. 2006; Zemojtel et al. 2006; Besson-Bard et al. 2008a, b; Wilson et al. 2008). The plant enzyme displaying NOS-like activity is structurally different from classical mammalian NOS (Guo et al. 2003). Crawford et al. (2006) suggested that the AtNOS1 from Arabidopsis might act as a GTPase.

Polyamines may also be involved in NO synthesis (Fig. 6.1; Besson-Bard et al. 2008a). Rapid production of NO was observed in primary leaves of *Arabidopsis* when the polyamines spermidine and spermine were added to the *Arabidopsis* seedlings (Tun et al. 2006). This conversion might be carried out by unknown enzymes or by polyamine oxidases (Yamasaki and Cohen 2006; Besson-Bard et al. 2008a, b).

NO may be formed also from nitrite by the action of nitrate reductase, which catalyzes the reduction of nitrate to nitrite using NAD(P)H as an electron donor and also generates NO from nitrite (Fig. 6.1; Yamasaki and Sakihama 2000; Desikan et al. 2002; Morot-Gaudry-Talarmain et al. 2002; Rockel et al. 2002; Bethke et al. 2004; Planchet and Kaiser 2006; Wilson et al. 2008). NO generation was reduced in ammonium –fed tobacco plants compared to $NO_3^-$ -fed plants and $NO_3^-$ -fed plants showed enhanced disease resistance against *Pseudomonas syringae* pv. *phaseolicola* (Gupta et al. 2013). NO production was completely absent in ammonium-grown tobacco cell suspensions totally devoid of nitrate (Planchet et al. 2005, 2006). It indicates that NO synthesis is catalyzed through a nitrate/nitrite-dependent pathway rather than an L-arginine-dependent NOS mediated pathway in these tobacco cells. Nitrate reductase transcript and protein levels increase in response to an elicitor in potato tubers, suggesting a role for nitrate reductase in the synthesis of NO during the defense response (Delledonne 2005). Treatment of protoplasts prepared from *Nicotiana benthamiana* leaves with the PAMP elicitin INF1 secreted by the oomycete pathogen *Phytophthora infestans* elevated NO production. INF1-induced NO generation was suppressed by an NO-specific scavenger. Silencing of nitrate reductase (NR) genes significantly decreased INF1-induced NO production (Yamamoto-Katou et al. 2006). These results suggest that nitrate reductase is involved in the PAMP-triggered NO generation.

NO may also be synthesized from nitrite in a nonenzymatic manner (Fig. 6.1; Yamasaki 2000). In this process nitrite is protonated to form nitrous acid ($HNO_2$). Two molecules of $HNO_2$ interact through a series of reactions and give rise to NO (Yamasaki and Sakihama 2000). The non-enzymatic conversion of nitrite to NO occurs in the apoplast at acidic pH in the presence of reductants such as ascorbic acid (Bethke et al. 2004).

## 6.3 Upstream Events in NO Production

### *6.3.1 $Ca^{2+}$ Influx into Cytosol May Be an Early Upstream Event in NO Production*

PAMP elicitor signals trigger a NO burst within minutes in plant cells after the elicitor treatment (Foissner et al. 2000; Lamotte et al. 2004; Tischner et al. 2010). The rapid NO production has been shown to be dependent on $Ca^{2+}$ signaling system (Lamotte et al. 2004; Ali et al. 2007; Courtois et al. 2008; Ma et al. 2008; Choi et al. 2009; Vatsa et al. 2011; Ma et al. 2012). Influx of extracellular $Ca^{2+}$ through the cell membrane appears to be one of the earliest events triggered by the PAMPs (Laohavisit et al. 2009; Vadassery and Oelmūller 2009; Kwaaitaal et al. 2011; Vincill et al. 2012). The massive influx of $Ca^{2+}$ occurs via different calcium ion channels within 15–30 min after PAMP/elicitor treatment (Lecourieux-Ouaked et al. 2000; Kwaaitaal et al. 2011). The generated $Ca^{2+}$ current leads to downstream NO production (Ali et al. 2007).

The different calcium ion channels involved in the calcium influx have also been found to be involved in NO production. Cyclic nucleotide-gated ion channels (CNGCs) have been reported to play important role in NO synthesis. Plants without functional CNGC2 lack the cell membrane $Ca^{2+}$ current and do not display immune responses induced by NO. The hypersensitive response impaired phenotype to an avirulent pathogen in *cngc2* mutant plants could be complemented by the addition of an NO donor (Ali et al. 2007). The results suggest the importance of the cyclic nucleotide gated channel in induction of NO in the immune signaling system (Ali et al. 2007). Another type of ion channels involved in $Ca^{2+}$influx is glutamate receptor (GLR)-like channels Ma et al. 2012). An oomycete PAMP elicitor has been shown to activate GLR calcium channels triggering NO production (Vatsa et al. 2011). The addition of the $Ca^{2+}$ channel blocker $Gd^{3+}$ or the $Ca^{2+}$ chelator EGTA abolished the PAMP-induced NO synthesis (Ali et al. 2007), suggesting that NO synthesis occurs downstream of cytosolic $Ca^{2+}$ elevation.

### 6.3.2 Role of Calmodulin in NO Production

NOS production in plants has been reported to be dependent on calmodulin (CaM), a $Ca^{2+}$ sensor protein (Delledonne et al. 1998; Courtois et al. 2008; Ma and Berkowitz 2011). The pepper CaM gene *CaCaM1* has been shown to be involved in NO generation (Choi et al. 2009). Upon treatment with the CaM antagonist, virulent *Pseudomonas syringae* pv. *tomato* – induced NO generation was also compromised in *CaCaM1* overexpressing plants (Choi et al. 2009). The calmodulin antagonist N-(6-aminohexyl)-5-chloro-1-naphthalene sulfonamide (W7) blocked NO generation. The effect of W7 was downstream from the $Ca^{2+}$ channel, as the elicitor activation of the channel was demonstrated in the presence of W7 (Ali et al. 2007).

NO synthase (NOS) is the key enzyme involved in NO production and NOS is a calmodulin (CaM)-binding protein. NOS contains CaM-binding motifs and full activation of the enzyme needs both $Ca^{2+}$ and CaM (Guo et al. 2003; Lamotte et al. 2004; Zeidler et al. 2004; Ma and Berkowitz 2007). The pathogen-induced $Ca^{2+}$ signals lead to CaM activation of NOS (Ma et al. 2008). Application of a CaM antagonist prevents NO generation and induction of immune responses. $Ca^{2+}$ chelation also abolishes NO generation (Ma et al. 2008). Collectively these results suggest that the initial pathogen recognition signal of $Ca^{2+}$ influx into the cytosol activates CaM, which then acts to induce downstream NO synthesis, leading to innate immune responses (Fig. 6.2).

### 6.3.3 ROS and ABA Act Upstream of NO Production

NO has been shown to act downstream of $H_2O_2$ in ABA signaling system (Lü et al. 2005). ABA-mediated NO generation has been shown to be dependent on ABA-induced $H_2O_2$ production in *Vicia faba* and *Arabidopsis* (Lü et al. 2005; Bright et al. 2006).

**Fig. 6.2** Induction of NO production through activation of $Ca^{2+}$ signaling system

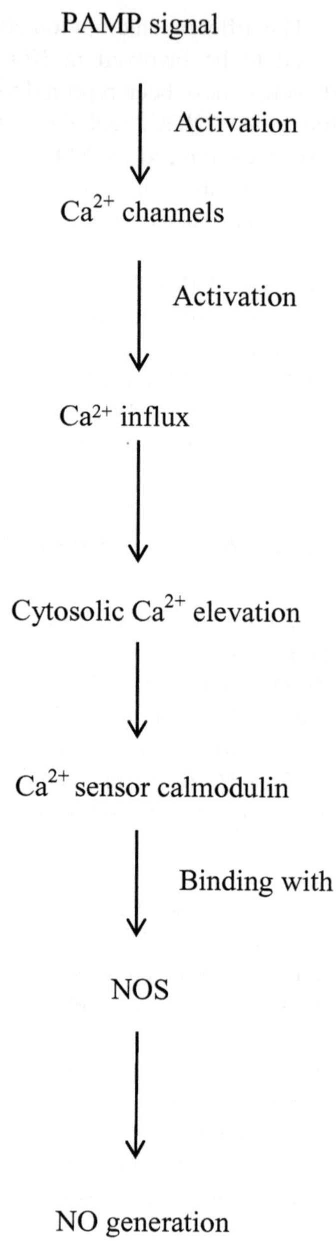


Nitrate reductase, but not nitric oxide synthase, was involved in the ABA-mediated NO production and stomatal closure (Bright et al. 2006). However, Yan et al. (2007) showed that ABA-induced NO is synthesized by a NOS-like enzyme in *V. faba*. ABA-induced NO is involved in ABA-induced stomatal closure; stomatal closure is part of plant immune response to restrict bacterial invasion (Melotto et al. 2006). It is suggested that both $H_2O_2$ and NO are synthesized in parallel, in response to ABA (Wang and Song 2008). Action of both ABI1-1 and ABI2-1 phosphatases occurs downstream of NO synthesis (Desikan et al. 2002).

## 6.4 Nitric Oxide-Target Proteins

Several proteomic and transcriptomic studies have led to the identification of numerous NO target proteins (Fig. 6.3; Delledonne et al. 2003; Polverari et al. 2003; Lindermayr et al. 2005, 2006; Grün et al. 2006; Belenghi et al. 2007; Besson-Bard et al. 2008b). NO targets metal- and thiol-containing proteins, such as catalase and peroxidase (Clark et al. 2000), glutathione S-transferase, superoxide dismutase, thioredoxin, and glutaredoxin and these proteins are involved in redox signaling system (Lindermayr et al. 2005). NO induced an increased expression of a cDNA

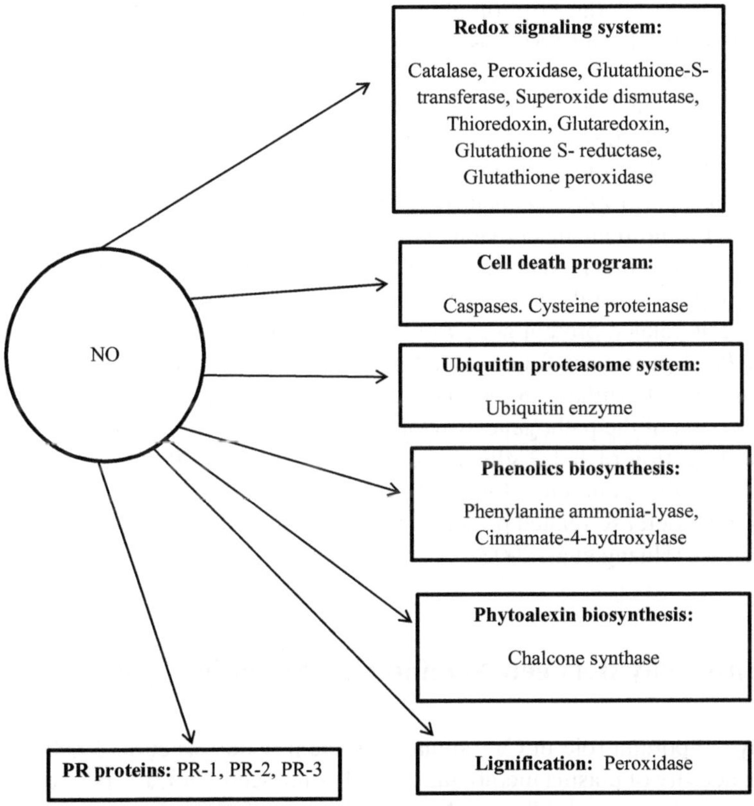


**Fig. 6.3** NO-target proteins

corresponding to fructose 1-6 bisphosphatase, one of the major targets of the thioredoxin system (Polverari et al. 2003). Its expression can lead to an increase in total cellular glutathione, which participates in the redox signaling system (Voehringer et al. 2000). NO activates glutathione-S-reductase and glutathione peroxidase which are involved in cell death process (Polverari et al. 2003).

NO triggers the activation of two ubiquitin conjugating enzymes UBC10 and Ahus5 (Polverari et al. 2003) and ubiquitin system plays an important role host defense (Bachmair et al. 1990, 2001a, b; Schenk et al. 2000). NO is involved in induction of proteins involved in hypersensitive cell death. ATAF2, a protein belonging to the NAC domain protein family, involved in cell death is induced by NO (Collinge and Boller 2001). Caspases are involved in plant cell death and a caspase-like activity is detectable in hypersensitively reacting cells following treatment with NO donors (Clarke et al. 2000). Specific caspase inhibitors suppress NO-induced cell death (Clarke et al. 2000). A gene encoding the cysteine proteinase RD21A is shown to be induced by NO at 10 min after treatment and the enzyme is involved in cell death activation in *Arabidopsis thaliana* (Swidzinski et al. 2002).

NO induces key enzymes in the phenylpropanoid pathway. It induces phenylalanine ammonia-lyase (PAL) which is the key enzyme in biosynthesis of phenolics (Neill et al. 2002). Transcription of the genes encoding the enzymes cinnamate-4-hydroxylase (C4H) and PAL has been shown to be induced by NO (Durner et al. 1998). C4H transcription increases following SNP (Sodium nitroprusside, a NO donor) infiltration (Delledonne et al. 1998). These results suggest that NO is involved in biosynthesis of phenolics. Chalcone synthase (CHS) is involved in the synthesis of isoflvonoid phytoalexins and its transcription is also modulated by NO (Delledonne et al. 1998). NO is involved in phytoalexin synthesis (Romero-Puertas et al. 2004). It up-regulates phytoalexin production (Modolo et al. 2002). NO activates peroxidases involved in lignification (Huang et al. 2002; Polverari et al. 2003).

NO triggers several pathogenesis-related protein genes. It also induces the pathogenesis-related protein gene *Pr1* in tobacco (Durner et al. 1998). Expression level of *PR-1* gene rises following administration of NO donors or expression of recombinant NO-synthase in tobacco (Levine et al. 1994). NO is capable of inducing expression of the pathogenesis-related protein PR-1 (Neill et al. 2002). NO can activate induction of β-1,3-glucanase, the PR-2 protein (Polverari et al. 2003). NO induces *Pr3* gene encoding a chitinase (Grün et al. 2006). It also induces various other defense-related proteins including alternative oxidase and glutathione S-transferase (Huang et al. 2002).

## 6.5 Interplay Between NO and $Ca^{2+}$ Signaling Systems

NO plays important role in $Ca^{2+}$ signaling system (Courtois et al. 2008). It modulates the activity of plasma membrane as well as intracellular $Ca^{2+}$-permeable channels (Besson-Bard et al. 2008a). Almost all types of $Ca^{2+}$ channels appear to be regulated by NO (Clementi 1998). NO released by NO donors induced a transient rise in $[Ca^{2+}]_{cyt}$ in *Vicia faba* guard cells (Garcia-Mata et al. 2003) and in tobacco suspension cells (Gould et al. 2003; Lamotte et al. 2004, 2006). Treatment of transgenic *Nicotiana plumbaginifolia* cells expressing the $Ca^{2+}$ reporter aequorin addressed in the cytosol with a NO donor resulted in a rapid and transient elevation in $[Ca^{2+}]_{cyt}$ (Besson-Bard et al. 2008a). NO scavengers and mammalian NOS inhibitors reduced the increase in $[Ca^{2+}]_{cyt}$ triggered by elicitors (Lamotte et al. 2006; Vandelle et al. 2006; Besson-Bard et al. 2008a). These observations suggest that NO is involved in influx of $Ca^{2+}$ into the cytosol. Inhibitors of plasma membrane and intracellular $Ca^{2+}$ permeable channels have been found to inhibit NO-induced increases in $[Ca^{2+}]_{cyt}$ (Gould et al. 2003; Lamotte et al. 2006; Vandelle et al. 2006), suggesting that NO might promote an influx of $Ca^{2+}$ from the extracellular space and/or mobilization of $Ca^{2+}$ sequestered in intracellular $Ca^{2+}$ stores. Ryanodine receptors (RYR) may be the main targets for NO (Durner et al. 1998; Lamotte et al. 2004).

The activation of intracellular $Ca^{2+}$ channels by NO may be due to the second messenger cGMP (guanosine-3′,5′-cyclic monophosphate), produced following the activation of soluble guanylate cyclase (GC) (Willmott et al. 1996; Hanafy et al. 2001). cGMP generation activated by NO in plants activates CNGCs, cytosolic $Ca^{2+}$ elevation, and downstream signaling (Ma and Berkowitz 2007). NO posttranslationally activates GC (Klessig et al. 2000). cGMP activates ADP-ribosylcyclase (ADPRC), through a cGMP-dependent protein kinase This results in elevated levels of another second messenger, cyclic ADP-ribose (cADPR) (Willmott et al. 1996; Durner et al. 1998; Klessig et al. 2000; Minorski 2003).

cADPR is a $Ca^{2+}$ mobilizing metabolite and activates intracellular $Ca^{2+}$ release channels (Ma and Berkowitz 2007). It has been suggested that cADPR mediates $Ca^{2+}$ release by activating the intracellular $Ca^{2+}$ channels ryanodine receptors (RYR) in animals and also in plants (Allen et al. 1995; Fliegert et al. 2007). The NO-mediated $Ca^{2+}$ transient influx is reduced by almost 40 % by the cADPR antagonist 8-bromo-cADPR (Besson-Bard et al. 2008a). cADPR is involved in NO signaling of various defense genes. NO-induced accumulation of *PR-1* transcripts in tobacco leaves was suppressed in the presence of the cADPR-selective antagonist 8-bromo-cADPR (Klessig et al. 2000). Vacuum infiltration of nanomolar concentrations of cADPR in tobacco leaf disks triggered the expression of the PR-1 gene, which was suppressed by RYR inhibitors (Durner et al. 1998). These results suggest that NO-induced cADPR is involved in downstream defense signaling system.

Together with cADPR, protein kinases may also be involved in mediating NO-induced changes in $[Ca^{2+}]_{cyt}$. Inhibitors of protein kinases, such as staurosporine and K252a reduced the $[Ca^{2+}]_{cyt}$ triggered by NO in *Vicia faba* guard cells and tobacco cell suspensions (Sokolovski et al. 2005; Lamotte et al. 2006). Treatment of *Nicotiana plumbaginifolia* with NO resulted in the activation of a protein kinase belonging to SNF1-related protein kinase type 2 (SnRK2) family and the mitogen-activated protein kinase SIPK (Besson-Bard et al. 2008a, b). Several other studies have demonstrated that artificially generated NO stimulated MAPKs including SIPK (Clarke et al. 2000; Pagnussat et al. 2004; Zhang et al. 2007). These observations suggest that NO triggers cellular events in plant cells by causing an increase in $[Ca^{2+}]_{cyt}$.

It has also been shown that a rise in $[Ca^{2+}]_{cyt}$ induces NO production through NOS-like enzyme (Lamotte et al. 2004; Corpas et al. 2006; Vandelle et al. 2006). $Ca^{2+}$-dependent NO production induced by an elicitor in *Arabidopsis thaliana* was brought about by the activation of the plasma membrane by cyclic nucleotide gated channel CNGC2 (Ali et al. 2007). Arabidopsis *dnd1* (defense no death 1) mutants without functional CNGC2 lack inward plasma membrane $Ca^{2+}$ currents and fail to produce NO in response to the bacterial elicitor treatment (Ali et al. 2007). These results suggest that a complex interaction between NO and $Ca^{2+}$ influx may exist and NO increases $[Ca^{2+}]_{cyt,}$ while $[Ca^{2+}]_{cyt}$ increases NO production.

It has also been reported that the generated NO can induce cytosolic $Ca^{2+}$ increase through activation of plasma membrane- and intracellular membrane-localized $Ca^{2+}$ channels during pathogen induced signaling cascades (Ali et al. 2007). The bacterial PAMP LPS could elicit NO generation in leaf guard cells and facilitate $Ca^{2+}$ influx into the cytosol (Ali et al. 2007). NO synthesis occurring during the plant-pathogen interactions causes elevation of cytosolic $Ca^{2+}$ level (Lamotte et al. 2004, 2006; Vandelle et al. 2006; Besson-Bard et al. 2008a, b). It is suggested that the NO generated downstream $Ca^{2+}$ influx may diffuse to neighboring cells and activate new $Ca^{2+}$ signals, which may amplify the NO generation process (Ma et al. 2007; Ma and Berkowitz 2011).

## 6.6 Interplay Between NO and ROS Signaling Systems

$H_2O_2$ and NO are known to operate together in several signaling cascades (Delledonne et al. 2002; Grennan 2007; Wang et al. 2010). Both an oxidative and a NO burst have been reported to occur prior to activation of the signal cascade that eventually activates the transcription of defense genes (Zaninotto et al. 2006). Potential target of $H_2O_2$ in *Arabidopsis* includes glyceraldehyde-3-phosphate dehydogenase (GAPDH) that is reversibly inhibited by $H_2O_{2.}$ GADPH has a role in mediating ROS signaling in plants as a target of $H_2O_{2.}$ GADPH is also a target of NO-mediated S-nitrosylation and is inhibited by NO (Lindermayr et al. 2005). Since GADPH has been identified as the protein that interacts with both $H_2O_2$ and NO, it is suggested that it may be the link between these two signaling pathways (Grennan 2007). Another direct target of $H_2O_2$ action is Met adenosyltransferase, which is inactivated by $H_2O_2$ through reversible and covalent oxidation of a Cys residue. The same Cys residue is also a target for NO, which similarly causes enzyme inactivation (Hancock et al. 2006).

MAPK cascade may be involved in production of both NO and ROS (Asai et al. 2008). The MEK2 – SIPK/NTF4 cascade activates NOS inducing production of NO and also activates NADPH oxidase inducing production of ROS (Asai et al. 2008). ROS have been shown to influence the transcription and activation of a number of mitogen-activated protein kinases (MAPKs) (Rentel et al. 2004). The MAPK signaling pathways are also potential targets for NO, which influences the activity of

MAP kinases (Neill et al. 2002). Death of host cells during HR results from the simultaneous, balanced production of NO and ROS (Delledonne et al. 2001, 2002). The cytotoxic effects of NO and ROS derive from the diffusion-limited reaction of NO with $O_2^-$ to form the peroxynitrite anion $ONOO^-$. Peroxynitrite causes oxidative damage and protein modifications such as Tyr nitration and oxidation of thiol residues (Radi 2004). These results suggest that NO and $H_2O_2$ work in strong partnership during induction of the defense genes (Zago et al. 2006). NO and $H_2O_2$ can also act independently in the same signaling pathways with similar downstream responses as a consequence (Zago et al. 2006).

## 6.7 Role of NO in SA, JA, and Ethylene Signaling Systems

NO is involved in the production of salicylic acid (Chamnongpol et al. 1998; Durner et al. 1998). NO triggers UDP-glucose:SA:glucosyltransferase that converts SA to SA β-glucoside, a conjugated and stable form of SA (Zago et al. 2006). SA in turn, activates nitric oxide synthesis in *A. thaliana* (Zottini et al. 2007). A tight interrelationship between NO and SA in plant defense has been reported. NO donors produce SA accumulation (Durner et al. 1998). NO is required for the full function of SA as a SAR (systemic acquired resistance) inducer (Song and Goodman 2001). Many NO-regulated enzymes, including aconitase or catalase, are regulated by SA (Durner et al. 1997; Clark et al. 2000). Thus NO may be involved in SA signaling system. Both NO and SA activated SIPK in tobacco (Kumar and Klessig 2000). Studies with transgenic *NahG* tobacco revealed that SA is required in the NO-mediated induction of SIPK. SIPK may function downstream of SA in the NO signaling pathway (Kumar and Klessig 2000).

NO is involved in induction of biosynthesis of oxylipins and JA. NO induces the key enzymes of the JA biosynthesis pathway (Fig. 6.4; del Rio et al. 2004; Palmieri et al. 2008). Transcripts encoding lipoxygenase (LOX), 12-oxophytodienoate reductase (12-OPR), and diacylglycerol kinase (DAGK), all involved in the biosynthesis of oxylipins and JA, are up-regulated by NO (Grün et al. 2006; Zago et al. 2006). Interaction between NO and JA signaling has been described (Palmieri et al. 2008). NO induces allene oxide synthase gene (*AOS*) and 12-oxophytodienoate reductase gene *OPR3* (Grün et al. 2006); both are involved in JA biosynthesis (Vidhyasekaran 2007a, b). NO induces the JA-regulated *PDF1.2* gene encoding PR12 protein (Grün et al. 2006).

NO induces S-Adenosyl-L-Met synthetase, which catalyzes the conversion of ATP and L-Met into the ethylene precursor *S*-adenosyl-L-Met (Fig. 6.5; Zago et al. 2006). NO induces ACC synthase involved in ethylene biosynthesis (Lamotte et al. 2004). Thus, NO is involved in ethylene biosynthesis (Lindermayr et al. 2005, 2006). NO activates *ein3*, a gene involved in ethylene perception and transduction (Chang and Stadler 2001). NO is known to influence several ethylene-dependent processes in the plant (Leshem et al. 1998).

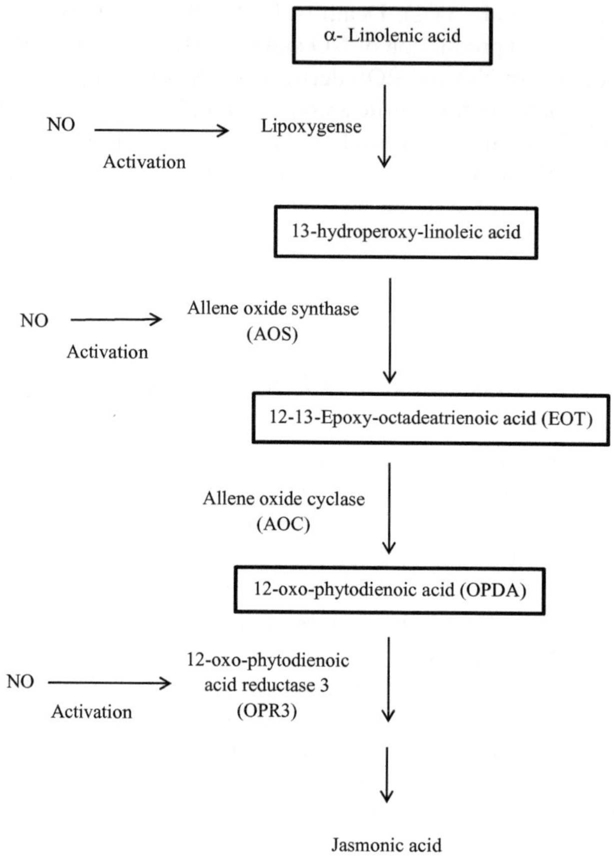


**Fig. 6.4** Role of NO in activation of jasmonic acid biosynthesis

## 6.8 Role of NO in Protein *S*-Nitrosylation

NO rapidly induces reversible *S*-nitrosylation of proteins involved in signal transduction. The addition of an NO moiety to a cysteine (Cys) thiol to form an S-nitrosothiol (SNO) is termed S-nitrosylation (Malik et al. 2011). This redox-based post-translational modification is a key regulator of protein function in plant immunity (Malik et al. 2011). Intracellular NO reacts with various proteins and nonprotein thiols to form nitrosothiols (Stamler et al. 2001). Most of the NO-modified proteins are regulated by *S*-nitrosylation. S-Nitrosylation refers to the incorporation of the NO moiety to a Cys sulfur atom to form a S-NO bond (Martinez-Ruiz and Lamas 2004). Protein S-nitrosylation occurs at a single critical Cys residue by oxygen-dependent chemical reactions or by the transfer of NO from a nitrosothiol to a protein sulfhydryl group (Lindermayr et al. 2005). NO can react with sulfhydryl groups on proteins, yielding SNOs, which lead to a change in protein function or

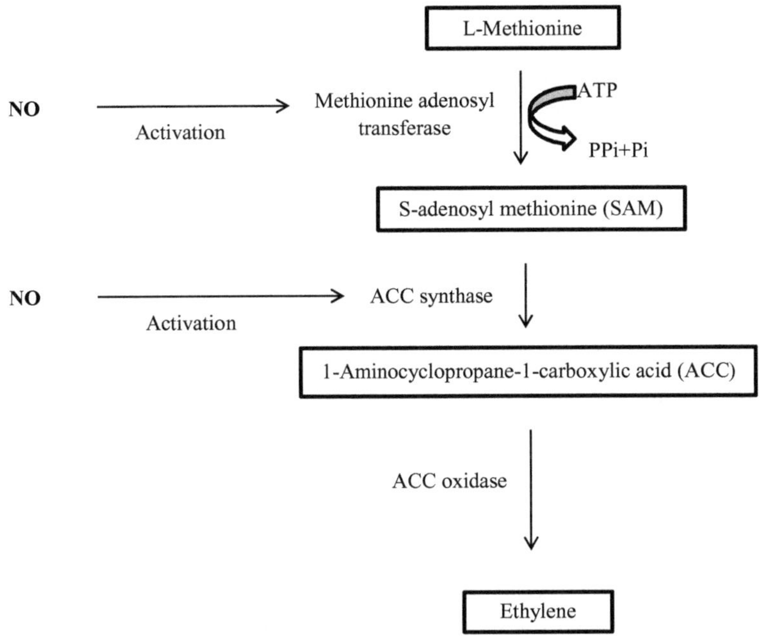


**Fig. 6.5** Role of NO in ethylene biosynthesis

activity (Grennan 2007). Since SNOs interact with intracellular reducing agents, such as ascorbic acid or glutathione (GSH), they are highly labile. This lability results in tissue half-lives of seconds to a few minutes and therefore provides a very sensitive mechanism for regulating cellular processes. Thus S-nitrosylation is considered as posttranslational modification which is similar to phosphorylation (Lindermayr et al. 2005). Like phosphorylation, S-nitrosylation is a precisely targeted and rapidly reversible posttranslational modification (Mannick and Schonhoff 2004; Astier and Lindermayr 2012). S-nitrosylated proteins can be easily denitrosylated, as the S-NO bond is labile in a reductive environment. Posttranslational modification of proteins by *S*-nitrosylation potentially alters function of the proteins. Rapidly reversible *S*-nitrosylation of proteins induced by NO is involved in signal transduction (Grennan 2007).

The proteins, which are regulated by nitrosylation, include catalase (Foster and Stamler 2004), superoxide dismutase, glutathione peroxidase and peroxiredoxin (Lindermayr et al. 2005) and these enzymes are involved in redox signaling system. S-Nitrosylation regulates Met adenosyltransferase involved in ethylene biosynthesis (Lindermayr et al. 2006) and metacaspase involved in HR-related cell death (Belenghi et al. 2007). NO reacts rapidly with glutathione (GSH), the major intracellular low-molecular-mass antioxidant to yield S-nitrosoglutathione (GSNO). GSNO is a bioactive, stable, and mobile reservoir of NO and it is an important player in plant defense responses against pathogens (Espunya et al. 2012). GSNO is considered to represent a functionally relevant signaling molecule that might act both as NO reservoir and NO donor (Lindermayr et al. 2005) or independently of homolytic cleavage to NO (Gaston 1999). GSNO acts synergistically with SA in SAR (Espunya et al. 2012). It has been suggested that GSNO would act as a long- distance phloematic signal in SAR (Durner and Klessig 1999).

GSNO may be broken down by GSNO reductase (GSNOR) (Liu et al. 2001; Malik et al. 2011). GSNOR reduces GSNO, an essential reservoir for NO activity (Wünsche et al. 2011). It is the main enzyme responsible for the in vivo control of intracellular levels of GSNO (Espunya et al. 2012). GSNOR controls not only the cellular levels of GSNO but also the levels of *S*-nitrosylated proteins (Grennan 2007). NO bioactivity is controlled by NO synthesis by the different routes and by NO degradation, which is mainly performed by the GSNOR (Liu et al. 2004). Mutation of the gene *AtGSNOR1* in *Arabidopsis* controls cellular S-nitrosothiols during plant-pathogen interactions (Feechan et al. 2005). GSNOR is encoded by a single-copy gene in *Arabidopsis thaliana* (Sakomoto et al. 2002).

GSNOR has been shown to play a role in plant defense response (Rustérucci et al. 2007). Transgenic *Arabidopsis* plants with decreased amounts of GSNOR (using antisense strategy) show enhanced basal resistance against *Peronopora parasitica*, which correlates with higher levels of intracellular SNOs and constitutive activation of *PR-1* gene (Rustérucci et al. 2007). SNOs also play important role in systemic acquired resistance (SAR). SAR is impaired in plants overexpressing GSNOR and enhanced in the antisense plants, and this correlated with changes in the S-nitrosothiol content both in local and systemic leaves. The loss of AtGSNOR1 function compromises defense responses in *A. thaliana* (Feechan et al. 2005). GSNOR was found to be localized in the phloem, suggesting that GSNOR would regulate SAR signal transport through the vascular system (Rustérucci et al. 2007). A reduction in NO accumulation leads to pathogen susceptibility (Delledonne et al. 1998; Zeidler et al. 2004), a decrease in SNOs promotes protection against microbial infection (Feechan et al. 2005). Collectively these results show that GSNOR controls SNO in vivo levels and the SNO content positively regulates plant defense responses.

NO mediates the S-nitrosylation of peroxiredoxin II E (PrxII E), a member of the peroxiredoxin family consisting of peroxidases that reduce $H_2O_2$ and alkyl hydroperoxides to $H_2O$ and the corresponding alcohol using equivalents from thioredoxin or glutaredoxins (Dietz 2005; Horling et al. 2003). During $H_2O_2$ reduction, the catalytic Cys residues of peroxiredoxins undergo oxidation and must be reduced by electron donors such as thioredoxins, glutaredoxins or cyclophins before the next catalytic cycle (Horling et al. 2003). S-nitrosylation severely inhibits the peroxidase activity of PrxII E, thus revealing a novel regulation mode for peroxiredoxins (Romero-Puertas et al. 2007). PrxII E possesses peroxynitrite reductase activity and S-nitrosylation inhibits this activity (Romero-Puertas et al. 2007).

Peroxynitrite ($ONOO^-$) is a toxic reactive nitrogen species generated by the interaction of ROS and NO during oxidative burst and PrxII E detoxifies $ONOO^-$(Romero-Puertas et al. 2007). S-Nitrosylation inhibits the peroxynitrite detoxification activity of PrxII E (Romero-Puertas et al. 2007). GSNO was found to severely inhibit PrxII E peroxidase activity in a concentration-dependent manner. This effect could be reversed by the thiol-specific reductant DTT, indicating that

GSNO affects PrxII E activity through S-nitrosylation of the Cys residues. GSNO has been shown to be able to S-nitrosylate PrxII E in vivo ((Romero-Puertas et al. 2007). $ONOO^-$is an important signal molecule involved in triggering accumulation of PR proteins (Durner et al. 1998).

## 6.9 Role of NO in Protein Nitration

NO can also perform posttranslational protein modifications through nitration, besides through S-nitrosylation (Zaninotto et al. 2006). Peroxynitrite ($ONOO^-$), the reactive nitrogen species is generated by the reaction of NO with superoxide anion ($O_2^-$) (Bryk et al. 2000; Neill et al. 2008; Vandelle and Delledonne 2011). During the hypersensitive reaction (HR), the formation of $ONOO^-$is promoted by the rate of NO reaction with $O_2^-$, which is approximately three times faster than the reaction of $O_2^-$with superoxide dismutase (SOD) forming $H_2O_2$ during the oxidative burst (Ichiropoulos and al-Mehdi 1995). $ONOO^-$generation in tobacco cells treated with an oomycete elicitor occurred within 1 h and reached a maximum level at 6–12 h after elicitor treatment (Saito et al. 2006). Urate, a $ONOO^-$scavenger, abolished elicitor-induced $ONOO^-$generation (Saito et al. 2006).

$ONOO^-$causes protein tyrosine nitration through the nitration of Tyr residues in Tyr kinase (Schopfer et al. 2002; Radi 2004; Romero-Puertas et al. 2008). $ONOO^-$donor SIN-1 [3-(4-morpholinyl) sydnonimine hydrochloride) treatment induced nitrotyrosine-containing proteins in tobacco cells (Saito et al. 2006). The number of nitrated proteins increased during disease resistance response in *Arabidopsis thaliana* (Romero-Puertas et al. 2008). Protein extracts from leaves of *A. thaliana* treated with $ONOO^-$showed a significant increase in nitrated proteins when pretreated with GSNO (Romero-Puertas et al. 2008). $ONOO^-$was found to induce protein nitration in soybean and tobacco (Delledonne et al. 2001; Saito et al. 2006).

*S*-nitrosylation inhibits the $ONOO^-$detoxification activity of peroxiredoxin II E (PrxII E), causing a dramatic increase of $ONOO^-$-dependent nitrotyrosine residue formation. The same increase was observed in prxII E mutant line after exposure to $ONOO^-$, indicating that PrxII E modulation of $ONOO^-$is important in the signaling system (Romero-Puertas et al. 2008). $ONOO^-$may have important signaling functions in plants. SIN, an NO donor that provides continuous source of $ONOO^-$was found to induce the accumulation of the transcript encoding PR-1 in tobacco leaves (Durner et al. 1998).

Protein nitration alters catalytic activity and interferes with cellular signaling processes (Schopfer et al. 2002). Protein nitration is a reversible and selective process associated with protein Tyr phosphorylation (Klotz et al. 2002). Tyr phosphorylation may mediate signaling events induced by nitrating agents like $ONOO^-$and depending on $ONOO^-$local concentrations, the nitration and phosphorylation of critical Tyr residues may be competing processes (Brito et al. 1999). Several protein phosphatases have been characterized in plants, implying that Tyr phosphorylation and dephosphorylation also serve important functions in signaling system (Luan 2002).

## 6.10 Role of NO in Salicylic Acid-Regulated Systemic Acquired Resistance

SA signaling system activates not only local resistance, but also systemic acquired resistance (SAR) observed in distal (systemic) tissues. SAR is a SA-dependent heightened defense to a broad spectrum of pathogens that is activated throughout a plant following local infection (Liu et al. 2011). Infection of plants by necrotizing pathogens, which induce the accumulation of SA, or treatment of plants with synthetic compounds, which are able to trigger SA signaling, causes the induction of a unique physiological state called "priming" (Slaughter et al. 2012). SAR is associated with priming of defense (Kohler et al. 2002; Luna et al. 2011) and the priming results in a faster and stronger induction of defense mechanisms after pathogen attack (Conrath 2011). The priming can be inherited epigenetically from disease-exposed plants (Pastor et al. 2013) and descendants of primed plants exhibit next-generation systemic acquired resistance (Luna et al. 2011; Slaughter et al. 2012). The transgenerational SAR has been recently reported (Luna et al. 2011).

NPR1 is an important regulator of SAR downstream of SA (Mou et al. 2003; Zhang et al. 2003). The events downstream of SA include an increase of NO (Zottini et al. 2007), which may then serve as signaling mediator itself (Krinke et al. 2007). NO is required for the full function of NPR1 in SA-triggered SAR (Song and Goodman 2001). Nuclear localization of NPR1 protein is essential for its function (Kinkema et al. 2000; Meur et al. 2006). Without induction, NPR1 protein forms an oligomer and is excluded from the nucleus. Redox changes cause monomeric NPR1 to emerge and accumulate in the nucleus and activate PR gene expression (Kinkema et al. 2000; Mou et al. 2003).

Plant immunity requires conformational changes of NPR1 via S-nitrosylation and thioredoxin (Tada et al. 2008). NPR1 is sequestered in the cytoplasm as an oligomer through intermolecular disulfide bonds. NO-mediated S-nitrosylation of NPR1 by S-nitrosoglutathione (GSNO) at Cys156 facilitates the NPR1 oligomerization, which maintains protein homeostasis upon SA induction. Conversely, the SA-induced NPR1 oligomer-to-monomer reaction is catalyzed by thioredoxins (TRX). Mutants in both NPR1 Cys156 and TRX compromised NPR1-mediated disease resistance response. Thus, the regulation of NPR1 is through opposing action of NO-dependent GSNO and ROS-dependent TRX (Tada et al. 2008). These results suggest that NO is involved in the action of NPR1 in triggering SAR. It also has been shown that NO-induced nitrosoglutathione could act as a long-distance phloematic signal in SAR (Durner and Klessig 1999).

## References

Ali R, Ma W, Lemtiri-Chlieh F, Tsaltas D, Leng Q, von Bodman S, Berkowitz GA (2007) Death don't have no mercy and neither does calcium: Arabidopsis CYCLIC NUCLEOTIDE GATED CHANNEL2 and innate immunity. Plant Cell 19:1081–1095

Allen GJ, Muir SR, Sanders D (1995) Release of $Ca^{2+}$ from individual plant vacuoles by both InsP3 and cyclic ADP-ribose. Science 268:735–737

Asai S, Ohta K, Yoshioka H (2008) MAPK signaling regulates nitric oxide and NADPH oxidase-dependent oxidative bursts in *Nicotiana benthamiana*. Plant Cell 20:1390–1406

Astier J, Lindermayr C (2012) Nitric oxide-dependent posttranslational modification in plants: an update. Int J Mol Sci 13:15193–15208

Bachmair A, Becker F, Masterson RV, Schell J (1990) Perturbation of the ubiquitin system causes leaf curling, vascular tissue alterations and necrotic lesions in a higher plant. EMBO J 9:4543–4549

Bachmair A, Novatchkova M, Potuschak T, Eisenhaber F (2001) Ubiquitylation in plants: a post-genomic look at a post-translational modification. Trends Plant Sci 6:463–470

Belenghi B, Romero-Puertas MC, Vercammen D, Brackenier A, Inzé D, Delledonne M, Van Breusegem F (2007) Metacaspase activity of *Arabidopsis thaliana* is regulated by S-nitrosylation of a critical cysteine residue. J Biol Chem 282:1352–1358

Bellin D, Asai S, Delledonne M, Yoshioka H (2013) Nitric oxide as a mediator for defense responses. Mol Plant Microbe Interact 26:271–277

Benhar M, Forrester MT, Hess DT, Stamler JS (2008) Regulated protein denitrosylation by cytosolic and mitochondrial thioredoxins. Science 320:1050–1054

Besson-Bard A, Courtois C, Gauthier A, Dahan J, Dobrowolska G, Jeandroz S, Pugin A, Wendehenne D (2008a) Nitric oxide in plants: production and cross-talk with $Ca^{2+}$ signaling. Mol Plant 1:218–228

Besson-Bard A, Pugin A, Wendehenne D (2008b) New insights into nitric oxide signaling in plants. Annu Rev Plant Biol 99:21–39

Bethke PC, Badger MR, Jones RL (2004) Apoplastic synthesis of nitric oxide by plant tissues. Plant Cell 16:332–341

Bright J, Desikan R, Hancock JT, Weir IS, Neill SJ (2006) ABA-induced NO generation and stomatal closure in *Arabidopsis* are dependent on $H_2O_2$ synthesis. Plant J 45:113–122

Brito C, Navillat M, Tiscomia AC, Vullier F, Gualco G, Dighiero G, Radi R, Cayota AM (1999) Peroxynitrite inhibits T lymphocyte activation and proliferation by promoting impairment of tyrosine phosphorylation and peroxynitrite-driven apoptotic death. J Immunol 162:3356–3366

Bryk R, Griffin P, Nathan C (2000) Peroxynitrite reductase activity of bacterial peroxiredoxins. Nature 407:211–215

Chamnongpol S, Willekens H, Moeder W, Langebartels C, Sandermann H Jr, Van Montagu M, Inzé D, Van Camp W (1998) Defense activation and enhanced pathogen tolerance induced by $H_2O_2$ in transgenic plants. Proc Natl Acad Sci U S A 95:5818–5823

Chang C, Stadler R (2001) Ethylene hormone receptor action in *Arabidopsis*. Bioessays 23:619–627

Choi HW, Lee DH, Hwang BK (2009) The pepper calmodulin gene CaCaM1 is involved in reactive oxygen species and nitric oxide generation required for cell death and the defense response. Mol Plant Microbe Interact 22:1389–1400

Chun HJ, Park HC, Koo SC, Lee JH, Park CY, Choi MS, Kang CH, Baek D, Cheong YH, Yun DJ, Chung WS, Cho MJ, Kim MC (2012) Constitutive expression of mammalian nitric oxide synthase in tobacco plants triggers disease resistance to pathogens. Mol Cells 34:463–471

Clark D, Durner J, Navarre DA, Klessig DF (2000) Nitric oxide inhibition of tobacco catalase and ascorbate peroxidase. Mol Plant Microbe Interact 13:1380–1384

Clarke JD, Volko SM, Ledford H, Ausubel FM, Dong X (2000) Roles of salicylic acid, jasmonic acid, and ethylene in *cpr*-induced resistance in Arabidopsis. Plant Cell 12:2175–2190

Clementi E (1998) Role of nitric oxide and its intracellular signaling pathways in the control of $Ca^{2+}$ homeostasis. Biochem Pharmacol 55:713–718

Collinge M, Boller T (2001) Differential induction of two potato genes, *Stprx2* and *StNAC* in response to infection by *Phytophthora infestans* and to wounding. Plant Mol Biol 46:521–529

Conrath U (2011) Molecular aspects of defence priming. Trends Plant Sci 16:524–531

Corpas FJ, Barroso JB, Carreras A, Valderrama R, Palma JM, León AM, Sandalio LM, del Rio LA (2006) Constitutive arginine-dependent nitric oxide synthase activity in different organs of pea seedlings during plant development. Planta 224:246–254

Courtois C, Besson A, Dahan J, Bourque S, Dobrowolska G, Pugin A, Wendehenne D (2008) Nitric oxide signaling in plants: interplays with $Ca^{2+}$ and protein kinases. J Exp Bot 59:155–163

Crawford NM, Galli M, Tischner R, Heimer YM, Okamoto M, Mack A (2006) Response to Zemojtel et al.: plant nitric oxide synthase: back to square one. Trends Plant Sci 11:526–527

Delledonne M (2005) NO news is good news for plants. Curr Opin Plant Biol 8:390–396

Delledonne M, Murgia I, Ederie D, Sbicego PF, Biondani A, Polverari A, Lamb C (2002) Reactive oxygen intermediates modulate nitric oxide signaling in the plant hypersensitive disease-resistance response. Plant Physiol Biochem 40:605–610

Delledonne M, Polverari A, Murgia I (2003) The functions of nitric oxide-mediated signaling and changes in gene expression during the hypersensitive response. Antioxid Redox Signal 5:33–41

Delledonne M, Xia Y, Dixon RA, Lamb C (1998) Nitric oxide functions as a signal in plant disease. Nature 394:585–588

Delledonne M, Zeier J, Marocco A, Lamb C (2001) Signal interactions between nitric oxide and reactive oxygen intermediates in the plant hypersensitive disease resistance response. Proc Natl Acad Sci U S A 98:13454–13459

del Rio LA, Corpas FJ, Barroso JB (2004) Nitric oxide and nitric oxide synthase activity in plants. Phytochemistry 65:783–792

Desikan R, Griffiths R, Hancock J, Neill S (2002) A new role for an old enzyme: nitrate reductase-mediated nitric oxide generation is required for abscisic acid-induced stomatal closure in *Arabidopsis thaliana*. Proc Natl Acad Sci U S A 99:16314–16318

Dietz KJ (2005) Plant peroxiredoxins. Annu Rev Plant Biol 54:93–107

Durner J, Klessig DF (1999) Nitric oxide as a signal in plants. Curr Opin Plant Biol 2:369–374

Durner J, Shah J, Klessig DF (1997) Salicylic acid and disease resistance in plants. Trends Plant Sci 2:266–274

Durner J, Wendehenne D, Klessig DF (1998) Defense gene induction in tobacco by nitric oxide, cyclic GMP, and cyclic ADP ribose. Proc Natl Acad Sci U S A 95:10328–10333

Espunya C, De Michele R, Gomez-Cadenas AA, Martinez MC (2012) S-Nitrosoglutathione is a component of wound- and salicylic acid-induced systemic responses in *Arabidopsis thaliana*. J Exp Bot 63:3219–3227

Feechan A, Kwon E, Yun B, Wang Y, Pallas JA, Loake GJ (2005) A central role for *S*-nitrosothiols in plant disease resistance. Proc Natl Acad Sci U S A 192:8054–8059

Fliegert R, Gasser A, Guse AH (2007) Regulation of calcium signalling by adenine-based second messengers. Biochem Soc Trans 35:109–114

Foissner L, Wendehenne D, Langebartels C, Durner J (2000) In vivo imaging of an elicitor-induced nitric oxide burst in tobacco. Plant J 23:817–824

Foster MW, Stamler JS (2004) New insights into protein S-nitrosylation: mitochondria as a model system. J Biol Chem 279:25891–25897

Garcia-Mata C, Gay R, Sokolovski S, Hills A, Lamattina L, Blatt MR (2003) Nitric oxide regulates $K^+$ and $Cl^-$ channels in guard cells through a subset of abscisic acid-evoked signaling pathways. Proc Natl Acad Sci U S A 100:11116–11121

Gaston B (1999) Nitric oxide and thiol groups. Biochim Biophys Acta 1411:323–333

Gould KS, Lamotte O, Klinguer A, Pugin A, Wendehenne D (2003) Nitric oxide production in tobacco leaf cells: a generalized stress response? Plant Cell Environ 26:1851–1862

Grennan AK (2007) Protein *S*-nitrosylation: protein targets and roles in signal transduction. Plant Physiol 144:1237–1239

Grün S, Lindermayr C, Sell S, Durner J (2006) Nitric oxide and gene regulation in plants. J Exp Bot 57:507–516

Guo F-Q, Okamoto M, Crawford NM (2003) Identification of a plant nitric oxide synthase gene involved in hormonal signaling. Science 302:100–103

Gupta KJ, Brotman Y, Segu S, Zeier T, Zeier J, Persijn ST, Cristescu SM, Harren FJM, Bauwe H, Fernie AR, Kaiser WM, Mur LAJ (2013) The form of nitrogen nutrition affects resistance against *Pseudomonas syringae* pv. *phaseolicola* in tobacco. J Exp Bot 64:553–568

Hanafy KA, Krumenacker JS, Murad F (2001) NO, nitrotyrosine, and cyclic GMP in signal transduction. Med Sci Monit 7:801–819

Hancock J, Desikan R, Harrison J, Bright J, Hooley R, Neill S (2006) Doing the unexpected: proteins involved in hydrogen peroxide perception. J Exp Bot 57:1711–1718

Horling F, Lamkemeyer P, Konig J, Finkemeier L, Kandlbinder A, Baier M, Dietz KJ (2003) Divergent light-, ascorbate- and oxidative stress-dependent regulation of expression of the peroxiredoxin gene family in *Arabidopsis*. Plant Physiol 131:317–325

Huang X, von Rad U, Durner J (2002) Nitric oxide induces transcriptional activation of the nitric oxide-tolerant alternative oxidase in Arabidopsis suspension cells. Planta 215:914–923

Ichiropoulos H, al-Mehdi AB (1995) Peroxynitrite-mediated oxidative protein modifications. FEBS Lett 364:279–282

Kinkema M, Fan W, Dong X (2000) Nuclear localization of NPR1 is required for activation of PR gene expression. Plant Cell 12:2339–2350

Klessig DF, Durner J, Noad R, Navarre DA, Wendehenne D, Kumar D, Zhou JM, Shah J, Zhang S, Kachroo P, Trifa Y, Pontier D, Lam E, Silva H (2000) Nitric oxide and salicylic acid signalling in plant defense. Proc Natl Acad Sci U S A 97:8849–8855

Klotz LO, Schroeder P, Sies H (2002) Peroxynitrite signaling: receptor tyrosine kinases and activation of stress-responsive pathways. Free Radic Biol Med 33:737–743

Kohler A, Schwindling S, Conrath U (2002) Benzothiadiazole-induced priming for potentiated responses to pathogen infection, wounding, and infiltration of water into leaves requires the *NPR1/NIM1* gene in *Arabidopsis*. Plant Physiol 128:1046–1056

Krinke O, Ruelland E, Valentová O, Vergnolle C, Renou J-P, Taconnat L, Flemr M, Burketova L, Zachowski A (2007) Phosphatidylinositol 4-kinase activation is an early response to salicylic acid in *Arabidopsis* suspension cells. Plant Physiol 144:1347–1359

Kumar D, Klessig DF (2000) Differential induction of tobacco MAP kinases by the defense signals nitric oxide, salicylic acid, ethylene, and jasmonic acid. Mol Plant Microbe Interact 13:347–351

Kwaaitaal M, Huisman R, Maintz J, Reinstädler A, Panstruga R (2011) Ionotropic glutamate receptor (iGluR)-like channels mediate MAMP-induced calcium influx in *Arabidopsis thaliana*. Biochem J 440:355–365

Lamotte O, Courtois C, Dobrowolska G, Besson A, Pugin A, Wendehenne D (2006) Mechanisms of nitric oxide-induced increase of free cytosolic $Ca^{2+}$ concentration in *Nicotiana plumbaginifolia* cells. Free Radic Biol Med 40:1369–1376

Lamotte O, Gould K, Lecourieux D, Sequeira-Legrand A, Lebrun-Garcia A, Durner J, Pugin A, Wendehenne D (2004) Analysis of nitric oxide signaling functions in tobacco cells challenged by the elicitor cryptogein. Plant Physiol 135:516–529

Laohavisit A, Mortimer JC, Demidchik V, Coxon KM, Stancombe MA, Macpherson N, Brownlee C, Hofmann A, Webb AAR, Miedema H, Battey NH, Davies JM (2009) *Zea mays* annexins modulate cytosolic free $Ca^{2+}$ and generate a $Ca^{2+}$-permeable conductance. Plant Cell 21:479–493

Leshem YY, Wills RBH, Veng-Va Ku V (1998) Evidence for the function of the free radical gas-nitric oxide (NO) – as an endogenous maturation and senescence regulating factor in higher plants. Plant Physiol Biochem 36:825–833

Levine A, Tenhaken R, Dixon R, Lamb C (1994) $H_2O_2$ from the oxidative burst orchestrates the plant hypersensitive disease resistance response. Cell 79:583–593

Lecourieux-Ouaked F, Pugin A, Lebrun-Garcia A (2000) Phosphoproteins involved in the signal transduction of cryptogein, an elicitor of defense reactions in tobacco. Mol Plant Microbe Interact 13:821–829

Lindermayr C, Saalbach G, Bahnweg G, Durner J (2006) Differential inhibition of *Arabidopsis* methionine adenosyltransferases by protein *S*-nitrosylation. J Biol Chem 281:4285–4291

Lindermayr C, Saalbach G, Durner J (2005) Proteomic identification of *S*-nitrosylated proteins in *Arabidopsis*. Plant Physiol 137:921–930

Lindermayr C, Sell S, Müller B, Leister D, Dumer J (2010) Redox regulation of NPR1-TGA1 system of Arabidopsis thaliana by nitric oxide. Plant Cell 22:2894–2907

Liu L, Hausladen A, Zeng M, Que L, Heitman J, Stamler JS (2001) A metabolic enzyme for *S*-nitrosothiol conserved from bacteria to humans. Nature 410:490–494

Liu P-P, von Dahl CC, Klessig DF (2011) The extent to which methyl salicylate is required for signaling systemic acquired resistance is dependent on exposure to light after infection. Plant Physiol 157:2216–2226

Liu L, Yan Y, Zeng M, Zhang J, Hanos MA, Ahearn G, McMahon TJ, Dickfeld T, Marshall HE, Quo LG, Stamler JS (2004) Essential roles of S-nitrosothiols in vascular homeostasis and endotoxic shock. Cell 116:617–628

Lü D, Zhang X, Jiang J, An GY, Zhang I, Song C-P (2005) NO may function in the downstream of $H_2O_2$ in ABA-induced stomatal closure in *Vicia faba* L. J Plant Physiol Mol Biol 31:62–70

Luan S (2002) Tyrosine phosphorylation in plant cell signaling. Proc Natl Acad Sci U S A 99:11567–11569

Luna E, Bruce TJA, Roberts MR, Flors V, Ton J (2011) Next generation systemic acquired resistance. Plant Physiol 158:844–853

Ma W, Berkowitz GA (2007) The grateful dead: calcium and cell death in plant innate immunity. Cell Microbiol 9:2571–2585

Ma W, Berkowitz GA (2011) $Ca^{2+}$ conduction by plant cyclic nucleotide gated channels and associated signaling components in pathogen defense signal transduction cascades. New Phytol 190:566–572

Ma W, Smigel A, Tsai YC, Braam J, Berkowitz GA (2008) Innate immunity signaling: cytosolic $Ca^{2+}$ elevation is linked to downstream nitric oxide generation through the action of calmodulin or a calmodulin-like protein. Plant Physiol 148:818–828

Ma W, Yoshioka K, Berkowitz GA (2007) Cyclic nucleotide gated channels and $Ca^{2+}$-mediated signal transduction during plant innate immune response to pathogens. Plant Signal Behav 2:548–550

Ma Y, Walker RK, Zhao Y, Berkowitz GA (2012) Linking ligand perception by PEPR pattern recognition receptors to cytosolic $Ca^{2+}$ elevation and downstream immune signaling in plants. Proc Nat Acad Sci U S A 109:19852–19857

Malik SI, Hussain A, Yun BW, Spoel SH, Loake GJ (2011) GSNOR-mediated de-nitrosylation in the plant defence response. Plant Sci 181:540–544

Mannick JB, Schonhoff CM (2004) NO means no and yes. Regulation of cell signaling by protein nitrosylation. Free Radic Res 38:1–7

Martinez-Ruiz A, Lamas S (2004) *S*-Nitrosylation: a potential new paradigm in signal transduction. Cadiovasc Res 62:43–52

Melotto M, Underwood W, Koczan J, Nomura K, He SY (2006) Plant stomata function in innate immunity against bacterial invasion. Cell 126:969–980

Meur G, Budatha M, Gupta AD, Prakash S, Kirti PB (2006) Differential induction of NPR1 during defense responses in *Brassica juncea*. Physiol Mol Plant Pathol 68:128–137

Minorski PV (2003) The hot and classic. Plant Physiol 131:1578–1579

Modolo LV, Augusto O, Almeida IMG, Pinto-Maglio CAF, Oliveira HC, Seligman K, Salgado I (2002) Decreased arginine and nitrite levels in nitrate reductase-deficient *Arabidopsis thaliana* plants impair nitric oxide synthesis and the hypersensitive response to *Pseudomonas syringae*. Plant Sci 171:34–40

Morot-Gaudry-Talarmain Y, Rockel P, Moureaux T, Quilleré I, Leydecker MT, Kaiser WM, Morot-Gaudry F (2002) Nitrite accumulation and NO emission in relation to cellular signalling in NiR antisense tobacco. Planta 215:708–715

Mou Z, Fan W, Dong X (2003) Inducers of plant systemic acquired resistance regulate NPR1 function through redox changes. Cell 113:935–944

Mur LA, Mandon J, Persijn S, Cristescu SM, Moshkov GV, Novikova GV, Hall MA, Harren FJ, Hebelstrup KH, Gupta KJ (2013) Nitric oxide in plants: an assessment of the current state of knowledge. AoB Plants 5:plso52. doi:10.1093/aobpla/pls052

Mur LAJ, Santosa IE, Laarhoven LJJ, Holton NJ, Harren FJM, Smith AR (2005) Laser photoacoustic detection allows in planta detection of nitric oxide in tobacco following challenge with avirulent and virulent *Pseudomonas syringae* pathovars. Plant Physiol 138:1247–1258

Neill SJ, Desikan R, Clarke A, Hurst RD, Hancock JT (2002) Hydrogen peroxide and nitric oxide as signaling molecules in plants. J Exp Bot 53:1237–1247

Neill S, Bright J, Desikan R, Hancock J, Harrison J, Wilson I (2008) Nitric oxide evolution and perception. J Exp Bot 59:25–35

Pagnussat GC, Lanteri ML, Lombardo MC, Lamattina L (2004) Nitric oxide mediates the indole acetic acid induction activation of a mitogen-activated protein kinase cascade involved in adventitious root development. Plant Physiol 135:279–286

Palmieri MC, Sell S, Huang X, Scherf M, Werner T, Durner J, Lindermayr C (2008) Nitric oxide-responsive genes and promoters in *Arabidopsis thaliana*: a bioinformatics approach. J Exp Bot 59:177–180

Pastor V, Luna E, Mauch-Mani B, Ton J, Flors V (2013) Primed plants do not forget. Environ Exp Bot 94:46–56

Perchepied L, Balagué C, Riou C, Claudel-Renard C, Riviére N, Grezes-Besset B, Roby D (2010) Nitric oxide participates in the complex interplay of defense-related signaling pathways controlling disease resistance to *Sclerotinia sclerotiorum* in *Arabidopsis thaliana*. Mol Plant Microbe Interact 23:846–860

Planchet E, Gupta KJ, Sonoda M, Zeier J, Kaiser WM (2005) Nitric oxide emission from tobacco leaves and cell suspensions: rate limiting factors and evidence for the involvement of mitochondrial electron transport. Plant J 41:732–743

Planchet E, Kaiser WM (2006) Nitric oxide production in plants. Plant Signal Behav 1:46–51

Planchet E, Sonoda M, Zeier J, Kaiser WM (2006) Nitric oxide (NO) as an intermediate in cryptogein-induced hypersensitive response: a critical evaluation. Plant Cell Environ 29:59–69

Polverari A, Molesini B, Pezzotti M, Buonaurio R, Marte M, Delledonne M (2003) Nitric oxide-mediated transcriptional changes in *Arabidopsis thaliana*. Mol Plant Microbe Interact 16:1094–1105

Radi R (2004) Nitric oxide, oxidants, and protein tyrosine nitration. Proc Natl Acad Sci U S A 101:4003–4008

Rentel MC, Lecourieux D, Ouaked F, Usher SL, Petersen L, Okamoto H, Knight H, Peck SC, Grierson CS, Hirt H, Knight MR (2004) OXI1 kinase is necessary for oxidative burst-mediated signaling in *Arabidopsis*. Nature 427:858–861

Rockel P, Strube F, Rockel A, Wildt J, Kaiser WM (2002) Regulation of nitric oxide (NO) production by plant nitrate reductase in vivo and in vitro. J Exp Bot 53:103–110

Romero-Puertas MC, Campostrini NN, Matté A, Righetti PG, Perazzolli M, Zolla L, Roepstorff P, Delledonne M (2008) Proteomic analysis of *S*-nitrosylated proteins in *Arabidopsis thaliana* undergoing hypersensitive response. Proteomics 8:1459–1469

Romero-Puertas MC, Laxa M, Matté A, Zaninotto F, Finkemeier I, Jones AME, Perazzolli M, Vandelle E, Dietz K-J, Delledonne M (2007) *S*-nitrosylation of peroxiredoxin II E promotes peroxynitrite-mediated tyrosine nitration. Plant Cell 19:4120–4130

Romero-Puertas MC, Perazzolli M, Zago ED, Delledonne M (2004) Nitric oxide signalling functions in plant-pathogen interactions. Cell Microbiol 6:795–803

Rustérucci C, Espunya MC, Diaz M, Chabannes M, Martinez MC (2007) *S*-Nitrosoglutathione reductase affords protection against pathogens in Arabidopsis, both locally and systemically. Plant Physiol 143:1282–1292

Saito S, Yamamoto-Katou A, Yoshioka H, Doke N, Kawakita K (2006) Peroxynitrite generation and tyrosine nitration in defense responses in tobacco BY-2 cells. Plant Cell Physiol 47:689–697

Sakomoto A, Ueda M, Morikawa H (2002) *Arabidopsis* glutathione-dependent formaldehyde dehydrogenase is an S-nitrosoglutathione reductase. FEBS Lett 515:20–24

Schenk PM, Kazan K, Wilson IW, Anderson JP, Richmond T, Somerville SC, Manners JM (2000) Coordinated plant defense responses in *Arabidopsis* revealed by microarray analysis. Proc Natl Acad Sci U S A 97:11655–11660

Schopfer FJ, Baker PR, Freeman BA (2002) NO-dependent protein nitration: a cell signaling event or an oxidative inflammatory response? Trends Biochem Sci 28:646–654

Slaughter A, Daniel X, Flors V, Luna E, Hohn B, Mauch-Mani B (2012) Descendants of primed *Arabidopsis* plants exhibit resistance to biotic stress. Plant Physiol 158:835–843

Sokolovski S, Hills A, Gay R, Garcia-Mata C, Lamattina L, Blatt MR (2005) Protein phosphorylation is a prerequisite for intracellular $Ca^{2+}$ release and ion channel control by nitric oxide and abscisic acid in guard cells. Plant J 43:520–529

Song F, Goodman RM (2001) Molecular biology of disease resistance in rice. Physiol Mol Plant Pathol 59:1–11

Stamler JS, Lamas S, Fang FC (2001) Nitrosylation: the prototypic redox-based signaling mechanism. Cell 106:675–683

Swidzinski JA, Sweetlove LJ, Leaver CJ (2002) A custom microarray analysis of gene expression during programmed cell death in *Arabidopsis thaliana*. Plant J 30:431–446

Tada Y, Spoel SH, Pajerowska-Mukhtar K, Mou Z, Song J, Dong X (2008) Plant immunity requires conformational changes of NPR1 via S-nitrosylation and thioredoxin. Science 321:952–956

Tischner R, Koltermann M, Haesse H, Plath M (2010) Early responses of *Arabidopsis thaliana* to infection by *Verticillium longisporum*. Physiol Mol Plant Pathol 74:419–427

Tun NN, Santa-Catarina C, Begum T, Silveira V, Handro W, Floh EIS, Scherer GFE (2006) Polyamines induce rapid biosynthesis of nitric oxide (NO) in *Arabidopsis thaliana* seedlings. Plant Cell Physiol 47:346–354

Vadassery J, Oelmüller R (2009) Calcium signaling in pathogenic and beneficial plant microbe interactions. What can we learn from the interaction between *Piriformospora indica* and *Arabidopsis thaliana*. Plant Signal Behav 411:1024–1027

Vandelle E, Delledonne M (2011) Peroxynitrite formation and function in plants. Plant Sci 181:534–539

Vandelle E, Poinssot B, Wendehenne D, Bentéjac M, Pugin A (2006) Integrated signaling network involving calcium, nitric oxide, active oxygen species but not mitogen-activated protein kinases in BcPG-1 elicited grapevine defenses. Mol Plant Microbe Interact 19:429–440

Vatsa P, Chiltz A, Bourque S, Wendehenne D, Garcia-Brugger A, Pugin A (2011) Involvement of putative glutamate receptors in plant defence signaling and NO production. Biochimie 93:2095–2101

Vidhyasekaran P (2007a) Fungal pathogenesis in plants and crops: molecular biology and host defense mechanisms, IIth edn. CRC Press/Taylor & Francis Group, Boca Raton, p 510

Vidhyasekaran P (2007b) Handbook of molecular technologies in crop disease management. Haworth Press, Binghamton, p 599

Vincill ED, Bieck AM, Spalding EP (2012) $Ca^{2+}$ conduction by an amino acid-gated ion channel related to glutamate receptors. Plant Physiol 159:40–46

Voehringer DW, Hirschberg DL, Xiao J, Lu Q, Roederer M, Lock CB, Herzenberg LA, Steinman L, Herzenberg LA (2000) Gene microarray identification of redox and mitochondrial elements that control resistance or sensitivity to apoptosis. Proc Natl Acad Sci U S A 97:2680–2685

Wang P, Du Y, Li Y, Ren D, Song C-P (2010) Hydrogen peroxide-mediated activation of MAP kinase 6 modulates nitric oxide biosynthesis and signal transduction in *Arabidopsis*. Plant Cell 22:2981–2998

Wang P, Song C-P (2008) Guard-cell signaling for hydrogen peroxide and abscisic acid. New Phytol 178:703–718

Willmott N, Sethi JK, Walseth TF, Lee H, White AM, Galione A (1996) Nitric oxide–induced mobilization of intracellular calcium via the cyclic ADP-ribose signalling pathway. J Biol Chem 271:3699–3705

Wilson ID, Neill SJ, Hancock JT (2008) Nitric oxide synthesis and signalling in plants. Plant Cell Environ 31:622–631

Wünsche H, Baldwin IT, Wu J (2011) *S*-Nitrosoglutathione reductase (GSNOR) mediates the biosynthesis of jasmonic acid and ethylene induced by feeding of the insect herbivore *Manduca sexta* and is important for jasmonate-elicited responses in *Nicotiana attenuata*. J Exp Bot 62:4605–4616

Xu YC, Cao YL, Guo P, Tao Y, Zhao BL (2004) Detection of nitric oxide in plants by electron spin resonance. Phytopathology 94:402–407

Yamamoto-Katou A, Katou S, Yoshioka H, Doke N, Kawakita K (2006) Nitrate reductase is responsible for elicitin-induced nitric oxide production in *Nicotiana benthamiana*. Plant Cell Rep 47:726–735

Yamasaki H (2000) Nitrite dependent nitric oxide production pathway: implications for involvement of active nitrogen species in photoinhibition in vivo. Philos Trans R Soc Lond 355:1477–1488

Yamasaki H, Cohen MF (2006) NO signal at the crossroads: polyamine-induced nitric oxide synthesis in plants? Trends Plant Sci 11:522–524

Yamasaki H, Sakihama Y (2000) Simultaneous production of nitric oxide and peroxynitrite by plant nitrate reductase: in vitro evidence for the NR-dependent formation of active nitrogen species. FEBS Lett 468:89–92

Yan J, Tsuichihara N, Etoh T, Iwai S (2007) Reactive oxygen species and nitric oxide are involved in ABA inhibition of stomatal opening. Plant Cell Environ 30:1320–1325

Zago E, Morsa S, Dat JF, Alard P, Ferrarini A, Inzé D, Delledonne M, Van Breusegem F (2006) Nitric oxide- and hydrogen peroxide-responsive gene regulation during cell death induction in tobacco. Plant Physiol 141:404–411

Zaninotto F, Camera SL, Polverari A, Delledonne M (2006) Cross talk between reactive nitrogen and oxygen species during the hypersensitive disease resistance response. Plant Physiol 141:379–383

Zeidler D, Zahringer U, Gerber I, Dubery I, Hartung T, Bors W, Hutzler P (2004) Innate immunity in *Arabidopsis thaliana*: Lipopolysaccharides activate nitric oxide synthase (NOS) and induce defense genes. Proc Natl Acad Sci U S A 101:15811–15816

Zemojtel T, Fröhlich A, Palmieri MC, Kolanczyk M, Mikula I, Wyrwicz LS, Wanker EE, Mundlos S, Vingron M, Martasek P, Durner J (2006) Plant nitric oxide synthase: a never-ending story? Trends Plant Sci 11:524–525

Zhang C, Czymmek KJ, Shapiro AD (2003) Nitric oxide does not trigger early programmed cell death events but may contribute to cell-to-cell signaling governing progression of the *Arabidopsis* hypersensitive response. Mol Plant Microbe Interact 16:962–972

Zhang X, Dai Y, Xiong Y, DeFraia C, Li J, Dong X, Mou Z (2007) Overexpression of Arabidopsis *MAP kinase kinase 7* leads to activation of plant basal and systemic acquired resistance. Plant J 52:1066–1079

Zhao XM, She XP, Yu W, Liang XM, Du YG (2007) Effects of oligochitosans on tobacco cells and role of endogenous nitric oxide burst in resistance of tobacco to *Tobacco mosaic virus*. J Plant Pathol 89:55–65

Zottini M, Costa A, De Michele R, Ruzzene M, Carimi F, Lo Schiavo F (2007) Salicylic acid activates nitric oxide synthesis in *Arabidopsis*. J Exp Bot 58:1397–1405

# Chapter 7
# Mitogen-Activated Protein Kinase Cascades in Plant Innate Immunity

**Abstract** Mitogen-activated protein kinase cascades are major signal transduction systems functioning downstream of pattern recognition receptors upon perception of PAMP elicitor signals. The MAPKs transduce extracellular stimuli into intracellular transcription factors through activation of $Ca^{2+}$, ROS, SA, JA, and ethylene-dependent signaling systems, and enhance expression of defense-related genes in plant innate immune system. A typical MAPK signaling module consists of three kinases: a MAP kinase kinase kinase (MAPKKK), a MAP kinase kinase (MAPKK), and a MAP kinase (MAPK). MAPKs function at the bottom of the three-kinase cascade and are activated by MAPKKs through phosphorylation. The activity of MAPKKs is, in turn, regulated by MAPKKKs via phosphorylation. MAPKKKs receive PAMP signals from upstream receptors/sensors to activate the MAPK signaling system. Several distinct MAPK pathways are involved in activating the plant immune system. Rather than linear pathways, multiple interconnected MAPK pathways may be required to transmit elicitor signals and integrate defense responses. Some MAP kinase cascades are involved in biosynthesis of SA, JA, and ethylene, which are key plant hormones modulating plant innate immune systems. MAP kinases play important role in interplay and crosstalk of the plant hormone signaling systems. Some MAP kinases have been found to be involved in priming defense responses in systemic acquired resistance. MAP kinases modulate phosphorylation of transcription factors to trigger transcription of defense genes. 14-3-3 proteins enhance the activity of MAPKKKs in triggering phosphorylation. MAP kinases modulate stomatal closure immune responses. Enhanced expression of some MAP kinases trigger enhanced defense responses against a wide spectrum of viral, bacterial, fungal, and oomycete pathogens. Potential pathogens secrete effectors to suppress MAPK signaling system and enhance plant susceptibility.

**Keywords** MAPK • MAPKK • MAPKKK • Three-kinase cascade • Priming • 14-3-3 proteins • Phosphorylation • Effectors

P. Vidhyasekaran, *PAMP Signals in Plant Innate Immunity: Signal Perception and Transduction*, Signaling and Communication in Plants 21,
DOI 10.1007/978-94-007-7426-1_7, © Springer Science+Business Media Dordrecht 2014

## 7.1 MAPK Signaling Three-Kinase Modules

Mitogen-activated protein kinase (MAPK) cascades are major pathways downstream of sensors/receptors that transduce extracellular stimuli into intracellular responses in plants (Liu et al. 2003; Pedley and Martin 2005; Tena et al. 2011; Hettenhausen et al. 2012; Zhang et al. 2012a). A typical MAPK signaling module consists of three interconnected protein kinases: a MAP kinase kinase kinase (MAPKKK or MEKK [for MAPK/Extracellular signal-regulated kinase Kinase Kinase]), a MAP kinase kinase (MAPKK or MKK), and a MAP kinase (MAPK or MPK) (Mészáros et al. 2006). MAP kinase cascade involves sequence of phosphorylation events (Hirt 2000). MAPKs function at the bottom of the three-kinase cascade and are activated by MAPKKs through phosphorylation on the Thr and Tyr residues in their activation motif between the kinase subdomain VII and VIII. The activity of MAPKKs is, in turn, regulated by MAPKKKs via phosphorylation of two Ser/Thr residues in the activation loop of MAPKKs. MAPKKKs receive signals from upstream receptors/sensors (Ichimura et al. 2002; Li et al. 2012).

MAPKs (MPKs) are serine/threonine protein kinases with two-lobed structure (Hirt 2000). The active site contains the MAPK-specific T-X--Y (threonine-X-tyrosine, where X denotes any amino acid) motif that is targeted by MAPKKS (MKKs). The MAPKKs are dual-specificity protein kinases that activate MAPKs by phosphorylation of both the threonine and tyrosine residue of the T-X-Y motif located between kinase subdomains VII and VIII (Ligterink and Hirt 2000; Liu et al. 2000). MAPKKs are activated themselves by phosphorylation of two conserved serine or threonine residues (S/TXXXS/T) by MAPKKKs (MEKKs) (Hirt 2000). MAPKKKs contain different regulatory motifs, proline-rich sequences involved in Src homology3 binding, zinc finger motifs, leucine zippers, and binding sites for G proteins (Hirt 2000). MAPKKKs can be activated by various elicitors, pathogens and a wide range of stress stimuli (Teige et al. 2004).

MAP kinase cascade components are abundant in plants. There are more than 80 putative MEKKS, 10 MKKs, and at least 20 MPKs in *Arabidopsis* (Ichimura et al. 2002; Jonak et al. 2002; Nakagami et al. 2005). MAPKs constitute a large gene family with 20 family members in *Arabidopsis*, 15 in rice and 21 in *Populus* spp. (MAPK Group 2002; Hamel et al. 2006). Several MAPKKKs, MAPKKs, and MAPKs have been identified in tobacco (Zhang and Klessig 1998a, b; Romeis et al. 1999; Yang et al. 2001), tomato (Mayrose et al. 2004; Kandoth et al. 2007; Stulemeijer et al. 2007), rice (He et al. 1999; Agrawal et al. 2003; Cheong et al. 2003; Xiong and Yang 2003), alfalfa (Cardinale et al. 2002), pea (Uppalapati et al. 2004), and parsley (Ligterink et al. 1997). These also include salicylic acid-induced protein kinase (SIPK) (Zhang and Klessig 1998b; Yang et al. 2001), wound-induced protein kinase (WIPK) (Seo et al. 1995; Zhang et al. 2000; Jin et al. 2003; Waller et al. 2006), elicitor-responsive MAPK (ERMK) (Ligterink et al. 1997), stress-activated MAPK (SAMK) (Nakagami et al. 2004), and salt-induced MAPK (SIMK) (Nakagami et al. 2004). Three MAPKKs, MKK1, PRKK, and SIMKK have been isolated from alfalfa (Cardinale et al. 2002; Nakagami et al. 2004), and two MAPKKs, MKK1

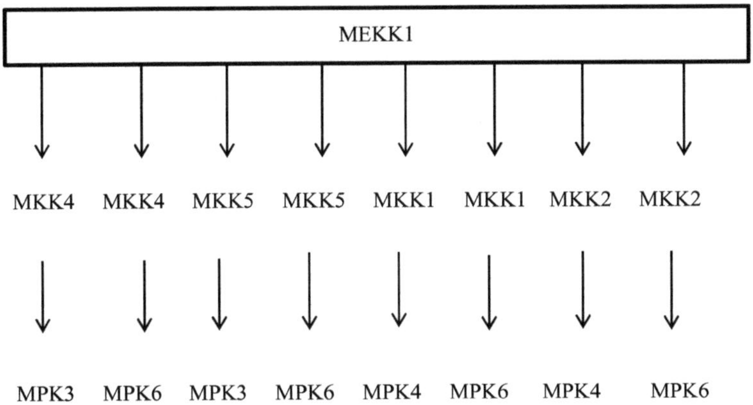


**Fig. 7.1** MAPK pathways with the common MAPKKK component MEKK1 in *Arabidopsis thaliana*

(also called as MEKI) and MKK2 have been detected in *Arabidopsis* (Ichimura et al. 1998; Nakagami et al. 2004). A MAPKK, NbMKK1, has been detected in *Nicotiana benthamiana* and the protein is localized to the nucleus (Takahashi et al. 2007b). The MAPKKKs MEKK1, ANP1 (Arabidopsis NPK1-related 1), and CTR1 (constitutive triple-response 1) in *Arabidopsis* and NPK1 (a homolog of ANP1) in tobacco have been reported to be the components of MAPK cascade (Kovtun et al. 2000; Nakagami et al. 2004). OMTK1 (oxidative stress-activated MAP triple kinase 1) is a MAPKKK from alfalfa (Nakagami et al. 2004).

MAPKKKs can feed into multiple MAPK pathways (Fig. 7.1). For example, in *Arabidopsis* the MAPKKK MEKK1 acts in the MEKK1 – MKK4 – MPK3, MEKK1 – MKK4 – MPK6, MEKK1 – MKK5 – MPK3, MEKK1 – MKK5 – MPK6, MEKK1 – MKK1 – MPK4, MEKK1 – MKK1 – MPK6, MEKK1 – MKK2 – MPK4, and MEKK1 – MKK2 – MPK6 pathways (Ichimura et al. 1998; Mizoguchi et al. 1998; Asai et al. 2002; Teige et al. 2004; Mészáros et al. 2006; Takahashi et al. 2007a). By contrast, MAPKKs usually have restricted substrate specificity, functioning mainly in a single cascade. Four MAPKKs – MKK2, MKK3, MKK4, and MKK5 – activate MPK6 in *Arabidopsis* (Takahashi et al. 2007a). MKK2, MKK3, MKK4, and MKK5 interact with MPK6 to constitute different MAPK complexes to transduce different signals and cross talk with each other (Takahashi et al. 2007a). The MKK3 activates four MAPKs, MPK1, MPK2, MPK7, and MPK14 (Dóczi et al. 2007). The MKK2 activates two MAP kinases, MPK4 and MPK6 (Brader et al. 2007). Another distinct MAP kinase module in *Arabidopsis* consists of MEKK1-MKK1 and MKK2-MPK4 (Ichimura et al. 1998; Mizoguchi et al. 1998; Zhang et al. 2012a). Although MKK1 and MKK2 form complexes with identical upstream and downstream kinases, they appear to have distinctly different functions. While MKK1 is activated by biotic elicitors, MKK2 is activated by abiotic stress signals (Teige et al. 2004). MPK4 is phosphorylated and activated by MKK1 (Huang et al. 2000; Matsuoka et al.

2002; Teige et al. 2004). MPK4 is activated by both biotic and abiotic stresses (Droillard et al. 2004; Teige et al. 2004). MAPKKs can also activate multiple MAPKs in *Medicago*, tobacco, and *Arabidopsis* (Yang et al. 2001; Asai et al. 2002; Cardinale et al. 2002; Jin et al. 2003). The tobacco MAPK kinase, NtMEK2, activates three different MAP kinases, SIPK, WIPK, and Ntf4 (Ren et al. 2006).

The MAPKK pair MKK4 and MKK5 seems to mediate elicitor-induced MPK3 and MPK6 activation (Asai et al. 2002). Another MAPKK pair of MKK1 and MKK2 was found to interact, phosphorylate, and activate MPK4, while only MKK2 was able to target also MPK6 (Teige et al. 2004). MAPKKs are multifunctional entry routes for upstream signal integration as well as bifurcation points for activation of downstream MAPKs (Teige et al. 2004). Different kinases are assembled into distinct modules by scaffold proteins. Scaffold proteins are important for preventing cross-talk between different cascades and allow a given kinase in more than one module without affecting the specificity of the response (Hirt 2000).

Several distinct MAPK pathways have been detected in plants (Cardinale et al. 2000; Pedley and Martin 2005; Mészáros et al. 2006). Various stimuli differentially induce the highly varying MAPK pathways. For example, the *Arabidopsis* MAPKK MKK2 is activated by salt and cold stress, but not by the MAMP elicitors flagellin, or laminarin. In contrast, the *Arabidopsis* MKK1 is activated by flagellin or laminarin, but not by salt or cold stress (Teige et al. 2004). SA strongly induced p48 and p44 MAPKs in pea, while these kinases were not activated by JA (Uppalapati et al. 2004). Four specific MAPK pathways involving MMK2 (for M*edicago* MAPK2), MMK3, SAMK (for stress-activated MAPK), and SIMK (for salicylate-induced protein kinase) in alfalfa were found to be activated to different levels and with different kinetics by four different elicitors, chitin, β-glucan, ergosterol, and yeast extract (Cardinale et al. 2000). Bacterial elicitor strongly activated MPK6 but resulted in poor activation of MPK7. MPK6 and MPK7 were both activated by $H_2O_2$ (Dóczi et al. 2007). The tomato MAPKs, LeMPK1 and LeMPK2, were activated in response to systemin, four different oligosaccharide elicitors, and UV-B radiation. However, another tomato MAPK LeMPK3 was only activated by UV-B radiation (Holley et al. 2003). The tobacco MAP kinase SIPK is activated both by the oomycete elicitor, β-megaspermin and the bacterial elicitor $hrpZ_{psph}$ (Hall et al. 2007). However, SIPK activation induced by the oomycete elicitor required external calcium influx, whereas that induced by the bacterial elicitor does not. It suggests that SIPK activation is involved in different elicitor-initiated signaling pathways (Hall et al. 2007).

## 7.2 MAP Kinases Involved in Plant Immune Responses

### 7.2.1 Arabidopsis thaliana *MPK3 and MPK6 Positively Regulate Plant Immune Responses*

The *Arabidopsis* MAP kinases, MPK3 and MPK6, have been implicated in positive plant immune responses (Petersen et al. 2000; Innes 2001; Zhang and Klessig 2001;

Asai et al. 2002; Menke et al. 2005; Takahashi et al. 2007a; Gao et al. 2008; Ren et al. 2008; Pitzschke et al. 2009b; Liu et al. 2011). MPK3, MPK6, MKK4, and MKK5 form a cascade that positively regulates plant defenses (Pitzschke et al. 2009b). MPK3 has been shown to be required for camalexin accumulation upon *Botrytis cinerea* infection (Ren et al. 2008). Inactivation of MPK3 and MPK6 by the *P. syringae* effector HopA/1 and inactivation of MKKs by the *P. syringae* effector HopF2 severely impair PAMP-induced defenses and render plants highly susceptible to nonpathogenic *P. syringae* bacteria (Zhang et al. 2007b; Wang et al. 2010). MPK3 and MPK6 have been shown to be required for priming of defense responses during induced resistance (Beckers et al. 2009).

*Arabidopsis* MPK3 and MPK6 have been shown to be positive regulators of plant immune responses. These MAPKs are activated by pathogens, PAMP elicitors and DAMPs (endogenous elicitors) (Asai et al. 2002; Bethke et al. 2012). The bacterial PAMP flg22 and the oligogalacturonides elicitor of host plant origin activated MPK3 and MPK6 in *Arabidopsis* (Galletti et al. 2011). Analysis of single *mapk* mutants revealed that lack of *MPK3* increased basal susceptibility to the fungal pathogen *Botrytis cinerea*, but did not significantly affect elicitors-induced disease resistance. By contrast, lack of MPK6 had no effect on basal resistance but suppresses flg22- and OGs- induced resistance to *B. cinerea*. Overexpression of the AP2C1 phosphatase led to impaired flg22- and OGS-induced phosphorylation of both MPK3 and MPK6 (Galletti et al. 2011). These results suggest that both MPK3 and MPK6 are involved in plant innate immunity, but their mode of action may vary.

Root treatment with *N*-3-oxo-tetradecanoyl-*L*-homoserine lactone (HSL) induced resistance against *Pseudomonas syringae* pv. *tomato* DC3000 in *Arabidopsis*. HSL treatment promoted a stronger activation of MPK3 and MPK6 when challenged with flg22, followed by a higher expression of the defense-related transcription factors *WRKY22* and *WRKY29*, and the *PR-1* gene (Schikora et al. 2011). These studies show that MPK3 and MPK6 are involved in triggering induced resistance.

### 7.2.2 Arabidopsis thaliana *MPK4 Negatively Regulates Plant Immune Responses*

MPK4, its upstream MAP kinase kinases MKK1 and MKK2, and the MAP kinase kinase kinase MEKK1 form a cascade that negatively regulates defenses in *Arabidopsis* because loss-of-function mutations in this cascade result in constitutive activation of defenses (Petersen et al. 2000; Mészáros et al. 2006; Suarez-Rodriguez et al. 2007; Gao et al. 2008; Qiu et al. 2008a; Pitzschke et al. 2009b). The *mpk4* plants exhibit constitutive systemic acquired resistance, including elevated salicylic acid levels and increased resistance to virulent pathogens (Petersen et al. 2000). The results suggest that MPK4 may negatively regulate SA signaling system. The induction of JA-responsive genes was blocked in *mpk4* mutant plants, suggesting that MPK4 positively regulates the JA pathway (Petersen et al. 2000; Gao et al. 2008;

Qiu et al. 2008a; Pitzschke et al. 2009a). Overexpression of an inactive form of MPK4 failed to complement *mpk4* phenotypes, indicating that kinase activity is required for MPK4 function (Petersen et al. 2000).

The 'Three Kinase Module' including MPK4 shows that MEKK1 and MKK1/2 act upstream of MPK4. The *mekk1/mkk1/mkk2* double mutants display similar constitutive defense responses, including elevated levels of SA and ROS, constitutive PR gene expression, and pathogen resistance (Petersen et al. 2000; Gao et al. 2008; Qiu et al. 2008b). It has also been shown that both MPK4 and MEKK1 interact with MKK1 and MKK2 in vivo (Gao et al. 2008). MEKK1 and MKK1/2 have also been shown to be essential for activation of MPK4 (Ichimura et al. 2006; Suarez-Rodriguez et al. 2007; Gao et al. 2008). Many defense–related genes were similarly deregulated in *mekk1*, *mkk1/2*, and *mpk4* mutants (Qiu et al. 2008b; Pitzschke et al. 2009a). Collectively these studies suggest the existence of three kinase module involving MEKK1-MKK1/2-MPK4 in *Arabidopsis thaliana.*

### *7.2.3* Arabidopsis thaliana *MPK11 in Plant Immune Responses*

Flg22 treatment is known to increase *MPK11* expression in *Arabidopsis thaliana* (Bethke et al. 2012). MPK11 is an MPK4 paralog and has overlapping function with MPK4 in regulating cell division (Bethke et al. 2012). MPK11 transcripts accumulate following pathogen infection or treatments with PAMPs, pathogen-derived molecules, or resistance-inducing chemicals such as benzothiadiazole. *MPK11* expression is rapidly induced within 10–15 min after flg22 treatment. MPK11 interacted with an ethylene response transcription factor (ERF104). MPK11 is activated during flg22 treatment (Bethke et al. 2012). MPK11 constitutes a fourth MAPK activated by flg22, in addition to MPK3, MPK4, and MPK6. MPK4 and MPK11 are nearly identical at the amino acid sequence level and similar in molecular weight (Bethke et al. 2012). MPK11 differed in its function from MPK4, although it was structurally similar to MPK4. In contrast to reports for *mpk4* (Petersen et al. 2000), no enhanced expression of *PR* genes such as *PR1*, *PR2*, or *PR5* was observed in the *mpk11*mutant. The levels of SA and its conjugates were not elevated in the *mpk11* mutant relative to wild-type plants, whereas levels in *mpk4* mutants were at least 20-fold higher (Bethke et al. 2012).

### *7.2.4 Arabidopsis MPK9 and MPK12 Positively Regulate ROS-Mediated ABA Signaling*

The MAP kinases MPK9 and MPK12 have been shown to be preferentially expressed in stomata guard cells and positively regulate ROS-mediated ABA signaling (Jammes et al. 2009). The MAPKs act upstream of anion channels in guard cell

ABA signaling. MPK9 and MPK12 function downstream of ROS to regulate ABA signaling (Jammes et al. 2009). ABA-induced stomatal closure plays an important role in bacterial disease resistance. The *mpk9/mpk12* double mutants have been shown to be highly susceptible to *P. syringae* pv. *tomato* DC3000 compared to wild-type *Arabidopsis thaliana* plants (Jammes et al. 2011). The results suggest that the regulation of stomatal apertures by MPK9 and MPK12 contributes to the first line of defense against pathogens.

### *7.2.5 Soybean GmMPK4 Negatively Regulates SA and ROS Signaling Systems*

Soybean homolog of MPK4, GmMPK4, negatively regulates defense responses. Silencing *GmMPK4* resulted in activation of immune responses and it also led to elevated levels of SA and $H_2O_2$. In soybean, GmMPK4 is a negative regulator of SA, ROS, and defense responses. Thus the functions of MPK4s from *Arabidopsis* and soybean are almost similar in negatively activating SA signaling system (Liu et al. 2011).

### *7.2.6 Rice OsMPK6 Positively Regulates Local Resistance and Negatively Regulates Systemic Acquired Resistance*

Rice, OsMPK6, an ortholog of AtMPK4, functions both as an activator and a repressor in conferring resistance against the bacterial blight pathogen *Xanthomonas oryzae* pv. *oryzae* (Shen et al. 2010). Activation of OsMPK6 resulted in the formation of lesion mimics and local resistance to *X. oryzae* pv. *oryzae*, accompanied by the accumulation of SA and JA, and the induced expression of SA- and JA-signaling genes. By contrast, the knock-out of *OsMPK6* results in increased accumulation of SA and enhanced resistance to *X. oryzae* pv. *oryzae* in systemic tissues. The knock-out of *OsMPK6* induces the expression of *PR1a*, the marker gene of systemic acquired resistance (SAR) when challenged with the pathogen (Shen et al. 2010). These results suggest that OsMPK6 positively regulates local resistance while it negatively regulates systemic acquired resistance against *X. oryzae* pv. *oryzae* by differentially modulating SA and JA signaling pathways.

### *7.2.7 Oilseed Rape BnMPK4 Positively Regulates JA-Mediated Defense Responses*

In oilseed rape (*Brassica napus*), overexpression of *BnMPK4* enhances resistance to *Sclerotinia sclerotiorum* (Wang et al. 2009). The transgenic plants inhibited growth of *S. sclerotiorum* and constitutively activated *PDF1.2*, the gene activated by JA signaling (Wang et al. 2009). The results suggest that BnMPK4 positively regulates JA-mediated defense response, which might play an important role in resistance to *S. sclerotiorum* in oilseed rape.

### *7.2.8 Cotton GhMPK2 Are Involved in Ethylene Biosynthesis - Mediated Plant Immune Responses*

A MAPK gene, *GhMPK2*, has been cloned from cotton and characterized. Transgenic tobacco plants overexpressing *GhMPK2* showed enhanced resistance to fungal and viral pathogens (Zhang et al. 2011). The transgenic plants showed enhanced expression of genes encoding ACC synthase (ACS) and ACC oxidase (ACO), which are involved in biosynthesis of ethylene. The expression of PR genes, *PR1*, *PR2*, *PR4*, and *PR5*, was upregulated in the transgenic plants expressing the *MAPK* gene (Zhang et al. 2011). The results suggest that the cotton MPK2 plays an important role in innate immune system.

### *7.2.9 Cotton GhMPK7 Triggers SA Signaling System*

The *MPK7* gene from cotton, *GhMPK7*, has been found to have a role in activating defense responses in plants. Transgenic *Nicotiana benthamiana* plants overexpressing *GhMPK7* showed broad-spectrum disease resistance. The transgenic plants displayed significant resistance against *Colletotrichum nicotianae* and *Potato virus Y* (Shi et al. 2010). The *GhMPK7* transcript was induced by pathogen infection and also by different defense-related signal molecules. Overexpression of the cotton *MAPK* gene in *N. benthamiana* induced rapid and strong expression of SA pathway genes (Shi et al. 2010). The results suggest that the GhMPK7 plays an important role in triggering SA signaling pathway in plant innate immunity.

### *7.2.10 GhMPK16 Activates ROS-Mediated Signaling System*

*GhMPK16* gene has been cloned from cotton and characterized. It showed high homology to *Arabidopsis AtMPK16*. The gene transcripts have been shown to accumulate following pathogen infection and treatment with several defense-related signal molecules (Shi et al. 2011). Transgenic *Arabidopsis* plants overexpressing *GhMPK16* showed enhanced resistance against the fungal pathogens *Colletotrichum nicotianae* and *Alternaria alternata* and also against the bacterial pathogen

*Pseudomonas solanacearum*. These *GhMPK16* overexpressing *Arabidopsis* plants showed rapid accumulation of ROS and enhanced expression of *PR* genes (Shi et al. 2011). The results suggest that MPK16 is another MAPK gene involved in plant immune responses.

### *7.2.11 SIPK and WIPK Activates Plant Immune Responses by Modulating SA and JA Signaling Systems*

Two mitogen-activated protein kinases, salicylic acid-induced protein kinase (SIPK) and wound-induced protein kinase (WIPK), are involved in plant immune responses (Ren et al. 2006; Kallenbach et al. 2010; Meldau et al. 2012). WIPK and SIPK play an important role in JA production and they function cooperatively to control SA biosynthesis (Seo et al. 2007). Activation of SIPK and WIPK by their upstream MAPK kinase (MAPKK), NtMEK2 leads to hypersensitive reaction-like cell death in tobacco (Jin et al. 2003). WIPK and SIPK have been implicated in TMV resistance in tobacco (Kobayashi et al. 2010). Silencing of WIPK/SIPK reduced TMV accumulation in tobacco and was correlated with an increase in SA and a decrease in JA. The reduction in viral accumulation was attenuated by expressing a gene for an SA-degrading enzyme or by exogenously applying JA (Kobayashi et al. 2010). These results suggest that WIPK and SIPK function to negatively regulate local resistance to TMV accumulation, partially through modulating accumulation of SA and JA.

## 7.3 MAPK Kinases (MAPKKs) in Plant Immune Responses

### *7.3.1 MKK1 in Plant Immune Responses*

*Arabidopsis* MAP kinase kinase MKK1 has been shown to be activated by PAMP elicitors. MKK1 is activated in cells treated with flg22, and it phosphorylates the MAPK MPK4. MKK1 negatively regulates the activity of flagellin-responsive genes. The *mkk1* mutant is compromised in resistance to both virulent and avirulent *Pseudomonas syringae* strains (Mészáros et al. 2006). The results suggest that MKK1 plays an important role in plant immune responses.

A MAPKK, NbMKK1, identified in *Nicotiana benthamiana* is a potent inducer of hypersensitive response (HR)-like cell death (Takahashi et al. 2007b). NbMKK1-mediated cell death was compromised in leaves where NbSIPK expression was silenced by virus-induced gene silencing. NbMKK1 and NbSIPK physically interact. *Phytophthora infestans* INF1 elicitor-mediated HR was delayed in NbMKK1-silenced plants, indicating NbMKK1 is involved in this immune response pathway (Takahashi et al. 2007a). NbMKK1 is also involved in inducing resistance against the bacterial pathogen *Pseudomonas cichorii* (Takahashi et al. 2007b).

### 7.3.2 *Arabidopsis* MKK2 Differentially Induces Defense Responses Against Different Pathogens

*AtMKK2* overexpressing transgenic *Arabidopsis thaliana* plants showed enhanced susceptibility to the fungal pathogen *Alternaria brassicicola*, whereas these plants were more resistant to the bacterial pathogens *Pseudomonas syringae* pv. *tomato* Dc3000 and *Erwinia carotovora* subsp. *carotovora* (Brader et al. 2007). To assess the downstream events in the MAPK cascade, transgenic *Arabidopsis* plants overexpressing the MAPKK MKK2 were developed (Teige et al. 2004). The MKK2 overexpressing plants showed altered expression of 152 genes involved in transcriptional regulation, signal transduction, and defense responses (Teige et al. 2004). The activated genes included transcription factors such as WRKY transcription factors, Myb DNA binding protein, and ethylene-responsive element binding factor. The activated genes in the MKK2 overexpressing plants also included several genes involved in signal transduction pathways. Several calmodulin and calcium binding proteins were strongly upregulated in the *MKK2* overexpressor lines. Genes involved in phosphorylation/dephosphorylation systems were also activated in the *MKK2* overexpressing plants. These included *MAPKK5* gene and protein phosphatase 2C genes. The MKK2 overexpressor lines also showed upregulation of several enzymes and targets of the ethylene and jasmonate pathways. The increased expression included JA-mediated defense gene *PDF1.2a* encoding defensin, ethylene biosynthesis gene *AtACS-6* encoding ACC synthase, and lipoxygenase gene involved in JA biosynthesis (Teige et al. 2004). Downstream events include activation of calcium/calmodulin signaling system (Teige et al. 2004), and JA-, SA-, ET-, and ABS- mediated signaling systems (Takahashi et al. 2007a). In *Arabidopsis*, the constitutively active AtMKK2 increases the expression levels of genes that encode enzymes in ET/JA signaling system (Brader et al. 2007).

### 7.3.3 *Arabidopsis* MKK3 Positively Regulates Immune Responses

MKK3 is an upstream activator of MPK1, MPK2, MPK7, and MPK14 (Dóczi et al. 2007). *Arabidopsis* plants overexpressing *AtMKK3* show increased expression of *PR* genes (Dóczi et al. 2007). These plants showed enhanced disease resistance against *Pseudomonas syringae* pv. *tomato* (Dóczi et al. 2007). MKK3 has been shown to play important role in plant innate immunity (Dóczi et al. 2007). The growth of virulent bacterial pathogen *P. syringae* pv. *tomato* DC3000 was increased in *mkk3* knockout plants and decreased in *MKK3*-overexpressing plants. *MKK3* overexpression lines showed increased expression of several PR genes (Dóczi et al. 2007). These results suggest that MKK3 positively regulate plant immune responses.

### 7.3.4 *Arabidopsis MKK7 Positively Regulates SA-Mediated SAR*

Another MAPKK gene detected in *Arabidopsis*, *MKK7*, positively regulates plant basal and systemic acquired resistance. MKK7 has been shown to trigger accumulation of SA and the increases in SA levels resulted in enhanced expression of PR genes. Overexpression of MKK7 induced *PR* gene expression and triggered resistance to the bacterial pathogen *Pseudomonas syringae* pv. *maculicola* and the oomycete pathogen *Hyaloperonospora parasitica* (Zhang et al. 2007c). The enhanced expression of disease resistance and PR gene expression induced by MKK7 was shown to be dependent on SA and NPR1 (Zhang et al. 2007c). The results suggest that the MAPKK is involved in triggering SA signaling system to confer disease resistance.

The expression of the *MKK7* gene in *Arabidopsis* was induced by pathogen infection. Reducing mRNA levels of *MKK7* by antisense RNA expression blocked the induction of SA-dependent systemic acquired resistance (SAR) in plants. Ectopic expression of *MKK7* in local tissues induced *PR* gene expression and disease resistance to the bacterial pathogen in systemic tissues (Zhang et al. 2007c). The results suggest that *MKK7* is involved in generating the mobile signal of SAR. The homologue of AtMKK7 in tomato, LeMKK4, is involved in *Pto*-mediated immune responses (Pedley and Martin 2004).

### 7.3.5 *Cotton GhMKK5 Triggers ROS-Mediated Signaling Systems*

A MAPKK encoding gene *GhMKK5* from cotton has been isolated and characterized. The expression of *GhMKK5* is induced by pathogen infection and defense-related signal molecules. The overexpression of *GhMKK5* in *Nicotiana benthamiana* showed enhanced resistance against the bacterial pathogen *Ralstonia solanacearum* but the transgenic plants were susceptible to the oomycete pathogen *Phytophthora parasitica* var. *nicotianae* (Zhang et al. 2012b). *GhMKK5*-overexpressing plants showed enhanced expression of *NtRbohA* gene encoding NADPH oxidase involved in accumulation of ROS. The transgenic plants showed accumulation of $H_2O_2$, suggesting that MKK5 may be involved in ROS-mediated immune signaling system (Zhang et al. 2012b). GhMKK5 is significantly induced by SA, methyl jasmonate and ethephon (Zhang et al. 2012b). The results suggest that GhMKK5 might be involved in the SA or JA/ET signaling pathways. The expression of SA signaling system-inducible *PR1a* and *PR5* and JA signaling system-inducible *PR-4* genes were greatly elevated in *GhMKK5*-overexpressing plants (Zhang et al. 2012b). Another SA signaling pathway gene *NPR1*, which is involved in SAR response, was also significantly increased in *GhMKK5*-overexpressing plants (Zhang et al. 2012b).

## 7.4 MAPKK Kinase EDR1 Modulates SA-JA-ET Signaling

EDR1 (ENHANCED DISEASE RESISTANCE 1) is a MAPKK Kinase (MAPKKKK), which functions at the top of a MAP kinase cascade. The *edr1* (*enhanced disease resistance 1*) gene encodes a putative MAPKKK, which negatively regulates SA signaling system. All *edr1*-associated phenotypes are suppressed by mutations that block SA perception (*nim1*) or reduce SA production (*pad4* and *eds1*). The *NahG* transgene, which lowers endogenous SA levels, also suppressed *edr1* (Frye et al. 2001). These results suggest that EDR1 plays an important role in SA-mediated defense responses. The *ein2* mutation did not suppress *edr1*-mediated disease resistance, suggesting that ethylene and JA-induced responses are not required for *edr1* resistance (Frye et al. 2001). *EDR1* gene has been isolated from *Arabidopsis* and putative orthologs of EDR1 have been detected in rice and barley (Frye et al. 2001).

The *edr1* mutant exhibits enhanced resistance against the powdery mildew pathogen *Golovinomyces cichoracearum* in *Arabidopsis thaliana* (Frye and Innes 1998; Frye et al. 2001), suggesting that EDR1 acts as a negative regulator of defense responses. Plant defensin *PDF* genes are downregulated in *edr1* mutants. PDF1.2 (PR-12; defensin), is an important pathogenesis-related protein involved in plant innate immune responses (Vidhyasekaran 2007) and its expression is triggered by the JA signaling system (Jung et al. 2007; Oñate-Sánchez et al. 2007; Pré et al. 2008). *MYC2/JIN1* encodes a basic helix-loop-helix leucine zipper transcription factor and differentially regulates JA-responsive defense genes (Lorenzo et al. 2004). *MYC2* is involved in repression of *PDF1.2* expression and *PDF1.2* was highly induced in *edr1myc2* double mutant (Hiruma and Takano 2011). It has been shown that EDR1 is critical for expression of plant defensin genes and the *MYC2-encoded* transcription factor represses defensin expression. Inactivation of *MYC2* fully restored defensin expression in *edr1* mutants (Hiruma and Takano 2011). It suggests that EDR1 cancels MYC2 function to regulate defensin expression.

The *edr1* mutant of *Arabidopsis* confers resistance against bacterial and fungal pathogens. When the *edr1* plants were inoculated with the powdery mildew pathogen *Golovinomyces cichoracearum*, the mutant plants showed increased expression of several defense-related genes (Christiansen et al. 2011). Many of the genes with elevated expression encoded WRKY transcription factors. EDR1 was found to be localized to the nucleus, suggesting that EDR1 could potentially interact with transcription factors in the nucleus. Elevated expression of ROS-related genes was also observed early during infection with the pathogen (Christiansen et al. 2011).

EDR1 kinase domain displays autophosphorylation activity and phosphorylates the common MAP kinase substrate myelin basic protein. The EDR1 kinase domain also phosphorylates a kinase-deficient EDR1 protein, indicating that EDR1 autophosphorylation can occur via an intermolecular mechanism (Tang and Innes 2002).

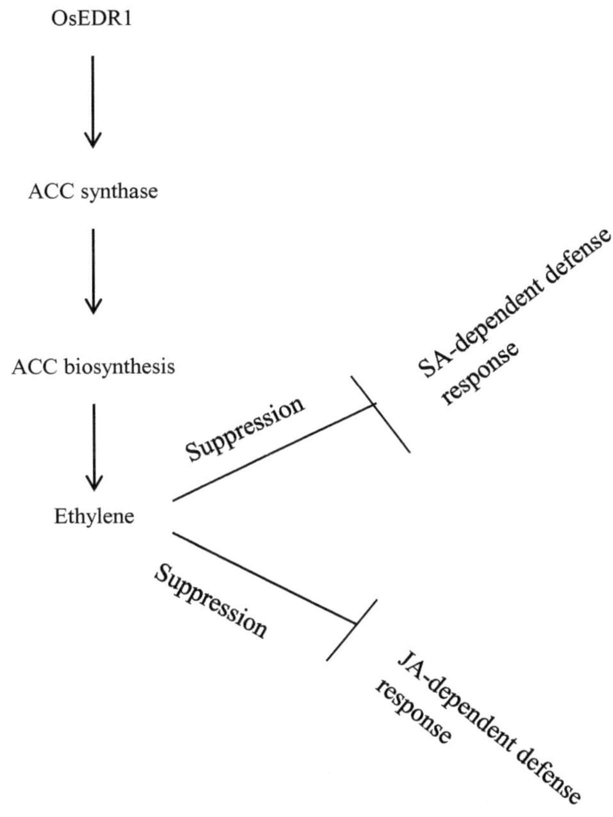


**Fig. 7.2** Role of a MAPKKK (EDR1) in modulation of SA-JA-ET signaling system in rice plant immune responses (Adapted from Shen et al. 2011)

OsEDR1 is a sequence ortholog of *Arabidopsis* EDR1. It has been shown that OsEDR1 negatively regulates plant defense responses via the activation of ethylene biosynthesis (Shen et al. 2011). OsEDR1-suppressing knockout (KO) rice plants showed enhanced resistance against the bacterial blight pathogen *Xanthomonas oryzae* pv. *oryzae*. This resistance was associated with increased accumulation of SA and JA, induced expression of SA- and JA-related defense genes and suppressed accumulation of 1-aminocyclopropane-carboxylic acid (ACC), the precursor of ethylene, and expression of ethylene-related genes. Knockout of OsEDR1suppressed the ACC synthase (ACS) gene family, which encodes the enzymes of ethylene biosynthesis by catalyzing the formation of ACC. The enhanced disease resistance of the OsEDR1-knockout plants was complemented by ACC treatment. ACC treatment decreased SA and JA biosynthesis in OsEDR1-knockout plants. In contrast, aminoethoxyvinylglycine, the inhibitor of ethylene biosynthesis promoted expression of SA and JA synthesis-related genes in OsEDR1-knockout plants (Shen et al. 2011). These studies show that OsEDR1 transcriptionally promotes the synthesis of ethylene that, in turn, suppresses SA- and JA-associated defense signaling (Fig. 7.2).

## 7.5 MAPK Pathways Involved in Defense Signal Transduction May Be Interconnected

Rather than linear pathways, multiple interconnected MAPK pathways may be required to transmit PAMP elicitor signals and integrate defense responses (Pedley and Martin 2005; Mészáros et al. 2006). Two separate MAPK pathways including MAPKKK – NtMEK2 – WIPK and MAPKKK – NtMEK2 – SIPK have been shown to be involved in defense response signaling in tobacco (Liu et al. 2003). Both these pathways are interconnected in induction of defense signals. SIPK, NtMEK2 and the upstream MAPKKK pre-exist in cells. Upon recognition of elicitor signals, NtMEK2 is activated by its upstream MAPKKK, which in turn activates the pre-existing SIPK. Activation of SIPK turns on the transcription of *WIPK* gene, which leads to the accumulation of WIPK protein. The newly synthesized WIPK protein is then activated by the NtMEK2 which is activated by a putative MAPKKK (Fig. 7.3; Liu et al. 2003). WIPK triggers HR-related cell death probably by the production of $H_2O_2$. SIPK besides activating synthesis of WIPK, activates transcription of various defense genes, particularly 3-hydroxy-3-methylglutaryl CoA reductase (*HMGR*) and phenylalanine ammonia lyase (*PAL*) genes encoding key enzymes in the phytoalexin and salicylic acid biosynthesis pathways (Yang et al. 2001; Liu et al. 2003).

Two independent MAPK pathways, MEKK1 – MKK1 – MPK4 and MEKK1 – MKK4/5 – MPK3/6, have been shown to be interconnected in defense signaling system in *Arabidopsis* (Mészáros et al. 2006). Both these pathways have common MAPKK, MEKK1, but their MAPKKs differ. The first pathway has MKK1 as MAPKK, while the second one has the MAPKK pair, MKK4 and MKK5. MKK1 specifically interacts with, and phosphorylates MPK4 (Teige et al. 2004), while it cannot phosphorylate the MAPK pair MPK3 and MPK4 (Mészáros et al. 2006). MKK4/5 can phosphorylate MPK3/4. It has been suggested that MEKK1 tethers MPK4 and MKK1 through its N- and C-terminal domains, respectively. This would prevent the association of MKK4/MKK5 until MEKK1 is activated by elicitor signals and dissociates the MKK1-MPK4 complex (Mészáros et al. 2006). The activated MEKK1 in the first pathway after sequential phosphorylation of MKK1 and MPK4 activates the second pathway. The MPK4 pathway positively regulates ethylene/JA signaling system and negatively regulates SA-mediated signaling system (Petersen et al. 2000; Liu and Zhang 2004). The MPK3/MPK6 pathway is involved in JA/ET signaling system and also in activation of WRKY22 and WRKY29 transcription factors (Asai et al. 2002; Takahashi et al. 2007a).

Another MAP kinase cascade in *Arabidopsis thaliana* involves MEKK1 – MKK2 – MPK4/MPK6 (Teige et al. 2004). Transgenic *Arabidopsis* plants overexpressing a MAPKK, MKK2, were developed (Teige et al. 2004). The MAPKK MKK5 also showed enhanced expression levels in MKK2 overexpressing plants. MKK5 is an activator of MPK3 and MPK6. Upregulation of *MKK5* in the *MKK2* overexpressor lines indicates some level of crosstalk between these two pathways (Teige et al. 2004).

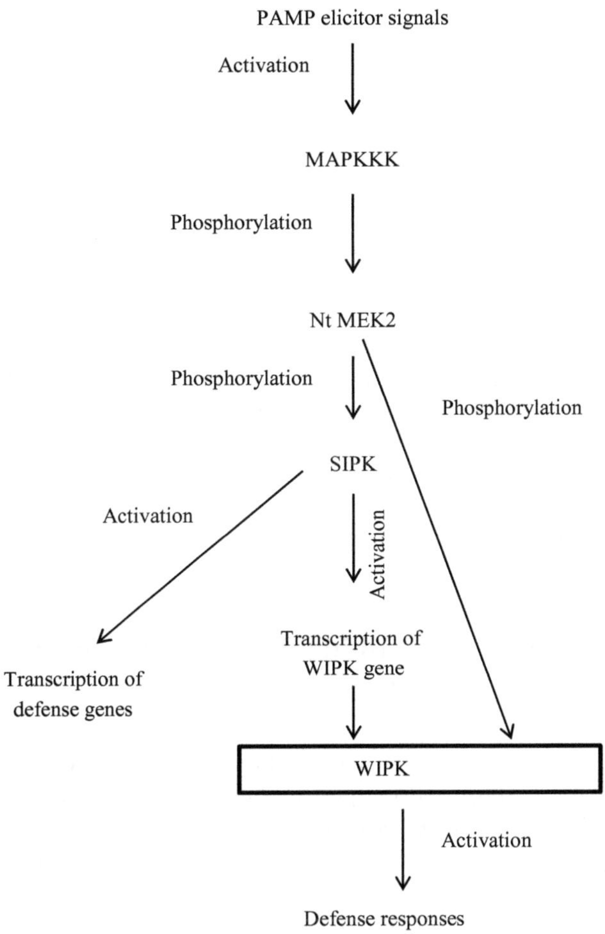


**Fig. 7.3** Interconnection of two different MAPK pathways to activate immune responses in tobacco (Adapted from Liu et al. 2003)

## 7.6 14-3-3 Protein Enhances Signaling Ability of MAPKKK in Activating Plant Innate Immune Responses

14-3-3 Proteins are a group of proteins that have been shown to be involved in regulation of protein kinases and phosphatases (Ferl 2004). The 14-3-3 proteins are given this nomenclature based on their chromatography and electroprofiles. The 14-3-3 proteins play a role in the completion of signal transduction events. Phosphorylation may tag the proteins for association with 14-3-3 and the subsequent binding of 14-3-3s may complete the signal-induced changes in the protein activity (Ferl 2004). 14-3-3 proteins occur as homo- and heterodimers in vitro and in vivo and these dimers may mediate interaction between pairs of associated proteins (Jones et al. 1995). It has been shown that 14-3-3 proteins bind to phosphorylated Ser residues present within one of a small number of consensus sequences found in many of the proteins with which they interact (Yaffe et al. 1997; Yaffe 2002).

MAPKKKα is a positive regulator of immunity-associated programmed cell death (PCD) in tomato and *Nicotiana benthamiana.* A 14-3-3 protein, TFI7, has been identified as a MAPKKKα – interacting protein in tomato (Oh et al. 2010). TFI7 protein contains a phosphopeptide binding motif, which was found to be essential for the interaction with MAPKKKα in vivo and also the PCD-enhancing activity of TFI7. A 14-3-3 binding motif, including a putative phosphorylated Ser-535, is present in the C-terminal region of MAPKKKα. An S535A substitution in MAPKKKα reduced interaction with TFI7 and both PCD-eliciting ability and stability of MAPKKKα. Coexpression of the 14-3-3 protein with tomato MAPKKKα enhanced MAPKKKα -mediated PCD. Coexpression TF17 with MAPKKKα in vivo caused increased accumulation of the kinase and enhanced phosphorylation of two MAP kinases (Oh et al. 2010). Collectively, these results suggest that the 14-3-3 protein enhances signaling ability of MAPKKKα in activating plant innate immune responses.

## 7.7 Role of MAPKs in Priming Plants for Augmented Defense Gene Activation

Some dormant MAPKs have been suggested to be important components required for priming in *Arabidopsis* and the prestress deposition of these inactive kinases may be a possible mechanism of priming during development of systemic acquired resistance (Beckers et al. 2009). MPK3, and functionally redundant MPK6, have been found to be important components for full priming in *Arabidopsis.* The resistance-inducing avirulent strains of *P. syringae* pv. *tomato* DC3000 and *P. syringae* pv. *phaseolicola* induced SA accumulation and MPK3 expression. Both SA and the SA-related compounds benzothiadiazole (BTH) and 4-chloro-SA activate *MPK3* gene expression and induce priming and SAR. In contrast, another SA-related compound 3-hydroxybenzoic acid did not induce *MPK3* gene expression, priming and disease resistance. This strong correlation between the ability of avirulent bacteria and various SA-related compounds to activate *MPK3* gene expression and their capacity to prime plants for augmented defense gene activation and induced resistance suggests that MPK3 plays a role in priming (Beckers et al. 2009). Similarly another MAPK gene, *MPK6* was also found to be involved in priming process. However, the BTH-induced accumulation of *MPK6* transcript and protein was less pronounced. Both MPK3 and MPK6 accumulate in an inactive form during priming of *Arabidopsis* with BTH (Beckers et al. 2009). Both MPK3 and MPK6 displayed greater activity in *Arabidopsis* plants which are primed and subsequently challenged with the virulent *P. syringae* pv. *maculicola.* These two enzymes were more strongly activated in primed plants than in nonprimed plants. Priming of defense gene expression and induced resistance were lost or reduced in *mpk3* or *mpk6* mutants (Beckers et al. 2009). These results suggest that prestress deposition of the signaling components MPK3 and MPK6 is a critical step in priming plants for full induction of defense responses during induced resistance.

## 7.8 PAMP Signals Activate MAP Kinases

PAMP and DAMP/HAMP elicitors activate several MAP kinases in plants. A typical array of early defense responses induced by PAMPs includes $Ca^{2+}$ influx and the generation of ROS, nitric oxide, and ethylene. Much of this follows the activation of mitogen-activated protein kinase cascades, leading to transcriptional changes of many defense-related genes (Aslam et al. 2009; Boller and He 2009; Boller and Felix 2009). PAMP-triggered immunity requires a signal transduction from receptors to downstream components via the MAPK cascade and many of the known PAMPs were shown to activate MAP kinases (Pitzschke et al. 2009b). Four specific MAPK pathways involving MMK2 (for M*edicago* MAPK2), MMK3, SAMK (for stress-activated MAPK), and SIPK (for salicylate-induced protein kinase) in alfalfa were found to be activated to different levels and with different kinetics by four different elicitors, chitin, β-glucan, ergosterol, and yeast extract (Cardinale et al. 2000). The tomato MAPKs, LeMPK1 and LeMPK2, were activated in response to four different oligosaccharide elicitors (Holley et al. 2003). The tobacco MAP kinase SIPK is activated both by the oomycete elicitor, β-megaspermin and the bacterial elicitor $hrpZ_{psph}$ (Hall et al. 2007). However, SIPK activation induced by the oomycete elicitor required external calcium influx, whereas that induced by the bacterial elicitor does not. It suggests that SIPK activation is involved in different elicitor-initiated signaling pathways (Hall et al. 2007).

The *Arabidopsis* MKK1 has been shown to be activated by flagellin or laminarin (Teige et al. 2004). *Arabidopsis* MPK3 and MPK6 have been shown to be positive regulators of plant immune responses. These MAPKs are activated by PAMP elicitors and DAMPs (endogenous elicitors) (Asai et al. 2002; Bethke et al. 2012). The bacterial PAMP flg22 and the oligogalacturonides elicitor of host plant origin activated MPK3 and MPK6 in *Arabidopsis* (Galletti et al. 2011). *Phytophthora infestans* INF1 elicitor activates NbMKK1 in *Nicotiana benthamiana* (Takahashi et al. 2007b).

The MAPK module MEKK1-MKK4/MKK5-MPK3/MPK6 has been proposed to be responsible for flg22 signal transmission (Asai et al. 2002; Bethke et al. 2009b). The PAMP flg22 triggers a rapid and strong activation of MPK3, MPK4 and MPK6 (Droillard et al. 2004). The PAMP flg22 treatment activated MPK11 in *Arabidopsis* and MPK11 constitutes a fourth MAPK activated by flg22, in addition to MPK3, MPK4, and MPK6 in *Arabidopsis* (Bethke et al. 2012). Flagellin-derived flg22 peptide strongly activated MPK6 but it only poorly activated MPK7 in *Arabidopsis* (Dóczi et al. 2007). *Arabidopsis* MAP kinase kinase MKK1 is activated in cells treated with flg22, and it phosphorylates the MAPK MPK4 (Mészáros et al. 2006). Thus several studies have shown that MAPK signaling is activated by flagellin in *Arabidopsis* (Tsuda et al. 2009; Wu et al. 2011). Although such studies have shown that MAPK cascades are important components in the PAMP flagellin signaling system, there are also reports that flg22 may trigger the plant immune responses independent of MAPK cascades in *Arabidopsis*. The *bik1* mutant is significantly compromised in PAMP-induced resistance, but not the flg22-induced MAPK activation (Zhang and Zhou 2010).

## 7.9 Signals and Signaling Systems Activating MAPK Cascades

Several MAP kinases have been shown to be involved in inducing resistance against bacterial, oomycete, and fungal pathogens (Menke et al. 2004; Brader et al. 2007; Zhang et al. 2007c). These MAP kinase cascades may induce different signaling systems and induce resistance or susceptibility to various types of pathogens (Xiong and Yang 2003). Various signals activate MAP kinases. Fungal elicitor, $H_2O_2$, SA, JA and ethylene activated BWMK1, the MAPK in rice cells (Cheong et al. 2003). The kinase activity of BWMK1 was rapidly and transiently activated by all these defense signals. The activity peaked 5–30 min after treatment. The transcript levels of *BWMK1* increased in a delayed manner after activation with the defense signals; the increase was observed only 1 to 6 h after treatment (Cheong et al. 2003). However, BWMK1 protein levels did not change. This indicates that the BWMK1 protein is maintained at steady-state levels in the cell, which would permit the plant to respond rapidly to the external stimuli. The response consumes BWMK1 protein; thus the plants produce new *BWMK1* transcripts to maintain the baseline level of protein. These observations suggest that BWMK1 activation is primarily achieved by post-translational modification (Cheong et al. 2003).

Several other MAPKs, including ERMK (for Elicitor-responsive MAPK) (Ligterink et al. 1997), SIPK (Zhang and Klessig 1997; Zhang et al. 1998), and WIPK (Seo et al. 1995) have also shown to be activated by elicitors, SA, and JA. The ERMK and WIPK are also activated by post-translational modification (Seo et al. 1995; Ligterink et al. 1997). LeMPK3 is a MAPK involved in defense response in tomato and *LeMPK3* was found to be transcriptionally up-regulated by both bacterial and fungal elicitors (Mayrose et al. 2004).

$H_2O_2$ activates a MAPKK kinase from alfalfa, OMTK1 (oxidative stress-activated MAP triple-kinase 1) (Nakagami et al. 2004). JA activates MKK3 – MPK6 cascade in *Arabidopsis* (Takahashi et al. 2007a) and OsBIMK1, a rice MAP kinase (Song and Goodman 2002). Salicylic acid activates a 48-kD MAP kinase and SIPK in tobacco (Zhang and Klessig 1997; Zhang and Liu 2001) and p48 and p44 MAPKs in pea (Uppalapati et al. 2004). ABA activates p48 MAPK in pea (Uppalapati et al. 2004) and rice MAPK gene OsMAPK5 (Xiong and Yang 2003). Both nitric oxide (NO) and SA activated SIPK in tobacco (Kumar and Klessig 2000). Studies with transgenic NahG tobacco revealed that SA is required in the NO-mediated induction of SIPK. These observations suggest that SIPK may function downstream of SA in the NO signaling pathway (Kumar and Klessig 2000).

## 7.10 MAPKs May Function Downstream of G-Proteins, $Ca^{2+}$, ROS, SA, ABA, and NO Signaling Pathways

Silencing of a small GTPase, *OsRac1*, by RNA interference or loss of function mutation of the heterotrimeric G-protein α-subunit gene resulted in strong reduction of the OsMAPK6 protein levels and of kinase activation by the elicitor in rice (Lieberherr et al. 2005). These results suggest that both small G protein and heterotrimeric G protein act upstream of MAPK in induction of defense genes such as *PAL* in rice (Lieberherr et al. 2005). It has been reported that Gα functions upstream of OsRac1 (Suharsono et al. 2002) and the MAPK may act downstream of OsRac1 in rice (Lieberherr et al. 2005).

The signaling cascade initiated by the DAMP/HAMP elicitor AtPep1 leads to expression of defense genes in a $Ca^{2+}$-dependent manner in *Arabidopsis* (Qi et al. 2010). The endogenous elicitor AtPep1 after binding with its PRR AtPepR1 activates plant membrane inwardly conducting $Ca^{2+}$ permeable channels in mesophyll cells, resulting in cytosolic $Ca^{2+}$ elevation (Qi et al. 2010). The resulting $Ca^{2+}$ signature triggers the expression of the MAPK MPK3. The results suggest that AtPep-dependent expression of *MPK3* is mediated by the $Ca^{2+}$ signaling pathway.

$H_2O_2$ activates a MAPKK kinase from alfalfa, OMTK1 (oxidative stress-activated MAP triple-kinase 1) (Nakagami et al. 2004). JA activates MKK3 – MPK6 cascade in *Arabidopsis* (Takahashi et al. 2007a) and OsBIMK1, a rice MAP kinase (Song and Goodman 2002). ABA activates p48 MAPK in pea (Uppalapati et al. 2004) and rice MAPK gene OsMAPK5 (Xiong and Yang 2003). Salicylic acid activates a 48-kD MAP kinase and SIPK in tobacco (Zhang and Klessig 1997; Zhang and Liu 2001) and p48 and p44 MAPKs in pea (Uppalapati et al. 2004). Both nitric oxide (NO) and SA activated SIPK in tobacco (Kumar and Klessig 2000). Studies with transgenic *NahG* tobacco revealed that SA is required in the NO-mediated induction of SIPK. These observations suggest that SIPK may function downstream of SA in the NO signaling pathway (Kumar and Klessig 2000). Collectively these studies suggest that MAPKs may function downstream of G-proteins, $Ca^{2+}$ influx, ROS, NO, SA, JA, ET, and ABA signaling.

## 7.11 Some MAPKs May Act Upstream of SA, JA, and ET Signaling Pathways

There are also reports that JA or SA or ethylene may not be able to induce expression of MAPKs and they may act only at downstream of MAPK. JA or SA could not activate transcription of *TIPK* (*Trichoderma*-induced protein kinase) in cucumber, even at high concentrations (Shoresh et al. 2005). JA was unable to induce expression of *LeMPK3* in tomato (Mayrose et al. 2004) or activate WIPK or its alfalfa homolog, SAMK (Bögre et al. 1997; Kumar and Klessig 2000). WIPK in tobacco is not activated by ethylene (Kumar and Klessig 2000). Ethylene did not affect expression of the tomato LeMPK3 (Mayrose et al. 2004). The tomato MAPKs LeMPK1 and LeMPK2 function upstream of JA biosynthesis (Kandoth et al. 2007). These studies suggest that MAPKs may act at upstream of various signaling systems in different plants.

## 7.12 Some MAP Kinases Act Downstream of Phosphoinositide (PI) Signal Transduction Pathway

Some MAP kinases have been shown to act downstream of phosphoinositide (PI) signal transduction pathway. In this pathway, PI is phosphorylated by phosphatidylinositol kinase to form phosphatidylinositol-4-phosphate (PIP), which is phosphorylated by phosphatidylinositol-4-monophosphate kinase to form phosphatidylinositol-4,5-bisphosphate (PIP2). The PIP2 is hydrolysed by phospholipase C into inositol trisphosphate (IP3) and diacylglycerol (DAG). It has been reported that a fungal elicitor induced a rapid and biphasic increase in levels of PIP2 and IP3 in pea that was apparent within 15 min (Toyoda et al. 1993). Neomycin, a known inhibitor of phospholipase C blocked the elicitor-induced accumulation both of IP3 and the phytoalexin pisatin. These results suggest that rapid changes in PI metabolism are indispensable in the defense signaling system in pea plants (Toyoda et al. 1993). In another related study, increased activation of two MAP kinases, p44 and p48 kinases, by a fungal pathogen elicitor within 15–30 min was observed (Uppalapati et al. 2004). The elicitor-induced p44 kinase activation was inhibited by a MAPKK inhibitor, PD098059 and this inhibition was correlated with the suppression of elicitor-induced expression of the defense gene *PAL* encoding phenylalanine ammonia-lyase. These observations suggest that the MAPK is involved in defense signaling system in pea (Uppalapati et al. 2004). Pre-treatment of epicotyls of pea plants with the phospholipase C inhibitor neomycin completely suppressed elicitor-induced p44 kinase activation and subsequent phytoalexin accumulation (Uppalapati et al. 2004). A DAG kinase which degrades the DAG was shown to activate the p44 kinase. This was demonstrated using a DAG kinase inhibitor, R59022, and this inhibitor potentiated the elicitor-induced MAP kinase activities (Uppalapati et al. 2004). Toyoda et al. (2000) reported that inhibition of DAG kinase potentiated the *PAL* gene activation and accumulation of phytoalexin in pea. These results suggest that DAG may act as a second messenger and inhibition of breakdown of DAG by DAG kinase may increase induction of the defense gene and defense compound. It demonstrates that the MAP kinase acts downstream of DAG in the PI metabolism (Fig. 7.4).

## 7.13 MAP Kinase Cascades May Act Either Upstream or Downstream of ROS Signaling System

ROS may act either upstream or downstream of MAPK pathways The overexpression of a constitutively active MAPK kinase (MAPKK), which activates endogenous SIPK and WIPK, induced the *NbrbohB* (*Nicotiana benthamiana respiratory burst*

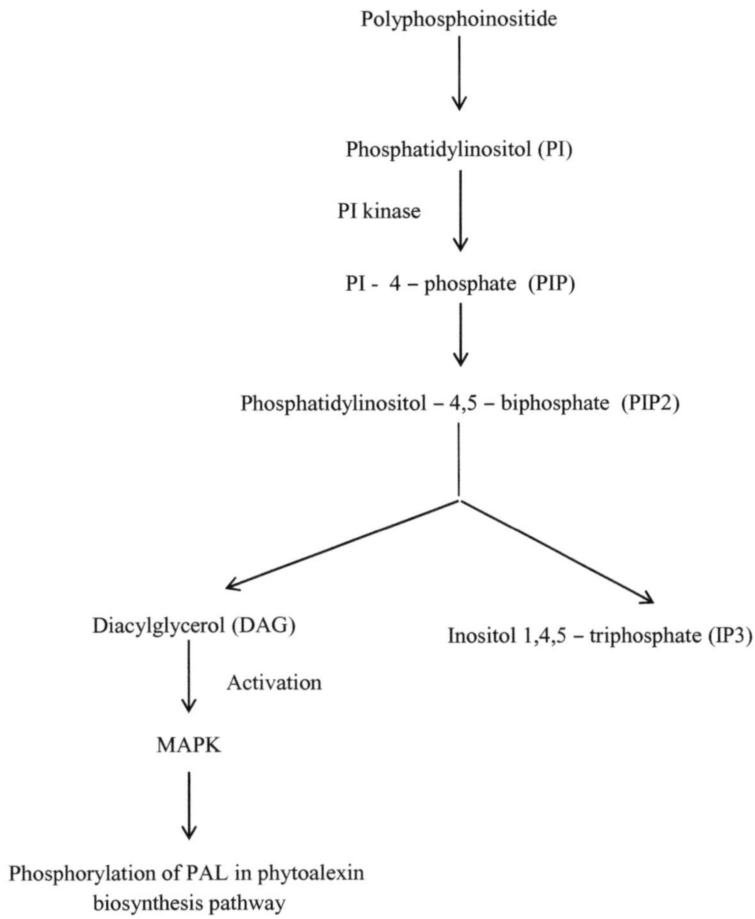


**Fig. 7.4** Activation of MAP kinase in phosphatidylinositol signaling pathway (Adapted from Toyoda et al. 2000; Uppalapati et al. 2004)

*oxidase homolog B*) gene encoding a NADPH oxidase inducing ROS in *N. benthamiana* (Yang et al. 2001). Ren et al. (2002) reported that the activation of endogenous *Arabidopsis* MAPKs by *MEK* transgenes encoding MAPK Kinase under induced conditions leads to the generation of $H_2O_2$. Elicitor signals have been shown to activate a MAPK (a SIPK ortholog of tobacco), StrbohB (*Solanum tuberosum* respiratory burst oxidase homolog B) and accumulation of ROS in potato tubers (Katou et al. 1999; Yoshioka et al. 2001).

The MAPK pathway might induce the NADPH oxidase at the gene transcriptional level and also by post-translational activation of the enzyme. Yoshioka et al. (2003) demonstrated that the *N. benthamiana* MAPKK, MEK, induced the NADPH oxidase gene *NbrbohB* at the transcriptional level. It has also been shown that the oxidative burst is controlled through phosphorylation activation by its upstream kinase and dephosphorylation inactivation by its negative regulator, phosphatase

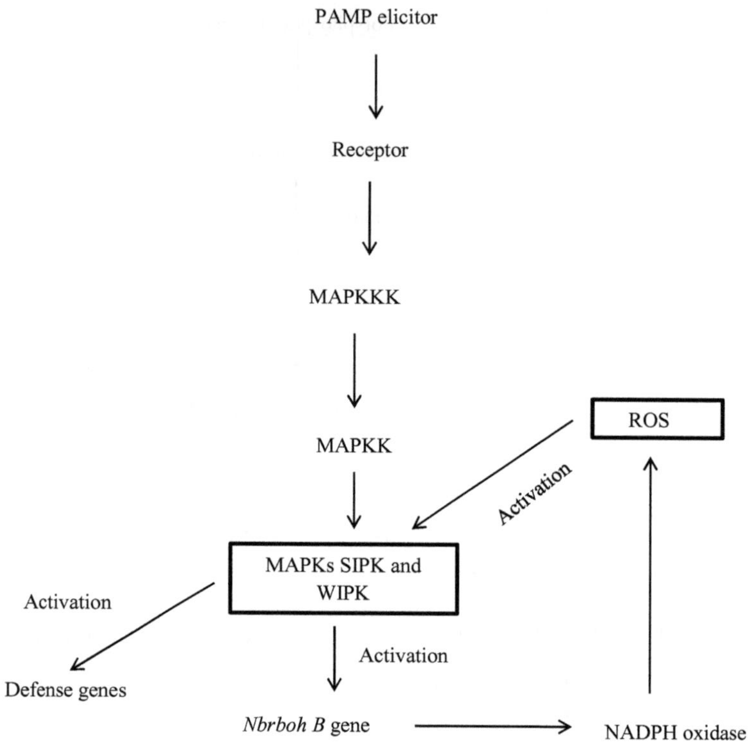


**Fig. 7.5** MAP kinase cascade acting both upstream and downstream of ROS signaling system (Yang et al. 2001; Yoshioka et al. 2003)

(Yoshioka et al. 2003). It suggests that post-translational modification of the NADPH oxidase by phosphorylation may also be involved in ROS production. The MAPK pathway may also act at downstream of $H_2O_2$, which activates MAPKs (SIPK and WIPK in tobacco). The activated MAPKs trigger defense gene activation (Fig. 7.5; Yang et al. 2001; Yoshioka et al. 2003).

The role of MAPK cascade in the generation of $H_2O_2$ has been demonstrated in *Arabidopsis* (Ren et al. 2002). Transgenic Arabidopsis plants expressing active mutants of two MAPK kinases, AtMEK4 and AtMEK5, were developed (Ren et al. 2002). The external signal stimulated the activation of the endogenous MAPKs and generation of $H_2O_2$ (Ren et al. 2002).

Some MAP kinase may control $H_2O_2$ accumulation by the action of catalase (Xing et al. 2008). A catalase (*CAT1*) transcript was induced in an abscisic acid (ABA)-dependent way in *Arabidopsis thaliana* and the induction was abolished in the T-DNA insertion mutant *mkk1*, a gene encoding a MAPK kinase. Overexpression of *AtMKK1* significantly enhanced ABA-dependent *CAT1* expression and $H_2O_2$ production (Xing et al. 2008). Another component in the MAPK cascade, MPK6 (a MAP kinase) was also involved in signal transduction. The *mpk6* mutant blocked and overexpressing AtMPK6 enhanced the ABA-dependent expression of *CAT1* and $H_2O_2$ production. The activity of AtMPK6 was increased by ABA in an AtMKK1-dependent manner. These results suggest an ABA-dependent signaling pathway connecting *CAT1* expression through a phosphorylation system including AtMKK1 and AtMPK6 (Xing et al. 2008).

In another study, Xing et al. (2007) showed that ABA-induced expression of *CAT1* catalase is mediated by an *Arabidopsis* MAPK kinase, AtMEK1, by triggering $H_2O_2$ signal production. The *mek1* mutant totally blocked stress-induced *CAT1* expression and $H_2O_2$ production. Overexpression of *AtMEK1* significantly induced *CAT1* expression and $H_2O_2$ production (Xing et al. 2007). These results suggest that MAPK pathway acts upstream of $H_2O_2$ signaling.

An *Arabidopsis* MAPKKK, MEKK1 has been shown to act downstream of $H_2O_2$. MEKK1 kinase activity and protein stability was regulated by $H_2O_2$ in a proteasome-dependent manner (Nakagami et al. 2006). $H_2O_2$ has been shown to activate two MAPKs, AtMPK6 and AtMPK3 in *Arabidopsis* (Grant et al. 2000; Kovtun et al. 2000; Desikan et al. 2001), through a MAPKKK, AN1 (Kovtun et al. 2000). In maize, $H_2O_2$ has been reported to activate a 46 kDa MAPK (Zhang et al. 2006). In tobacco, ROS induced SIPK and WIPK (Samuel and Ellis 2002).

## 7.14 MAP Kinases Positively or Negatively Regulate SA Signaling System

Several MAP kinases are known to positively or negatively regulate SA-signaling systems in plants. The *Arabidopsis MKK7* gene encoding MAP kinase kinase 7 positively regulates SA-signaling system (Zhang et al. 2007c). The activation-tagged *bud1* mutant, in which the expression of MKK7 is increased, accumulates SA, exhibits constitutive PR gene expression, and displays enhanced resistance to both the oomycete pathogen *Hyaloperonospora parasitica* and the bacterial pathogen *Pseudomonas syringae* pv. *maculicola* (Zhang et al. 2007c). Expression of a WIPK-activated transcription factor results in increase of endogenous salicylic acid in tobacco (Waller et al. 2006). Salicylic acid levels were 50-fold higher in the transgenic plants than those in wild-type plants. The levels of JA did not significantly differ (Waller et al. 2006). SA pathway genes were more rapidly and strongly induced in plants overexpressing a MAPK gene. These tobacco plants showed enhanced defense responses against fungal and viral pathogens (Shi et al. 2010). Collectively these results suggest that the MAPKs may act upstream of SA biosynthesis.

MPK4 acts as a negative regulator of defense responses through a salicylic acid-dependent signaling system (Petersen et al. 2000). The *mpk4* knockout mutant shows elevated SA levels and constitutively expresses pathogenesis-related (PR) genes (Petersen et al. 2000). Expression of the bacterial NahG salicylate hydroxylase in *mpk4* plants abolishes PR gene expression, indicating the role of the MAPK in SA-mediated signaling system (Petersen et al. 2000; Brodersen et al. 2006).

A substrate for MPK4 has been identified and it was designated MKS1 (for MAP Kinase 4 Substrate 1) (Andreasson et al. 2005). MPK4 interacts with the nuclear protein MKS1 that in turn interacts with two WRKY transcription factors, WRKY25 and WRKY33 (Andreasson et al. 2005). The molecular phenotypes of plants over- or under-expressing MKS1 indicate that it mediates some effects of MPK4 on SA-mediated resistance responses. The results suggest that the MKS1 is required for SA-dependent resistance in *Arabidopsis* (Andreasson et al. 2005). The transcription factors WRKY25 and WRKY33 may function as downstream components of the MPK4-mediated signaling pathway and contribute to repression of SA-dependent disease resistance response (Andreasson et al. 2005).

An *Arabidopsis* MAPKKK, EDR1 (Enhanced Disease Resistance1), negatively regulates SA-mediated defense responses (Frye et al. 2001). Inactivation of EDR1 interacting receptors by fungal pathogen – derived elicitor signals activated SA-inducible defense responses. The *Arabidopsis* mutant *edr1* showed enhanced disease resistance response. These results suggest that the mitogen-activated protein kinase is involved in SA-mediated resistance response (Frye et al. 2001).

## 7.15 MAP Kinase Cascades Activate JA Signaling System

Some MAP kinase cascades are involved in jasmonic acid (JA) signaling system. The *Arabidopsis* MKK3-MPK6 cascade is involved mainly in JA signaling system. This cascade negatively regulates *ATMYC2* function (Takahashi et al. 2007a). *Arabidopsis JIN1* (*JASMONATE INSENSITIVE1*, also known as *MYC2* [*MYELOCYTOMATOSIS2*]) encodes a basic helix-loop-helix (bHLH) – type transcription factor. MYC2 involved in the transcriptional regulation of JA-responsive gene expression (Lorenzo et al. 2004; Chini et al. 2007). *ATMYC2* plays a predominant role in JA pathway (Boter et al. 2004). *ATMYC2* was shown to function as a downstream factor of the MKK3-MPK6 cascade in JA signaling (Takahashi et al. 2007a).

JAZ (for JASMONATE ZIM [Zinc-finger protein expressed in Inflorescence Meristem]-domain) family of transcriptional repressors has been identified as an important component in a receptor complex involved in the JA perception process (Sheard et al. 2010). JAZ proteins have been identified as key regulators of jasmonate signaling (Chini et al. 2007; Thines et al. 2007). JAZ proteins negatively regulate the key transcriptional activator of jasmonate responses, MYC2 (Chini et al. 2007). JAZ proteins interact with MYC2 involved in the transcriptional regulation of JA-responsive gene expression (Lorenzo et al. 2004; Chini et al. 2007). It is suggested that, in the absence of a JA signal, JAZ proteins repress MYC2. Upon sensing of the JA signals, JAZ repressors are recruited to the SCF E3 complex for ubiquitination and subsequent degradation by the proteasome. The removal of these repressors then paves the way for MYC2 to regulate JA-dependent gene expression.

MYC2 is a master regulator of the JA signaling pathway. MYC2 is required for induced systemic resistance (ISR) triggered by beneficial soil microbes (Kazan and

Manners 2013). MYC2 function is targeted by pathogens during effector-mediated suppression of innate immunity in roots. MYC2 regulates crosstalk between the signaling pathways of JA and those of other phytohormones such as ABA, SA, GA, and auxin (Kazan and Manners 2013). MYC2 orthologs act as 'master switches' that regulate JA-mediated biosynthesis of secondary metabolites (Kazan and Manners 2013).

The MKK3-MPK6 cascade affects gene expression controlled by JA. The JA-responsive *PDF1.2* encoding the PR-12 defensin protein is activated by the MKK3-MPK6 cascade (Takahashi et al. 2007a). The tobacco orthologue of the *Arabidopsis* MPK4, NtMPK4, is activated by SIPKK, a SIPK-interacting MAPK kinase. In NtMPK4-silenced tobacco plants, the induction of JA-responsive genes was inhibited, suggesting that NtMPK4 activates JA-responsive genes (Gomi et al. 2005).

The MAPK WIPK-overproducing tobacco plants showed 3- to 4-fold higher levels of JA and methyl jasmonate than in the wild type plants. These WIPK-overproducing plants showed constitutive accumulation of PI-II (Proteinase Inhibitor-II) transcript accumulation and *PI-II* is a JA-inducible PR gene (Seo et al. 1999). Rice plants transformed with MK1, a homolog of WIPK from *Capsicum annuum*, showed a three-fold higher level of JA than the wild type (Lee et al. 2004). These observations suggest a role for the MAPK in the production of JA and in the activation of JA-mediated signaling system.

MAP kinase cascades may positively or negatively regulate JA signaling system. It has been reported that the induction of jasmonate-responsive genes was blocked in *mpk4 Arabidopsis* plants (Petersen et al. 2000), suggesting that it positively regulates the JA pathway. AP2C1, an *Arabidopsis* Ser/Thr phosphatase of type 2C is a defense signal regulator that inactivates MPK4. The *ap2C1* mutants produce significantly higher amounts of JA upon wounding (Schweighofer et al. 2007). The result provides additional evidence to show that *Arabidopsis thaliana* MPK4 positively regulates JA pathway. By contrast, the MPK4 from soybean (*Glycine max*) GmMPK4 negatively regulates JA pathway. JA signaling is activated in *GmMPK4-silenced* plants. Significantly, a hallmark JA-responsive gene encoding defensin had 315-fold change in *GmMPK4*-silenced plants (Liu et al. 2011).

## 7.16 Some MAP Kinase Cascades Are Involved in Biosynthesis of Ethylene and Ethylene-Mediated Signaling Systems

Some of the MAP kinase cascades have been shown to be involved in biosynthesis of ethylene and ethylene (ET)-mediated signaling systems. Activation in tobacco of MAP kinase cascades, which include the NtMEK2 MAPKK and the SIPK and WIPK MAP kinases, has been reported to cause an increase of ethylene levels (Kim et al. 2003). The MKK4/5-MPK6 cascade in *Arabidopsis* is involved in ET signaling system (Kim et al. 2003; Liu and Zhang 2004). MPK6 is phosphorylated by MKK4/MKK5. MPK6 is involved in defense signaling system. It has been demonstrated that silencing of MPK6 compromised disease resistance response (Menke et al. 2004). The *Arabidopsis* CTR1 (Constitutive triple Response 1) is a Raf-like kinase considered as a MAPKKK and it controls MPK3/6 activation via MKK9 (MAPK kinase9). It negatively regulates ET signaling (Guo and Ecker 2004; Yoo et al. 2008). The *Arabidopsis* MAP kinase MPK4 has been shown to be a requirement for induction of a subset of ET-regulated genes (Brodersen et al. 2006).

The ethylene-dependent defense signaling pathway begins with the induction of ethylene biosynthesis. 1-Amino-cyclopropane-1-carboxylic acid (ACC) synthase (ACS) is the key enzyme involved in ethylene biosynthesis. S-adenosyl-methionine (SAM) is converted to ACC by ACC synthase (Wang et al. 2002). ACS is PAMP elicitor-inducible enzyme involved in induction of biosynthesis of ethylene (Li et al. 2012). Another enzyme involved in biosynthesis of ethylene is ACC oxidase. The enzyme oxidatively cleaves the ACC resulting in generation of ethylene (Wang et al. 2002).

Several *ACS* genes have been identified and transcriptional activation of these genes contributes to the increase in ACS isozymes in plants (Vogel et al. 1998; Chae et al. 2003; Skottke et al. 2011). Phosphorylation of the ACC synthases appears to be essential for the activation of these enzymes. In the absence of phosphorylation by mitogen-activated protein kinase, the newly synthesized ACC synthases are rapidly degraded through ubiquitin-proteasome pathway, resulting in no net increase in the ACS proteins (Liu and Zhang 2004; Joo et al. 2008). ACC synthase isozymes have been shown to be substrates for E3 ligases (Dreher and Callis 2007).

The *Arabidopsis* ACS isoforms ACS2 and ACS6 have been shown to be substrates of MPK3 and MPK6 (Liu and Zhang 2004; Han et al. 2010). The ACS isoforms were shown to be phosphorylated and stabilized by MPK3 and MPK6 functioning in the MAPK cascade consisting of MEKK1-MKK4/MKK5-MPK3/MPK6 (Han et al. 2010; Li et al. 2012). Phosphorylation of the two ACC synthases, ACS2 and ACS6 by MPK3 and MPK6 prevents rapid degradation of ACS2/ACS6 by the 26S proteasome pathway, resulting in an increase in cellular ACS activity (Han et al. 2010). The phosphorylation resulted in ACS stability and accumulation of ACS isozymes, which led to increased synthesis of ACC (Liu and Zhang 2004). Another enzyme involved in biosynthesis of ethylene, ACC oxidase, oxidatively cleaves the accumulated ACC resulting in generation of ethylene (Wang et al. 2002). In addition to direct phosphorylation modification and stabilization of ACS proteins, MPK3 and MPK6 also regulate the expression of ACS genes through another MPK3/MPK6 substrate, the WRKY transcription factor WRKY 33 (Fig. 7.6; Li et al. 2012). WRKY33 binds to the W-boxes in the *ACS2/ACS6* promoters in vivo and is directly involved in MPK3/MPK6-induced *ACS2/ACS6* gene expression. Regulations of ACS isoforms at both transcriptional and post-translational levels by MPK3 and MPK6 seem to contribute to the high-level ethylene production in plants challenged by invading pathogens (Li et al. 2012). These studies suggest that MPK3 and MPK6 not only function in the phosphorylation-induced stabilization of ACS2/ACS5 proteins, but also signal the *ACS2* and *ACS6* gene activation after *Botrytis cinerea* in *Arabidopsis* (Li et al. 2012).

**Fig. 7.6** Role of MAPK signaling cascade in triggering ethylene biosynthesis in *Arabidopsis* (Adapted from: Li et al. 2012)

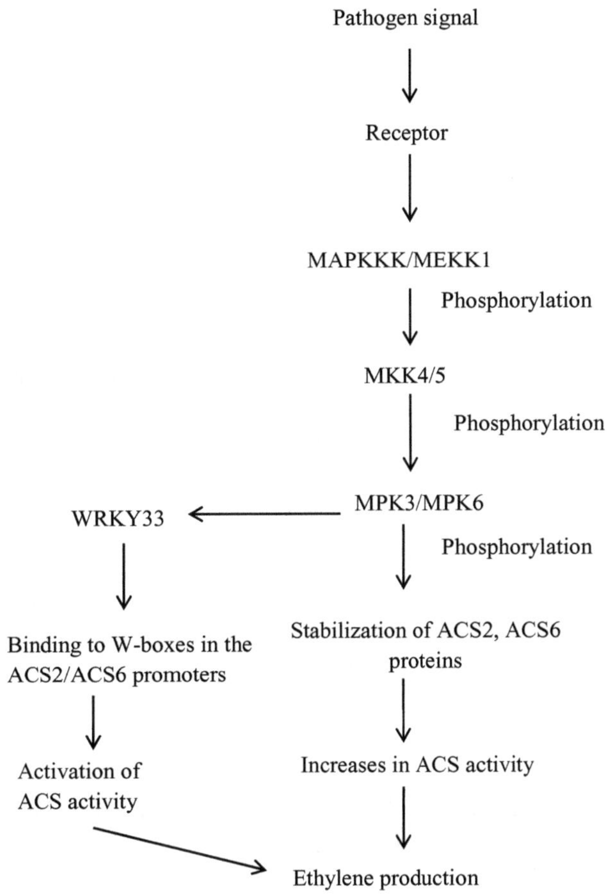


## 7.17 Involvement of MAP Kinase in Crosstalk Between SA and JA/ET Signaling Systems

Crosstalk between SA and JA/ET signaling systems has been reported in different plant-pathogen interactions. The tobacco basic PR protein gene *NtPRB1b* responds positively to JA and ET signaling systems, but negatively to SA (Li et al. 2012). The MAP kinase MPK4 plays a role in the antagonism between SA and JA signaling systems in *Arabidopsis thaliana*. The *mpk4* knockout plants exhibit constitutive activation of SA-dependent signaling system, but fail to induce JA-dependent signaling system (Petersen et al. 2000; Brodersen et al. 2006).

EDS1 and PAD4 are defense regulators and they modulate SA/JA signal antagonism as activators of SA but repressors of JA signaling (Wiermer et al. 2005; Brodersen et al. 2006). The function of EDS1 and PAD4 in the antagonistic interaction between SA and JA depends on the MAP kinase, MPK4 (Brodersen et al. 2006).

The *mpk4* mutant is impaired in the induction of JA- and ET-responsive genes. The SA-repressing and the ET/JA-activating functions depend on the defense regulators EDS1 and PAD4 (Brodersen et al. 2006). EDS1 and PAD4 participate in defense amplification loop that responds to SA and reactive oxygen species (ROS)-derived signals (Rustérucci et al. 2001). Mutations in *EDS1/PAD4* affect SA-ET/JA signal antagonism as activators of SA but as repressors of ET/JA defenses, and MPK4 negatively regulates both of these functions (Brodersen et al. 2006).

MAP kinase acts as a negative regulator of SA signaling and a positive regulator of JA signaling in Arabidopsis (Petersen et al. 2000). Inactivation of MPK4 in mutant *mpk4 Arabidopsis* plants resulted in increased expression of SA-responsive genes and suppression of JA-responsive genes (Petersen et al. 2000). MAP KINASE4 SUBSTRATE 1 (MKS1) is the target of MPK4 and phosphorylation of MKS1 by MPK4 has been reported to repress SA signaling. MSK1 interacts with the WRKY transcription factors WRKY25 and WRKY33 and both of them can be phosphorylated by MPK4 (Andreasson et al. 2005). Overexpression of both WRKY25 and WRKY33 results in repression of SA signaling responses, suggesting that these transcription factors, after activation by phosphorylation by the action of MPK4, suppress SA signaling system (Zheng et al. 2006, 2007). By contrast, *wrky33* mutant plants showed reduced expression of JA-responsive genes, suggesting that WRKY33 after phosphorylation by MPK4 activates JA signaling system (Zheng et al. 2006). These studies suggest that MPK4 suppresses SA signaling system, while it activates JA signaling system and it is involved in cross-talk between SA and JA signaling systems.

## 7.18 MAPK Phosphatases as Negative Regulators of MAP Kinases

MAPK cascades include both phosphorylation and dephosphorylation events resulting in transient increases in MAPK activity. MAP kinases are dephosphorylated and inactivated by protein phosphatases, including tyrosine-specific phosphatases, serine/threonine-specific phosphatases, and dual-specificity MAPK phosphatases (MKPs), which are highly specific to MAPKs (Keyse 2000; Theodosiou and Ashworth 2002). MAPK phosphatases (MKPs) are negative regulators of MAPKs (Katou et al. 2005). Overexpression of the tobacco MKP NtMKP1 compromises wound-induced SIPK and WIPK, suggesting a role for the MKP in regulating these MAPKs in vivo (Yamakawa et al. 2004). In *Arabidopsis*, the transmission of ROS and pathogen signaling by MAPKs involves the coordinated activation of MPK6 and MPK3. MAPK phosphatase 2 (MKP2) regulates plant immune responses and functionally interacts with MPK3 and MPK6. Plants lacking a functional *MKP2* gene show defense response against the bacterial pathogen *Ralstonia solanacearum* and by contrast, these plants show enhanced susceptibility against the fungal pathogen *Botrytis cinerea* (Lumbreras et al. 2010). This MKP2 function appears to be linked to differential regulation of MPK3 and MPK6 networks by different types of pathogens (Lumbreras et al. 2010).

In contrast to many members of the MAPK family, plant MKPs form only a small gene family. Only five MKPs have been reported in *Arabidopsis* (Kerk et al. 2002). This disproportionate ratio of MAPK to MKP suggests that one MKP regulates multiple MAPKs in plants. The Arabidopsis MAPK phosphatase AtMKP1 specifically interacts with AtMPK3, AtMPK4, and AtMPK6 (Ulm et al. 2002). The tobacco MKP NtMKP1 inactivates salicylic acid-induced protein kinase (SIPK) through dephosphorylation of the TEY motif of SIPK (Katou et al. 2005). The phosphatase activity of NtMKP1 was increased strongly by the binding of SIPK and only weakly by another MAPK, WIPK, revealing the specificity of NtMKP1 (Katou et al. 2005).

MKPs have been shown to negatively regulate JA and ET signaling systems in *Arabidopsis thaliana* (Schweighofer et al. 2007). An *A. thaliana* Ser/Thr phosphatase of type 2C, AP2C1, inactivates the MAPKs MPK4 and MPK6. Mutant *ap2c1* plants produce significantly higher amounts of JA in response to external stimulus. Plants with increased AP2C1 levels display lower activation of MAPKs, reduced ethylene production, and compromised innate immunity against *Botrytis cinerea.* These results suggest that the phosphatase negatively regulates the MAPK pathway and ET and JA signaling system (Schweighofer et al. 2007).

## 7.19 MAP Kinase Cascades Modulate Phosphorylation of Transcription Factors to Trigger Transcription of Defense Genes

MAP kinase cascades have been shown to be involved in phosphorylation of transcription factors, which are involved in transcription of defense genes activated through SA- or JA-, or ET- dependent signaling systems. BWMK1 (Blast- and wounding- activated MAPK 1), a rice mitogen-activated protein kinase is targeted to the nucleus. This protein phosphorylates the rice transcription factor OsEREBP1 (Oryza sativa ethylene-responsive element-binding protein 1). EREBPs are known to bind to the GCC box DNA motif (AGCCGCC) that is located in the promoter of several *PR* genes. In vitro phosphorylation of OsEREBP1 by BWMK1 enhanced its ability to bind to the GCC box. Ectopic expression of the BWMK1 in tobacco plant induced the expression of a broad spectrum of *PR* genes (Cheong et al. 2003).

Ethylene response factor6 (ERF6) is another substrate of MPK3/MPK6 and it regulates *Arabidopsis* defense gene expression and resistance against *Botrytis cinerea.* Phosphorylation of ERF6 by MPK3/MPK6 in either the gain-of-function transgenic plants or in response to *B. cinerea* infection increases ERF6 protein stability in vivo (Meng et al. 2013). Chitin elicitors induced the kinase activity of two MAPK genes, *AtMPK6* and *AtMPK3* in *Arabidopsis* (Wan et al. 2004). In addition, *WRKY22*, *WRKY29*, *WRKY33*, and *WRKY53*, which encode four WRKY transcription factors that recognize TTGAC(C/T) W-box elements in promoters of several defense-related genes, were up-regulated by the elicitor treatment (Wan et al. 2004). WRKY33 is a pathogen-inducible transcription factor, whose expression is regulated by the MPK3/MPK6 cascade. WRKY33 is a substrate of MPK3/MPK6. It has been demonstrated that WRKY33 is phosphorylated by MPK3/MPK6 in vivo in response to *Botrytis cinerea* infection in *Arabidopsis*. WRKY33 is required for MPK3/MPK6-induced camalexin biosynthesis (Mao et al. 2007).

*Nicotiana benthamiana* WRKY8 transcription factor has been shown to be a physiological substrate of the MAPKs, SIPK, NTF4, and WIPK (Ishihama et al. 2011). Clustered Pro-directed Ser residues, which are conserved in group 1 WRKY proteins, in the N-terminal region of WRKY8 were phosphorylated by these MAPKs in vitro. The interaction of WRKY8 with MAPKs depended on its D domain, which is a MAPK-interacting motif, and this interaction was required for effective phosphorylation of WRKY8 in plants. Phosphorylation of WRKY8 increased its DNA binding activity to the cognate W-box sequence (Ishihama et al. 2011). Ectopic expression of WRKY8 induced defense-related genes. By contrast, silencing of WRKY8 decreased the expression of defense-related genes and increased susceptibility to the oomycete pathogen *Phytophthora infestans* and the fungal pathogen *Colletotrichum orbiculare* (Ishihama et al. 2011). These results suggest that MAPK-mediated phosphorylation of WRKY8 has an important role in triggering downstream immune responses.

The tobacco MAP kinase WIPK phosphorylates and activates NtWIF, a transcription factor. The transgenic tobacco plants overexpressing *NtWIF* exhibited constitutive accumulation of transcripts for *PR* genes, *PR-1a* and *PR-2* (Waller et al. 2006). MPK3 phosphorylates a plant VirE2-interacting protein 1 (VIP1), a bZIP transcription factor (Liu et al. 2010). VIP1 is a direct target of the PAMP-induced MPK3. Upon phosphorylation by MPK3, VIP1 relocalizes from the cytoplasm to the nucleus and regulates the expression of the *PR1* pathogenesis-related gene (Djamei et al. 2007). Collectively, these results suggest that phosphorylation of transcription factors by MAPKs is an important event in triggering expression of defense-related genes.

## 7.20 MAPKs Regulate Defense Gene Expression by Releasing Transcription Factors in the Nucleus

Transcription factor release may be a common theme after MAPK activation to control downstream gene expression. MAPKs activate expression of defense-related genes. MKS1 (for MAP kinase 4 substrate 1) is a substrate for the *Arabidopsis* MAPK MPK4. MKS1 interacts with the transcription factors WRKY 25 and WRKY33 (Andreasson et al. 2005). The interaction of MKS1 with WRKY33 has been shown to be dependent on the phosphorylation status of MKS1 induced by MPK4 (Qiu et al. 2008b). In the absence of pathogens, inactivated MPK4 forms a ternary complex with MKS1 and WRKY33 in the nucleus, which prevents WRKY33 from functioning as a transcription factor. MPK4 is activated by the PAMP treatment or pathogen inoculation in *Arabidopsis*. Upon activation of MPK4, MKS1 is phosphorylated by MPK4. Subsequently, phosphorylated MKS1 and WRKY33 proteins are released from MPK4. The unbound WRKY33 targets the promoter of a defense-related gene *PAD3* (*PHYTOALEXIN DEFICIENT3*) for transcriptional activation (Qiu et al. 2008a, b).

Bethke et al. (2009b) identified a transcription factor of the Ethylene Response Factor (ERF) family, ERF104, which interacted with the *Arabidopsis* MAPK MPK6. The ERF104 was found to be a nuclear substrate involved in plant defense and MPK6 binds with ERF104. The continued binding of MPK6 to ERF104 might constrain physical interactions with subsequent ERF104 targets and impringe on its role in transcription activation (Bethke et al. 2009b). The release of ERF104 from MPK6 in the nucleus required rapid ET signaling (Bethke et al. 2009b). Bethke et al. (2009a) suggested that the PAMP flg22 signal network includes one pathway for MPK6 to target ERF104 directly through phosphorylation and on a separate branch, to stimulate ET production, which triggers a yet unknown mechanism (that is dependent on EIN2 and the EIN3/EIL members) for the release of ERF104 from MPK6 in the nucleus. The released transcription factor ERF104 may activate transcription of defense genes. These results suggest that transcription factor release after MAPK activation in the nucleus controls the downstream gene expression.

## 7.21 Role of MAPK Signaling Cascade in Triggering Phytoalexin Biosynthesis

Phytoalexins are key components in plant defense responses and several elicitors are known to trigger production of phytoalexins (Vidhyasekaran 2007). MAPK signaling cascades have been shown to activate the phytoalexin camalexin biosynthesis (Ren et al. 2008). The *Arabidopsis* MAPKKKα/MEKK1-MKK4/MKK5-MPK3/MPK6 cascade has been shown to trigger the camalexin biosynthesis (Fig. 7.7). Both MPK3 and MPK6 play important role in triggering biosynthesis of camalexin. MPK3/MPK6 cascade coordinates the induction of multiple genes in the camalexin biosynthetic pathway. The camalexin biosynthetic genes include the genes encoding anthranilate synthase α and β subunits (ASA and ASB), phosphoribosylanthranilate transferase (PAT), indole-3-glycerolphosphate synthase (IGPS), tryptophan synthase α and β subunits (TSA) and TSB), and the P450 enzymes CYP79B2, CYP79B3, and CYP71B15 (PAD3). The induction of all of these genes was partially compromised in *mpk3*/*mpk6* mutant plants (Ren et al. 2008).

## 7.22 Role of MAPK Signaling Cascade in Stomatal Immune Response

Stomata serve as passive ports of bacterial entry during infection. They constitute one entry point for bacteria, which need to reach apoplastic spaces to multiply and cause disease (Nicaise et al. 2009). The stomata in the *Arabidopsis* leaf epidermis

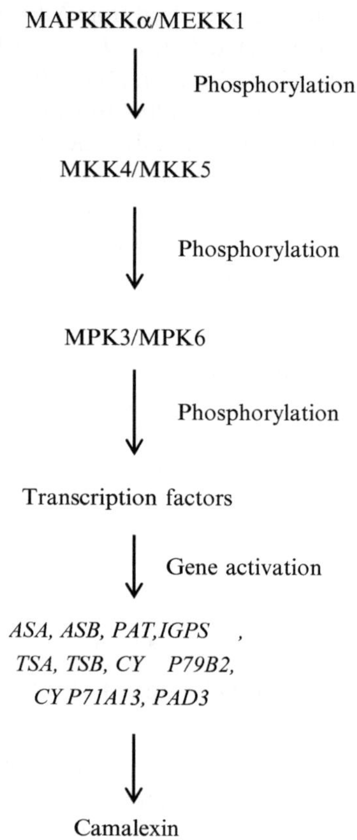


**Fig. 7.7** Role of MPK3 and MPK6 signaling cascade in triggering biosynthesis of camalexin in *Arabidopsis* (Adapted from Ren et al. 2008)

have been shown to act as innate immunity gates to actively prevent bacteria from entering the plant leaf (Melotto et al. 2006). The PAMP flg22 triggers closure of stomata which occurs within the first hour of contact with plant tissue (Melotto et al. 2006). The PAMP-triggered stomatal closure was dependent on several signals and signaling systems. The PAMP-induced ABA signaling system has been reported to be involved in stomatal closure (Hubbard et al. 2010). ABA increase was the critical early event in stomatal closure induced by flg22 (Melotto et al. 2006). ROS signaling system is also involved in the stomatal closure. ABA and $H_2O_2$ treatments induced inhibition of stomatal opening or the promotion of stomatal closure (Gudesblat et al. 2007; Jammes et al. 2009, 2011; Hettenhausen et al. 2012). NO production, activation of OST1 (for OPEN STOMATA1) kinase, $Ca^{2+}$ influx, and modulation of S-type anion channel have been found to be important for induction of stomatal closure responses (Melotto et al. 2006; Vahisalu et al. 2008; Kim et al. 2010). Flg22 triggered ABA synthesis, NO production, and OST1 (for OPEN STOMATA1) kinase, which are required for stomatal closure (Melotto et al. 2006).

MAPKs also have been found to play important role in the stomatal closure immune response. AtMPK9 and AtMPK12 are mainly localized in guard cells, and plants silenced in both MAPKs have strong defects in ABA-induced stomatal closure (Jammes et al. 2009). AtMPK9 and AtMPK12 redundantly and positively control stomatal closure in response to ABA and $H_2O_2$ treatment (Jammes et al. 2009, 2011). Silencing AtMPK3 in *Arabidopsis* guard cells reduces $H_2O_2$-induced inhibition of stomatal opening or the promotion of stomatal closure (Gudesblat et al. 2007).

A *Nicotiana attenuata* MAPK, NaMPK4, plays an important role in guard cell-mediated defense against the bacterial pathogen *Pseudomonas syringae* pv. *tomato* DC3000. NaMPK4 appears to act downstream of ABA in regulating stomatal closure (Hettenhausen et al. 2012). NaMPK4 also has been reported to function downstream of ROS to mediate stomatal closure. Knocking down *NaMPK4* of *N. attenuata* compromises the stomatal closure response after supplying $H_2O_2$ (Hettenhausen et al. 2012). The results suggest that NaMPK4 acts downstream of ROS in stomatal closure response. The S-type anion channel-associated 1 is important for stomatal closure in response to ABA, $H_2O_2$, and $Ca^{2+}$ (Vahisalu et al. 2008; Kim et al. 2010). Downstream of ABA/ROS, NaMPK4 modulates the activity of S-type anion channels, which regulate the stomatal closure response (Hettenhausen et al. 2012). Supplying irNaMPK4 guard cells with $Ca^{2+}$ induced the normal stomatal closure, suggesting the importance of $Ca^{2+}$ in stomatal closure (Hettenhausen et al. 2012). Collectively these studies suggest that ABA- and $H_2O_2$-induced cytosolic $Ca^{2+}$ ($[Ca^{2+}]$cyt) signature activates the anion channels, which regulate stomatal closure. *Arabidopsis thaliana* MPK3 also has been shown to be involved in stomatal guard cell signaling. It probably acts in signaling downstream of $H_2O_2$ in the signaling system. MPK3 is activated by abscisic acid and $H_2O_2$, which control stomatal closure (Gudesblat et al. 2007). These studies suggest that MAP kinase signaling cascades are involved in PAMPs-triggered stomatal closure immune responses.

## 7.23 Effectors Inhibit PAMP-Triggered MAPK Signaling to Suppress Plant Immune Responses

Plant innate immune systems have high potential to fight against a wide range of viral, bacterial, oomycete, and fungal pathogens (Lacombe et al. 2010; Hwang and Hwang 2011; Alkan et al. 2012). However, potential pathogens produce several effectors to nullify the defense responses induced by the innate immune system (Wu et al. 2011; Cheng et al. 2012). To avoid or suppress or delay the expression of the defense gene-activating signaling systems, the pathogens secrete several effectors into the host cell (Göhre et al. 2008; Kim et al. 2010; Wu et al. 2011; Cheng et al. 2012). MAP kinase signaling cascades constitute a major immune response system. Activation of the MAPK signaling system confers resistance against viral (Shi et al. 2010, 2011; Zhang et al. 2011), bacterial (Brader et al. 2007; Dóczi et al. 2007;

Zhang et al. 2007c, 2012c; Shen et al. 2010; Jammes et al. 2011; Schikora et al. 2011; Hettenhausen et al. 2012), fungal (Wang et al. 2009; Shi et al. 2010, 2011), and oomycete (Zhang et al. 2007c) diseases.

Pathogens secrete effectors to suppress the immune responses activated by PAMP elicitors. The bacterial effectors AvrPto and AvrPtoB act as suppressors of early-defense gene transcription and MAPK signaling. These effectors intercept multiple PAMP-mediated signaling upstream of MAPKKK at the plasma membrane linked to the receptor (He et al. 2006). The *Pseudomonas syringae* effector HopF2 shows mono-ADP- ribosyltransferase activity and it inhibits the MAPKK MKK5 preventing the phosphorylation of MPK3 and MPK6 in response to PAMP treatment (Wang et al. 2010). HopPtoD2 is the effector secreted by *Pseudomonas syringae* pv. *tomato* DC3000 and it possessed tyrosine phosphatase activity (Espinosa et al. 2003). A constitutively active MAPK kinase, NtMEK2, is involved in triggering hypersensitive responses. The effector HopPtoD2 suppressed the action of NtMEK2 in eliciting defense responses. It has been suggested that inactivation of MAPK pathways is a virulence strategy by the bacterial pathogen (Bretz et al. 2003; Espinosa et al. 2003).

HopF2 has been found to be a potent suppressor of early immune gene transcription and mitogen-activated protein kinase signaling activated by multiple PAMPs, including bacterial flagellin, ef-Tu, peptidoglycan, lipopolysaccharide and HrpZ1 harpin, and fungal chitin (Wu et al. 2011). The conserved surface-exposed residues of HopF2 may be essential for its PAMP suppression activity. HopF2 is targeted to the plant plasma membrane through a putative myristoylation site, and the membrane association appears to be required for its PAMP-suppression function (Wu et al. 2011). These results suggest that HopF2 likely intercepts PAMP signaling at the plasma membrane immediately of signal perception. Expression of HopF2 in transgenic plants compromised plant nonhost immunity to bacterial pathogen *P. syringae* pv. *phaseolicola* and plant immunity to the fungal pathogen *Botrytis cinerea* (Wu et al. 2011). HopF2 severely impairs PAMP-induced defenses and render plants highly susceptible to nonpathogenic *P. syringae* bacteria (Wang et al. 2010). These results suggest that HopF2 plays important role in suppression of function of multiple PAMP signaling.

A *Pseudomonas syringae* effector HopAI1 inactivates the MAPKs MPK3 and MPK6 to suppress PAMP-induced immunity in plants (Zhang et al. 2007b). HopAI1 inactivates MAPKs by removing the phosphate group from phosphothreonine through phosphothreonine lyase activity. The inhibition of MAPKs by HopI1 suppresses transcriptional activation of PAMP response genes (Zhang et al. 2012c). The HopAI1 has been shown to inhibit MPK4, a negative regulator of defense responses. The MEKK1-MKK1/MKK2-MPK4 cascade negatively regulates plant immune responses (Kong et al. 2012). However, when the MAPK cascade is targeted by HopAI1, it positively regulates the basal immunity, probably due to inhibition of MPK4 by the effector (Zhang et al. 2012c). *P. syringae* effector AvrB interacts with and stimulates the activity of MPK4, a negative regulator of plant defense responses, thereby perturbing hormone signaling and enhancing plant susceptibility (Cui et al. 2010). Collectively, these studies suggest that pathogens secrete effectors, which suppress the action of MAPKs in triggering plant immune responses.

## References


Agrawal GK, Iwahashi H, Rakwal R (2003) Rice MAPKs. Biochem Biophys Res Commun 302:171–180

Alkan N, Fluhr R, Prusky D (2012) Ammonium secretion during *Colletotrichum coccodes* infection modulates salicylic and jasmonic acid pathways of ripe and unripe tomato fruit. Mol Plant Microbe Interact 25:85–96

Andreasson E, Jenkins T, Brodersen P, Thorgrimsen S, Petersen NHT, Zhu S, Qiu J-L, Micheelsen P, Rocher A, Petersen M, Newman A, Nielsen HB, Hirt H, Somssich I, Mattsson O, Mundy J (2005) The MAP kinase substrate MKS1 is a regulator of plant defense responses. EMBO J 24:2579–2589

Asai T, Tena G, Plotnikova J, Willmann MR, Chiu W-L, Gómez-Gómez L, Boller T, Ausubel FM, Sheen J (2002) MAP kinase signalling cascade in *Arabidopsis* innate immunity. Nature 415:977–983

Asai S, Ohta K, Yoshioka H (2008) MAPK signaling regulates nitric oxide and NADPH oxidase-dependent oxidative bursts in *Nicotiana benthamiana*. Plant Cell 20:1390–1406

Aslam SN, Erbs G, Morrissey KL, Newman M-A, Chinchilla D, Boller T, Molinaro A, Jackson RW, Cooper RM (2009) Microbe-associated molecular pattern (MAMP) signatures, synergy, size, and charge: influences on perception or mobility and host defense responses. Mol Plant Pathol 10:375–387

Beckers GJM, Jaskiewicz M, Liu Y, Underwood WR, He Y, Zhang S, Conrath U (2009) Mitogen-activated protein kinases 3 and 6 are required for full priming of stress responses in *Arabidopsis thaliana*. Plant Cell 21:944–953

Bethke G, Scheel DK, Lee J (2009a) Sometimes new results raise new questions: the question marks between mitogen-activated protein kinase and ethylene signaling. Plant Signal Behav 4:672–674

Bethke G, Unthan T, Uhrig JF, Poschl Y, Gust AA, Scheel DK, Lee J (2009b) Flg22 regulates the release of an ethylene response factor substrate from MAP kinase 6 in *Arabidopsis thaliana* via ethylene signaling. Proc Natl Acad Sci U S A 106:8067–8072

Bethke G, Pecher P, Eschen-Lippold L, Tsuda K, Katagiri F, Glazebrook J, Scheel D, Lee J (2012) Activation of the *Arabidopsis thaliana* mitogen-activated protein kinase MPK11 by the flagellin-derived elicitor peptide, flg22. Mol Plant Microbe Interact 25:471–480

Bögre L, Ligterink W, Meskiene I, Barker PJ, Heberle-Bors E, Huskisson NS, Hirt H (1997) Wounding induces the rapid and transient activation of a specific MAP kinase pathway. Plant Cell 9:75–83

Boller T, Felix G (2009) A renaissance of elicitors: perception of miceobe-associated molecular patterns and danger signals by pattern-recognition receptors. Annu Rev Plant Biol 60:379–406

Boller T, He SY (2009) Innate immunity in plants: an arms race between pattern recognition receptors in plants and effectors in microbial pathogens. Science 324:742–744

Boter M, Ruiz-Rivero O, Abdeen A, Prat S (2004) Conserved MYC transcription factors play a key role in jasmonate signalling both in tomato and *Arabidopsis*. Genes Dev 18:1577–1591

Brader G, Djamei A, Teige M, Palva ET, Hirt H (2007) The MAP kinase kinase MKK2 affects disease resistance in *Arabidopsis*. Mol Plant Microbe Interact 20:589–596

Bretz JR, Mock NM, Charity JC, Zeyad S, Baker CJ, Hutcheson SW (2003) A translocated protein tyrosine phosphatase of *Pseudomonas syringae* pv. *tomato* DC3000 modulates plant defense response to infection. Mol Microbiol 49:389–400

Brodersen P, Petersen M, Bjorn Nielsen H, Zhu S, Newman MA, Shokat KM, Rietz S, Parker J, Mundy J (2006) *Arabidopsis* MAP kinase 4 regulates salicylic acid- and jasmonic acid/ethylene-dependent responses via EDS1 and PAD4. Plant J 47:532–546

Cardinale F, Jonak C, Ligterink W, Niehaus K, Boller T, Hirt H (2000) Differential activation of four specific MAPK pathways by distinct elicitors. J Biol Chem 275:36734–36740

Cardinale F, Meskiene I, Ouaked F, Hirt H (2002) Convergence and divergence of stress induced mitogen-activated protein kinase signaling pathways at the level of two distinct mitogen-activated protein kinase kinases. Plant Cell 14:703–711

Chae HS, Faure F, Kieber JJ (2003) The *eto1*, *eto2*, and *eto3* mutations and cytokinin treatment increase ethylene biosynthesis in Arabidopsis by increasing the stability of ACS protein. Plant Cell 15:545–559

Cheng B, Yu X, Ma Z, Dong S, Dou D, Wang Y, Zheng X (2012) *Phytophthora sojae* effector Avh331 suppresses the plant defence response by disturbing the MAPK signalling pathway. Physiol Mol Plant Pathol 77:1–9

Cheong YH, Moon BC, Kim JK, Kim CY, Kim MC, Kim IH, Park CY, Kim JC, Park BO, Koo SC, Yoon HW, Chung WS, Lim CO, Lee SY, Cho MJ (2003) BWMK1, a rice mitogen-activated protein kinase, locates in the nucleus and mediates pathogenesis-related gene expression by activation of a transcription factor. Plant Physiol 132:1961–1972

Chini A, Fonseca S, Fernández G, Adie B, Chico JM, Lorenzo O, Garcia-Casado G, López-Vidriero I, Lozano FM, Ponce MR, Micol JL, Solano R (2007) The JAZ family of repressors is the missing link in jasmonate signaling. Nature 448:666–671

Christiansen KM, Gu Y, Rodibaugh N, Innes RW (2011) Negative regulation of defence signalling pathways by the EDR1 protein kinase. Mol Plant Pathol 12:746–758

Cui H, Wang Y, Xue L, Chu J, Yan C, Fu J, Chen M, Innes RW, Zhou JM (2010) *Pseudomonas syringae* effector protein AvrB perturbs *Arabidopsis* hormone signaling by activating MAP kinase 4. Cell Host Microbe 7:164–175

Desikan R, Hancock JT, Ichimura K, Shinozaki K, Neill SJ (2001) Harpin induces activation of the *Arabidopsis* mitogen-activated protein kinases AtMPK4 and AtMPK6. Plant Physiol 126:1579–1587

Djamei A, Pitzschke A, Nakagami H, Rajh I, Hirt H (2007) Trojan horse strategy in *Agrobacterium* transformation: abusing MAPK defense signaling. Science 318:453–456

Dóczi R, Brader G, Pettkó-Szandtner A, Rajh I, Djamei A, Pitzschke A, Teige M, Hirt H (2007) The *Arabidopsis* mitogen-activated protein kinase kinase MKK3 is upstream of group C mitogen-activated protein kinases and participates in pathogen signaling. Plant Cell 19:3266–3279

Dreher K, Callis J (2007) Ubiquitin, hormones and biotic stress in plants. Ann Bot 99:787–822

Droillard MJ, Boudsocq M, Barbier-Brygoo H, Laurière C (2004) Involvement of MPK4 in osmotic stress response pathways in cell suspensions and plantlets of *Arabidopsis thaliana* : activation by hypoosmolarity and negative role in hyperosmolarity tolerance. FEBS Lett 574:42–48

Espinosa A, Guo M, Tam VC, Fu ZQ, Alfano JR (2003) *Pseudomonas syringae* type III-secreted protein HopPtoD2 possesses protein tyrosine phosphatase activity and suppresses programmed cell death in plants. Mol Microbiol 49:377–387

Ferl RJ (2004) 14-3-3 proteins: regulation of signal-induced events. Physiol Plant 120:173–178

Frye CA, Innes RW (1998) An *Arabidopsis* mutant with enhanced resistance to powdery mildew. Plant Cell 10:947–956

Frye CA, Tang D, Innes RW (2001) Negative regulation of defense responses in plants by a conserved MAPKK kinase. Proc Natl Acad Sci U S A 98:373–378

Galletti R, Ferrari S, De Lorenzo G (2011) *Arabidopsis* MPK3 and MPK6 play different roles in basal and oligogalacturonide- or flagellin-induced resistance against *Botrytis cinerea*. Plant Physiol 157:804–814

Gao M, Liu J, Bi D, Zhang Z, Cheng F, Chen S, Zhang Y (2008) MEKK1, MKK1/MKK2 and MPK4 function together in a mitogen-activated protein kinase cascade to regulate innate immunity in plants. Cell Res 18:1190–1198

Göhre V, Spallek T, Häweker H, Mersmann S, Mentzel T, Boller T, de Torres M, Mansfield JW, Robatzek S (2008) Plant pattern-recognition receptor FLS2 is directed for degradation by the bacterial ubiquitin ligase AvrPtoB. Curr Biol 18:1824–1832

Gomi K, Ogawa D, Katou S, Kamada H, Nakajima N, Saji H, Soyano T, Sasabe M, Machida Y, Mitsuhara I, Ohashi Y, Seo S (2005) A mitogen-activated protein kinase NtMPK4 activated by SIPKK is required for jasmonic acid signaling and involved in ozone tolerance via stomatal movement in tobacco. Plant Cell Physiol 46:1902–1914

Grant JJ, Yun B-W, Loake GJ (2000) Oxidative burst and cognate redox signaling reported by luciferase imaging: identification of a signal network that functions independently of ethylene, SA, and Me-JA but is dependent on MAPKK activity. Plant J 24:569–582

Gudesblat GE, Iusem ND, Morris PC (2007) Guard cell-specific inhibition of *Arabidopsis* MPK3 expression causes abnormal stomatal responses to abscisic acid and hydrogen peroxide. New Phytol 173:713–721

Guo H, Ecker JR (2004) The ethylene signaling pathway; new insights. Curr Opin Plant Biol 7:40–49

Hall HC, Samuel MA, Ellis BE (2007) SIPK conditions transcriptional responses unique to either bacterial or oomycete elicitation in tobacco. Mol Plant Pathol 8:581–594

Hamel LP, Nicole MC, Sritubtim S, Morency MJ, Ellis M, Ehiting J, Beaudoin N, Barbazuk B, Klessig D, Lee J, Martin G, Mundy J, Ohashi Y, Scheel D, Sheen J, Xing T, Zhang S, Sequin A, Ellis BE (2006) Ancient signals: comparative genomics of plant MAPK and MAPKK gene families. Trends Plant Sci 11:192–198

Han L, Li GJ, Yang KY, Mao G, Wang R, Liu Y, Zhang S (2010) Mitogen-activated protein kinase 3 and 6 regulate *Botrytis cinerea*-induced ethylene production in *Arabidopsis*. Plant J 64:114–127

He Z-H, Cheeseman I, He D, Kohorn BD (1999) A cluster of live cell wall-associated receptor kinases, Wak1-5 are expressed in specific organs of *Arabidopsis*. Plant Mol Biol 39:1189–1999

He P, Shan L, Lin NC, Martin GB, Kemmerling B, Nurnberger T, Sheen J (2006) Specific bacterial suppressors of MAMP signaling upstream of MAPKKK in *Arabidopsis* innate immunity. Cell 125:563–575

Hettenhausen C, Baldwin IT, Wu J (2012) Silencing *MPK4* in *Nicotiana attenuata* enhances photosynthesis and seed production but compromises abscisic acid-induced stomatal closure and guard cell-mediated resistance to *Pseudomonas syringae* pv. *tomato* DC3000. Plant Physiol 158:759–776

Hirt H (2000) Connecting oxidative stress, auxin, and cell cycle regulation through a plant mitogen-activated protein kinase pathway. Proc Natl Acad Sci U S A 97:2405–2407

Hiruma K, Takano Y (2011) Roles of EDR1 in non-host resistance of Arabidopsis. Plant Signal Behav 6:1831–1833

Holley SR, Yalamanchili RD, Moura DS, Ryan CA, Stratmann JW (2003) Convergence of signaling pathways induced by systemin, oligosaccharide elicitors, and ultraviolet-B radiation at the level of mitogen-activated protein kinases in *Lycopersicon peruvianum* suspension-cultured cells. Plant Physiol 132:1728–1738

Huang Y, Li H, Gupta R, Morris PC, Luan S, Kieber JJ (2000) AtMPK4, an Arabidopsis homolog of mitogen-activated protein kinase, is activated *in vitro* by AtMEK1 through threonine phosphorylation. Plant Physiol 122:1301–1310

Hubbard KE, Nishimura N, Hitoni K, Getzoff ED, Schroeder JI (2010) Early abscisic acid signal transduction mechanisms: newly discovered components and newly emerging questions. Genes Dev 24:1695–1708

Hwang IS, Hwang BK (2011) The pepper mannose-binding lectin gene *CaMBL1* is required to regulate cell death and defense responses to microbial pathogens. Plant Physiol 155:447–463

Ichimura K, Mizoguchi T, Irie K, Morris P, Giraudat J, Matsumoto K, Shinozaki K (1998) Isolation of ATMEKK1 (a MAP kinase kinase kinase)-interacting proteins and analysis of a MAP kinase cascade in *Arabidopsis*. Biochem Biophys Res Commun 253:532–543

Ichimura K, Shinozaki K, Tena G, Sheen J, Henry Y (2002) Mitogen-activated protein kinase cascades in plants: a new nomenclature. Trends Plant Sci 7:301–308

Ichimura K, Casais C, Peck SC, Shinozaki K, Shirasu k (2006) MEKK1 is required for MPK4 activation and regulates tissue-specific and temperature-dependent cell death in Arabidopsis. J Biol Chem 281:36969–36976

Innes RW (2001) Mapping out the roles of MAP kinases in plant defense. Trends Plant Sci 6:392–394

Ishihama N, Yamada R, Yoshioka M, Katou S, Yoshioka H (2011) Phosphorylation of the *Nicotiana benthamiana* WRKY8 transcription factor by MAPK functions in the defense response. Plant Cell 23:1153–1170

Jammes F, Song C, Shin D, Munemasa S, Takeda K, Gu D, Cho D, Lee S, Giordo R, Sritubtim S, Leonhardt N, Ellis BE, Murata Y, Kwak JM (2009) MAP kinases MPK9 and MPK12 are preferentially expressed in guard cells and positively regulate ROS-mediated ABA signaling. Proc Natl Acad Sci U S A 106:20520–20525

Jammes F, Yang X, Xiao S, Kwak JM (2011) Two Arabidopsis guard cell-preferential MAPK genes, *MPK9* and *MPK12*, function in biotic stress response. Plant Signal Behav 6:1875–1877

Jin H, Liu Y, Yang K-Y, Kim CY, Baker B, Zhang S (2003) Function of a mitogen-activated protein kinase pathway in *N*-gene mediated resistance in tobacco. Plant J 33:719–731

Jonak C, Okrész L, Bogre L, Hirt H (2002) Complexity, cross-talk and integration of plant MAP kinase signaling. Curr Opin Plant Biol 5:415–424

Jones DH, Ley S, Aitken A (1995) Isoforms of 14-3-3 protein can form homo- and heterodimers in vivo and in vitro: implications for function as adapter proteins. FEBS Lett 368:55–58

Joo S, Liu Y, Leuth A, Zhang S (2008) MAPK phosphorylation-induced stabilization of ACS6 protein is mediated by the non-catalytic C-terminal domain, which also contains the *cis-determinant* for rapid degradation by the 26S proteasome pathway. Plant J 54:129–140

Jung J, Won SY, Suh SC, Kim H, Wing R, Jeong Y, Hwang I, Kim M (2007) The barley ERF-type transcription factor HvRAF confers enhanced pathogen resistance and salt tolerance in *Arabidopsis*. Planta 225:575–585

Kallenbach M, Alagna F, Baldwin IT, Bonaventure G (2010) *Nicotiana attenuata* SIPK, WIPK, NPR1, and fatty acid-amino acid conjugates participate in the induction of jasmonic acid biosynthesis by affecting early enzymatic steps in the pathway. Plant Physiol 152:98–106

Kandoth PK, Ranf S, Pancholi SS, Jayanty S, Walla MD, Miller W, Howe GA, Lincoln DE, Stratmann JW (2007) Tomato MAPKs LeMPK1, LeMPK2, and LeMPK3 function in the systemin-mediated defense response against herbivorous insects. Proc Natl Acad Sci U S A 104:12205–12210

Katou S, Senda K, Yoshioka H, Doke N, Kawakita K (1999) A 51 kDa protein kinase of potato activated with hyphal wall components from *Phytophthora infestans*. Plant Cell Physiol 40:825–831

Katou S, Karita E, Yamakawa H, Seo S, Mitsuhara I, Kuchitsu K, Ohashi Y (2005) Catalytic activation of the plant MAPK phosphatase NtMKP1 by its physiological substrate salicylic acid-induced protein kinase but not by calmodulins. J Biol Chem 280:39569–39581

Kazan K, Manners JM (2013) MYC2: the master in action. Mol Plant 6:686–703

Kerk D, Bulgrien J, Smith DW, Barsam B, Veretnik S, Gribskov M (2002) The complement of protein phosphatase catalytic subunits encoded in the genome of *Arabidopsis*. Plant Physiol 129:908–925

Keyse SM (2000) Protein phosphatases and the regulation of mitogen-activated protein kinase signaling. Curr Opin Cell Biol 12:186–192

Kim CY, Liu Y, Thorne ET, Yang H, Fukushige H, Gassmann W, Hildebrand D, Sharp RE, Zhang S (2003) Activation of a stress-responsive mitogen-activated protein kinase cascade induces the biosynthesis of ethylene in plants. Plant Cell 15:2707–2718

Kim TH, Böhmer M, Hu H, Nishimura N, Schroeder JI (2010) Guard cell signal transduction network: advances in understanding abscisic acid, $CO_2$, and $Ca^{2+}$ signaling. Annu Rev Plant Biol 61:561–591

Kobayashi M, Seo S, Hirai K, Yamamoto-Katou A, Katou S, Seto H, Meshi T, Mitsuhara I, Ohashi Y (2010) Silencing of WIPK and SIPK mitogen-activated protein kinases reduces *Tobacco mosaic virus* accumulation but permits systemic viral movement in tobacco possessing the *N* resistance gene. Mol Plant Microbe Interact 23:1032–1041

Kong G, Qu N, Gao M, Zhang Z, Ding X, Yang F, Li Y, Dong OX, Chen S, Li X, Zhang Y (2012) The MEKK1-MKK1/MKK2-MPK4 kinase cascade negatively regulates immunity mediated by a mitogen-activated protein kinase kinase kinase in *Arabidopsis*. Plant Cell 24:2225–2236

Kovtun Y, Chiu W-L, Tena G, Sheen J (2000) Functional analysis of oxidative stress-activated mitogen activated protein kinase cascade in plants. Proc Natl Acad Sci U S A 97:2940–2945

Kumar D, Klessig DF (2000) Differential induction of tobacco MAP kinases by the defense signals nitric oxide, salicylic acid, ethylene, and jasmonic acid. Mol Plant Microbe Interact 13:347–351

Lacombe S, Rougon-Cardoso A, Sherwood E, Peeters N, Dahlbeck D, van Esse HP, Smoker M, Rallapalli G, Thomma BPHJ, Stakawicz B, Jones JDG, Zipfel C (2010) Interfamily transfer of a plant pattern-recognition receptor confers broad-spectrum bacterial resistance. Nat Biotechnol 28:365–369

Lee DE, Lee IJ, Han O, Baik MG, Han SS, Back K (2004) Pathogen resistance of transgenic rice expressing mitogen-activated protein kinase 1, MK1, from *Capsicum anuum*. Mol Cells 17:81–85

Li G, Meng X, Wang R, Mao G, Han L, Liu Y, Zhang S (2012) Dual-level regulation of ACC synthase activity by MPK3/MPK6 cascade and its downstream WRKY transcription factor during ethylene induction in Arabidopsis. PLoS Genet 8(6):e1002767. doi:10.1371/journal.pgen.1002767

Lieberherr D, Thao NP, Nakashima A, Umemura K, Kawasaki T, Shimamoto K (2005) A sphingolipid elicitor-inducible mitogen-activated protein kinase is regulated by the small GTPase OsRac1 and heterotrimeric G-protein in rice. Plant Physiol 138:1644–1652

Ligterink W, Hirt T (2000) MAP kinase pathways in plants: versatile signaling tools. Int Rev Cytol 201:209–258

Ligterink W, Kroj T, Hirt H, Scheel D, zur Nieden U (1997) Receptor-mediated activation of a MAP kinase in pathogen defense of plants. Science 276:2054–2057

Liu Y, Zhang S (2004) Phosphorylation of 1-aminocyclopropane-1-carboxylic acid synthase by MPK6, a stress-responsive mitogen-activated protein kinase, induces ethylene biosynthesis in *Arabidopsis*. Plant Cell 16:3386–3399

Liu Y, Zhang S, Klessig DF (2000) Molecular cloning and characterization of a tobacco MAP kinase kinase that interacts with SIPK. Mol Plant Microbe Interact 13:118–124

Liu Y, Jin H, Yang K, Kim C, Baker B, Zhang S (2003) Interaction between two mitogen-activated protein kinases during tobacco defense signaling. Plant J 34:149–160

Liu Y, Kong X, Pan J, Li D (2010) VIP1: linking *Agrobacterium*-mediated transformation to plant immunity? Plant Cell Rep 29:805–812

Liu J-Z, Horstman HD, Braun E, Graham MA, Zhang C, Navarre D, Qiu W-L, Lee Y, Nettleton D, Hill JH, Whitham SA (2011) Soybean homologs of MPK4 negatively regulate defense responses and positively regulate growth and development. Plant Physiol 157:1363–1378

Lorenzo O, Chico JM, Sánchez-Serrano JJ, Solano R (2004) JASMONATE-INSENSITIVE1 encodes a MYC transcription factor essential to discriminate between different jasmonate-regulated defense responses in *Arabidopsis*. Plant Cell 16:1938–1950

Lumbreras V, Vilela B, Irar S, Solé M, Capellades M, Valls M, Coca M, Pagés M (2010) MAPK phosphatase MKP2 mediates disease responses in *Arabidopsis* and functionally interacts with MPK3 and MPK6. Plant J 63:1017–1030

Mao P, Duan M, Wei C, Li Y (2007) WRKY62 transcription factor acts downstream of cytosolic NPR1 and negatively regulates jasmonate responsive gene expression. Plant Cell Physiol 48:833–842

MAPK Group (2002) Mitogen-activated protein kinase cascades in plants: a new nomenclature. Trends Plant Sci 7:301–308

Matsuoka D, Nanmori T, Sato K, Fukami Y, Kikkawa U, Yasuda T (2002) Activation of AtMEK1, an Arabidopsis mitogen-activated protein kinase kinase, *in vitro* and *in vivo*: analysis of active mutants expressed in *E. coli* and generation of the active form in stress response in seedlings. Plant J 29:637–647

Mayrose M, Bonshtien A, Sessa GJ (2004) LeMPK3 is a mitogen-activated protein kinase with dual specificity induced during tomato defense and wounding responses. J Biol Chem 279:14819–14827

Meldau S, Ullman-Zeunert L, Govind G, Bartram S, Baldwin IT (2012) MAPK-dependent JA and SA signalling in *Nicotiana attenuata* affects plant growth and fitness during competition with conspecifics. BMC Plant Biol 12:213

Melotto M, Underwood W, Koczan J, Nomura K, He SY (2006) Plant stomata function in innate immunity against bacterial invasion. Cell 126:969–980

Meng X, Xu J, He Y, Yang K-Y, Mordorski B, Liu Y, Zhang S (2013) Phosphorylation of an ERF transcription factor by *Arabidopsis* MPK3/MPK6 regulates plant defense gene induction and fungal resistance. Plant Cell 25:1126–1142

Menke FLH, van Pelt JA, Pieterse CMJ, Klessig DF (2004) Silencing of the mitogen-activated protein kinase MPK6 compromises disease resistance in *Arabidopsis*. Plant Cell 16:897–907

Menke FLH, Kang H-G, Chen Z, Mee Park J, Kumar D, Klessig DF (2005) Tobacco transcription factor WRKY1 is phosphorylated by the MAP kinase SIPK and mediates HR-like cell death in tobacco. Mol Plant Microbe Interact 10:1027–1034

Mészáros T, Helfer A, Hatzimasoura E, Magyar Z, Serazetdinova L, Rios G, Bardóczy V, Teige M, Koncz C, Peck S, Bögre L (2006) The Arabidopsis MAP kinase kinase MKK1 participates in defence responses to the bacterial elicitor flagellin. Plant J 48:485–495

Mizoguchi T, Ichimura K, Irie K, Morris P, Giraudat J, Matsumoto K, Shinozaki K (1998) Identification of a possible MAP kinase cascade in *Arabidopsis thaliana* based on pairwise yeast two-hybrid analysis and functional complementation tests of yeast mutants. FEBS Lett 56–60

Nakagami H, Kiegerl S, Hirt H (2004) OMTK1, a novel MAPKKK, channels oxidative stress signaling through direct MAPK activation. J Biol Chem 279:26959–26966

Nakagami H, Pitzschke A, Hirt H (2005) Emerging MAP kinase pathways in plant stress signalling. Trends Plant Sci 10:339–346

Nakagami H, Soukupová H, Schikora A, Zársk V, Hirt H (2006) A mitogen-activated protein kinase kinase kinase mediates reactive oxygen species homeostasis in *Arabidopsis*. J Biol Chem 281:38697–38704

Nicaise V, Roux M, Zipfel C (2009) Recent advances in PAMP-triggered immunity against bacteria: pattern recognition receptors watch over and raise the alarm. Plant Physiol 150:1638–1647

Oh C-S, Pedley KF, Martin GB (2010) Tomato 14-3-3 protein 7 positively regulates immunity-associated programmed cell death by enhancing protein abundance and signaling ability of MAPKKKα. Plant Cell 22:260–272

Oñate-Sánchez L, Anderson JP, Young J, Singh KB (2007) AtERF14, a member of the ERF family of transcription factors, plays a nonredundant role in plant defense. Plant Physiol 143:400–409

Pedley KF, Martin GB (2004) Identification of MAPKs and their possible MAPKK activators involved in the Pto-mediated defense response of tomato. J Biol Chem 279:49229–49235

Pedley KF, Martin GB (2005) Role of mitogen-activated protein kinases in plant immunity. Curr Opin Plant Biol 8:541–547

Petersen M, Brodersen P, Naested H, Andreasson E, Lindhart U, Johansen B, Nielsen HB, Lacy M, Austin MJ, Parker JE, Sharma SB, Klessig DF, Martienssen R, Mattsson O, Jensen AB, Mundy J (2000) *Arabidopsis* MAP kinase 4 negatively regulates systemic acquired resistance. Cell 103:1111–1120

Pitzschke A, Djamei A, Bitton F, Hirt H (2009a) A major role of the MEKK1-MKK1/2-MPK4 pathway in ROS signaling. Mol Plant 2:120–137

Pitzschke A, Schikora A, Hirt H (2009b) MAPK cascade signalling networks in plant defense. Curr Opin Plant Biol 12:421–426

Pré M, Atallah M, Champion A, De Vos M, Pieterse CMJ, Memelink J (2008) The AP2/ERF domain transcription factor ORA59 integrates jasmonic acid and ethylene signals in plant defense. Plant Physiol 147:1347–1357

Qi Z, Verma R, Gehring C, Yamaguchi Y, Zhao Y, Ryan CA, Berkowitz GA (2010) $Ca^{2+}$ signaling by plant *Arabidopsis thaliana* Pep peptides depends on AtPepR1, a receptor with guanylyl cyclase activity, and cGMP-activated $Ca^{2+}$ channels. Proc Natl Acad Sci U S A 107:21193–21198

Qiu JL, Fiil BK, Petersen K, Nielsen HB, Botanga CJ, Thorgrimsen S, Palma K, Suarez-Rodriguez MC, Sandbech-Clausen S, Lichota J, Brodersen P, Grasser KD, Mattson O, Glazebrook J,

Mundy J, Petersen M (2008a) *Arabidopsis* MAP kinase 4 regulates gene expression through transcription factor release in the nucleus. EMBO J 27:2214–2221

Qiu JL, Zhou L, Yun B, Nielsen H, Fiil BK, Petersen K, MacKinlay J, Loake GJ, Mundy J, Morris PC (2008b) *Arabidopsis* mitogen-activated protein kinase kinases MKK1 and MKK2 have overlapping functions in defense signaling mediated by MEKK1, MPK4, and MKS1. Plant Physiol 148:212–222

Ren D, Yang H, Zhang S (2002) Cell death mediated by MAPK is associated with hydrogen peroxide production in *Arabidopsis*. J Biol Chem 277:559–565

Ren D, Yang K-Y, Li G-J, Liu Y, Zhang S (2006) Activation of Ntf4, a tobacco mitogen-activated protein kinase, during plant defense and its involvement in hypersensitive response-like cell death. Plant Physiol 141:1482–1493

Ren D, Liu Y, Yang KY, Han L, Mao G, Glazebrook J, Zhang S (2008) A fungal responsive MAPK cascade regulates phytoalexin biosynthesis in *Arabidopsis*. Proc Natl Acad Sci U S A 105:5638–5643

Romeis T, Piedras P, Zhang S, Klessig DF, Hirt H, Jones JD (1999) Rapid Avr9- and Cf-9 – dependent activation of MAP kinases in tobacco cell cultures and leaves: convergence of resistance gene, elicitor, wound, and salicylate responses. Plant Cell 11:273–287

Rustérucci C, Aviv DH, Holt BF, Dangl JL, Parker JE (2001) The disease resistance signaling components *EDS1* and *PAD4* are essential regulators of the cell death pathway controlled by *LSD1* in *Arabidopsis*. Plant Cell 13:2211–2224

Samuel MA, Ellis BE (2002) Double jeopardy: both overexpression and suppression of a redox-activated protein kinase render tobacco plants ozone sensitive. Plant Cell 14:2059–2069

Schikora A, Schenk ST, Stein E, Molitor A, Zuccaro A, Kogel K-H (2011) *N*-acyl-homoserine lactone confers resistance toward biotrophic and hemibiotrophic pathogens via altered activation of AtMPK6. Plant Physiol 157:1407–1418

Schweighofer A, Kazanaviciute V, Scheikl E, Teige M, Doczi R, Hirt H, Schwanninger M, Kant M, Schuurink R, Mauch F, Buchala A, Cardinale F, Meskiene I (2007) The PP2C-type phosphatase AP2C1, which negatively regulates MPK4 and MPK6, modulates innate immunity, jasmonic acid, and ethylene levels in *Arabidopsis*. Plant Cell 19:2213–2224

Seo S, Okamoto M, Settee H, Ishizuka K, Sano H, Ohashi Y (1995) Tobacco MAP kinase: a possible mediator in wound signal transduction pathways. Science 270:1988–1992

Seo S, Sano H, Ohashi Y (1999) Jasmonate-based wound signal transduction requires activation of WIPK, a tobacco mitogen-activated protein kinase. Plant Cell 11:289–298

Seo S, Katou S, Seto H, Gomi K, Ohashi Y (2007) The mitogen-activated protein kinases WIPK and SIPK regulate the levels of jasmonic and salicylic acids in wounded tobacco plants. Plant J 49:899–909

Sheard LB, Tan X, Mao H, Withers J, Ben-Nissan G, Hinds TR, Kobayashi Y, Hsu EF, Sharon M, Browse J, He SY, Rizo J, Howe GA, Zheng N (2010) Jasmonate perception by inositol-phosphate-potentiated COI1-JAZ co-receptor. Nature 468:400–405

Shen X, Yuan B, Liu H, Li X, Xu C, Wang S (2010) Opposite functions of a rice mitogen-activated protein kinase during the process of resistance against *Xanthomonas oryzae*. Plant J 64:86–99

Shen X, Liu H, Yuan B, Li X, Xu C, Wang S (2011) OsEDR1 negatively regulates rice bacterial resistance via activation of ethylene biosynthesis. Plant Cell Environ 34:179–191

Shi J, An H, Zhang L, Gao Z, Guo X (2010) GhMPK7, a novel multiple stress-responsive cotton group C MAPK gene, has a role in broad spectrum disease resistance and plant development. Plant Mol Biol 74:1–17

Shi J, Zhang L, An H, Wu C, Guo X (2011) *GhMPK16*, a novel stress-responsive group D *MAPK* gene from cotton, is involved in disease resistance and drought sensitivity. BMC Mol Biol 12:22

Shoresh M, Yedidia I, Chet I (2005) Involvement of jasmonic acid/ethylene signaling pathway in the systemic resistance induced in cucumber by *Trichoderma asperellum* T203. Phytopathology 95:76–84

Skottke KR, Yoon GM, Kieber JJ, DeLong A (2011) Protein phosphatase 2A controls ethylene biosynthesis by differentially regulating the turnover of ACC synthase isoforms. PLoS Genet 7:e1001370. doi:10.1371/journal pgen.1001370

Song F, Goodman RM (2002) OsBIMK1, a rice MAP kinase gene involved in disease resistance responses. Planta 215:997–1005

Stulemeijer IJE, Stratmann JW, Joosten MHAJ (2007) Tomato mitogen-activated protein kinases LeMPK1, LeMPK2, and LeMPK3 are activated during the Cf-4/Avr4-induced hypersensitive response and have distinct phosphorylation specificities. Plant Physiol 144:1481–1494

Suarez-Rodriguez MC, Adams-Phillips L, Liu Y, Wang H, Su S-H, Jester PJ, Zhang S, Bent AF, Krysan PJ (2007) MEKK1 is required for flg22-induced MPK4 activation in Arabidopsis plants. Plant Physiol 143:661–669

Suharsono U, Fujisawa Y, Kawasaki T, Iwasaki Y, Satoh H, Shimamoto K (2002) The heterotrimeric G protein subunit acts upstream of the small GTPase Rac in disease resistance of rice. Proc Natl Acad Sci U S A 99:13307–13312

Takahashi F, Yoshida R, Ichimura K, Mizoguchi T, Seo S, Yonezawa M, Maruyama K, Yamaguchi-Shinozaki K, Shinozaki K (2007a) The mitogen-activated protein kinase cascade MKK3-MPK6 is an important part of the jasmonate signal transduction pathway in *Arabidopsis*. Plant Cell 19:805–818

Takahashi Y, Nasir KHB, Ito A, Kanzaki H, Matsumura H, Saitoh H, Fujisawa S, Kamoun S, Terauchi R (2007b) A high-throughput screen of cell-death-inducing factors in *Nicotiana benthamiana* identifies a novel MAPKK that mediates INF1-induced cell death signaling and non-host resistance to *Pseudomonas cichorii*. Plant J 49:1030–1040

Tang D, Innes RW (2002) Overexpression of a kinase-deficient form of the *EDR1* gene enhances powdery mildew resistance and ethylene-induced senescence in *Arabidopsis*. Plant J 32:975–983

Teige M, Scheikl E, Eulgem T, Doczi R, Ichimura K, Shinozaki K, Dangl JL, Hirt H (2004) The MKK2 pathway mediates cold and salt stress signaling in *Arabidopsis*. Mol Cell 15:141–152

Tena G, Boudsocq M, Sheen J (2011) Protein kinase signalling networks in plant innate immunity. Curr Opin Plant Biol 14:519–529

Theodosiou A, Ashworth A (2002) MAP kinase phosphatases. Genome Biol 3:1–10

Thines B, Katsir L, Melotto M, Niu Y, Mandaokar A, Liu G, Nomura K, He SY, Howe GA, Browse J (2007) JAZ repressor proteins are targets of the $SCF^{COI1}$ complex during jasmonate signaling. Nature 448:661

Toyoda K, Shiraishi T, Yamada T, Ichinose Y, Oku H (1993) Rapid changes in phosphoinositide metabolism in pea in response to fungal signals. Plant Cell Physiol 34:729–735

Toyoda K, Kawahara T, Ichinose Y, Yamada T, Shiraishi T (2000) Potentiation of phytoalexin accumulation in elicitor-treated epicotyls of pea (*Pisum sativum*) by a diacylglycerol kinase inhibitor. J Phytopathol 148:1–5

Tsuda K, Sato M, Stoddard T, Glazebrook J, Katagiri F (2009) Network properties of robust immunity in plants. PLoS Genet 5(12):e1000772, pp16

Ulm R, Ichimura K, Mizoguchi T, Peck SC, Zhu T, Wang X, Shinozaki K, Paszkowski J (2002) Distinct regulation of salinity and genotoxic stress responses by *Arabidopsis* MAP kinase phosphatase 1. EMBO J 21:6483–6493

Uppalapati SR, Toyoda K, Yasuhiro I, Ichinose Y, Shiraishi T (2004) Differential regulation of MBP kinases by a glycoprotein elicitor and a polypeptide suppressor from *Mycosphaerella pinodes* in pea. Physiol Mol Plant Pathol 64:17–25

Vahisalu T, Kollist H, Wang YF, Nishimura N, Chan WY, Valerio G, Lamminmäki A, Brosché M, Moldau H, Desikan R, Schroeder JI, Kangäsjarvi J (2008) SLAC1 is required for plant guard cell S-type anion channel function in stomatal signalling. Nature 452:487–491

Vidhyasekaran P (2007) Fungal pathogenesis in plants and crops: molecular biology and host defense mechanisms, IIth edn. CRC Press/Taylor & Francis Group, Boca Raton, 510pp

Vogel JP, Woeste KE, Theologis A, Kicher JJ (1998) Recessive and dominant mutations in the ethylene biosynthetic gene *ACS5* of Arabidopsis confer cytokinin insensitivity and ethylene overproduction, respectively. Proc Natl Acad Sci U S A 95:4766–4771

Waller F, Müller A, Chung K-M, Yap Y-K, Nakamura K, Weiler E, Sano H (2006) Expression of a WIPK-activated transcription factor results in increase of endogenous salicylic acid and pathogen resistance in tobacco plants. Plant Cell Physiol 47:1169–1174

Wan J, Zhang S, Stacey G (2004) Activation of a mitogen-activated protein kinase pathway in *Arabidopsis* by chitin. Mol Plant Pathol 5:125–135

Wang KLC, Li H, Ecker IP (2002) Ethylene biosynthesis and signaling networks. Plant Cell 14:S131–S155

Wang Z, Mao H, Dong C, Ji R, Cai L, Fu H, Liu S (2009) Overexpression of *Brassica napus* MPK4 enhances resistance to *Sclerotinia sclerotiorum* in oilseed rape. Mol Plant Microbe Interact 22:235–244

Wang Y, Li J, Hou S, Wang X, Li Y, Ren D, Chen S, Tang X, Zhou J-M (2010) A *Pseudomonas syringae* ADP-ribosyltransferase inhibits *Arabidopsis* mitogen-activated protein kinase kinases. Plant Cell 22:2033–2044

Wiermer M, Feys BJ, Parker JE (2005) Plant immunity: the EDS1 regulatory node. Curr Opin Plant Biol 8:383–389

Wu S, Lu D, Kabbage M, Wei H-L, Swingle B, Records AR, Dickman M, He P, Shan L (2011) Bacterial effector HopF2 suppresses *Arabidopsis* innate immunity at the plasma membrane. Mol Plant Microbe Interact 24:585–593

Xing Y, Jia W, Zhang J (2007) AtMEK1 mediates stress-induced gene expression of CAT1 catalase by triggering $H_20_2$ production in *Arabidopsis*. J Exp Bot 58:2969–2981

Xing Y, Jia W, Zhang J (2008) AtMKK1 mediates ABA-induced *CAT1* expression and $H_2O_2$ production via AtMPK6-coupled signaling in *Arabidopsis*. Plant J 54:440–451

Xiong I, Yang Y (2003) Disease resistance and abiotic stress tolerance in rice are inversely modulated by an abscisic acid-inducible mitogen-activated protein kinase. Plant Cell 15:745–759

Yaffe MB (2002) How do 14-3-3 proteins work? – Gatekeeper phosphorylation and the molecular anvil hypothesis. FEBS Lett 513:53–57

Yaffe MB, Rittinger K, Volinia S, Caron PR, Aitken A, Leffers H, Gamblin SJ, Smerdon SJ, Cantley LC (1997) The structural basis for 14-3-3 phosphopeptide binding specificity. Cell 91:961–971

Yamakawa H, Katou S, Seo S, Mitsuhara I, Kamada H, Ohashi Y (2004) Plant mitogen-activated protein kinase phosphatase interacts with calmodulin. J Biol Chem 279:928–936

Yang KY, Liu YD, Zhang SQ (2001) Activation of a mitogen-activated protein kinase pathway is involved in disease resistance in tobacco. Proc Natl Acad Sci U S A 98:741–746

Yoo SD, Cho Y-H, Tena G, Xiong Y, Sheen J (2008) Dual control of nuclear EIN3 by bifurcate MAPK cascades in $C_2H_4$ signalling. Nature 451:789–795

Yoshioka K, Sugie K, Park HJ, Maeda H, Tsuda N, Kawakita K, Doke N (2001) Induction of plant gp91 phox homolog by fungal cell wall, arachidonic acid, and salicylic acid in potato. Mol Plant Microbe Interact 14:725–736

Yoshioka H, Numata N, Nakajima K, Katou S, Kawakita K, Rowland O, Jones JD, Doke N (2003) *Nicotiana benthamiana* gp91$^{phox}$ homologs *NbrbohA* and *NbrbohB* participate in $H_2O_2$ accumulation and resistance to *Phytophthora infestans*. Plant Cell 15:706–718

Zhang SQ, Klessig DF (1997) Salicylic acid activates a 48-kD MAP kinase in tobacco. Plant Cell 9:809–824

Zhang SQ, Klessig DF (1998a) Resistance gene *N*-mediated de novo synthesis and activation of a tobacco mitogen-activated protein kinase by tobacco mosaic virus infection. Proc Natl Acad Sci U S A 95:7225–7230

Zhang SQ, Klessig DF (1998b) The tobacco wounding-activated mitogen-activated protein kinase is encoded by *SIPK*. Proc Natl Acad Sci U S A 95:7225–7230

Zhang SQ, Klessig DF (2001) MAP cascades in plant defense signaling. Trends Plant Sci 6:5320–5327

Zhang S, Liu Y (2001) Activation of salicylic acid-induced protein kinase, a mitogen-activated protein kinase, induces multiple defense responses in tobacco. Plant Cell 13:1877–1889

Zhang J, Zhou J-M (2010) Plant immunity triggered by microbial molecular signatures. Mol Plant 3:783–793

Zhang SQ, Du H, Klessig DF (1998) Activation of the tobacco SIP kinase by both a cell wall-derived carbohydrate elicitor and purified proteinaceous elicitins from *Phytophthora* spp. Plant Cell 10:435–449

Zhang S, Liu Y, Klessig DF (2000) Multiple levels of tobacco WIPK activation during the induction of cell death by fungal elicitins. Plant J 23:339–347

Zhang AY, Jiang MY, Zhang JH, Tan M, Hu XL (2006) Mitogen-activated protein kinase is involved in abscisic acid-induced antioxidant defense and acts downstream of reactive oxygen species production in leaves of maize plants. Plant Physiol 141:475–487

Zhang A, Jiang MY, Zhang J, Ding H, Xu S, Hu X, Tan M (2007a) Nitric oxide induced by hydrogen peroxide mediates abscisic acid-induced activation of the mitogen-activated protein kinase cascade involved in antioxidant defense in maize leaves. New Phytol 175:36–50

Zhang J, Shao F, Li Y, Cui H, Chen L, Li H, Zou Y, Long C, Lan L, Chai J, Chen S, Tang X, Zhou J-M (2007b) A *Pseudomonas syringae* effector inactivates MAPKs to suppress PAMP-induced immunity in plants. Cell Host Microbe 1:175–185

Zhang X, Dai Y, Xiong Y, DeFraia C, Li J, Dong X, Mou Z (2007c) Overexpression of *Arabidopsis MAP kinase kinase 7* leads to activation of plant basal and systemic acquired resistance. Plant J 52:1066–1079

Zhang L, Xi D, Luo L, Meng F, Li Y, Wu CA, Guo X (2011) Cotton GhMPK2 is involved in multiple signaling pathways and mediates defense responses to pathogen infection and oxidative stress. FEBS J 278:1267–1378

Zhang H, Li D, Wang M, Liu J, Teng W, Cheng B, Huang Q, Wang M, Song W, Dong S, Zheng X, Zhang Z (2012a) The *Nicotiana benthamiana* mitogen-activated protein kinase cascade and WRKY transcription factor participate in $Nep1_{Mo}$-triggered plant responses. Mol Plant Microbe Interact 25:1639–1653

Zhang L, Li Y, Lu W, Meng F, Wu C, Guo X (2012b) Cotton *GhMKK5* affects disease resistance, induces HR-like cell death, and reduces the tolerance to salt and drought stress in transgenic *Nicotiana benthamiana*. J Exp Bot 63:3935–3952

Zhang Z, Wu Y, Gao M, Zhang J, Kong Q, Liu Y, Ba H, Zhou J, Zhang Y (2012c) Disruption of PAMP-induced MAP kinase cascade by a *Pseudomonas syringae* effector activates plant immunity mediated by the NB-LRR protein SUMM2. Cell Host Microbe 11:253–263

Zheng Z, Qamar SA, Chen Z, Mengiste T (2006) Arabidopsis WRKY33 transcription factor is required for resistance to necrotrophic fungal pathogens. Plant J 48:596–605

Zheng Z, Mosher SL, Fan B, Klessig DF, Chen Z (2007) Functional analysis of Arabidopsis WRKY25 transcription factor in plant defense against *Pseudomonas syringae*. BMC Plant Biol 7:2

# Chapter 8
# Phosphololipids Signaling System in Plant Innate Immunity

**Abstract** Phospholipids are the sources for production of the second messengers phosphatidic acid (PA), diacylglycerol (DAG), and inositol 1,4,5-trisphosphate ($IP_3$), which are involved in defense signaling system. Phospholipase C (PLC) and phospholipase D (PLD) are the key enzymes involved in generation of the phospholipid second messengers. G-proteins, $Ca^{2+}$ influx, nitric oxide (NO), and reactive oxygen species (ROS) are involved in PAMP elicitors-triggered activation of PLC and PLD. $IP_3$ is involved in activation of $Ca^{2+}$ signaling system. PA is an important second messenger in activating ROS, jasmonate (JA), abscisic acid (ABA) systems and it also activates phosphorylation/dephosphorylation in various signaling systems. DAG is involved in JA biosynthesis and ROS signaling system. Biphasic production of PA and ROS through distinctly different phospholipase pathways has been reported. Protein kinases and phosphatases play key roles in phospholipid signaling system.

**Keywords** Phosphatidic acid (PA) • Diacetylglycerol (DAG) • Inositol 1,4,5-trisphosphate (IP3) • Phospholipases • Protein kinases • Phosphorylation • Second messengers

## 8.1 Biosynthesis of Phospholipids-Derived Second Messengers

Phospholipids are the structural components of cell membranes and they are also sources of second messengers involved in defense signaling system. Many different phospholipids can be cleaved by phospholipases to generate second messengers (Berridge 1984; Chasan 1995). Phospholipase C (PLC) and phospholipase D (PLD) are the key enzymes involved in generation of various phospholipid second messengers Munnik et al. 1998b). In the phospholipase C-mediated pathway for the generation of second messengers (Fig. 8.1), PLC hydrolyzes the signaling phospholipids (PLs) phosphatidylinositol 4-phosphate (PIP) or phosphatidylinositol

P. Vidhyasekaran, *PAMP Signals in Plant Innate Immunity: Signal Perception and Transduction*, Signaling and Communication in Plants 21,
DOI 10.1007/978-94-007-7426-1_8, © Springer Science+Business Media Dordrecht 2014

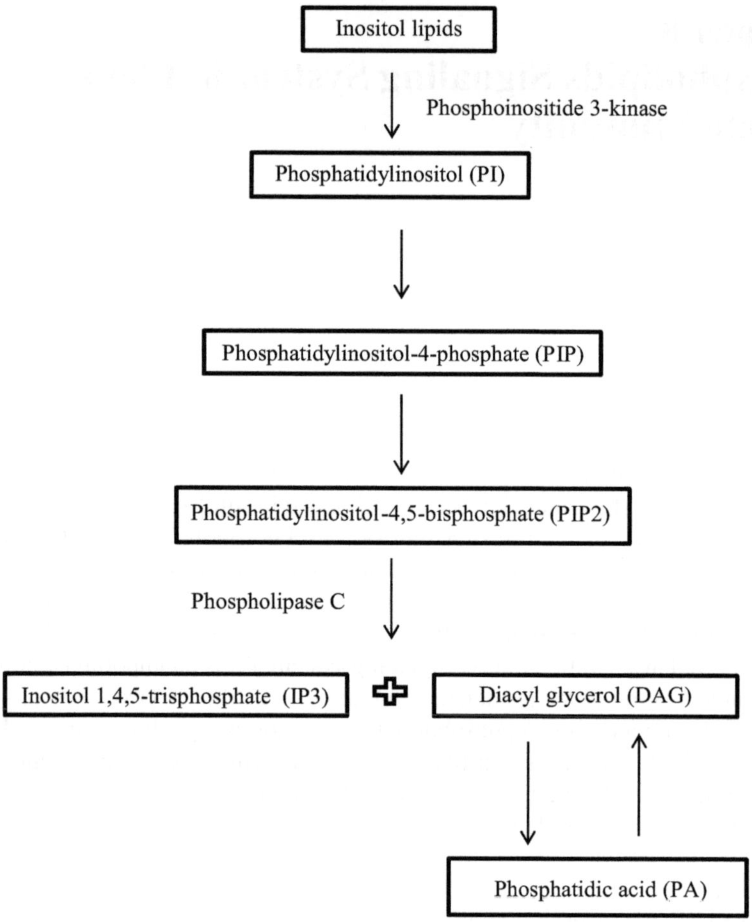


**Fig. 8.1** Phospholipase C-mediated pathway in generation of the second messengers IP3, DAG, and PA (Adapted from Sang et al. 2001; Anthony et al. 2006)

4,5-bisphosphate ($PIP_2$) to generate the second messengers diacylglycerol (DAG) and inositol 1,4-bisphosphate (IP2) or inositol 1,4,5-trisphosphate ($IP_3$), respectively. $IP_2$ can be subsequently phosphorylated to $IP_3$ (Lanteri et al. 2008). The second product of PLC activity, DAG, can be phosphorylated to phosphatidic acid (PA) through the action of DAG kinase (Munnik 2001; Lanteri et al. 2008). PA can be dephosphorylated to DAG (Munnik et al. 1995).

PA can also be synthesized by the action of PLD (Fig. 8.2). PLD hydrolyzes structural phospholipids such as phosphatidylcholine (PC) or phosphatidylethanolamine (PE) at the terminal phosphodiester bond to produce PA and free head groups such as choline (Wang 2001). Several specific elicitors or pathogen-associated molecular pattern (PAMPs) are known to activate PLC/DAG kinase- enzymatic pathway and biosynthesis of PA (Van der Luit et al. 2000; de Jong et al. 2004; Yamaguchi et al. 2005).

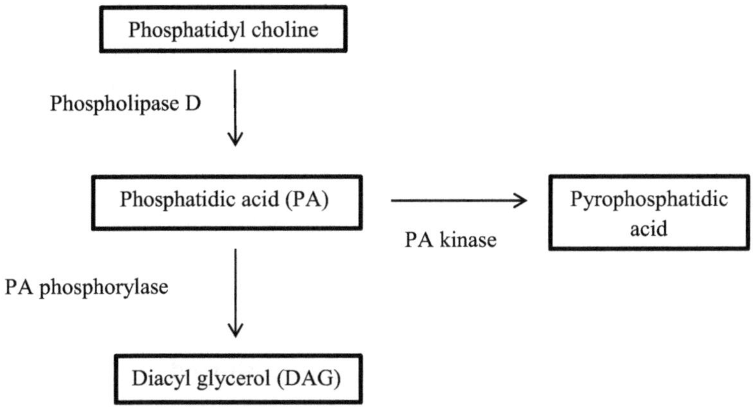


**Fig. 8.2** Phospholipase D-mediated phosphatidic acid and diacyl glycerol generation pathway (Adapted from Sang et al. 2001; Anthony et al. 2006)

Laxalt et al. (2007) demonstrated that NO is required for the production of the lipid second messenger PA via the activation of the phospholipase C (PLC) and DAG kinase pathway. Treatment of tomato cell cultures with the fungal elicitor xylanase resulted in a rapid NO accumulation (Laxalt et al. 2007). NO donor S-nitroso N-acetyl penicillamine (SNAP) treatment induced PA, PIP, and $PIP_2$ accumulation within 1 min after treatment in cucumber, probably through activation of PLC (Lanteri et al. 2008). NO triggers PA formation also through PLD activation (Lanteri et al. 2008).

ROS signaling system is also involved in accumulation of PA (Desikan et al. 2004; Mittler et al. 2004). $H_2O_2$ induced rapid and transient accumulation of PA in suspension-cultured rice cells (Yamaguchi et al. 2004). $H_2O_2$ directly induced PLD in vitro (Yamaguchi et al. 2004). $H_2O_2$ might act upstream of PA and or even upstream of NO (Laxalt et al. 2007). There are also reports that $H_2O_2$ is required for NO production (Lum et al. 2002; de Pinto et al. 2006).

G proteins may activate PLC and PLD (Munnik et al. 1995; Ritchie and Gilroy 2000). Heterotrimeric G protein α-subunit regulates PLD through a motif analogous to the DRY motif in G-protein-coupled receptors in *Arabidopsis* (Zhao and Wang 2004). The activated PLD may hydrolyze phospholipids to produce the lipid second messenger PA (Zhao and Wang 2004). PLD is regulated by $Ca^{2+}$ (Zheng et al. 2000). Downstream, $Ca^{2+}$ influx activates PLC, DAG kinase, and PLD. $Ca^{2+}$ activates NADPH oxidase, which is involved in ROS production (Munnik et al. 1998a).

## 8.2 Phospholipids in $Ca^{2+}$ Signaling System

The phospholipids-derived second messenger $IP_3$ releases $Ca^{2+}$ from intracellular compartments into the cytosol (Meijer and Munnik 2003; Lanteri et al. 2008; Munnik and Testerink 2009). A number of $Ca^{2+}$ release channels have been found in the vacuolar membranes. $IP_3$ releases calcium through an intact intracellular plant membrane by activating a $Ca^{2+}$ channel (Alexandre et al. 1990). It operates through receptors which resemble ryanodine receptors of human muscle (Berridge et al. 2000). This calcium channel is voltage-dependent and opened only on depolarization of the vacuoles (Alexandre et al. 1990). The calcium released through this channel induces calcium waves and oscillations in the cytosol (Berridge 1993; Hisatsune et al. 2005). The calcium ion influx into the cytosol may activate $Ca^{2+}$-dependent protein kinases (Munnik et al. 1995) and $Ca^{2+}$ signaling system (Huang et al. 2001; Luan et al. 2002).

## 8.3 Phosphatidic Acid in G Proteins-Mediated Signaling System

The second messenger PA has been shown to be involved in activation of ROS signaling system. PA induces ROS-induced hypersensitive cell death through G proteins-mediated signaling system (Park et al. 2004). The plant Rac-like GTPases, named ROPs (Rho-related small G proteins) are involved in ROS generation (Yang 2002). PA activates Rho-related small G protein GTPase-mediated pathway of ROS generation (Park et al. 2004). The induced ROS generation may be due to activation of NADPH oxidase by PA (Park et al. 2004).

## 8.4 Phosphatidic Acid in ROS Signaling System

PA has been shown to be able to trigger an oxidative burst (Fig. 8.3; Sang et al. 2001; de Jong et al. 2004; Park et al. 2004). PA is involved in the activation of NADPH oxidase in macrophages (McPhail et al. 1999) and it has been suggested that similar activation of NADPH oxidase may also occur in plants and ROS is generated through the action of NADPH oxidase (Laxalt et al. 2007). PA and DAG directly activate NADPH oxidase by interacting with enzyme components (Palicz et al. 2001). PA has been shown to induce ROS in tomato cells. Scavenging of NO or inhibition of either the PLC or the DAG kinase enzyme diminished elicitor-induced ROS production (Laxalt et al. 2007). PA promotes superoxide-generating activity in plants through the activation of NADPH oxidase (Sang et al. 2001). DAG is also able to generate NADPH-dependent superoxide synthesis (Sang et al. 2001).

PA has been shown to bind a protein kinase, 3′-phosphoinositide-dependent kinase 1 (PDK1) in *Arabidopsis* and to activate protein kinase AGC2-1 in a PDK-dependent manner (Deak et al. 1999; Anthony et al. 2004). PDK1 is specifically activated by PLD-generated PA in Arabidopsis cells treated with a fungal elicitor (Anthony et al. 2006). AGC2-1 is identical to OXI1, a protein kinase implicated in oxidative burst-mediated signaling in *Arabidopsis* (Rentel et al. 2004).

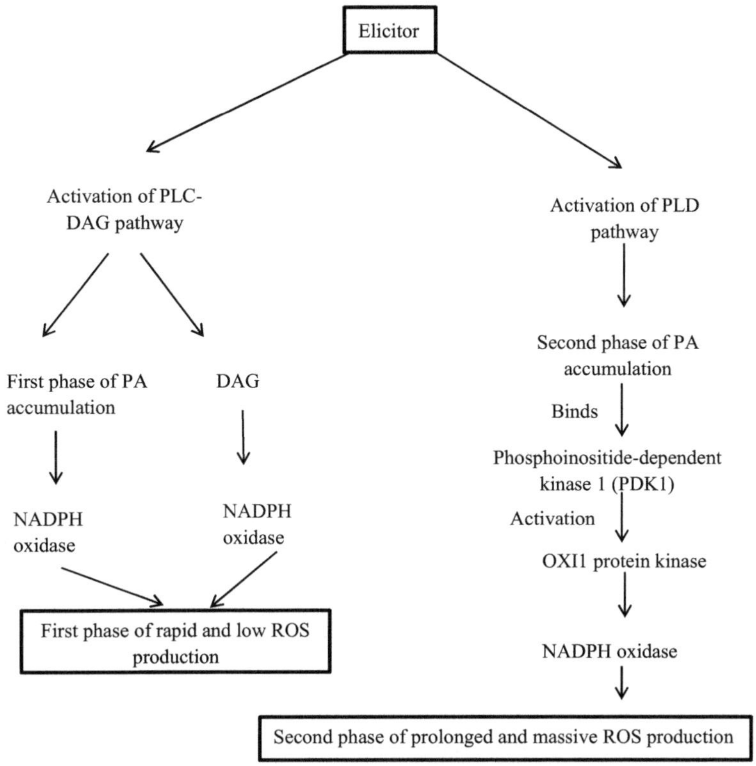


**Fig. 8.3** Role of phospholipase pathways in biphasic PA and ROS generation in *Arabidopsis* (Adapted from Yamaguchi et al. 2005; Andersson et al. 2006)

The oxidative burst mediated by elicitors occurs in two phases (Fig. 8.4). The first phase shows a rapid (5 min onward) and low ROS production whereas the second phase shows a prolonged (3–6 h) and massive ROS production (Lamb and Dixon 1997). The biphasic ROS generation in rice cells induced by an elicitor was associated with the activation of PLC and PLD. The activation of both enzymes was shown for the first phase of ROS generation, whereas for the second phase only the activation of PLD was observed (Yamaguchi et al. 2005). A biphasic accumulation of PA has been reported in *Arabidopsis* on recognition of elicitors (Fig. 8.3). The first wave was attributed to the PLC/DAG kinase pathway and the second to PLD. Both the phospholipase pathways acted upstream of ROS formation (Andersson et al. 2006). PA accumulated rapidly via PLC/DAG kinase pathway in tobacco cells treated with an elicitor and PLC activity was required for the rapid ROS accumulation (de Jong et al. 2004). The NO-dependent, PLC/DAG kinase-generated PA is involved in the induction of ROS production (Laxalt et al. 2007).

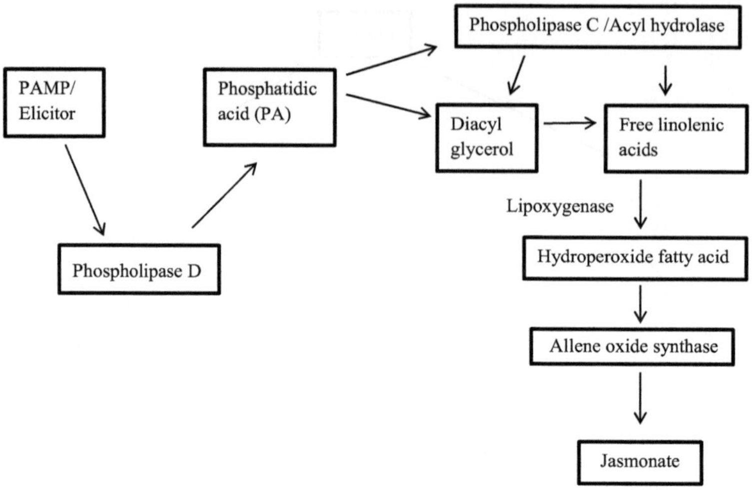


**Fig. 8.4** Role of phospholipases in jasmonate biosynthesis pathway (Adapted from Wang et al. 2000)

## 8.5 Phospholipids in JA Signaling System

PLD-mediated formation of PA may initiate a lipolytic pathway, consisting of PLD, PA phosphatase, and acyl-hydrolyzing enzymes. In this pathway, phospholipids are converted sequentially into PA, DAG, and free linoleic acids (Ryu and Wang 1998; Wang et al. 2000). The free linolenic acid is the substrate for JA synthesis (Fig. 8.4).

## 8.6 Phospholipid Signaling System in ABA Signaling Network

Phospholipid signaling system involving phospholipase C, phospholipase D, and inositol (1,4,5) trisphosphate (IP3) have all been implicated in ABA signaling. Phospholipase D-mediated phosphatidic acid (PA) production has been shown to promote ABA-induced gene expression (Zhang et al. 2005). Phospholipase Dα-1-derived PA regulates a protein phosphatase 2C, ABI1, which is a negative regulator of ABA responses in *Arabidopsis* (Zhang et al. 2004). ABA treatment promotes an increase in PA from phosphatidylcholine. The PA binds to ABI1 and arginine 73 in ABI1 is essential for PA-ABI1 binding. Binding of PA to ABI1 protein results in anchorage of ABI1 to the plasma membrane and a decrease in ABI1 PP2C activity. This membrane tethering reduces the movement of ABI1 from the cytosol into the nucleus. The lack of ABA-induced production of PA in *PLDα1*-null cells results in a decrease in the association of ABI1 with the plasma membrane in response to ABA (Zhang et al. 2004). These results suggest that PA produced by PLDα1 inhibits the function of the negative regulator ABI1, thus promoting ABA signaling. It is also shown that ABI1 is a direct target of PA and there is a functional link between the two signaling enzymes, phospholipase and phosphatase (Zhang et al. 2004).

Phospholipase D (PLD) has been shown to be involved in ABA signaling-mediated stomatal closure immune response (Uraji et al. 2012). Two *Arabidopsis* PLDs (PLDα1and PLDδ) are involved in ABA stomatal closure signaling in guard cells. ABA induced stomatal closure was suppressed in the *pldα1 pldδ* double mutant but not in the *pld* single mutants. The *pldα1* and *pldδ* mutations reduced ABA-induced phosphatidic acid production. During ABA-induced stomatal closure, wild-type guard cells ROS and NO and showed increased cytosolic alkalization. These changes were reduced in guard cells of the *pldα1pldδ* double mutant plants. Inward-rectifying $K^+$ channel currents of guard cells were inhibited by ABA in the wild-type but not in *pldα1pldδ* mutant plants (Uraji et al. 2012). ABA signaling system has been shown to involve ROS, NO, and phosphatidic acid, the product of PLD activity, in *Arabidopsis*. PLD seems to act downstream of NO and ROS (Distéfano et al. 2012). These studies suggest that ABA induces production of phosphatidic acid through the action of phospholipase D and the phosphatidic acid acting as a second messenger triggers ROS, NO, and $Ca^{2+}$ signaling systems triggering stomatal closure immune response.

## 8.7 Phosphatidic Acid in Phosphorylation/Dephosphorylation System

PA activates a calcium-dependent protein kinase (CDPK) and a mitogen-activated kinase (MAPK) (Munnik et al. 1995; Farmer and Choi 1999; Lee et al. 2001). Together with DAG, IP3 activates protein kinase C (Munnik et al. 1995; Chasan 1995). A phosphatidic acid-activated protein kinase C isoform phosphorylates $p22^{phox}$, an NADPH oxidase component (Regier et al. 1999). PA binds with ABI1 (ABA insensitive1), a protein phosphatase 2C (PP2C) that is a negative regulator of abscisic acid (ABA) in *Arabidopsis*. The PA binding decreases PP2C activity and also appears to reduce its translocation to nuclei in response to ABA by tethering ABI1 to the plasma membrane (Zhang et al. 2004; Wang 2005; Mishra et al. 2006). These results suggest that activation of PLD inhibits the function of the negative regulator ABI1, thus promoting ABA signaling (Wang 2005).

## References

Alexandre J, Lassales JP, Kado RT (1990) Opening of $Ca^{2+}$ channels in isolated red beet vacuole membrane by inositol-1,4,5-trisphosphate. Nature 343:567–570

Andersson MX, Kourtchenko O, Dangl JL, Mackey D, Ellerstrom M (2006) Phospholipase-dependent signaling during the AvrRpm1- and AvrRpt2-induced disease resistance responses in *Arabidopsis thaliana*. Plant J 47:947–959

Anthony RG, Henriques R, Helfer A, Meszaros T, Rios G, Testerink C, Munnik T, Deak M, Koncz C, Rogre L (2004) A protein kinase target of a PDK1 signalling pathway is involved in root hair growth in *Arabidopsis*. EMBO J 23:572–581

Anthony RG, Khan S, Costa J, Pais MS, Bögre L (2006) The *Arabidopsis* protein kinase PTI1-2 is activated by convergent phosphatidic acid and oxidative stress signaling pathways downstream of PDK1 and OXI1. J Biol Chem 281:37536–37546

Berridge MJ (1984) Inositol trisphosphate and diacylglycerol as second messengers. Biochem J 220:345–360

Berridge MJ (1993) Inositol trisphosphate and calcium signaling. Nature 361:315–325

Berridge MJ, Lipp P, Bootman MD (2000) The versatility and universatility of calcium signaling. Nat Rev Mol Cell Biol 1:11–21

Chasan R (1995) Signaling a role for phospholipids-derived compounds in plants. Plant Cell 7:1971–1974

de Jong CF, Laxalt AM, Bargmann BO, de Wit PJ, Joosten MH, Munnik T (2004) Phosphatidic acid accumulation is an early response in the *Cf-4/Avr4* interaction. Plant J 39:1–12

de Pinto MC, Paradiso A, Leonetti P, De Gara L (2006) Hydrogen peroxide, nitric oxide and cytosolic ascorbate peroxidase at the crossroad between defence and cell death. Plant J 48:784–795

Deak M, Casamayor A, Currie RA, Downes CP, Alessi DR (1999) Characterisation of a plant 3-phosphinositide-dependent protein kinase-1 homologue which contains a pleckstrin homology domain. FEBS Lett 451:220–226

Desikan R, Cheung M-K, Bright J, Henson D, Hancock JT, Neill SJ (2004) ABA, hydrogen peroxide and nitric oxide signaling in stomatal guard cells. J Exp Bot 55:205–212

Distéfano AM, Scuffi D, Garcia-Mata C, Lamattina L, Laxalt AM (2012) Phospholipase Dδ is involved in nitric oxide-induced stomatal closure. Planta 236:1899–1907

Farmer PK, Choi JH (1999) Calcium and phospholipids activation of a recombinant calcium-dependent protein kinase (DcCPK1) from carrot (*Daucus carota* L). Biochim Biophys Acta 1434:6–17

Hisatsune C, Nakamura K, Kuroda Y, Nakamura T, Mikoshiba K (2005) Amplification of $Ca^{2+}$ signaling by diacylglycerol-mediated inositol 1,4,5-trisphosphate production. J Biol Chem 280:11723–11730

Huang JZ, Hardin SC, Huber JC (2001) Identification of a novel phosphorylation motif for CDPKs: phosphorylation of synthetic peptides lacking basic residues at P-3/P-4. Arch Biochem Biophys 393:61–66

Lamb CJ, Dixon RA (1997) The oxidative burst in plant disease resistance. Annu Rev Plant Physiol Plant Mol Biol 48:251–275

Lanteri ML, Laxalt AM, Lamattina L (2008) Nitric oxide triggers phosphatidic accumulation via phospholipase D during auxin-induced adventitious root formation in cucumber. Plant Physiol 147:188–198

Laxalt AM, Raho N, ten Have A, Lamattina L (2007) Nitric oxide is critical for inducing phosphatidic acid accumulation in xylanase-elicited cells. J Biol Chem 282:21160–21168

Lee S, Hirt H, Lee Y (2001) Phosphatidic acid activates a wound-activated MAPK in *Glycine max*. Plant J 479–486

Luan S, Kudla J, Rodriguez-Concepcion M, Yalovsky S, Gruissem W (2002) Calmodulins and calcineurin B-like proteins: calcium sensors for specific signal response coupling in plants. Plant Cell 14 Suppl: S389–S400

Lum HK, Butt YK, Lo SC (2002) Hydrogen peroxide induces a rapid production of nitric oxide in mung bean (*Phaseolus aureus*). Nitric Oxide 6:205–213

McPhail LC, Waite KA, Regier DS, Nixon JB, Qualliotine-Mann D, Zhang WX, Wallin R, Sergeant S (1999) A novel protein kinase target for the lipid second messenger phosphatidic acid. Biochim Biophys Acta 1439:277–290

Meijer HJ, Munnik T (2003) Phospholipid-based signaling in plants. Annu Rev Plant Biol 54:265–306

Mishra G, Zhang W, Deng F, Zhao J, Wang X (2006) A bifurcating pathway directs abscisic acid effects on stomatal closure and opening in *Arabidopsis*. Science 312:264–266

Mittler R, Vanderauwera S, Gollery M, Van Breusegem F (2004) Reactive oxygen network of plants. Trends Plant Sci 9:490–498

Munnik T (2001) Phosphatidic acid: an emerging plant lipid second messenger. Trends Plant Sci 6:227–233

Munnik T, Testerink C (2009) Plant phospholipid signaling – 'in a nutshell'. J Lipid Res 50:S260–S265

Munnik T, Arisz SA, de Vrije T, Musgrave A (1995) G protein activation stimulates phospholipase D signaling in plants. Plant Cell 7:2197–2210

Munnik T, Irvine RF, Musgrave A (1998a) Phospholipid signalling in plants. Biochim Biophys Acta 1389:222–272

Munnik T, van Himbergen JAJ, ter Riet B, Braun F-J, Irvine RF, van den Ende H, Musgrave A (1998b) Detailed analysis of the turnover of polyphosphoinositides and phosphatidic acid upon activation of phospholipase C and –D in *Chlamydomonas* cells treated with non-permeabilizing concentrations of mastoparan. Planta 207:133–145

Palicz A, Foubert TR, Jesaitis AJ, Marodi L, McPhail LC (2001) Phosphatidic acid and diacylglycerol directly activate NADPH oxidase by interacting with enzyme components. J Biol Chem 276:3090–3097

Park J, Gu Y, Lee Y, Yang Z, Lee Y (2004) Phosphatidic acid induces leaf cell death in *Arabidopsis* by activating the Rho-related small G protein GTPase-mediated pathway of reactive oxygen species generation. Plant Physiol 134:129–136

Regier DS, Waite KA, Wallin R, McPhail LC (1999) A phosphatidic acid activated protein kinase and conventional protein kinase C isoforms phosphorylate $p22^{phox}$, an NADPH oxidase component. J Biol Chem 274:36601–36608

Rentel MC, Lecourieux D, Ouaked F, Usher SL, Petersen L, Okamoto H, Knight H, Peck SC, Grierson CS, Hirt H, Knight MR (2004) OXI1 kinase is necessary for oxidative burst-mediated signaling in *Arabidopsis*. Nature 427:858–861

Ritchie S, Gilroy S (2000) Abscisic acid stimulation of phospholipase D in the barley aleurone is G-protein-mediated and localized to the plasma membrane. Plant Physiol 124:693–702

Ryu SB, Wang X (1998) Increase in free linolenic and linoleic acids associated with phospholipase D-mediated hydrolysis of phospholipids in wounded castor bean leaves. Biochim Biophys Acta 1393:193–202

Sang Y, Cui D, Wang X (2001) Phospholipase D and phosphatidic acid-mediated generation of superoxide in *Arabidopsis*. Plant Physiol 126:1449–1458

Uraji M, Katagiri T, Okuma E, Te W, Hossain M-A, Masuda C, Miura A, Nakamura Y, Mori I-C, Shinozaki K, Murata Y (2012) Cooperative function of PLDδ and PLDα1in abscisic acid-induced stomatal closure in Arabidopsis. Plant Physiol 159:450–460

van der Luit AH, Piatti T, van Doorn A, Musgrave A, Felix G, Boller T, Munnik T (2000) Elicitation of suspension-cultured tomato cells triggers the formation of phosphatidic acid and diacylglycerol pyrophosphate. Plant Physiol 123:1507–1515

Wang X (2001) Plant phospholipases. Annu Rev Plant Physiol Plant Mol Biol 52:211–231

Wang X (2005) Regulatory functions of phospholipase D and phosphatidic acid in plant growth, development, and stress responses. Plant Physiol 139:566–573

Wang C, Zien C, Afitlhile M, Welti R, Hildebrand DF, Wang X (2000) Involvement of phospholipase D in wound-induced accumulation of jasmonic acid in *Arabidopsis*. Plant Cell 12:2237–2246

Yamaguchi T, Tanabe S, Minami E, Shibuya N (2004) Activation of phospholipase D induced by hydrogen peroxide in suspension-cultured rice cells. Plant Cell Physiol 45:1261–1270

Yamaguchi T, Minami E, Ueki J, Shibuya N (2005) Elicitor-induced activation of phospholipases plays an important role for the induction of defense responses in suspension-cultured rice cells. Plant Cell Physiol 46:579–587

Yang Z (2002) Small GTPases: versatile signaling switches in plants. Plant Cell 14(Suppl): S375–S388

Zhang W, Qin C, Zhao J, Wang X (2004) Phospholipase Dα1-derived phosphatidic acid interacts with ABI1 phosphatase 2C and regulates abscisic acid signaling. Proc Natl Acad Sci U S A 101:9508–9513

Zhang WH, Yu LJ, Zhang YY, Wang XM (2005) Phospholipase D in the signaling networks of plant response to abscisic acid and reactive oxygen species. Biochim Biophys Acta 1736:1–9

Zhao J, Wang X (2004) *Arabidopsis* phospholipase Dα1 interacts with the heterotrimeric G-protein α-subunit through a motif analogous to the DRY motif in G-protein-coupled receptors. J Biol Chem 279:1794–1800

Zheng L, Krishnamoorthi R, Zolkiewski M, Wang X (2000) Distinct $Ca^{2+}$ binding properties of the novel C2 domains of plant phospholipase Dα and β. J Biol Chem 275:19700–19706

# Chapter 9
# Protein Phosphorylation and Dephosphorylation in Plant Immune Signaling Systems

**Abstract** Protein phosphorylation plays a central role in the plant immune response signaling. PAMP signals induce rapid and transient phosphorylation of several proteins/enzymes involved in defense signaling system. The proteins that make up the signal transduction pathway are present in the cell prior to the perception of PAMP elicitor signal. On perception of the signal these proteins are activated by post-translational modifications and conformational changes induced by phosphorylation. Protein phosphorylation is carried out by different protein kinases. PAMP signals are perceived by plant pattern recognition receptors (PRRs), which belong to the family of receptor-like kinases (RLKs). The PAMPs have been shown to activate the RLKs by autophosphorylation by their own serine/threonine kinase. The autophosphorylation of the receptor kinases takes place within few seconds to few minutes after PAMP treatment and the autophosphorylated RLKs have been shown to be essential for PAMP signaling in plants. Calcium-dependent protein kinases (CDPKs) and mitogen-activated protein kinases (MAPKs) regulate expression of various enzymes involved in ROS, salicylate, jasmonate, ethylene, and abscisic acid signaling systems by inducing protein/enzyme phosphorylation. Protein kinase C is involved in phosphorylation of some transcription factors. His kinase family protein kinase takes part in ethylene signaling system. Protein dephosphorylation may also be involved in defense signaling and the phosphatases negatively regulate innate immune responses.

**Keywords** Receptor-like kinases (RLKs) • Calcium-dependent protein kinases (CDPKs) • Mitogen-activated protein kinases (MAPKs) • Protein kinase C • His kinase family protein kinase • Autophosphorylation • Phosphatases

P. Vidhyasekaran, *PAMP Signals in Plant Innate Immunity: Signal Perception and Transduction*, Signaling and Communication in Plants 21,
DOI 10.1007/978-94-007-7426-1_9, © Springer Science+Business Media Dordrecht 2014

## 9.1 Protein Phosphorylation Plays Key Roles in Plant Immune Signal Transduction

Phosphoproteomic studies have established that the information signal is transferred to a protein by phosphorylation and this simple modification of the protein causes a change in its activity and localization. The incoming signal is transduced to the target protein by a protein kinase and the signal alone is responsible for the change in activity of the target protein (Ferl 2004). Phosphatase activity may remove the signal and return the protein to its original state, providing only transitional activity of the target protein/enzyme. The activated kinases in turn may activate other kinases such that cascades of phosphorylation events propagate to enzymatic or structural proteins, where their phosphorylation may result in several changes in the activities of these proteins (Ferl 2004; Vidhyasekaran 2007). It has been demonstrated that signals initiate cascades while the protein kinases propagate the signaling processes (Ferl 2004).

Rapid and transient phosphorylation of several proteins involved in defense signaling system has been reported. Phosphorylation of various protein kinases, either by autophosphorylation or by other related kinases appears to be the crucial factor in triggering phosphorylation of various defense signaling-related proteins. Elicitor signals transiently activate various protein kinases within a few minutes after application (Romeis et al. 2000; Vitart et al. 2000). This type of transition from nonelicited to elicited form of protein kinase is caused by a phosphorylation event.

Protein phosphorylation has been shown to play a central role in the plant immune response signaling (Peck et al. 2001; Zipfel et al. 2004; Benschop et al. 2007; Tena et al. 2011). Protein kinases and protein phosphatases and their corresponding protein substrates play key roles in the pathogen-associated molecular pattern (PAMP)–plant pattern recognition receptor (PRR) mediated defense signal transduction (Benschop et al. 2007; Tischner et al. 2010).

## 9.2 Protein Phosphorylation Is an Early PAMP/Elicitor-Triggered Event

The proteins that make up the signal transduction pathway are present in the cell prior to the perception of elicitor (Benschop et al. 2007). These proteins are activated by post-translational modifications and conformational changes. The most widely recognized post-translational modification involved in signal transduction is protein phosphorylation (Benschop et al. 2007; Tischner et al. 2010). Early signaling and induction of defense responses are likely mediated through plasma membrane-associated proteins. Many membrane-associated proteins have one or more phosphorylation sites (Nühse et al. 2004). Very early signaling appears to be transient protein phosphorylation. The phosphorylation pattern of proteins was transiently changed within 10 min in *Arabidopsis thaliana* after contact with *Verticillium longisporum* conidia (Tischner et al. 2010). At least 30 proteins were differentially phosphorylated within first 4 min after the bacterial PAMP flg22 or the fungal

PAMP chitin treatment in *Arabidopsis* (Peck et al. 2001). Perception of flg22 activates MAP kinase activity within minutes and it peaks within 5–10 min and diminishes after 60 min in *Arabidopsis* suspension-cultured cells (Nühse et al. 2000). Transient protein phosphorylation is involved in various defense signal transduction systems triggered by PAMPs (Dietrich et al. 1990; Felix et al. 1991; Lecourieux-Ouaked et al. 2000; Nühse et al. 2000, 2003; Peck et al. 2001; Romeis et al. 2001; Lecourieux et al. 2002, 2006).

## 9.3 Protein Phosphorylation Is Carried Out by Different Protein Kinases

Protein phosphorylation is carried out by different protein kinases. Protein kinase superfamily has been classified into five main groups. These include "AGC" group (protein kinase A, G, and C group), CaMK group (calcium and calmodulin-dependent protein kinase group), CMGC group (cyclin-dependent kinase group), PTK group (protein Tyr kinase group), and "other" group. The AGC group is represented by the cyclic nucleotide-dependent kinases (protein kinase A [PKA] and protein kinase G [PKG]) and the calcium-phospholipid-dependent kinases (protein kinase C [PKC]). This group is regulated by cAMP (cyclic adenosine monophosphate), cGMP (cyclic guanosine monophosphate), diacylglycerol, and $Ca^{2+}$ (Stone and Walker 1995). Protein kinase C has been reported to induce phosphorylation in soybean (Dröge-Laser et al. 1997) and potato (Desprès et al. 1995; Subramaniam et al. 1997).

The CaMK group of protein kinases includes the calcium-/calmodulin-dependent and SNF1 (Suc nonfermenting1)/AMP-activated protein kinase families. Calcium-dependent protein kinases (CDPKs) are the most common protein kinases in plants (Ellard-Ivey et al. 1999; Harmon et al. 2000; Romeis et al. 2000, 2001; Cheng et al. 2002; Zhang et al. 2002; Harmon 2003; Hrabak et al. 2003; Boudsocq et al. 2010). The CDPKs constitute one of the largest families of protein kinases. Thirty-four different genes encoding CDPKs have been detected in *A. thaliana* (Harmon et al. 2000; Cheng et al. 2002; Hrabak et al. 2003).

The CMGC group contains MAPK (mitogen-activated protein kinase), CDK (cyclin-dependent kinase), GSK-3 (glycogen synthase kinase-3), and CKII (casein kinase II) families. MAPKs are also known as ERKs (extracellular-regulated protein kinases) and these are Ser-Thr protein kinases activated by dual phosphorylation. The enzyme responsible for this dual phosphorylation, MAPKK (MAPK kinase), represents an unusual class of protein kinases that will phosphorylate on Ser, Thr, and Tyr residues (Stone and Walker 1995). A MAPK cascade generally involves MAP kinase kinase kinase (MAPKK) – MAPKK – MAPK module that transduces extracellular signals through the receptors into a wide range of intracellular responses (He et al. 2007). In this module, a MAPKKK phosphorylates and activates a MAPK. Activated MAPK is imported into the nucleus, where it phosphorylates and activates specific downstream signaling components such as transcription factors (Ligterink and Hirt 2000). In Arabidopsis, 68 MAPKKKs (MTKs), 10 MAPKKs (MKKs), and 20 MAPKs (MPKs) have been identified (He et al. 2007). Activation of the MAPK cascades triggers massive transcript changes and confers resistance to multiple pathogens (He et al. 2007).

The CDK (cyclin-dependent kinase) family protein kinases have a regulatory subunit, cyclin, and a catalytic subunit, CDK. In addition to their interaction with cyclins, CDKs are themselves regulated by protein phosphorylation. Phosphorylation of a Thr residue is required for kinase activation, whereas phosphorylation of a Tyr residue serves an inhibitory function (Stone and Walker 1995). The GSK-3 family includes ASKs (Apoptosis signal regulating kinases). ASKs autophosphorylate on Ser, Thr, and Tyr. The CKII family protein kinases phosphorylate and promote the DNA-binding activity of G-box binding factor 1, a transcription factor that binds to the plant G-box promoter element (Klimczak et al. 1995).

PTK group includes Tyr-specific protein kinases, which have not been detected in plants (Stone and Walker 1995). Several other protein kinases, which do not fall into any of the above four groups have been detected in plants. These include protein kinases belonging to His kinase family. His kinase family protein kinase takes part in ethylene signaling system. In *Arabidopsis*, ethylene is perceived by a family of five membrane-bound receptors (ETR1, ERS1, ETR2, EIN4, and ERS2), which transmit the signal to downstream effectors Among them the ETR1 receptor shows His kinase activity, while others show Ser/Thr kinase activity (Wang et al. 2006).

The other group of protein kinases includes receptor-like kinases (RLKs). These RLKs are serine/threonine kinases (Shiu and Bleecker 2001a, b). Large numbers of RLKs have been detected in plants (Wang et al. 2005a, b; Benschop et al. 2007) and at least 340 genes encoding putative RLKs have been detected in *Arabidopsis* genome (Montesano et al. 2003). RLKs are transmembrane proteins that recognize an extracellular signal, in the form of a polypeptide ligand. Ligand-binding leads to autophosphorylation on the cytoplasmic kinase domain, a requirement for propagation (Stone and Walker 1995).

## 9.4 PAMPs/Elicitors Activate Receptor-Like Kinases

Plants rely on recognition of elicitors to activate defense signaling. Defense signaling is initiated through interaction of the elicitor with a plasma membrane-localized receptor (Zipfel et al. 2004; Benschop et al. 2007). Most of the receptors belong to the huge family of receptor-like kinases/receptor protein kinases (RLKs/RPKs; Tena et al. 2011). About 610 receptor-like kinases (RLKs) have been identified in the *Arabidopsis thaliana* genome (Shiu and Bleecker 2001b). Several receptor-like kinases (RLKs) have been found to be phosphorylated in response to elicitors in *A. thaliana* (Benschop et al. 2007). For example, the general elicitor flagellin is recognized in *Arabidopsis* through a conserved 22-amino acid sequence (flg22). Recognition involves the receptor-like kinase FLS2, which activates a downstream response that includes the production of ROS, ethylene biosynthesis, activation of a MAPK cascade, and activation of defense gene expression (Navarro et al. 2004; Zipfel et al. 2004). The PAMP has been shown to activate the receptor kinase by autophosphorylation by its own serine/threonine kinase (Gómez-Gómez et al. 2001; Wang et al. 2001; Robatzek et al. 2006). Mutation of the threonine residue 867 hampers FLS2 response, suggesting that autophosphorylation of the general elicitor occurs at threonine residue 867 (Robatzek et al. 2006).

Activation of other receptor-like kinases by induced autophosphorylation has been reported (Nasrallah 2000; Gómez-Gómez et al. 2001; Wang et al. 2005b). Arabidopsis CERK1 is the receptor of the PAMP chitin. It is involved in the perception of the chitin oligosaccharide elicitor at the cell surface and the transduction of the signal into the cytoplasm via its intracellular serine/threonine kinase activity (Miya et al. 2007). It contains an intracellular serine/threonine kinase domain (Eckardt 2008; Lohmann et al. 2010). The rice homolog of CERK1, OsCERK1 encoded a receptor-like kinase consisting of 624 amino acid residues, containing a signal peptide, an extracellular domain, a transmembrane region and an intracellular Ser/Thr kinase domain. The expression of *OsCERK1* was up-regulated by elicitor treatment (Shimizu et al. 2010). CERK1 is autophosphorylated *in vitro* (Iizasa et al. 2010). The autophosphorylated CERK1 has been shown to be essential for chitin signaling in plants (Wan et al. 2008a, b). The PRR for the PAMP elicitin INF1 of *Phytophthora infestans* has been identified as a lectin-like receptor kinase and it was designated NbLRK1. NbLRK1 is a typical RD kinase (Kanzaki et al. 2008). The 31 amino acids fragment of NbLRK1 kinase domain within VIb subdomain has been shown to interact with INF1 *in vitro*. The VIb subdomain of Ser/Thr kinase is known to contain the catalytic loop with an invariant Asp serving as the catalytic base necessary for the kinase function. This site is close to the VII and VIII domains where the activation loop is located, which is necessary for autophosphorylation of kinases (Dardick and Ronald 2006; Kanzaki et al. 2008). It is suggested that INF1 binding to the VIb subdomain of NbLRK1 alters its kinase activity presumably by autophosphorylation (Kanzaki et al. 2008). INF1 treatment induced autophosphorylation of NbLRK1 *in vivo* (Kanzaki et al. 2008).

The importance of phosphorylation of receptor-like kinases in signal transduction has been demonstrated by developing mutants with impaired kinase activity (Wang et al. 2005a). Mutation studies with BRI1, a receptor – like kinase in *Arabidopsis* have revealed that mutations in the kinase domain activation loop nearly abolished kinase activity, with respect to both autophosphorylation and protein substrate phosphorylation. The results suggest that autophosphorylation of the activation loop of BRI1 is required for downstream signaling events (Wang et al. 2005a).

In *Arabidopsis*, BAK1 belongs to the LRR-receptor-like kinase (RLK). BAK1 is a positive regulator of PAMP-triggered plant immunity and it acts as an adaptor of multiple LRR-RKs that act in defense signaling, including the PRRs FLS2, EFR, PEPR1 and PEPR2 (Chinchilla et al. 2007b; Ryan and Pearce 2003; Gao et al. 2009; Postel et al. 2010; Schulze et al. 2010). It also acts as an adaptor of the receptor kinases BIR1 and SOBIR1, which seem to act as part of a presumed PRR complex(es) and/or at a downstream step in the signaling cascade (Saijo 2010). Flg22 perception by the PRR FLS2 triggers an interaction between FLS2 and BAK1 (Chinchilla et al. 2007a, b; Heese et al. 2007). The flg22 induced FLS2-BAK1 association occurs within seconds and is accompanied by increased phosphorylation on both FLS2 and

BAK1 (Schulze et al. 2010). The *de novo* phosphorylation of both FLS2 and BAK1 has been detected within 15 s of stimulation with flg22. It is suggested that several LRR-RKs form tight complexes with BAK1 almost instantaneously after ligand binding and the subsequent phosphorylation events are key initial steps in signal transduction (Schulze et al. 2010). Collectively, these studies suggest that autophosphorylation of receptor-like kinases is an important event in PAMP-activated defense response signal transduction system.

## 9.5 PAMP/Elicitor Induces Phosphorylation of Calcium-Dependent Protein Kinases

Phosphorylation of calcium-dependent protein kinases plays an important role in plant immune responses. Romeis et al. (2000) identified a membrane-bound calcium-dependent protein kinase (CDPK) that showed a shift in electrophoretic mobility from 68 to 70 kDa within 5 min after an elicitor was added in tobacco cell cultures. The interconversion of the corresponding CDPK forms could be induced in vitro in both directions by treatments with either phosphatase or ATP. CDPK activity of the phosphorylated 70-kDa CDPK form was greater than that of nonelicited 68-kD form (Romeis et al. 2000). The conversion of the nonelicited CDPK into active form was not due to autophosphorylation (Romeis et al. 2000). The results suggest that phosphorylation of the CDPK results in activation of the kinase to trigger the downstream events in the signal transduction system (Romeis et al. 2001).

Autophosphorylation of CDPK due to elicitor action has also been reported. CDPK is autoinhibited by an interaction of a pseudosubstrate site within its junction domain that blocks the active site of the kinase domain. Binding of $Ca^{2+}$ to the calmodulin-like domain of the CDPK causes a conformational change that extends to the adjacent junction domain and finally disengages the autoinhibitor of the active site (Huang et al. 1996).

## 9.6 PAMP/Elicitor Triggers Phosphorylation of MAP Kinases

Mitogen-activated protein kinase cascade involves sequence of phosphorylation events. MAPKKK (MEKK) phosphorylates and activates a particular MAPKK (MKK) by the phosphorylation of serine/threonine residues in the SXXXS/T motif. As a dual-specificity kinase, MAPKK then activates MAPK through the phosphorylation of threonine and tyrosine residues in the TXY motif located between kinase subdomains VII and VIII (Ligterink and Hirt 2000; Liu et al. 2000). The MAP kinase itself may be autophosphorylated (Mayrose et al. 2004). The tomato MAP kinase *LeMPK3* is specifically induced at the mRNA level upon treatment with a fungal elicitor. The transcript accumulation was followed by an increase in LeMPK3 kinase activity. The LeMPK3 autophosphorylates in vitro mainly on tyrosine and less so on threonine and serine, whereas it phosphorylates myelin basic protein on serine and threonine (Mayrose et al. 2004). The autophosphorylation of LeMPK3 may not be sufficient for its full activation, which requires an upstream MAPKK (Mayrose et al. 2004). *Arabidopsis* AtMPK4 MAP kinase autophosphorylates in vitro on tyrosine residues and is activated by the AtMEK1 MAPKK through phosphorylation of threonine residues (Huang et al. 2000). Phosphorylation of the MAP kinase promotes its homodimerization and nuclear translocation (Khokhlatchev et al. 1998). The activated MAPK phosphorylates transcription factors (Ligterink and Hirt 2000).

The MAP kinases WIPK and SIPK have been shown to be involved in defense signaling system (Liu et al. 2003). Reversible phosphorylation/dephosphorylation events were involved in the activation of WIPK and SIPK in tobacco cells. The protein kinase inhibitor staurosporine inhibited the activation of SIPK and WIPK (Zhang et al. 2000; Liu et al. 2003) and the phosphatase inhibitors calyculin A and okadaic acid activated SIPK and induced WIPK expression (Liu et al. 2003). These results suggest the importance of transient phosphorylation/dephosphorylation events in activation of the MAPK pathway.

A protein that interacts with SIPK has been identified as a member of the MAP kinase kinase family and named as SIPKK. SIPKK phosphorylates myelin basic protein in vitro (Liu et al. 2000). SIPK is activated exclusively at the posttranslational level by phosphorylation (Liu et al. 2000). Fungal elicitors transiently activated a 47-kD putative MAPK via tyrosine phosphorylation in tobacco cells (Suzuki and Shinshi 1995). Activation of this 47-kD kinase was inhibited by staurosporine, a protein kinase inhibitor staurosporine and the $Ca^{2+}$ channel blocker $Gd^{2+}$ (gadolinium) suggesting that upstream kinases and $Ca^{2+}$ might be involved in the activation of this kinase (Suzuki and Shinshi 1995).

## 9.7 Role of 14-3-3 Proteins in Protein Phosphorylation

In some cases, besides the kinases, 14-3-3 proteins complete a multiplex signal-induced change in the target protein (Roberts 2000; Yaffe 2002; Ferl 2004). 14-3-3 proteins physically interact with other protein families by binding with the phosphorylated proteins (Ferl 2004). 14-3-3 proteins can bind with a large number of proteins, since 14-3-3 proteins require relatively simple amino acid sequence for their binding (Yaffe et al. 1997). 14-3-3 proteins have been detected in different cellular compartments. They are found in cytoplasm, inside chloroplasts and found associated with mitochondria and some of them have been detected in plasma membrane (Fuglsang et al. 1999; Roberts and Bowles 1999). It appears that phosphorylation cannot cause a change in protein activity, but phosphorylation is the only means to connect the signal to the target protein (Ferl 2004). With the assistance of 14-3-3 proteins, phosphorylation signals can reach many additional targets (Ferl 2004).

14-3-3 proteins exist as dimers and hence they can bring together two different proteins, or two different domains within one protein, showing a direct interaction between the clients (Yaffe et al. 1997).

## 9.8 PAMP/Elicitor Triggers Phosphorylation of PEN Proteins

Three PEN (for PENETRATION) proteins, PEN1, PEN2, and PEN3 have been reported to be involved in penetration resistance by limiting pathogen entry into host cells (Collins et al. 2003; Assad et al. 2004; Stein et al. 2006). PEN1, also called SYP121, is a syntaxin, while PEN2 is a glycosyl hydrolase and PEN3 is an ABC (ATP binding cassette) transporter (Collins et al. 2003; Lipka et al. 2005). Expression of the genes encoding these proteins, *PEN1*, *PEN2*, and *PEN3*, were found to be induced in response to elicitor perception in *A. thaliana*. Both PEN1 and PEN3 were found to be phosphorylated upon elicitor treatment (Benschop et al. 2007). PEN3 was phosphorylated on two residues in response to the elicitors and PEN1 was phosphorylated on the N terminus (Ser-7). These phosphorylation sites may be involved in the activation of the PEN proteins (Benschop et al. 2007). The importance of PEN proteins in defense response was demonstrated by developing mutants. *Arabidopsis* mutants, *pen1*, *pen2*, and *pen3*, supported higher frequency of penetration of *Blumeria graminis* f. sp. *hordei* into leaves (Stein et al. 2006). Elicitor-induced phosphorylation of PEN proteins is involved in early defense responses (Collins et al. 2003; Stein et al. 2006).

## 9.9 Protein Phosphorylation Involved in Early Defense Signaling Events Triggered by PAMPs/Elicitors

Protein phosphorylation/dephosphorylation plays an important role in early signaling events (Fig. 9.1). An oomycete elicitor induces calcium ion influx, anion efflux, and activation of a plasma membrane NADPH oxidase responsible for a transient production of ROS in tobacco. All these effects were inhibited by staurosporine, a protein kinase inhibitor, indicating that phosphorylation reactions occurred upstream from these effects (Viard et al. 1994; Tavernier et al. 1995). Phosphorylation of proteins involved in G-protein coupled signaling, $Ca^{2+}$/calmodulin-dependent signaling pathways, redox signaling system, and $H^+$-ATPase regulation of intracellular pH has been reported in tobacco cells treated with a bacterial elicitor (Fig. 9.1; Gerber et al. 2006).

PAMP-triggered early events include G proteins modulated protein phosphorylation/dephosphorylation systems which trigger $Ca^{2+}$ influx. Phosphorylation of proteins involved in G-protein coupled signaling has been reported in tobacco cells treated

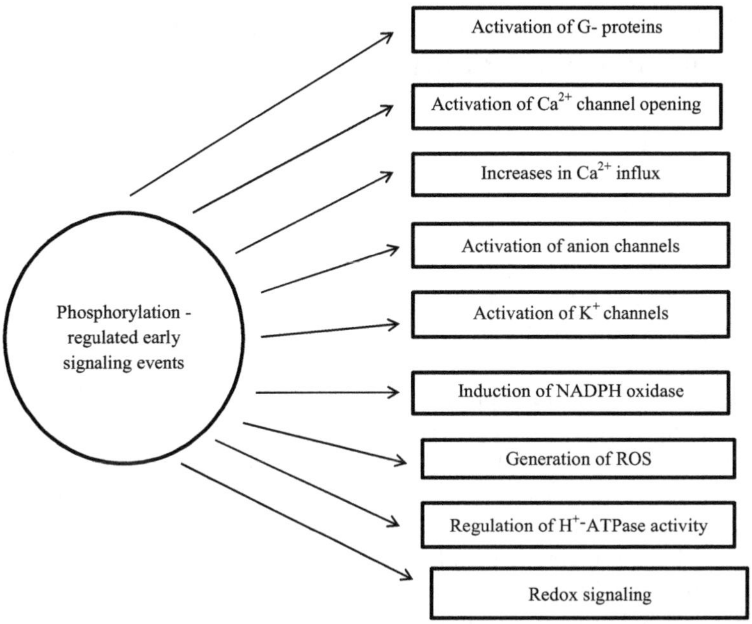


**Fig. 9.1** Phosphorylation/dephosphorylation – modulated PAMP-triggered early signaling events

with a bacterial PAMP (Gerber et al. 2006). G-proteins may be involved in $Ca^{2+}$ channel opening (Gelli et al. 1997). The activation of the $Ca^{2+}$ channel by PAMPs was modulated by a heterotrimeric G-protein–dependent phosphorylation of the channel protein in tomato, probably by activating protein kinase, and inhibiting protein phosphatase (Gelli et al. 1997). Protein phosphorylation precedes $Ca^{2+}$ influx in tobacco cells treated with a PAMP isolated from the oomycete pathogen *Phytophthora cryptogea* (Tavernier et al. 1995). The G-proteins modulate the phosphorylation/dephosphorylation system in the plasma membrane of tomato cells and transduce the $Ca^{2+}$ influx signal (Vera-Estrella et al. 1994a, b).

An elicitor induced increase in $H^+$ ATPase activity was shown to be activated by the G-proteins in tomato (Vera-Estrella et al. 1994a). G-protein has been shown to activate a membrane-bound phosphatase that mediates the dephosphorylation of the host plasma membrane $H^+$-ATPase in tomato (Xing et al. 1997). The dephosphorylation of $H^+$-ATPase was followed by rephosphorylation by protein kinase and $Ca^{2+}$-dependent kinase (Xing et al. 1996).

Anion channels are also activated by protein phosphorylation. Protein kinases act as positive regulators, while phosphatases negatively regulate anion channel activity (Pei et al. 1996; Wendehenne et al. 2002). The activated anion channels trigger $Ca^{2+}$-mediated signaling system (Ward et al. 1995). Activation of $K^+$ channel by phosphorylation has also been reported (Li et al. 1998).

## 9.10 Phosphorylation of Proteins Involved in $H^+$ Fluxes Induced by PAMP/Elicitor

An important target protein of CDPK for phosphorylation is plasma membrane $H^+$-ATPase (Piedras et al. 1998). Elicitation of tobacco cell cultures with an elicitor resulted in changes of $H^+$ fluxes, detectable as media alkalinization. These fluxes are accomplished by way of inactivation of an $H^+$-ATPase by CDPK (Piedras et al. 1998). Reversible phosphorylation of an $H^+$-ATPase in tobacco has been reported (Xing et al. 1996). CDPKs are known to regulate $H^+$-ATPase (Camoni et al. 1998b; Schaller and Oecking 1999). A CDPK phosphorylates $H^+$-ATPase of oat roots (Schaller et al. 1992).

Regulation of $H^+$-ATPases appear to depend on the presence or absence of 14-3-3 proteins (Chung et al. 1999; Fuglsang et al. 1999). An in vitro interaction between a phosphorylated CDPK and 14-3-3 isoforms from *Arabidopsis* has been reported (Camoni et al. 1998a). There may be a functional link among phosphorylated CDPK, $H^+$-ATPase and 14-3-3 protein in defense signaling (Romeis et al. 2000). Binding of 14-3-3 proteins to the plasma membrane $H^+$-ATPase involves the three C-terminal residues Tyr-Thr-Val and requires phosphorylation of Thr (Fuglsang et al. 1999). 14-3-3 proteins recognize phosphate-bearing amino acids and regulate the $H^+$-ATPase enzyme activity (Romeis et al. 2000).

## 9.11 Phosphorylation of Proteins Involved in ROS Signaling System

Several and different protein kinases induce a sequence of phosphorylation events in the production of reactive oxygen species (ROS) and downstream signaling events (Fig. 9.2; Anthony et al. 2006). Elicitors induce very rapid production of ROS resulting in the oxidative burst in plant cells. The oxidative burst is mostly mediated by NADPH oxidases (Torres et al. 2002; Torres and Dangl 2005; Davies et al. 2006). In response to elicitors, *Arabidopsis* leaves produce ROS within minutes and this oxidative burst peaks around 10 min. During this period, phosphorylation on several different residues in a NADPH oxidase (RBOHD [respiratory burst oxidase protein D]) was observed (Benschop et al. 2007). CDPK has been shown to phosphorylate NADPH oxidase (Xing et al. 1997; Blumwald et al. 1998). Accumulation of ROS requires $Ca^{2+}$ influx and protein kinase activity (Keller et al. 1998; Piedras et al. 1998; Romeis et al. 1999).

Anthony et al. (2006) reported the participation of 3-phosphoinositide-dependent protein kinase 1 (PDK1), the AGC family protein kinase called oxidative signal-inducible 1 (OXI1) protein kinase, the serine/Thr kinase (PTI1-2 for Pto kinase interactor1-2), and the mitogen-activated kinase MPK6 in ROS signaling system in *A. thaliana*. OXI1 was identified as a downstream signaling component to the PDK1 (Rentel et al. 2004). OXI1 is activated by PDK1-mediated phosphorylation

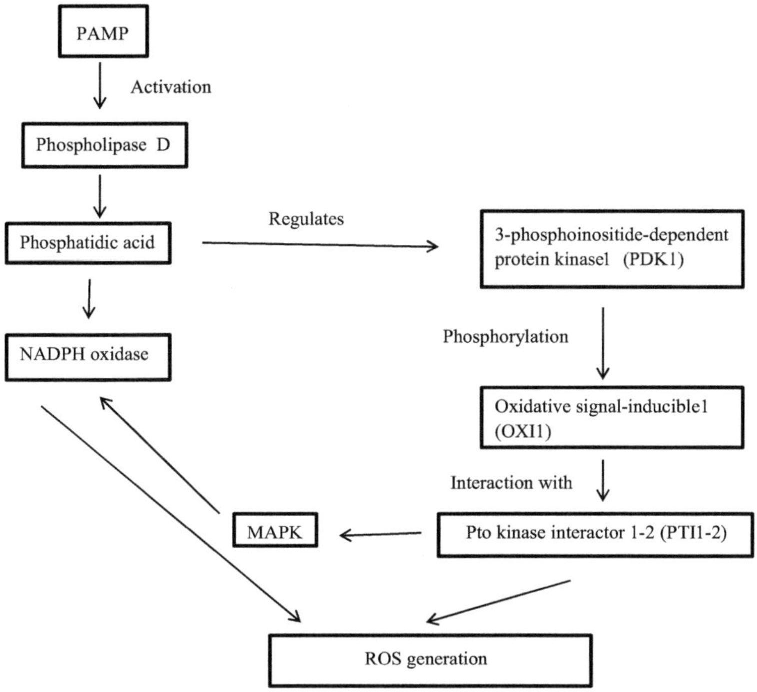


**Fig. 9.2** Role of PDK1 – OXI1 – PTI1 – MAPK mediated phosphorylation events in ROS generation (Adapted from Rentel et al. 2004; Anthony et al. 2004, 2006)

(Devarenne et al. 2006; Zegzouti et al. 1999). PTI1-2 has been found to be an interacting partner of OXI1, which is downstream of OXI1. Its sequence closely resembled the tomato Pto kinase, which is involved in triggering programmed cell death-mediated disease resistance (Anthony et al. 2006). $H_2O_2$ is produced via OXI1-PTI1-2 pathway (Anthony et al. 2006). PDK1 enzyme activity is regulated by phosphatidic acid (PA) (Anthony et al. 2004). PA is generated via two distinct phospholipase pathways, either directly by phospholipase D (PLD) or the sequential action of phospholipase C (PLC) and diaceylglycerol kinase (Testernik and Munnik 2005). PDK1 is specifically activated by PLD-generated PA ($PA_{PLD}$) (Anthony et al. 2004). The lipids signaling pathways converge via the PDK1-OXI1 axis (Anthony et al. 2006). Thus, three protein kinases (PDK1, OXI1, and PTI1-2) are involved in phosphorylation-mediated ROS signaling system in *A. thaliana* (Anthony et al. 2006).

Kinase active PTI1-2 is able to increase the expression of ROS promoters indicating that PTI1-2 functions in specific ROS signaling pathways (Anthony et al. 2006). Oxi1 null mutants are impaired in the activation of the MAPKs MPK3 and MPK6 upon oxidative stress, suggesting that OXI1 functions downstream of ROS but upstream of the MAPK module (Rentel et al. 2004).

MPK3, MPK4, and MPK6 are all activated by fungal and bacterial elicitors and ROS (Kovtun et al. 2000; Nühse et al. 2000). To assess the potential involvement of PDK1 in OXI1-MPK signaling, the expression of PDK1 in OXI1-MPK signaling was ablated. The results showed that activation of MPK6 by the elicitor was PDK1-dependent (Anthony et al. 2006). It suggests that MAPK signaling cascades function downstream of OXI1 and PTI1-2, resulting in the eventual activation of PR genes. MAPK cascades may also play a role upstream and have been implicated in the activation of the NADPH oxidase genes (Yoshioka et al. 2003). PA has the potential to activate targets such as PDK1 and NADPH oxidase simultaneously; and both of them trigger ROS by phosphorylation (Anthony et al. 2006).

The possible role of various phosphorylation events in ROS signaling system is presented in Fig. 9.2.

## 9.12 Phosphorylation of Proteins Involved in Ethylene-Signaling System

Protein phosphorylation and dephosphorylation have been shown to be involved in the biosynthesis of ethylene. S-adenosylmethionine and 1-aminocyclopropane-1-carboxylic acid (ACC) are the precursors of ethylene. S-adenosylmethionine is converted to ACC by ACC synthase (ACS). ACC is oxidized by ACC oxidase (ACO) to form ET (Vidhyasekaran 2007). Treatment with the protein kinase inhibitors staurosporine or K-252 inactivated the ACS activity, whereas protein phosphatase inhibitor calyculin A stimulated ACS activity (Spanu et al. 1991). These results suggest that phosphorylation/dephosphorylation of ACS is involved in the increased activity. Wang et al. (2002) suggested that ACS is unstable *in vivo* and present at low abundance, and phosphorylation of ACS may increase its stability to sustain the elevated activity.

An oomycete elicitor induced a dramatic increase in ACS activity in parsley cell cultures (Chappell et al. 1984). Similar increase in ACS activity coincided with activation of the MAPK SIPK followed by an increase in ET production in tobacco (Kim et al. 2003a). The protein kinase inhibitor H-7 (1-[5-isoquinolinylsulfonyl]-2-methylpiperazine) and the protein phosphatase inhibitors vanadate and okadaic acid inhibited the induction of ET in pea (Kwak and Lee 1997). These results suggest that protein phosphorylation/dephosphorylation plays an important role in ET signaling system.

MAPK cascade has been shown to negatively regulate ethylene signaling by constitutive phosphorylation of downstream components that ultimately repress the accumulation of EIN3 and its relatives (Ouaked et al. 2003). Binding of ethylene to the receptors (ETR1, ETR2, EIN4, ERS1, and ERS2) blocks the MAPK cascade by failing to activate the MAPKKK protein (CTR1). This in turn allows EIN3 and its relatives to accumulate. The accumulated transcription factors induce the transcription of a variety of ethylene-response genes such as the ERFs (Ouaked et al. 2003).

ERFs (ethylene-responsive transcription factors) have been shown to bind specifically to the GCC box of the defense genes (Yamamoto et al. 1999). *Trichoderma viride* xylanase elicitor induced accumulation of mRNA for ERF2 in tobacco cells. The elicitor-inducible accumulation of the transcription factor was inhibited by staurosporine (an inhibitor of serine/threonine protein kinase) and calcyculin A (an inhibitor of protein phosphatase). These results suggest that ERF2 might play a major role in the elicitor-induced GCC box-mediated transcription of defense genes and that both protein kinase and protein phosphatase might be involved, as positive regulators in the signal transduction pathway that leads to expression of ERF2 and subsequent GCC box-mediated transcription of defense genes (Yamamoto et al. 1999).

## 9.13 Phosphorylation of Proteins Involved in Salicylic Acid Signaling System

Protein phosphorylation has been shown to be involved in SA-dependent signaling system. Okadaic acid blocked SA-induced PR-1 gene expression (Conrath et al. 1987). Okadaic acid is an inhibitor of phosphorylation of protein kinase and hence it suggests that phosphorylation is a component in SA-dependent *PR* gene expression.

## 9.14 Protein Phosphorylation in ABA Signaling System

Protein phosphorylation is the important mechanism for ABA signaling (He and Li 2008). Specific protein kinases are activated in response to ABA and they play a positive role in ABA signaling (Li et al. 2000; Lu et al. 2002; Fujii et al. 2009). ABA response element binding factors (ABFs) have been shown to be activated by phosphorylation by protein kinases (Uno et al. 2000). SnRK (for SNF1-related protein kinase) family of protein kinases, belonging to calcium-dependent protein kinases (CDPKs) are involved in phosphorylation of ABF proteins (Fujii et al. 2009). The *Arabidopsis* genome contains 38 SnRKs, of which 10 are SnRK2s (Hrabak et al. 2003). Five of these SnRK2s were shown to be activated by ABA and they phosphorylated ABI1, ABF2 and ABF4 (Furihata et al. 2006; Yoshida et al. 2006). SnRK2.2 and SnRK2.3 are the two protein kinases and the effect of these protein kinases is mediated by phosphorylating ABFs and regulating ABA-responsive genes (Fujii et al. 2009). In rice, SAPK8, SAPK9, and SAPK10, which are homologous with SnRK2.2, SnRK2.3, and SnRK2.6 of *Arabidopsis*, were activated by ABA (Kobayashi et al. 2004). These SAPKs phosphorylated TRAB1, which is a rice ortholog of the *Arabidopsis* ABFs (Kobayashi et al. 2005). ABA induced the wheat SnRK, PKABA1, at the transcript level. The ABA-induced PKAB1 phosphorylated TaABF, the wheat bZIP transcription factor (Johnson et al. 2002). Several other CDPKs have been shown to be involved in ABA signaling (Sheen 1996; Romeis et al. 2001; Choi et al. 2005). Two Arabidopsis CDPKs, AtCPK10 and AtCPK30 activate an ABA-inducible promoter in maize leaf protoplasts (Sheen 1996). Another CDPK in *Arabidopsis*, AtCPK32, interacts with C2-C3 conserved region of ABF4 (Choi et al. 2005). AtCPK32 has autophosphorylation activity and phosphorylates ABF4 in vitro. The CDPK has been shown to be an ABA signaling component that positively modulates ABF4 function (Choi et al. 2005).

A calcineurin B-like (CBL) protein kinase CIPK15 interacts with the calcium-modulated protein phosphatases ABI1 and ABI2 and a CBL $Ca^{2+}$-binding protein ScaBP5. CIPK15 and one of its homologs CIPK2 are involved in ABA signaling as negative regulators (Guo et al. 2002; Kim et al. 2003b). The kinase substrate of CIPK15 has been identified as an AP2 transcription factor AtERT7 that negatively regulates ABA signaling (Song et al. 2005).

Another SNF1-related protein kinase, AAPK (for ABA-activated protein kinase) was detected in broad bean (*Vicia faba*). AAPK is stimulated by ABA and it is a positive regulator of ABA response (Li et al. 2000, 2002; Johnson et al. 2002). Takahashi et al. (2007) showed that ABA induced binding of a 14-3-3 protein to proteins with molecular masses of 61, 43 and 39 kDa. Autophosphorylation of AAPK, which mediates anion channel activation and ABA-induced phosphorylation of the 61 kDa protein showed similar time courses. AAPK elicits the binding of the 14-3-3 protein to the 61-kDa protein in vitro when AAPK was activated by ABA. It has also been suggested that the 61 kDa protein may be a substrate for AAPK and the 61 kDa protein is located upstream of $H_2O_2$ and $Ca^{2+}$, or on $Ca^{2+}$-independent signaling pathway (Takahashi et al. 2007).

An ABA-insensitive *Vicia faba* mutant, *fia* (faba bean impaired in ABA-induced stomatal closure) had been isolated. Unlike ABA, $H_2O_2$, and nitric oxide (NO) induced stomatal closure in the *fia* mutant. ABA did not induce production of either reactive oxygen species or NO in the mutant. ABA also did not suppress inward-rectifying $K^+$ currents or activate AAPK in mutant guard cells. These results suggest that FIA functions as an early signal component upstream of AAPK activation in ABA signaling in guard cells of *Vicia faba* (Sugiyama et al. 2012).

Arabidopsis OST1/SRK2E, an Arabidopsis ortholog of *Vicia faba* AAPK, which is an ABA-activated, $Ca^{2+}$-independent protein kinase, is involved in ABA signaling and it acts upstream of ROS production (Mustilli et al. 2002; Yoshida et al. 2002). Phosphorylation has been shown to function in ABA-induced ROS production and $I_{ca}$ channel activation (Murata et al. 2001; Köhler and Blatt 2002).

ABA activates mitogen activated protein kinase (MAPK)-mediated signaling system (Gomi et al. 2005; Wang and Song 2008). An *Arabidopsis* MAPK, AtMAPK3 and a rice MAPK, OsMAPK5, have been identified as ABA-activated MAPKs (Lu et al. 2002; Xiong and Yang 2003). It has been found that ABA and $H_2O_2$

can activate the same MAPK (Desikan et al. 2004). ROS may act upstream of the MAPK cascade in the ABA signaling system in maize leaves (Zhang et al. 2006). ABA-induced activation of MAPK was almost fully arrested by pretreatment with inhibitors of ROS production, suggesting that ABA-induced $H_2O_2$ production activates MAPK (Zhang et al. 2006).

## 9.15 Phosphorylation of Transcription Factors

Protein kinases may phosphorylate transcription factors involved in transcription of defense genes. Protein kinase C-mediated phosphorylation activates a basic leucine zipper transcription factor G/HBF-1, enabling its binding to the chalcone synthase *Ch15* promoter from soybean (Dröge-Laser et al. 1997). An ERF transcription factor, ERF6, regulates *Arabidopsis* defense gene expression and resistance to the necrotrophic fungal pathogen *Botrytis cinerea*. Phosphorylation of ERF6 by the mitogen-activated protein kinases MPK3 and MPK6 has been shown to increase ERF6 stability in vivo. The phosphorylated ERF6 activates defense-related genes (Meng et al. 2013). The WRKY33 transcription factor is phosphorylated by MPK3/MPK6 in vivo in response to *B. cinerea* infection in *Arabidopsis*. The phosphorylated transcription factor induces the phytoalexin camalexin biosynthesis (Mao et al. 2011).

Phosphorylation of WRKY8 by the MAPKs SIPK, NTF4, and WIPK increased the DNA binding activity of WRKY8 to the cognate W-box sequence in *Nicotiana benthamiana* (Ishihama et al. 2011). The ectopic expression of phosphorylated WRKY8 induced defense-related genes (Ishihama et al. 2011). The bZIP transcription factor VIP1is phosphorylated by MPK3 and the phosphorylated transcription factor regulates the expression of pathogenesis-related genes (Djamei et al. 2007; Liu et al. 2010).

## 9.16 Phosphorylation Events Induced by MAP Kinases in Various Signaling Systems

Phosphorylation induced by MAP kinases have been shown to be involved in ROS signaling (Yang et al. 2001; Ren et al. 2002; Yoshioka et al. 2001, 2003; Nakagami et al. 2006;cv Zhang et al. 2006; Xing et al. 2007, 2008), SA signaling (Petersen et al. 2000; Frye et al. 2001; Brodersen et al. 2006; Zhang et al. 2007); JA signaling (Seo et al. 1999; Gomi et al. 2005; Takahashi et al. 2007), and ethylene signaling systems (Kim et al. 2003a; Guo et al. 2002; Liu and Zhang 2004; Menke et al. 2004; Brodersen et al. 2006). Phosphorylation of various transcription factors involved in transcription of defense genes by MAP kinases has also been described (Asai et al. 2002; Cheong et al. 2003; Wan et al. 2004; Waller et al. 2006).

## 9.17 Dephosphorylation Induced by Phosphatases May Negatively Regulate Innate Immune Responses

Dephosphorylation may also be involved in defense signaling. CESA1 and CESA3 are two cellulose synthase proteins detected in *A. thaliana* and both of them were dephosphorylated in response to elicitor treatment (Benschop et al. 2007). A reduction in CESA3 levels or CESA3 activity was shown to induce JA and ethylene accumulation and invoke defense gene expression in *A. thaliana* (Ellis et al. 2002; Cano-Delgado et al. 2003). Staurosporine, a protein kinase inhibitor, inhibited $Ca^{2+}$ uptake, extracellular alkalinization, ROS production, and protein phosphorylation induced by an elicitor in tobacco (Lecourieux-Ouaked et al. 2000). In contrast, calyculin A, a protein phosphatase inhibitor triggered all the above effects. Protein phosphatase inhibitors calyculin A and okadaic acid stimulated the inducible defenses in the absence of elicitors. The results suggest that continuous phosphorylation of proteins may occur in the non-elicited cells and the inhibition of dephosphorylation may be sufficient to initiate signal transduction (Lecourieux-Ouaked et al. 2000). Phosphatase may negatively regulate protein kinase in the signaling system. Phosphatases may function as negative regulators in defense signaling system (Lecourieux-Ouaked et al. 2000).

## References

Anthony RG, Henriques R, Helfer A, Meszaros T, Rios G, Testerink C, Munnik T, Deak M, Koncz C, Rogre L (2004) A protein kinase target of a PDK1 signalling pathway is involved in root hair growth in *Arabidopsis*. EMBO J 23:572–581

Anthony RG, Khan S, Costa J, Pais MS, Bögre L (2006) The *Arabidopsis* protein kinase PTI1-2 is activated by convergent phosphatidic acid and oxidative stress signaling pathways downstream of PDK1 and OXI1. J Biol Chem 281:37536–37546

Asai T, Tena G, Plotnikova J, Willmann MR, Chiu W-L, Gómez-Gómez L, Boller T, Ausubel FM, Sheen J (2002) MAP kinase signalling cascade in *Arabidopsis* innate immunity. Nature 415:977–983

Assad FF, Qiu JL, Youngs H, Ehrhardt D, Zimmerli L, Kalde M, Wanner G, Peck SC, Edwards H, Ramonell K, Somerville CR, Thordal-Christensen H (2004) The PEN1 syntaxin defines a novel cellular compartment upon fungal attack and is required for the timely assembly of papillae. Mol Biol Cell 15:5118–5129

Benschop JJ, Mohammed S, O'Flaherty M, Heck AJR, Slijper M, Menke FLH (2007) Quantitative phosphoproteomics of early elicitor signaling in *Arabidopsis*. Mol Cell Proteomics 6:1198–1214

Blumwald E, Aharon GS, Lam BC-H (1998) Early signal transduction pathways in plant-pathogen interactions. Trends Plant Sci 3:342–346

Boudsocq M, Willmann MR, McCormack M, Lee H, Shan L, He P, Bush J, Cheng SH, Sheen J (2010) Differential innate immune signalling via $Ca^{2+}$ sensor protein kinases. Nature 464:418–422

Brodersen P, Petersen M, Bjorn Nielsen H, Zhu S, Newman MA, Shokat KM, Rietz S, Parker J, Mundy J (2006) *Arabidopsis* MAP kinase 4 regulates salicylic acid- and jasmonic acid/ethylene-dependent responses via EDS1 and PAD4. Plant J 47:532–546

Camoni L, Fullone MR, Marra M, Aduccia P (1998a) The plasma membrane $H^+$-ATPase from maize roots is phosphorylated in the C-terminal domain by a calcium-dependent protein kinase. Physiol Plant 104:549–555

Camoni L, Harper JF, Palmgren MG (1998b) 14-3-3 proteins activate a plant calcium-dependent protein kinase (CDPK). FEBS Lett 430:381–384

Cano-Delgado A, Penfield S, Smith C, Catley M, Bevan M (2003) Reduced cellulose synthesis invokes lignification and defense responses in *Arabidopsis thaliana*. Plant J 34:351–362

Chappell J, Hahlbrock K, Boller T (1984) Rapid induction of ethylene biosynthesis in cultured parsley cells by fungal elicitor and its relationship to the induction of phenylalanine ammonia-lyase. Planta 161:475–480

Cheng S-H, Willmann MR, Chen H-C, Sheen J (2002) Calcium signaling through protein kinases: the Arabidopsis calcium-dependent protein kinase gene family. Plant Physiol 129:469–485

Cheong YH, Moon BC, Kim JK, Kim CY, Kim MC, Kim IH, Park CY, Kim JC, Park BO, Koo SC, Yoon HW, Chung WS, Lim CO, Lee SY, Cho MJ (2003) BWMK1, a rice mitogen-activated protein kinase, locates in the nucleus and mediates pathogenesis-related gene expression by activation of a transcription factor. Plant Physiol 132:1961–1972

Chinchilla D, Boller T, Robatzek S (2007a) Flagellin signaling in plant immunity. Adv Exp Med Biol 598:358–371

Chinchilla D, Zipfel C, Robatzek S, Kemmerling B, Nürnberger T, Jones JDG, Felix G, Boller T (2007b) A flagellin-induced complex of the receptor FLS2 and BAK1 initiates plant defence. Nature 448:497–500

Choi MS, Kim MC, Yoo JH, Moon BC, Koo SC, Park BO, Lee JH, Koo YD, Han HJ, Lee SY, Chung WS, Lim CO, Cho MJ (2005) Isolation of a calmodulin-binding transcription factor from rice (*Oryza sativa* L.). J Biol Chem 280:40820–40831

Chung H-J, Sehnke PC, Fert RJ (1999) The 14-3-3 proteins: cellular regulation of plant metabolism. Trends Plant Sci 4:367–371

Collins NC, Thordal-Christensen H, Lipka V, Bau S, Kombrink E, Qiu JL, Huckelhoven R, Stein M, Frelaldenhoven A, Somerville SC, Schulze-Lefert P (2003) SNARE-protein-mediated disease resistance at the plant cell wall. Nature 425:973–977

Conrath U, Silva A, Klessig DF (1987) Protein dephosphorylation mediates salicylic acid-induced expression of *PR1* genes in tobacco. Plant J 11:747–757

Dardick C, Ronald P (2006) Plant and animal pathogen recognition receptors signal through non-RD kinases. PLoS Pathog 2:e2

Davies DR, Bindschedler LV, Strickland TS, Bolwell GP (2006) Production of reactive oxygen species in *Arabidopsis thaliana* cell suspension cultures in response to an elicitor from *Fusarium oxysporum*: implications for basal resistance. J Exp Bot 57:1817–1827

Desikan R, Cheung M-K, Bright J, Henson D, Hancock JT, Neill SJ (2004) ABA, hydrogen peroxide and nitric oxide signaling in stomatal guard cells. J Exp Bot 55:205–212

Després C, Subramaniam R, Matton DPR, Brisson N (1995) The activation of the potato *PR-10a* gene requires the phosphorylation of the nuclear factor, PBF-1. Plant Cell 7:589–598

Devarenne TP, Ekengren SK, Pedley KF, Martin GB (2006) Adi3 is a Pdk1-interacting AGC kinase that negatively regulates plant cell death. EMBO J 25:255–265

Dietrich A, Mayer JE, Hahlbrock K (1990) Fungal elicitor triggers rapid, transient, and specific protein phosphorylation in parsley cell suspension cultures. J Biol Chem 265:6360–6368

Djamei A, Pitzschke A, Nakagami H, Rajh I, Hirt H (2007) Trojan horse strategy in *Agrobacterium* transformation: abusing MAPK defense signaling. Science 318:453–456

Dröge-Laser W, Kaiser A, Lindsay WP, Halker BA, Loake GJ, Doerner P, Dixon RA, Lamb C (1997) Rapid stimulation of a soybean protein-serine kinase which phosphorylates a novel bZIP DNA-binding protein, G/HBF-1 during the induction of early transcription-dependent defenses. EMBO J 16:726–738

Eckardt NA (2008) Chitin signaling in plants: insights into the perception of fungal pathogens and rhizobacterial strains. Plant Cell 20:241–243

Ellard-Ivey M, Hopkins RB, White TJ, Lomax TL (1999) Cloning, expression and N-terminal myristoylation of CpCPK1, a calcium-dependent protein kinase from zucchini (*Cucurbita pepo* L.). Plant Mol Biol 39:199–208

Ellis C, Karafyllidis I, Wasternack C, Turner JG (2002) The Arabidopsis mutant *cev1* links cell wall signaling to jasmonate and ethylene responses. Plant Cell 14:1557–1566

Felix G, Grosskopf DG, Regenass M, Boller T (1991) Rapid changes of protein phosphorylation are involved in transduction of the elicitor signal in plant cells. Proc Natl Acad Sci U S A 88:8831–8834

Ferl RJ (2004) 14-3-3 proteins: regulation of signal-induced events. Physiol Plant 120:173–178

Frye CA, Tang D, Innes RW (2001) Negative regulation of defense responses in plants by a conserved MAPKK kinase. Proc Natl Acad Sci U S A 98:373–378

Fuglsang AT, Visconti S, Drumm K, Jahn T, Stensballe A, Mattei B, Jensen ON, Aducci P, Palmgren MG (1999) Binding of 14-3-3 protein to the plasma membrane $H^+$-ATPase AHA2 involves the three C-terminal residues $Tyr^{946}$-Thr-Val and requires phosphorylation of $Thr^{947}$. J Biol Chem 274:36774–36780

Fujii H, Chinnusamy V, Rodrigues A, Rubio S, Antoni R, Park SY, Cutler SR, Sheen J, Rodriguez PL, Zhu JK (2009) *In vitro* reconstitution of an abscisic acid signalling pathway. Nature 462:660–664

Furihata T, Manuyama K, Fujita Y, Umezawa T, Yoshida R, Shinozaki K, Yamaguchi-Shinozaki K (2006) Abscisic acid-dependent multisite phosphorylation regulates the activity of a transcription activator AREB1. Proc Natl Acad Sci U S A 103:1988–1993

Gao M, Wang X, Wang D, Xu F, Ding X, Zhang Z, Bi D, Cheng YT, Chen S, Li X, Zhang Y (2009) Regulation of cell death and innate immunity by two receptor-like kinases in *Arabidopsis*. Cell Host Microbe 23:34–44

Gelli A, Higgins VJ, Blumwald E (1997) Activation of plant plasma membrane $Ca^{2+}$-permeable channels by race-specific fungal elicitors. Plant Physiol 113:269–279

Gerber IB, Laukens K, Witters E, Dubery IA (2006) Lipopolysaccharide-responsive phosphoproteins in *Nicotiana tabacum* cells. Plant Physiol Biochem 44:369–379

Gómez-Gómez L, Bauer Z, Boller T (2001) Both the extracellular leucine-rich repeat domain and the kinase activity of FLS2 are required for flagellin binding and signaling in *Arabidopsis*. Plant Cell 13:1155–1163

Gomi K, Ogawa D, Katou S, Kamada H, Nakajima N, Saji H, Soyano T, Sasabe M, Machida Y, Mitsuhara I, Ohashi Y, Seo S (2005) A mitogen-activated protein kinase NtMPK4 activated by SIPKK is required for jasmonic acid signaling and involved in ozone tolerance via stomatal movement in tobacco. Plant Cell Physiol 46:1902–1914

Guo Y, Xiong L, Song CP, Gong D, Halfter U, Zhu JK (2002) A calcium sensor and its interacting protein kinase are global regulators of abscisic acid signaling in *Arabidopsis*. Dev Cell 3:233–244

Harmon AC (2003) Calcium-regulated protein kinases of plants. Gravit Space Biol Bull 16(2):83–90

Harmon AC, Gribskov M, Harper JF (2000) CDPKs – a kinase for every $Ca^{2+}$ signal? Trends Plant Sci 5:154–159

He H, Li J (2008) Proteomic analysis of phosphoproteins regulated by abscisic acid in rice leaves. Biochem Biophys Res Commun 371:883–888

He P, Shan L, Sheen J (2007) Elicitation and suppression of microbe-associated molecular pattern-triggered immunity in plant-microbe interactions. Cell Microbiol 9:1385–1396

Heese A, Hann DR, Gimenez-Ibanez S, Jones AME, He K, Li J, Schroeder JI, Peck SC, Rathjen JP (2007) The receptor-like kinase SERK3/BAK1 is a central regulator of innate immunity in plants. Proc Natl Acad Sci U S A 104:12217–12222

Hrabak EM, Chan CW, Gribskov M, Harper JF, Choi JH, Halford N, Kudla J, Luan S, Nimmo HG, Sussman MR, Thomas M, Walker-Simmons K, Zhu J-K, Harmon AC (2003) The Arabidopsis CDPK-SnRK superfamily of protein kinases. Plant Physiol 132:666–680

Huang J-F, Teyton L, Harper JF (1996) Activation of a $Ca^{2+}$-dependent protein kinase involves intramolecular binding of a calmodulin-like regulatory domain. Biochemistry 35:13222–13230

Huang Y, Li H, Gupta R, Morris PC, Luan S, Kieber JJ (2000) AtMPK4, an Arabidopsis homolog of mitogen-activated protein kinase, is activated *in vitro* by AtMEK1 through threonine phosphorylation. Plant Physiol 122:1301–1310

Iizasa E, Mitsutomi M, Nagano Y (2010) Direct binding of a plant LysM receptor-like kinase, LysM RLK1/CERK1, to chitin in vitro. J Biol Chem 285:2996–3004

Ishihama N, Yamada R, Yoshioka M, Katou S, Yoshioka H (2011) Phosphorylation of the *Nicotiana benthamiana* WRKY8 transcription factor by MAPK functions in the defense response. Plant Cell 23:1153–1170

Johnson RR, Wagner RL, Verhey SD, Walker-Simmons MK (2002) The abscisic acid-responsive kinase PKABA1 interacts with a seed-specific abscisic acid response element-binding factor, TaABF, and phosphorylates TaABF peptide sequences. Plant Physiol 130:837–846

Kanzaki H, Saitoh H, Takahashi Y, Berberich T, Ito A, Kamoun S, Terauchi R (2008) NbLRK1, a lectin-like receptor kinase protein of *Nicotiana benthamiana*, interacts with *Phytophthora infestans* INF1 elicitin and mediates INF1-induced cell death. Planta 228:977–987

Keller T, Damude HG, Werner D, Doerner P, Dixon RA, Lamb CA (1998) A plant homolog of the neutrophil NADPH oxidase gp91phox subunit gene encodes a plasma membrane protein with $Ca^{2+}$ binding motifs. Plant Cell 10:255–266

Khokhlatchev AV, Canagarajah B, Wilsbacher J, Robinson M, Atkinson M, Goldsmith E, Cobb MH (1998) Phosphorylation of the MAP kinase ERK2 promotes its homodimerization and nuclear translocation. Cell 93:605–615

Kim CY, Liu Y, Thorne ET, Yang H, Fukushige H, Gassmann W, Hildebrand D, Sharp RE, Zhang S (2003a) Activation of a stress-responsive mitogen-activated protein kinase cascade induces the biosynthesis of ethylene in plants. Plant Cell 15:2707–2718

Kim K-N, Cheong YH, Grant JJ, Pandey GK, Luan S (2003b) CIPK3, a calcium sensor-associated protein kinase that regulates abscisic acid and cold signal transduction in *Arabidopsis*. Plant Cell 15:411–423

Klimczak LJ, Collinge MA, Farini D, Guiliano G, Walker JC, Cashmore AR (1995) Reconstitution of *Arabidopsis* casein kinase II from recombinant subunits and phosphorylation of transcription factor GBF1. Plant Cell 7:105–115

Kobayashi Y, Yamamoto S, Minami H, Kagaya Y, Hattori T (2004) Differential activation of the rice sucrose nonfermenting 1-related protein kinase2 family by hyperosmotic stress and abscisic acid signaling. Plant Cell 16:1163–1177

Kobayashi Y, Murata M, Minami H, Yamamoto S, Kagaya Y, Hobo T, Yamamoto A, Hattori T (2005) Abscisic acid-activated SNRK2 protein kinases function in the gene-regulation pathway of ABA signal transduction by phosphorylation of ABA response element –binding factors. Plant J 44:939–949

Köhler B, Blatt MR (2002) Protein phosphorylation activates the guard cell $Ca^{2+}$ channel and is a prerequisite for gating by abscisic acid. Plant J 32:185–194

Kovtun Y, Chiu WL, Tena G, Sheen J (2000) Functional analysis of oxidative stress-activated-mitogen-activated protein kinase cascade in plants. Proc Natl Acad Sci U S A 97(6):2940–2945

Kwak SH, Lee SH (1997) The requirements for $Ca^{2+}$, protein phosphorylation, and dephosphorylation for ethylene signal transduction in *Pisum sativum* L. Plant Cell Physiol 38:1142–1149

Lecourieux – Ouaked D, Pugin A, Lebrun-Garcia A (2000) Phosphoproteins involved in the signal transduction of cryptogein, an elicitor of defense reactions in tobacco. Mol Plant Microbe Interact 13:821–829

Lecourieux D, Mazars C, Pauly N, Ranjeva R, Pugin A (2002) Analysis and effects of cytosolic free calcium elevations in response to elicitors in *Nicotiana plumbaginifolia* cells. Plant Cell 14:2627–2641

Lecourieux D, Ranjeva R, Pugin A (2006) Calcium in plant defence-signalling pathways. New Phytol 171:249–269

Li J, Lee Y-R, Assmann SM (1998) Guard cells possess a calcium-dependent protein kinase that phosphorylates the KAT1 potassium channel. Plant Physiol 116:785–795

Li J, Wang XQ, Watson MB, Assman SM (2000) Regulation of abscisic acid-induced stomatal closure and anion channels by guard cell AAPK kinase. Science 287:300–303

Li J, Wen J, Lease KA, Doke JT, Tax FE, Walker JC (2002) BAK1, an Arabidopsis LRR receptor-like protein kinase, interacts with BRI1 and modulates brassinosteroid signaling. Cell 110:213–222

Ligterink W, Hirt T (2000) MAP kinase pathways in plants: versatile signaling tools. Int Rev Cytol 201:209–258

Lipka V, Dittgen J, Bednarek P, Bhat R, Wiermer M, Stein M, Landtag J, Brandt W, Rosahl S, Scheel D, Llorente F, Molina A, Parker J, Somerville S, Schulze-Lefert P (2005) Pre- and post-invasion defenses both contribute to nonhost resistance in Arabidopsis. Science 310:1180–1183

Liu Y, Zhang S (2004) Phosphorylation of 1-aminocyclopropane-1-carboxylic acid synthase by MPK6, a stress-responsive mitogen-activated protein kinase, induces ethylene biosynthesis in *Arabidopsis*. Plant Cell 16:3386–3399

Liu Y, Zhang S, Klessig DF (2000) Molecular cloning and characterization of a tobacco MAP kinase kinase that interacts with SIPK. Mol Plant Microbe Interact 13:118–124

Liu Y, Jin H, Yang K, Kim C, Baker B, Zhang S (2003) Interaction between two mitogen-activated protein kinases during tobacco defense signaling. Plant J 34:149–160

Liu Y, Kong X, Pan J, Li D (2010) VIP1: linking *Agrobacterium*-mediated transformation to plant immunity? Plant Cell Rep 29:805–812

Lohmann GV, Shimoda Y, Nielsen W, Jørgensen FG, Grossmann C, Sandal N, Sørensen K, Thirup S, Madsen LH, Tabata S, Sato S, Stougaard J, Radutoiu S (2010) Evolution and regulation of the *Lotus japonica* LysM receptor gene family. Mol Plant Microbe Interact 23:510–521

Lu C, Han MH, Guevara-Garcia A, Fedoroff V (2002) Mitogen-activated protein kinase signaling in postgermination arrest of development by abscisic acid. Proc Natl Acad Sci U S A 99:15812–15817

Mao G, Meng X, Liu Y, Zheng Z, Chen Z, Zhang S (2011) Phosphorylation of a WRKY transcription factor by two pathogen-responsive MAPKs drives phytoalexin biosynthesis in *Arabidopsis*. Plant Cell 23:1639–1653

Mayrose M, Bonshtien A, Sessa GJ (2004) LeMPK3 is a mitogen-activated protein kinase with dual specificity induced during tomato defense and wounding responses. J Biol Chem 279:14819–14827

Meng X, Xu J, He Y, Yang K-Y, Mordorski B, Liu Y, Zhang S (2013) Phosphorylation of an ERF transcription factor by *Arabidopsis* MPK3/MPK6 regulates plant defense gene induction and fungal resistance. Plant Cell 25:1126–1142

Menke FLH, van Pelt JA, Pieterse CMJ, Klessig DF (2004) Silencing of the mitogen-activated protein kinase MPK6 compromises disease resistance in *Arabidopsis*. Plant Cell 16:897–907

Miya A, Albert P, Shinya T, Desaki Y, Ichimura K, Shirasu K, Narusaka Y, Kawakami N, Kaku H, Shibuya N (2007) CERK1, a LysM receptor kinase, is essential for chitin elicitor signaling in *Arabidopsis*. Proc Natl Acad Sci U S A 104:19613–19618

Montesano M, Brader G, Palva ET (2003) Pathogen-derived elicitors: searching for receptors in plants. Mol Plant Pathol 4:73–78

Murata Y, Pei ZM, Mori IC, Schroeder J (2001) Abscisic acid activation of plasma membrane $Ca^{2+}$ channels in guard cells requires cytosolic NAD(P)H and is differentially disrupted upstream and downstream of reactive oxygen species production in *abi1-1* and *abi2-1* protein phosphatase 2C mutants. Plant Cell 13:2513–2523

Mustilli A-C, Merlot S, Vavasseur A, Fenzi F, Giraudat J (2002) *Arabidopsis* OST1 protein kinase mediates the regulation of stomatal aperture by abscisic acid and acts upstream and downstream of reactive oxygen species production. Plant Cell 14:3089–3099

Nakagami H, Soukupová H, Schikora A, Zársk V, Hirt H (2006) A mitogen-activated protein kinase kinase kinase mediates reactive oxygen species homeostasis in *Arabidopsis*. J Biol Chem 281:38697–38704

Nasrallah JB (2000) Cell-cell signalling in the self-incompatible response. Curr Opin Plant Biol 3:368–373

Navarro L, Zipfel C, Rowland O, Keller I, Robatzek S, Boller T, Jones JD (2004) The transcriptional innate immune response to flg22. Interplay and overlap with Avr gene-dependent defense responses and bacterial pathogenesis. Plant Physiol 135:1113–1128

Nühse TS, Peck SC, Hirt H, Boller T (2000) Microbial elicitors induce activation and dual phosphorylation of the *Arabidopsis thaliana* MAPK 6. J Biol Chem 275:7521–7526

Nühse TS, Boller T, Peck SC (2003) A plasma membrane syntaxin is phosphorylated in response to the bacterial elicitor flagellin. J Biol Chem 278:45248–45254

Nühse TS, Stenballe A, Jensen ON, Peck SC (2004) Phosphoproteomics of the Arabidopsis plasma membrane and a new phosphorylation site database. Plant Cell 16:2394–2405

Ouaked F, Rozhon W, Lecourieux D, Hirt H (2003) A MAPK pathway mediates ethylene signaling in plants. EMBO J 22:1282–1288

Peck SC, Nühse TS, Hess D, Iglesias A, Meins F, Boller T (2001) Directed proteomics identifies a plant-specific protein rapidly phosphorylated in response to bacterial and fungal elicitors. Plant Cell 13:1467–1475

Pei ZM, Ward JM, Harper JF, Schroeder JI (1996) A novel chloride channel in *Vicia faba* guard cell vacuoles activated by the serine/threonine kinase, CDPK. EMBO J 15:6564–6574

Petersen M, Brodersen P, Naested H, Andreasson E, Lindhart U, Johansen B, Nielsen HB, Lacy M, Austin MJ, Parker JE, Sharma SB, Klessig DF, Martienssen R, Mattsson O, Jensen AB, Mundy J (2000) *Arabidopsis* MAP kinase 4 negatively regulates systemic acquired resistance. Cell 103:1111–1120

Piedras P, Hammond-Kosack KE, Harrison K, Jones JDG (1998) Rapid, *Cf-9-* and *Avr9*-dependent production of active oxygen species in tobacco suspension cultures. Mol Plant Microbe Interact 11:1155–1166

Postel S, Kufner I, Beueter C, Mazzotta S, Schwedt A, Borlotti A, Halter T, Kemmerling B, Nürnberger T (2010) The multifunctional leucine-rich repeat receptor kinase BAK1 is implicated in *Arabidopsis* development and immunity. Eur J Cell Biol 89:169–174

Ren D, Yang H, Zhang S (2002) Cell death mediated by MAPK is associated with hydrogen peroxide production in *Arabidopsis*. J Biol Chem 277:559–565

Rentel MC, Lecourieux D, Ouaked F, Usher SL, Petersen L, Okamoto H, Knight H, Peck SC, Grierson CS, Hirt H, Knight MR (2004) OXI1 kinase is necessary for oxidative burst-mediated signaling in *Arabidopsis*. Nature 427:858–861

Robatzek S, Chinchilla D, Boller T (2006) Ligand-induced endocytosis of the pattern recognition receptor FLS2 in *Arabidopsis*. Genes Dev 20:537–542

Roberts MR (2000) Regulatory 14-3-3 protein-protein interactions in plant cells. Curr Opin Plant Biol 3:400–405

Roberts MR, Bowles DJ (1999) Fusicoccin, 14-3-3 proteins, and defense responses in tomato plants. Plant Physiol 119:1243–1250

Romeis T, Piedras P, Zhang S, Klessig DF, Hirt H, Jones JD (1999) Rapid Avr9- and Cf-9 – dependent activation of MAP kinases in tobacco cell cultures and leaves: convergence of resistance gene, elicitor, wound, and salicylate responses. Plant Cell 11:273–287

Romeis T, Piedras P, Jones JD (2000) Resistance gene-dependent activation of a calcium-dependent protein kinase in the plant defense response. Plant Cell 12:803–815

Romeis T, Ludwig AA, Martin R, Jones JD (2001) Calcium-dependent protein kinases play an essential role in a plant defence response. EMBO J 20:5556–5567

Ryan CA, Pearce G (2003) Systemins: a functionally defined family of peptide signals that regulate defensive genes in Solanaceae species. Proc Natl Acad Sci U S A 100:14577–14580

Saijo Y (2010) ER quality control of immune receptors and regulators in plants. Cell Microbiol 12:716–724

Schaller A, Oecking C (1999) Modulation of plasma membrane $H^+$-ATPase activity differentially activates wound and pathogen defense responses in tomato plants. Plant Cell 11:263–272

Schaller GE, Harmon AC, Sussman MR (1992) Characterization of a calcium- and lipid-dependent protein kinase associated with the plasma membrane of oat. Biochemistry 31:1721–1727

Schulze B, Mentzel T, Jehle A, Mueller K, Beeler S, Boller T, Felix G, Chinchilla D (2010) Rapid heteromerization and phosphorylation of ligand-activated plant transmembrane receptors and their associated kinase BAK1. J Biol Chem 285:9444–9451

Seo S, Sano H, Ohashi Y (1999) Jasmonate-based wound signal transduction requires activation of WIPK, a tobacco mitogen-activated protein kinase. Plant Cell 11:289–298

Sheen J (1996) $Ca^{2+}$-dependent protein kinase and stress signal transduction in plants. Science 274:1900–1902

Shimizu T, Nakano T, Takamizawa D, Desaki Y, Ishi-Minami N, Nishizawa Y, Minami E, Okada K, Yamane H, Kaku H, Shibuya N (2010) Two LysM receptor molecules, CEBiP and OsCERK1, cooperatively regulate chitin elicitor signaling in rice. Plant J 64:204–214

Shiu S-H, Bleecker AB (2001a) Plant receptor-like kinase gene family: diversity, function, and signaling. Sci STKE 113:RE22

Shiu S-H, Bleecker AB (2001b) Receptor-like kinases from Arabidopsis form a monophyletic gene family related to animal receptor kinases. Proc Natl Acad Sci U S A 98:10763–10768

Song CP, Agarwal M, Ohta M, Guo Y, Halfter U, Wang P, Zhu JK (2005) Role of an *Arabidopsis* AP2/EREBP-type transcriptional repressor in abscisic acid and drought stress responses. Plant Cell 17:2384–2396

Spanu P, Grosskopf DG, Felix G, Boller T (1991) The apparent turnover of 1-aminocyclopropane-1-carboxylate synthase in tomato cells is regulated by protein-phosphorylation and dephosphorylation. Plant Physiol 106:529–535

Stein M, Dittgen J, Sanchez-Rodriquez C, Hou BH, Molina A, Schulze-Lefert P, Lipka V, Somerville S (2006) *Arabidopsis* PEN3/PDR8, an ATP binding cassette transporter, contributes to non-host resistance to inappropriate pathogens that enter by direct penetration. Plant Cell 18:731–746

Stone JM, Walker JC (1995) Plant protein kinase families and signal transduction. Plant Physiol 108:451–457

Subramaniam R, Desprěs C, Brisson N (1997) A functional homolog of mammalian protein kinase C participates in the elicitor-induced defense response in potato. Plant Cell 9:653–664

Sugiyama Y, Uraji M, Watanabe-Sugimoto M, Okuma E, Munemasa S, Shimoishi Y, Nakamura Y, Mori IC, Iwai S, Murata Y (2012) FIA functions as an early signal component of abscisic acid signal cascade in *Vicia faba* guard cells. J Exp Bot 63:1357–1365

Suzuki K, Shinshi H (1995) Transient activation and tyrosine phosphorylation of a protein kinase in tobacco cells treated with fungal elicitor. Plant Cell 7:639–647

Takahashi F, Yoshida R, Ichimura K, Mizoguchi T, Seo S, Yonezawa M, Maruyama K, Yamaguchi-Shinozaki K, Shinozaki K (2007) The mitogen-activated protein kinase cascade MKK3-MPK6 is an important part of the jasmonate signal transduction pathway in Arabidopsis. Plant Cell 19:805–818

Tavernier E, Wendehenne J-P, Blein J-P, Pugin A (1995) Involvement of free calcium in action of cryptogein, a proteinaceous elicitor of a hypersensitive reaction in tobacco cells. Plant Physiol 109:1025–1031

Tena G, Boudsocq M, Sheen J (2011) Protein kinase signalling networks in plant innate immunity. Curr Opin Plant Biol 14:519–529

Testernik C, Munnik T (2005) Phosphatidic acid: a multifunctional stress signaling lipid in plants. Trends Plant Sci 10:368–375

Tischner R, Koltermann M, Haesse H, Plath M (2010) Early responses of *Arabidopsis thaliana* to infection by *Verticillium longisporum*. Physiol Mol Plant Pathol 74:419–427

Torres MA, Dangl JL (2005) Functions of the respiratory burst oxidase in biotic interactions, abiotic stress and development. Curr Opin Plant Biol 8:397–403

Torres MA, Dangl JL, Jones JDG (2002) Arabidopsis gp91phox homologues *AtrbohD* and *AtrbohF* are required for accumulation of reactive oxygen intermediates in plant defense response. Proc Natl Acad Sci U S A 99:517–522

Uno Y, Furihata T, Abe H, Yoshida R, Shinozaki K, Yamaguchi-Shinozaki K (2000) *Arabidopsis* basic leucine zipper transcription factors involved in abscisic acid-dependent signal transduction pathway under drought and high-salinity conditions. Proc Natl Acad Sci U S A 97:11632–11637

Vera-Estrella R, Barkla BJ, Higgins VJ, Blumwald E (1994a) Plant defense response to fungal pathogens. Activation of host-plasma membrane $H^+$-ATPase by elicitor-induced enzyme dephosphorylation. Plant Physiol 104:209–215

Vera-Estrella R, Higgins VJ, Blumwald E (1994b) Plant defense responses to fungal pathogens. II. G protein mediated changes in host plasma membrane redox reactions. Plant Physiol 106:97–103

Viard MP, Martin F, Pugin A, Blein J-P (1994) Protein phosphorylation is induced in tobacco cells by the elicitor cryptogein. Plant Physiol 104:1245–1249

Vidhyasekaran P (2007) Fungal pathogenesis in plants and crops: molecular biology and host defense mechanisms, IIth edn. CRC Press/Taylor & Francis Group, Boca Raton, p510

Vitart V, Christodoulou J, Huang JF, Chazin WJ, Harper JF (2000) Intramolecular activation of a $Ca^{2+}$-dependent protein kinase is disrupted by insertions in the tether that connects the calmodulin-like domain to the kinase. Biochemistry 39:4004–4011

Waller F, Müller A, Chung K-M, Yap Y-K, Nakamura K, Weiler E, Sano H (2006) Expression of a WIPK-activated transcription factor results in increase of endogenous salicylic acid and pathogen resistance in tobacco plants. Plant Cell Physiol 47:1169–1174

Wan J, Zhang S, Stacey G (2004) Activation of a mitogen-activated protein kinase pathway in *Arabidopsis* by chitin. Mol Plant Pathol 5:125–135

Wan J, Zhang X-C, Neece D, Ramonell KM, Clough S, Kim S-Y, Stacey MG, Stacey G (2008a) A LysM receptor-like kinase plays a critical role in chitin signaling and fungal resistance in *Arabidopsis*. Plant Cell 20:471–481

Wan J, Zhang X-C, Stacey G (2008b) Chitin signaling and plant disease resistance. Plant Signal Behav 3:831–833

Wang P, Song C-P (2008) Guard-cell signaling for hydrogen peroxide and abscisic acid. New Phytol 178:703–718

Wang XQ, Ullah H, Jones AM, Assmann SM (2001) G protein regulation of ion channels and abscisic acid signaling in *Arabidopsis* guard cells. Science 292:2070–2072

Wang KLC, Li H, Ecker IP (2002) Ethylene biosynthesis and signaling networks. Plant Cell 14:S131–S155

Wang X, Goshe MB, Soderblom EJ, Phinney BS, Kuchar JA, Li J, Asami T, Yoshida S, Huber SC, Clouse SD (2005a) Identification and functional analysis of in vivo phosphorylation sites of the *Arabidopsis* BRASSINOSTEROID-INSENSITIVE 1 receptor kinase. Plant Cell 17:1685–1703

Wang X, Li X, Meisenhelder J, Hunter T, Yoshida S, Asami T, Chory J (2005b) Autoregulation and homodimerization are involved in the activation of the plant steroid receptor BRI1. Dev Cell 8:855–865

Wang W, Esch JJ, Shiu S-H, Agula H, Binder BM, Chang C, Patterson S, Bleecker AB (2006) Identification of important regions for ethylene binding and signaling in the transmembrane domain of the ER1 ethylene receptor of *Arabidopsis*. Plant Cell 18:3429–3442

Ward JM, Pei Z-M, Schroeder JI (1995) Roles of ion channels in initiation of signal transduction in higher plants. Plant Cell 7:833–844

Wendehenne D, Lamotte O, Frachisse J-M, Barbier-Brygoo H, Pugin A (2002) Nitrate efflux is an essential component of the cryptogein signaling pathway leading to defense responses and hypersensitive cell death in tobacco. Plant Cell 14:1937–1951

Xing T, Higgins VJ, Blumwald E (1996) Regulation of plant defense response to fungal pathogens: two types of protein kinases in the reversible phosphorylation of the host plasma membrane $H^+$-ATPase. Plant Cell 8:555–564

Xing T, Higgins VJ, Blumwald E (1997) Identification of G proteins mediating fungal elicitor-induced dephosphorylation of host plasma membrane $H^+$-ATPase. J Exp Biol 48:229–237

Xing Y, Jia W, Zhang J (2007) AtMEK1 mediates stress-induced gene expression of CAT1 catalase by triggering $H_2 0_2$ production in *Arabidopsis*. J Exp Bot 58:2969–2981

Xing Y, Jia W, Zhang J (2008) AtMKK1 mediates ABA-induced *CAT1* expression and $H_2O_2$ production via AtMPK6-coupled signaling in *Arabidopsis*. Plant J 54:440–451

Xiong I, Yang Y (2003) Disease resistance and abiotic stress tolerance in rice are inversely modulated by an abscisic acid-inducible mitogen-activated protein kinase. Plant Cell 15:745–759

Yaffe MB (2002) How do 14-3-3 proteins work? – Gatekeeper phosphorylation and the molecular anvil hypothesis. FEBS Lett 513:53–57

Yaffe MB, Rittinger K, Volinia S, Caron PR, Aitken A, Leffers H, Gamblin SJ, Smerdon SJ, Cantley LC (1997) The structural basis for 14-3-3 phosphopeptide binding specificity. Cell 91:961–971

Yamamoto S, Suzuki K, Shinshi H (1999) Elicitor-responsive, ethylene-independent activation of GCC box-mediated transcription that is regulated by both protein phosphorylation and dephosphorylation in cultured tobacco cells. Plant J 20:571–579

Yang KY, Liu YD, Zhang SQ (2001) Activation of a mitogen-activated protein kinase pathway is involved in disease resistance in tobacco. Proc Natl Acad Sci U S A 98:741–746

Yoshida R, Hobo T, Ichimura K, Mizoguchi T, Takahashi S, Takahashi F, Alonso J, Ecker JR, Shinozaki K (2002) ABA-activated SnRK2 protein kinase is required for dehydration stress signaling in *Arabidopsis*. Plant Cell Physiol 43:1473–1483

Yoshida R, Umezawa T, Mizoguchi T, Takahashi S, Takahashi F, Shinozaki K (2006) The regulatory domain of SRK2E/OST1/SnRK2.6 interacts with ABI1 and integrates abscisic acid (ABA) and osmotic stress signals controlling stomatal closure in *Arabidopsis*. J Biol Chem 281:5310–5318

Yoshioka K, Sugie K, Park HJ, Maeda H, Tsuda N, Kawakita K, Doke N (2001) Induction of plant gp91 phox homolog by fungal cell wall, arachidonic acid, and salicylic acid in potato. Mol Plant Microbe Interact 14:725–736

Yoshioka H, Numata N, Nakajima K, Katou S, Kawakita K, Rowland O, Jones JD, Doke N (2003) *Nicotiana benthamiana* gp91$^{phox}$ homologs *NbrbohA* and *NbrbohB* participate in $H_2O_2$ accumulation and resistance to *Phytophthora infestans*. Plant Cell 15:706–718

Zegzouti H, Jones B, Frasse P, Marty C, Maitre B, Latch A, Pech JC, Bouzayen M (1999) Ethylene-regulated gene expression in tomato fruit: characterization of novel ethylene-responsive and ripening-related genes isolated by differential display. Plant J 589–600

Zhang S, Liu Y, Klessig DF (2000) Multiple levels of tobacco WIPK activation during the induction of cell death by fungal elicitins. Plant J 23:339–347

Zhang L, Liu BF, Liang S, Jones RL, Lu YT (2002) Molecular and biochemical characterization of a calcium/calmodulin-binding protein kinase from rice. Biochem J 368:145–167

Zhang AY, Jiang MY, Zhang JH, Tan M, Hu XL (2006) Mitogen-activated protein kinase is involved in abscisic acid-induced antioxidant defense and acts downstream of reactive oxygen species production in leaves of maize plants. Plant Physiol 141:475–487

Zhang X, Dai Y, Xiong Y, DeFraia C, Li J, Dong X, Mou Z (2007) Overexpression of Arabidopsis *MAP kinase kinase 7* leads to activation of plant basal and systemic acquired resistance. Plant J 52:1066–1079

Zipfel C, Robatzek S, Navarro L, Oakeley EJ, Jones JDG, Felix G, Boller T (2004) Bacterial disease resistance in *Arabidopsis* through flagellin perception. Nature 428:764–767

# Chapter 10
# Ubiquitin-Proteasome System-Mediated Protein Degradation in Defense Signaling

**Abstract** Ubiquitin-proteasome system plays important role in the complex PAMP signal transduction systems involved in plant innate immunity. JAZ proteins, which are activated by jasmonate (JA) signals, act as repressors of JA-dependent transcription factors. COI1 protein, a receptor of JA signal, forms a functional E3 ubiquitin ligase and is required for removal of repressors of the JA signaling pathway. Ubiquitin proteins-cullin-RING ligases negatively regulate biosynthesis of ethylene. Ethylene signal transduction terminates in a transcription cascade involving the EIN3/EIL and ERF families of transcription factors and ubiquitin ligases regulate the stability and expression of these transcription factors. Ubiquitin proteasome may positively or negatively regulate SA biosynthesis. A 26S proteosome is involved in triggering SA accumulation, probably by removing/degrading an inhibitor of SA biosynthesis. Ubiquitin ligases are also involved in regulation of R proteins-mediated defense responses. Major function of ubiquitin ligases may be in conferring stability to R proteins, probably by degrading the proteins involved in reducing the stability of R proteins. Small ubiquitin-like modifier (SUMO) plays a significant role in SA-mediated systemic acquired resistance. Ubiquitin-proteasome is involved in triggering defense responses and virulent pathogens may subvert ubiquitin-proteasome system to cause disease.

**Keywords** JAZ proteins • COI1 protein • Ubiquitin ligases • 26S proteasome • SUMO

## 10.1 Ubiquitin-Proteasome System in Plants

Ubiquitin- and proteasome-mediated degradation of proteins plays an important role in plant defense signaling system (Dreher and Callis 2007; van den Burg et al. 2008; Trujillo et al. 2008; Sahana et al. 2012; Yao and Ndoja 2012; Yao et al. 2012; Zhang et al. 2012). More than 1,300 genes identified in the *Arabidopsis thaliana* genome have been shown to be involved in the ubiquitin-proteasome pathway,

P. Vidhyasekaran, *PAMP Signals in Plant Innate Immunity: Signal Perception and Transduction*, Signaling and Communication in Plants 21,
DOI 10.1007/978-94-007-7426-1_10, 

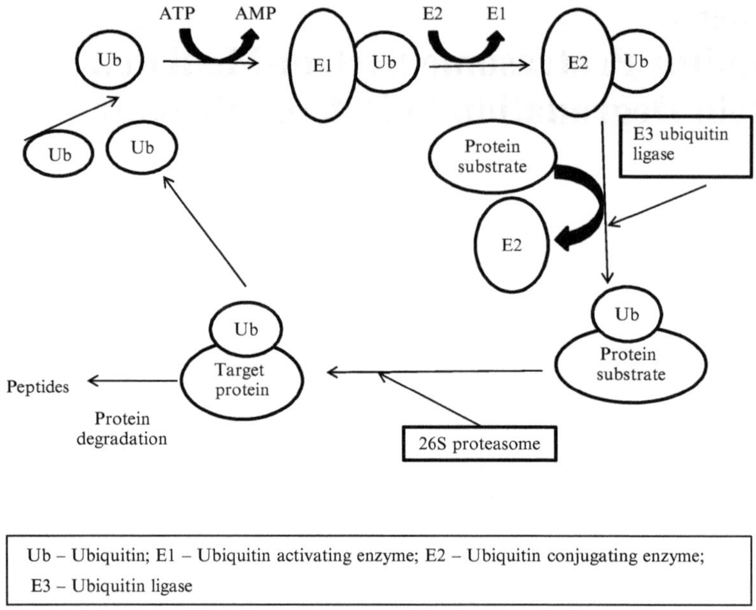


**Fig. 10.1** Ubiquitin–proteasome pathway

suggesting that this pathway is one of the most elaborate regulatory mechanisms in plants (Vierstra 2003; Serrano et al. 2006). Proteasomes are large protein complexes located in the nucleus and the cytoplasm (Peters et al. 1994). The main function of the proteasome is to degrade unneeded or damaged proteins by proteolysis. Proteasomes regulate the concentration of particular proteins and degrade misfolded proteins. Proteins are tagged for degradation by a small protein called ubiquitin. The ubiquitin protein is 76 amino acids long and was named due to its ubiquitous nature (Pickart and Eddins 2004).

Ubiquitin acts as a covalent molecular tag and its attachment requires three distinct enzymatic activities. Proteins are targeted for degradation by the proteasome by covalent modification of a lysine residue that requires the coordinated reactions of three enzymes. In the first step, a ubiquitin-activating enzyme, E1, activates ubiquitin C-terminal carboxyl group by adenylation, and then forms a thioester bond with cysteinyl sulfhydryl residue on the E1 protein itself. The ubiquitin-activating enzyme E1 activates ubiquitin in an ATP-dependent manner (Fig. 10.1; Lee et al. 2011). The adenylated ubiquitin is then transferred to a cysteine of a second enzyme, ubiquitin-conjugating enzyme (E2). In the last step, a member of a highly diverse class of enzymes known as ubiquitin ligases (E3) recognizes the specific protein to be ubiquitinated and catalyzes the transfer of ubiquitin from E2 to this target protein (Haas et al. 1982; Clechanover 1998; Pickart and Eddins 2004; Smalle and Vierstra 2004; Dreher and Callis 2007). A target protein must be labeled with at least four ubiquitin monomers in the form of a polyubiquitin chain before it is recognized by the proteasome lid (Thrower et al. 2000). Ubiquitin-ubiquitin linkages may serve as proteolytic signals (Kirkpatrick et al. 2006).

The common 26S proteasome contains one 20S core particle structure and two 19S regulatory caps. The core is hollow and provides an enclosed cavity in which proteins are degraded; openings at the two ends of the core allow the target protein to enter. Each end of the core particle associates with a 19S regulatory subunit that contains multiple ATPase active sites and ubiquitin binding sites. This structure recognizes polyubiquitinated proteins and transfers them to the catalytic core. After delivery to the proteasome mediated in part by ubiquitin binding proteins, the polyubiquitylated substrate can be deubiquitylated by the proteasome's regulatory cap or associated proteases. The deubiquitylated substrate is fed into the proteolytic core of the proteasome where it is cleaved into small peptides (Book et al. 2005; Zhu et al. 2005; Dreher and Callis 2007).

E3 ligases play a key role in the ubiquitin-proteasome system (UPS). Two mechanistic classes of E3 ligases have been recognized. In case of HECT (for Homologous to E6-AP COOH terminus) domain E3 ligases, ubiquitin forms a covalent thioester linkage with a cysteinyl sulfhydryl group on HECT protein before being transferred to a lysine on the substrate (Downes et al. 2003). The other E3 ligase class non-covalently interacts with an E2 protein carrying ubiquitin. There are two groups within this class, the U-box (UFD2-homology) domain- and RING (for Really Interesting New Gene) domain-containing proteins (Zheng et al. 2000; Pickart 2001; Andersen et al. 2004). Several proteins containing U-box and RING domains have been reported in plants. Approximately 61 proteins containing U-box domain and more than 450 proteins with one or more RING domains have been identified in *Arabidopsis thaliana* (Stone et al. 2005).

The RING domain has a consensus sequence containing Cys and His residues (Cys-$X_2$-Cys-$X_{9\text{-}39}$-Cys-$X_{1\text{-}3}$-His-$X_{2\text{-}3}$-Cys/His-$X_2$-Cys-$X_{4\text{-}48}$-Cys-$X_2$-Cys), which functions as a binding site for the ubiquitin-conjugating enzyme (E2) intermediate that has a zinc-binding domain formed by conserved Cys and His residues (Lee et al. 2011). Based on the presence of Cys or His in the fifth position, the RING domains of RING finger type proteins can be divided into two types (Borden and Freemont 1996).

Major types of E3 ligases belong to the Skp, Cullin, F-box containing complex (SCF complex), which are composed of four primary subunits: SKP1 (for S-phase Kinase-associated Protein 1); a cullin family member protein (e.g. CUL1); a RING finger protein; and an F-box protein (Guo and Ecker 2003; Potuschak et al. 2003; Risseeuw et al. 2003; Dreher and Callis 2007). Within this complex, the F-box protein directly binds the substrate through protein-protein interaction domains, the cullin binds the RING finger protein, together they recruit E2 ubiquitin-conjugating enzyme, and SKP1 helps to link the F-box protein and cullin (Deshaies 1999). The cullins are modified covalently by NEDDB/RUB1, a ubiquitin-like protein, in a process called neddylation. This modification stimulates SCF ubiquitin ligase activity in plants (Kawakami et al. 2001; del Pozo et al. 2002).

A nucleus-enriched multisubunit protein complex, called "COP9 signalosome (CSN)" is known to regulate ubiquitin-proteasome-mediated protein degradation

(Feng et al. 2003). CSN is commonly present in plants. Several molecular studies have shown the presence of six loci, *COP9*, *FUS/COP11*, *FUS5*, *FUS4/COP8*, *FUS11*, and *FUS12*, encoding subunits of the COP9 signalosome, which are CSN8, CSN1, CSN7, CSN4, CSN3, and CSN2, respectively in plants (Feng et al. 2003). CSN is associated with multiple SCF-type E3 ubiquitin ligases (Wang et al. 2003). CSN shows deneddylation activity toward the neddylated cullin subunit of SCF complexes, which is important for SCF ubiquitin ligase activity (Yang et al. 2002).

## 10.2 Ubiquitin-Proteasome in Jasmonate Signaling System

### *10.2.1 JAZ Proteins Act as Repressors of JA Signaling Pathway*

Jasmonate (JA) signaling system plays important role in plant innate immunity (Lozano-Durán et al. 2011; Qi et al. 2011; Zhang et al. 2012). In the absence of elicitor signals, the JA signaling pathway is generally repressed by a family of jasmonate ZIM (for ZING FINGER PROTEIN EXPRESSED IN INFLORESCENCE MERISTEM) domain (JAZ) proteins (Chini et al. 2007; Thines et al. 2007), which recruit the corepressor TOPLESS through the linker NOVEL INTERACTOR OF JAZ (Pauwels et al. 2010). JAZ family of proteins and related JAI3 (for JASMONATE-INSENSITIVE 3) protein have been identified as key suppressors of jasmonate signaling (Chini et al. 2007; Thines et al. 2007). The JAI3 protein contains a ZIM domain and hence it was renamed as JASMONATE ZIM DOMAIN3 (JAZ3; Chini et al. 2007). Within 30 min of JA treatment, several genes encoding individual members of the JAZ protein family showed strong induction in *Arabidopsis* (Mandaokar et al. 2006). JAZ1 and JAZ3/JAI3 are the repressors of the JA signaling pathway (Chini et al. 2007; Thines et al. 2007). Both JAZ1 and JAZ3/JAI3 each interact with JIN1 (JASMONATE INSENSITIVE1, also known as MYC2 [MYELOCYTOMATOSIS2]) (Chini et al. 2007). *JIN1/MYC2* encodes a basic helix-loop-helix-type transcription factor involved in the transcriptional regulation of JA-responsive gene expression (Lorenzo et al. 2004). It is suggested that, in the absence of a JA signal, JAZ1 and JAZ3 repress JIN1/MYC2. The *Arabidopsis* bHLH transcription factors MYC3 and NYC4 are targets of JAZ repressors and act additively with MYC2 in the activation of jasmonate response (Fernández-Calvo et al. 2011).

### *10.2.2 JA Signaling Pathway Is Activated by the Removal of the JAZ Repressor Proteins by Ubiquitination*

Once activated by stress signals, JA is rapidly synthesized and further converted into numerous conjugates, including the highly bioactive (+)-7-iso-jasmonoyl-L-isoleucine (JA-Ile) (Thines et al. 2007; Fonseca et al. 2009). Synthesized JA and its bioactive conjugates are perceived by a receptor complex consisting of CORONATINE INSENSITIVE1 (COI1), JAZs, and inositol pentakisphosphate (Katsir et al. 2008; Sheard et al. 2010). *COI1* (*coronatine insensitive 1*) gene has been shown required for jasmonate-regulated defense signaling in *Arabidopsis* (Xie et al. 1998). The *Arabidopsis* null mutant for *COI1-1* is fully insensitive to jasmonates, and the COI1 protein is required for all JA-dependent responses (Feys et al. 1994; Xie et al. 1998).

The COI1 protein has been identified as an F-box protein, suggesting the involvement of ubiquitin-proteasome-mediated protein degradation in JA signaling (Xie et al. 1998). *COI1* forms a functional E3 ubiquitin ligase, $SCF^{COI1}$, in plants (Devoto et al. 2002; Xu et al. 2002). COI1 is present in the functional SKIP-CULLIN-F-box-type E3 ubiquitin ligase complex (Devoto et al. 2002). COI1 associates physically with CUL1, RBX1, and SKP1-like proteins to assemble $SCF^{COI1}$ complexes (Xu et al. 2002). A single amino acid substitution in the F-box motif of COI1 abolishes the formation of $SCF^{COI1}$ complexes and affects JA-inducible gene expression. These observations suggest that $SCF^{COI1}$ complex is important for JA signaling (Xu et al. 2002).

The SCF complex is involved in marking proteins with ubiquitin tags to facilitate their degradation by the 26S proteasome (Stone and Callis 2007). JA-related phenotypes have been identified in plants with mutations in components of the ubiquitin proteasome system and these mutations suggested a linkage between JA signaling and ubiquitin-mediated protein degradation (Lorenzo and Solano 2005; Moon et al. 2007). It is suggested that COI1 is required for removal of repressors of the JA signaling pathway (Xie et al. 1998; Balbi and Devoto 2008).

The mode of action of $SCF^{COI1}$ in the JA-regulated defense signaling pathway is still not known. It has been shown that several genes such as *Cev*, *Cet*, *Cex* encode proteins, which negatively regulate JA pathway. The recessive mutants of these genes constitutively activate JA pathway (Ellis and Turner 2001; Hilpert et al. 2001; Xu et al. 2001). It is suggested that these genes might encode negative regulators, which might act as putative repressors to negatively regulate the expression of their downstream target genes that are involved in the JA response (Xu et al. 2002). In response to external signals, the JA signal might be activated to modify some or all of these regulatory proteins, conceivably through phosphorylation or dephosphorylation. COI1 could recruit the modified proteins to $SCF^{COI1}$ for ubiquitination. Subsequent degradation of the ubiquitynated substrates by the 26S proteasome would result in removal of the putative repressors, leading to the expression of the downstream defense genes (Xu et al. 2002).

Histone acetylation has been demonstrated to be involved in ubiquitin-proteasome-mediated proteolytic pathway. Regulation of histone deacetylases (HDACs) by ubiquitination has been demonstrated (Gaughan et al. 2005). It was suggested that COI1 may form a functional E3-type ubiquitin ligase in plants to regulate expression of a HDAC (Devoto and Turner 2003). The interaction of COI1 with SKP1-like proteins and HDAC *in planta* has been demonstrated (Vierstra 2003).

Arabidopsis HDA6, was shown to interact with COI1, an F-box protein which interacts with SKP1 and cullin proteins to form E3 ubiquitin ligases known as the SCF complexes that selectively recruit regulatory proteins targeted for ubiquitination

(Devoto et al. 2002; Vierstra 2003). It suggests that COI1 may form a functional E3-type ubiquitin ligase in plants to regulate expression of jasmonate responsive genes by targeted ubiquitination of a histone deacetylase (Devoto and Turner 2003; Zhou et al. 2005).

Ubiquitin-proteasome-mediated proteolytic pathway may activate JA-inducible transcription factors. The COI1 activated four JA-inducible transcription factors, which include WRKY18, At1g74930, At3g53600, and AtMYC2 in Arabidopsis. These transcription factors positively regulated the jasmonate-mediated signaling system (Wang et al. 2008; Kazan and Manners 2013). Both JAZ1 and JAZ3/JAI3 each interact with JIN1 (JASMONATE INSENSITIVE1, also known as MYC2 [MYELOCYTOMATOSIS2]) (Chini et al. 2007). *JIN1/MYC2* encodes a basic helix-loop-helix-type transcription factor involved in the transcriptional regulation of JA-responsive gene expression (Lorenzo et al. 2004). It is suggested that, in the absence of a JA signal, JAZ1 and JAZ3 repress JIN1/MYC2. Upon sensing of the JA signals, JAZ repressors are recruited to the SCF E3 complex for ubiquitination and subsequent degradation by the proteasome. The removal of these repressors then paves the way for JIN1/MYC2 to regulate JA-dependent gene expression.

JA-Ile, but not jasmonic acid itself or MeJA, promotes the interaction between $SCF^{COI1}$-JAZ complexes (Thines et al. 2007). This finding suggests that jasmonic acid may not be the signal directly responsible for the activation of the JA signaling pathway, but possibly it undergoes further modifications to be converted to a biologically active signal. *JAR1* (*JASMONATE RESISTANT1*) encodes a JA-amino synthetase, which activates conjugation of JA to Ile (Staswick and Tiryaki 2004). JA-Ile produced by JAR1 promotes the interaction between JAZ and $SCF^{COI1}$ and takes part in the removal of repressors of JA signaling pathway. $SCF^{COI1}$ targets key regulators of JA signaling pathway for ubiquitination and subsequent degradation by the 26S proteasome (Kazan and Manners 2008; Melotto et al. 2008; Kawamura et al. 2009).

The COP9 signalasome (CSN) has been shown to be necessary for the function of $SCF^{COI1}$ ubiquitin ligase. COP9 signalosome associates physically with $SCF^{COI1}$ (Feng et al. 2003). Most of the COI1-dependent JA-responsive genes also required CSN function, and CSN abundance was shown to be important for JA responses (Feng et al. 2003). These observations suggest that both CSN and $SCF^{COI1}$ work together to control genome expression and promote JA responses.

### *10.2.3 Ubiquitin-Proteasome–Mediated Proteolysis in JA Signaling System*

Downstream events in the JA signaling pathway are controlled by repressors of downstream transcriptional networks and the degradation of these repressors is the early downstream event (Kazan and Manners 2008). Ubiquitin-proteasome-mediated proteolysis has been shown to be involved in jasmonate signaling system (Xu et al. 2002; Feng et al. 2003). Proteome studies have revealed the participation of a ubiquitin-conjugating protein in the JA-signal transduction pathway (Hondo et al. 2007). A cell wall protein fraction (CWP) elicitor derived from the biocontrol agent *Pythium oligandrum* induced expression of *LeATL6* gene encoding ubiquitin-ligase enzyme E3 and triggered the synthesis of PR-6 and TPI-1 defense-related proteins in tomato via JA-dependent signaling system (Hondo et al. 2007). The role of the ubiquitin ligase3 enzyme in the JA signaling system was demonstrated by overexpressing *LeATL6* under the control of the *Cauliflower mosaic virus* 35S promoter in tomato plants. Overexpression of the gene induced the defense genes *PR-6* and *TPI-1* in wild tomato but not in the *jai-1* mutant in which the JA-mediated signaling pathway was impaired (Hondo et al. 2007). It suggests that LeATL6 may be a part of the JA- signal transduction system. *LeATL6* expression was induced by elicitor treatment in *jai-1* mutant tomato cells; however, JA-dependent expression of the basic PR-6 and TPI-1 genes was not induced in elicitor-treated *jai-1* mutants. These results indicated that ubiquitin ligase E3 (LeATL6) may act upstream of JA signaling (Hondo et al. 2007).

The expression of *LeATL6*, which encodes RING-H2 zinc finger ubiquitin ligase E3 was highly induced in tomato roots treated with the elicitor from *P. oligandrum* (Takahashi et al. 2010). The target protein of *LeATL6* was identified as S-adenosylmethionine decarboxylase (SAMDC) (Takahashi et al. 2010), which is involved in biosynthesis of polyamines (Kresge et al. 2007). Polyamines are known to act in the JA-signaling system (Chen et al. 2004). The elicitor suppressed the activity of SAMDC in treated tomato roots. The interaction of SAMDC with LeATL6 and the decreased SAMDC activity may be associated with JA-dependent induced resistance in tomato treated with *P. oligandrum* (Takahashi et al. 2010).

Two RING-type ubiquitin ligases, RGLG3 and RGLG4, have been found to be essential for JA-mediated responses in *Arabidopsis*. Both RGLG3 and RGLG4 possessed ubiquitin ligase activities (Zhang et al. 2012). Altered expression of *RGLG3* and *RGLG4* affected JA-inducible gene expression. The ubiquitin ligases have been found to act as upstream modulators of JA signaling (Zhang et al. 2012).

## 10.3 Ubiquitin-Proteasome in Ethylene Signaling System

Ethylene signaling system is an important component in defense signaling (Iwai et al. 2006; Binder et al. 2007; Ralph et al. 2007; Dreher and Callis 2007). Ubiquitin-proteasome proteolytic pathway has been shown to be involved in both regulation of ethylene biosynthesis and downstream activation of transcription factors, leading to transcription of defense genes. ACC synthase (ACS) isozymes have been shown to be substrates for E3 ligases (Dreher and Callis 2007). Arabidopsis ethylene overproduction mutants (*eto2* and *eto3*) are shown to bear mutations in the ACS genes, *ACS5* and *ACS9*, respectively (Vogel et al. 1998; Chae et al. 2003).

Bostick et al. (2004) showed that silencing of two ubiquitin-related proteins, RUB1 and RUB2, which modify cullin RING ligases and regulate their activity, caused overproduction of ethylene in *Arabidopsis*. Similarly, mutation of the RUB-conjugating enzyme RCE1 also increased ethylene production (Larsen and Cancel 2004). These results suggest that ubiquitin proteins-cullin RING ligases negatively regulate biosynthesis of ethylene.

Ethylene signal transduction terminates in a transcription cascade involving the EIN3/EIL (ETHYLENE INSENSITIVE 3/EIN3-LIKE) and ERF (Ethylene Responsive Factor) families of plant-specific transcription factors (Potuschak et al. 2003; van Loon et al. 2006). EIN3 is expressed constitutively, but is unable to accumulate because it is subjected to permanent proteolysis mediated by two Arabidopsis SCF complex F box proteins called EBF1 and EBF2 (for EIN3-Binding F box protein 1 and 2). The F-box proteins specifically recruit the target EIN proteins to an SCF ubiquitin ligase for degradation by the proteasome (Potuschak et al. 2003). EIN3 increases in response to increased levels of ethylene (Guo and Ecker 2003). EIN3 is rapidly degraded by ubiquitylation by the $SCF^{EBF1/EBF2}$through a proteasome-mediated pathway, but is stabilized upon ethylene treatment (Guo and Ecker 2003; Kepinski and Leyser 2003). EIN3 becomes stabilized after perception of ethylene and acts on its target promoters (Potuschak et al. 2003). These observations suggest that ethylene signaling action depends on EIN3 protein stabilization and proteolytic regulation of this protein may affect transcription of the defense-related genes (Potuschak et al. 2003).

ERFs are plant-specific transcription factors detected in tobacco, tomato, and Arabidopsis. They have been shown to bind nucleotide sequences containing the GCC box, the core sequence of an ethylene-responsive element of defense genes and regulate the expression of GCC box-mediated transcriptions (Fujimoto et al. 2000). ERF proteins are grouped into three classes based on amino acid sequence identities within the ERF domain. Class I and class III ERFs act as activators, whereas class II ERFs act as repressors (Ohta et al. 2000; Koyama et al. 2003). Class II ERF repressors down-regulate the transactivation activity of class I and class II ERFs (Fujimoto et al. 2000). Ubiquitin-proteasome system has been shown to be involved in the repression of class II ERFs (Koyama et al. 2003).

In tobacco, the ERF3 gene coding for a class II repressor as well as genes for activators such as ERF2 and ERF4 were transcriptionally upregulated in response to ethylene (Ohme-Takagi and Shinshi 1995; Kitajima et al. 2000). These ERF genes for both activators and a repressor were both rapidly induced by a fungal elicitor treatment (Yamamoto et al. 1999). A ubiquitin-conjugating enzyme (NtUBC2) was found to be involved in the repression activity of ERF3 (Koyama et al. 2003). The ubiquitin-conjugation activity of NtUBC2 may be involved in the regulation of repression activity of ERF3. The NtUBC2 interacted with ERF3 but not with ERF2 or ERF4. This suggests that the mechanism of regulation of the repression activity of ERF3 is distinct from that of the activation activity of ERF2 and ERF4. Since ERF repressor can suppress transactivation activity of ERF activators (Koyama et al. 2003), down-regulation of the repression activity of ERF3 by NtUBC2 may be operating for the induction of the GCC box-mediated transcription of defense genes (Koyama et al. 2003). Thus the interaction between ERF3 and NtUBC2 may be a critical step in activating transcription of various defense genes.

# 10.4 Ubiquitin-Proteasome in SA Signaling System

## *10.4.1 Ubiquitin-Proteasome May Be Involved in Regulation of SA Levels in the SA Signaling System*

Protein degradation through ubiquitin-proteasome pathway has been shown to play an important role in SA-regulated defense signaling. Ubiquitin proteasome may positively or negatively regulate SA accumulation. SGT1 (for Suppressor of the G2 allele of SKP1) associates with SKP1, a component of the SCF-type E3 complexes (Kitagawa et al. 1999; Liu et al. 2002; Peart et al. 2002). SGT1 plays key role in ubiquitin-proteasome-mediated proteolytic pathway (Seo et al. 2008). *SGT1* is involved in basal defense response besides effector triggered immunity (ETI). Silencing *GmSGT1-2* impaired resistance to virulent bacterial pathogens and systemic acquired resistance (SAR) in soybean (Fu et al. 2009). Overexpression of *OsSGT1* in rice significantly induced basal resistance to both the rice bacterial blight pathogen *Xanthomonas oryzae* pv. *oryzae* and the fungal pathogen *Magnaporthe oryzae* (Wang et al. 2008). *SGT1* genes have been shown to be required for SA accumulation in *Arabidopsis* for induction of disease resistance (Zhou et al. 2008). SGT1 triggers expression of various SA-regulated defense-related genes including *PR-1*, *PR-2*, *PR-5*, *RPW8.1*, *RPW8.2*, *WRKY6*, *WRKY29*, and *EDS1* (Zhou et al. 2008). The results suggest that ubiquitin-proteasome mediated proteolytic pathway positively regulates SA levels and triggers SA signaling system triggering defense responses.

RPN1a, a 26S proteasome subunit has been shown to be required for SA – mediated innate immunity in *Arabidopsis* (Yao et al. 2012). EDR2 (ENHANCED DISEASE RESISTANCE2) negatively regulates SA-based defense response against the powdery mildew pathogen *Golovinomyces cichoracearum* (Vorwerk et al. 2007). Loss-of-function mutations in *EDR2* lead to enhanced resistance against the pathogen (Yao et al. 2012). Mutations in the gene encoding RPN1a, a subunit of the 26S proteasome, suppressed *edr2*-associated disease resistance. RPN1a also has been shown to be required for *edr1* and *pmr4* (recessive *R* gene)-mediated powdery mildew resistance. The *rpn1a* mutant displayed enhanced susceptibility against both *G. cichoracearum* and the bacterial pathogen *Pseudomonas syringae* pv. *tomato*. The *rpn1a* mutant showed defects in SA accumulation upon *P. syringae* pv. *tomato* infection (Yao et al. 2012). The results suggest that the 26S proteosome is involved in triggering SA accumulation, probably by removing/degrading an inhibitor of SA biosynthesis.

The pepper E3 ubiquitin ligase RING1 gene, *CaRING1*, has been shown to regulate SA accumulation in pepper plants (Lee et al. 2011). Overexpression of *CaRING1* gene in *Arabidopsis thaliana* showed enhanced accumulation of SA, while in pepper plants, virus-induced gene silencing of *CaRING1* lowered SA levels. The results suggest that *CaRING1* modulates SA levels. The E3 ubiquitin ligase activity of *CaRING1* modulated SA signaling in innate immune system. Overexpression of *CaRING1* in *A. thaliana* conferred enhanced resistance to the bacterial pathogen

*Pseudomonas syringae* pv. *tomato* and to the oomycete pathogen *Hyaloperonospora arabidopsidis* infections. In pepper plants, silencing of *CaRING1* conferred enhanced susceptibility to avirulent *Xanthomonas campestris* pv. *vesicatoria* infection accompanied by reduced expression of SA-dependent *PR-1* gene expression (Lee et al. 2011). These results suggest that the E3 ubiquitin ligase positively regulates SA accumulation and SA-mediated defense responses.

By contrast, ubiquitin-proteasome system has been shown to negatively regulate SA levels in tobacco. Transgenic tobacco plants expressing an inhibitor of ubiquitin-dependent protein degradation have been developed (Conrath et al. 1998). These plants constitutively accumulated enhanced levels of salicylic acid and/or its glucoside. These transgenic plants showed enhanced resistance to *Tobacco mosaic virus* (Conrath et al. 1998). The results suggest that ubiquitin –proteasome pathway negatively regulates SA accumulation in the defense signaling in tobacco.

### 10.4.2 Role of an E3 Ubiquitin Ligase, OsRHC1, in SA-Dependent NPR1 Signaling

A novel RING zinc finger protein (OsRHC1) was detected in rice and it was identified as an E3 ubiquitin ligase. Its function was dependent on the ubiquitin-mediated protein degradation via the 26S proteasome (Cheung et al. 2007). *OsRHC1* cDNA from rice in transgenic *Arabidopsis thaliana* enhanced the defense response toward *Pseudomonas syringae* pv. *tomato*, suggesting its role in defense signaling in rice (Cheung et al. 2007). The defense response effects were neutralized in an *npr1* mutation background, suggesting that the function of OSRHC1 is dependent on presence of the key defense regulator NPR1 (Cheung et al. 2007). NPR1 is master regulator of SA signaling and an important regulator of responses downstream of SA (Mou et al. 2003; Zhang et al. 2003; Chern et al. 2008). The activity of NPR1 has been shown as a prerequisite for the functioning of OsRHC1. Therefore, it is suggested that OsRHC1 acts either upstream from the NPR1 in the signal transduction pathway or it acts on a negative regulator of the NPR1 pathway (Cheung et al. 2007).

### 10.4.3 Role of SON1 (F-Box Protein in E3 Ubiquitin-Ligase Complex) in SA–Mediated Immune Responses

The gene SON1 (for Suppressor of NIM1-1) was cloned from *Arabidopsis thaliana*. It was found to encode a protein containing an F-box motif, an element found in the E3 ubiquitin-ligase complex. The gene negatively regulates defense signaling system (Kim and Delaney 2002). The *son1* mutant exhibited resistance response. The *son1* plants that contain a functional *NPR1* gene revealed a constitutive increase in PR genes expression. In contrast, *son1 npr1* double mutant plants do not show induction of PR gene expression (Kim and Delaney 2002). These results suggest that SON1 represses NPR1-dependent PR gene expression. SON1, the ubiquitin-ligase, may target for degradation of different regulators of PR gene expression. Candidates for such a regulatory factor would be proteins that displace negative transcriptional regulators of PR genes whose expression may be associated with *son1*-mediated defense responses. The target of SON1 may be NPR1 itself (Kim and Delaney 2002).

## 10.5 Ubiquitin-Proteasome in *R*-Gene Mediated Early Signaling System

Several *R* genes encode receptor proteins (Kawchuck et al. 2001) and on perception of the external stimuli by the receptor, a well-orchestrated signaling system is activated. Ubiquitin-proteasome-mediated proteolytic pathway plays an important role in the rice R protein XA21-mediated signaling system (Wang et al. 2006; Yang et al. 2006). TheXA21 protein carries both a leucine-rich repeat motif and a serine-threonine kinase-like domain, suggesting a role in cell surface recognition of a pathogen ligand and subsequent activation of an intracellular defense response (Song et al. 1995). Lee et al. (2009) reported that XA21 is a pattern recognition receptor (PRR) and it recognizes a 194-amino acid protein designated Ax21 as a pathogen-associated molecular pattern (PAMP).

A RING domain-containing protein named as XB3 (for XA21 binding protein 3) interacts with the rice bacterial blight resistance gene Xa21-encoded protein (Wang et al. 2006). The RING domain ubiquitinates XB3 protein, indicating that XB3 is an E3 ubiquitin ligase. XB3 is specifically transphosphorylated by the kinase domain of XA21 (Wang et al. 2006). The activated XB3 may ubiquitinate a third protein and target its degradation. The degraded protein may be a negative regulator of the defense signaling (Wang et al. 2006). It is suggested that XB3 protein may be involved in conferring stability to the R protein XA21; otherwise, the XA21 protein would have been degraded by a ubiquitin-proteasome-mediated proteolytic pathway. Thus the ubiquitin ligase activity of XB3 protein is involved in the *R* gene-mediated signaling system.

Cf-9 is a disease resistance gene in tomato conferring resistance against *Cladosporiun fulvum*. Transgenic tobacco plants expressing this gene were developed. Upon an elicitor treatment, the gene ACRE276 (Avr9/Cf-9 Rapidly Elicited gene 276) was upregulated and the gene encoded an E3 ubiquitin ligase requiring an intact U-box domain (Yang et al. 2006). *ACRE276* RNA interference (RNAi) silencing in tobacco resulted in loss of defense response induced by *Cf* resistance genes. *ACRE276* RNAi plants also lost defense responses induced by a viral elicitor (Yang et al. 2006). Another ACRE gene, ACRE74, was also found to be induced upon elicitor treatment in tobacco (González-Lamothe et al. 2006). ACRE74 encodes a U-box E3 ligase homolog, highly related to parsley CMPG1 (Kirsch et al. 2001) and *Arabidopsis thaliana* PLANT U-BOX20 (PUB20) and PUB21 proteins (Azevedo et al. 2001), and was called NtCMPG1. The NtCMPG1 was shown to be involved in induction of HR (hypersensitive response) in Cf9 tobacco after AVR9 elicitor infiltration (González-Lamothe et al. 2006). A homolog of CMPG1 was detected in tomato and it conferred resistance to the fungal pathogen *Cladosporium fulvum*. It was shown to be involved in the Pto/AvrPto-mediated bacterial disease resistance and Infl–mediated oomycete resistance responses in tomato

(González-Lamothe et al. 2006). *CMPG1* gene was also rapidly induced after elicitation with the oomycete elicitor Pep-13 in parsley (Kirsch et al. 2001). Arabidopsis PUB20 expression is induced after treatment with the oomycete pathogen elicitor Pmg and bacterial elicitor flg22 (Heise et al. 2002; Navarro et al. 2004). These results suggest the involvement of U-box E3 ligases in defense signaling system.

Another U-box protein was detected in *Arabidopsis* and it was called PUB17 (PLANT U-BOX17) and it is a homolog of tobacco ARCE276 (Yang et al. 2006). *Arabidopsis* PUB17 knockout plants lost the *R* genes *RPM1*- and *RPS4*-mediated resistance against *Pseudomonas syringae* pv. *tomato*. Transiently expressing PUB17 in Cf-9 tobacco silenced for *ACRE276* restored defense-related hypersensitive response (Yang et al. 2006). These observations suggest the U-box E3 ubiquitin ligase PUB17 is also involved in defense signaling system. Another U-box protein involved in ubiquitin- proteasome proteolytic pathway is ARC1 in *Brassica napus* (Yang et al. 2006).

SGT1 (for Suppressor of the G2 allele of SKP1) is a plant disease resistance response protein required for the function of multiple *R* genes (Austin et al. 2002; Azevedo et al. 2002; Peart et al. 2002; Tör et al. 2002; Fu et al. 2009; Kud et al. 2013). It is a co-chaperone in the HSP90-SGT1-RAR1 molecular chaperone complex, a core modulator in plant immunity (Seo et al. 2008). SGT1 associates with SKP1, a component of the SCF-type E3 complexes (Liu et al. 2002). It is required for the function of an SCF complex (Kitagawa et al. 1999; Peart et al. 2002). It suggests that SGT1 plays key role in ubiquitin-proteasome-mediated proteolytic pathway. SGT1 plays an important regulatory role in early *R*-gene-mediated plant defense responses (Austin et al. 2002; Azevedo et al. 2002; Bieri et al. 2004; Wang et al. 2006; Seo et al. 2008). SGT1 functions in R protein accumulation in disease resistance (Azevedo et al. 2006). Silencing of SGT1 in *Nicotiana benthamiana* results in reduced steady-state levels of R proteins (Azevedo et al. 2006). It suggests that the major function of SGT1 may be in conferring stability to R proteins, probably by degrading the proteins involved in reducing the stability of R proteins. It has been shown that SGT1 contributes to the *Prf*-mediated defense responses by stabilizing Prf protein via its co-chaperone activity (Kud et al. 2013).

## 10.6 Small Ubiquitin-Like Modifier (SUMO) in Plant Immunity

### *10.6.1 Role of SUMOylation in SA Biosynthesis*

SUMO is a post-translational modification that can be reversibly conjugated to target proteins, similar to its well-known cousin Ubiquitin (Miura and Hasegawa 2010). In contrast to ubiquitinylation, reversible SUMO conjugation only requires the actions of the SUMO activating enzymes SAE1/SAE2, the SUMO conjugating enzyme SCE1 and SUMO proteases. SUMO conjugation is promoted by SUMO E3 ligases, such as SIZ1 or HPY2 (NSE1) (Ishida et al. 2009; Miura and Hasegawa 2010).

SUMOylation plays an important role in SA signaling system. It alters the levels of SA in the immune response pathway. Several regulatory proteins are involved in upstream of SA signaling and involved in SA production in *Arabidopsis*. Mutations in *eds1* (for *enhanced disease susceptibility1*), or *pad4* (for *phytoalexin-deficient4*) lead to reduced SA levels in infected leaves (Zhou et al. 1998; Gupta et al. 2000; Feys et al. 2001). EDS1 is required for SA production and it controls SA production to amplify defense signals (Rustérucci et al. 2001; Eulgem et al. 2004; Song et al. 2004). PAD4 is a key regulator acting at upstream of SA (Lippok et al. 2007). *Arabidopsis* plants carrying *pad4* mutations have a defect in accumulation of SA upon pathogen infection (Zhou et al. 1998). PAD4 is required for amplification of weak signals to a level sufficient for activation of SA signaling (Jirage et al. 1999). The PAD4 protein sequence displays similarity to triacyl glycerol lipases and other esterases (Jirage et al. 1999). It is suggested that EDS1 and PAD4 transduce ROS-derived signals leading to SA production (Rustérucci et al. 2001; Wiermer et al. 2005). EDS1 and PAD4 may have a fundamental role in transducing redox signals. EDS1 forms several molecularly and spatially distinct complexes with PAD4 (Wiermer et al. 2005; Xing and Chen 2006). Another gene *EDS5* encodes a protein, which transports precursors for SA biosynthesis. EDS5 exhibits homology to multidrug and toxin extrusion (MATE) transporter proteins from animals (Nawrath et al. 2002). EDS5 expression requires PAD4, placing EDS5 downstream of PAD4 (Nawrath et al. 2002). SIZ1 gene, which encodes an Arabidopsis SUMO E3 ligase, regulates SA-mediated plant immunity. SIZ1 interacts epistatically with PAD4 and EDS1 and inhibits the SA biosynthesis pathway (Lee et al. 2006).

### 10.6.2 Role of SUMOylation in SA-Mediated Systemic Acquired Resistance

SUMO conjugation suppresses defense signaling in unstressed healthy plants without any exposure to PAMP signals or pathogen invasion. Upon pathogen/PAMP perception plant innate immune receptors activate various signaling pathways that trigger host defenses. SUMO (Small ubiquitin-like modifier) conjugation is essential to suppress defense signaling in non-infected plants. SUMO conjugation can transform transcription activators into repressors, thereby preventing defense induction in the absence of a pathogen signal (van der Burg and Takken 2010).

SAR is activated upon recognition of pathogens and activation of SAR requires SA, which induces SA-responsive gene expression. The SA-induced changes in gene expression have been found to have a link to chromatin remodeling, such as histone modifications and histone replacement. The recruitment of chromatin-modifying complexes to SA-responsive loci controls their basal and SA-induced expression (March-Diaz et al. 2008; van den Burg and Takken 2009, 2010; Jaskiewicz et al. 2011). Basal repression of these loci may require the post-translational modifier SUMO. SUMO conjugation has been reported to control the activity, assembly and disassembly of chromatin-modifying complexes to transcription complexes (van den Burg and Takken 2009). SUMO conjugation determines recruitment and activity of chromatin-modifying enzymes, and thereby indirectly controls gene expression (van der Burg and Takken 2010).

The *SIZ1* gene, which encodes an *Arabidopsis* SUMO E3 ligase, regulates SAR. Mutant *siz1* plants exhibit constitutive SAR characterized by elevated accumulation of salicylic acid and increased resistance to *Pseudomonas syringae* pv. *tomato*. Transfer of the *NahG* gene to *siz1* plants results in reversal of these phenotypes back to wild-type. Analyses of the double mutants, *npr1 siz1*, *pad4 siz1*, *ndr1 siz1* revealed that *SIZ1* controls SA signaling (Lee et al. 2007). The results suggest that SUMOylation suppresses SA signaling at the level of transcription. It is suggested that SUMO conjugation of the SA-dependent transcription factors may transform the transcription factors into repressors. Pathogen invasion/PAMP application, which activates phosphorylation, may revert back the transcription repressors into transcription activators (van den Burg and Takken 2010). Sumoylation transforms an activator into a repressor and phosphorylation induced by MAP kinases may convert the repressor into activator. The interaction between sumoylation and phosphorylation may determine the immune response. Plants with disturbed SUMOylation levels exhibit constitutive expression of early and late defense genes, increased accumulation of SA, and increased disease resistance (Lee et al. 2007; van den Burg and Takken 2009).

## 10.7 Pathogens May Subvert Ubiquitin-Proteasome System to Cause Disease

Ubiquitin-proteasome system has been shown to be involved in plant immune system. Many E3 ligase class RING finger proteins have been found to be induced in plants by pathogen attack and play an important role in plant defense (Zeng et al. 2006). The fungal PAMP chitin up-regulated the *ATL2* and *ATL6* genes encoding RING proteins in *Arabidopsis* (Salinas-Mondragón et al. 1999). The bacterial PAMP flg22 upregulated the genes encoding ten putative RING finger E3 ligases in *Arabidopsis* (Navarro et al. 2004). The fungal PAMP N-acetyloligosaccharide has been shown to induce the expression of EL5 RING E3 ligase (Takai et al. 2002). Overexpression of the pepper E3 ubiquitin ligase RING1 gene confers enhanced resistance against *Pseudomonas syringae* pv. *tomato* and *Hyaloperonospora arabidopsidis* in *Arabidopsis*, which was accompanied by rapid production of SA and various PR genes (Lee et al. 2011). U-box E3 ligases have also been implicated in plant defense response (Zeng et al. 2004; González-Lamothe et al. 2006; Yang et al. 2006). Some E3 ubiquitin ligases have been shown to negatively regulate plant defense responses (Hong et al. 2004; Trujillo et al. 2008; Zhang et al. 2012).

Recently it has been reported that some pathogens interfere with the ubiquitin-proteasome system and subvert the defense responses induced by the proteolysis to cause disease. Park et al. (2012) showed that the rice blast pathogen *Magnaporthe oryzae* produces an effector called AvrPiz-t. The effector targets the RING E3 ubiquitin ligase APIP6 to suppress PAMP-triggered immunity in rice. AvrPiz-1 accumulates in the specialized structure called the biotrophic interfacial complex and is then translocated into rice cells. AvrPiz-t suppressed the ubiquitin ligase activity of the rice RING E3 ubiquitin ligase. Silencing of the ubiquitin ligase in transgenic rice enhanced susceptibility of rice plants to *M. oryzae*, suggesting the role of the ubiquitin ligase in disease resistance (Park et al. 2012). The results suggest that the fungal pathogen causes the disease by suppressing the ubiquitin-proteasome.

DNA viruses selectively interfere with CUL1-based SCF ubiquitin E3 ligases to cause infection in plants (Lozano-Durán and Bejarano 2011; Lozano-Durán et al. 2011). The DNA viruses redirect ubiquitination by interfering with the activity of the COP9 signalosome (CSN) complex. The geminiviral C2 protein interacts with CSN5 (COP9 signalosome5), and its expression in transgenic plants compromises CSN activity on CUL1. Several responses regulated by the CUL1-based SCF ubiquitin E3 ligases (including responses to jasmonates, auxins, gibberellins, ethylene, and abscisic acid) are altered in these plants (Lozano-Durán et al. 2011). Transcriptomic analysis of the transgenic plants showed that the response to jasmonates is the main SCF-dependent process affected by geminiviral C2 protein. Exogenous JA treatment of *Arabidopsis* plants disrupts geminivirus infection suggesting that the suppression of the JA response might be crucial for infection. SCFs are key regulators of JA signaling. The capability of viruses to selectively interfere with or hijack the activity of these complexes may be a powerful strategy in viral infection (Lozano-Durán et al. 2011).

Sahana et al. (2012) showed that a cell permeable proteasomal inhibitor facilitated an increase in *Papaya ringspot virus* (PRSV) accumulation in the host papaya plants. The PRSV viral protein HcPro was found to interact with the α1 subunit of the 20S proteasome, inhibiting the action of the 20S proteasome. The results suggest that inhibition of the host proteasome facilitates the virus accumulation in the host plant and the proteosomal catalytic activity is modulated by the viral protein. Collectively, these studies suggest that proteasome is involved in virus disease resistance and the potential viral pathogens inhibit the proteasome activity and induce the disease incidence infection.

## References

Andersen P, Kragelund BB, Olsen AN, Larsen FH, Chua NH, Poulsen FM, Skriver K (2004) Structure and biochemical function of a prototypical Arabidopsis U-box domain. J Biol Chem 279:40053–40061

Austin MJ, Muskett P, Kahn K, Feys BJ, Jones JD, Parker JE (2002) Regulatory role of *SGT1* in early *R* gene-mediated plant defenses. Science 295:2077–2080

Azevedo C, Santos-Ross MJ, Shirasu K (2001) The U-box protein family in plants. Trends Plant Sci 6:354–358

Azevedo C, Sadanandom A, Kitagawa K, Freialdenhoven A, Shirasu K, Schulze-Lefert P (2002) The RAR1 interactor SGT1, an essential component of *R* gene-triggered disease resistance. Science 295:2073–2076

Azevedo C, Betsuyaku S, Peart J, Takahashi A, Noel L, Sadanandom A, Casals C, Parker J, Shirasu K (2006) Role of SGT1 in resistance protein accumulation in plant immunity. EMBO J 25:2007–2016

Balbi V, Devoto A (2008) Jasmonate signalling network in *Arabidopsis thaliana*: crucial regulatory nodes and new physiological scenarios. New Phytol 177:301–318
Bieri S, Mauch S, Shen QH, Peart J, Devoto A, Casals C, Ceron F, Schulze S, Steinbiss HH, Shirasu K, Schulze-Lefert P (2004) RAR1 positively controls steady state levels of barley MLA resistance proteins and enables sufficient MLA6 accumulation for effective resistance. Plant Cell 16:3480–3495
Binder BM, Walker JM, Gagne JM, Emborg TJ, Hemmann G, Bleecker AB, Vierstra RD (2007) The Arabidopsis EIN3 binding F-Box proteins EBF1 and EBF2 have distinct but overlapping roles in ethylene signaling. Plant Cell 19:509–523
Book AJ, Yang P, Scalf M, Smith LM, Viersta RD (2005) Tripeptidyl peptidase II. An oligomeric protease complex from *Arabidopsis*. Plant Physiol 138:1046–1057
Borden KL, Freemont PS (1996) The RING finger domain: a recent example of a sequence-structure family. Curr Opin Struct Biol 55:395–401
Bostick M, Lochhead SR, Honda A, Palmer S, Callis J (2004) Related to ubiquitin 1 and 2 are redundant and essential and regulate vegetative growth, auxin signaling, and ethylene production in *Arabidopsis*. Plant Cell 16:2418–2432
Chae HS, Faure F, Kieber JJ (2003) The *eto1*, *eto2*, and *eto3* mutations and cytokinin treatment increase ethylene biosynthesis in Arabidopsis by increasing the stability of ACS protein. Plant Cell 15:545–559
Chen H, McCaig BC, Melotto M, He SY, Howe GA (2004) Regulation of plant argininase by wounding, jasmonate, and the phytotoxin coronatine. J Biol Chem 279:45998–46007
Chern M, Canlas PE, Ronald PC (2008) Strong suppression of systemic acquired resistance in *Arabidopsis* by NRR is dependent on its ability to interact with NPR1 and its putative repressive domain. Mol Plant 1:552–559
Cheung M-Y, Zeng N-Y, Tong S-W, Li FW-Y, Zhao K-J, Zhang Q, Sun SS-M, Lam H-M (2007) Expression of a RING-HC protein from rice improves resistance to *Pseudomonas syringae* pv. *tomato* DC3000 in transgenic *Arabidopsis thaliana*. J Exp Bot 58:4147–4159
Chini A, Fonseca S, Fernández G, Adie B, Chico JM, Lorenzo O, Garcia-Casado G, López-Vidriero I, Lozano FM, Ponce MR, Micol JL, Solano R (2007) The JAZ family of repressors is the missing link in jasmonate signaling. Nature 448:666–671
Clechanover A (1998) The ubiquitin-proteasome pathway: on protein death and cell life. EMBO J 17:7151–7160
Conrath U, Klessig DF, Bachmair A (1998) Tobacco plants perturbed in the ubiquitin-dependent protein degradation system accumulate callose, salicylic acid, and pathogenesis-related protein 1. Plant Cell Rep 17:876–880
del Pozo JC, Dharmasiri S, Hellmann H, Walker L, Gray WM, Estelle M (2002) AXR1-ECR1-dependent conjugation of RUB1 to the *Arabidopsis* cullin AtCUL1 is required for auxin response. Plant Cell 14:421–433
Deshaies RJ (1999) SCF and cullin/ring H2-based ubiquitin ligases. Annu Rev Cell Dev Biol 15:435–467
Devoto A, Turner JC (2003) Regulation of jasmonate-mediated plant responses in Arabidopsis. Ann Bot 92:329–337
Devoto A, Nieto-Rostro M, Xie D, Ellis C, Harmston R, Patrick E, Davis J, Sherratt L, Coleman M, Turner JG (2002) COI1 links jasmonate signaling and fertility to the SCF ubiquitin-ligase complex in *Arabidopsis*. Plant J 32:457–466
Downes BP, Stupar RM, Gingerich DJ, Vierstra RD (2003) The HECT ubiquitin-protein ligase (UPL) family in *Arabidopsis*: UPL3 has a specific role in trichome development. Plant J 35:729–742
Dreher K, Callis J (2007) Ubiquitin, hormones and biotic stress in plants. Ann Bot 99:787–822
Ellis C, Turner JG (2001) The Arabidopsis mutant *cev1* has constitutively active jasmonate and ethylene signal pathways and enhanced resistance to pathogens. Plant Cell 13:1025–1033
Eulgem T, Weighman VJ, Chang H-S, McDowell JM, Holub EB, Glazebrook J, Zhu T, Dangl JL (2004) Gene expression signatures from three genetically separable resistance gene signaling pathways for downy mildew resistance. Plant Physiol 135:1129–1144

Feng S, Ma L, Wang X, Xie D, Dinesh-Kumar SP, Wei N, Deng XW (2003) The COP9 signalosome interacts physically with $SCF^{COI1}$ and modulates jasmonate responses. Plant Cell 15:1083–1094

Fernández-Calvo P, Chini A, Fernández-Barbero G, Chico J-M, Gimenez-Ibanez S, Geerinck J, Eeckhout D, Schweizer F, Godoy M, Franco-Zorrilla JM, Pauwels L, Witters E, Puga MI, Paz-Ares J, Goossens A, Reymond P, De Jaeger G, Solano R (2011) The *Arabidopsis* bHLH transcription factors $MYC_3$ and $MYC_4$ are targets of JAZ repressors and act additively with $MYC_2$ in the activation of jasmonate responses. Plant Cell 23:701–715

Feys BJF, Benedetti CE, Penfold CN, Turner JG (1994) Arabidopsis mutants selected for resistance to the phytotoxin coronatine are male sterile, insensitive to methyl jasmonate, and resistant to a bacterial pathogen. Plant Cell 6:751–759

Feys BJ, Moisan LJ, Newman MA, Parker JE (2001) Direct interaction between the Arabidopsis disease resistance signaling proteins, EDS1 and PAD4. EMBO J 20:5400–5411

Fonseca S, Chini A, Hamberg M, Adie B, Porzel A, Kramell R, Miersch O, Wasternack C, Solano R (2009) (+)-7-iso-Jasmonoyl-L-isoleucine is the endogenous bioactive jasmonate. Nat Chem Biol 5c:344–350

Fu DQ, Ghabrial S, Kachroo A (2009) *GmRAR1* and *Gm SGT1* are required for basal, *R* gene-mediated and systemic acquired resistance in soybean. Mol Plant Microbe Interact 22:86–95

Fujimoto SY, Ohta M, Usui A, Shinshi H, Ohme-Takagi M (2000) Arabidopsis ethylene-responsive element binding factors act as transcriptional activators or repressors of GCC box-mediated gene expression. Plant Cell 12:393–404

Gaughan L, Logan IR, Neal DE, Robson CN (2005) Regulation of androgen receptor and histone deacetylase 1 by Mdm2-mediated ubiquitylation. Nucleic Acids Res 33:13–26

González-Lamothe R, Tsitsigiannis DJ, Ludwig AA, Panicot M, Shirasu K, Jones JD (2006) The U-box protein CMPG1 is required for efficient activation of defense mechanisms triggered by multiple resistance genes in tobacco and tomato. Plant Cell 18:1067–1083

Guo H, Ecker JR (2003) Plant responses to ethylene gas are mediated by $SCF^{(EBF1/EBF2)}$-dependent proteolysis of EIN3 transcription factor. Cell 115:667–677

Gupta V, Willits MG, Glazebrook J (2000) *Arabidopsis thaliana* EDS4 contributes to salicylic acid (SA)-dependent expression of defense responses: evidence for inhibition of jasmonic acid signaling by SA. Mol Plant Microbe Interact 13:503–511

Haas AL, Warms JV, Hershko A, Rose IA (1982) Ubiquitin-activating enzyme: mechanism and role in protein-ubiquitin conjugation. J Biol Chem 257:2543–2548

Heise A, Lippok B, Kirsch C, Hahlbrock K (2002) Two immediate-early pathogen-responsive members of the *AtCMPG* gene family in *Arabidopsis thaliana* and the W-box-containing elicitor-response element of *AtCMPG1*. Proc Natl Acad Sci U S A 99:9049–9054

Hilpert B, Bohlmann H, op den Camp R, Przybyla D, Miersch O, Cuchala A, Apel K (2001) Isolation and characterization of signal transduction mutants of *Arabidopsis thaliana* that constitutively activate the octadecanoid pathway and form necrotic microlesions. Plant J 26:1–14

Hondo D, Hase S, Kanayama Y, Yoshikawa N, Takenaka S, Takahashi H (2007) The *LeATL6-associated* ubiquitin/proteasome system may contribute to fungal elicitor-activated defense response via the jasmonic acid-dependent signaling pathway in tomato. Mol Plant Microbe Interact 20:72–81

Hong JK, Jung HW, Lee BK, Lee SC, Lee YK, Hwang BK (2004) An osmotin-like protein gene, *CAOSM1*, from pepper: differential expression and in situ localization of its mRNA during pathogen infection and abiotic stress. Physiol Mol Plant Pathol 64:301–310

Ishida T, Fujiwara S, Miura K, Stacey N, Yoshimura M, Schneider K, Adachi S, Minamisawa K, Umeda M, Sugimoto K (2009) E3 ligase HIGH PLOIDY2 regulates endocycle onset and meristem maintenance in *Arabidopsis*. Plant Cell 21:2284–2297

Iwai T, Miyasaka A, Seo S, Ohashi Y (2006) Contribution of ethylene biosynthesis for resistance to blast fungus infection in young rice plants. Plant Physiol 142:1202–1215

Jaskiewicz M, Conrath U, Peterhänsel C (2011) Chromatin modification acts as a memory for systemic acquired resistance in the plant stress response. EMBO Rep 12:50–55

Jirage D, Tootle TL, Reuber TL, Frost LN, Feys BJ, Parker JE (1999) *Arabidopsis thaliana* PAD4 encodes a lipase-like gene that is important for salicylic acid signaling. Proc Natl Acad Sci U S A 96:13583–13588

Katsir L, Schilmiller A, Staswick P, He SY, Howe GA (2008) COI1 is a critical component of a receptor for jasmonate and the bacterial virulence factor coronatine. Proc Natl Acad Sci U S A 105:7100–7105

Kawakami T, Chiba T, Suzuki T, Iwai K, Yamanaka K (2001) NEDD8 recruits E2-ubiquitin to SCF E3 ligase. EMBO J 20:4003–4012

Kawamura Y, Hase S, Takenaka S, Kanayama Y, Yoshioka H, Kamoun S, Takahashi H (2009) INF1 elicitin activates jasmonic acid- and ethylene-mediated signaling pathways and induces resistance to bacterial wilt disease in tomato. J Phytopathol 157:287–297

Kawchuck LM, Hachey J, Lynch DR, Kulcsar F, van Rooijen G, Waterer DR, Robertson A, Kokko E, Byers R, Howard RJ, Fischer R, Prufer D (2001) Tomato *Ve* disease resistance genes encode cell surface-like receptors. Proc Natl Acad Sci U S A 98:6511–6515

Kazan K, Manners JM (2008) Jasmonate signaling: toward an integrated view. Plant Physiol 146:1459–1468

Kazan K, Manners JM (2013) MYC2: the master in action. Mol Plant 6:686–703

Kepinski S, Leyser O (2003) SCF-mediated proteolysis and negative regulation in ethylene signaling. Cell 115:647–648

Kim HS, Delaney TP (2002) Arabidopsis SON1 is an F-box protein that regulates a novel induced defense response independent of both salicylic acid and systemic acquired resistance. Plant Cell 14:1469–1482

Kirkpatrick DS, Hathaway NA, Hanna J, Elsasser S, Rush J, Finley D, King RW, Gygi SP (2006) Quantitative analysis of in vitro ubiquitinated cyclin B1 reveals complex chain topology. Nat Cell Biol 8:700–710

Kirsch C, Logemann E, Lippok B, Schmelzer E, Hahlbrock K (2001) A highly specific pathogen-responsive promoter element from the immediate-early activated *CMPG1* gene in *Petroselinum crispum*. Plant J 26:217–227

Kitagawa K, Skowyra D, Elledge SJ, Harper JW, Hieter P (1999) SGT1 encodes an essential component of the yeast kinetochore assembly pathway and a novel subunit of the SCF ubiquitin ligase complex. Mol Cell 4:21–33

Kitajima S, Koyama T, Ohme-Takagi M, Shinshi H, Sato F (2000) Characterization of gene expression of NsERFs, transcription factors of basic PR genes from *Nicotiana sylvestris*. Plant Cell Physiol 41:817–824

Koyama T, Okada T, Kitajima S, Ohme-Takagi M, Shinshi H, Sato F (2003) Isolation of tobacco ubiquitin-conjugating enzyme cDNA in a yeast two-hybrid system with tobacco ERF3 as bait and its characterization of specific interaction. J Exp Bot 54:1175–1181

Kresge N, Simoni RD, Hill RL (2007) The biosynthesis of polyamines: the work of Herbert Tabor and Celia White Tabor. J Biol Chem 282:e26–e28

Kud J, Zhao Z, Du X, Liu Y, Zhao Y, Xiao F (2013) SGT1 interacts with the Prf resistance protein and is required for Prf accumulation and Prf-mediated defense signaling. Biochem Biophys Res Commun 431:501–505

Larsen PB, Cancel JD (2004) A recessive mutation in the RUB1-conjugating enzyme, RCE1, reveals a requirement for RUB modification for control of ethylene biosynthesis and proper induction of basic chitinase and PDF1.2 in *Arabidopsis*. Plant J 38:626–638

Lee S-W, Han S-W, Bartley LE, Ronald PC (2006) Unique characteristics of *Xanthomonas oryzae* pv. *oryzae* AvrXa21 and implications for plant innate immunity. Proc Natl Acad Sci U S A 103:18395–18400

Lee J, Nam J, Park HC, Na G, Miura K, Jin JB, Yoo CY, Baek D, Kim DH, Jeong JC, Kim D, Lee SY, Salt DE, Mengiste T, Gong Q, Ma S, Bohnert HJ, Kwak S-S, Bressan RA, Hasegawa PM, Yun D-J (2007) Salicylic acid-mediated innate immunity in *Arabidopsis* is regulated by SIZ1 SUMO E3 ligase. Plant J 49:79–90

Lee S-W, Han S, Sririyanum M, Park C, Seo Y, Ronald PC (2009) A type 1-secreted, sulfated peptide triggers XA21-mediated innate immunity. Science 326:850–853

Lee DH, Choi HW, Hwang BK (2011) The pepper E3 ubiquitin ligase *RING1* gene, *CaRING1*, is required for cell death and the salicylic acid-dependent defense response. Plant Physiol 156:2011–2025

Lippok B, Birkenbihl RP, Rivory G, Brummer J, Schmelzer E, Logemann E, Somssich IE (2007) Expression of *AtWRKY33* encoding a pathogen- or PAMP-responsive WRKY transcription factor is regulated by a composite DNA motif containing W box elements. Mol Plant Microbe Interact 20:420–429

Liu Y, Schiff M, Serino G, Deng X-W, Dinesh-Kumar SP (2002) Role of SCF ubiquitin-ligase and COP9 signalosome in the *N* gene-mediated resistance response to *Tobacco mosaic virus*. Plant Cell 14:1483–1496

Lorenzo O, Solano R (2005) Molecular players regulating the jasmonate signaling network. Curr Opin Plant Biol 8:532–540

Lorenzo O, Chico JM, Sánchez-Serrano JJ, Solano R (2004) JASMONATE-INSENSITIVE1 encodes a MYC transcription factor essential to discriminate between different jasmonate-regulated defense responses in *Arabidopsis*. Plant Cell 16:1938–1950

Lozano-Durán R, Bejarano ER (2011) Geminivirus C2 protein might be the key player for geminiviral co-option of SCF-mediated ubiquitination. Plant Signal Behav 6:999–1001

Lozano-Durán R, Rosas-Diaz T, Gusmaroli G, Luna AP, Taconnat L, Deng XW, Bejarano ER (2011) Geminiviruses subvert ubiquitination by altering CSN-mediated derubylation of SCF E3 ligase complexes and inhibit jasmonate signaling in *Arabidopsis thaliana*. Plant Cell 23:1014–1032

Mandaokar A, Thines B, Shin B, Lange BM, Choi G, Koo YJ, Yoo YJ, Choi YD, Choi G, Browse J (2006) Transcriptional regulators of stamen development in *Arabidopsis* identified by transcriptional profiling. Plant J 46:984–1008

March-Diaz R, Garcia-Domínguez M, Lozano-Juste J, León J, Florencio FJ, Reyes JC (2008) Histone H2A.Z and homologues of components of the SWR1 complex are required to control immunity in *Arabidopsis*. Plant J 53:475–487

Melotto M, Underwood W, He SY (2008) Role of stomata in plant innate immunity and foliar bacterial diseases. Annu Rev Phytopathol 46:101–122

Miura K, Hasegawa PM (2010) Sumoylation and other ubiquitin-like post-translational modifications in plants. Trends Cell Biol 20:223–232

Moon J, Zhao Y, Dai X, Zhang W, Gray WM, Huq E, Estelle M (2007) A new *CULLIN1* mutant has altered responses to hormones and light in *Arabidopsis*. Plant Physiol 143:684–696

Mou Z, Fan W, Dong X (2003) Inducers of plant systemic acquired resistance regulate NPR1 function through redox changes. Cell 113:935–944

Navarro L, Zipfel C, Rowland O, Keller I, Robatzek S, Boller T, Jones JD (2004) The transcriptional innate immune response to flg22. Interplay and overlap with Avr gene-dependent defense responses and bacterial pathogenesis. Plant Physiol 135:1113–1128

Nawrath C, Heck S, Parinthawong N, Metraux JP (2002) EDS5, an essential component of salicylic acid-dependent signaling for disease resistance in *Arabidopsis*, is a member of MATE transporter family. Plant Cell 14:275–286

Ohme-Takagi M, Shinshi H (1995) Ethylene-inducible DNA binding proteins that interact with an ethylene-responsive element. Plant Cell 7:173–182

Ohta M, Ohme-Takagi M, Shinshi H (2000) Three ethylene-responsive transcription factors in tobacco with distinct transactivation functions. Plant J 22:29–38

Park CH, Chen S, Shirsekar G, Zhou B, Khang CH, Songkumam P, Afzal AJ, Ning Y, Wang R, Bellizzi M, Valent B, Wang GL (2012) The *Magnaporthe oryzae* effector AvrPiz-t targets the RING E3 ubiquitin ligase APIP6 to suppress pathogen-associated molecular pattern-triggered immunity in rice. Plant Cell 24:4748–4762

Pauwels L, Barbero GF, Geerinck J, Tilleman S, Grunewald W, Pérez AC, Chico JM, Bossche RV, Sewell J, Gill E, Garcia-Casado G, Witters E, Inzé D, Long JA, De Jaeger G, Solano R, Goossens A (2010) NINJA connects the co-repressor TOPLESS to jasmonate signaling. Nature 464:788–791

Peart JR, Lu R, Sadanandom A, Malcuit I, Moffett P, Brice DC, Schauser L, Jaggard DAW, Xiao S, Coleman MJ, Dow M, Jones JDG, Shirasu K, Baulcombe DC (2002) Ubiquitin ligase-associated protein SGT1 is required for host and nonhost disease resistance in plants. Proc Natl Acad Sci U S A 99:10865–10869

Peters JM, Franke WW, Kleinschmidt JA (1994) Distinct 19S and 20S subcomplexes of the 26S proteasome and their distribution in the nucleus and the cytoplasm. J Biol Chem 269:7709–7718

Pickart CM (2001) Mechanisms underlying ubiquitination. Annu Rev Biochem 70:503–533

Pickart CM, Eddins MJ (2004) Ubiquitin: structures, functions, mechanisms. Biochim Biophys Acta 1695:55–72

Potuschak T, Lechner E, Parmentier Y, Yanagisawa S, Grava S, Koncz C, Genschik P (2003) EIN3-dependent regulation of plant ethylene hormone signaling by two *Arabidopsis* F box proteins EBF1 and EBF2. Cell 115:679–689

Qi RC, Song SS, Ren QC, Wu DW, Huang H, Chen Y, Fan M, Peng W, Ren CM, Xie DX (2011) The jasmonate-ZIM domain proteins interact with the WD-repeat/bHLH/MYB complexes to regulate jasmonate-mediated anthocyanin accumulation and trichome initiation in *Arabidopsis thaliana*. Plant Cell 23:1795–1814

Ralph SG, Hudgins JW, Jancsik S, Franceschi VR, Bohlmann J (2007) Aminocyclopropane carboxylic acid synthase is a regulated step in ethylene-dependent induced conifer defense. Full-length cDNA cloning of a multigene family, differential constitutive, and wound- and insect-induced expression, and cellular and subcellular localization in spruce and Douglas Fir. Plant Physiol 143:410–424

Risseeuw EP, Daskalchuk TE, Banks TW, Liu E, Cotelesage J, Hellmann H, Estelle M, Somers DE, Crosby WI (2003) Protein interaction analysis of SCF ubiquitin E3 ligase subunits from *Arabidopsis*. Plant J 34:753–767

Rustérucci C, Aviv DH, Holt BF, Dangl JL, Parker JE (2001) The disease resistance signaling components *EDS1* and *PAD4* are essential regulators of the cell death pathway controlled by *LSD1* in *Arabidopsis*. Plant Cell 13:2211–2224

Sahana N, Kaur H, Basavaraj H, Tena F, Jain RK, Palukaitis P, Canto T, Praveen S (2012) Inhibition of the host proteasome facilitates papaya ringspot virus accumulation and proteasomal catalytic activity is modulated by viral factor HcPro. PLoS One 7:e52546

Salinas-Mondragón RE, Garcidueñas-Piña C, Guzmán P (1999) Early elicitor induction in members of a novel multigene family coding for highly related RING-H2 proteins in *Arabidopsis thaliana*. Plant Mol Biol 40:579–590

Seo W-S, Lee S-K, Song M-Y, Suh J-P, Hahn T-R, Ronald P, Jeon J-S (2008) The HSP90-SGT1-RAR1 molecular chaperone complex: a core modulator in plant immunity. J Plant Biol 51:1–10

Serrano M, Paara S, Alcaraz LD, Guzman P (2006) The ATL gene family from *Arabidopsis thaliana* and *Oryza sativa* comprises a large number of putative ubiquitin ligases of the RING-H2 type. J Mol Evol 62:434–445

Sheard LB, Tan X, Mao H, Withers J, Ben-Nissan G, Hinds TR, Kobayashi Y, Hsu EF, Sharon M, Browse J, He SY, Rizo J, Howe GA, Zheng N (2010) Jasmonate perception by inositol-phosphate-potentiated COI1-JAZ co-receptor. Nature 468:400–405

Smalle J, Vierstra RD (2004) The ubiquitin 26S proteasome proteolytic pathway. Annu Rev Plant Biol 55:555–590

Song W-Y, Wang G-L, Chen L-L, Kim H-S, Pi L-Y, Holsten T, Gardner J, Wang B, Zhai W-X, Zhu L-H, Fauquet C, Ronald P (1995) A receptor kinase-like protein encoded by the rice disease resistance gene *Xa21*. Science 270:1804–1806

Song JT, Lu H, McDowell JM, Greenberg JT (2004) A key role for *ALD1* in activation of local and systemic defenses in *Arabidopsis*. Plant J 40:200–212

Staswick PE, Tiryaki I (2004) The oxylipin signal jasmonic acid is activated by an enzyme that conjugates it to isoleucine in *Arabidopsis*. Plant Cell 16:2117–2127

Stone SL, Callis J (2007) Ubiquitin ligases mediate growth and development by promoting protein death. Curr Opin Plant Biol 10:624–632

Stone SL, Hauksdottir H, Troy A, Herschleb J, Kraft E, Callis J (2005) Functional analysis of the RING-type ubiquitin ligase family of Arabidopsis. Plant Physiol 137:13–30

Takahashi H, Hase S, Kanayama Y, Takenaka S (2010) Identification of a protein that interacts with LeATL6 ubiquitin-protein ligases E3 upregulated in tomato treated with elicitin-like cell wall proteins of *Pythium oligandrum*. J Phytopathol 158:132–136

Takai R, Matsuda N, Nakano A, Hasegawa K, Akimoto C, Shibuya N, Minami E (2002) EL5, a rice N-acetylchitooligosaccharide elicitor-responsive RING-H2 finger protein, is a ubiquitin ligase which functions *in vitro* in co-operation with an elicitor-responsive ubiquitin-conjugating enzyme, OsUBC5b. Plant J 30:447–455

Thines B, Katsir L, Melotto M, Niu Y, Mandaokar A, Liu G, Nomura K, He SY, Howe GA, Browse J (2007) JAZ repressor proteins are targets of the SCF$^{COI1}$ complex during jasmonate signaling. Nature 448:661

Thrower JS, Hoffmann L, Rechsteiner M, Pickart CM (2000) Recognition of the polyubiquitin proteolytic signal. EMBO J 19:94–102

Tör M, Gordon P, Cuzick A, Eulgem T, Sinapidou E, Mert-Turk F, Can C, Dangl JL, Holub EB (2002) Arabidopsis SGT1b is required for defense signaling conferred by several downy mildew resistance genes. Plant Cell 14:993–1003

Trujillo M, Ichimura K, Casais C, Shirasu K (2008) Negative regulation of PAMP-triggered immunity by an E3 ubiquitin ligase triplet in *Arabidopsis*. Curr Biol 18:1396–1401

van den Burg HA, Takken FLW (2009) Does chromatin remodeling mark systemic acquired resistance? Trends Plant Sci 14:286–294

van den Burg HA, Takken FLW (2010) SUMO-, MAPK- and resistance protein-signaling converge at transcription complexes that regulate plant innate immunity. Plant Signal Behav 5:1507–1601

van den Burg HA, Tsitsigiannis DI, Rowland O, Lo J, Rallapalli G, Maclean D, Takken FL, Jones JD (2008) The F-box protein ACRE189/ACIF1 regulates cell death and defense responses activated during pathogen recognition in tobacco and tomato. Plant Cell 20:697–719

Van Loon LC, Rep M, Pieterse C (2006) Significance of inducible defense-related proteins in infected plants. Annu Rev Phytopathol 44:135–162

Vierstra RD (2003) The ubiquitin/26S proteasome pathway, the complex last chapter in the life of many plant proteins. Trends Plant Sci 8:135–142

Vogel JP, Woeste KE, Theologis A, Kicher JJ (1998) Recessive and dominant mutations in the ethylene biosynthetic gene *ACS5* of Arabidopsis confer cytokinin insensitivity and ethylene overproduction, respectively. Proc Natl Acad Sci U S A 95:4766–4771

Vorwerk S, Schiff C, Santamaria M, Koh S, Nishimura M, Vogel J, Somerville C, Somerville S (2007) EDR2 negatively regulates salicylic acid-based defenses and cell death during powdery mildew infections of *Arabidopsis thaliana*. BMC Plant Biol 7:35

Wang X, Feng S, Nakayama N, Crosby WL, Irish V, Deng XW, Wei N (2003) The COP9 signalosome interacts with SCF$^{UFO}$ and participates in Arabidopsis flower development. Plant Cell 15:1071–1082

Wang D, Amonsiripanitch N, Dong X (2006) A genomic approach to identify regulatory nodes in the transcriptional network of systemic acquired resistance in plants. PLoS Pathog 2:e123

Wang Y, Gao M, Li Q, Wang L, Jeon J-S, Qu N, Zhang Y, He Z (2008) *OsRAR1* and *OsSGT1* physically interact and function in rice basal disease resistance. Mol Plant Microbe Interact 21:294–303

Wiermer M, Feys BJ, Parker JE (2005) Plant immunity: the EDS1 regulatory node. Curr Opin Plant Biol 8:383–389

Xie DX, Feys BF, James S, Nieto-Rostro M, Turner JG (1998) *COI1*: an *Arabidopsis* gene required for jasmonate-regulated defense and fertility. Science 280:1091–1094

Xing D, Chen Z (2006) Effects of mutations and constitutive overexpression of *EDS1* and *PAD4* on plant resistance to different types of microbial pathogens. Plant Sci 171:251–262

Xu LH, Liu FQ, Wang ZL, Peng W, Huang RF, Huang RF, Huang DF, Xie DX (2001) An Arabidopsis mutant *cex1* exhibits constant accumulation of jasmonate-regulated *AtVSP*, *Thi2.1* and *PDF1.2*. FEBS Lett 494:161–164

Xu L, Liu F, Lechner E, Genschik P, Crosby WL, Ma H, Peng W, Huang D, Xie D (2002) The SCF(COI1) ubiquitin-ligase complexes are required for jasmonate response in *Arabidopsis*. Plant Cell 14:1919–1935

Yamamoto S, Suzuki K, Shinshi H (1999) Elicitor-responsive, ethylene-independent activation of GCC box-mediated transcription that is regulated by both protein phosphorylation and dephosphorylation in cultured tobacco cells. Plant J 20:571–579

Yang X, Menon S, Lykke-Andersen K, Tsuge T, Di X, Wang X, Rodriguez-Suarez-Suarez RJ, Zhang H, Wei N (2002) The COP9 signalosome inhibits $p27^{klp1}$ degradation and impedes G1-S phase progression via deneddylation of SCF Cul1. Curr Biol 12:667–672

Yang CW, Gonzalez-Lamothe R, Ewan RA, Rowland O, Yoshioka H, Shenton M, Ye H, O'Donnell E, Jones JD, Sadanandom A (2006) The E3 ubiquitin ligase activity of *Arabidopsis* PLANT U-BOX17 and its functional tobacco homolog ACRE276 are required for cell death and defense. Plant Cell 18:1084–1098

Yao T, Ndoja A (2012) Regulation of gene expression by the ubiquitin-proteasome system. Semin Cell Dev Biol 23:523–529

Yao C, Wu Y, Nie H, Tang D (2012) RPN1a, a 26S proteasome subunit, is required for innate immunity in Arabidopsis. Plant J 71:1015–1028

Zeng LR, Qu S, Bordeos A, Yang C, Baraoidan M, Yan H, Xie Q, Nahm BH, Leung H, Wang GL (2004) *Spotted leaf11*, a negative regulator of plant cell death and defense, encodes a U-box/armadillo repeat protein endowed with E3 ubiquitin ligase activity. Plant Cell 16:2795–2808

Zeng LR, Vega-Sanchez ME, Zhu T, Wang GL (2006) Ubiquitination-mediated protein degradation and modification: an emerging theme in plant-microbe interactions. Cell Res 16:413–426

Zhang Y, Goritschnig S, Dong X, Li X (2003) A gain of function mutation in a plant disease resistance gene leads to constitutive activation of downstream signal transduction pathways in suppressor of *npr1-1*, constitutive 1. Plant Cell 15:2636–2646

Zhang X, Wu Q, Ren J, Qian W, He S, Huang K, Yu XC, Gao Y, Huang P, An C (2012) Two novel RING-type ubiquitin ligases, RGLG4 and RGLG4, are essential for jasmonate-mediated responses in Arabidopsis. Plant Physiol 160:808–822

Zheng N, Wang P, Jeffrey PD, Pavletich NP (2000) Structure of a c-Cbi-UbcH7 complex: RING domain function in ubiquitin-protein ligases. Cell 102:533–539

Zhou N, Tootle T, Tsui F, Klessig D, Glazebrook J (1998) *PAD4* functions upstream from salicylic acid to control defense responses to *Arabidopsis*. Plant Cell 10:1021–1030

Zhou C, Zhang L, Duan J, Miki B, Wu K (2005) Histone deacetylase19 is involved in jasmonic acid and ethylene signaling of pathogen response in *Arabidopsis*. Plant Cell 17:1196–1204

Zhou F, Mosher S, Tian M, Sassi G, Parker J, Klessig DF (2008) The *Arabidopsis* gain-of-function mutant *ssi4* requires *RAR1* and *SGT1b* differentially for defense activation and morphological alterations. Mol Plant Microbe Interact 21:40–49

Zhu Q, Wani G, Wang QE, El-mahdy M, Snapka RM, Wani AA (2005) Deubiquitination by proteasome is coordinated with substrate translocation for proteolysis in vivo. Exp Cell Res 307:436–451

# Index



P. Vidhyasekaran, *PAMP Signals in Plant Innate Immunity: Signal Perception and Transduction*, Signaling and Communication in Plants 21,
DOI 10.1007/978-94-007-7426-1, © Springer Science+Business Media Dordrecht 2014

**R**

FSC
www.fsc.org
MIX
Papier aus verantwortungsvollen Quellen
Paper from responsible sources
FSC® C105338

If you have any concerns about our products,
you can contact us on
**ProductSafety@springernature.com**

In case Publisher is established outside the EU,
the EU authorized representative is:
**Springer Nature Customer Service Center GmbH**
**Europaplatz 3, 69115 Heidelberg, Germany**

Printed by Libri Plureos GmbH
in Hamburg, Germany